Lecture Notes in Computer Science 9678

Commenced Publication in 1973
Founding and Former Series Editors:
Gerhard Goos, Juris Hartmanis, and Jan van Leeuwen

More information about this series at http://www.springer.com/series/7407

Harald Sack · Eva Blomqvist
Mathieu d'Aquin · Chiara Ghidini
Simone Paolo Ponzetto · Christoph Lange (Eds.)

The Semantic Web

Latest Advances and New Domains

13th International Conference, ESWC 2016
Heraklion, Crete, Greece, May 29 – June 2, 2016
Proceedings

 Springer

Editors

Harald Sack
Universität Potsdam
Potsdam
Germany

Eva Blomqvist
Linköping University
Linköping
Sweden

Mathieu d'Aquin
The Open University
Milton Keynes
UK

Chiara Ghidini
Fondazione Bruno Kessler
Trento
Italy

Simone Paolo Ponzetto
Universität Mannheim
Mannheim
Germany

Christoph Lange
Universität Bonn
Bonn
Germany

ISSN 0302-9743   ISSN 1611-3349   (electronic)
Lecture Notes in Computer Science
ISBN 978-3-319-34128-6   ISBN 978-3-319-34129-3   (eBook)
DOI 10.1007/978-3-319-34129-3

Library of Congress Control Number: 2016938379

LNCS Sublibrary: SL1 – Theoretical Computer Science and General Issues

Printed on acid-free paper

This Springer imprint is published by Springer Nature
The registered company is Springer International Publishing AG Switzerland

Preface

The goal of the Semantic Web is to create a Web of knowledge and services in which the semantics of content is made explicit and content is linked to both other content and services, enabling novel applications that combine content from heterogeneous sources in unforeseen ways and support enhanced matching between users' needs and content. This network of knowledge-based functionality weaves together a large network of human knowledge and distributed linked data, and makes this knowledge machine-processable to support intelligent behavior by machines. Creating such an interlinked Web of Knowledge that spans unstructured, RDF, as well as multimedia content and services requires the collaboration of many disciplines, including but not limited to: artificial intelligence, natural language processing, database and information systems, information retrieval and data mining, machine learning, multimedia, distributed systems, social networks, Web engineering, and Web science. These complementarities are reflected in the outline of the technical program of ESWC 2016.

The ESWC Conference is established as a yearly major venue for discussing the latest scientific results and technology innovations related to the Semantic Web and linked data. At ESWC, international scientists, industry specialists, and practitioners meet to discuss the future of applicable, scalable, user-friendly, as well as potentially game-changing solutions. This 13th edition took place from May 29 to June 2, 2016, in Heraklion, Crete, Greece. Building on its past success, ESWC is seeking to broaden its focus to span other relevant research areas in which Web semantics plays an important role. Thus, the chairs of ESWC 2016 decided to broaden the scope to span further emerging relevant research areas with two special tracks putting particular emphasis on inter-disciplinary research topics and areas that show the potential of exciting synergies for the future, namely: "Trust and Privacy" and "Smart Cities and GeoSpatial Data."

This choice also resulted in three exciting invited keynotes. Jim Hendler (Rensselaer Polytechnic Institute) is well known as one of the originators of the Semantic Web. In his keynote, he explored some of the uses and needs of ontologies on the Web in data integration, emerging technologies, and linked data applications. In particular, he pointed out deficiencies in OWL's design that have hindered its application, and suggested directions for making OWL more relevant to the modern Web. Ernesto Damiani (Università degli Studi di Milano) discussed the idea that techniques used for semantic enrichment of big data can be seen as non-linear leakage and privacy risk boosters. Semantic technologies might increase leakage risk by increasing the value for the attacker per unit of information leaked. Furthermore, they might increase intrusion risk, making injection attacks more effective per unit of poisoned information injected. Eleni Pratsini (IBM Research Ireland) discussed the typical challenges of intelligent semantic systems that often prevent a business from even starting to look at the information and make sense of it. On the other hand there are novel business opportunities enabled by advances in cognitive computing that offer new possibilities in analyzing unstructured information for richer insights.

The main scientific program of the conference comprised 47 papers: 39 research and eight in-use, selected out of 204 submissions, which corresponds to an acceptance rate of 21 % for the 184 research papers submitted and of 40 % for the 20 in-use papers submitted. This program was completed by a demonstration and poster session, in which researchers had the chance to present their latest results and advances in the form of live demos. In addition, the PhD Symposium program included 10 contributions, selected out of 21 submissions.

This year's edition of ESWC's main scientific program presented a significant number of research papers with a focus on solving typical Semantic Web problems, such as entity linking, discoverability, etc., by using methods and techniques borrowed from other areas like machine learning and natural language processing. Likewise, research problems from those related areas, in particular also from smart cities and geospatial data-related problems, are tackled by adapting typical approaches to incorporate Semantic Web resources as well as technologies.

To have an open, multidisciplinary, and cross-fertilizing event, we complemented the conference program with 15 workshops, nine tutorials, as well as the EU Project Networking session. This year, an open call for challenges also allowed us to select and support eight challenges.

As general and Program Committee chairs, we would like to thank the many people who were involved in making ESWC 2016 a success.

First of all, our thanks go to the 24 track chairs and 378 reviewers including 83 external reviewers for ensuring a rigorous blind review process that led to an excellent scientific program and an average number of four reviews per article. This was also completed by an inspiring selection of posters and demos chaired by Nadine Steinmetz and Giuseppe Rizzo.

Special thanks go to the PhD symposium chairs, Chiara Ghidini and Simone Paolo Ponzetto, who proposed and managed a very constructive organization of this ESWC key event ensuring a real mentoring to all the brilliant students who participated.

We had a great selection of workshops and tutorials thanks to the dynamism and commitment of our workshop chairs, Dunja Mladenic and Sören Auer, and tutorial chairs, H. Sofia Pinto and Tommaso di Noia.

Thanks to our EU Project Networking session chairs, Erik Mannens, Mauro Dragoni, Lyndon Nixon, and Oscar Corcho, we had the opportunity to arrange meetings and exciting discussions between the contributors of the currently leading European research projects.

We are grateful for the work and commitment of Anna Tordai, Stefan Dietze, and all the challenges chairs, who successfully established a challenge track with an open-call leading to a very useful comparison of the latest solutions for eight challenge areas.

Thanks to STI International for supporting the conference organization. YouVivo GmbH and in particular Katharina Haas deserve special thanks for the professional support of the conference organization.

We are very grateful to Heiko Paulheim, our publicity chair, who kept our community informed at every stage, and Venislav Georgiev, who administered the website.

Our sponsorship chairs, Steffen Lohmann and Freddie Lecue, played an extremely important role in collecting sponsorships for the conference, the awards, and the grants.

And of course we also thank our sponsors listed on the next pages, for their vital support to this edition of ESWC.

We also want to stress the huge work achieved by the Semantic Technologies coordinators Anna Lisa Gentile and Andrea Giovanni Nuzzolese, who maintained and updated our "ESWC Conference Live" mobile app.

A special thanks also to our proceedings chair, Christoph Lange, who did a remarkable job in preparing this volume with the kind support of Springer.

March 2016

Harald Sack
Eva Blomqvist
Mathieu d'Aquin

Organization

Organizing Committee

General Chair

Harald Sack — Hasso Plattner Institute (HPI), Germany

Program Chairs

Mathieu d'Aquin — Knowledge Media Institute KMI, UK
Eva Blomqvist — Linköping University, Sweden

Workshops Chairs

Dunja Mladenic — Jožef Stefan Institute, Slovenia
Sören Auer — University of Bonn and Fraunhofer IAIS, Germany

Poster and Demo Chairs

Giuseppe Rizzo — Istituto Superiore Mario Boella, Italy
Nadine Steinmetz — Technische Universität Ilmenau, Germany

Tutorials Chairs

Tommaso di Noia — Politecnico di Bari, Italy
H. Sofia Pinto — INESC-ID, Instituto Superior Técnico, Universidade de Lisboa, Portugal

PhD Symposium Chairs

Simone Paolo Ponzetto — Universität Mannheim, Germany
Chiara Ghidini — FBK-IRST, Italy

Challenge Chairs

Stefan Dietze — L3S, Germany
Anna Tordai — Elsevier, The Netherlands

Semantic Technologies Coordinators

Andrea Giovanni Nuzzolese — University of Bologna/STLab ISTC-CNR, Italy
Anna Lisa Gentile — University of Mannheim, Germany

EU Project Networking Session Chairs

Erik Mannens	Data Science Lab – iMinds – Ghent University, Belgium
Mauro Dragoni	Fondazione Bruno Kessler, Italy
Lyndon Nixon	Modul Universität Vienna, Austria
Oscar Corcho	Universidad Politécnica de Madrid, Spain

Publicity Chair

Heiko Paulheim	University of Mannheim, Germany

Sponsor Chairs

Steffen Lohmann	Fraunhofer IAIS, Germany
Freddie Lecue	IBM, Ireland

Web Presence

Venislav Georgiev	STI International, Austria

Proceedings Chair

Christoph Lange	University of Bonn and Fraunhofer IAIS, Germany

Treasurer

Ioan Toma	STI International, Austria

Local Organization and Conference Administration

Katharina Haas	YouVivo GmbH, Germany

Program Committee

Program Chairs

Mathieu d'Aquin	Knowledge Media Institute KMI, UK
Eva Blomqvist	Linköping University, Sweden

Track Chairs

Vocabularies, Schemas, Ontologies

Krzysztof Janowicz	University of California, Santa Barbara, USA
Rinke Hoekstra	VU University Amsterdam, The Netherlands

Reasoning

Uli Sattler	University of Manchester, UK
Thomas Schneider	Universität Bremen, Germany

Linked Data

Monika Solanki   University of Oxford, UK
Aidan Hogan   Universidad de Chile, Santiago de Chile, Chile

Social Web and Web Science

Claudia Müller-Birn   Freie Universitat Berlin, Germany
Steffen Staab   Universität Koblenz, Germany

Semantic Data Management, Big Data, Scalability

Philippe Cudré-Mauroux University of Fribourg, Switzerland
Katja Hose   Aalborg University, Denmark

Natural Language Processing and Information Retrieval

Nathalie Aussenac   IRIT - Université Toulouse, France
  Gilles
Pablo N. Mendes   IBM, USA

Machine Learning

Claudia d'Amato   University of Bari, Italy
Jens Lehmann   Universität Leipzig, Germany

Mobile Web, Sensors and Semantic Streams

Raúl García Castro   Universidad Politécnica de Madrid, Spain
Jean-Paul Calbimonte   École Polytechnique Fédérale de Lausanne, Switzerland

Services, APIs, Processes and Cloud Computing

Maria Maleshkova   AIFB, Karlsruhe Institute of Technology, Germany
Karthik Gomadam   Accenture Technology Labs, USA

In-Use and Industrial Track

Mike Lauruhn   Elsevier Labs, The Netherlands
Jacco van Ossenbruggen   Centrum Wiskunde & Informatica (CWI), Amsterdam,
                The Netherlands

Trust and Privacy

Sabrina Kirrane   Wirtschaftsuniversität Wien, Austria
Pompeu Casanovas   Universidad Autónoma de Barcelona, Spain

Smart Cities, Urban and Geospatial Data

Carsten Kessler   Hunter College, CUNY, New York, USA
Vanessa Lopez   IBM, Ireland

Program Committee (All Tracks)

Maribel Acosta
Nitish Aggarwal
Guadalupe Aguado-De-Cea
Carlo Allocca
Bernd Amann
Pramod Anantharam
Lora Aroyo
Manuel Atencia
Martin Atzmueller
Sören Auer
Nathalie Aussenac-Gilles
Michele Barbera
Payam Barnaghi
Pierpaolo Basile
Zohra Bellahsene
Bettina Berendt
Chris Biemann
Antonis Bikakis
Peter Bloem
Eva Blomqvist
Fernando Bobillo
Kalina Bontcheva
Stefano Borgo
Johan Bos
Gosse Bouma
Alessandro Bozzon
Charalampos Bratsas
Joseph Busch
Elena Cabrio
Jean-Paul Calbimonte
Nicoletta Calzolari
Erik Cambria
Amparo E. Cano
David Carral
Marco Antonio Casanova
Pompeu Casanovas
Michele Catasta
Irene Celino
Pierre-Antoine Champin
Jean Charlet
Vinay Chaudhri
Paolo Ciccarese
Pieter Colpaert

Marco Combetto
Bonaventura Coppola
Oscar Corcho
Gianluca Correndo
David Corsar
Fabio Cozman
Michael Crandall
Danilo Croce
Philippe Cudré-Mauroux
Olivier Curé
Claudia d'Amato
Mathieu d'Aquin
Aba-Sah Dadzie
Enrico Daga
Danica Damljanovic
Jérôme David
Ernesto William De Luca
Thierry Declerck
Luciano Del Corro
Emanuele Della Valle
Gianluca Demartini
Elena Demidova
Tommaso Di Noia
Stefan Dietze
Djellel Eddine Difallah
Dejing Dou
Mauro Dragoni
Anca Dumitrache
Vadim Ermolayev
Jérôme Euzenat
Nicola Fanizzi
Catherine Faron Zucker
Miriam Fernandez
Besnik Fetahu
Fabian Flöck
Flavius Frasincar
Fred Freitas
Johannes Fürnkranz
Fabien Gandon
Aldo Gangemi
Roberto Garcia
José María García
Nuria García Santa

Raúl García-Castro
Daniel Garijo
Dragan Gasevic
Anna Lisa Gentile
Chiara Ghidini
Alain Giboin
Fausto Giunchiglia
François Goasdoué
Karthik Gomadam
Jose Manuel Gomez-Perez
Jorge Gracia
Michael Granitzer
Alasdair Gray
Paul Groth
Tudor Groza
Alessio Gugliotta
Ramanathan Guha
Giancarlo Guizzardi
Kalpa Gunaratna
Peter Haase
Ollivier Haemmerlé
Andreas Harth
Olaf Hartig
Oktie Hassanzadeh
Tom Heath
Benjamin Heitmann
Sebastian Hellmann
Pascal Hitzler
Rinke Hoekstra
Aidan Hogan
Laura Hollink
Matthew Horridge
Katja Hose
Geert-Jan Houben
Eero Hyvönen
Antoine Isaac
Krzysztof Janowicz
Frederik Janssen
Mustafa Jarrar
Ernesto Jimenez-Ruiz
Lalana Kagal
Pavan Kapanipathi
Tomi Kauppinen
Takahiro Kawamura
C. Maria Keet
Carsten Kessler

Sabrina Kirrane
Friederike Klan
Szymon Klarman
Matthias Knorr
Spyros Kotoulas
Manolis Koubarakis
Markus Krause
Adila A. Krisnadhi
Udo Kruschwitz
Kaushik Kumar Ram
Oliver Kutz
Birgitta König-Ries
Manuel Lama Penin
Dave Lambert
Patrick Lambrix
Christoph Lange
Mike Lauruhn
Nico Lavarini
Agnieszka Lawrynowicz
Danh Le Phuoc
Jens Lehmann
Domenico Lembo
Maurizio Lenzerini
Juanzi Li
Wenwen Li
Jean Lieber
Nuno Lopes
Vanessa Lopez
Markus Luczak-Roesch
Yue Ma
Frederick Maier
Maria Maleshkova
Vincenzo Maltese
Maria Vanina Martinez
Diana Maynard
Suvodeep Mazumdar
John P. McCrae
Fiona McNeill
Parvathy Meenakshy
Alexander Mehler
Pablo N. Mendes
Nandana Mihindukulasooriya
Peter Mika
Riichiro Mizoguchi
Dunja Mladenic
Pascal Molli

Alexandre Monnin
Mikolaj Morzy
Alessandro Moschitti
Claudia Müller-Birn
Roberto Navigli
Maximilian Nickel
Nadeschda Nikitina
Andriy Nikolov
Malvina Nissim
Olaf Noppens
Andrea Giovanni Nuzzolese
Andreas Nürnberger
Leo Obrst
Alessandro Oltramari
Jacco Van Ossenbruggen
Matteo Palmonari
Jeff Z. Pan
Patrick Paroubek
Heiko Paulheim
Terry Payne
Carlos Pedrinaci
Tassilo Pellegrini
Silvio Peroni
Dimitris Plexousakis
Axel Polleres
Livia Predoiu
Valentina Presutti
Yuzhong Qu
Achim Rettinger
Chantal Reynaud
Mikko Rinne
Carlos R. Rivero
Giuseppe Rizzo
Mariano Rodriguez-Muro
Víctor Rodríguez Doncel
Haggai Roitman
Dumitru Roman
Camille Roth
Marie-Christine Rousset
Matthew Rowe
Edna Ruckhaus
Marta Sabou
Harald Sack
Hassan Saif
Felix Sasaki
Uli Sattler

Marco Luca Sbodio
Ansgar Scherp
Stefan Schlobach
Jodi Schneider
Thomas Schneider
Stefan Schulte
Juan F. Sequeda
Luciano Serafini
Bariş Sertkaya
Amit Sheth
Pavel Shvaiko
Gerardo Simari
Elena Simperl
Monika Solanki
Steffen Staab
Yannis Stavrakas
Thomas Steiner
Armando Stellato
Giorgos Stoilos
Umberto Straccia
Heiner Stuckenschmidt
Gerd Stumme
Fabian Suchanek
Vojtiěch Svátek
Marcin Sydow
Pedro Szekely
Valentina Tamma
Kunal Taneja
Kerry Taylor
Dhaval Thakker
Keerthi Thomas
Thanassis Tiropanis
Ioan Toma
Alessandra Toninelli
Farouk Toumani
Thanh Tran
Volker Tresp
Raphaël Troncy
Giovanni Tummarello
Anni-Yasmin Turhan
Jürgen Umbrich
Christina Unger
Alejandro A. Vaisman
Herbert Van De Sompel
Marieke Van Erp
Frank Van Harmelen

Pierre-Yves Vandenbussche
Joaquin Vanschoren
Paola Velardi
Ruben Verborgh
Maria Esther Vidal
Serena Villata
Boris Villazón-Terrazas
Holger Wache
Claudia Wagner
Haofen Wang

Kewen Wang
Shenghui Wang
Erik Wilde
Cord Wiljes
Gregory Todd Williams
Josiane Xavier Parreira
Ondřej Zamazal
Ziqi Zhang
Antoine Zimmermann

Additional Reviewers

Azad Abad
Markus Ackermann
Mohamed Ahmed Sherif
Albin Ahmeti
Saud Aljaloud
Stephan Baier
Gianni Barlacchi
Ciro Baron Neto
Valerio Basile
Wouter Beek
Mohamed Ben Ellefi
Tarek Richard Besold
Stefan Bischof
Emanuele Bottazzi
Alessander Botti Benevides
Janez Brank
Carlos Buil Aranda
Jose Camacho Collados
Alessio Carenini
Nilesh Chakraborty
Vinay Chaudhri
Jiaoyan Chen
Lu Chen
Mihai Codescu
Diego Collarana
Olivier Corby
Fabrizio Cucci
Minh Dao-Tran
Fariz Darari
Tom De Nies
Cristhian Ariel David Deagustini
Jeremy Debattista

Elena Demidova
Zlatan Dragisic
Andreas Ecke
Steffen Eger
Kemele M. Endris
Cristobal Esteban
Lorena Etcheverry
Nicola Fanizzi
Nazli Farajidavar
Catherine Faron Zucker
Giorgos Flouris
Nuno Freire
Aldo Gangemi
Jie Gao
Jhonatan Garcia
Tatiana Gossen
Simon Gottschalk
Kalpa Gunaratna
Lavdim Halilaj
Tom Hanika
Tim Hill
Yingjie Hu
Steffen Hölldobler
Ignacio Iacobacci
Valentina Ivanova
Hailong Jin
Amit Joshi
Pavan Kapanipathy
Fariba Karimi
Evgeny Kharlamov
Sefki Kolozali
Olga Kovalenko

Alja Košmerljž
Markus Krötzsch
Tobias Käfer
Sarasi Lalithsena
Maxime Lefrançois
Mingyang Li
Rinaldo Lima
Michael Luggen
Andy Lücking
Zide Meng
Isabelle Mirbel
Aditya Mogadala
Alexandre Monnin
Gabriela Montoya
Kay Mueller
Raghava Mutharaju
Andriy Nikolov
Nikolay Nikolov
Andreas Nolle
Inna Novalija
Theodore Patkos
Bianca Pereira
Sujan Perera
Niklas Petersen
Johann Petrak
Patrick Philipp
Danae Pla Karidi
Valentina Presutti
Sambhawa Priya
Daniel Puschmann
Ashequl Qadir
Ella Rabinovich
Steffen Remus
Ryan Ribeiro De Azevedo
Cleyton Rodrigues

Oscar Rodríguez Rocha
Eugen Ruppert
Emilio Sanfilippo
Domenico Fabio Savo
Viachaslau Sazonau
Stefan Schlobach
Andreas Schmidt
Patrik Schneider
Joerg Schoenfisch
Hala Skaf-Molli
Panayiotis Smeros
Martin Stephenson
Annette Ten Teije
Andrea Tettamanzi
Steffen Thoma
Veronika Thost
Konstantin Todorov
Andrei Tolstikov
Pierpaolo Tommasi
Riccardo Tommasini
Despoina Trivela
Kateryna Tymoshenko
Ricardo Usbeck
Tassos Venetis
Andrew Walker
Xin Wang
Zhe Wang
Philip Webster
Yinchong Yang
Seid Muhie Yimam
Ran Yu
Fadi Zaraket
Chrysostomos Zeginis
Jiangtao Zhang
Lei Zhang

PhD Symposium Program Committee

Chairs

Simone Paolo Ponzetto	Universität Mannheim, Germany
Chiara Ghidini	FBK-IRST, Italy

Members

Sören Auer	University of Bonn and Fraunhofer IAIS, Germany
Chris Biemann	TU Darmstadt, Germany
Stefano Borgo	ISTC, National Research Council, Italy
Elena Cabrio	Inria, France
Oscar Corcho	Universidad Politécnica de Madrid, Spain
Claudia d'Amato	University of Bari, Italy
Chiara Di Francescomarino	Fondazione Bruno Kessler-IRST, Italy
Anna Lisa Gentile	University of Mannheim, Germany
Chiara Ghidini	FBK-IRST, Italy
Paul Groth	Elsevier Labs, The Netherlands
Tudor Groza	The Garvan Institute of Medical Research
Pascal Hitzler	Wright State University, USA
Hajira Jabeen	FAST-National University of Computer and Emerging Sciences
Patrick Lambrix	Linköping University, Sweden
Bernardo Magnini	FBK-IRST, Italy
Pablo Mendes	IBM Research Almaden, USA
Matteo Palmonari	University of Milano-Bicocca, Italy
Heiko Paulheim	University of Mannheim, Germany
Simone Paolo Ponzetto	University of Mannheim, Germany
Sebastian Rudolph	Technische Universität Dresden, Germany
Stefan Schlobach	Vrije Universiteit Amsterdam, The Netherlands
Guus Schreiber	VU University Amsterdam, The Netherlands
Monika Solanki	University of Oxford, UK
Heiner Stuckenschmidt	University of Mannheim, Germany
Vojtěch Svátek	University of Economics, Prague, Czech Republic
Danai Symeonidou	INRA, France
Valentina Tamma	University of Liverpool, UK
Serena Villata	CNRS, I3S Laboratory, France

Additional Reviewers

Terry Payne

Steering Committee

Chair

John Domingue   The Open University, UK and STI International, Austria

Members

Claudia d'Amato   Universià degli Studi di Bari, Italy
Philipp Cimiano   Bielefeld University, Germany
Oscar Corcho   Universidad Politécnica de Madrid, Spain
Fabien Gandon   Inria, W3C, Ecole Polytechnique de l'Université de Nice
                         Sophia Antipolis, France
Axel Polleres   Vienna University of Economics and Business, Austria
Valentina Presutti   CNR, Italy
Marta Sabou   Vienna University of Technology, Austria
Elena Simperl   University of Southampton, UK

Sponsoring Institutions

Gold Sponsors

http://www.eutravelproject.eu/

Silver Sponsors

http://byte-project.eu/

http://entropy-project.eu/

http://project-hobbit.eu/

http://www.springer.com/lncs

Abstracts of Invited Talks

Wither OWL in a Knowledge-Graphed, Linked-Data World?

Jim Hendler

Rensselaer Institute for Data Exploration and Applications
hendler@cs.rpi.edu

Abstract. The need for ontologies in the real world is manifest and increasing. On the Web, ontologies are increasingly needed — but OWL isn't being used in many of these applications. This talk explores some of the use and needs for ontologies on theWeb in data integration, emerging technologies, and linked data applications. It focuses on deficiencies in OWL's design that have hindered its application, and suggests some directions for making OWL more relevant to the modern Web, rather than the Web of the early 2000's. The talk ends with some challenges to the OWL, and greater ontology, community needed to be addressed if we are to see more use of ontologies on the Web.

Semantic Technologies in Business: Are We There Yet?

Eleni Pratsini

Smarter Cities Technology Center, IBM Research, Ireland
elenip@ie.ibm.com

Abstract. Developing intelligent solutions requires a comprehensive under-standing and management of the data. Intelligent semantic systems provide the smart technologies to harvest large amounts of data and insight in order to find solutions to the problems in various application areas. Typical challenges are: data acquisition from different types of sources; establishing links among different data types using both structure and content; dynamic, real-time processing of data; scalability for analytics and query processing, just to name a few. These challenges often prevent a business from even starting to look at the information and make sense out of it. At the same time, advances in cognitive computing offer new possibilities in analyzing unstructured information for richer insights. In this talk, we will use applications to discuss the use of semantic technologies, point out the research challenges, and highlight the business benefit from these technologies. We will conclude with a view on future research directions.

Controlling Leakage and Disclosure Risk in Semantic Big Data Pipelines

Ernesto Damiani

Università degli Studi di Milano, Italy
ernesto.damiani@unimi.it

Abstract. Ernesto Damiani is the Director of the Information Security Research Center at Khalifa University, Abu Dhabi, and the leader of the Big Data Initiative at the Etisalat British Telecom Innovation Center (EBTIC). Ernesto is on extended leave from the Department of Computer Science, Università degli Studi di Milano, Italy, where he leads the SESAR research lab and coordinates several large scale research projects funded by the European Commission, the Italian Ministry of Research and by private companies such as British Telecom, Cisco Systems, SAP, Telecom Italia and many others. Ernesto's research interests include business process analysis and privacy-preserving Big Data analytics. Ernesto is the Principal Investigator of the TOREADOR H2020 project on models and tools for Big data-as-a-service.

Contents

Mobile Web, Sensors and Semantic Streams Track

Natural Language Processing and Information Retrieval Track

Reasoning Track

Semantic Data Management, Big Data, Scalability Track

Services, APIs, Processes and Cloud Computing Track

Smart Cities, Urban and Geospatial Data Track

Trust and Privacy Track

Vocabularies, Schemas, Ontologies Track

In-Use & Industrial Track

PhD Symposium

Linked Data Track

Detecting Similar Linked Datasets Using Topic Modelling

Michael Röder[1], Axel-Cyrille Ngonga Ngomo[1(✉)], Ivan Ermilov[1], and Andreas Both[2]

[1] AKSW, Leipzig University, Leipzig, Germany
{roeder,ngonga}@informatik.uni-leipzig.de
[2] Mercateo AG, Leipzig, Germany

Abstract. The Web of data is growing continuously with respect to both the size and number of the datasets published. Porting a dataset to five-star Linked Data however requires the publisher of this dataset to link it with the already available linked datasets. Given the size and growth of the Linked Data Cloud, the current mostly manual approach used for detecting relevant datasets for linking is obsolete. We study the use of topic modelling for dataset search experimentally and present TAPIOCA, a linked dataset search engine that provides data publishers with similar existing datasets automatically. Our search engine uses a novel approach for determining the topical similarity of datasets. This approach relies on probabilistic topic modelling to determine related datasets by relying solely on the metadata of datasets. We evaluate our approach on a manually created gold standard and with a user study. Our evaluation shows that our algorithm outperforms a set of comparable baseline algorithms including standard search engines significantly by 6 % F1-score. Moreover, we show that it can be used on a large real world dataset with a comparable performance.

1 Introduction

The Web of Data and the Linked Open Data Cloud have grown considerably over the last years and are continuing to grow steadily. Following the statistics of LODStats,[1] several thousands of RDF datasets can already be found online. With the growth of the number of datasets available as well as the growth of their size comes the problem of effectively detecting not only the links between the datasets (as studied in previous works [12]) but also of determining the datasets with which a novel dataset should be linked. A naive approach to linking these datasets would choose two datasets and check, whether they can be linked with each other. Such an approach would need $\mathcal{O}(n^2)$ pairwise comparisons of datasets to find possible candidates of linking, which is clearly impracticable. Addressing the problem of finding relevant datasets for linking is however of crucial importance to facilitate the integration of novel datasets into the Linked Data Clouds as well as the discovery of relevant data sources in enterprise Linked Data [12].

---

[1] http://stats.lod2.eu/.

© Springer International Publishing Switzerland 2016
H. Sack et al. (Eds.): ESWC 2016, LNCS 9678, pp. 3–19, 2016.
DOI: 10.1007/978-3-319-34129-3_1

In this paper, we study the search for similar datasets given an input dataset. In this context, we define two datasets as being similar if they cover the same topics and should thus be linked to each other. In particular, we aim to elucidate the question whether topic modelling (in particular LDA [3]) can be used to improve the search of similar datasets. To address this research question, we present different approaches pertaining to how datasets can be modelled for dataset search. We then compare these different modelling possibilities against the state of the art. Our findings are implemented into TAPIOCA, a search engine that takes a description of a dataset and searches for topically similar datasets that could be candidates for link discovery. Our engine learns topics of datasets by analysing their ontologies and uses these topics to map datasets to domains in a fuzzy manner. Based on this representation, TAPIOCA can compare the topic vector of an input dataset to datasets in its index so as to suggest topically similar datasets, which are assumed to be good candidates for linking. Note that we do not study the link discovery problem herein and address exclusively the search for data for linking under the assumption that datasets should be linked if they describe similar topics.

Our contributions are thus as follows:

- We present six combinations of approaches for modelling data in RDF datasets that can be used for dataset search.
- We apply topic modelling to these combinations, compare them with state-of-the-art baselines and show that topic modelling does lead to significant improvements over several baseline methods.
- We provide a gold standard for dataset search and make it available for future research on the topic.

The rest of this paper is structured as follows: In Sect. 2 we present other approaches related to our work. Section 3 introduces Latent Dirichlet Allocation— a model from the probabilistic topic modelling domain. In Sect. 4, our novel approach for a dataset search engine is presented and subsequently evaluated in Sect. 5. Section 6 concludes this paper. More information on TAPIOCA, the data we used for the evaluation and a demo can be found at http://aksw.org/projects/tapioca.

2   Related Work

Link discovery is a task of central importance when publishing Linked Data [12]. While a large number of approaches have been devised for discovering links between datasets, the task at hand is a precursor of link discovery and can be regarded as a similarity computation task. The usage of document similarities that are based on topic modelling is well known and have been widely studied in previous works, e.g., in [14]. Especially for information retrieval applications, topic modelling has been used for documents containing natural language. Buntime et al. [4] developed an information retrieval system that is based on an hierarchical topic modelling algorithm to retrieve documents topically related to a given query. Lu et al. [10] analysed the effect of topic modelling for information retrieval. Their results show that while its performance is not good for a keyword search, it has a good performance for clustering and classification tasks in which only a coarse matching is needed and training data is sparse. We think that the task of retrieving similar linked datasets matches this task description.

The Semantic Web is already used for information retrieval tasks. For example, Hogan et al. [8] as well as Tummarello et al. [15] published approaches for semantic web search engines retrieving single entities and consolidated information about them given a keyword query. One of the problems that have to be solved for this task is the consolidation of retrieved entities. Since inside different datasets a single entity could have different URIs, the workflow of such a search engine has to have a consolidation step identifying URIs mentioning the same entity. In both approaches two resources are assumed to mention the same entity if (1) they are connected by an **owl:sameAs** property[2] or (2) both resources have an OWL inverse functional property with the same value. The values of such inverse functional properties are typically assumed to be unique, e.g., an e-mail address. This problem is further studied in [7]. These approaches differ from our topical search engine, since they can't be used to identify topical similar datasets for linkage, because the entities must have been already linked—directly or indirectly by inverse functional properties.

The search engine proposed in [15] has an additional consolidation step summarising properties that are assumed to describe the same fact. This summary is created by using the name of the property, i.e., the last part of its URI. Additionally, the authors wrote that they want to use the labels of the properties in a future release of their search engine. This usage of labels or names of properties to decide whether they stand for a similar fact overlaps with our approach to detect topically similar datasets based on the labels of their properties or classes.

Kunze and Auer [9] proposed a search engine for RDF datasets that is mainly based on filters that work similar to a faceted search. For ranking, the authors use a similarity function that comprises different aspects. One of these aspects is called topical aspect and is based on the vocabularies, that are used inside the different datasets. We will use this aspect as a baseline for comparison and explain it in more detail in Sect. 5.1.

Recently, Sleeman et al. [13] proposed an approach to use topic modelling with RDF data. While their work has a similar basis it differs in many ways since it aims at other use cases. Their approach generates a single document for every entity described in a dataset while our approach creates documents that describe a complete dataset. Thus, their documents are based on a different set of triples and on different textual data gathered from the dataset.

3 Latent Dirichlet Allocation

3.1 Overview

Our approach uses Latent Dirichlet Allocation (LDA) to identify the topics of RDF datasets. LDA is a generative model for the creation of natural language documents [3]. This process is based on probabilistic sampling rules [14] and the following assumptions [3]:

---

[2] **owl** is the abbreviation for http://www.w3.org/2002/07/owl.

- Every topic is defined as a distribution over words ϕ with higher probabilities for words that are essential for the topic.
- A document is a mixture of topics. Thus, it has a distribution over topics θ.

The generation of a corpus based on a given vocabulary as well as the hyper parameters α and β is defined as follows:

1. Create the set of topics T by sampling a distribution over words ϕ for every topic t using a dirichlet distribution and a prior β.
2. Create every single document d of the corpus using the following steps.
   (a) Create a distribution over topics using a dirichlet distribution and the prior α.
   (b) For every word w in the document, choose a topic that creates it by sampling a topic index z from $\theta^{(d)}$.
   (c) Sample a word from the $\phi^{(z)}$ distribution of the topic t_z.

$$\phi^{(z)} \sim Dir(\beta) \quad \theta^{(d)} \sim Dir(\alpha) \quad z \sim Discrete(\theta^{(d)}) \quad w \sim Discrete(\phi^{(z)}) \quad (1)$$

Figure 1 shows the generative model using plate notation and Eq. 1 contains the relations between the elements. It can be seen that only the word tokens w are observable. All other elements are hidden and have to be derived from the observed word tokens. Therefore, several inference algorithms have been developed that try to estimate all hidden distributions [3,6]. In our work, we use an inference algorithm that is based on Gibbs sampling [6]. Additionally, we use hyper parameter optimisation to automatically determine α and β during inference [16]. The inference algorithm generates the topics as distributions over words and the document's distribution over topics.

3.2   Number of Topics

An important parameter of LDA inference is the number of topics. If this number is too low, the topic model is not able to describe the complexity of the training data. If it is too high, one of the model's main assumption, i.e., the orthogonality of the topics, will not hold anymore. Thus, picking a good number of topics has

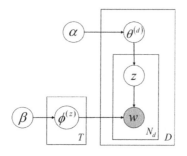

Fig. 1. LDA in plate notation [14].

a high influence on the model's performance. Unfortunately, there is no general applicable method to determine a good number of topics for a given corpus. In this section, we present two different methods that we will apply to our use case during the evaluation. Both methods suggest to determine topic models with different numbers of topics. After the generation, the single models are evaluated regarding their quality using different approaches [1,6].

The first approach is the calculation of $P(\mathbf{w}|T)$ proposed by Griffiths and Steyvers [6]. $\mathbf{w}$ is the set of all tokens present inside the corpus while T is the set of topics of the model. Thus, this probability shows how likely it is that the model could generate the corpus on which it has been trained. Since this probability is intractable, Griffiths and Steyvers presented an approximation by calculating the harmonic mean of a set of $P(\mathbf{w}|\mathbf{z}, T)$ where $\mathbf{z}$ are topic assignments that are sampled from the posterior $P(\mathbf{z}|\mathbf{w}, T)$.

The second approach has been proposed by Arun et al. [1] and is based on the observation that LDA can be regarded as a non-negative matrix factorisation. This factorisation takes the corpus Matrix M of order $|D| \times |V|$ into two matrices M_1 of order $|D| \times |T|$ and M_2 of order $|T| \times |V|$ where D is the set of documents, V is the vocabulary and T is the set of topics. The proposed measure—which we will call $\mathcal{A}$ throughout the paper—is based on the idea that the sum of assignments to the single topics have to be the same in both matrices. But since the rows of both matrices represent probability distributions and are thus normalised, these sums cannot be used directly. Hence, $\mathcal{A}$ is defined as

$$\mathcal{A}(M_1, M_2) = KL(v_1||v_2) + KL(v_2||v_1) \tag{2}$$

where KL is the Kullback-Leibler divergence, v_1 is the distribution of singular values of M_1, $v_2 = L \times M_2$ and L is a vector containing the lengths of the single documents. [1] predicts that with an increasing number of topics the values of $\mathcal{A}$ will decrease until a certain point and start to increase from that point on. They argue that the lowest point inside this dip is created by the model with the best number of topics [1].

4 Our Approach

The goal of TAPIOCA is to detect topically similar datasets with the aim of supporting the link discovery process. Ergo, given a dataset D and a set of datasets $\mathcal{U} = \{D_1, D_2, \ldots, D_n\}$, our aim is to is to rank the datasets by their likelihood of containing resources that should be linked to resources in D. The basic assumption behind our approach towards this goal is that datasets that should be linked should have similar topics. Hence, we adopt a topic-based modelling of the problem.

The TAPIOCA search engine comprises three major components:

1. An index that contains known datasets,
2. A way to formulate a query and
3. A method to calculate the similarity between a given query and the indexed datasets.

Of these three, the most challenging component is the definition of *topical similarity* between datasets. A definition of a similarity automatically results in requirements for the indexing and querying components. Therefore, we concentrate on this similarity calculation and present our new probabilistic topic-modelling-based approach. We will use the two example datasets esd-columbia-gorge and esd-south-coast to explain our approach. These examples are derived from real RDF datasets generated from open government data published by the State of Oregon. They contain contracts that have been concluded by different education service districts in 2013. The Listings 1.1 and 1.2 show two example entities of these datatsets.[3]

An RDF dataset contains two types of information that are relevant for our purposes: The first ones are the *instances* that are described inside a dataset. However, instance data is not a good starting point for finding topically similarities between two datasets, since there would have to be at least one instance both datasets have in common. This would be like comparing the two example datasets, i.e., names, titles, keywords and numbers, but without knowing that the data comprises contract data. In such a case, we could only be sure that the two datasets are similar if we were able to find instances that occur in both datasets.

```
1  @prefix cg: <http://data.oregon.gov/resource/i3bn-rwu4/> .
2
3  cg:1
4      a cg:Contract ,
5      cg:type_of_contract_subcontract    "Material"  ,
6      cg:esd_name  "Columbia Gorge Education Service District" ,
7      cg:award_title  "Technology Equipment" ,
8      cg:award_type    "Price Agreement" ,
9      cg:contractor_name  "TelCompany" ,
10     cg:original_start_amendment_date  "03-07-12" ,
11     cg:original_award_value  32456.92 ,
12     cg:total_award_value_amendments   32456.92 .
```

Listing 1.1. Example entity of the esd-columbia-gorge dataset.

```
1  @prefix sc: <http://data.oregon.gov/resource/qhct-wumz/> .
2
3  sc:1
4      a sc:Contract ,
5      sc:esd_name  "South Coast ESD" ,
6      sc:award_title  "Server" ,
7      sc:award_type    "Lease" ,
8      sc:contractor_information  "computer company" ,
9      sc:start_date_expiration_date "7/1/10-6/30/14" ,
10     sc:award_amount  5181.87 .
```

Listing 1.2. Example entity of the esd-south-coast dataset.

---

[3] The original datasets can be found at http://catalog.data.gov/dataset/contracts-esd-columbia-gorge-fiscal-year-2013-c3848   and   http://catalog.data.gov/dataset/contracts-esd-south-coast-fiscal-year-2013-3cb8d. For a better explanation of our approach, we made minor changes, e.g., we added two contract classes.

A much more promising approach is to look at the *structure* of the datasets. By doing so, we would know that both datasets contain a class and properties related to contracts. Following these assumptions, our approach is based on extracting this structural metadata from a dataset and transform it into a description of the topically content of the dataset.

Our approach is thus based on three different steps as can be seen in Fig. 2. At first, the metadata of every single dataset is extracted. In the second step, the metadata is used to create a document describing the dataset. In the last step, a topic model is created based on the documents of the datasets. The resulting topic model and distributions enable a similarity calculation between single datasets based on their topic distribution. Additionally, the topic model can be used to determine the topic distribution of documents derived from new, unseen datasets. Thus, our approach is able to handle user input containing datasets that where not known during model inference. The steps underlying TAPIOCA are explained in more detail in the following subsections.

4.1 Metadata Extraction

Our approach for finding topical similarities between datasets is based on the metadata of these datasets and the RDF and OWL semantics[4] which underlie the Linked Data Web. The metadata comprises the classes and properties used or defined inside a dataset. To every URI of a class or property a frequency count c is assigned, i.e., the number of entities of an extracted class or the number of triples of an extracted property. If a dataset contains metadata, i.e., triples with elements of the VOID Vocabulary, these information are extracted as well. After the extraction, classes and properties of the well-known vocabularies RDF, RDFS, OWL, SKOS and VOID are removed because these vocabularies do not contain any information about the topic of a dataset. Table 1 contains the URIs that would have been extracted from the two example datasets. Note, that the

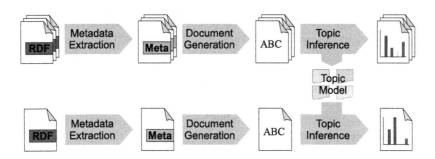

Fig. 2. The single steps of our approach. The upper part shows the index phase in which the topic model is generated while the lower part shows the handling of a query dataset.

---

[4] https://www.w3.org/RDF/ and http://www.w3.org/TR/owl-semantics/.

Table 1. Example URIs extracted from the two example datasets.

URI	Type
cg:Contract	Class
cg:type_of_contract_subcontract	Property
cg:esd_name	Property
cg:award_title	Property
cg:award_type	Property
cg:contractor_name	Property
cg:original_start_amendment_date	Property
cg:original_award_value	Property
cg:total_award_value_amendments	Property
sc:Contract	Class
sc:esd_name	Property
sc:award_title	Property
sc:award_type	Property
sc:contractor_information	Property
sc:start_date_expiration_date	Property
sc:award_amount	Property

table does not contain the rdf:type property, because it has been removed as part of the RDF vocabulary.

4.2   Document Generation

The generation of a document describing a certain dataset is based on the metadata extracted from this dataset. First, URIs and their frequency counts c are selected from the metadata. After that, the labels of the URIs are retrieved. The last step comprises the generation of the document corresponding to the dataset at hand by filtering stop words and determining the frequency of the single words.

There are three different possibilities to use the URIs contained in the metadata of a dataset, leading to three different variants V. Variant V_C uses only the class URIs of the dataset, while V_P uses its property URIs. V_{CP} uses both URI types—classes and properties. Depending on the variant, the URIs and their counts are selected for the next step.

The labels of each of the selected URIs are retrieved and tokenized. This label retrieval is based on the list of URIs that have been identified as label containing properties by Ell et al. [5]. If there are no labels available, the vocabulary part of the URI is removed and the remaining part is used as label. If this generated label is written in camel case or contains symbols like underscores, it is split into multiple words. The derived words inherit the counts c of their URI. If more than one URI created the same word, their counts are summed up.

After generating a list of words all stop words are removed[5]. After that the words are inserted into the document based on their frequency counts. Since LDA uses the bag-of-words assumption only the frequency of the words matters while their order makes no difference. However, using the extracted counts directly could result in large documents, because a dataset can contain millions of triples. Therefore, we tested two different variants to reduce the counts c to a manageable frequency f of a word inside the document. The first variant V_U inserts every word only once, therewith creating a list of unique words with $f = 1$. The second variant V_L uses the logarithm of the counts leading to $f = r(log(c) + 1)$ where r is the rounding function which results the next integer value.

Thus, the whole document generation has six different variants—the product of three different URI selections and two different word frequency definitions. Throughout this paper we will use their abbreviations – V_{CU}, V_{PU} and V_{CPU} for the variants that are using lists of unique words as well as V_{CL}, V_{PL} and V_{CPL} for the logarithm based variants.

At the end of the Document Generation every dataset is represented by a single document. With the variant V_{CPU}, the following two documents would have been created for the two example datasets.

contract type subcontract esd name award title contractor
original start amendment date value amendments

contract esd name award title type contractor information
start date expiration amount

4.3   Topic Model Inference

At this stage of our approach, there is a corpus containing a single document for every dataset. This corpus is used to generate a topic model using the LDA inference algorithm of the Mallet library [11]. The model comprises a distribution over topics for every document of the corpus ($\theta^{(d)}$) and a distribution over words for every topic of the model ($\phi^{(t)}$). The second type of distribution allows the inference of a θ distribution for a new document not contained inside the training corpus.

In our simple example, there might be three topics. While the words *subcontract, original, amendment, amendments* and *value* are marked with the first topic, the second topic could contain the words *information, expiration* and *amount*. The third topic contains the remaining words.

---

[5] The stop word list used can be found at https://github.com/AKSW/topicmodeling/ blob/master/topicmodeling.lang/src/main/resources/english.stopwords.

contract type subcontract *esd name award title contractor*
original *start* amendment *date* value amendments

contract esd name award title type contractor **information**
start date **expiration amount**

4.4   Similarity Calculation

The similarity of two datasets d_1 and d_2 is defined as the similarity of their topic distributions $\theta^{(d_1)}$ and $\theta^{(d_2)}$. Since the topic distributions can be seen as vectors, we are using the cosine similarity of these vectors [14][6].

$$sim(d_1, d_2) = \frac{\theta^{(d_1)} \cdot \theta^{(d_2)}}{\left|\theta^{(d_1)}\right| \times \left|\theta^{(d_2)}\right|} \qquad (3)$$

The **esd-columbia-gorge** document of our example would have $\theta = \{\frac{5}{14}, 0, \frac{9}{14}\}$ while the **esd-south-coast** document has $\theta = \{0, \frac{3}{12}, \frac{9}{12}\}$. Thus, the similarity of our example datasets would be 0.829.

5   Evaluation

The aim of our evaluation was threefold. In the first experiment, we focused on the evaluation of the different possible combinations of the features for topic modelling against several baselines. In our second experiment, we evaluated the two approaches for detecting the best number of topics presented in Sect. 3.2 to test whether they can be applied to dataset search. In the third experiment, we repeated the first two experiments at a larger scale to show that our approach works with larger data as well.

The dataset used for the evaluation is based on RDF datasets that have been indexed by LODStats. We removed those datasets that had no English description or not at least one class URI or one property URI of a vocabulary, that is not filtered out by our approach. The remaining evaluation dataset contained 1680 RDF datasets with 776 213 346 triples.

5.1   Baselines

We compare our approach with three baselines from the field of Information Retrieval as well as the Semantic Web. The first baseline is *tf-idf* [2] for which we extracted the metadata and generate a document for every dataset as described in Sects. 4.1 and 4.2. Let D be the set of known documents and V the vocabulary containing all known words w. Let $tf(d, w)$ be the number of times the

---

[6] Since we are comparing distributions, it would be possible to use the well-known Jensen-Shannon divergence instead of the cosine. However, during the evaluation of our approach both similarity calculations had a similar performance.

word w occurs inside the document d and let D_w be the set of documents that contain w at least once. Then, a vector can be generated for every document d by calculating a *tf-idf* value for every word w using

$$tf\text{-}idf(d, w) = tf(d, w) * idf(d, w) \quad \text{with} \quad idf(d, w) = \log \frac{|D|}{|D_w|}. \qquad (4)$$

Since *tf-idf* uses term frequencies and an instantiation of the single words is not needed, we used the pure frequencies instead of the logarithm or unique variant. After generating a vector for every document, the cosine similarity can be calculated.

The second baseline is the topical aspect (BL_T) used by Kunze and Auer [9] as part of their RDF search engine described in Sect. 2. The main idea of this topical aspect is to identify topically similar datasets based on the vocabularies that are used inside the datasets, i.e., the datasets contain URIs of the same vocabularies. Let D be the set of all known datasets and $d_1, d_2 \in D$. Let V be the union of the vocabularies used in d_1 or d_2 and let D_v be the set of all known datasets that are using the vocabulary v. Than, the BL_T is defined as

$$BL_T(d_1, d_2) = \sum_{v \in V} w(v)g(d_1, v)g(d_2, v) \qquad (5)$$

$$\text{with} \quad w(v) = -\log q(v) \quad \text{and} \quad q(v) = \frac{|D_v|}{|D|}, \qquad (6)$$

where $g(d, v)$ is a function that returns 1 if the vocabulary v is used inside the dataset d or 0 otherwise. The weighting function $w(v)$ is inspired by the *idf* term of the *tf-idf* function. Thus, the more datasets are using the vocabulary, the less important it is for the topical similarity and the lower its weighting [9].

The last baseline is using Apache Lucene[7]. The generated documents are indexed using the standard analysis of Lucene, i.e., the documents are tokenized, the tokens are transformed into their lower-cased form and Lucene's stop word filter is applied. For every dataset, its document is used to generate a weighted boolean query containing the words of the documents and their counts as weights. This query is used to retrieve similar documents from the index together with Lucene's similarity score for them.

5.2 Experiment I

For the first experiment, we randomly selected 100 RDF datasets to generate a gold standard. Two researchers independently determined topically similar datasets. For solving this task, they got the description of those datasets as well as the possibility to take a deeper look inside the data itself. The ratings of both researchers were compared and showed an inter-rater agreement of 97.58 %. Cases in which the ratings differed were discussed to compile a final rating. With

---

[7] http://lucene.apache.org/.

this approach 86 dataset pairs could be identify as topically similar.[8] Table 2 shows the features of the corpora that have been created by the different variants of our approach based on these 100 datasets (3 659 152 triples).

For all six approaches presented above, we calculated the similarities of every dataset to every other dataset using the leave-one-out method: One dataset was used as query while the topic model was trained using the other 99 datasets of the gold standard. The result of this step was a ranked and scored list of corresponding datasets for each of the datasets in our gold standard. We then searched for a similarity threshold that led to a maximal F1-score over all datasets. For every variation of our approach, we run experiments in the range of [2, 200] topics. Since the F1-score of the variant V_{AL} was still rising near 200 topics, we further increased the number of topics for this variant until 500.[9]

The best F1-scores that were achieved by the different variants and the different baselines are shown in Table 3. Based on this data, our approach clearly outperforms all baselines if the document generation is based only on properties

Table 2. Features of the corpora generated by the different variants.

Variant	Words	Tokens
V_{AL}	10 182	252 406
V_{AU}	10 182	34 264
V_{CL}	9 500	239 108
V_{CU}	9 500	32 020
V_{PL}	1 173	14 078
V_{PU}	1 173	2 501

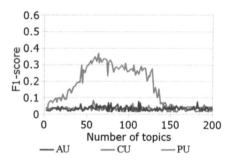

Fig. 3. The F1-scores of the three unique word based variants.

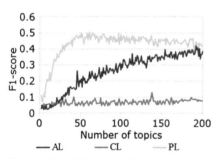

Fig. 4. The F1-scores of the three logarithm based variants.

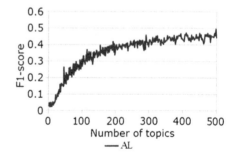

Fig. 5. The F1-scores of V_{AL} for different numbers of topics in the range [2, 500].

---

[8] The gold standard can be found at the project's web page.

[9] For all topic numbers, the inference was carried out with 1040 iterations, $\alpha = 0.1$, $\beta = 0.01$ and a hyper parameter optimisation after every 50 iterations starting after iteration 200.

Table 3. Best F1-scores achieved by the different variants and the baselines. In the most left column, there are the results for the variants V_{CL}, V_{PL} and V_{AL} while the results of V_{CU}, V_{PU} and V_{AU} are in the second most left column.

URIs used	TAPIOCA (log.)	TAPIOCA (unique)	tf-idf	BL_T	Lucene
Class	0.128	0.083	0.103	0.292	0.096
Properties	**0.505**	0.350	0.436	0.356	0.418
Both	0.495	0.078	0.444	0.333	0.241

and logarithmic counts. Moreover, our approach performs much better with logarithmic counts than with unique word frequencies. In Fig. 3, we also see that with varying numbers of topics V_{AU} and V_{CU} stay at a low level. Only V_{PU} achieves competitive F1-scores. We think that this has two causes. First, the unique-based variants do not assign a weight to the labels regarding the importance that a class or a property has inside a dataset. Secondly, it has already been shown that LDA does not perform well on short documents in which many different words appear rarely, e.g., messages of short messaging services [17].

Regarding the URIs used for the document creation, it can be seen that all approaches show a poor performance if they are only based on classes. These variants are only able to find similar datasets if the similarity is very obvious, e.g., different eagle-i[10] datasets that are using the same vocabularies. Additionally, they have the drawback, that only 88 out of the 100 datasets define or use classes which makes them unable to calculate similarities for 12 datasets.

Another observation of the experiment is that BL_T does not perform well. Thus, the assumption that topically similar datasets are using the same vocabularies does not hold in reality. One core reason might be that many of the datasets we consider have been generated automatically from tables or CSV files. Every generated dataset has an own, generated vocabulary URI like the two example datasets in Sect. 4.

The Figs. 3, 4 and 5 show the influence of the number of topics on the models performance. For V_{PL}, V_{AL} and V_{PU}, there is a range of numbers of topics in which the F1-score is maximised. Models with too few topics have a much worse performance while—especially for V_{PL} and V_{AL}—the performance deterioration caused by too many topics is rather small. Thus, we can summarise that finding a good number of topics is important for our approach. However, in case an exact number cannot be determined, a high number of topics should be preferred.

5.3 Experiment II

Based on the results of the first experiment, we evaluated whether the two approaches for determining a good number of topics presented in Sect. 3.2 are useful in the present use case. Thus, for the topic range [2,200] we generated topic models using all documents of the gold standard datasets that have been

---

[10] The gold standard contains datasets of the eagle-i project. https://www.eagle-i.net.

generated by the V_{PL} variant of our approach. For every number of topics we generated five models, calculated $P(\mathbf{w}|T)$ as well as $\mathcal{A}$ and determined the average values of these five runs.

Figure 6 shows the average logarithm of $P(\mathbf{w}|T)$ and reveals that the probability increases steadily with an increasing number of topics. Thus, this method would recommend a much higher number of topics than the 61 topics with which the V_{PL} variant performed best. The average value of $\mathcal{A}$ is shown in Fig. 7. The curve shows a dip as described by Arun et al. [1]. But the minimum value of this dip has been achieved by models with 11 topics with which V_{PL} has only an F1-score of 0.21.

From this experiment, we can summarise that none of these approaches seems to be appropriate to determine a good number of topics for our use case. Therefore, we have to fall back on a simple alternative that we will present during the third experiment.

5.4 Experiment III

To evaluate whether our approach can handle a larger number of datasets, we repeated the first two experiments but trained the model on the complete LOD-

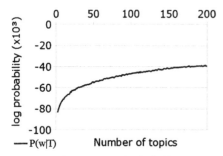

Fig. 6. Average $\log(P(\mathbf{w}|T))$ calculated on the gold standard corpus of the V_{PL} variant.

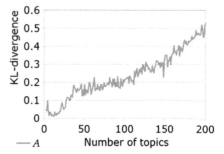

Fig. 7. Average values of $\mathcal{A}$ calculated on the gold standard corpus of the V_{PL} variant.

Table 4. Best F1-scores achieved by Tapioca and the baselines for the complete LODStats corpus.

Approach	Classes	Properties	Both
Tapioca (log.)	—	**0.538**	—
tf-idf	0.103	0.436	0.444
BL_T	0.014	0.014	0.014
Lucene	0.214	0.241	0.385

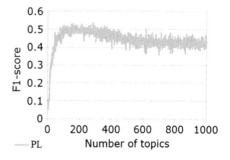

Fig. 8. The F1-scores of V_{PL} calculated on the complete LODStats corpus for different numbers of topics.

Stats dataset. In detail, this means that for every dataset of the gold standard, we removed it from the set of all 1 680 LODStats datasets. We trained the variant V_{PL} of our approach on the 1 679 remaining datasets and calculated the similarity between the removed dataset and the other 99 datasets contained in the model. After that we compared the similarities with the gold standard and searched the similarity threshold that maximised the F1-score. Using the V_{PL} document creation, the 1 680 documents of the complete LODStats dataset comprise 175 080 tokens of 5 816 different words.

Figure 8 shows the F1-score achieved by V_{PL}. The maximum F1-score of 0.538 was achieved by a model with 284 topics. The results in Table 4 show that even with a much larger input our approach is able to achieve an F1-score that is higher than the scores of the baselines and comparable to the score achieved in the first experiment.

We repeated the calculation of $P(\mathbf{w}|T)$ and $\mathcal{A}$ for the complete LODStats corpus (for the sake of space, we do not show the resulting figures since they are similar to the results of the second experiment). While the average value of $P(\mathbf{w}|T)$ increases steadily with a larger number of topics, the minimum of the average value of $\mathcal{A}$ is at 33 where the F1-score is only 0.336. But since the gold standard is part of the dataset our search engine indexes, we can use it as a pragmatic way to determine a good number of topics. This pragmatic method assumes that a good topic model that has been trained on the datasets of the gold standard and additional datasets should give a high F1-score if it is compared to the gold standard. Thus, in practice we shall train multiple models with different numbers of topics on the same large dataset that comprises the gold standard datasets and use the model that achieves the highest F1-score compared to the gold standard.

6   Conclusion

The aim of this work was to present TAPIOCA—a search engine that tackles the problem of finding topically similar linked datasets inside the LOD cloud. With this search engine we address the gap between creating an RDF dataset and linking it to other datasets. Our evaluation shows that our approach is better than several baselines and performs well on a large number of datasets. We could identify different parts of a datasets metadata and show that the properties are most important for determining the datasets topic. Additionally, we created a gold standard for this task that can be downloaded from the projects web page[11].

The most challenging future task is the search for a good number of topics that can be used to generate the topic model and that is not bound to the gold standard created by us. Besides this, another challenge is the handling of classes and properties that only have labels in foreign languages instead of English.

---

[11] http://aksw.org/Projects/tapioca.html.

Additionally, we want to increase the search engine's usability, including a tool with which a user can extract the metadata from its own dataset easily.

References

1. Arun, R., Suresh, V., Veni Madhavan, C.E., Narasimha Murthy, M.N.: On finding the natural number of topics with latent Dirichlet allocation: some observations. In: Zaki, M.J., Yu, J.X., Ravindran, B., Pudi, V. (eds.) PAKDD 2010, Part I. LNCS, vol. 6118, pp. 391–402. Springer, Heidelberg (2010)
2. Baeza Yates, R.A., Neto, B.R.: Modern Information Retrieval. Addison-Wesley Longman Publishing Co., Inc., Boston (1999)
3. Blei, D.M., Ng, A.Y., Jordan, M.I.: Latent Dirichlet allocation. J. Mach. Learn. Res. **3**, 993–1022 (2003)
4. Buntine, W., Lofstrom, J., Perkio, J., Perttu, S., Poroshin, V., Silander, T., Tirri, H., Tuominen, A., Tuulos, V.: A scalable topic-based open source search engine. In: Proceedings of the WI 2004, pp. 228–234, September 2004
5. Ell, B., Vrandečić, D., Simperl, E.: Labels in the web of data. In: Aroyo, L., Welty, C., Alani, H., Taylor, J., Bernstein, A., Kagal, L., Noy, N., Blomqvist, E. (eds.) ISWC 2011, Part I. LNCS, vol. 7031, pp. 162–176. Springer, Heidelberg (2011)
6. Griffiths, T.L., Steyvers, M.: Finding scientific topics. Proc. Nat. Acad. Sci. **101**(suppl. 1), 5228–5235 (2004)
7. Herzig, D.M., Mika, P., Blanco, R., Tran, T.: Federated entity search using on-the-fly consolidation. In: Alani, H., Kagal, L., Fokoue, A., Groth, P., Biemann, C., Parreira, J.X., Aroyo, L., Noy, N., Welty, C., Janowicz, K. (eds.) ISWC 2013, Part I. LNCS, vol. 8218, pp. 167–183. Springer, Heidelberg (2013)
8. Hogan, A., Harth, A., Umrich, J., Kinsella, S., Polleres, A., Decker, S.: Searching and browsing linked data with swse: the semantic web search engine. Web Semant. Sci. Serv. Agents World Wide Web **9**(4), 365–401 (2011)
9. Kunze, S., Auer, S.: Dataset retrieval. In: IEEE Seventh International Conference on Semantic Computing (ICSC), pp. 1–8, September 2013
10. Lu, Y., Mei, Q., Zhai, C.: Investigating task performance of probabilistic topic models: an empirical study of PLSA and LDA. Inf. Retrieval **14**(2), 178–203 (2011)
11. McCallum, A.K.: Mallet: A machine learning for language toolkit (2002). http://mallet.cs.umass.edu
12. Ngomo, A.-C.N., Auer, S., Lehmann, J., Zaveri, A.: Introduction to linked data and its lifecycle on the web. In: Koubarakis, M., Stamou, G., Stoilos, G., Horrocks, I., Kolaitis, P., Lausen, G., Weikum, G. (eds.) Reasoning Web 2014. LNCS, vol. 8714, pp. 1–99. Springer, Heidelberg (2014)
13. Sleeman, J., Finin, T., Joshi, A.: Topic modeling for rdf graphs. In: 3rd International Workshop on Linked Data for Information Extraction, 14th International Semantic Web Conference (2015)
14. Steyvers, M., Griffiths, T.: Probabilistic topic models. Handb. Latent Semant. Anal. **427**(7), 424–440 (2007)
15. Tummarello, G., Cyganiak, R., Catasta, M., Danielczyk, S., Delbru, R., Decker, S.: Sig.ma: live views on the web of data. Web Semant. Sci. Serv. Agents World Wide Web **8**(4), 355–364 (2010)
16. Wallach, H.M., Mimno, D.M., McCallum, A.: Rethinking LDA: why priors matter. In: Advances in Neural Information Processing Systems, vol. 22, pp. 1973–1981 (2009)

17. Zhao, W.X., Jiang, J., Weng, J., He, J., Lim, E.-P., Yan, H., Li, X.: Comparing twitter and traditional media using topic models. In: Clough, P., Foley, C., Gurrin, C., Jones, G.J.F., Kraaij, W., Lee, H., Mudoch, V. (eds.) ECIR 2011. LNCS, vol. 6611, pp. 338–349. Springer, Heidelberg (2011)

Heuristics for Connecting Heterogeneous Knowledge via FrameBase

Jacobo Rouces[1(✉)], Gerard de Melo[2], and Katja Hose[1]

[1] Aalborg University, Aalborg, Denmark
jrg@es.aau.dk, khose@cs.aau.dk
[2] Tsinghua University, Beijing, China
gdm@demelo.org

Abstract. With recent advances in information extraction techniques, various large-scale knowledge bases covering a broad range of knowledge have become publicly available. As no single knowledge base covers all information, many applications require access to integrated knowledge from multiple knowledge bases. Achieving this, however, is challenging due to differences in knowledge representation. To address this problem, this paper proposes to use linguistic frames as a common representation and maps heterogeneous knowledge bases to the FrameBase schema, which is formed by a large inventory of these frames. We develop several methods to create complex mappings from external knowledge bases to this schema, using text similarity measures, machine learning, and different heuristics. We test them with different widely used large-scale knowledge bases, YAGO2s, Freebase and WikiData. The resulting integrated knowledge can then be queried in a homogeneous way.

1 Introduction

In the past decades, numerous large-scale knowledge bases (KBs) have become available and are now essential both in research and in the commercial world, e.g., for IBM's Jeopardy!-winning question answering system Watson [16] and for Google's Knowledge Graph-driven search results. The Web of Linked Data has grown to the point that the numerous different KBs that have been published can no longer easily be visualized in a single cloud image.

Since numerous stakeholders are publishing separate KBs focusing on different domains and sources, a given application often needs to combine knowledge from multiple KBs. Hence, there is a clear need for methods to *integrate* such knowledge. A substantial body of work has aimed to address this problem by automatically aligning individual entries across KBs, both at the schema level [9] and at the level of entity instances [6]. These methods often produce a list of binary links using properties such as `owl:sameAs`. Unfortunately, different KBs often model the world in quite distinct ways. Despite the adoption of standards such as the use of subject-predicate-object triples in RDF [15], the same piece of information can be represented in ways such that a one-to-one alignment is no longer possible.

© Springer International Publishing Switzerland 2016
H. Sack et al. (Eds.): ESWC 2016, LNCS 9678, pp. 20–35, 2016.
DOI: 10.1007/978-3-319-34129-3_2

Consider, for instance, a marriage between two people. The YAGO KB [25] captures this using a binary predicate (`isMarriedTo`) between two persons. The Freebase KB [1], in contrast, relies on a special entity called a mediator or Compound Value Type (CVT) to describe the marriage, as well as several subject-predicate-object triples to list properties of the marriage, such as involved people, time, location, etc. In cases like this, which are not uncommon, neither `owl:sameAs`, `rdfs:subClassOf`, `owl:equivalentProperty`, nor any other individual property or binary relation can fully express the complex n-ary relationships between these resources.

In this paper, we propose to address this problem by integrating heterogeneous data into the FrameBase schema [21], which consists of a large inventory of *frames* that homogeneously represent n-ary relations. Frame structures are used in linguistics to describe the meaning of a sentence as scenarios with multiple participants and properties filling specific semantic roles. A marriage frame involves two partners, a time and a place, among other things. This is similar to Freebase's CVTs. However, in contrast to the few hundreds of CVTs in Freebase, FrameBase uses a larger number of frames ($\sim 20,000$) organized in a dense hierarchy [11].

While FrameBase offers a flexible system for representing knowledge from existing knowledge sources [21,22], there has not been any research showing how to automatically or semi-automatically integrate heterogeneous knowledge under its schema. In this paper, we develop a generic algorithm to create complex integration rules from external KBs into this schema. These rules go beyond existing alignment mechanisms designed for binary mappings between elements of different KBs. In our experiments, we show results on three particularly heterogeneous sources: Freebase [1] and WikiData [8] are KBs with an especially large schema. YAGO2s [25], in contrast, uses only a small number of properties, but relies heavily on reification to describe phenomena such as time and locations.

2 Related Work

Connecting knowledge sources is a long-standing problem. At the level of individual records in databases, this has variously been addressed as record linkage, entity resolution, and data de-duplication [7]. In KBs, this roughly corresponds to the problems of ontology alignment, data linking [26], and instance matching [6].

For KBs, there has been substantial work on ontology alignment [9] to identify matching classes from different sources, and in some cases also instances and properties across different sources [24]. A closely related task is that of canonicalizing or reconciliating knowledge from open information extraction [12,18], which focuses on aligning names of entities and predicates by clustering synonymous entries. To achieve this, the knowledge extracted from each text source has to be reconciled, sometimes using complex graph matching algorithms [18]. But as the same extraction tool is used for each text source, the resulting graphs are constructed in similar ways and therefore follow a common model. Hence, the applied techniques for reconciliation are different from the ones necessary to reconcile ontologies created by completely independent parties and tools.

Only very little work has considered scenarios in which the same type of ontological knowledge is modeled in entirely different ways. In these cases, alignment by means of binary properties such as equivalence or subsumption is no longer sufficient, because a KB may not have a direct counterpart for an element of another KB. The EDOAL (Expressive and Declarative Ontology Alignment Language) format [3] has been proposed to express complex relationships between properties. It defines a way to describe complex correspondences but it does not address how to create them. Similarly, complex correspondence patterns between ontologies – or ontologies and databases – have been described and classified in an ontology [23]. However, this approach does not provide any method to create the correspondence patterns, neither fully nor semi-automatically. The iMAP tool [4] explores a search space of possible complex relationships between the values of entries in two databases, e.g., `room-price = room-rate * (1 + tax-rate)`, but these are limited to specific types of attribute combinations. The S-Match tool [13] uses formal ontological reasoning to prove possible matches between ontology classes, involving union and intersection operators, but does not address complex matching of properties beyond this. Ritze et al. [20] use a rule-based approach to detect specific kinds of complex alignment patterns between entries in small ontologies.

Unlike previous work, the approach presented in this paper does not focus on matching individual entities but provides techniques to match knowledge that can also be expressed with complex patterns involving multiple entities.

3   Frames for Data Integration

FrameBase [21] relies on the concept of linguistic frames as provided by FrameNet [11]. Such frames represent events or situations with characteristics denoted as Frame Elements (FEs). As FrameNet's original purpose is semantic annotation of natural language, many frames have associated Lexical Units (LUs), i.e., terms that, when appearing in a text, may evoke a frame, which may be connected via FEs to some other parts of the text.

FrameBase represents the information about *"John's 7-year marriage to Mary"* by creating an entity e that is an instance of FrameNet's `Personal_relationship` frame (or a more specific one for marriages, as we describe later on). Relevant FEs such as the marriage partners and the duration are then captured by adding triples with e as subject. For instance, properties `Partner_1` and `Partner_2` connect e to entities representing John and Mary, respectively, while the property `Duration` is used for the time their marriage lasted.

FrameBase thus repurposes FrameNet frames, originally intended to represent natural language semantics, for knowledge representation with subject-predicate-object triples, using what is also called *neo-Davidsonian representation*: One first introduces an entity e that is an instance of a frame

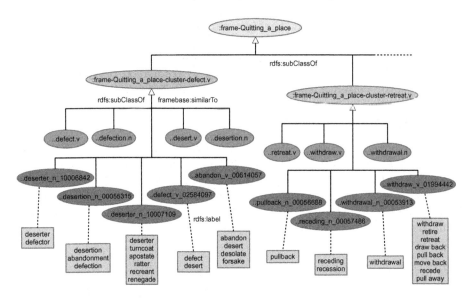

Fig. 1. Example of a hierarchy with a macroframe :`frame-Quitting_a_place`, two cluster-microframes that are direct subclasses of the macroframe, and several LU- and synset-microframes that are direct subclasses of the cluster-microframe. All the microframes under a given synset-microframe are also connected via the symmetric property `framebase:similarTo` (for clarity, the transitive closure is omitted). The synset-microframes also have labels extracted from WordNet. The microframe identifiers have a shared prefix that has been abbreviated.

class, and hence represents a particular event or situation. This entity is then connected to other entities (for example other frame instances, literals, or named entities) by means of properties representing the frame elements.

To adapt FrameNet for knowledge representation, FrameBase extends the inventory of frames defined by FrameNet in a hierarchy consisting of the following levels (Fig. 1):

– *Macroframes* are very coarse-grained and correspond to regular frames in FrameNet. The `Personal_relationship` frame class, for example, subsumes `spouse`, `marriage`, `girlfriend`, and `divorced`.
– *Microframes* inherit the general semantics and FE properties from their parent macroframes. They can be classified into 3 types:
  1. *LU-micoframes* are based on a frame's LUs and are represented as subframes in FrameNet, and therefore as subclasses in FrameBase. `Personal_relationship-married.a` and `Personal_relationship-divorced.a`, for example, are subclasses of `Personal_relationship`.
  2. *Synset-microframes* are created for synsets (sense-disambiguated synonymous words) in WordNet [10] that LUs can be mapped to. For instance, the two LU-microframes `Personal_relationship-suitor.n`

and ...`Personal_relationship-court.v` are connected to each other by means of synset-microframes.

3. *Cluster-microframes* are created to cluster sets of LU- and synset-microframes with similar meaning. `Personal_relationship`, for instance, clusters (encoded as subclasses) `Personal_relationship-married.a`, `Personal_relationship-divorced.a`, `Personal_relation-ship-suitor.n`, and `Personal_relationship-court.v`.

To enable more efficient querying without involving frame instances, FrameBase also provides Direct Binary Predicates (DBPs) that directly connect pairs of FEs. For instance, the two partners involved in a marriage are directly connected by a triple with `marriedTo` as property. The schema provides both reification and dereification (ReDer) rules to convert knowledge between the two representations (frame and DBPs). Two example ReDer rules are presented in Fig. 2.

```
?S :...isSplitIntoParts ?O
↕
?R a :frame-Separating-split.v ,
?R :fe-Separating-Whole_1 ?S ,
?R :fe-Separating-Parts_2 ?O .
```

```
?S :dbp-Motion-movesInCarrier ?O
↕
?R a :frame-Motion-move.v ,
?R :fe-Motion-Theme_1 ?S ,
?R :fe-Motion-Carrier_2 ?O .
```

Fig. 2. Two example ReDer rules. The direct binary predicate is the property in the dereified pattern, on the top. The reified pattern is at the bottom.

Overall, the FrameBase RDFS schema currently contains 19,376 frames, including 11,939 frames for specific lexical units and 6,418 frames for Word-Net's sets of synonyms. In addition to ReDer rules, the schema uses efficient RDFS+ inference (RDFS extended with a transitive, symmetrical, and recipro-cal property used to link elements of a cluster).

4   Knowledge Base Integration

We now outline our approach for integrating heterogeneous knowledge bases using the FrameBase schema. Although the techniques can be applied to a wide range of KBs, we focus in particular on YAGO2s [25], Freebase [1], and Wiki-Data [8].

Our integration algorithm produces integration rules describing how to trans-form knowledge from a KB into FrameBase. These rules do not connect individ-ual instances but are defined at the schema level and therefore resemble Global-As-View mappings in relational database systems [5]. Formally speaking, the produced integration rules can be expressed in first-order logic – with triples represented as 3-ary predicates (Fig. 3). Nevertheless, we implement these rules using SPARQL CONSTRUCT queries [14] because SPARQL is a widely sup-ported standard for KBs available in RDF format. Non-RDF KBs can also be

---

Algorithm 1. FrameBase Integration Algorithm.

Require: K ▷ input knowledge base
1: $R \leftarrow \emptyset$ ▷ set of SPARQL CONSTRUCT rules
2: **for all** classes C in K **do**   ▷ create class-frame rules
3:     **for all** frames $F \in \mathrm{mappings}_{C-F}(C)$ **do**
4:        $M \leftarrow \emptyset$ ▷ property mappings
5:        **for all** properties P such that $\exists\, s, o : \langle s\ P\ o \rangle \in K$ **and** $s \in C$ **do**
6:           **for all** $E \in \mathrm{mappings}_{PF-E}(P, F)$: E is not in R **do**
7:              $M \leftarrow M \cup (P, E)$
8:        $R \leftarrow R \cup \mathrm{ClassFrameRule}(C, F, M)$
9: **for all** properties P in K **do**   ▷ create core property-frame rules
10:     **if** the domain of P is not `rdf:Statement` **then**
11:        **for all** $(F, E_s, E_o) \in \mathrm{mappings}_{P-FEE}(P)$ **do**
12:           $R \leftarrow R \cup \mathrm{PropertyFrameRule}(P, F, E_s, E_o)$
13: **for all** properties P' in K **do**   ▷ extend property-frame rules
14:     **if** the domain(P')=`rdf:Statement` **then**
15:        **for all** properties P in K satisfying $\langle$P `^rdf:property/P'` y$\rangle$ **do**
16:           **for all** property-frame rules r in R **do**
17:              **if** r matches PropertyFrameRule(P, F, E_s, E_o) **then**
18:                 **for all** frame elements $E_{P'} \in \mathrm{mappings}_{PF-E}(P', F)$ **do**
19:                    Extend$(r, P', E_{P'})$
20: **return** R ▷ final set of integration rules

---

integrated by either using an alternative rule formalism or invoking off-the-shelf or custom-purpose RDF converters[1].

Algorithm 1 sketches our approach, which relies on three mapping functions that are discussed in Sect. 4.3 and three rule instantiation functions given in Fig. 3. The mapping functions relate entities from the source KB with entities from FrameBase into which they can be translated, but they do not provide the structure of the integration rules. The structure is specified by the instantiation functions, which take elements from the source KB and FrameBase, and return structured integration rules.

The instantiation functions are used to create two kinds of integration rules: (i) class-frame rules, which convert classes and properties from the original KB into similar elements in FrameBase (Sect. 4.1) and (ii) property-frame rules, which convert properties from the source KB into frames (Sect. 4.2).

4.1 Class-Frame Rules

The process of creating class-frame rules starts in line 2 in Algorithm 1, relying on mapping functions $\mathrm{mappings}_{C-F}$ and $\mathrm{mappings}_{PF-E}$. Class-frame rules are produced by the rule instantiation function ClassFrameRule(C, F, M) from Fig. 3. They convert a class C into a frame F that represents an event, situation

---

[1] http://www.w3.org/wiki/ConverterToRdf.

ClassFrameRule(C, F, M)
given $M = \{(P_1, E_1), ...(P_n, E_n)\}$

$\forall v_1...v_n v_{n+1}($

$\quad \exists e_1 ($

$\qquad t_f(e_1, \texttt{rdf:type}, F) \wedge$

$\qquad t_f(e_1, E_1, v_1) \wedge$

$\qquad ...$

$\qquad t_f(e_1, E_n, v_n)$

$\quad) \leftarrow ($

$\qquad t_s(v_{n+1}, \texttt{rdf:type}, C) \wedge$

$\qquad t_s(v_{n+1}, P_1, v_1) \wedge$

$\qquad ...$

$\qquad t_s(v_{n+1}, P_n, v_n)$

$\quad)$

$)$

PropertyFrameRule(P, F, E_s, E_o)

$\forall v_1 v_2 (\exists e_1 ($

$\quad t_f(e_1, \texttt{rdf:type}, F) \wedge$

$\quad t_f(e_1, E_s, v_1) \wedge$

$\quad t_f(e_1, E_o, v_2) \wedge$

$\quad) \leftarrow t_s(v_1, P, v_2)$

$)$

Extend($r, P', E_{P'}$)
given $r = $ PropertyFrameRule(P, F, E_s, E_o)

Add inside $\exists e_1 (...)$ in r

$... \wedge \forall v_3 (t_f(e_1, E_{P'}, v_3) \leftarrow \exists e_2 ($

$\quad t_s(e_2, \texttt{rdf:type}, \texttt{rdf:Statement}) \wedge$

$\quad t_s(e_2, \texttt{rdf:subject}, v_1) \wedge$

$\quad t_s(e_2, \texttt{rdf:predicate}, P) \wedge$

$\quad t_s(e_2, \texttt{rdf:object}, v_2)$

$\quad t_s(e_2, P', v_3)$

$\quad)$

$)$

Fig. 3. Instantiation functions for the integration rules used by Algorithm 1. $t_s(s, p, o)$ stands for a triple in a source KB and $t_f(s, p, o)$ for a triple in FrameBase. v_i and e_i are variables (universally and existentially quantified, respectively) over entities in the source KB.

or state of affairs, given $M = \{(P_1, E_1), \ldots (P_n, E_n)\}$ maping properties P_i for C to frame elements E_i of F. Figure 4 provides an example of a class-frame rule automatically generated for integrating Freebase.

```
CONSTRUCT {
    _:e a :frame-Win_prize-win.v      ; :fe-Win_prize-Time ?y
    ; :fe-Win_prize-Prize ?a          ; :fe-Win_prize-Competitor ?aw
    ; :fe-Win_prize-Explanation ?hf ; :fe-Win_prize-Competition ?c
    ; :fe-Win_prize-Rank ?al          ; :fe-Win_prize-Event_description ?ed .
} WHERE {
    ?m a fb:award.award_honor .
    OPTIONAL { ?m fb:award.award_honor.year ?y }
    ...honor.award ?a }                ...honor.award_winner ?aw }
    ...honor.honored_for ?hf }         ...honor.ceremony ?c }
    ...honor.achievement_level ?al }   ...honor.notes_description ?ed } }
```

Fig. 4. Class-frame rule, automatically generated rule for integrating Freebase.

```
#SOURCE_PROPERTY_NAME='depicts'
#SOURCE_PROPERTY_DESCR='depicted person, place, object or event'
CONSTRUCT {
 _:r a :frame-Communicate_categorization-depict.v .
 _:r :fe-Communicate_categorization-Speaker ?S .
 _:r :fe-Communicate_categorization-Item ?O .
} WHERE { ?S <http://www.wikidata.org/entity/P180> ?O }
```

Fig. 5. Property-frame rule, automatically generated for integrating Wikidata.

4.2   Property-Frame Rules

In general, the purpose of a property-frame rule is to translate a property in a source KB as an instance of a frame with at least two properties. These rules are built in two steps.

Creation of Core Property-Frame Rules. The process of creating core property-frame rules starts in line 9 in Algorithm 1, relying on the mapping function mappings$_{P-FEE}$. Core property-frame rules are produced by the instantiation function PropertyFrameRule(P, F, E_s, E_o) from Fig. 3. Each RDF triple in the source KB matching pattern ?x P ?y, is transformed into a frame instance of type F with two frame-element properties E_s and E_o whose values are ?x and ?y, respectively. Figure 5 provides an example of a core property-frame rule automatically generated for integrating Wikidata.

Extending Core Property-Frame Rules to Capture RDF Reification. Additional clauses may be added by Algorithm 1 in the loop starting in line 13. This process relies on the mapping function mappings$_{PF-E}$. It uses the instantiation function Extend(r, P', E) from Fig. 3, which takes a property-frame rule r = PropertyFrameRule(P, F, E_s, E_o) as argument and returns an extended version of it to capture knowledge attached to triples by means of RDF reification [21]. KBs such as YAGO use this to represent n-ary relationships, but the FrameBase model is more efficient for this purpose. Figure 6 provides an example of an extended property-frame rule generated for integrating YAGO.

4.3   Mapping Functions

The mapping functions use an automatic general technique meant to be used with big and dynamic source KBs, extended with heuristics that apply for common patterns across large source KBs or cover most small source KBs.

P-FEE Mapping Function. Given property P from the source KB, mappings$_{P-FEE}(P)$ returns 3-tuples of a frame F, and frame element properties

```
CONSTRUCT { _:event a :frame-Ride_vehicle-flight.n          core
  ; :fe-Ride_vehicle-Source ?s ; :fe-Ride_vehicle-Goal ?o core
  ; :fe-Ride_vehicle-Vehicle ?objTransp .                  extension
} WHERE { ?s yago:isConnectedTo ?o .                       core
  OPTIONAL { ?sid rdf:type rdf:Statement .                 extension
    ?sid rdf:subject ?s ; rdf:object ?o                    extension
      ; rdf:predicate yago:isConnectedTo .                 extension
    OPTIONAL { ?sid yago:byTransport ?objTransp }}}         extension
```

Fig. 6. Extended property-frame rule, generated for integrating YAGO.

E_s, E_o associated with P. Informally, it means that property P from the source KB can be substituted with a path $\char94 E_s/E_o$ in FrameBase.

General Method. For the general variant of mappings$_{P-FEE}(P, F)$, we exploit the fact that the direct binary predicates built into FrameBase, which allow us to directly connect two frame elements, are directly mappable to external properties that should evoke a frame and two frame elements. Since the direct binary predicates were created with labels that follow the prevailing conventions in other LOD KBs [21], we can use a text similarity measure to find equivalent direct binary predicates, and for those found, use the frame and FEs in the associated reification rule. For example, if a property in a source KB is named "is split in", it turns out to be similar to the direct binary property "is split into parts" from the first example in Fig. 2, which can be used to create an integration rule that translates that source KB property into the reified pattern of FrameBase's ReDer rule.

To compare direct binary predicates with external ones, the text similarity we use is cosine distance of bag-of-words vectors. We split predicate names into tokens using capitalization, use proper lemmatization (with Stanford CoreNLP 3.6.0 [17]) instead of stemming, and do not filter stop-words, since in this case certain closed-set parts of speech such as prepositions are very important. The use of this measure significantly improved the results compared to using ADW [19], arguably because the latter is not tuned for our kind of text. Besides, our method was much faster.

For each external KB property, we run the similarity function against all existing DBPs in FrameBase, and we take the best candidate if it has a score higher than a threshold of 0.8. The threshold value was chosen empirically to balance precision and recall.

Additional Heuristics. Our system admits manually crafted heuristics to be added to mappings$_{P-FEE}(P, F)$. When one of the heuristics fire, they take preference over the general method. The vast majority of datasets in the Linked Open Data cloud rely on very small hand-crafted ontologies and vocabularies. In this case, relying on the heuristics is particularly useful, because they can cover most of the elements of the source KB. In particular, we do this for YAGO2s, which is not a small ontology per se (it has a rather big class hierarchy and millions of instances) but uses just 77 different non-metadata properties. The heuristics can be expressed using an RDF ontology that is loaded by the system at startup.

PF-E Mapping Function. Given a property P from the source KB and a frame F associated with P, mappings$_{PF-E}(P, F)$ returns frame element properties E with domain F, and associated with P. Informally, this means that property P from the source KB can be substituted with property E in Frame-Base.

General Method. The implementation of mappings$_{PF-E}(P, F)$ computes the text similarity between the name of P concatenated with the name of its range, and the names of the FEs whose domain is F, using the ADW similarity measure [19]. It chooses the candidate with the maximum score for each FE. Note that our algorithm only considers these mappings in restricted settings, e.g. when a frame F has already been chosen. This greatly reduces the set of candidates in practice and enables this approach to deliver good results.

Additional Heuristics. To the general method, we add a heuristic that increases similarity to 1 if the following condition is met: endsWith(P, X) $\wedge$ endsWith(FE, Y). The possible values required for X and Y can also be loaded from the heuristic ontology. For Freebase, we use the following two pairs: $(X, Y) \in \{(\text{from}, \text{time}), (\text{place}, \text{place})\}$. For YAGO, 4 pairs are required: $(X, Y) \in \{(\text{happenedIn}, \text{place}), (\text{happenedOnDate}, \text{time}), (\text{endedOnDate}, \text{time}), (\text{startedOnDate}, \text{time})\}$.

C-F Mapping Function. Given a class C from the source KB, mappings$_{C-F}(C)$ returns frames F associated with C. Informally, this means that class C from the source KB can be substituted with class F in FrameBase.

General Method. We let $F(C)$ denote a candidate set of relevant frames F. In order to filter out noisy and incomplete parts of the source KB, mappings$_{C-F}(C)$ returns $\emptyset$ for classes from the origin KB that do not have at least 10 instances and at least 3 outgoing properties with text annotations. Otherwise, $F(C)$ is defined to include all LU-microframes with non-zero lexical overlap (some word in common in the text labels) between C's name and the set of text labels for the synonymous frames from the cluster that F belongs to. Clusters of synonymous frames are formed by LU-microframes that are deemed equivalent via links through synset-microframes. To disambiguate and choose the best frame F among all candidates $F(C)$, we train (and later test, c.f. Sect. 5.1) logit and SVM classifiers over (C, F) pairs of this form, taken from a gold standard. The (C, F) pairs are considered true when there is a class-frame rule in the gold standard with C in the antecedent and F in the consequent, and false otherwise. Then, for each source KB item, we choose the frame whose pair has the highest score. Although this entails an implicit assumption of functionality, in practice, this results in very significant gains in precision. As input to the model, we use the following four features:

1. The lexical overlap between (i) C's name and (ii) the lexical labels of the cluster of synonymous LU-microframes for F.
2. The lexical overlap between (i) the syntactic head of C's name determined iteratively using the Collins Algorithm [2] and (ii) the lexical labels of the cluster of synonymous LU-microframes for f.
3. The lexical overlap between the descriptions (the longer text labels sometimes identified as comments).
4. If C is a class, the lexical overlap between the union of labels and descriptions of the outgoing properties, upweighting the labels by a factor of 10. When available, the labels and descriptions of the ranges are added too.

For all features, we lemmatize and filter out stop words (closed word classes) and use TF-IDF to compute the feature values (although the second feature is boolean in practice).

In Sect. 5.1, we test this method using a gold standard manually created for Freebase [1], which is a typical case of a large, open-ended schema, where a fully automatic approach becomes more necessary.

Additional Heuristics. A high-accuracy heuristic can be applied for those source KBs that are linked to WordNet, leveraging that FrameBase includes a significant part of WordNet synset as synset-microframes, which are linked to FrameNet-based LU-microframes.

The heuristic works as follows. If a given source KB class C is associated with a WordNet synset, the synset-microframe based on this synset is looked up in FrameBase. If found, this is the match, and if it is not found, a class C' is selected that is the next most specific WordNet-based parent of C. That is, $C' \supset C \wedge (C'' \supset C \rightarrow C'' = C')$. Now a synset-microframe is searched for C'. If it is not found, the process is repeated until a match is found, or a maximum number of steps is reached (e.g., 6), in order to avoid overly general rules. With this method, a sound rule can still be created, even if it loses some specificity, and it accounts for the fact that not all synsets are mapped in FrameBase.

This heuristic is particularly relevant for YAGO2s, whose upper class hierarchy is based on WordNet nouns, which makes the mapping obvious. However, it also applies to any other KB for which a mapping to WordNet exists, even if this is an external or a-posteriori one. Since WordNet is a very commonly used linguistic resource, this is reasonably common in LOD KBs.

5   Evaluation

5.1   Integration Rules Created

In this section we present examples of creating integration rules with Algorithm 1 for the test cases of YAGO2s, Freebase (2014-09-21 version), and WikiData (2015-09-28 version).

Table 1. Evaluation of mapping external classes to FrameBase classes.

	Baseline-1	Baseline-2	Our method	
			Logit	SVM
Recall	0.21	0.60	0.50	0.77
Precision	0.12	0.15	0.88	0.77
F1	0.15	0.24	0.63	0.77

Table 2. Evaluation of mapping external properties to FrameBase properties.

Metric	Score
Precision	0.81
Recall	0.30
Accuracy	0.36

Creation of Class-Frame Integration Rules

Freebase. To evaluate the results of this method on an arbitrary KB, we produced a manual gold standard consisting of 31 classes and 141 external properties from Freebase [1], paired with their candidate frames or FE properties, respectively. The gold standard is available at http://framebase.org/data. Using two independent annotators, we obtained a Cohen's kappa (inter-annotator agreement) $k = 0.69$ for class to macroframe mappings, and $k = 0.38$ for property to frame element mappings. The second is lower because it accumulates the errors from the first, which illustrates how difficult it is to create a gold standard for structured knowledge integration. The classes were randomly chosen from Freebase, disregarding classes whose candidate set did not include a valid match in FrameBase. Freebase was chosen for testing this method because it features Compound Value Types (CVTs), which have a similar role to frames, but we are also able to map some non-CVT classes. Out of a total of 155 outgoing properties for the randomly chosen Freebase classes, 141 could successfully manually be matched to frame elements in the gold standard.

Table 1 shows the results for automatic class mappings, averaging over 10 random training/test partitions of ratio 2:1. We compare three different methods.

- Baseline-1 takes the frame class with maximum lexical overlap in names (as in feature 1 of our method) and for which the candidate set $F(x)$ consists of all FrameBase classes (which is a sort of metric that can be configured with the Link Specification language in Silk [26], a state-of-the-art ontology alignment system). However ontology alignment systems alone cannot produce complex mappings.
- Baseline-2 uses the same measure as above, but applying the candidate set $F(C)$ chosen in our method, described in Sect. 4.3.
- Our method described in Sect. 4.3, using a logistic regression (logit) classifier and the functional assumption in conjunction with a fixed acceptance threshold of $p > 0.5$, where p is the probability obtained from the logit.
- Our method described in Sect. 4.3, using a support vector machine (SVM) with radial kernel, selecting, for each source KB class, the candidate frame whose score is highest, given by the distance to the frontier (functional assumption).

Table 2 provides the results obtained by our method for properties, averaging over 10 random training/test partitions of the ground truth data, each of ratio 2:1. Precision and recall are calculated with respect to the gold standard. We obtain higher precision with the logit method because we use the output probabilities to apply a condition that filters out false positives at the cost of a lower recall. Both classifier-based methods outperform the baseline.

Note that in general, word sense disambiguation is considered a hard and yet unsolved problem in natural language processing. This is particular relevant when matching properties that come with little or no metadata. For example, the Freebase classes *education.academic_post* and *base.banned.exiled* must be mapped to the *Employing* and *Residence-reside.v* frames, respectively, for which there is no obvious lexical connection. The same applies when mapping, for example, Freebase properties *education.academic_post.institution* and *geography.river.length* to frame elements *Employing-Employer* and *Natural_features-Descriptor*, respectively. A complete high-precision integration of Freebase into another knowledge base thus requires a larger community effort with additional manual revisions. Our system can be used to automatically propose suggestions to speed up this process.

YAGO. 450 class-frame integration rules were automatically created for YAGO2s. The results are given in Table 3. It shows how the number of matches decreases as n increases and the WordNet-based heuristic for mappings$_{C-F}(C)$ moves up the WordNet hierarchy. For $n > 6$ the results are negligible. The ratio of correctly matched entities is 0.789, which is equivalent to the precision of the WordNet-FrameNet mapping used for creating the schema [21] – via clustering of near-equivalent microframes, which uses other links in FrameNet and WordNet that are annotated by experts and therefore expected to be nearly error-free. Figure 7 provides an example of a class-frame integration rule created for YAGO2s.

```
CONSTRUCT { ?s a :frame-Change_of_leadership-revolt.n ;
            :fe-Change_of_leadership-Place ?o .
} WHERE {   ?s a/rdfs:subClassOf yago:wordnet_rebellion_100962129 .
            OPTIONAL { ?s yago:happenedIn ?o }  }
```

Fig. 7. Class-frame rule, automatically generated for integrating YAGO2s.

Property-Frame Integration Rules
State-of-the-art ontology alignment systems cannot produce something comparable to property-frame integration rules because the binary links produced by these systems (equality, subsumption, etc.) cannot reflect the complex 4-ary nature of property-frame integration rules.

Table 3. Number of created class-frame rules for YAGO2s. Matches(n) denotes the number of matches obtained for n being the maximum number of generalization steps. For each column, the left side shows the number of created rules and the right side the number of triples in YAGO2s matching these rules. `endedOnDate` has no significant occurrence in YAGO2s and was therefore omitted.

	happenedIn		happenedOnDate		startedOnDate	
	Rules	Triples	Rules	Triples	Rules	Triples
Matches(0)	38	11,149	86	16,836	4	13
Matches(1)	25	944	83	3,579	5	5
Matches(2)	24	469	58	14,329	1	1
Matches(3)	15	1,232	39	2,315	1	2
Matches(4)	9	540	30	986	0	0
Matches(5)	5	42	14	121	0	0
Matches(6)	2	2	11	39	0	0
All matches	118	14,378	321	38,205	11	21
No match	42	633	148	13,195	0	0
Total	160	15,011	469	51,400	11	21
% Match	73 %	95 %	68 %	74 %	100 %	100 %

WikiData. We use the general method to automatically extract property-frame rules from WikiData. We evaluate it on YAGO2s, as we can re-use the manually created FrameBase mappings for YAGO2s (described below) as a ground truth. Evaluating the results directly we obtain a precision of 0.80, and using the YAGO ground truth we obtain a recall of 0.21. Figure 5 shows an example of a rule extracted from WikiData.

YAGO. Using the RDF ontology with manually specified heuristics mentioned in Sect. 4.3, 62 out of the 77 non-metadata properties in YAGO2s (i.e., 81 %) could be perfectly integrated into FrameBase using simple property-frame rules.

6   Conclusion

In this paper, we have shown that knowledge base heterogeneity is a problem that goes beyond just the use of different identifiers that need to be aligned. We provide a general analysis of declarative constructs – integration rules – that can also achieve kinds of mappings other than basic entity alignments. We further show that FrameBase is able to incorporate multiple broad-coverage knowledge sources, despite their structural heterogeneity, opening up the possibility for it to serve as a hub for semantic integration of other KBs.

We also provide practical methods to produce these rules, combining general methods with heuristics. The quality of the output is certainly not perfect, but while traditional ontology alignment is already a difficult task, complex mappings have combinatorially more possible candidates and are thus much harder. Our results constitute a first step towards a more comprehensive linking of knowledge.

The total size of the instance data obtained from these source KBs is 40,411,393 statements, which renders it the largest collection of facts linked to FrameNet.

All FrameBase data (schema, ReDer rules, integration rules, instance data, and gold standards) is published under a CC–BY 4.0 International license at http://framebase.org.

Acknowledgments. The research leading to these results has received funding from the European Union Seventh Framework Programme under grant agreement No. FP7-SEC-2012-312651. Additional funding was received from National Basic Research Program of China Grants 2011CBA00300, 2011CBA00301, NSFC Grants 61033001, 61361136003, 61550110504, as well as from the Danish Council for Independent Research (DFF) under grant agreement No. DFF-4093-00301.

References

1. Bollacker, K., Evans, C., Paritosh, P., Sturge, T., Taylor, J.: Freebase: a collaboratively created graph database for structuring human knowledge. In: SIGMOD 2008, pp. 1247–1250 (2008)
2. Collins, M.: Head-driven statistical models for natural language parsing. Comput. Linguist. **29**(4), 589–637 (2003)
3. David, J., Euzenat, J., Scharffe, F., dos Santos, C.T.: The alignment API 4.0. Semant. Web J. **2**(1), 3–10 (2011)
4. Dhamankar, R., Lee, Y., Doan, A.H., Halevy, A., Domingos, P.: iMAP: discovering complex semantic matches between database schemas. SIGMOD **2004**, 383–394 (2004)
5. Doan, A., Halevy, A.Y., Ives, Z.G.: Principles of Data Integration. Morgan Kaufmann, Burlington (2012)
6. Dragisic, Z., et al.: Results of the ontology alignment evaluation initiative 2014. In: OM 2014, pp. 61–104 (2014)
7. Elmagarmid, A.K., Ipeirotis, P.G., Verykios, V.S.: Duplicate record detection: a survey. IEEE TKDE **19**(1), 1–16 (2007)
8. Erxleben, F., Günther, M., Krötzsch, M., Mendez, J., Vrandečić, D.: Introducing Wikidata to the linked data web. In: Mika, P., et al. (eds.) ISWC 2014, Part I. LNCS, vol. 8796, pp. 50–65. Springer, Heidelberg (2014)
9. Euzenat, J., Shvaiko, P.: Ontology Matching. Springer, Berlin (2007)
10. Fellbaum, C.: WordNet: An Electronic Lexical Database. MIT Press, Cambridge (1998)
11. Fillmore, C.J., Johnson, C.R., Petruck, M.R.L.: Background to framenet. Int. J. Lexicogr. **16**(3), 235–250 (2003)
12. Galárraga, L., Heitz, G., Murphy, K., Suchanek, F.M.: Canonicalizing open knowledge bases. In: CIKM 2014, pp. 1679–1688 (2014)
13. Giunchiglia, F., Shvaiko, P., Yatskevich, M.: S-match: an algorithm and an implementation of semantic matching. In: Bussler, C.J., Davies, J., Fensel, D., Studer, R. (eds.) ESWS 2004. LNCS, vol. 3053, pp. 61–75. Springer, Heidelberg (2004)

14. Harris, S., Seaborne, A.: SPARQL 1.1 Query Language. W3C Recommendation, W3C Consortium, March 2013
15. Hayes, P.: RDF Semantics. Technical report, W3C Consortium (2004). http://www.w3.org/TR/2004/REC-rdf-mt-20040210/
16. Kalyanpur, A., et al.: Structured data, inference in DeepQA. IBM J. Res. Dev. **56**(3.4), 10:1–10:14 (2012)
17. Klein, D., Manning, C.D.: Accurate unlexicalized parsing. In: ACL 2003, pp. 423–430 (2003)
18. Mongiovì, M., Recupero, D.R., Gangemi, A., Presutti, V., Nuzzolese, A.G., Consoli, S.: Semantic reconciliation of knowledge extracted from text through a novel machine reader. In: K-CAP 2015, pp. 25:1–25:4 (2015)
19. Pilehvar, M.T., Jurgens, D., Navigli, R.: Align, disam-biguate and walk: a unified approach for measuring semantic similarity. In: ACL 2013, pp. 1341–1351 (2013)
20. Ritze, D., Meilicke, C., Svb-Zamazal, O., Stuckenschmidt, H.: A pattern-based ontology matching approach for detecting complex correspondences. In: OM 2010 (2008)
21. Rouces, J., de Melo, G., Hose, K.: FrameBase: representing n-ary relations using semantic frames. In: Gandon, F., Sabou, M., Sack, H., d'Amato, C., Cudré-Mauroux, P., Zimmermann, A. (eds.) ESWC 2015. LNCS, vol. 9088, pp. 505–521. Springer, Heidelberg (2015)
22. Rouces, J., de Melo, G., Hose, K.: Representing specialized events with FrameBase. In: DeRiVE 2015 (2015)
23. Scharffe, F., Fensel, D.: Correspondence patterns for ontology alignment. In: Gangemi, A., Euzenat, J. (eds.) EKAW 2008. LNCS (LNAI), vol. 5268, pp. 83–92. Springer, Heidelberg (2008)
24. Suchanek, F.M., Abiteboul, S., Senellart, P.: PARIS: probabilistic alignment of relations, instances, and schema. PVLDB **5**(3), 157–168 (2011)
25. Suchanek, F.M., Hoffart, J., Kuzey, E., Lewis-Kelham, E.: YAGO2s: modular high-quality information extraction with an application to flight planning. In: BTW, pp. 515–518 (2013)
26. Volz, J., Bizer, C., Gaedke, M., Kobilarov, G.: Discovering and maintaining links on the web of data. In: Bernstein, A., Karger, D.R., Heath, T., Feigenbaum, L., Maynard, D., Motta, E., Thirunarayan, K. (eds.) ISWC 2009. LNCS, vol. 5823, pp. 650–665. Springer, Heidelberg (2009)

Dataset Recommendation for Data Linking: An Intensional Approach

Mohamed Ben Ellefi[1]([✉]), Zohra Bellahsene[1], Stefan Dietze[2], and Konstantin Todorov[1]

[1] LIRMM, University of Montpellier, Montpellier, France
{benellefi,bella,todorov}@lirmm.fr
[2] L3S Research Center, Leibniz University Hannover, Hannover, Germany
dietze@l3s.de

Abstract. With the growing quantity and diversity of publicly available web datasets, most notably Linked Open Data, recommending datasets, which meet specific criteria, has become an increasingly important, yet challenging problem. This task is of particular interest when addressing issues such as entity retrieval, semantic search and data linking. Here, we focus on that last issue. We introduce a dataset recommendation approach to identify linking candidates based on the presence of schema overlap between datasets. While an understanding of the nature of the content of specific datasets is a crucial prerequisite, we adopt the notion of dataset profiles, where a dataset is characterized through a set of schema concept labels that best describe it and can be potentially enriched by retrieving their textual descriptions. We identify schema overlap by the help of a semantico-frequential concept similarity measure and a ranking criterium based on the *tf*idf* cosine similarity. The experiments, conducted over all available linked datasets on the Linked Open Data cloud, show that our method achieves an average precision of up to 53 % for a recall of 100 %. As an additional contribution, our method returns the mappings between the schema concepts across datasets – a particularly useful input for the data linking step.

1 Introduction

With the emergence of the Web of Data, in particular Linked Open Data (LOD) [1], an abundance of data has become available on the web. Dataset recommendation is becoming an increasingly important task to support challenges such as entity interlinking [2], entity retrieval or semantic search [3]. Particularly with respect to interlinking, the current topology of the LOD cloud underlines the need for practical and efficient means to recommend suitable datasets: currently, only very few, well established knowledge graphs show a high amount of inlinks, with DBpedia being the most obvious target [4], while a large amount of datasets is largely ignored.

This is due in part to the challenge to identify suitable linking candidates without prior knowledge of the available datasets and their characteristis. Linked

© Springer International Publishing Switzerland 2016
H. Sack et al. (Eds.): ESWC 2016, LNCS 9678, pp. 36–51, 2016.
DOI: 10.1007/978-3-319-34129-3_3

datasets vary significantly with respect to represented resource types, current-ness, coverage of topics and domains, size, used languages, coherence, acces-sibility [5] or general quality aspects [6]. This heterogeneity poses significant challenges for data consumers when attempting to find useful datasets. Hence, a long tail of datasets from the LOD cloud[1] has hardly been reused and adopted, while the majority of data consumption, linking and reuse focuses on established knowledge graphs such as DBpedia [7] or YAGO [8].

In line with [9], we define dataset recommendation as the problem of com-puting a rank score for each of a set of datasets D_T (for Target Dataset) so that the rank score indicates the relatedness of D_T to a given dataset, D_S (for Source Dataset). The rank scores provide information of the likelihood of a D_T dataset to contain linking candidates for D_S.

We adopt the notion of a dataset profile, defined as a set of concept labels that describe the dataset. By retrieving the textual descriptions of each of these labels, we can map the label profiles to larger text documents. This representa-tion provides richer contextual and semantic information and allows to compute efficiently and inexpensively similarities between profiles.

Although different types of links can be defined across datasets, here we focus on the identity relation given by the statement "owl:sameAs". Our working hypothesis is simple: datasets that share at least one concept, i.e., at least one pair of semantically similar concept labels, are likely to contain at least one potential pair of instances to be linked by a "owl:sameAs" statement. We base our recommendation procedure on this hypothesis and propose an approach in two steps: (1) for every D_S, we identify a cluster[2] of datasets that share schema concepts with D_S and (2) we rank the datasets in each cluster with respect to their relevance to D_S.

In step (1), we identify concept labels that are semantically similar by using a similarity measure based on the frequency of term co-occurence in a large corpus (the web) combined with a semantic distance based on WordNet without relying on string matching techniques [10]. For example, this allows to recommend to a dataset annotated by "school" one annotated by "college". In this way, we form clusters of "comparable datasets" for each source dataset. The intuition is that for a given source dataset, any of the datasets in its cluster is a potential target dataset for interlinking.

Step (2) focuses on ranking the datasets in a D_S-cluster with respect to their importance to D_S. This allows to evaluate the results in a more meaningful way and of course to provide quality results to the user. The ranking criterium should not be based on the amount of schema overlap, because potential to-link instances can be found in datasets sharing 1 class or sharing 100. Therefore, we need a similarity measure on the *profiles* of the comparable datasets. We have proceeded by building a vector model for the document representations of the profiles and computing cosine similarities.

---

[1] http://datahub.io/group/lodcloud.

[2] We note that we use the term "cluster" in its general meaning, referring to a set of datasets grouped together by their similarity and not in a machine learning sense.

To evaluate the approach, we have used the current topology of the LOD as evaluation data (ED). As mentioned in the beginning, the LOD link graph is far from being complete, which complicates the interpretation of the obtained results—many false positives are in fact missing positives (missing links) from the evaluation data—a problem that we discuss in detail in the sequel. Note that as a result of the recommendation process, the user is not only given candidate datasets for linking, but also pairs of classes where to look for identical instances. This is an important advantage allowing to run more easily linking systems like SILK [11] in order to verify the quality of the recommendation and perform the acutal linking. Our experimental tests with SILK confirm the hypothesis on the incompleteness of the ED.

To sum up, the paper contains the following contributions: (1) new definitions of dataset profiles based on schema concepts, (2) a recommendation framework allowing to identify the datasets sharing schema with a given source dataset, (3) an efficient ranking criterium for these datasets, (4) an output of additional metadata such as pairs of similar concepts across source and target datasets, (5) a large range of reproducible experiments and in depth analysis with all of our results made available.

We proceed to present the theoretical grounds of our technique in Sect. 2. Section 3 defines the evaluation framework that has been established and reports on our experimental results. Related approaches are presented and discussed in Sect. 4 before we conclude in Sect. 5.

2   A Dataset Interlinking Recommendation Framework

Our recommendation approach relies on the notion of a dataset profile, providing comparable representations of the datasets by the help of characteristic features. In this section, we first introduce the definitions of a dataset profile that we are using in this study. Afterwards, we describe the profile-based recommendation technique that we apply.

2.1   Intensional Dataset Profiles

A dataset profile is seen as a set of dataset characteristics that allow to describe in the best possible way a dataset and that separate it maximally from other datasets. A feature-based representation of this kind allows to compute distances or measure similarities between datasets (or for that matter profiles), which unlocks the dataset recommendation procedure. These descriptive characteristics, or features, can be of various kinds (statistical, semantic, extensional, etc.). As we observe in [12], a dataset profile can be defined based on a set of types (schema concepts) names that represent the topic of the data and the covered domain. In line with that definition, we are interested here in intensional dataset characteristics in the form of a set of keywords together with their definitions that best describe a dataset.

Definition 1 (Dataset Label Profile). *The* **label profile** *of a dataset* D, *denoted by* $\mathcal{P}_l(D)$, *is defined as the set of* n *schema concept labels corresponding to* D: $\mathcal{P}_l(D) = \{L_i\}_{i=1}^n$.

Note that the representativity of the labels in $\mathcal{P}_l(D)$ with respect to D can be improved by filtering out certain types. We rely on two main heuristics: (1) remove too popular types (such as foaf : Person), (2) remove types with too few instances in a dataset. These two heuristics are based on the intuition that the probability of finding identical instances of very popular or underpopulated classes is low. We support (1) experimentally in Sect. 3 while we leave (2) for future work.

Each of the concept labels in $\mathcal{P}_l(D)$ can be mapped to a text document consisting of the label itself and a textual description of this label. This textual description can be the definition of the concept in its ontology, or any other external textual description of the terms composing the concept label. We define a document profile of a dataset in the following way.

Definition 2 (Dataset Document Profile). *The* **document profile** *of a dataset* D, $\mathcal{P}_d(D)$, *is defined as a text document constructed by the concatenation of the labels in* $\mathcal{P}_l(D)$ *and the textual descriptions of the labels in* $\mathcal{P}_l(D)$.

Note that there is no substantial difference between the two definitions given above. The document profile is an extended label profile, where more terms, coming from the label descriptions, are included. This allows to project the profile similarity problem onto a vector space by indexing the documents and using a term weighting scheme of some kind (e.g., *tf*idf*).

By the help of these two definitions, a profile can be constructed for any given dataset in a simple and inexpensive way, independent on its connectivity properties on the LOD. In other words, a profile can be easily computed for datasets that are already published and linked, just as for datasets that are to be published and linked, allowing to use the same representation for both kinds of datasets and thus allowing for their comparison by the help of feature-based similarity measures.

As stated in the introduction, we rely on the simple intuition that datasets with similar intension have extensional overlap. Therefore, it suffices to identify at least one pair of semantically similar types in the schema of two datasets in order to select these datasets as potential linking candidates. We are interested in the semantic similarity of concept labels in the dataset label profiles. There are many off-the-shelf similarity measures that can be applied, known from the ontology matching literature. We have focused on the well known semantic measures Wu and Palmer [13] and Lin's [14], as well as the UMBC [10] measure that combines semantic distance in WordNet with frequency of occurrence and co-occurrence of terms in a large external corpus (the web). We provide the definition of that measure, since it is less well-known and it showed to perform best in our experiments. For two labels, x and y, we have

$$sim_{\text{UMBC}}(x,y) = sim_{\text{LSA}}(x,y) + 0.5e^{-\alpha D(x,y)}, \tag{1}$$

where $sim_{\text{LSA}}(x, y)$ is the Latent Semantic Analysis (LSA) [15] word similarity, which relies on the words co-occurrence in the same contexts computed in a three billion words corpus[3] of good quality English. $D(x, y)$ is the minimal WordNet [16] path length between x and y. According to [10], using $e^{-\alpha D(x,y)}$ to transform simple shortest path length has shown to be very efficient when the parameter α is set to 0.25.

With a concept label similarity measure at hand, we introduce the notion of dataset comparability, based on the existence of shared intension.

Definition 3 (Comparable Datasets). *Two datasets D' and D'' are comparable if there exists L_i and L_j such that $L_i \in \mathcal{P}_l(D')$, $L_j \in \mathcal{P}_l(D'')$ and $sim_{UMBC}(L_i, L_j) \geq \theta$, where $\theta \in [0, 1]$.*

2.2   Recommendation Process: The CCD-CosineRank Approach

A dataset recommendation procedure for the linking task returns, for a given source dataset, a set of target datasets ordered by their likelihood to contain instances identical to those in the source dataset.

Let D_S be a source dataset. We introduce the notion of a *cluster of comparable datasets* related to D_S, or $CCD(D_S)$ for short, defined as the set of target datasets, denoted by D_T, that are comparable to D_S according to Definition 3. Thus, D_S is identified by its CCD and all the linking candidates D_T for this dataset are found in its cluster, following our working hypothesis.

Finally, we need a ranking function that assigns scores to the datasets in $CCD(D_S)$ with respect to D_S expressing the likelihood of a dataset in $CCD(D_S)$ to contain identical instances with those of D_S. To this end, we need a similarity measure on the dataset profiles.

We have worked with the document profiles of the datasets (Definition 2). Since datasets are represented as text documents, we can easily build a vector model by indexing the documents in the corpus formed by all datasets of interest – the ones contained in one single CCD. We use a *tf*idf* weighting scheme, which allows to compute the cosine similarity between the document vectors and thus assign a ranking score to the datasets in a CCD with respect to a given dataset from the same CCD. Note that this approach allows to consider the information of the intensional overlap between datasets prior to ranking and indexing – we are certain to work only with potential linking candidates when we rank, which improves the quality of the ranks. For a given dataset D_S, the procedure returns datasets from $CCD(D_S)$, ordered by their cosine similarity to D_S.

Finally, an important outcome of the recommendation procedure is the fact that, along with an ordered list of linking candidates, the user is provided the pairs of types of two datasets—a source and a target—where to look for identical instances. This information facilitates considerably the linking process, to be performed by an instance matching tool, such as SILK.

---

[3] http://ebiquity.umbc.edu/resource/html/id/351.

2.3   Application of the Approach: An Example

We illustrate our approach by an example. We consider *education-data-gov-uk*[4] as a source dataset (D_S). The first step consists in retrieving the schema concepts from this dataset and constructing a clean label profile (we filter out noisy labels, as discussed above), as well as its corresponding document profile (Definitions 1 and 2, respectively). We have $\mathcal{P}_l(education\text{-}data\text{-}gov\text{-}uk)$ = {London Borough Ward, School, Local Learning Skills Council, Adress}. We perform a semantic comparison between the labels in $\mathcal{P}_l(education\text{-}data\text{-}gov\text{-}uk)$ and all labels in the profiles of the accessible LOD datasets. By fixing θ = 0.7, we generate $CCD(education\text{-}data\text{-}gov\text{-}uk)$ containing the set of comparable datasets D_T, as described in Definition 3. The second step consists of ranking the D_T datasets in $CCD(education\text{-}data\text{-}gov\text{-}uk)$ by computing the cosine similarity between their document profiles and $\mathcal{P}_d(education\text{-}data\text{-}gov\text{-}uk)$. The top 5 ranked candidate datasets to be linked with *education-data-gov-uk* are (1) *rkb-explorer-courseware*[5], (2) *rkb-explorer-courseware*[6], (3) *rkb-explorer-southampton*[7], (4) *rkb-explorer-darmstadt*[8], and (5) *oxpoints*[9].

Finally, for each of these datasets, we retrieve the pairs of shared (similar) schema concepts extracted in the comparison part:

- *education-data-gov-uk* and *statistics-data-gov-uk* share two labels "London Borough Ward" and "LocalLearningSkillsCouncil".
- *education-data-gov-uk* and *oxpoints* contain similar labels which are, respectively, "School" and "College", for the SILK results see Sect. 3.5.

3   Experiments and Results

We proceed to report on the experiments conducted in support of the proposed recommendation method.

3.1   Evaluation Framework

The quality of the outcome of a recommendation process can be evaluated along a number of dimensions. Ricci *et al.* [17] provide a large review of recommender systems evaluation techniques and cite three common types of experiments: (i) offline experiments, where recommendation approaches are compared without user interaction, (ii) user studies, where a small group of subjects experiments with the system and reports on the experience, and (iii) online experiments, where real user populations interact with the system.

---

[4] http://education.data.gov.uk/.
[5] http://courseware.rkbexplorer.com/.
[6] http://courseware.rkbexplorer.com/.
[7] http://southampton.rkbexplorer.com/.
[8] http://darmstadt.rkbexplorer.com/.
[9] https://data.ox.ac.uk/sparql/.

For the task of dataset recommendation, the system suggests to the user a list of n target datasets candidates to be linked to a given source dataset. There does not exist a common evaluation framework for the datasets recommendation, thus, we evaluate our method with an offline experiment by using a pre-collected set of linked data considered as evaluation data (ED). The most straightforward, although not unproblematic (see below) choice of evaluation data for the data linking recommendation task is the existing link topology of the current version of the LOD cloud.

In our recommendation process, for a given source dataset D_S, we identify a cluster of target datasets, D_T, that we rank with respect to D_S (*cf.* Sect. 2.2). To evaluate the quality of the recommendation results given the ED of our choice, we compute the common evaluation measures for recommender systems, precision and recall, defined as functions of the true positives (TP), false positives (FP), true negatives (TN) and false negatives (FN) as follows:

$$Pr = \frac{TP}{TP + FP}; \quad Re = \frac{TP}{TP + FN}. \tag{2}$$

The number of potentially useful results that can be presented to the user has to be limited. Therefore, to assess the effectiveness of our approach, we rely on the measure of precision at rank k denoted by $P@k$. Complementarily, we evaluate the precision of our recommendation when the level of recall is 100 % by using the mean average precision at $Recall = 1$, MAP@R, given as:

$$MAP@R = \frac{\sum_{q=1}^{Total_{D_S}} Pr@R(q)}{Total_{D_S}}, \tag{3}$$

where $R(q)$ corresponds to the rank, at which recall reaches 1 for the qth dataset and $Total_{D_S}$ is the entire number of source datasets in the evaluation.

3.2 Experimental Setup

We started by crawling all available datasets in the LOD cloud group on the Data Hub[10] in order to extract their profiles. In this crawl, only 90 datasets were accessible via endpoints or via dump files. In the first place, for each accessible dataset, we extracted its implicit and explicit schema concepts and their labels, as described in Definition 1. The explicit schema concepts are provided by resource types, while the implicit schema concepts are provided by the definitions of a resource properties [18]. As noted in Sect. 2, some labels such as "Group", "Thing", "Agent", "Person" are very generic, so they are considered as noisy labels. To address this problem, we filter out schema concepts described by generic vocabularies such as VoID[11], FOAF[12] and SKOS[13]. The dataset document profiles, as defined in Definition 2, are constructed by extracting the textual

---

[10] http://datahub.io/group/lodcloud.
[11] http://rdfs.org/ns/void.
[12] http://xmlns.com/foaf/0.1/.
[13] http://www.w3.org/2004/02/skos/core.

descriptions of labels by querying the Linked Open Vocabularies[14] (LOV) with each of the concept labels per dataset.

To form the clusters of comparable datasets from Definition 3, we compute the semantico-frequential similarity between labels (given in Eq. (1)). We apply this measure via its available web API service[15]. In addition, we tested our system with two more semantic similarity measures based on WordNet: Wu Palmer and Lin's. For this purpose, we used the 2013 version of the $WS4J$[16] java API.

The evaluation data (ED) corresponds to the outgoing and incoming links extracted from the generated VoID file using the *datahub2void* tool[17]. It is made available on http://www.lirmm.fr/benellefi/void.ttl. We note that out of 90 accessible datasets, only those that are linked to at least one accessible dataset in the ED are evaluated in the experiments.

3.3   Evaluation Results

We started by considering each dataset in the ED as an unlinked source (newly published) dataset D_S. Then, we ran the *CCD-CosineRank* workflow, as described in Sect. 2.2. The first step is to form a $CCD(D_S)$ for each D_S. The CCD construction process depends on the similarity measure on dataset profiles. Thus, we evaluated the CCD clusters in terms of recall for different levels of the threshold θ (*cf.* Definition 3) for the three similarity measures that we apply. We observed that the recall value remains 100 % in the following threshold intervals per similarity measure: **Wu Palmer**: $\theta \in [0, 0.9]$; **Lin**: $\theta \in [0, 0.8]$; **UMBC**: $\theta \in [0, 0.7]$.

The CCD construction step ensures a recall of 100 % for various threshold values, which will be used to evaluate the ranking step of our recommendation process by the Mean Average Precision (MAP@R) at the maximal recall level, as defined in Definition 3. The results in Fig. 1 show highest performance of the UMBC's measure with a $MAP@R \cong 53\%$ for $\theta = 0.7$, while the best MAP@R values for Wu Palmer and Lin's measures are, respectively, 50 % for $\theta = 0.9$ and 51 % for $\theta = 0.8$. Guided by these observations, we evaluated our ranking in terms of precision at ranks $k = \{5, 10, 15, 20\}$, as shown in Table 1. Based on these results, we choose UMBC at a threshold $\theta = 0.7$ as a default setting for *CCD-CosineRank*, since it performs best for three out of four k-values and it is more stable than the two others especially with MAP@R.

3.4   Baselines and Comparison

To the best of our knowledge, there does not exist a common benchmark for dataset interlinking recommendation. Since our method uses both label profiles and document profiles, we implemented two recommendation approaches to be

---

[14] http://lov.okfn.org/dataset/lov/.
[15] http://swoogle.umbc.edu/SimService/.
[16] https://code.google.com/p/ws4j/.
[17] https://github.com/lod-cloud/datahub2void.

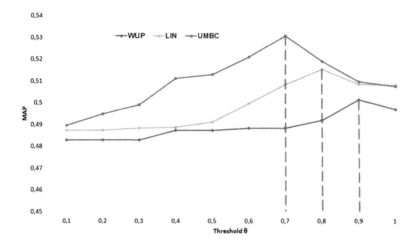

Fig. 1. The MAP@R of our recommender system by using three different similarity measures for different similarity threshold values

considered as baselines – one using document profiles only, and another one using label profiles:

Doc-CosineRank: All datasets are represented by their *document profiles*, as given in Definition 2. We build a vector model by indexing the documents in the corpus formed by all available LOD datasets (no CCD clusters). We use a *tf*idf* weighting scheme, which allows us to compute the cosine similarity between the document vectors and thus assign a ranking score to each dataset in the entire corpus with respect to a given dataset D_S.

UMBCLabelRank: All datasets are represented by their *label profiles*, as given in Definition 1. For a source dataset D_S, we construct its $CCD(D_S)$ according to Definition 3 using UMBC with $\theta = 0.7$. Thus, D_S is identified by its CCD and all target datasets D_T are found in its cluster. Let AvgUMBC be a ranking function that assigns scores to each D_T in $CCD(DS)$, defined by:

$$AvgUMBC(D', D'') = \frac{\sum_{i=1}^{|\mathcal{P}_l(D')|} \sum_{j=1}^{|\mathcal{P}_l(D'')|} \max \text{sim}_{UMBC}(L_i, L_j)}{\max(|\mathcal{P}_l(D')|, |\mathcal{P}_l(D'')|)}, \quad (4)$$

where L_i in $\mathcal{P}_l(D')$ and L_j in $\mathcal{P}_l(D'')$.

Figure 2 depicts a detailed comparison of the precisions at recall 1 obtained by the three approaches for each D_S taken as source dataset. It can be seen that the CCD-*CosineRank* approach is more stable and largely outperforms the two other approaches by an MAP@R of up to 53 % as compared to 39 % for *UMBCLabelRank* and 49 % for *CCD-CosineRank*. However, the *UMBCLabel-Rank* approach produces better results than the other ones for a limited number

Table 1. Precision at 5, 10, 15 and 20 of the *CCD-CosineRank* approach using three different similarity measures over their best threshold values based on Fig. 1

Measure\P@k	P@5	P@10	P@15	P@20
WU Palmer ($\theta = 0.9$)	0.56	0.52	0.53	0.51
Lin ($\theta = 0.8$)	0.57	**0.54**	**0.55**	0.51
UMBC ($\theta = 0.7$)	**0.58**	**0.54**	0.53	**0.53**

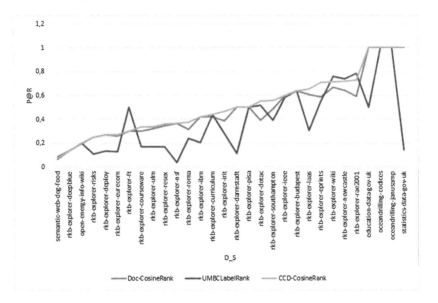

Fig. 2. Precisions at recall = 1 of the *CCD-CosineRank* approach as compared to *Doc-CosineRank* and *UMBCLabelRank*

of source datasets, especially in the case when D_S and D_T share a high number of identic labels in their profiles.

The performance of the *CCD-CosineRank* approach demonstrates the efficiency and the complementarity of combining in the same pipeline (i) the *semantic* similarity on labels for identifying recommendation candidates (*CCD* construction process) and (ii) the *frequential* document cosine similarity to rank the candidate datasets. We make all of the ranking results of the *CCD-CosineRank* approach available to the community on http://www.lirmm.fr/benellefi/CCD-CosineRank_Result.csv.

3.5 Discussion

We begin by a note on the vocabulary filtering that we perform (Sect. 3.2). We underline that we have identified the types which improve/decrease the performance empirically. As expected, vocabularies, which are very generic and wide-spread have a negative impact, acting like hub nodes, which dilute the results. For comparison, the results of the recommendation before removal are made available on http://www.lirmm.fr/benellefi/RankNoFilter.csv.

The different experiments described above show a high performance of the introduced recommendation approach with an average precision of 53 % for a recall of 100 %. Likewise, it may be observed that this performance is completely independent of the dataset size (number of triples) or the schema cardinality (number of schema concepts by datasets). However, we note that better performance was obtained for datasets from the *geographic* and *governmental* domains with precision and recall of 100 %. Naturally, this is due to the fact that a recommender system in general and particularly our system performs better with datasets having high quality schema description and datasets reusing existing vocabularies (the case for the two domains cited above), which is considered as linked data modeling best practice. An effort has to be made for improving the quality of the published dataset [19].

We believe that our method can be given a more fair evaluation if better evaluation data in the form of ground truth are used. Indeed, our results are impacted by the problem of false positives overestimation. Since data are not collected using the recommender system under evaluation, we are forced to assume that the false positive items would have not been used even if they had been recommended, i.e., that they are uninteresting or useless to the user. This assumption is, however, generally false, for example when the set of unused items contains some interesting items that the user did not select. In our case, we are using declared links in the LOD cloud as ED, which is certain but far from being complete for it to be considered as ground truth. Thus, in the recommendation process the number of false positives tends to be overestimated, or in other words an important number of missing positives in the ED translates into false positives in the recommendation process.

To further illustrate the effect of false positives overestimation, we ran SILK as an instance matching tool to discover links between D_S and their corresponding D_Ts that have been considered as false positives in our ED. SILK takes as an input a *Link Specification Language* file, which contains the instance matching configuration. We recall that our recommendation procedure provides pairs of shared or similar types between D_S and every D_T in its corresponding CCD, which are particularly useful to configure SILK. However, all additional information, such as the datatype properties of interest, has to be given manually. This makes the process very time consuming and tedious to perform over the entire LOD. Therefore, as an illustration, we ran the instance matching tool on two flagship examples of false positive D_Ts:

Semantically Similar Labels: We choose *education-data-gov-uk*[18] as a D_S and its corresponding false positive D_T *oxpoints*[19]. The two datasets contain in their profiles, respectively, the labels "School" and "College", detected

---

[18] http://education.data.gov.uk/.
[19] https://data.ox.ac.uk/sparql.

as highly similar labels by the UMBC measure, with a score of 0.91. The instance matching gave as a result 10 accepted "owl:sameAs" links between the two datasets.

Identical Labels: We choose *rkb-explorer-unlocode*[20] as a D_S and its corresponding D_Ts, which are considered as FP: *yovisto*[21] *datos-bcn-uk*[22] *datos-bcn-cl*[23]. All 4 datasets share the label "Country" in their corresponding profiles. The instance matching process gave as a result a set of accepted "owl:sameAs" links between *rkb-explorer-unlocode* and each of the three D_T.

We provide the set of newly discovered linksets to be added to the LOD topology and we made the generated linksets and the corresponding SILK configurations available on http://www.lirmm.fr/benellefi/Silk_Matching.

It should be noted that the recommendation results provided by our approach may contain some broader candidate datasets with respect to the source dataset. For example, two datasets that share schema labels such as books and authors are considered as candidates even when they are from different domains like science vs. literature. This outcome can be useful for predicting links such as "rdfs:seeAlso" (rather than "owl:sameAs"). We have chosen to avoid the inclusion of instance-related information in order to keep the complexity of the system as low as possible and still provide reasonable precision by guaranteeing a 100 % recall.

As a conclusion, we outline three directions of work in terms of dataset quality that can considerably facilitate the evaluation of any recommender system in that field: (1) improving descriptions and metadata; (2) improving accessibility; (3) providing a reliable ground truth and benchmark data for evaluation.

4 Related Work

With respect to finding relevant datasets on the Web, we cite briefly several studies on discovering relevant datasets for query answering. Based on well-known data mining strategies, [20, 21] present techniques to find relevant datasets, which offer contextual information corresponding to the user queries. A feedback-based approach to incrementally identify new datasets for domain-specific linked data applications is proposed in [22]. User feedback is used as a way to assess the relevance of the candidate datasets.

In the following, we cite approaches that have been devised for the datasets interlinking candidates recommendation task and which are, therefore, directly relevant to our work. Nikolov and d'Aquin [23] propose a keyword-based search approach to identify candidate sources for data linking consisting of two steps: (i) searching for potentially relevant entities in other datasets using as keywords

---

[20] http://unlocode.rkbexplorer.com/sparql/.
[21] http://sparql.yovisto.com/.
[22] http://data.open.ac.uk/query.
[23] http://data.open.ac.cl/query.

randomly selected instances over the literals in the source dataset, and (ii) filtering out irrelevant datasets by measuring semantic concept similarities obtained by applying ontology matching techniques.

Mehdi *et al.* [24] propose a method to automatically identify relevant public SPARQL endpoints from a list of candidates. First, the process needs as input a set of domain-specific keywords, which are extracted from a local source or can be provided manually by an expert. Then, using natural languages processing techniques and queries expansion techniques, the system generates a set of queries that seek to exact literal matches between the introduced keywords and the target datasets, i.e., for each term supplied to the algorithm, the system runs a comparison to a set of eight queries: {original-case, proper-case, lower-case, upper-case} × {no-lang-tag, @en-tag}. Finally, the produced output consists of a list of potentially relevant SPARQL endpoints of datasets for linking. In addition, an interesting contribution of this technique is the bindings returned for the subject and predicate query variables, which are recorded and logged when a term match is found on some particular SPARQL endpoint. The records are useful in the linking step.

Leme *et al.* [25] present a ranking method for datasets with respect to their relevance for the interlinking task. The ranking is based on Bayesian criteria and on the popularity of the datasets, which affects the generality of the approach. The authors extend this work and overcome this drawback in [9] by exploring the correlation between different sets of features—properties, classes and vocabularies—and the links to compute new rank score functions for all the available linked datasets.

None of the studies outlined above have evaluated the ranking measure in terms of Precision/Recall, except for [9] which, according to the authors, achieves a mean average precision of around 60 % and an excepted recall of 100 % with rankings over all LOD datasets. However, a direct comparison to our approach seems unfair since the authors did not provide the list of the datasets and their rank performance by datasets considered as source.

In comparison to the work discussed above, our approach has the potential of overcoming a series of complexity related problems, precisely, considering the complexity to generate the matching in [23], to produce the set of domain-specific keywords as input in [24] and to explore the set of features of all the network datasets in [9]. Our recommendation results are much easier to obtain since we only manipulate the schema part of the dataset. They are also easier to interpret and apply since we automatically recommend the corresponding schema concept mappings together with the candidate datasets.

5   Conclusion and Future Work

Following the linked data best practices, metadata designers reuse and build on, instead of replicating, existing RDF schema and vocabularies. Motivated by this observation, we propose the *CCD-CosineRank* interlinking candidate dataset recommendation approach, based on concept label profiles and schema overlap across datasets. Our approach consists of identifying clusters of comparable datasets, then, ranking the datasets in each cluster with respect to a given dataset. We discuss three different similarity measures, by which the relevance of our recommendation can be achieved. We evaluate our approach on real data coming from the LOD cloud and compare it two baseline methods. The results show that our method achicves a mean average precision of around 53 % for recall of 100 %, which reduces considerably the cost of dataset interlinking. In addition, as a post-processing step, our system returns sets of schema concept mappings between source and target datasets, which decreases considerably the interlinking effort and allows to verify explicitly the quality of the recommendation.

In the future, we plan to improve the evaluation framework by developing a more reliable and complete evaluation data for dataset recommendation. We plan to elaborate a ground truth based on certain parts of the LOD, possibly by using crowdsourcing techniques, in order to deal with the false positives overestimation problem. Further work should go into btaining high quality profiles, in particular by considering the population of the schema elements. We also plan to investigate the effectiveness of machine learning techniques, such as classification or clustering, for the recommendaiton task. One of the conclusions of our study shows that the recommendation approach is limited by the lack of accessibility, explicit metadata and quality descriptions of the datasets. As this can be given as an advice to data publishers, in the future, we will work on the development of recommendation methods for datasets with noisy and incomplete descriptions.

Acknowledgements. This research has been partially funded under the Datalyse project-FSN-AAP Big Data n3- (http://www.datalyse.fr/), by the European Commission-funded DURAARK project (FP7 Grant Agreement No. 600908) and the COST Action IC1302 (KEYSTONE).

References

1. Bizer, C., Heath, T., Berners-Lee, T.: Linked data - the story so far. Int. J. Semant. Web Inf. Syst. **5**(3), 1–22 (2009)
2. Pereira Nunes, B., Dietze, S., Casanova, M.A., Kawase, R., Fetahu, B., Nejdl, W.: Combining a co-occurrence-based and a semantic measure for entity linking. In: Cimiano, P., Corcho, O., Presutti, V., Hollink, L., Rudolph, S. (eds.) ESWC 2013. LNCS, vol. 7882, pp. 548–562. Springer, Heidelberg (2013)
3. Blanco, R., Mika, P., Vigna, S.: Effective and efficient entity search in RDF data. In: Aroyo, L., Welty, C., Alani, H., Taylor, J., Bernstein, A., Kagal, L., Noy, N., Blomqvist, E. (eds.) ISWC 2011, Part I. LNCS, vol. 7031, pp. 83–97. Springer, Heidelberg (2011)
4. Schmachtenberg, M., Bizer, C., Paulheim, H.: Adoption of the linked data best practices in different topical domains. In: Mika, P., et al. (eds.) ISWC 2014, Part I. LNCS, vol. 8796, pp. 245–260. Springer, Heidelberg (2014)
5. Buil-Aranda, C., Hogan, A., Umbrich, J., Vandenbussche, P.-Y.: SPARQL web-querying infrastructure: ready for action? In: Alani, H., et al. (eds.) ISWC 2013, Part II. LNCS, vol. 8219, pp. 277–293. Springer, Heidelberg (2013)

6. Guéret, C., Groth, P., Stadler, C., Lehmann, J.: Assessing linked data mappings using network measures. In: Simperl, E., Cimiano, P., Polleres, A., Corcho, O., Presutti, V. (eds.) ESWC 2012. LNCS, vol. 7295, pp. 87–102. Springer, Heidelberg (2012)

7. Auer, S., Bizer, C., Kobilarov, G., Lehmann, J., Cyganiak, R., Ives, Z.G.: DBpedia: a nucleus for a web of open data. In: Aberer, K., et al. (eds.) ASWC 2007 and ISWC 2007. LNCS, vol. 4825, pp. 722–735. Springer, Heidelberg (2007)

8. Suchanek, F.M., Kasneci, G., Weikum, G.: Yago: a core of semantic knowledge. In: Proceedings of WWW, pp. 697–706 (2007)

9. Rabello Lopes, G., Paes Leme, L.A.P., Pereira Nunes, B., Casanova, M.A., Dietze, S.: Two approaches to the dataset interlinking recommendation problem. In: Benatallah, B., Bestavros, A., Manolopoulos, Y., Vakali, A., Zhang, Y. (eds.) WISE 2014, Part I. LNCS, vol. 8786, pp. 324–339. Springer, Heidelberg (2014)

10. Han, L., Kashyap, A.L., Finin, T., Mayfield, J., Weese, J.: Umbc_ebiquity-core: semantic textual similarity systems. In: Proceedings of *SEM. Association for Computational Linguistics (2013)

11. Volz, J., Bizer, C., Gaedke, M., Kobilarov, G.: Silk - a link discovery framework for the web of data. In: Proceedings of WWW, LDOW (2009)

12. Ellefi, M.B., Bellahsene, Z., Scharffe, F., Todorov, K.: Towards semantic dataset profiling. In: Proceedings of Dataset PROFIling and fEderated Search for Linked Data Workshop Co-located with the 11th ESWC (2014)

13. Wu, Z., Palmer, M.: Verbs semantics and lexical selection. In: Proceedings of 32nd ACL, pp. 133–138 (1994)

14. Lin, D.: An information-theoretic definition of similarity. In: Proceedings of ICML, pp. 296–304 (1998)

15. Deerwester, S.C., Dumais, S.T., Landauer, T.K., Furnas, G.W., Harshman, R.A.: Indexing by latent semantic analysis. In: JASIS 1990, vol. 41, pp. 391–407 (1990)

16. Miller, G.A.: Wordnet: a lexical database for English. Commun. ACM **38**, 39–41 (1995)

17. Ricci, F., Rokach, L., Shapira, B., Kantor, P.B.: Recommender Systems Handbook, vol. 1. Springer, Heidelberg (2011)

18. Gottron, T., Knauf, M., Scheglmann, S., Scherp, A.: A systematic investigation of explicit and implicit schema information on the linked open data cloud. In: Cimiano, P., Corcho, O., Presutti, V., Hollink, L., Rudolph, S. (eds.) ESWC 2013. LNCS, vol. 7882, pp. 228–242. Springer, Heidelberg (2013)

19. Ellefi, M.B., Bellahsene, Z., Todorov, K.: Datavore: a vocabulary recommender tool assisting linked data modeling. In: Proceedings of ISWC Posters & Demonstrations Track a Track (2015)

20. Wagner, A., Haase, P., Rettinger, A., Lamm, H.: Discovering related data sources in data-portals. In: Proceedings of 1st IWSS (2013)

21. Wagner, A., Haase, P., Rettinger, A., Lamm, H.: Entity-based data source contextualization for searching the web of data. In: Proceedings of Dataset PROFIling & fEderated Search for Linked Data Workshop Co-located with the 11th ESWC, pp. 25–41 (2014)

22. de Oliveira, H.R., Tavares, A.T., Lóscio, B.F.: Feedback-based data set recommendation for building linked data applications. In: Procedings of 8th ISWC, pp. 49–55. ACM (2012)

23. Nikolov, A., d'Aquin, M.: Identifying relevant sources for data linking using a semantic web index. In: WWW, LDOW (2011)

24. Mehdi, M., Iqbal, A., Hogan, A., Hasnain, A., Khan, Y., Decker, S., Sahay, R.: Discovering domain-specific public SPARQL endpoints, a life-sciences use-case. In: Proceedings of 18th IDEAS, pp. 39–45 (2014)
25. Leme, L.A.P.P., Lopes, G.R., Nunes, B.P., Casanova, M.A., Dietze, S.: Identifying candidate datasets for data interlinking. In: Daniel, F., Dolog, P., Li, Q. (eds.) ICWE 2013. LNCS, vol. 7977, pp. 354–366. Springer, Heidelberg (2013)

From Queriability to Informativity, Assessing "Quality in Use" of DBpedia and YAGO

Tong Ruan$^{(\boxtimes)}$, Yang Li, Haofen Wang, and Liang Zhao

Department of Computer Science and Engineering,
East China University of Science and Technology, Shanghai 200237, China
{ruantong,whfcarter}@ecust.edu.cn, marine1ly@163.com, 252007913@qq.com

Abstract. In recent years, an increasing number of semantic data sources have been published on the web. These sources are further inter-linked to form the Linking Open Data (LOD) cloud. To make full use of these data sets, it is necessary to learn their data qualities. Researchers have proposed several metrics and have developed numerous tools to measure the qualities of the data sets in LOD from different dimensions. However, there exist few studies on evaluating data set quality from the users' usability perspective and usability has great impacts on the spread and reuse of LOD data sets. On the other hand, usability is well studied in the area of software quality. In the newly published standard ISO/IEC 25010, usability is further broadened to include the notion of "quality in use" besides the other two factors, namely, internal and external. In this paper, we first adapt the notions and the methods used in software quality to assess the data set quality. Second, we formally define two quality dimensions, namely, Queriability and Informativity from the perspective of quality in use. The two proposed dimensions correspond to querying and answering, respectively, which are the most frequent usage scenarios for accessing LOD data sets. Then we provide a series of metrics to measure the two dimensions. Last, we apply the metrics to two representative data sets in LOD (i.e., YAGO and DBpedia). In the evaluating process, we select dozens of questions from both QALD and WebQuestions and ask a group of users to construct queries as well as to check the answers with the help of our usability testing tool. The findings during the assessment not only illustrate the capability of our method and metrics but also give new insights on data quality of the two knowledge bases.

1 Introduction

In recent years, an increasing number of semantic data sources are published on the web. These sources are further interlinked to form Linking Open Data (LOD).

This work was partially supported by the 863 plan of China Ministry of Science and Technology (project No: 2015AA020107), and Software and Integrated Circuit Industry Development Special Funds of Shanghai Economic and Information Commission (project No: 140304).

© Springer International Publishing Switzerland 2016
H. Sack et al. (Eds.): ESWC 2016, LNCS 9678, pp. 52–68, 2016.
DOI: 10.1007/978-3-319-34129-3_4

They include not only encyclopedic knowledge bases (KBs) such as DBpedia and YAGO, which serve as data hubs, but also domain-specific LOD data sets such as DrugBank[1] and DailyMed.[2]

For a user who wants to utilize existing KBs, it is a demanding task to know their qualities. Their quality can be measured in various ways. A systematic review of the different approaches for assessing the data quality of LOD can be referred to [1]. The authors summarized 68 metrics and categorized them into four dimensions, namely, *Availability, Intrinsic, Contextual* and *Representational*. Glenn and Dave[3] listed 15 metrics to assess the quality of a data set. The metrics include *Accuracy, Completeness, Typing,* and *Currency,* etc.

While the metrics proposed in literature could measure different characteristics of a data set, these metrics neither take enough users' point of view into consideration, nor do they measure the "usability" of the data set. Despite that most studies [1–3] agree with the opinion that data quality is "fitness for use in special application context," no research works have proposed quality models or metrics related to this definition. In contrast to the state of the art of LOD usability research, software usability is well studied and has mature models and metrics. Since the definitions of software usability in traditional research do not distinguish different usage contexts, the ISO/IEC 25010 (2011) broadened the concept of software usability with *quality in use.* In the new standard, software quality contains three factors, which are internal quality, external quality, and quality in use. Quality in use is measured from the users' point of view and is obtained from using the software in the working environment.

In this paper, we propose metrics and methods to evaluate quality in use of data sets. We use the concept *quality in use* instead of usability. The reason is that usability is usually used to measure the user interface design, and a data set may not provide any user interface. However, a data set without a user interface can still be utilized in different contexts so that we call how easy a user utilizes a data set *quality in use.* The most common usage scenario of LOD data set is to access the information returned by executing a query. Therefore, we propose two quality dimensions, namely, Queriability and Informativity. Queriability measures how easily an end user can construct a correct query on a data set. Informativity shows how informative a data set is under a particular usage context. Furthermore, we define three metrics *Query Construction Time, Number of Attempts,* and *Difficulty Rating* to measure Queriability, and we also use *Precision, Recall, Comprehensive Informativity,* and *Informativeness Rating* to measure Informativity.

To investigate the effectiveness of our method, we carried out a few evaluations on DBpedia and YAGO. The two encyclopedic knowledge bases contain a large amount of instances or entities distributed in multiple domains. They are not designed for specific purposes of a particular group of users and are widely applicable theoretically. Therefore, the quality in use in different usage contexts

---

[1] http://www.drugbank.ca/.
[2] http://dailymed.nlm.nih.gov.
[3] http://lists.w3.org/Archives/Public/public-lod/2011Apr/0145.html/.

is very important for the spreading of these knowledge bases. We choose questions from two standard Q&A (questions and answers) test sets, namely, QALD and WebQuestions as query contexts and ask a group of users to construct queries complying with these questions and check the results with the answers in the test sets. We also develop a GUI tool to help users to construct queries in case they are not familiar with the SPARQL syntax. Our evaluation leads to a few interesting findings. For example, DBpedia has too many similar properties with different names, which greatly degrade the Queriability and Informtivity of DBpedia. The number of classes in YAGO is so large, which makes it difficult for evaluators to find the suitable domain in the query.

The paper is organized as follows: Sect. 2 introduces related work. Section 3 provides a quality model on LOD as well as the definitions of the metrics. Section 4 proposes our quality assessment process and related tools. Section 5 analyzes the results of the evaluation. Section 6 gives a conclusion and points out the future direction of our work.

2   Related Work

Our work focuses on devising new metrics and assessing KBs such as DBpedia and YAGO with these new metrics. Therefore, we survey literature regarding *metrics on LOD* and *Quality Evaluation on DBpedia and YAGO*. Since we learned usability assessment methods and metrics from the area of software usability, we would also give a brief introduction on software usability and quality in use.

2.1   Metrics on LOD

Besides the systematic review of approaches for accessing the data quality of LOD, there exist a lot of researches focusing on evaluating particular aspects of LOD quality. Labels are considered as an important quality factor of LOD in [4], and the authors introduced a number of related metrics to measure the completeness, accessibility, and other quality aspects of labels. Zhang et al. [5] designed a few complexity metrics on web ontologies. Gueret et al. [6] focused on assessing the quality of links in LOD. They assumed that unsuitable network structures are related to the low quality of links. Farber et al. [7] gave a survey on major cross-domain data sets of LOD cloud. They compared DBpedia, Freebase, OpenCyc, Wikidata, and YAGO from 35 aspects including schema constraints, data types, LOD linkages, and so on. However, they used natural languages and checklists instead of quantitative metrics to describe the special characteristics of data sets. While there are a lot of metrics on LOD, they do not measure the quality from the user's usability point of view.

2.2   Quality Evaluation on DBpedia and YAGO

Quality evaluations and quality improvements on encyclopedic KBs became a hot research topic recently. Zaveri et al. [8] classified the errors in DBpedia into four dimensions, including Accuracy, Relevancy, Representational-Consistency and Interlinking. YAGO2 (a version of YAGO) [9] used statistic sampling with *Wilson score interval* to reduce human efforts when evaluating the correctness of the YAGO2 manually. Wienand et al. [10] detected incorrect numerical data in DBpedia using unsupervised numerical outlier detection methods. Paulheim and Bizer [11] added missing type statements and removed faulty statements in both DBpedia and NELL using statistical distributions. Kontokostas et al. [12] presented a methodology for test-driven quality assessment which automatically generated test cases based on predefined test patterns. While there is quality evaluation work on encyclopedic KBs, the assessment work on YAGO and DBpedia is mostly focused on internal qualities such as correctness.

2.3 Software Usability and Quality in Use

There are various standards and models defining software usability [13,14]. The GE model is one of the earliest work working on software quality by Mccall et al. This hierarchical quality model consists of 11 quality factors, 25 quality criteria, and 41 quality metrics. Usability is a quality factor in the GE model and relates to 3 quality criteria. Usability is the second factor in the FURPS+ quality model adopted by Rational Software.

While usability is defined as a characteristic of products in traditional approaches, more recent researches find that the required quality attributes vary according to usage scenarios [15]. As the example in [15], the required quality attributes of a text editor for a programmer should be different from those for a casual user. Therefore, the quality model in the revised ISO/IEC 9126 [16] and later in ISO/IEC 25010 [17] distinguishes three quality approaches: internal quality, external quality, and quality in use. Internal quality concerns the static properties of the code. External quality can be obtained by executing the software system. Quality in use is measured from the users' view and shows whether the user is satisfied with the software in the working environment. The three components are interrelated. The internal software attributes will determine the quality of a software product in use in a particular context.

In this paper, we adapt the notions of quality in use to the research area of data quality. Our metrics such as *Query Construction Time* and *Number of Incorrect Tempts* are highly inspired by the task-oriented metrics in [18].

3 Metrics Design

3.1 Quality Model

Inspired by the software quality model in ISO/IEC 9261, we also divide the quality of LOD into three factors, namely, internal quality, external quality, and quality in use. We further map the quality dimensions of linked data mentioned in [1] into internal and external factors in our model. The Queriability and Informativity dimensions are put into quality in use, as shown in Fig. 1.

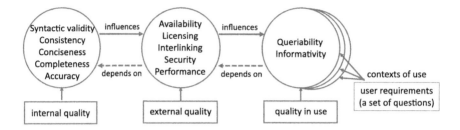

Fig. 1. The quality model of linked data

- Quality in use relates to the notion "context of use", which sometimes refers as usage context or usage scenario. Context of use implies user requirements. Users provide their data requirements in the form of a set of questions in our paper, just like user requirements are modeled as a set of use cases in software engineering. In the evaluation process, user-oriented questions are converted into data set oriented queries, the queries are executed on the data set, and the results are returned. Both the converting process and the results are measured by quality in use metrics. The whole quality evaluation process is requirement-oriented, conforming to the data quality definition "fitness for use in special application context". As a result, the metrics results depend on the requirements.
- The internal quality of a KB influences the quality in use of a KB. For example, the *accuracy* in Fig. 1 is categorized as internal quality. If there are errors in the triples, users may find the query results less informative.
- The external quality of a KB influences the quality in use of a KB. For example, the availability in Fig. 1 is categorized as external quality. Both DBpedia and YAGO have SPARQL endpoints, but the exploring service of YAGO[4] on the web can be accessed easily, which makes it easier for users who are not familiar with SPARQL to find the data.

3.2   Analyzing the Process of Constructing a Query

We designed an experiment to investigate the processes of constructing SPARQL queries for different evaluators on different data sets with different questions. In the experiment, evaluators assessed Queriability and Informativeness manually with the source files of data sets. We selected ten questions from WebQuestions and QALD and asked five graduate students to construct queries of the ten questions on both DBpedia and YAGO. Each evaluator wrote down his steps in constructing the query. After we checked all the step flows in their reports, we give a summarization, and the detail of the experiments can be accessed via our web sites[5].

---

[4] https://gate.d5.mpi-inf.mpg.de/webyago3spotlx/Browser.
[5] http://kbeval.nlp-bigdatalab.com/qiu.html.

1. Analyze each question and find the patterns of the question. For example, the question "what was Abe lincoln's wife name?" contains a subject and its property, and the object is the answer. But the question "Give me all female Russian astronauts" is much more complex. The pattern may contain a target domain (astronauts or Russian astronauts) and one or two constraints (female and Russian).
2. Find suitable vocabularies in the KBs. The vocabularies include domain names, property names, property values as well as instance names. We call the step to find domain names the "domain selection" step. As property names and properties values are related, the sub-step to find them are adjacent in our experiment records. They are combined together as the "property constraint selection" step. We find "domain selection" step is usually before "property constraint selection" step. Among 100 SPARQL constructing records (10 questions*5 evaluators*2 data sets), there is only one record in which a property name instead of a domain name is first found. The two steps are time-consuming (15 min on average) and evaluators may try many times before success. For example, the "wife" may not be the property name in a data set. It could be "married to" or something else. What's worse, the domain could be "Russian astronauts" instead of "astronauts," and its property names in the question are vacant.
3. Construct the query using SPARQL syntax and execute the query.
4. Repeat step 1–3 if the results are not desirable.

While there exist questions which are so simple that some steps may not be required, the above steps are unavoidable in general. The above steps contain all the possible steps in constructing a query, which give us guidelines on developing the evaluation tool. Since major difficulties arise from the "domain selection" step and the "property constraint selection" step, we design special metrics for the two steps in Sect. 3.3. Since we target at evaluating KBs and we do not want to bother evaluators on question understanding and syntax of SPARQL, we should eliminate the difficulties in steps 1 and 3. We develop a GUI tool which help users construct a SPARQL query interactively. The functions in the tool have direct mappings with the steps above in the manual construction process. Thus, evaluators can construct queries in the same process with the same results manually as in the tool and vice versa, as described in Sect. 4.3.

3.3 Metrics

The Queriability and Informativity focus on the process of query and the results of the query, and the corresponding metrics are shown in Table 1. Queriability measures how easily an end user can construct a correct query on a data set. Informativity shows how informative a data set is under a particular usage context.

3.3.1 Queriability

We design two kinds of Queriability metrics: one is the subjective metrics which are collected from direct feedbacks of evaluators, and the other is the objective

Table 1. Metrics definitions

Dimension	Metric	Description
Queriability	Query construction time on domain	Time (T_a) spent on setting the domain of the query $T_a = \frac{1}{NOA}\sum_{i=1}^{NOA} T_{ai}$ (T_{ai} is the time spent on setting the domain of the attempt i)
	Query construction time on property constraint	Time (T_b) spent on setting the properties and property values of the query $T_b = \frac{1}{NOA}\sum_{i=1}^{NOA} T_{bi}$ (T_{bi} is the time spent on property constraint setting of the attempt i)
	Query construction time	Time (T) spent on constructing the query $T = NOA(T_a + T_b)$
	Number of attempts	Times (NOA) tried for constructing the query
	Difficulty rating	Users' rating on the difficulties on constructing the query
Informativity	Precision	The precision (P) of the results for the query $P = NCA/A$ (NCA is the number of correct results, and A is the number of query results)
	Recall	The recall (R) of the results for the query $C = NCA/NA$ (NA is the number of standard answer for the query)
	Comprehensive informativity	Comprehensive Informativity (CI) is a comprehensive metric to measure the informativity of the answer
	Informativeness rating	Users' rating on the information that the results contain

metrics which are collected in the process of constructing a query. *Difficulty Rating* in Table 1 is a subjective metric. After evaluators finish (may be success or fail) constructing a query, they give ratings on how difficulty the process was. The rating has five levels: (1) very easy, (2) easy, (3) average, (4) difficult, and (5) very difficult. Objective metrics are designed according to the process of constructing a query. There are two important aspects in query constructing: time spent in constructing a query and how many times a user has tried. We use *Query Construction Time On Domain* T_a and *Query Construction Time On Property Constraint* T_b to measure the former, and we use *Number of Attempts* NOA to measure the latter. Then, *Query Construction Time* $T = NOA(T_a+T_b)$.

Based on the analysis of the query construction process in Sect. 3.2, constructing a query typically consists of two steps. One step is to set the domain of the question, and the other step is to set the property constraints. T_a is closely linked to the taxonomy system of the KB. Both the complexity of the taxonomy system and the specificity of the vocabularies of the classes may lead to larger T_a. T_b is closely linked to properties used in the KB. The redundancy of the properties and the ambiguity of the properties may influence the time T_b. In general, the longer the time spent, the harder the process is.

A user may have tried many incorrect queries before he/she successfully constructed a query. If a user finds nothing returned or the returned results are wrong after he executes a query, he may reconstruct a query. For example, for the question "Which presidents were born in 1945?" if the property is set as "wasBornIn" on YAGO, nothing will be returned because the range of the property "wasBornIn" is the city, and the correct property should be "wasBornOn-Date." There may be extreme conditions that whatever the user tries, he/she does not get the intended answer since the KB does not contain the answer. The whole process fails. In whatever situation, the *Number of Attempts* reflects the quality of the KBs. The larger the number is, the more difficult to construct a query. Since we limit the maximum number to 10 so that the value of *Number of Attempts* is an integer between 1 to 10.

3.3.2   Informativity

We also design two kinds of Informativity metrics as the Queriability metrics above. *Informativeness Rating* in Table 1 is the subjective metric, and it is the users' rating on the informativeness that the results contain. The *Informativeness Rating* has five levels: (1) very little information, (2) little information, (3) some information, (4) a fair amount of information, and (5) lots of information. The objective metrics are computed according to the standard answers of the questions, and we use *Precision, Recall, Comprehensive Informativity* to measure the query results.

Precision and *Recall* are used to measure the query results in the Informativity dimension. Precision is the fraction of returned results that are relevant, and it measures the correctness of the query results. Recall is the fraction of relevant instances that are returned with querying, and it measures the completeness of the query results. *Comprehensive Informativity* (CI) is a metric which integrates different factors that influence the users' comprehension of information. From our understanding of information comprehension, the factors should include not only precision and recall of a query results but also the accuracy of data returned as well as the understandability of the data. Therefore, the formula is

$$CI = \frac{NCA}{NA} * (\frac{NCA}{A})^2 * \alpha * \beta \qquad (1)$$

NCA is the number of correct answers of the query results. NA is the number of standard answers for the question. A is the number of query results. The range of CI is [0,1]. The square function is used to punish the irrelevant answers. The data in the KBs may be inaccurate. In our previous work [19], we use the *correctness ratio of facts* metric to measure the accuracy. Here, we use α to denote the metric result. In the process of calculating the CI, we just directly reuse results in [19] to α. β is the understandability of the data in the KB, and it measures whether the data of the KBs is human-readable. We do not measure the dimension in this paper, so we just set it to a constant 0.8.

4 Experiment Design

4.1 Evaluated Knowledge Bases

We applied our metrics to DBpedia and YAGO in our experiment. DBpedia and YAGO are two major general-purpose knowledge bases serving as the hubs of Linked Open Data. In particular, we chose two representative versions (DBpedia2014 and YAGO2S) for DBpedia and YAGO, respectively.

4.2 Questions and Patterns

Data requirements of users are modeled as question sets, as mentioned in Sect. 3.1. Here we use the following criteria to collect questions which meet the requirements for a good "wikipedia like data set".

1. Biases on either data set should be avoided. We utilized two question sets in the KB-based QA area. One source is the Question Answering over Linked Data[6] (QALD), and the other is WebQuestions[7] from the NLP laboratory of Standford. We choose QALD-2 whose questions are cross-domain. Because the number of questions in WebQuestions is huge, we chose 50 questions from the beginning, 50 questions in the middle, and 50 questions at the end.
2. Questions having no answers in either DBpedia or YAGO are avoided.
3. The diversity of question patterns is considered. For example, "what are the official languages in spain?" and "what is the official language spoken in mexico?" are similar, therefore, we only choose one of them.

At last, we chose 13 questions from QALD and 13 questions from WebQuestions. A full list of 26 questions, 150 questions from WebQuestions, and 100 questions from QALD could be found on our website.[8]

As explained before, our intention is to assess data set quality instead of SPARQL syntax. We should try our best to eliminate the time that the evaluators spend on SPARQL syntax. To meet that goal, we analyze the questions in QALD and WebQuestions and find most of them (the detail statistics are also on our website mentioned above) can be categorized to special patterns shown in Table 2. The patterns in Table 2 cover all the 26 questions in the list. We build a tool to support these patterns so that users will just fill in the appropriate vocabularies and operators to instantiate the corresponding pattern, and the executable SPARQL query is generated automatically.

---

[6] http://greententacle.techfak.uni-bielefeld.de/~cunger/qald/index.php?
x=home\&q=home.
[7] http://nlp.stanford.edu/software/sempre/.
[8] http://kbeval.nlp-bigdatalab.com/qiu.html.

Table 2. Question patterns and SPARQL query patterns

Pattern	Transformation	Example
DomainPattern	Find all sub classes	Give me all school types
PropertyPattern	Select * where { ?s propertyname ?o }	Population of cities
InstancePattern	Select * where { instancename ?p ?o}	Information of Google
ValuePattern	Select * where { instancename propertyname ?o}	What was abe lincoln's wife name?
AttributeEqual	Select * where { ?s attributename attributevalue }	Which presidents were born in 1945?
AttributeRange	Select * where{ ?s attributename ?o filter(?o operator value) }	The cities whose population bigger than 3 million
AggreExpression	Select ?s where {?s attributename ?o} group by ?s having(aggreateFuncname(?o) operator value)	Which countries have more than two official languages?
OrderedTop	Select ?s where {?s attributename ?o} order by ?o	What is the highest mountain?
OrderedTop + AggreExpression...	Select ?s where {?s attributename ?o} order by aggreateFuncname(?o)	Who produced the most films?

4.3   Evaluation Process and Evaluation Tool

We design a tool to support the manual process in Sect. 3.2, as Fig. 2 shown. Each evaluator first chooses a question from the question list and selects a target KB (YAGO or DBpedia). He then sets the domain of the question and inputs property constraints to construct a corresponding query. The tool is shown in Fig. 3. After clicking the "execute" button, the tool constructs and executes the SPARQL query automatically, and the results are returned to the evaluator.

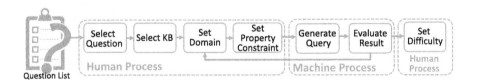

Fig. 2. Evaluation process

The vocabulary finding step in manual constructing process requires too much human repeated laboring work, for example, searching for a property name in an editor with the search function while the occurrences of the name is large, or find all sub domains. Since these repeated laboring work may obscure our mission of finding usability problems in the data sets, we provide some user-friendly functions to relieve the evaluators from these work. All the functions

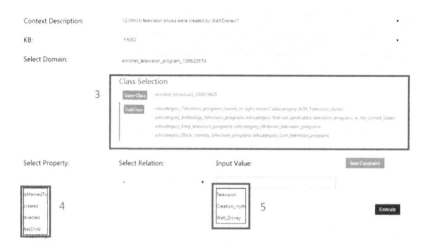

Fig. 3. Our evaluation tool

are just suggestions to help user find vocabularies, and it is the evaluator who ultimately chooses the vocabularies and determines the SPARQL query to be executed.

1. Autofill functions. When users or evaluators input the name of a domain or a property name, all the class names or property names containing the input letter sequence would be popped up as suggestions. The function is a replacement of the general search function in the editor in the manual construction process.

2. Show the subclasses and superclasses of a domain. In the manual construction process, the evaluator may find appropriate domain names by searching the subclasses or superclasses of existing class. For example, a user needs to select the domain of the question "Give me all presidents of the United States." The user may first come up with "President," and then he may find a subclass "Presidents_of_the_United_States" which may be closer to the question. Therefore, we provide the function which shows both subclasses and superclasses of a domain.

3. PATTY [20] is integrated in our tool to obtain the candidates of a property. We can use the relation patterns provided by PATTY to get the properties contained in the sentence. For example, the question is "Which television shows were created by Walt Disney?" and the candidates are "isMarriedTo," "created," "directed," "hasChild." While candidates provided by PATTY often have nothing to do with the question, the function broadens users' conjectures on possible properties names.

4. DBpedia Spotlight [21] is integrated in our tool to annotate the instances in the question. For the above question, DBpedia Spotlight can annotate the

instance "http://dbpedia.org/resource/Walt_Disney." This function can be considered as an advanced search function for instances.

After executing the query, the results may be empty or are not the ones as the users expect. Users need to construct another new query until the results are satisfactory or the number of attempts achieves the maximum limit. Each question in the question list has a standard answer set as ground truths, so the metrics in the Informativity dimension can be computed. Finally, users should set the *Difficulty Rating* according to their evaluation process and set the *Informativeness Rating* according to the query results after the evaluation for each question. Since the entire evaluation process includes several manual steps, in order to reduce errors caused by the subjectivity of the users, we employ eight people, and each of them tested all the 26 questions.

5   Evaluation Results

5.1   Queriability of DBpedia and YAGO

The comparison results of two KBs with respect to the average *Query Construction Time*, *Query Construction Time on Domain*, as well as the *Query Construction Time on Property Constraints* for every question are shown in Fig. 4(a–c), respectively. The means and variations for these metrics are shown in Fig. 4(d).

We find in Fig. 4(a) that it costs evaluators more time on YAGO than on DBpedia to find a satisfactory query. We look into detail Fig. 4(b) and Fig. 4(c). From Fig. 4(b), we find that when users select domains, they spend much more time on YAGO than on DBpedia. This is because YAGO contains a huge number of classes, and some class names are excessively long. As mentioned in Sect. 4.1, YAGO has 451k classes. The length of some class names may exceed 50 characters. For example, the category name wikicategory_Failed_assassins_of_United_States_presidents is really difficult for users to read or input. On the contrary, classes of DBpedia are fewer, and the names of the classes are easier to understand. From Fig. 4(c), we find that when considering properties selection, it takes longer on DBpedia than on YAGO. Compared with the number of properties defined in YAGO (i.e., 75), DBpedia has more than 55,000 properties. Moreover, there exist a lot of nearly duplicated properties, which lead to more effort on selecting properties. For example, the property names "dateOfBirth," "birthDate," "birth," "birthdate," and "birthday" in DBpedia are all related to the notion "birthday."

Figure 4(d) shows that when users query on YAGO, *Query Construction Time On Domain* and *Query Construction Time On Property Constraint* have a larger fluctuation. The fluctuation of domain selection results from the differences of the classes in YAGO. Classes with relatively simple names which are close to the root of the taxonomy hierarchy are easier to be found, such as "wordnet_actor_109765278." Classes with long names which are close to the leaf level of taxonomy hierarchy are difficult to find. The reason for the fluctuation of

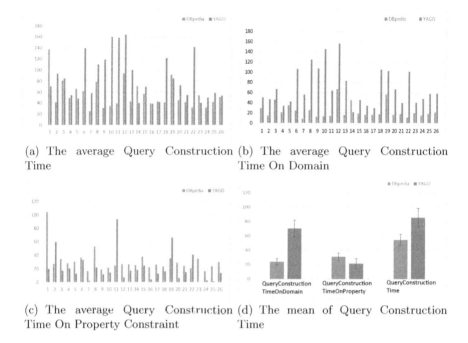

(a) The average Query Construction Time

(b) The average Query Construction Time On Domain

(c) The average Query Construction Time On Property Constraint

(d) The mean of Query Construction Time

Fig. 4. Results of query construction time

property selection is that, for simple queries, users only need to set the domain, and then the *Query Construction Time On Property Constraint* is 0. However, if the selected domain is not suitable, users have to spend a lot of time to switch between similar properties.

The comparison results of two KBs with respect to the average *Number of Attempts* are shown in Fig. 5(a). In general, the *Number of Attempts* in DBpedia is more than that in YAGO. The conclusion becomes more obvious for the second half of the questions. After investigation, we find that in the beginning, users are not familiar with YAGO classes. They find wrong classes and try to change the properties to construct a query, which leads to a number of failed attempts. In the later stages, users have a better understanding of the YAGO taxonomy and are aware that they should rely more on it.

The comparison results of two KBs with respect to the average *Difficulty Rating* are shown in Fig. 5(b). Even with the help of our tool, no question is rated by users as easy on average, whether for DBpedia or for YAGO. Figure 5(b) also shows a tendency that in the first-half questions, DBpedia is easier, while in the second half, YAGO is easier.

We also calculate the correlation between the *Difficulty Rating* and the other Queriability metrics. We find that *Difficulty Rating* closely correlates with *Number of Attempts* in DBpedia and closely correlates with the sum of *Query Construction Time on Domain* and *Query Construction Time On Property Constraint* in YAGO. For DBpedia, it does not cost much time to set the

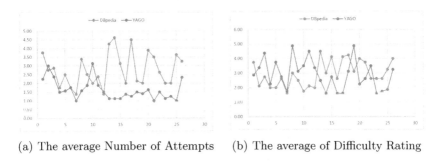

(a) The average Number of Attempts   (b) The average of Difficulty Rating

Fig. 5. Results of number of attempts and difficulty rating

domain and property. However, duplication exists in the property names. When users enter wrong property names with no query result, they have to try other property names. In this case, users have to try many times before being successful, and they feel confused and have a difficult time. That's the reason the *Difficulty Rating* correlates with *Number of Attempts* in DBpedia. For YAGO, it costs time to select the appropriate domain and property, especially when the target class is complex. That's the reason why *Difficulty Rating* correlates with the total time.

5.2   Informativity of DBpedia and YAGO

In terms of Informativitiy, we focus on the amount of information a user can gain under a certain context, and we do not care about how many incorrect queries are constructed by users. Therefore, we choose the best result achieved by all evaluators for each question to assess Informativity. In general, there are 20 questions that return at least one answer in YAGO and 14 questions in DBpedia. The comparison results of two KBs with respect to the average *Precision* and *Recall* as well as the *CI* for every question are shown in Fig. 6. All of them show that YAGO has better Informativity than DBpedia.

From Fig. 6(a), we find DBpedia has a rather low precision. One reason is the polysemy of type words. For example, presidents can be presidents of countries or presidents of organizations. So for the question "Give me all presidents of the United States," we found that the results include *the Chairman of the Federal Reserve* and other types of presidents. The low precision of YAGO arises from our misinterpretation of the vocabulary in YAGO. For example, the class "wikicategory_German_actors" means actors whose nationality, not birthplace, is German.

From Fig. 6(b), we find DBpedia and YAGO have low recalls. The reason for DBpedia is that it has duplicated properties as mentioned before. The reason for YAGO is that it really does not contain abundant properties with enough facts. For the question "Which actors were born in Germany?" if the users set the domain as "wordnet_actor_109765278," set the property as "wasBornIn," and set its value as "Germany," no results are returned. For the question "who is number 5 on the Boston Celtics?" evaluators cannot find the property in YAGO to describe "number" in the context.

Informativeness Rating before and after the user checked the correct answers are different. The former relates to how many times users attempt to construct a query and how many results are returned. The latter relates to the precision and the recall, and relates to CI as a whole.

In summary, the results on Queriability and Informativity relate to internal characteristics of KBs, namely, the schema design and the richness of the data. For YAGO, the huge class hierarchy increases difficulties in finding the classes, the small number of properties reduces the "expressiveness" of the data set, and the smaller number of facts makes query results less informative. While DBpedia seems to have a better balance between the number of classes and the number of properties, the property duplication problem largely decreases the quality in use of DBpedia.

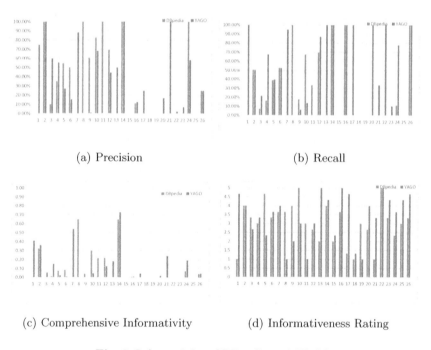

(a) Precision

(b) Recall

(c) Comprehensive Informativity

(d) Informativeness Rating

Fig. 6. Informativity of DBpedia and YAGO

The lessons learned from our assessment work include: (1) Naming convention for classes, objects and properties are required in the LOD world, similar to that in software (2) Duplicate property names should be avoided since they will mislead the users. (3) There should be a tradeoff between the number of classes and the number of properties. Users may encounter difficulties when both of the numbers become too large. In general, our experiments show that a well-designed schema is very important for people to utilize a data set. We wish someone could run a "linked open schema (LOS)" website which contains more schema data, vocabularies and constraints than existing schema web sites such as schema.org. Every data set should be linked and registered to the LOS websites before the data set is published. There could be facilities such as Q&A engines on this LOS site so that every data set can be accessed via natural language interfaces. In this way, the data sets published will be of higher quality, and end users could utilize the data set immediately.

6   Conclusion and Future Work

In this paper, we designed two metric sets, namely, Queriablity and Informativity with respect to the "quality in use" factor on LOD. The metric results on YAGO and DBpedia not only show that users have experienced difficulties in utilizing these KBs, but also give many hints on where the difficulties arise as well as how to improve these KBs. In the future, we plan to assess other usage scenarios of quality in use, such as search and browsing or Q&A, and assess more cross-domain KBs such as Wikidata. We also plan to collect data requirements on the medical domain and evaluate the quality in use of medical data sets of the LOD cloud.

References

1. Zaveri, A., Rula, A., Maurino, A., Pietrobon, R., Lehmann, J., Auer, S.: Quality assessment methodologies for linked open data. Semant. Web J. (2012)
2. Strong, D.M., Lee, Y.W., Wang, R.Y.: Data quality in context. Commun. ACM **40**(5), 103–110 (1997)
3. Tayi, G.K., Ballou, D.P.: Examining data quality. Commun. ACM **41**(2), 54–57 (1998)
4. Ell, B., Vrandečić, D., Simperl, E.: Labels in the web of data. In: Aroyo, L., Welty, C., Alani, H., Taylor, J., Bernstein, A., Kagal, L., Noy, N., Blomqvist, E. (eds.) ISWC 2011, Part I. LNCS, vol. 7031, pp. 162–176. Springer, Heidelberg (2011)
5. Zhang, H., Li, Y.F., Tan, H.B.K.: Measuring design complexity of semantic web ontologies. J. Syst. Softw. **83**(5), 803–814 (2010)
6. Guéret, C., Groth, P., Stadler, C., Lehmann, J.: Assessing linked data mappings using network measures. In: Simperl, E., Cimiano, P., Polleres, A., Corcho, O., Presutti, V. (eds.) ESWC 2012. LNCS, vol. 7295, pp. 87–102. Springer, Heidelberg (2012)
7. Färber, M., Ell, B., Menne, C., Rettinger, A.: A comparative survey of DBpedia, freebase, opencyc, wikidata, and yago
8. Zaveri, A., Kontokostas, D., Sherif, M.A., Bühmann, L., Morsey, M., Auer, S., Lehmann, J.: User-driven quality evaluation of DBpedia. In: Proceedings of the 9th International Conference on Semantic Systems, pp. 97–104. ACM (2013)
9. Hoffart, J., Suchanek, F.M., Berberich, K., Weikum, G.: Yago2: A spatially and temporally enhanced knowledge base from wikipedia. In: Proceedings of the Twenty-Third International Joint Conference on Artificial Intelligence, pp. 3161–3165. AAAI Press (2013)

10. Wienand, D., Paulheim, H.: Detecting incorrect numerical data in DBpedia. In: Presutti, V., d'Amato, C., Gandon, F., d'Aquin, M., Staab, S., Tordai, A. (eds.) ESWC 2014. LNCS, vol. 8465, pp. 504–518. Springer, Heidelberg (2014)
11. Paulheim, H., Bizer, C.: Improving the quality of linked data using statistical distributions. Int. J. Semant. Web Inf. Syst. (IJSWIS) **10**(2), 63–86 (2014)
12. Kontokostas, D., Westphal, P., Auer, S., Hellmann, S., Lehmann, J., Cornelissen, R., Zaveri, A.: Test-driven evaluation of linked data quality. In: Proceedings of the 23rd International Conference on World Wide Web, pp. 747–758. ACM (2014)
13. Al-Qutaish, R.E.: Quality models in software engineering literature: an analytical and comparative study. J. Am. Sci. **6**(3), 166–175 (2010)
14. Seffah, A., Donyaee, M., Kline, R.B., Padda, H.K.: Usability measurement and metrics: a consolidated model. Softw. Qual. J. **14**(2), 159–178 (2006)
15. Bevan, N., Azuma, M.: Quality in use: incorporating human factors into the software engineering lifecycle. In: Third IEEE International Software Engineering Standards Symposium and Forum, Emerging International Standards, ISESS 1997, pp. 169–179. IEEE (1997)
16. ISO/IEC: ISO/IEC 9126-4 Software engineering -Product quality- part4: Quality In Use metrics (2002)
17. ISO/IEC25010: Systems and software engineering – Systems and software Quality Requirements and Evaluation (SQuaRE) – System and software quality models (2011)
18. Albert, W., Tullis, T.: Measuring the User Experience: Collecting, Analyzing, and Presenting Usability Metrics. Newnes, Oxford (2013)
19. Ruan, T., Dong, X., Wang, H., Li, Y.: Kbmetrics - a multi-purpose tool for measuring quality of linked open data sets. In: The 14th International Semantic Web Conference, Poster and Demo Session (2015)
20. Nakashole, N., Weikum, G., Suchanek, F.: Patty: a taxonomy of relational patterns with semantic types. In: Proceedings of the 2012 Joint Conference on Empirical Methods in Natural Language Processing and Computational Natural Language Learning, Association for Computational Linguistics, pp. 1135–1145 (2012)
21. Daiber, J., Jakob, M., Hokamp, C., Mendes, P.N.: Improving efficiency and accuracy in multilingual entity extraction. In: Proceedings of the 9th International Conference on Semantic Systems (I-Semantics) (2013)

Normalized Semantic Web Distance

Tom De Nies[1]([✉]), Christian Beecks[2], Fréderic Godin[1], Wesley De Neve[1,3],
Grzegorz Stepien[2], Dörthe Arndt[1], Laurens De Vocht[1], Ruben Verborgh[1],
Thomas Seidl[2], Erik Mannens[1], and Rik Van de Walle[1]

[1] iMinds – Data Science Lab, Ghent University, Ghent, Belgium
{tom.denies,frederic.godin,wesley.deneve,dorthe.arndt,laurens.devocht,
ruben.verborgh,erik.mannens,rik.vandewalle}@ugent.be
[2] DME Group, RWTH Aachen University, Aachen, Germany
{beecks,grzegorz.stepien,seidl}@informatik.rwth-aachen.de
[3] IVY Lab, KAIST, Daejeon, Republic of Korea

Abstract. In this paper, we investigate the Normalized Semantic Web
Distance (NSWD), a semantics-aware distance measure between two concepts in a knowledge graph. Our measure advances the Normalized Web
Distance, a recently established distance between two textual terms, to
be more semantically aware. In addition to the theoretic fundamentals
of the NSWD, we investigate its properties and qualities with respect to
computation and implementation. We investigate three variants of the
NSWD that make use of all semantic properties of nodes in a knowledge
graph. Our performance evaluation based on the Miller-Charles benchmark shows that the NSWD is able to correlate with human similarity assessments on both Freebase and DBpedia knowledge graphs with
values up to 0.69. Moreover, we verified the semantic awareness of the
NSWD on a set of 20 unambiguous concept-pairs. We conclude that the
NSWD is a promising measure with (1) a reusable implementation across
knowledge graphs, (2) sufficient correlation with human assessments, and
(3) awareness of semantic differences between ambiguous concepts.

1 Introduction

The goal of semantic distance and/or similarity measures is to mimic the human
assessment of distance between two concepts. However, humans usually do not
explicitly quantify the distance between concepts – at least not consciously –
which makes the development and evaluation of semantic distances a challenging
task. These measures play an important role on the Web, particularly when it
comes to indexing, semantic search, and information retrieval. Despite being
an intensely researched field for the past decades, traditional, plain-text-based
similarity measures are still the dominant norm in practical scenarios [9]. This
is unfortunate because most of these have little to no semantic awareness since
they work with syntactic information instead of machine-interpretable data.

We argue that increasing the semantic awareness of similarity measures is
possible by making use of machine-interpretable concepts that are unambiguously linked to a resource on the Web [6]. Determining the distance between

© Springer International Publishing Switzerland 2016
H. Sack et al. (Eds.): ESWC 2016, LNCS 9678, pp. 69–84, 2016.
DOI: 10.1007/978-3-319-34129-3_5

individual concepts facilitates the application of more complex similarity measures such as Earth Mover's Distance or Signature Quadratic Form Distance [1] on Web documents that are modeled by the concepts they contain. To this end, we investigate the properties of the Normalized Semantic Web Distance (NSWD), an inter-concept distance we introduced [7] based on the statistics of the context of the entities in the knowledge graph where they are represented. In this paper, we provide additional theoretical fundamentals of the NSWD, as well as insights into its computation and implementation. Additionally, we further evaluate our approach on the DBpedia knowledge graph as well as Freebase, and verify its semantic awareness on a set of 20 unambiguous concept pairs.

2   Background Knowledge and Motivation

The distance measure we introduce in this paper is based on the Normalized Information Distance (NID), introduced in [15]. The NID is defined using the so called *Kolmogorov Complexity*. Formally, the Kolmogorov Complexity of a binary string x is defined as the length of the shortest program p with $U(p) = x$ for a fixed universal prefix Turing machine U [14]. Informally, it can be seen as the length of the maximally compressed version of x and in a sense reflects the amount of information encoded in x. Unfortunately, the Kolmogorov Complexity is non-computable, and thus needs to be approximated. This approximation is strongly affected by the representation of objects, either in *literal* or *non-literal* form. The literal form contains the object itself. A song or a novel, for example, can be provided as a binary file containing the actual song or novel. The non-literal form consists of a (possibly ambiguous) name referencing the actual object. Abstract concepts like "love" or "beauty" can only be provided by their non-literal name. This corresponds to the vision of information and non-information resources described in the original World Wide Web architecture [12].

If the input for the NID is given in **literal form**, we can approximate the Kolmogorov Complexity by the size of the output of a state of the art compression algorithm. This principle was introduced as the Normalized Compression Distance (NCD) [3]. However, in this paper we focus on **non-literal** objects, and we need a different approximation. In [4], it was shown that for a frequency function f, the NID can be approximated as in Definition 1.

Definition 1. *For two binary strings x and y, let $f(x, y) \in \mathbb{N}_0$ be a frequency function for which $N := \sum_{x,y} f(x, y) < \infty$ and $f(x) := f(x, x)$ holds. f resembles how common x and y occur together in a certain context. The NID can now be approximated as:*

$$NID_f(x, y) = \frac{\max\{\log f(x), \log f(y)\} - \log f(x, y)}{\log N - \min\{\log f(x), \log f(y)\}}$$

The challenge is to define a reasonable frequency function f. A well known instance is the so called Normalized Web Distance (NWD) [4]. It employs the statistics returned by an arbitrary Web search engine to calculate f, where the frequency $f(x)$ of a term x is the number of indexed pages mentioning x. The basic principle of the NWD is: if two terms occur together almost as often as they do separately, their semantic distance is likely to be low. More formally:

Definition 2. *Let W be the set of pages indexed by an arbitrary search engine able to return the (approximate) number of indexed pages containing a certain search term. For each search term x, let $\mathbf{X} \subseteq W$ denote the set of pages containing x. For two search terms x and y, the* Normalized Web Distance *corresponds to the NID_f from Definition 1 and the following frequency function f [4]:*

$$f(x) := |\mathbf{X}|$$
$$f(x, y) := |\mathbf{X} \cap \mathbf{Y}|$$

Note that search engines usually only estimate $f(x)$ and do not explicitly exclude duplicate pages from the search result. Furthermore, due to the volume of stored and indexed Web pages, a search engine cannot compute N as exactly as in Definition 1. However, since it merely serves as a scaling factor in the NID_f equation, it can be set to an arbitrary value $\geq |W|$ (this constraint ensures that the NWD is always non-negative).

Any search engine can be used for the NWD and *different engines* usually return *different results*. This approach is semantically unaware of the meaning of the input terms, as well as of the (human-understandable) Web pages returned by the search engine. The word "Java", for example, can either refer to the Indonesian island, or to the programming language. This means that calculating the NWD from the word "Java" to "Indonesia" leads to the same distance for either meaning, regardless of which the user intended. This is exactly the problem we are tackling in this paper: comparing unambiguous concepts instead of ambiguous lexical terms. While other approaches exist to achieve this (as seen in Sect. 3), none are based on the NID, which makes them less related to the way we represent knowledge. Since the essential foundations of the Semantic Web include an accurate representation of knowledge, it also makes sense to create a semantic distance that has roots in the same foundations. In Sect. 3, we highlight a number of known implementations of the NWD, as well as various other semantic distance or similarity measures relevant to our approach.

3   Related Work

The first instance of the NWD was the Normalized Google Distance (NGD), which made use of the Google search engine, and achieved up to 87 % agreement with human assessments [5]. However, due to changes to the functionality of the AND operator in the Google search engine [24], the returned page counts are no longer trustworthy. In our evaluation, we use the Bing search engine as an alternative, resulting in the Normalized Bing Distance (NBD) [10]. In fact, the NGD paper clearly states that other data sources than the World Wide Web, such as a dictionary, could be used to calculate the NGD [5]. This further reinforces our intuition of adapting the NWD to a graph-structured database.

The NGD has been used as an inspiration for a link-based similarity measure before, namely the Wikipedia Link-based Measure (WLM) [18]. WLM adapts the traditional TF-IDF-vector-based approach as well as the NGD approach to exploit the link structure of Wikipedia. When only considering the incoming-links-part of WLM, the principle is comparable to that of the NSWD, with the exception of WLM being specific to Wikipedia, whereas the NSWD is more generally applicable.

In previous work, we conducted a preliminary experiment with an approach we named the *Normalized Freebase Distance (NFD)* [10]. The NFD served as a first specific implementation of the concept of the NSWD. The promising results inspired us to generalize the concept to all knowledge graphs on the Semantic Web in [7]. We expand significantly upon the evaluation provided in these previous papers, as well as on the theoretical foundations for the NSWD.

The Jaccard similarity is one of the most efficient measures for semantic relatedness [2,13]. Unlike the Jaccard similarity, the Jaccard distance (inverse similarity) is a valid heuristic for e.g., a pathfinding algorithm. For example, the "Everything is Connected Engine" (EiCE) [8] uses a distance metric based on the Jaccard similarity for pathfinding. It applies the measure to estimate the similarity between two nodes and to assign a random-walk based weight, which ranks less popular resources higher, thereby guaranteeing that paths between resources prefer specific relations over general ones [19].

The Linked Data Semantic Distance (LDSD) [22] also partially relies on shared links. Similar to our proposed approach, the LDSD is extensible for specific domains, as shown in [25], where it is extended to model the similarity of human behavior processes.

An important category of semantic relatedness measures contains those that measure the distance between two concepts in the context of their concept hierarchy or ontology. Hliaoutakis et al. [11] provide a comparison of 11 such semantic relatedness measures, tested using WordNet[1] and MeSH[2]. More recently, Sanchez et al. [23] provide an overview over ontology-based semantic similarity measures, including a newly proposed one of their own. The correlation of these measures with popular benchmarks for word relatedness ranges between 0.7 and 0.86. For more details, we refer to the aforementioned surveys.

The Semantic Connectivity Score [21] is a measure to quantify the connectivity between two concepts. It considers the total number of paths between two concepts up to a user-specified maximum length. Authors from the same groups also introduced a co-occurence-based measure (CBM) [20], based on the same page counts as the NGD. Their measure does not require an estimation of the total number of Web pages. Note that this also means that the CBM of disjoint pairs of concepts are non-comparable. They argue for a combined connectivity/co-occurrence measure, since this would allow to find semantic relations between concepts that do not necessarily co-occur (with the connectivity-based measure),

---

[1] http://wordnet.princeton.edu/.
[2] http://www.nlm.nih.gov/mesh/.

while still emphasizing concept relations without the necessity of a strong connection in the semantic graph (with the CBM) [20].

A hybrid similarity metric for Linked Data was proposed in [16]. It considers the *information content* or *informativeness* of the features shared by two resources. The features of a resource are composed of the resources it links to, and the links themselves. In order for a feature to be shared, both the linked resource and the link type must be the same. Like the NSWD, the proposed metric requires an estimation of the number of concepts in the entire graph. It was specifically developed for a resource recommendation scenario and was preliminarily evaluated in such a scenario.

To sum up, most of the aforementioned instances of the NID are focused on comparison of purely textual, possibly ambiguous terms. Of those approaches mentioned above that do provide means to measure the distance between unambiguous concepts in a knowledge graph, none are based on the NID. Therefore, to the best of our knowledge, we provide the first NID-based (dis)similarity measure within the context of a knowledge graph.

4   Normalized Semantic Web Distance

The basic principle of the NSWD is to use the degree of co-occurrence of edges from and to two concepts in a knowledge graph to reflect their semantic dissimilarity. Instead of considering the human-understandable Web (accessed through a search engine), we consider a machine-understandable knowledge graph on the Semantic Web (accessed through a query client, e.g., a SPARQL endpoint).

The NSWD advances from possibly ambiguous natural language terms to unambiguous concepts, which are identified by URIs, as input. For example, when using the semantic dataset DBpedia.org, the island "Java" is uniquely identified by the URI dbpedia:Java[3], whereas the programming language "Java" is identified by dbpedia:Java_(programming_language). We design the NSWD to result in a lower distance between dbpedia:Java and dbpedia:Indonesia than between dbpedia:Java_(programming_language) and dbpedia:Indonesia.

A knowledge graph consists of a set of nodes V and a set T of directed triples $t \in V \times P \times V$, where P is a set of predicates. A node represents a certain real-world object or concept and is identified via an URI. A triple $(u, p, v) \in T$ indicates a subject-predicate-object relation. The starting node u is the subject, v is the object and the predicate p carries some additional information regarding the exact nature of the relation between those nodes. For example, the triple *"dbpedia:Statue_of_Liberty dbpedia-owl:location dbpedia:New_York"* signifies the semantic relationship that the Statue of Liberty is located in New York.

To model the semantic relationship between two nodes in a knowledge graph, we make use of these triples. We consider the semantic relationship between two *subjects* to be stronger the more concepts they share a triple with. Additionally, the semantic relationship between two *objects* is considered stronger the more

---

[3] We choose the prefix dbpedia: for convenience, which resolves to http://dbpedia. org/resource/.

often they *occur* in a triple with the same subject, similar to a term occurring on a Web page. Nodes for which both of this is true, will have an even stronger relationship. We formalize these sets of linked nodes in Definition 3.

Definition 3. *We define the following sets of nodes $V_\lambda \subseteq V$ for $\lambda \in \{in, out, all\}$ in a knowledge graph (V, T) with respect to a node $x \in V$:*

$$V_{in}(x) := \{v \in V \,|\, (v, p, x) \in T\}$$
$$V_{out}(x) := \{v \in V \,|\, (x, p, v) \in T\}$$
$$V_{all}(x) := V_{in}(x) \cup V_{out}(x)$$

In other words, the set $V_{in}(x)$ comprises distinct nodes with at least one link – i.e., predicate – pointing to node x, whereas the set $V_{out}(x)$ contains all distinct nodes where node x points to. The set $V_{all}(x)$ is the union of all nodes that link to, or are linked to from node x (not necessarily with the same predicate). Based on these sets, we can now define three variations of frequency functions f_λ with respect to parameter $\lambda \in \{in, out, all\}$, to be used to calculate three variations of our proposed distance.

Definition 4. *For two nodes x and y in V, we define f_λ as:*

$$f_\lambda(x) := |V_\lambda(x)|$$
$$f_\lambda(x, y) := |V_\lambda(x) \cap V_\lambda(y)|$$

We then define the Normalized Semantic Web Distance *between two nodes $x, y \in V$ from a knowledge graph (V, T) as follows:*

$$\text{NSWD}_\lambda(x, y) := \frac{\max\{\log f_\lambda(x), \log f_\lambda(y)\} - \log f_\lambda(x, y)}{\log N - \min\{\log f_\lambda(x), \log f_\lambda(y)\}}$$

As can be seen in the definition of the sets V_λ, the NSWD_λ makes use of all information available within the direct semantic context of the nodes in the corresponding knowledge graph. The parameter $\lambda \in \{in, out, all\}$ models the semantic context that is taken into account when determining the dissimilarity of two nodes. We investigate the question of which parameter λ performs best in practice in Sect. 8 and continue with and example of the NSWD_{in}.

Example 1. Consider the sub-graph of a graph with $|V| = 1000$, as depicted in Fig. 1. We want to calculate $\text{NSWD}_{in}(x, y)$. Here, $V_{in}(x) = \{a, d, e, f, y\}$ is the set of concepts containing a link to x. Similarly, $V_{in}(y) = \{a, b, c, f\}$ is the set of concepts containing a link to y. Finally, the set of concepts with a link to both x and y is $V_{in}(x) \cap V_{in}(y) = \{a, f\}$. This means that $f(x) = |V_{in}(x)| = 5$, $f(y) = |V_{in}(y)| = 4$, and $f(x, y) = |V_{in}(x) \cap V_{in}(y)| = 2$. We therefore have $\text{NSWD}_{in}(x, y) = \frac{log 5 - log 2}{log 1000 - log 4} \approx 0.16595$.

Note that the NSWD_{in} corresponds most closely to the original definition of the NWD (if we consider Web pages to be subjects and the search terms

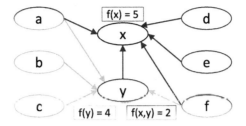

Fig. 1. Illustration of the values for the functions $f(x)$, $f(y)$, and $f(x, y)$ needed to calculate the NSWD$_{in}$ in Example 1.

as part of its description). On the other hand, the NSWD$_{out}$ corresponds more to the principles of the Jaccard Distance, since it relies on the shared outgoing links. Finally, the NSWD$_{all}$ corresponds more closely to the principles used in the Wikipedia Link-based Measure [18], as it combines outgoing and incoming links, albeit using a different formula. Which of these definitions performs best in a real-world scenario can only be determined through empirical evaluation, which we describe in Sect. 8. First, however, we discuss some of the properties of the NSWD in Sect. 5, and how to implement it in Sect. 6.

5   Properties

In order to fully understand its potential, applicability, and implementation, we discuss a number of important properties of the NSWD, such as its mathematical properties, a number of special cases, and its minimum and maximum values.

Mathematical Properties. Although the NSWD is based on the NID, it is not a metric. For two arbitrary non-equal concepts x and y, $NSWD(x, x) = 0$ and $NSWD(x, y) \geq 0$ always hold, but not necessarily $NSWD(x, y) > 0$, since both concepts could be connected to exactly the same nodes. The NSWD also does not fulfill the triangle inequality. Two concepts can have very few links in common, making their NSWD high, while there could be a third term having many links in common with both concepts, making the NSWD between this third concept and each of the other two low. As highlighted by the authors of the NWD, which is also not a metric [4], this is not necessarily a drawback since human knowledge consists of many concepts which, intuitively, do not fulfill a triangle inequality. For example, while "Paris" and "White House" intuitively have a low semantic similarity, both have a high semantic similarity to "Capital".

Special Cases. A first special case occurs when $f_\lambda(x) = |V| = f_\lambda(y)$. In this case, $f_\lambda(x, y) = |V|$, so the numerator and denominator of NSWD$_\lambda(x, y)$ both become 0, and NSWD$_{(}x, y)$ does not exist. Note that in reality, this situation will never occur, since this would mean that both x and y are linked to by every element in the dataset, and thus do not contribute any useful information. However, for the sake of completeness, we define the NSWD to be 0 in this case.

A second special case occurs when $f_\lambda(x,y) = 0$. In case of $\lambda = in$, this means there are no concepts in the dataset that link to both x and y. In case of $\lambda = out$, it means that x and y do not link to any common concept. In these cases, $log f_\lambda(x,y)$ does not exist, and thus a value for the NSWD must be defined. For the Normalized Google Distance, the authors chose $NGD(x,y) = 1$ in case there were no search results that contained both x and y [5]. However, that leads to a – in our view – very counter-intuitive situation where it is possible for two terms that have a low, but strictly positive number of search results for both terms, to have a higher NGD than two terms that do not. In fact, the first special case adds to this argument, since the distance is 1, whereas there is clearly some relationship between x and y (however abstract it may be). A much more elegant solution is to determine the upper bound $NSWD_{max}$ for $NSWD_\lambda$, and define $NSWD_\lambda(x,y) = NSWD_{max}$ in the case that $f_\lambda(x,y) = 0$.

A final special case occurs if $f_\lambda(x) = 0$ (or $f_\lambda(y) = 0$), in which case the $log f_\lambda(x)$ (or $log f_\lambda(y)$) does not exist. In this case, $f_\lambda(x,y)$ will automatically equal 0 as well, bringing us back to the second special case. Therefore, we will define $NSWD_\lambda(x,y) = NSWD_{max}$ in this case as well.

Maximum Value. With the values for the special cases as defined above in mind, we can rely on our knowledge of the behavior of $f_\lambda(x,y)$ with respect to $f_\lambda(x)$ and $f_\lambda(y)$ to calculate $NSWD_{max}$. It can be shown that

$$\forall x,y \in V : f_\lambda(x) + f_\lambda(y) - |V| \le f_\lambda(x,y) \le |V| - 1$$

Keeping these constraints in mind, we can iterate over all possible values of $NSWD_\lambda(x,y)$ for a given cardinality of V. This way, we determined that:

$$\forall x,y \in V : NSWD_{max} = \frac{log(\lfloor \frac{|V|}{2} \rfloor + 1)}{log|V| - log\lceil \frac{|V|}{2} \rceil}$$

For example, considering the same dataset V as in Example 1, the upper bound for $NSWD_\lambda(x,y)$ for any concepts x and y in V is $NSWD_{max} = \frac{log 501}{log 1000 - log 500} \approx 8.9686$. We can now use this value to determine the NSWD in the special case where $f_\lambda(x,y) = 0$, as well as to calculate a similarity score normalized between 0 and 1, as further explained in Sect. 7.

6   Implementation

The NSWD is not bound to one specific technology for its implementation. Any method that can calculate the frequency functions as described in Definition 4, as well as a count of the number of nodes in a knowledge graph is suitable. This makes the NSWD a very flexible measure, allowing it to be implemented using the most suitable technology for any specific use case. Additionally, it can be tailored to any knowledge domain by using a specialized dataset. However, typically, a knowledge graph on the Semantic Web is modeled in RDF, and it is made accessible through SPARQL. For any knowledge graph with a SPARQL endpoint, the following query can be used to calculate $f_\lambda(x)$ and $f_\lambda(x,y)$: `SELECT (COUNT(DISTINCT ?a) AS ?f) WHERE`, with the matching `WHERE`-clause from Table 1, and x and y the URIs of two concepts in the knowledge graph.

Table 1. `WHERE`-clauses for SPARQL queries to calculate the NSWD.

Frequency function	WHERE-clause
$f_{in}(x)$	`{ ?a ?p <x> }`
$f_{in}(x,y)$	`{?a ?p1 <x>  . ?a  ?p2  <y>  }`
$f_{out}(x)$	`{ <x> ?p  ?a  }`
$f_{out}(x,y)$	`{ <x> ?p1 ?a  .  <y> ?p2 ?a }`
$f_{all}(x)$	`{{ ?a ?p1 <x> } UNION { <x> ?p2 ?a }}`
$f_{all}(x,y)$	`{{?a ?p1 <x>} UNION {<x> ?p2 ?a} . {?a ?p3 <y>} UNION {<y> ?p4 ?a}}`

To calculate the total number of distinct concepts in the knowledge graph, a `COUNT DISTINCT` query could be used. However, due to the `DISTINCT` constraint, such a query – while conceptually straightforward – does pose a problem for larger knowledge graphs, as most SPARQL endpoints are not optimized for this kind of count. In these cases, it is advised to store a cached result for an earlier execution of this query, a value from the dataset's metadata (if available), or an estimate of the number of nodes in the knowledge graph, and use this for the calculations. Optimizing for these types of queries would be a must to guarantee the usability of the NSWD on a large scale.

7   Calculating Similarity

The NSWD is a *distance*, meaning that the more semantically related two concepts are, the smaller their distance is. However, in many cases the opposite is desired, i.e. when *similarity* must be measured – e.g., in the Miller-Charles benchmark [17], which we use for our evaluation. If the NSWD were normally distributed in the range $[0, \text{NSWD}_{max}]$, we could just scale it linearly using NSWD_{max} and substract it from 1. However, the values that occur most frequently are in the $[0,1]$ range. These are also the distance values that are most interesting in practical scenarios, such as recommendation systems. Scaling these values linearly using the NSWD_{max} would lead to a situation where the majority of distances would be in the range of $[0, \frac{1}{\text{NSWD}_{max}}]$, which is not very useful. Keeping this in mind, we define the NSWD-based similarity Sim_{NSWD} as follows:

Definition 5.

$$Sim_{NSWD_\lambda}(x,y) := \begin{cases} 1 - d(x,y) \times (1-c), & \text{if } d(x,y) \in [0,1] \\ (1 - \frac{d(x,y)}{NSWD_{max}}) \times c, & \text{if } d(x,y) \in]1, NSWD_{max}] \end{cases}$$

$$\text{with } d(x,y) = NSWD_\lambda(x,y) \text{ and } c = \frac{1}{NSWD_{max}}$$

This way, the most semantically significant distances – those between 0 and 1 – get mapped to the similarity range $[c, 1]$ with minimal scaling, and the distances higher than 1 get mapped to the similarity range $[0, c[$ with significant scaling. Note that if NSWD_{max} is accurately calculated, $Sim_{\text{NSWD}_\lambda}(x, y)$ is normalized between 0 and 1.

8   Evaluation

In this section, we first compare the NSWD-based similarity measures defined in Sect. 7 to human judgment of similarity, using a standard set of term-pairs. Next, we verify whether the NSWD and its variants accomplish what they were designed to do: increase semantic awareness w.r.t traditional distance measures.

To evaluate our approach, we chose to use the Miller-Charles dataset [17], which consists of 30 term-pairs that were judged for similarity by 38 people. While using lexical terms for evaluation of a distance in knowledge graphs is not ideal, and this is a relatively small dataset, it offers an insight to how humans judge the similarity between these terms, and more importantly, it gives us a number of related approaches for direct comparison, as it is very commonly used in this field. Therefore, the Miller-Charles benchmark provides the best starting point for external validation compared to established approaches. However, before we can use it, a disambiguation strategy has to be decided upon, as many of the terms are highly ambiguous, in the sense that they can correspond to more than one resource URI in the knowledge graph. To choose one to use for the NSWD calculation, we used three disambiguation strategies:

Manual: manually pick a disambiguated resource URI, or suggest an alternative URI (human judgment);
Count-Based: use the resource URI with the highest V_{in}, V_{out} or V_{all} (depending on whether the NSWD, NSWD_{out} or NSWD_{all} is calculated, respectively);
Similarity-Based: use the resource URI leading to the smallest distance (only possible in the context of a pairwise comparison);

Note that it is possible that the correct disambiguation cannot be determined due to the non-completeness of the dataset. In our evaluation, we calculated the distances using all aforementioned disambiguation strategies, to see which leads to the best results. For each of the 30 term-pairs, our evaluation process consists of the steps below.

1. Both terms are disambiguated, using the manual and automatic approaches. This results in 3 URI disambiguation options for each term: (a) manually selected, (b) based on the highest link-count, and (c) based on the highest similarity with the other term.
2. For each of the three URI disambiguation options, the NSWD_{in}, NSWD_{out}, and NSWD_{all} are calculated.
3. The above results in 9 distances (three for each variant of the NSWD), which are converted NSWD-based similarities, as defined in Definition 5.

These steps result in 9 similarity assessments for each of the 30 term-pairs, each value calculated with a different combination of disambiguation option and NSWD variant. These values are compared to the human-assessed scores from the Milled-Charles dataset by calculating the Pearson correlation coefficient.

As a baseline, we added a similarity score based on the Normalized Bing Distance [10] to the evaluation results. This NBD-based similarity was calculated as $1 - NWD(x, y)$, with $NWD(x, y)$ calculated as specified in Definition 2, using the Microsoft Bing Search API[4] as a search engine.

We calculated the Pearson correlation coefficient between the Miller-Charles scores and the NBD baseline, as well as the three NSWD variants. Each NSWD-based similarity measure was tested with three disambiguation strategies: manual (M), count-based (C), or similarity-based (S), using two widely used knowledge graphs: Freebase and DBpedia. The results are shown in Table 2, along with the reported correlations on the same benchmark for the Wikipedia Link-based Measure* [18], and Jaccard similarity as calculated in [13]. Higher correlation indicates a stronger positive relationship between the human-assessed scores and calculated similarities. To enable reproducibility of the results, we provide online access to the JSON files generated by our evaluation software, including all disambiguated URIs, Miller-Charles scores, and similarity scores. The file for Freebase can be accessed at http://semweb.mmlab.be/nswd/evaluation/mc30_results_freebase.json, the file for DBpedia at http://semweb.mmlab.be/nswd/evaluation/mc30_results_dbpedia.json, and the file for Bing at http://semweb.mmlab.be/nswd/evaluation/mc30_results_bing.json.

Table 2. Pearson correlation coefficient on the Miller-Charles benchmark for the NSWD similarity variants on the Freebase and DBpedia knowledge graphs, as well as the Normalized Bing Distance, Wikipedia Link-based Measure, and Jaccard similarity.

NBD	**0.23**								
	$Sim_{\text{NSWD}_{in}}$			$Sim_{\text{NSWD}_{out}}$			$Sim_{\text{NSWD}_{all}}$		
	M	C	S	M	C	S	M	C	S
Freebase	0.42	0.25	0.29	0.57	0.43	0.57	0.55	0.24	**0.58**
DBPedia	0.60	0.44	0.55	0.56	**0.69**	0.62	0.66	0.58	0.68
WLM[a]	**0.70**								
Jaccard[a]	**0.882**								

[a]Using a different disambiguation strategy

Note that for all distance and disambiguation options, the NSWD-based similarities achieved a higher correlation than the NBD-based similarity at the time of writing, with a maximum of 0.58 for Freebase and 0.69 for DBpedia. There is no consistent trend in which disambiguation strategy performed best. Overall,

---

[4] https://datamarket.azure.com/dataset/bing/search.

the NSWD$_{all}$ seemed to perform best, taking most of the semantic context of a node into account. None of the NSWD variants was able to perform better than the reported results of the WLM and Jaccard similarity. However, note that for these reported results, a different disambiguation strategy was applied. For the WLM as reported in [18], the disambiguation of the Miller-Charles terms was performed using a weighted combination of commonness, relatedness, and occurrence together in a sentence. However, the authors of [18] did not disclose the exact weighting scheme they used, nor the disambiguated terms. Do note that commonness and relatedness of the terms in a term-pair are factors that we also consider, by applying the disambiguation strategies using the highest link-count and highest similarity, respectively. Therefore, we can safely assume that the reported correlation of 0.70 is useful to compare with our results. In case of the Jaccard similarity as calculated in [13], disambiguation was left ad-hoc to a search engine, which makes it impossible for us to reproduce.

While the aforementioned results show the external validity of our approach, they do not highlight semantic awareness – the primary aspect it was designed for. To highlight this aspect of the NSWD, we created a small additional evaluation set of our own, consisting of 10 pairs of concept-pairs with the same plain-text representation, yet with the first concept in both concept-pairs remaining the same, and the second concept disambiguated to a more divergent meaning. While this is a limited set of concepts, the results clearly illustrate that the NSWD and its variants are capable of recognizing these differences in semantics, and assign smaller distances to concepts that are close in semantics than to concepts whose semantics diverge.

The evaluation set for the DBpedia knowledge graph is shown in Table 3, along with the results. We observe that in all 10 cases, all three variants of the NSWD maintain a semantically correct ordering w.r.t. the distance. Note that the impact of the lower link-density of DBpedia is seen here as well, with many distances being equal to 21.2, which is the NSWD$_{max}$ for DBpedia. This means that in all these cases there were no common links, and thus $f(x,y)$ was zero.

To use the same evaluation set for the Freebase graph, we mapped every DBpedia resource from Table 3 to a Freebase resource using the `owl:sameAs` link to Freebase included in the description of the resource[5]. As shown in Table 4, the NSWD and its variants also maintain a semantically correct ordering w.r.t. the distance when applied to Freebase. We observed one small discrepancy for the resource-pairs `:Automobile-:Bus` and `:Automobile-:Bus_(computing)`, where the ordering is (very slightly) reversed for the NSWD.

The quality of the knowledge graph greatly affects the performance and applicability of the NSWD. For example, during the disambiguation of the Miller-Charles dataset, we found that DBpedia often lacks a simple description of various concepts. For example, the concepts "journey" and "voyage" – resulting in the resources `dbpedia:Journey` and `dbpedia:Voyage`, repectively – both link to many disambiguation options, but none of these options capture the most straightforward meaning of the concepts. When inspecting the

---

[5] The full mapping is available at http://semweb.mmlab.be/nswd/evaluation/ resourcemapping_dbpedia_freebase.ttl.

Table 3. Results illustrating the semantic awareness of the NSWD variants using the DBpedia graph. The prefix : resolves to http://dbpedia.org/resource/. The 'order'-column indicates whether $d(c1, c2) < d(c1, c3)$, which is desired in these cases.

concept 1 (c1)	concept 2 (c2)	concept 1 (c3)	$d = $ NSWD			$d = $ NSWD$_{out}$			$d = $ NSWD$_{all}$		
			$d(c1,c2)$	$d(c1,c3)$	order	$d(c1,c2)$	$d(c1,c3)$	order	$d(c1,c2)$	$d(c1,c3)$	order
:Pear	:Apple	:Apple_Inc.	0.27	21.2	✓	0.16	21.2	✓	0.21	21.2	✓
:Trout	:Bass_(fish)	:Bass_guitar	21.2	21.2	-	0.31	21.2	✓	0.33	21.2	✓
:Cat	:Jaguar	:Jaguar_Cars	21.2	21.2	-	0.13	0.40	✓	0.20	0.50	✓
:Cat	:Mouse	:Mouse_(computing)	0.41	21.2	✓	0.13	21.2	✓	0.21	21.2	✓
:Automobile	:Bus	:Bus_(computing)	0.33	21.2	✓	21.2	21.2	-	0.34	21.2	✓
:Indonesia	:Java	:Java_(programming_language)	0.39	21.2	✓	0.24	0.49	✓	0.39	1.01	✓
:Lion	:Tiger	:Tiger_(Danish_store)	0.39	21.2	✓	0.14	21.2	✓	0.23	21.2	✓
:Musical_theatre	:Broadway_(play)	:Broadway_(Manhattan)	21.2	21.2	-	0.34	0.37	✓	0.51	0.58	✓
:Bird	:Crane_(bird)	:Crane_(machine)	0.55	21.2	✓	0.26	21.2	✓	0.58	21.2	✓
:Bass_guitar	:String_(music)	:String_(physics)	0.53	21.2	✓	0.33	21.2	✓	0.56	21.2	✓

Table 4. Results illustrating the semantic awareness of the NSWD variants using the Freebase graph. For clarity, the DBpedia concept names are shown instead of the Freebase hash codes. For each DBpedia resource, the actual Freebase URI used in the calculations is the object of the owl:sameAs relation to the corresponding Freebase resource.

concept 1 (c1)	concept 2 (c2)	concept 1 (c3)	$d = $ NSWD			$d = $ NSWD$_{out}$			$d = $ NSWD$_{all}$		
			$d(c1,c2)$	$d(c1,c3)$	order	$d(c1,c2)$	$d(c1,c3)$	order	$d(c1,c2)$	$d(c1,c3)$	order
:Pear	:Apple	:Apple_Inc.	0.13	0.37	✓	0.17	0.46	✓	0.17	0.44	✓
:Trout	:Bass_(fish)	:Bass_guitar	0.15	0.57	✓	0.32	0.67	✓	0.32	0.68	✓
:Cat	:Jaguar	:Jaguar_Cars	0.25	0.35	✓	0.31	0.46	✓	0.32	0.47	✓
:Cat	:Mouse	:Mouse_(computing)	0.23	0.33	✓	0.31	0.46	✓	0.31	0.47	✓
:Automobile	:Bus	:Bus_(computing)	0.34	0.32	×	0.38	0.46	✓	0.38	0.46	✓
:Indonesia	:Java	:Java_(programming_language)	0.30	0.58	✓	0.33	0.53	✓	0.36	0.60	✓
:Lion	:Tiger	:Tiger_(Danish_store)	0.11	0.28	✓	0.22	0.42	✓	0.21	0.43	✓
:Musical_theatre	:Broadway_(play)	:Broadway_(Manhattan)	0.42	0.43	✓	0.42	0.46	✓	0.44	0.50	✓
:Bird	:Crane_(bird)	:Crane_(machine)	0.22	0.33	✓	0.33	0.51	✓	0.33	0.52	✓
:Bass_guitar	:String_(music)	:String_(physics)	0.46	0.57	✓	0.46	0.73	✓	0.46	0.73	✓

corresponding human-understandable Wikipedia pages, it becomes clear that both "journey" and "voyage" are supposed to be disambiguated to the concept "travel", with resource URI dbpedia:Travel. Unfortunately, these links are not currently included in DBpedia. As a result, automatic disambiguation methods (such as the count-based and similarity-based disambiguation) that only follow links included in the knowledge graph will never find the correct result, leaving manual disambiguation by a human as the only correct option in these cases. In a number of other cases, no resource exists to represent a concept, as was the case with the terms "lad" and "madhouse". The lower connectivity between concepts in DBpedia also resulted in many of the distances defaulting to NSWD$_{max}$ during the evaluation. Freebase was found to be richer in this regard, as we found less zero-scores, and smaller variances in the similarities than in the DBpedia results. Concepts in Freebase were missing for fewer terms than in DBpedia, and there were less cases where two terms in a term-pair corresponded to the same URI. Still, terms such as "lad" and "madhouse" have no direct equivalent on Freebase.

9   Conclusion and Future Work

In this paper, we investigated the Normalized Semantic Web Distance: a semantically aware adaptation of the Normalized Information Distance, relying on links in a knowledge graph. We described three variations, taking into account incoming links, outgoing links, or both. We discussed the properties and special cases, and we proposed a conversion of the NSWD to measure similarity, using a customized normalization scheme. We extensively evaluated our approach, ensuring external validity by choosing an established benchmark: the Miller-Charles dataset of 30 human-assessed term-pairs. When applied to the Freebase knowledge graph, the NSWD and its variants exhibit a correlation of up to 0.58 with human similarity assessments, and when applied to DBpedia, the correlation was even higher at 0.69, albeit with less fine-grained similarity scores due to DBpedia's smaller size. We also verified that the NSWD maintains semantic awareness when confronted with ambiguous concept-pairs.

In future work, we will further illustrate the merit of the NSWD variants by applying them on more domain-specific knowledge graphs. We suspect that if the domain knowledge of the graph is high, the NSWD variants should be aware of these semantics and perform better than traditional approaches. Additionally, we aim to gather a larger set of concepts with an ambiguous plain-text representation as used to illustrate the semantic awareness of the NSWD variants, using a more systematic approach, supported by human assessments.

Acknowledgments. The research activities in this paper were funded by Ghent University, iMinds (by the Flemish Government), IWT Flanders, FWO-Flanders, the European Union, and RWTH Aachen University.

References

1. Beecks, C., Uysal, M.S., Seidl, T.: Signature quadratic form distance. In: Proceedings of the ACM International Conference on Image and Video Retrieval, pp. 438–445. ACM (2010)
2. Ceccarelli, D., Lucchese, C., Orlando, S., Perego, R., Trani, S.: Learning relatedness measures for entity linking. In: 22nd ACM International Conference on Information and Knowledge Management, pp. 139–148. ACM (2013)
3. Cilibrasi, R., Vitanyi, P.: Clustering by compression. IEEE Trans. Inf. Theory **51**(4), 1523–1545 (2005)
4. Cilibrasi, R., Vitányi, P.M.B.: Normalized web distance and word similarity. CoRR abs/0905.4039 (2009)
5. Cilibrasi, R.L., Vitanyi, P.M.B.: The Google similarity distance. IEEE Trans. Knowl. Data Eng. **19**(3), 370–383 (2007)
6. De Nies, T., Beecks, C., De Neve, W., Seidl, T., Mannens, E., Van de Walle, R.: Towards named-entity-based similarity measures: challenges and opportunities. In: ESAIR, pp. 9–11. ACM(2014)

7. De Nies, T., Beecks, C., Godin, F., De Neve, W., Stepien, G., Arndt, D., De Vocht, L., Verborgh, R., Seidl, T., Mannens, E., Van de Walle, R.: A distance-based approach for semantic dissimilarity in knowledge graphs. In: Proceedings of the 10th International Conference on Semantic Computing (ICSC, TBP). IEEE (2016)
8. De Vocht, L., Coppens, S., Verborgh, R., Vander Sande, M., Mannens, E., Van de Walle, R.: Discovering meaningful connections between resources in the web of data. In: Proceedings of the 6th Workshop on Linked Data on the Web (2013)
9. Eskevich, M., Jones, G.J., Aly, R., Ordelman, R., Chen, S., Nadeem, D., Guinaudeau, C., Gravier, G., Sébillot, P., De Nies, T., et al.: Multimedia information seeking through search and hyperlinking. In: Proceedings of the 3rd ACM Conference on International Conference on Multimedia Retrieval, pp. 287–294 (2013)
10. Godin, F., De Nies, T., Beecks, C., De Vocht, L., De Neve, W., Mannens, E., Seidl, T., Van de Walle, R.: The normalized Freebase distance. In: 11th Extended Semantic Web Conference (2014)
11. Hliaoutakis, A., Varelas, G., Voutsakis, E., Petrakis, E.G., Milios, E.: Information retrieval by semantic similarity. Int. J. Semant. Web Inf. Syst. (IJSWIS) 2(3), 55–73 (2006)
12. Jacobs, I., Walsh, N., et al. (eds.): Architecture of the World Wide Web, Volume One. W3C Recommendation 15 December 2004
13. Kulkarni, S., Caragea, D.: Computation of the semantic relatedness between words using concept clouds. In: Proceedings of the International Conference on Knowledge Discovery and Information Retrieval, pp. 183–188 (2009)
14. Li, M.: An Introduction to Kolmogorov Complexity and Its Applications. Springer, Heidelberg (1997)
15. Li, M., Chen, X., Li, X., Ma, B., Vitanyi, P.: The similarity metric. IEEE Trans. Inf. Theory 50(12), 3250–3264 (2004)
16. Meymandpour, R., Davis, J.G.: Recommendations using linked data. In: Proceedings of the 5th Ph.D. workshop on Information and knowledge, pp. 75–82 (2012)
17. Miller, G.A., Charles, W.G.: Contextual correlates of semantic similarity. Lang. Cogn. Process. 6(1), 1–28 (1991)
18. Milne, D., Witten, I.: An effective, low-cost measure of semantic relatedness obtained from Wikipedia links. In: Proceedings of the AAAI Workshop on Wikipedia and Artificial Intelligence: An Evolving Synergy, pp. 25–30, Chicago, USA (2008)
19. Moore, J.L., Steinke, F., Tresp, V.: A novel metric for information retrieval in semantic networks. In: García-Castro, R., Fensel, D., Antoniou, G. (eds.) ESWC 2011. LNCS, vol. 7117, pp. 65–79. Springer, Heidelberg (2012)
20. Nunes, B.P., Dietze, S., Casanova, M.A., Kawase, R., Fetahu, B., Nejdl, W.: Combining a co-occurrence-based and a semantic measure for entity linking. In: Cimiano, P., Corcho, O., Presutti, V., Hollink, L., Rudolph, S. (eds.) The Semantic Web: Semantics and Big Data. LNCS, vol. 7882, pp. 548–562. Springer, Heidelberg (2013)
21. Nunes, B.P., Herrera, J., Taibi, D., Lopes, G.R., Casanova, M.A., Dietze, S.: SCS connector-quantifying and visualising semantic connections between entity pairs. In: Presutti, V., Blomqvist, E., Troncy, R., Sack, H., Papadakis, I., Tordai, A. (eds.) The Semantic Web: ESWC 2014 Satellite Events. LNCS, vol. 8798, pp. 461–466. Springer, Heidelberg (2014)
22. Passant, A.: Measuring semantic distance on linking data and using it for resources recommendations. In: AAAI Spring Symposium: Linked Data Meets Artificial Intelligence (2010)

23. Sánchez, D., Batet, M., Isern, D., Valls, A.: Ontology-based semantic similarity: a new feature-based approach. Expert Syst. Appl. **39**(9), 7718–7728 (2012)
24. Schwartz, B.: Google removes the + search command (2011). http://searchengineland.com/google-sunsets-search-operator-98189
25. Zuo, Z., Huang, H.H., Kawagoe, K.: Similarity search of human behavior processes using extended linked data semantic distance. In: 25th International Workshop on Database and Expert Systems Applications (DEXA), pp. 178–182. IEEE (2014)

Gleaning Types for Literals in RDF Triples with Application to Entity Summarization

Kalpa Gunaratna[1]($\boxtimes$), Krishnaprasad Thirunarayan[1], Amit Sheth[1], and Gong Cheng[2]

[1] Kno.e.sis, Wright State University, Dayton, OH, USA
{kalpa,tkprasad,amit}@knoesis.org
[2] National Key Laboratory for Novel Software Technology,
Nanjing University, Nanjing, China
gcheng@nju.edu.cn

Abstract. Associating meaning with data in a machine-readable format is at the core of the Semantic Web vision, and typing is one such process. Typing (assigning a class selected from schema) information can be attached to URI resources in RDF/S knowledge graphs and datasets to improve quality, reliability, and analysis. There are two types of properties: object properties, and datatype properties. Type information can be made available for object properties as their object values are URIs. Typed object properties allow richer semantic analysis compared to datatype properties, whose object values are literals. In fact, many datatype properties can be analyzed to suggest types selected from a schema similar to object properties, enabling their wider use in applications. In this paper, we propose an approach to glean types for datatype properties by processing their object values. We show the usefulness of generated types by utilizing them to group facts on the basis of their semantics in computing diversified entity summaries by extending a state-of-the-art summarization algorithm.

Keywords: Type inference · Datatype properties · RDF triples · Feature grouping and ranking · Entity summarization · Dataset enrichment

1 Introduction

The rise of open data initiatives (e.g., Linking Open Data) has encouraged large-scale publication of data on the Web. Resource Description Framework (RDF) has been used extensively to encode information and publish as Semantic Web datasets and knowledge graphs. The direct consumers of these datasets, most of the time, are machines or software such as search and rank services. Therefore, it is imperative to enrich the datasets with additional information so that machines can interpret them properly. Assigning ontology classes as types to resources (typing) in RDF via the `rdf:type` property (which we refer to as semantic types) is one example of a data enrichment process. This can help machines to

© Springer International Publishing Switzerland 2016
H. Sack et al. (Eds.): ESWC 2016, LNCS 9678, pp. 85–100, 2016.
DOI: 10.1007/978-3-319-34129-3_6

identify similar or related resources by analyzing their types. Such enrichment can be exploited to improve the quality and reliability of datasets and facilitate analytics. Typing is performed on URI resources and, hence, only applies to object properties. Datatype property values are literals, and are usually associated with types by virtue of their syntactic data representation (which we refer to as syntactic types). For example, one assigns `datatypes` such as `xsd:string`, `xsd:integer,` and `xsd:date` to literals.

Syntactic types do not make explicit information that can be exploited for data analysis. However, the amount of information that datatype properties represent compared to object properties is significant in some real-world datasets. For instance, a recent version of DBpedia [10], which is one of the largest and most comprehensive encyclopedic datasets on the Web, has 1,600 datatype properties compared to 1,079 object properties[1]. Many of the literal values (other than noise) can be associated with types selected from a set of ontology classes that can promote proper semantic interpretation and use. For example, the property http://dbpedia.org/property/location has about 1,05,047 unique and simple literals that can be mapped to entities to infer the types (e.g., "California", "United States"). An entity is a thing (e.g., person, book, place) at the data level that encapsulates facts and is represented by a URI.

The importance of type information has been demonstrated by researchers in the Semantic Web community, including for inferring missing types for entities [13], ranking types for entities [17], and generating summaries using type graphs [18]. All these approaches make use of existing type information of "entities" or infer additional/missing types from them. There has not been any work related to inferring or computing types for literals in RDF/S datasets. In this work, we propose to address the issue of computing "semantic types" whenever possible for literal values of datatype properties. When types are available for datatype properties, they can be used in many interesting applications such as data integration, property alignment, and entity summarization. For example, in property alignment [5,7], we can utilize types to prune the candidates for alignment. Further, type prediction on datatype properties provides benefits similar to the works on type prediction for entities as in SDType [13]. We demonstrate the application of generated semantic types by extending the FACES entity summarization algorithm [6] and show how to group and rank features based on datatype properties. Our contributions in this work are twofold:

1. We analyze the object value of datatype properties and select a suitable class as the type from a given set of ontology classes.
2. We extend FACES (FACES-E) to group and rank both object and datatype properties to create entity summaries and demonstrate the usefulness of types to generate comprehensive entity summaries.

The rest of the paper is organized as follows. In Sect. 2, we analyze the problem of typing literals in datatype properties, and in Sect. 3, we define the problem and notions related to typing and summarization. In Sect. 4, we present

---

[1] http://wiki.dbpedia.org/About.

our typing algorithm and feature ranking for datatype properties, followed by evaluation in Sect. 5. We present a general discussion with future directions in Sect. 6, and present related work in Sect. 7. Finally, we conclude in Sect. 8.

2 Problem Analysis

Web Ontology Language (OWL) defines the types of properties: object properties that connect individuals to individuals and datatype properties that connect individuals to data values (literals)[2]. The object value of an object property is a URI that can be assigned an ontology class as its type via `rdf:type` property. But object values of datatype properties do not have ontology classes assigned as types and the only types available for them are the syntactic types referring to primitive, low-level implementation types. In this work, we try to suggest a class from a given set of classes as the type of the literal of the datatype property. See Fig. 1 which shows two triples (2) and (3) having datatype properties of the entities `dbr:Barack_Obama` and `dbr:Calvin_Coolidge` taken from DBpedia. The dotted boxes show the types for the two literals that we intend to compute, supplementing the syntactic type `xsd:string` which is already available. Whenever a semantic type can be computed for a literal, it can be used in practical applications for inferencing, grouping, and matching. For example, both computed types are `rdfs:subClassOf dbo:Politician` class in the DBpedia ontology and hence can be utilized for matching or grouping of related literals.

Datatype properties may have been provided instead of object properties for entities in datasets for various reasons, such as (i) the creator was unable to find a suitable entity URI for the object value, and hence chose to use a literal instead, (ii) the creator of the triple did not want to attach more details to the value and hence represented it in plain text, (iii) the value contains only basic implementation types like integer, boolean, and date, and hence not meaningful to create an entity, or (iv) the value has a lengthy description spanning several sentences (e.g., `dbo:abstract` property in DBpedia) that covers a diverse set of entities and facts. We attempt to assign a semantic type by analyzing cases

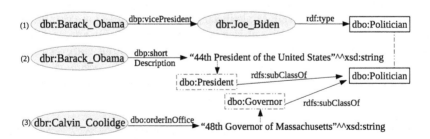

Fig. 1. Two triples corresponding to datatype properties and one triple corresponding to an object property taken from DBpedia. Computed types are shown in dashed boxes.

---

[2] http://www.w3.org/TR/owl-ref/#Property.

(i) and (ii) for text (up to a sentence long, delimited by a period) and avoid assigning a "type" to lengthy literal values as mentioned in (iv) because its focus is not clear (and multiple conflicting types can result).

3 Problem Statement

3.1 Typing Datatype Properties

Problem Statement: Let S P O be an RDF triple specifying subject (S), property (P), and object (O), and C be all classes (in the schema) for the set of triples D. If P is an object property, then O is an entity (i.e., individual). The type of O is a class assignment to O via the RDF triple, O `rdf:type` c, where $c \in C$. We refer to this as "semantic typing" in this work. Since the value of a datatype property instance is a literal, no semantic typing can be found for the value. We want to suggest a class $\bar{c} \in C$ for the value of a datatype property in addition to the syntactic type. As explained earlier in Sect. 2, we focus on text that is up to one sentence long. This realistic restriction has been imposed to ensure entity and type coherence. That is, deriving a unique type, which may not apply for longer texts, say a paragraph. Figure 1 illustrates types suggested for the object values of the triples (2) and (3).

For clarity of presentation, we define Type Set ($TS(v)$) for property value v as the set of classes that are assigned (via `rdf:type`) or inferred (via `rdfs:subClass Of`) from the class set C. If p is an object property, then $|TS(v))| > 0$, otherwise $|TS(v)| = 0$.

3.2 Ranking and Grouping Datatype Properties

Preliminaries: Let E, P, and L be sets of all entities, properties, and literals, respectively, in the triple set D (capturing a knowledge graph or dataset). V is the set of all property values and $V \subseteq E \cup L$. E and P are represented by URIs. An entity $e \in E$ is described using property-value pairs of the form (p, v) where $p \in P$ and value $v \in V$. Each of these property-value pairs of an entity is called a *feature f* of the entity. Furthermore, an entity and one of its features together correspond to an RDF triple in D. $prop(f)$ and $val(f)$ are two functions that return the property and its object value, respectively, for a feature f. The *feature set* of an entity e, denoted by $FS(e)$, is the set of all features that can be found for e in D. Using these notions, a *faceted entity summary* of entity e, which is a subset of all features that can be associated with e, is defined as follows (reproduced from [6] for completeness).

Definition 1 (Facet). *Given an entity e, a set of facets F(e) of e is a partition of the feature set FS(e). That is, $F(e) = \{A_1, A_2, \ldots A_n\}$ such that F(e) satisfies the following criteria: (i) Non-empty: $\emptyset \notin F(e)$. (ii) Collectively exhaustive:* $\bigcup_{A \in F(e)} A = FS(e)$. *(iii) Mutually (pairwise) disjoint: if $A_i, A_j \in F(e)$ and $A_i \neq A_j$ then $A_i \cap A_j = \emptyset$. Each A_i is called a facet of e.*

Faceted entity summary computation requires the identification of facets. Note that there can be multiple partitions for a given feature set, but in faceted entity summary generation, we compute a unique desirable partition by grouping related features together using a clustering algorithm that employs property name expansion (using WordNet[3]) and the "type information" associated with property values [6]. Informally, if the number of facets is n and the size of the summary is k, *at least one feature* from each facet is included in the summary when $k > n$. This can be achieved by iteratively picking features from each facet based on their ranking. If $k \leq n$, then *at most one feature* from each facet is included in the summary to improve diversity. Definition 2 formalizes the aforementioned idea about a faceted entity summary in set-theoretic terms.

Definition 2 *(Faceted Entity Summary)*. *Given an entity e and a positive integer $k < |FS(e)|$, faceted entity summary of e of size k, FSumm(e, k), is a collection of features such that FSumm$(e, k) \subset FS(e), |FSumm(e, k)| = k$. Further, either (i) $k > |F(e)|$ and $\forall X \in F(e)$, $X \cap FSumm(e, k) \neq \emptyset$ or (ii) $k \leq |F(e)|$ and $\forall X \in F(e), |X \cap FSumm(e, k)| \leq 1$ holds, where $F(e)$ is a set of facets of $FS(e)$.*

The ranking of features in each facet is computed using the informativeness $Inf(f)$ of feature f (Eq. 1) and popularity $Po(v)$ of value v, where $v = val(f)$ (Eq. 2). $N = |E|$ is the total number of entities in triple set D. The final rank score $Rank(f)$ is computed using Eq. 3. In our prior work [6], these ranking features have only been defined for object properties. In Sect. 4.2, we discuss how to adapt them for datatype properties.

$$Inf(f) = log(\frac{N}{|\{e \in E | f \in FS(e)\}|}) \tag{1}$$

$$Po(v) = log|\{triple\ t \in D | \exists\ e, f : t \equiv (e\ prop(f)\ v)\}| \tag{2}$$

$$Rank(f) = Inf(f) * Po(val(f)) \tag{3}$$

Problem Statement: Ranking and grouping features of object properties can be achieved using the entities that their values represent. For ranking, we can utilize Eqs. 1 to 3, and for grouping semantically related features, we could use their types [6]. Ranking features belonging to datatype properties cannot be done similarly because their literal values do not have a unique representation across the dataset (which URIs do provide for object properties) as the same entity may be referred to using minor variants of a literal. Therefore, we need to reflect related entities in ranking datatype properties by modifying Eqs. 1 to 3. For example, consider the third triple of Fig. 1 where we can spot: dbr:Governor and dbr:Massachusetts; here, we can use their frequency as opposed to checking the frequency of the entire literal value of the property.

---

[3] https://wordnet.princeton.edu/.

Grouping (conceptually similar) datatype properties is non-trivial compared to grouping object properties where types are available. Note that multiple entities and/or classes spotted in a datatype property value can confuse the groupings. For example, the second triple in Fig. 1 has the entity dbr:United States having the type dbo:Country but eventually results in the class dbo: President as the type. In general, this requires recognizing multiple types (e.g., country and president) and then resolving them suitably (e.g., to president) to enable the grouping of similar triples (among both object and datatype properties). See triple (1) and (2) in Fig. 1, where the first represents an object property and the second represents a datatype property. In fact, both values convey information about a person, while for the datatype property value, it is not explicit. The object property clearly has a type assigned to its value and if we compute the type for the datatype property as dbo:President, then we can abstract their values to type dbo:Politician which can be inferred for the datatype property value using rdfs:subClassOf.

4   Approach

First, we will investigate how to compute types for the values of datatype properties and then utilize them in grouping related properties and values based on the computed types. We will also discuss how to generate entity summaries based on new ranking measures for the datatype properties and groupings.

4.1   Typing Datatype Property Values

Determining the relevant type for a datatype property value is challenging due to several reasons. First, picking some term used in the literal value to determine the entity or class for the datatype property value does not work. For example, in triple (3) in Fig. 1, if we select the term "Massachusetts" as the entity to represent the entire value and use its type dbo:PopulatedPlace as the type, we obtain an incorrect interpretation. The main focus of the text is the term "Governor" and not "Massachusetts", as it conveys information about the governor. Therefore, we propose identifying this term, which we call the *focus term*, by analyzing the grammatical structure of the text. Then, we match the identified focus term to a suitable entity or a class in deriving the type for the value.

We utilize the Collins head word detection technique [2] to identify the focus terms. The Stanford CoreNLP[4] API offers an implementation of this technique using parse trees. We use the UMBC semantic similarity service [9] to compare the suggested class and the set of given classes. It facilitates the computation of phrase similarity, which generalizes and improves upon WordNet-based similarity. The algorithm for generating a type set $(TS(v))$ for a datatype property value v is presented in Algorithm 1.

Algorithm 1 shows details of the method *getTypesForText*, which generates types for the input text (datatype property value). We avoid processing if the

---

[4] http://nlp.stanford.edu/software/corenlp.shtml.

Algorithm 1. getTypesForText(*Text v*)

```
 1: initialize Set types to {} and pre-determined Integer n
 2: Set X ← getPhrases(v)
 3: for each Phrase x ∈ X do
 4:    if isNumeric(x) then
 5:       Set cls ← predefined date/numeric type
 6:    else
 7:       Set ngrams ← getNGrams(x, n)
 8:       Text focusTerm ← parseHeadWord(x) {head word identifier}
 9:       Set cls ← getTypeFromLabel(focusTerm)
10:       if isEmpty(cls) then
11:          cls ← getTypesFromNGrams(focusTerm, ngrams)
12:       end if
13:       if isEmpty(cls) then
14:          cls ← getMatchedType(focusTerm) {semantic matching}
15:       end if
16:    end if
17:    types ← cls
18: end for
19: return types
```

---

input text is more than one sentence long (segmented by "period"). If the identified sentence has phrases delimited by comma, we segment them (lines 2–3) and generate types for them. This is because these phrases normally align for the same abstract meaning (e.g., "Austrian-American bodybuilder, actor", "Denison, Texas"). We identify numeric or date values using simple regular expressions (lines 4–5). If the value is not numeric, we start the type computation process for the phrase by identifying n-grams associated with the phrase up to the maximum token length of n (line 7). Then, we retrieve the focus term by parsing the phrase using the head word identifier (line 8). Next, we check whether there is an exact match of the focus term and any of the types (via **rdfs:label** of classes) in the dataset. If a match is found, we take the class as the type of the phrase (line 9). Otherwise, we further analyze all the generated n-grams with the focus term to infer a type in the *getTypesFromNGrams* method (line 11). If there is still no match, we compute the similarity scores of the focus term against all the types in the dataset (via **rdfs:label** of classes) and get the highest match (>0) as the type of the phrase (line 14). Finally, we aggregate types generated for each phrase to obtain the set of types for the input text.

The method *getTypesFromNGrams* processes the n-grams set to allow for a maximal match of entity labels. It processes n-grams to extract types only if they contain the focus term. For each of those n-grams that contain the focus term, we check to see whether there is an exact match of the n-gram to a type. If no match is found, we spot entities for the n-gram and then get the types of those entities. We spot entities by exact matching of their labels (**rdf:label**) to n-grams. Looking for n-grams that contain the focus term (in descending order of n-gram token lengths from n to 1) can improve the quality of the identified types. For example, consider "Harvard Law School" as a datatype property value in DBpedia. The identified focus term for this phrase is "School." When we start processing n-grams in descending order of n, we encounter "Harvard Law School" as the first candidate for typing. This matches the entity **dbr:Harvard_Law_School** whose type `dbo:Educational_Institution` is then taken as the type of the phrase. We do not generate types for long text[5] (e.g., paragraphs) because they cannot be unambiguously typed as they can represent many different entities (with contrasting descriptions) and need further analysis to pick the correct type.

4.2 Grouping and Ranking Datatype Property Features for Summaries

The approach to rank and group object properties to create faceted entity summaries is presented in [6]. The idea of generating faceted summaries focuses on grouping related features (i.e., property-value pairs) and then picking the highest ranked ones from each group. Next, we discuss how to group and then rank datatype property based features.

Grouping Datatype Property Features: Grouping of features can be done at two levels: exact/ syntactic similarity and semantic/abstract similarity. Exact/ syntactic similarities can result in very fine-grained groups, while we are interested in groups based on their abstractions in FACES [6]. For

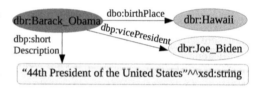

Fig. 2. Grouping both property features.

example, triples (2) and (3) in Fig. 1 present two literal values that do not share any common token (no/less syntactic similarity). However, when we compute a type for each, they are sub-types of the class `dbo:Politician`. The clustering algorithm that uses such type information of the values as in FACES can group them together. Further, it can also group features of both object and datatype properties which was hitherto not possible. For example, It can group features represented by triples (1) and (2) in Fig. 1 because the values are indirect instances of the type `dbo:Politician`. Figure 2 illustrates grouping of similar features with same color using the clustering algorithm presented in [6].

Ranking Datatype Property Features: We discuss ranking measures for datatype properties usable in the context of entity summarization. Recall that Eqs. 1 to 3 can be used only to rank object properties [6]. If we compute $Inf(f)$ as in Eq. 1 for datatype properties, it will have an artificially high value because the exact literal denoting an entity appears infrequently compared to URI references of the entities for object properties. As a consequence, every datatype property will have a high ranking score. To fix this discrepancy, we spot entities in datatype property values and get the frequency of entity URIs as a measure of informativeness and popularity. We first spot entities in the datatype property values by analyzing the n-grams generated in Algorithm 1. Let $ES(v)$ be the

---

[5] Note that we can still run the algorithm for each sentence to generate types.

set of all entities that can be spotted for value $v = val(f)$. We process all the n-grams generated for v (for a pre-defined n-gram token length n) and match them against the entity labels (`rdfs:label`) in the dataset to obtain $ES(v)$. Then, we choose the most popular (frequent) entity in the dataset from this set as the representative entity for v. The intuition is that humans spot and identify popular entities in text phrases and, hence, they can provide identifiable facts in the summary for datatype properties. Let $max(ES(v))$ be a function that returns the most popular entity e_{max} from the set $ES(v)$ based on the frequency of appearance of each entity (using Eq. 2). For example, for triple (2) in Fig. 1, function max identifies `dbr:President` and `dbr:United_States` as the entities and picks the latter to be the most popular entity. Hence, it is used to calculate the informativeness of the feature and popularity of the phrase.

We compute the informativeness of a feature f of a datatype property, $Inf(f)'$, using Eq. 4. We check for occurrences of features in entities "similar" to f (as opposed to checking the same feature as in object properties) by matching datatype property names and values that contain the most popular entity e_{max}. Then, we count those entities to compute informativeness of the feature. That is, informativeness is inversely proportional to the number of entities that are associated with overlapping values containing e_{max}. N is the total number of entities.

$$Inf(f)' = log\left(\frac{N}{\left|\left\{^{e \in E | \exists\, f' \in FS(e)\, :\, prop(f) = prop(f')\, and}_{max(ES(val(f))) \in ES(val(f'))}\right\}\right|}\right) \quad (4)$$

Similarly, for measuring the popularity $Po(v)'$ of a datatype property value v, we take the frequency of the most popular entity $e_{max} = max(ES(v))$ in v, as specified in Eq. 5. Then, the ranking score of feature f that belongs to datatype properties, $Rank(f)'$, is calculated using Eq. 6. When $ES(v) = \emptyset$, we take the denominator as the number of property instances in Eq. 4 and $Po(v)' = 1$ in Eq. 5, effectively ranking them low.

$$Po(v)' = log\left|\left\{^{triple\ t \in D | \exists\, e, f\, :\, t \equiv (e\ prop(f)\ e_{max})and}_{e_{max} = max(ES(v))}\right\}\right| \quad (5)$$

$$Rank(f)' = Inf(f)' * Po(val(f))' \quad (6)$$

Faceted Entity Summaries Using Object and Datatype Properties: Eqs. 1–6 are used to rank features within each facet (cluster partition). We further extend the FACES approach by ranking facets using the average of feature ranking scores ($FacetRank(F(e))$) generated by Eqs. 3 and 6, as shown in Eq. 7 for a facet $F(e)$. $R(f)$ is a function that selects the proper ranking method depending on whether $prop(f)$ is an object or datatype property. Then, facets are ordered from the highest to the lowest FacetRank score and we iterate over them in that order to pick individual features for the summary.

$$R(f) = \begin{cases} Rank(f), & \text{if } prop(f) \text{ is an object property} \\ Rank(f)', & \text{otherwise} \end{cases}$$

$$FacetRank(F(e)) = \frac{\sum_{f \in F(e)} R(f)}{n}, \; where \; n = |F(e)| \tag{7}$$

Given the feature set $FS(e)$ of an entity e and a positive integer $k < |FS(e)|$, the adapted process for the faceted entity summary creation is as follows. (1) The feature set $FS(e)$ is partitioned into facets. The algorithm yields a dendrogram (hierarchical tree) for $FS(e)$ and it is cut at an empirically determined level to get the facet set $F(e)$ of $FS(e)$. (2) Features in each facet are ranked using the ranking algorithms (Eqs. 3 and 6). (3) Then the feature ranking scores of features in each facet are aggregated and averaged to get the facet ranking score (Eq. 7). (4) The top ranked features, from highest to lowest ranked facet, are picked (Definition 2) to form the faceted entity summary of length k.

5   Evaluation and Results Discussion

We evaluate our contributions in two steps: (1) evaluation of types generated for literal values of datatype properties and (2) evaluation of faceted entity summarization using features belonging to both types of properties. We empirically set $n = 3$ (n-grams) for both evaluations. More details on approach, evaluation, and examples are available at http://wiki.knoesis.org/index.php/FACES.

5.1   Evaluating Datatype Property Types

Generating types for all the available datatype property values is not meaningful because there are labeling properties that simply represent human readable names for entities. The RDFS (RDF Schema) standard defines the `rdfs:label` property to provide such information, but in practice, there exist many such labeling properties (e.g., foaf:name). Ell et al. [3] studied the characteristics of these properties by manually inspecting the properties and their instance data. Similarly, we created a list of labeling properties for our data sample and filtered them out. We extracted a sample of unique datatype property-value pairs from DBpedia (version 3.9 and 2015-04). *Precision* for the identified types of a property value v is defined in terms of $TS(v)$ in Eq. 8. Then, we define the *Mean Precision* (MP) of property values as in Eq. 9, where n is the number of property values in the sample that have $|TS(v)| > 0$.

$$Precision(TS(v)) = \frac{\#correct \; types \; in \; TS(v)}{|TS(v)|} \tag{8}$$

$$MeanPrecision = \frac{\sum_{i=1}^{n} Precision(TS(val(f_i)))}{n} \tag{9}$$

Mean Precision is the average of precision over the property values in the feature sample. When the MP value is higher, the algorithm generates many correct types over different property values. It is important to know how often the algorithm can generate at least one correct type. Therefore, we define *Any Mean Precision* (AMP) as in Eq. 10, where n is the number of property values

Table 1. Type generation evaluation.

	Mean precision (MP)	Any mean precision (AMP)	Coverage
Our approach	0.8290	0.8829	0.8529
Baseline	0.4867	0.5825	0.5533

in the sample that have $|TS(v)| > 0$. It computes the average of all the ceiling values of Precision($TS(v)$). If the algorithm generates at least one correct type for a value, it counts the precision as 1 in averaging. When AMP is higher, the algorithm generates at least one correct type often. When both the MP and AMP values are higher, the algorithm can be considered reliable.

$$AnyMeanPrecision = \frac{\sum_{i=1}^{n} \lceil Precision(TS(val(f_i))) \rceil}{n} \qquad (10)$$

Table 1 shows the evaluation results performed by one evaluator. We constructed a baseline using a state-of-the-art tool to identify entities in the values and retrieved their types and super types (except `owl:Thing`). Specifically, we used DBpedia Spotlight [11] for this purpose and configured it with default parameters including the confidence of 0.5. We had a total of 1,117 unique property-value pairs after filtering out 118 labeling and noisy pairs. Coverage is the fraction of features that had a type generated. Our approach performed better compared to the baseline because we identify types using a combination of focus terms and matching entities and types. We did not measure recall because it is hard to produce an exhaustive list of all correct types for each value.

5.2   Evaluating Faceted Entity Summarization Use Case

We evaluated the proposed extended faceted entity summarization approach FACES-E against another state-of-the-art algorithm called RELIN [1]. It has been shown before that FACES outperformed RELIN for object properties [6]. RELIN has been the only tool to generate entity summaries for both datatype and object properties. We evaluate FACES-E against RELIN for the full range of features and show the benefits of the datatype property typing which enabled FACES-E to group features belonging to object and datatype properties in the partition algorithm. We randomly selected 20 entities from the FACES evaluation (DBpedia 3.9) [6] and another random sample of 60 entities from DBpedia version 2015-04 for a total of 80 unique entities. We retrieved object properties as earlier [6] and added datatype properties to each entity, filtering labeling properties (including date and numeric as in Sect. 5.1). We created a new gold standard for the entity samples by asking 17 human users to create summaries of length 5 and 10 for each of the 80 entities as the "ideal summaries" (total of 900 user-generated ideal summaries for both summary lengths). Each entity received at least 4 different ideal summaries and this comprises the gold standard. We use the same evaluation metrics as in [1,6]. When there are n ideal

Table 2. Evaluation of the summary quality (average for 80 entities) and % ↑ = 100 * (FACES-E avg. quality - Other system avg. quality) / (Other system avg. quality) for $k = 5$ and $k = 10$, where k is the summary length.

System	$k = 5$		$k = 10$	
	Avg. Quality	% ↑	Avg. Quality	% ↑
FACES-E	1.5308	–	4.5320	–
RELIN	0.9611	59 %	3.0988	46 %
RELINM	1.0251	49 %	3.6514	24 %
Avg. Agreement	2.1168		5.4363	

summaries denoted by $Summ_i^I(e)$ for $i = 1, .., n$ and an automatically generated summary denoted by $Summ(e)$ for entity e, the agreement on ideal summaries is measured by Eq. 11 and the quality of the automatically generated summary is measured by Eq. 12. In other words, the quality of an entity summary is its average overlap with the ideal summaries for the entity in the gold standard.

$$Agreement = \frac{2}{n(n-1)} \sum_{i=1}^{n} \sum_{j=i+1}^{n} |Summ_i^I(c) \cap Summ_j^I(e)| \quad (11)$$

$$Quality(Summ(e)) = \frac{1}{n} \sum_{i=1}^{n} |Summ(e) \cap Summ_i^I(e)| \quad (12)$$

We used previously determined thresholds for both FACES-E and RELIN. RELINM is the modified version of RELIN where it discards duplicate properties in the summary [6]. For FACES-E, we cut cluster hierarchies at level 3 and set the cut-off threshold of the clustering algorithm (Cobweb) to 5. For RELIN and RELINM, we set the jump probability to 0.85 and the number of iterations to 10. The evaluation results of entity summarization are presented in Table 2. Note that the *summary quality* is better when it is closer to the *agreement* value. Agreement is low for this evaluation, as was the case with previous evaluations [6], because the number of features per entity was relatively high (on average 44 features per entity). Our intuition for ranking datatype property features by giving precedence to popular entity mentions in the value, grouping features based on types, including the generated semantic types for datatype property values, and ranking facets to select faceted summaries, have been validated by the high summary quality in this evaluation. We conducted a *paired t-test* to confirm the significance of the FACES-E's mean summary quality improvements over RELINM. For $k = 5$ and $k = 10$, p_values for FACES against RELINM are 8.24E-11 and 9.42E-12. When p_values are less than 0.05, the results are statistically significant. According to results shown in Table 2 and the paired t-test, FACES-E performed better and benefited from datatype property typing.

6   General Discussion and Future Directions

Our typing algorithm enabled FACES-E to process both types of properties, a limitation of FACES, improving coverage. Likewise, typing for datatype properties can facilitate a wide range of data processing applications. Property alignment [7] can be one such instance where similarly typed property values can be used to limit the properties analyzed for equivalence and relatedness.

In the first evaluation, our algorithm showed MP and AMP values of 0.82 and 0.88 compared to 0.48 and 0.58, respectively of the baseline. This aligns with our claim that typing for values needs to identify the correct and appropriate focus. Furthermore, our algorithm showed good coverage. Getting both MP and AMP values to a relatively high level is desirable for faceted entity summaries. This is because when the algorithm can generate a correct type most of the time, also with high precision, it helps to group semantically similar features together. This, in turn, facilitates the creation of high quality, "diversified" entity summaries evidenced by the second evaluation in Sect. 5.2. However, we also note that datatype properties in our entity sample have labeling and noisy properties (due to incorrect or missing details) which is normal for real-world datasets on the Web. We manually filtered such labeling properties; however, this can be a challenging and important problem to solve in the future.

It is possible to use the meaning of the property names and word sequence relationships of values for type generation. For this, a machine learning model similar to Conditional Random Fields (CRF) could be utilized whereas now, focus term detection drives the type computation. This can facilitate predicting whether a value can be typed or not and filtering out noisy and labeling property values. The absence of matching entities in DBpedia can stymie the generation of type information. These are common dataset quality and completeness issues orthogonal to our problem. For missing information, we can: (1) use type inferencing approaches [13] to generate missing types and (2) use a comprehensive set of ontology classes and entities (e.g., from LOD) in our approach.

Our approach does not generate "semantic types" for numeric and date value properties, and challenges exist for measuring their popularity in the dataset for ranking. These properties will be investigated in the future for grouping and ranking in faceted entity summary generation. Furthermore, a formal model/approach needs to be adapted for RDF semantics to encapsulate type generations for datatype properties and then they can be encoded in the datasets similar to object properties whereas now, we keep computed values separate.

7   Related Work

Type assignment to text fragments is known as Named Entity Recognition (NER) [12]. NER consists of two subtasks: segmenting and classifying segmented text blocks into pre-defined categories (types). Entity Linking (EL) [8] maps entity mentions in text to their corresponding entities in knowledge bases. NER produces types for segments of the input text whereas EL identifies entities from which types can be inferred. NER and EL, however, differ from our problem in that they do not try to suggest a type based on the focus of the text but rather try to determine the types of all the entities present, causing ambiguity from the perspective of our problem (e.g., our evaluation with DBpedia Spotlight [11]).

Finding missing types for entities [13,14] is important for the reliability of datasets and reasoning. Paulheim and Bizer [13] infer types for entities in DBpedia and Sleeman and Finin [14] predict types of entities for efficient co-reference resolution. TRank [17] ranks entity types based on the context in which they appear (disambiguation). Fang et al. [4] use type information for search, and Tylenda et al. [18] generate summaries by analyzing type graphs. These approaches work on entities and/or object properties or infer types for entities, whereas we focus on typing datatype properties where no semantic types are available.

FACES [6] proposed a concise, comprehensive, and diversified approach for entity summarization for object properties and showed superior results over the state-of-the-art techniques and methods [1,15,16]. SUMMARUM [16] employs a PageRank based algorithm to rank and compute summaries for only object properties whereas RELIN [1] utilizes PageRank to compute summaries for both object and datatype properties. We developed FACES-E to utilize both object and datatype properties and showed superior results compared to RELIN.

8   Conclusion

We have investigated the problem of computing types for datatype property values. We generate types for a property value by: (1) exact and semantic matching of the focus term to class labels, and (2) spotting entities related to the focus term and retrieving their types. Our contributions in this work span over two significant problems: (1) enhancing datatype property values with type metadata, and (2) proposing FACES-E, which extends FACES to generate more comprehensive entity summaries using both types of properties. We evaluated both type generation and the extended entity summarization approach using the DBpedia encyclopedic dataset on the Web and showed improvement over the state-of-the-art. Our novel typing algorithm for datatype property values enhances data with additional semantics and, hence, is useful in applications beyond entity summarization, such as property alignment, data integration, and dataset profiling.

Acknowledgments. We acknowledge partial support from the National Science Foundation (NSF) award: EAR 1520870: Hazards SEES: Social and Physical Sensing Enabled Decision Support for Disaster Management and Response. Any opinions, findings, and conclusions/recommendations expressed in this material are those of the author(s) and do not necessarily reflect the views of the NSF.

References

1. Cheng, G., Tran, T., Qu, Y.: RELIN: relatedness and informativeness-based centrality for entity summarization. In: Aroyo, L., Welty, C., Alani, H., Taylor, J., Bernstein, A., Kagal, L., Noy, N., Blomqvist, E. (eds.) ISWC 2011, Part I. LNCS, vol. 7031, pp. 114–129. Springer, Heidelberg (2011)
2. Collins, M.: Head-driven statistical models for natural language parsing. Comput. Linguist. **29**(4), 589–637 (2003)
3. Ell, B., Vrandečić, D., Simperl, E.: Labels in the web of data. In: Aroyo, L., Welty, C., Alani, H., Taylor, J., Bernstein, A., Kagal, L., Noy, N., Blomqvist, E. (eds.) ISWC 2011, Part I. LNCS, vol. 7031, pp. 162–176. Springer, Heidelberg (2011)
4. Fang, Y., Si, L., Somasundaram, N., Al-Ansari, S., Yu, Z., Xian, Y.: Purdue at TREC 2010 entity track: a probabilistic framework for matching types between candidate and target entities. In: TREC (2010)
5. Gunaratna, K., Lalithsena, S., Sheth, A.: Alignment and dataset identification of linked data in semantic web. Wiley Interdisc. Rev.: Data Min. Knowl. Disc. **4**(2), 139–151 (2014)
6. Gunaratna, K., Thirunarayan, K., Sheth, A.: FACES: Diversity-aware entity summarization using incremental hierarchical conceptual clustering. In: Proceedings of the Twenty-Ninth AAAI Conference on Artificial Intelligence, pp. 116–123. AAAI (2015). http://www.aaai.org/ocs/index.php/AAAI/AAAI15/paper/view/9562/9233
7. Gunaratna, K., Thirunarayan, K., Jain, P., Sheth, A., Wijeratne, S.: A statistical and schema independent approach to identify equivalent properties on linked data. In: Proceedings of the 9th International Conference on Semantic Systems, pp. 33–40. ACM (2013)
8. Hachey, B., Radford, W., Nothman, J., Honnibal, M., Curran, J.R.: Evaluating entity linking with wikipedia. Artif. Intell. **194**, 130–150 (2013)
9. Han, L., Kashyap, A., Finin, T., Mayfield, J., Weese, J.: UMBC ebiquity-core: semantic textual similarity systems. In: Proceedings of the Second Joint Conference on Lexical and Computational Semantics, vol. 1, pp. 44–52 (2013)
10. Lehmann, J., Isele, R., Jakob, M., Jentzsch, A., Kontokostas, D., Mendes, P.N., Hellmann, S., Morsey, M., van Kleef, P., Auer, S., et al.: Dbpedia-a large-scale, multilingual knowledge base extracted from wikipedia. Seman. Web J. **5**, 1–29 (2014)
11. Mendes, P.N., Jakob, M., García-Silva, A., Bizer, C.: DBpedia spotlight: shedding light on the web of documents. In: Proceedings of the 7th International Conference on Semantic Systems, pp. 1–8. ACM (2011)
12. Nadeau, D., Sekine, S.: A survey of named entity recognition and classification. Lingvisticae Investigationes **30**(1), 3–26 (2007)
13. Paulheim, H., Bizer, C.: Type inference on noisy RDF data. In: Alani, H., et al. (eds.) ISWC 2013, Part I. LNCS, vol. 8218, pp. 510–525. Springer, Heidelberg (2013)
14. Sleeman, J., Finin, T.: Type prediction for efficient coreference resolution in heterogeneous semantic graphs. In: 2013 IEEE Seventh International Conference on Semantic Computing (ICSC), pp. 78–85. IEEE (2013)
15. Thalhammer, A., Knuth, M., Sack, H.: Evaluating entity summarization using a game-based ground truth. In: Cudré-Mauroux, P., et al. (eds.) ISWC 2012, Part II. LNCS, vol. 7650, pp. 350–361. Springer, Heidelberg (2012)

16. Thalhammer, A., Rettinger, A.: Browsing DBpedia entities with summaries. In: Presutti, V., Blomqvist, E., Troncy, R., Sack, H., Papadakis, I., Tordai, A. (eds.) ESWC Satellite Events 2014. LNCS, vol. 8798, pp. 511–515. Springer, Heidelberg (2014)
17. Tonon, A., Catasta, M., Demartini, G., Cudré-Mauroux, P., Aberer, K.: TRank: ranking entity types using the web of data. In: Alani, H., et al. (eds.) ISWC 2013, Part I. LNCS, vol. 8218, pp. 640–656. Springer, Heidelberg (2013)
18. Tylenda, T., Sozio, M., Weikum, G.: Einstein: physicist or vegetarian? summarizing semantic type graphs for knowledge discovery. In: Proceedings of the 20th International Conference Companion on World Wide Web, pp. 273–276. ACM (2011)

TermPicker: Enabling the Reuse of Vocabulary Terms by Exploiting Data from the Linked Open Data Cloud

Johann Schaible[1]($\boxtimes$), Thomas Gottron[2], and Ansgar Scherp[3]

[1] GESIS – Leibniz Institute for the Social Sciences, Cologne, Germany
`johann.schaible@gesis.org`
[2] Innovation Lab, SCHUFA Holding AG, Wiesbaden, Germany
`thomas.gottron@schufa.de`
[3] ZBW – Leibniz Information Center for Economics, Kiel University, Kiel, Germany
`asc@informatik.uni-kiel.de`

Abstract. Deciding which RDF vocabulary terms to use when modeling data as Linked Open Data (LOD) is far from trivial. In this paper, we propose *TermPicker* as a novel approach enabling vocabulary reuse by recommending vocabulary terms based on various features of a term. These features include the term's *popularity*, whether it is from an already used vocabulary, and the so-called *schema-level pattern* (SLP) feature that exploits which terms other data providers on the LOD cloud use to describe their data. We apply Learning To Rank to establish a ranking model for vocabulary terms based on the utilized features. The results show that using the SLP-feature improves the recommendation quality by 29–36 % considering the Mean Average Precision and the Mean Reciprocal Rank at the first five positions compared to recommendations based on solely the term's popularity and whether it is from an already used vocabulary.

1 Introduction

When modeling Linked Open Data (LOD), engineers employ Resource Description Framework (RDF) vocabularies—a collection of (unique) vocabulary terms, i.e., RDF types (also referred to as "classes") and properties—to represent their data. It is considered best practice to reuse terms from existing vocabularies before defining proprietary ones, since this reduces heterogeneity in the data representation [1]. However, finding vocabulary terms for reuse that are commonly used for representing similar data is far from trivial. Prominent vocabulary search services such as LOV [2] or LODstats [3] can be used to find specific terms based on string search. Vocabulary term recommendation tools, like CORE [4], suggest entire vocabularies based on a string search. They also provide statistics on vocabulary terms such as their total number of occurrence and which datasets use them. However, none of the services provide information on how data providers on the LOD cloud actually combine the RDF types and properties from the different vocabularies to model their data (cf. Sect. 2).

© Springer International Publishing Switzerland 2016
H. Sack et al. (Eds.): ESWC 2016, LNCS 9678, pp. 101–117, 2016.
DOI: 10.1007/978-3-319-34129-3_7

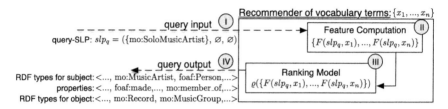

Fig. 1. *Example Run of TermPicker.* The engineer started modeling the data and chose to reuse mo:SoloMusicArtist from the Music Ontology. TermPicker uses the corresponding query-SLP (slp_q) (step (I)) and calculates feature values for each recommendation candidate x_i from the set of all terms published on the LOD cloud ($\{x_1, ..., x_n\}$) (step (II)). The ranking model ϱ uses the values $F(slp_q, x_i)$ (step (III)) to provide the engineer ranked lists of vocabulary terms (step (IV)).

In this paper, we introduce *TermPicker*[1] - a novel vocabulary term recommendation approach enabling the reuse of vocabulary terms by exploiting the following information: given that the engineer already generated a part of her LOD model, TermPicker suggests further RDF types and properties (from other vocabularies) that other LOD providers use together with the part the engineer has already modeled. We capture such information by using so-called *schema-level patterns* (SLPs), which are tuples describing the connection between two sets of RDF types via a set of properties (cf. Sect. 3).

Figure 1 shows TermPicker's workflow via an example: The input is an SLP describing a part of the data model the engineer has already designed, e.g. the query-SLP $slp_q = (\{mo:SoloMusicArtist\}, \varnothing, \varnothing)$. TermPicker allows both: adding further RDF types or properties to the query-SLP, and replacing already used terms in slp_q with better fitting ones (cf. Sect. 4). This applies for RDF types for resources in the subject or object position of an RDF triple, as well as properties between the resources. TermPicker's ranking model uses SLPs and other features, such as the popularity of a recommendation candidate or if the candidate is from an already used vocabulary, for presenting three ranked lists. These lists contain RDF type and property recommendations as output. In this way, the engineer can use TermPicker to find vocabulary terms for reuse that other data providers used together with the terms in slp_q.

To train the ranking model, we use a *Learning To Rank* approach. Learning to Rank approaches constitute a family of supervised machine learning algorithms which establish a ranking over a set of items by observing a general correlation between the utilized features and the relevance of items in the training data. In our case the items to be ranked are RDF vocabulary terms. To evaluate the benefit of using SLPs, we conduct a 10-fold leave-one-out evaluation for different situations where engineers need to select a vocabulary term for reuse (cf. Sect. 5). The recommendation quality is assessed using the Mean Average Precision (MAP) and the Mean Reciprocal Rank at the first five positions

---

[1] http://termpicker.lodrec.org, last access 3/4/16.

(MRR@5). Summarizing, the contributions of this paper are (cf. Sects. 6 and 7 for results and discussion): (i) Evaluation of Learning To Rank algorithms when calculating a ranking model for TermPicker's recommendations. (ii) Evaluation of the SLP-feature's impact on the recommendation quality. (iii) Evaluation of RDF type vs. property recommendations as this reflects different real-world LOD modeling scenarios.[2] For the interested reader, formalizations regarding the SLPs and the features for the L2R algorithm can be found in [5].

2   Related Work

To motivate the use of a recommendation service like TermPicker, we first illustrate current approaches that aid Linked Data engineers in reusing RDF vocabulary terms. Subsequently, we motivate the use of SLPs compared to current approaches for inducing schema information from Linked Open Data.

Vocabulary Search Engines. Services like LOV [2], LODstats [3], Watson [6], or Falcon's concept Search [7] aid the engineer in finding vocabulary terms based on a keyword-based approach. The input is a string describing the desired vocabulary term and the output is a set of vocabulary terms that are similar to the search-string. Each service provides various meta-information on the vocabulary terms and their vocabularies, such as the term's number of usages on the LOD cloud or which datasets use them. In addition, most services exploit schema-information encoded in the vocabularies (within the T-Box specification of vocabularies) such as sub-class, sub-property, or equivalence relations between vocabulary terms (in LOV also via SPARQL queries). However, sometimes engineers cannot express the needed vocabulary term via keywords, and not every relationship between vocabulary terms is explicitly defined in a vocabulary (especially across vocabularies). TermPicker does not use a string-based approach for search but a structure-based approach. This enables it to recommend specific vocabulary terms unknown to the engineer based on other usages of the term on the LOD cloud. Furthermore, for exploiting schema-information, TermPicker does not rely on T-Box information but rather induces the schema-information directly from the datasets on the LOD cloud, i.e. from the encoded A-Box specification of datasets.

Vocabulary Recommender Systems. Services that recommend RDF types and properties are generally based on syntactic and semantic similarity measures. Prominent examples are [4,6,8] and the "data2Ontology" module of the Datalift platform [9]. The input is a set of initial keywords describing a vocabulary term. The services determine a ranked list of domain-specific ontologies considerable for reuse based on string similarity and semantic similarity measures, such as synonyms (in [4] also on manual user evaluations of suggested ontologies). Falcons' Ontology Search [10] also identifies which vocabulary terms might

---

[2] https://www.w3.org/TR/ld-bp/#MODEL, last access 2/7/16.

express similar semantics, but it is rather designed to specify that different vocabularies contain terms describing similar data. Two other approaches [11,12], both embedded in the data integration tool Karma [13], use NLP techniques to recommend an RDF type in conjunction with a data type property for a column of a table, or use a user's previously designed RDF model for recommendations. Whereas these services provide recommendations based on string analysis, or on previously modeled data, they do not exploit any structural information on how other data providers on the LOD cloud use RDF vocabulary terms to describe the connection between objects. By using the schema-information induced directly from the datasets on the LOD cloud, TermPicker is able to alleviate this situation and recommend vocabulary terms with the goal to reduce the heterogeneity in data representation.

Inducing Schema Information. There are various existing concepts to induce schema information from data on the LOD cloud. Prominent examples are the Knowledge Patterns (KPs) [14], Statistical Knowledge Patterns (SKPs) [15], or the RDF graph summary [16]. KPs identify *PathElements* between all RDF types in the data. Statistical Knowledge Patterns (SKPs) find synonymous properties from different RDF types, and the RDF graph summary describes a Linked Data set via several layers, of which the Node-Collection Layer represents the schema-information. However, KPs do not contain object properties to resources that do not have an RDF type, nor do they contain data type properties to some literal values. SKPs only find synonymous vocabulary terms for a given term and not further vocabulary terms that other LOD providers have used together with the given term. Finally, besides the Node-Collection Layer, the graph summary also contains an Entity Layer describing the data on instance level. SPARQL queries over such a graph summary might be too costly, if the information on the RDF instance level is not needed. Therefore, we introduce the notion of *schema-level patterns* (SLPs) and utilize them to induce the schema information from datasets on the LOD cloud (cf. Sect. 3). SLPs have a rather flat structure and do not contain any information from the Entity Layer making them useful for fast queries. They also enable recommending data type properties and object properties that do not have an RDF type as rdfs:range encoded in the data. Statistical Knowledge Patterns can rather be used in addition to SLPs to suggest further synonymous terms. Other tools like LODSight [17] provide visualizations of a dataset's schema information, but are rather useful for exploring relations within and across datasets instead of providing recommendations.

3   Schema-Level Patterns (SLPs)

When reusing vocabularies with the goal to preferably reduce heterogeneity in data representation, one must investigate how other Linked Data providers modeled their data. To this end, we introduce the notion of schema-level patterns (SLPs). They illustrate how resources from a dataset on the LOD cloud are connected with each other. For example, the schema-level pattern

$$(\{\mathsf{foaf{:}Person}, \mathsf{dbo{:}ChessPlayer}\}, \{\mathsf{foaf{:}knows}\}, \{\mathsf{foaf{:}Person}, \mathsf{dbo{:}Coach}\}) \tag{1}$$

```
1  @prefix rdf: <http://www.w3.org/1999/02/22-rdf-syntax-ns#>
2  @prefix foaf: <http://xmlns.com/foaf/0.1/>
3  @prefix dbo: <http://dbpedia.org/ontology/>
4
5  <http://ex1.org/sports_001>
6       rdf:type foaf:Person;
7       rdf:type dbo:ChessPlayer;
8       foaf:knows <http://ex1.org/sports_002>.
9
10 <http://ex1.org/sports_002>
11      rdf:type foaf:Person;
12      rdf:type dbo:Coach.
```

Listing 1.1. Fictive RDF triples in Turtle syntax. The RDF triples specify that a resource of types Person and ChessPlayer knows a resource of types Person and Coach

illustrates that resources of types foaf:Person and dbo:ChessPlayer are connected to resources of types foaf:Person and dbo:Coach via the property foaf:knows and is calculated from the fictive RDF triples in Listing 1.1. We compute such SLPs based on two hash maps. The first map $M1$ contains a resource as key and the set of the resource's RDF types as value. For example, after iterating over the triples in Listing 1.1, $M1$ includes the entries [http://ex1.org/sports_001, {foaf:Person, dbo:ChessPlayer}] and [http://ex1.org/sports_002, {foaf:Person, dbo:Coach}]. The second map $M2$ contains a pair of the subject and object resource as key and the set of all properties between these resources as value. After iterating over the triples in Listing 1.1, $M2$ would contain the entry [(http://ex1.org/sports_001, http://ex1.org/sports_002), {foaf:knows}]. Subsequently, iterating over $M2$, we construct the SLPs using the RDF type information of every resource in $M1$. For the remainder of the paper, we refer to the set of SLPs that are calculated this way from all datasets on the LOD cloud as $\mathbb{SLP}$.

Formally, an SLP is a tuple $slp = (sts, ps, ots)$ (we use the generic term "tuple" to avoid a misunderstanding with the common Semantic Web RDF triple), where sts is the set of RDF types describing resources in subject position of a triple, ots the set of RDF types describing resources in object position of a triple, and ps the set of properties interlinking the resources of types in sts and ots. To operate with SLPs, we define the two operators $\oplus$ and $\ominus$. The operator $\oplus$ can be used for extending an SLP with a further vocabulary term by adding it either to the set sts, ps, or ots. The operator $\oplus_{sts}$ adds an RDF type to the set sts, operator $\oplus_{ots}$ adds a RDF type to ots and the operator $\oplus_{ps}$ adds a property to the set of properties ps. The operation to remove or extract terms from an SLP via the $\ominus$ is defined accordingly.

To compare two or more SLPs, we define the relationship "$\leq$" (*inclusion*) between two schema-level patterns slp_i and slp_j, which defines if an SLP is a subset of or equal to another SLP. It is defined as

$$slp_i \leq slp_j, \quad \text{iff} \quad (sts_i \subseteq sts_j) \wedge (ps_i \subseteq ps_j) \wedge (ots_i \subseteq ots_j) \tag{2}$$

4   Recommending Vocabulary Terms Using SLPs

TermPicker's recommendations are based on utilizing SLPs. Its input is the query-SLP slp_q, which defines a part of the already modeled dataset. Via the $\oplus$ operator it is first extended with a recommendation candidate x from the set of terms from all data sets on the LOD cloud. Subsequently, TermPicker calculates various feature values for each candidate x in conjunction with the query-SLP slp_q. These feature values are then used by a ranking model (calculated via Learning To Rank) to provide an ordered list of vocabulary terms. Optimally, all relevant candidates are ranked at the top of the list.

Features for Recommendation Candidates. Table 1 presents the features we have used to calculate the recommendations. They represent the popularity of each recommendation candidate, whether it is from an already used vocabulary, and the number of SLPs in $\mathbb{SLP}$ that contain all terms of the query-SLP together with the candidate.

Features f_1 to f_4 were derived from [18], which presents that the most common strategies and influencing factors to choose a vocabulary term for reuse is its *popularity* (features $f_1 - f_3$) and whether or not it is from a vocabulary that is already used (feature f_4). Reusing popular vocabulary terms is supposed to enable an easier consumption of the data, as many Linked Data tools provide support for popular vocabularies [1].[3] Whereas the total number of occurrences of a recommendation candidate x show its overall usage, the amount of data sources using x and its vocabulary specify whether the usage of x is spread across many datasets on the LOD cloud or concentrates on only a few ones. Reusing terms from the same vocabulary (feature f_4) is considered as important strategy as well, because of the following reason: When searching for vocabularies covering the domain of interest and subsequently using the vocabulary's RDF

Table 1. Overview of the utilized features. The features are computed for every recommendation candidate x from the set of terms from all data sets on the LOD cloud

Feature	Definition of the feature
f_1	Number of datasets on the LOD cloud using the recommendation candidate x
f_2	Number of datasets on the LOD cloud using the vocabulary V_x of recommendation candidate x
f_3	Total number of occurrences of recommendation candidate x on the LOD cloud
f_4	Whether the recommendation candidate x is from a vocabulary that is already used in query-SLP slp_q
f_5	Number of SLPs in $\mathbb{SLP}$ that contain recommendation candidate x together with all terms in slp_q

---

[3] http://www.w3.org/TR/ld-bp/#VOCABULARIES, last access 12/12/15.

types and properties for particular needs, it seems quite likely that one specific domain vocabulary contains many RDF types or properties that can be reused for describing data from that specific domain. For example, the geo[4] vocabulary contains many terms on specifying geographical information.

In addition, we use the introduced SLPs to generate the so-called SLP-feature (feature f_5) which calculates how many SLPs $slp_i \in \mathbb{SLP}$ exist with $slp_q \oplus x \lesssim slp_i$. Using recommendations based on this feature are likely to result in reducing heterogeneity in the data representation by relying on ontological agreement. The more SLPs in $\mathbb{SLP}$ use the recommendation candidate x together with all terms in the query-SLP, the more it seems to be appropriate to reuse it. Using the SLP-feature is basically analogous to a collaborative filtering approach, as it suggest to use terms that have been used similarly by other data providers on the LOD cloud.

To provide the engineer with relevant recommendation candidates, we must utilize the features $f_1 - f_5$ such that the recommendations are ranked from most relevant to least relevant. To this end, we must assign a weight to each feature, but doing so manually is time-consuming and error-prone. Thus, we make use of the machine learning approach Learning to Rank that calculates the weights for the features automatically.

Learning To Rank. (L2R) refers to a class of supervised machine learning algorithms for inducing a ranking model computed from a set of given features [19]. In detail, a ranking model ϱ is derived from some training data by observing correlations between the feature values and the relevance of a recommendation candidate. Ideally, the derived model lists all relevant vocabulary terms high and above less relevant vocabulary terms.

In our case, the training data is a set of query-SLPs with existing relevance information on each recommendation candidate. It contains SLPs such as $slp_q = (\{$mo:SoloMusicArtist$\}, \varnothing, \varnothing)$ with the relevance information that, e.g., for recommending properties solely the terms foaf:made and mo:member_of are considered relevant. Using this information, the L2R algorithm iterates over the training data multiple times — defined by the user — and adjusts the correlation between the features, such that the relevant recommendation candidates appear as far as possible at the top of the result list. This way, the learned ranking model can be used in new and previously unknown situations with new and unknown query-SLPs. For example, a query-SLP that was not part of the training set using terms from the Creative Commons[5] ontology and from an ontology for managing presentations at W3C[6] $slp_q = (\{$cc:Work$\}, \{$w3:presenter$\}, \varnothing)$ can get recommendations, such as the RDF types foaf:Person and/or dc:Agent to reuse for resources in object position.

L2R algorithms are categorized into three different groups according to their method for learning a ranking model [19]: (A) *point-wise* L2R algorithms,

---

[4] https://www.w3.org/2003/01/geo/wgs84_pos, last access 2/7/16.
[5] http://creativecommons.org/ns, last access 09/06/15.
[6] http://www.w3.org/2004/08/Presentations.owl, last access 09/06/15.

(B) *pair-wise* L2R algorithms, and (C) *list-wise* L2R algorithms. A point-wise approach ranks vocabulary terms directly by allocating a ranking score to each recommendation candidate individually. Pair-wise methods rank vocabulary terms solely in a given pair of two recommendation candidates. This way, a term is considered a better recommendation compared to the terms in a lower ranking position. List-wise approaches rank recommendation candidates by optimizing the quality measure of the result list, such as the Mean Average Precision (MAP).

5   Evaluation

We evaluate the proposed approach by using a 10-fold leave-one-out evaluation. Each fold comprises a *training set* to induce the ranking model, a *test set* to evaluate the ranking model, and a set which simulates the data sets that are already published on the LOD cloud to calculate features f_1 to f_5.[7]

We evaluate TermPicker's recommendation quality by investigating the following aspects: (i) Which Learning To Rank algorithm provides the best recommendation quality? (ii) To which extent does the SLP-feature (using features $f_1 - f_5$) enhance the recommendation quality compared to the baseline of reusing only popular vocabulary terms [18] (using features $f_1 - f_3$) and to the baseline of reusing popular vocabularies from the same vocabulary [20] (using features $f_1 - f_4$)? (iii) Is the recommendation quality better for recommending RDF types for resources in subject position of a triple, for recommending RDF types describing resources in object position, or for recommending properties?

To evaluate different L2R algorithms, we chose to use the RankLib[8] library, as it contains various L2R algorithms and provides an entire framework to train and evaluate an algorithm's ranking model. The recommendation quality is measured using the Mean Average Precision (MAP) and the Mean Reciprocal Rank at the first five positions (MRR@5). In the following, Sect. 5.1 describes the evaluation design in detail and we formalize the quality measures MAP and MRR@5 in Sect. 5.2.

5.1   Evaluation Design

TermPicker's recommendations are evaluated by simulating a search for a vocabulary term that can be reused. Thus, the training set and test set, which are used to induce and evaluate the ranking model, are disjoint sets of distinct SLPs. Before providing TermPicker with an SLP as input, i.e., the query-SLP, one or more vocabulary terms from that SLP are randomly selected. The selected terms are extracted from the query-SLP using the $\ominus$ operator and represent the set of *relevant* recommendation candidates, since they are the ones that have been initially used. All other recommendation candidates that are not contained in

---

[7] Evaluation data and results available at: https://github.com/WanjaSchaible/l2r_eval_material, last access 3/7/16.

[8] http://sourceforge.net/p/lemur/wiki/RankLib/, last access 12/12/15.

the set of the selected terms are considered as irrelevant recommendations. This way, for each query-SLP, the L2R algorithm is provided a set of recommendation candidates, five features categorizing each candidate, and the relevance information for each recommendation candidate. For example, given an SLP slp_j from the training or test set with

$$slp_j = (\{\mathsf{mo:SoloMusicArtist}\}, \{\mathsf{foaf:made}\}, \{\mathsf{mo:Record}\})$$

The property foaf:made is randomly selected and extracted via the $\ominus_{ps}$ operator.

$$slp_q = slp_j \ominus_{ps} \mathsf{foaf:made} = (\{\mathsf{mo:SoloMusicArtist}\}, \varnothing, \{\mathsf{mo:Record}\})$$

The query-SLP slp_q is now provided as input for TermPicker, and the property foaf:made is the single relevant recommendation candidate. This makes it possible to induce and evaluate a ranking model by interpreting a ranked list of recommendations

$$< \mathsf{foaf:name}, \mathsf{mo:remixed}, \mathbf{foaf:made}, ... >$$

in the following way: the first two recommendations are irrelevant, and the first relevant recommendation is at the third rank of the result list.

We validate TermPicker's recommendation quality by performing one evaluation on the DyLDO [21][9] dataset and a second evaluation on the Billion Triple Challenge (BTC) 2014 dataset [22][10] (crawl no. 1).

DyLDO comprises a considerably large amount of data from the LOD cloud with about 10.8 million triples from 382 different pay-level domains[11] (PLDs). In total there are about 2.3 million distinct vocabulary terms from about 600 vocabularies. In this context, we regard a vocabulary simply by its URI namespace, which is either a *hash namespace* or a *slash namespace* as specified by the W3C.[12] The BTC 2014 dataset contains about 1.4 billion triples, of which we use the first 34 millions, to reduce the memory requirements for the SLP computation. These 34 million triples are provided by $3,493$ pay-level domains. Within these triples there are about 5.5 million distinct RDF types and properties from about $1,530$ different vocabularies. The datasets differ by their size and the seed lists, i.e., the set of URIs that form the core of the data crawling, thus containing different data.

The evaluation dataset is split by PLDs such that the data from ten PLDs is used as training and test set and the data from the remaining PLDs is used to simulate the datasets published on the LOD cloud. The split could not be

---

[9] http://swse.deri.org/dyldo/, last access 12/12/15.
[10] http://km.aifb.kit.edu/projects/btc-2014/, last access 12/12/15.
[11] A pay-level domain (PLD) is a sub-domain of a top-level domain, such as *.org* or *.com*, or of a second-level country domain, such as *.de* or *.uk*. To calculate the PLD, we use the Google guava library: https://code.google.com/p/guava-libraries/, last access 2/3/16.
[12] http://www.w3.org/2001/sw/BestPractices/VM/http-examples/2006-01-18/# naming, last access 9/25/15.

Table 2. *PLDs that were selected for training and testing in our evaluation.* The selection is based on (*C*1) - PLDs that provide the highest number of distinct vocabulary terms - and (*C*2) - PLDs with the highest ratio between the reused vocabulary terms and all RDF types and properties.

DyLDO				BTC 2014			
PLD	(*C*1)	(*C*2)	# of SLPs	PLD	(*C*1)	(*C*2)	# of SLPs
bbc.co.uk	146	1.00	522	b4mad.net	291	1.00	393
bblfish.net	82	0.99	150	derby.ac.uk	137	1.00	197
bl.uk	102	0.46	246	heppnetz.de	121	1.00	199
data.gov.uk	258	0.93	920	ivan-herman.net	196	1.00	303
fundacionctic.org	110	0.97	390	jones.dk	164	1.00	155
kanzaki.com	176	0.99	294	ldodds.com	115	1.00	125
kasei.us	100	1.00	121	lmco.com	128	1.00	204
taxonconcept.org	139	0.92	424	mfd-consult.dk	192	1.00	315
thefigtrees.net	89	1.00	102	mit.edu	174	0.96	293
wikier.org	96	1.00	133	nickshanks.com	100	0.97	164

performed randomly, as we needed to make sure that the training and test data contained enough SLPs to train and evaluate the ranking model. Thus, to generate representative results, we selected the ten pay-level domains for training and testing based on two criteria.

(*C*1) A high number of distinct vocabulary terms within a PLD
(*C*2) A high ratio between the reused vocabulary terms and all used RDF types and properties within a PLD

(*C*1) indicates that resources of various RDF types are interlinked via several different properties. This way, it is very likely to calculate a high number of distinct SLPs from that data. (*C*2) indicates that most resources and their connections via properties are described by reused and not self-defined vocabulary terms. This ensures that the relevant recommendation candidates (the selected terms from the SLPs in the training and test sets) are not self-defined terms. Otherwise, all feature values for the relevant candidates would be zero which makes inducing a ranking model impossible. In our evaluation, we define a reused vocabulary term as follows: if a vocabulary term does not contain the PLD in its namespace, then it is considered as a reused vocabulary term. In summary, using (*C*1) and (*C*2) enables us to calculate SLPs that most likely contain many reused terms, which is important to generate valuable recommendations.

Table 2 provides an overview of the selected PLDs used for the evaluations based on the DyLDO (left half of the table) and BTC 2014 (right half of the table) dataset as well as the numbers considering (*C*1) and (*C*2). Furthermore, it displays the number of distinct SLPs that are calculated from the data of the selected pay-level domains. Of course, SLPs that are used to train the ranking model are different from the SLPs that are used to evaluate the model. The data from the remaining PLDs that is used for calculating the feature values contains $117,776$ (DyLDO) and $227,010$ (BTC 2014) SLPs, respectively.

5.2 Evaluation Metrics

As an engineer who searches for possible RDF types and properties for reuse is likely to browse only through the top-k vocabulary terms (where k is generally a small number such as 5 or 10), it is important to evaluate the ranking model by measures that use ordered sets of vocabulary terms. We use the Mean Average Precision (MAP) and the Mean Reciprocal Rank to the k-th position (MRR@k). Both measures present the quality of the ranking model well as they compute values using such ordered sets of vocabulary terms (in contrast to basic measures such as precision and recall).

On the one hand, MAP provides a measure of quality across recall levels. It illustrates the quality of the entire result list in which the ranking position of the relevant vocabulary term is considered. The higher the MAP value, the more relevant vocabulary terms are ranked to the top positions of the result list. On the other hand, MRR@k investigates the result list only to the rank position of the first relevant vocabulary term or to the k-th position. It either returns the reciprocal of the ranking position of the first relevant term, or zero, if no relevant term is contained in the first k results.

We define the set of query-SLPs as $Q = \{slp_{q_1}, ..., slp_{q_n}\}$. If the set of relevant vocabulary terms for a query $slp_{q_j} \in Q$ is $\{rt_1, \ldots, rt_{m_j}\}$ and R_{jh} $(1 \leq h \leq m_j)$ is the set of ranked retrieval results from the top result until one gets to the relevant vocabulary term rt_h, then the **Mean Average Precision** and the **Mean Reciprocal Rank** of Q defined as

$$\text{MAP}(Q) = \frac{1}{|Q|} \sum_{j=1}^{|Q|} \frac{1}{m_j} \sum_{h=1}^{m_j} \text{Precision}(R_{jh}) \qquad \text{MRR}(Q) = \frac{1}{|Q|} \sum_{j=1}^{|Q|} \frac{1}{|R_{jh}|} \qquad (3)$$

In the following, we use $k = 5$. For MRR@5 it means that one relevant terms must be in the first five results $(|R_{jh}| \leq 5)$. Furthermore, MRR@5 uses solely the ranked retrieval results R_{jh} with the minimum amount of vocabulary terms, such that iterating over h is not necessary.

6   Results

The results of our evaluation are presented in Fig. 2 as recommendation quality via box-plots based on MAP. The recommendation quality based on the MRR@5 measure is very much identical compared to the MAP values. The figure depicts the measurements of the recommendation quality considering the aspects (i), (ii), and (iii) mentioned in Sect. 5, where the most competitive L2R algorithms in the RankLib library are: *Coordinate Ascent* [23], *LambdaMART* [24], and *Random Forests* [25]. Both reusing solely popular vocabulary terms (using features $f_1 - f_3$

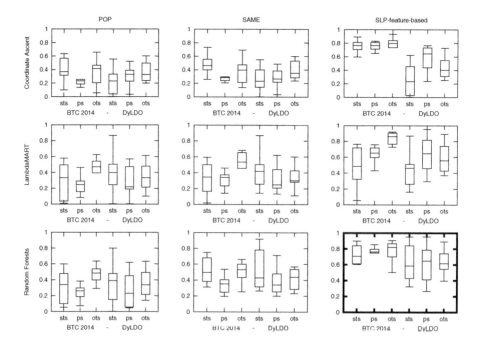

Fig. 2. Evaluation results based on MAP. The plots show the recommendation quality for RDF types for resources in subject position *sts*, properties *ps*, and RDF types for resources in object position *ots* (BTC on the left, DyLDO on the right) based on the different set of features (POP: $f_1 - f_3$; SAME: $f_1 - f_4$; SLP-feature-based: $f_1 - f_5$). The plot in bold font depicts the overall best results, which is the Random Forests algorithm using the SLP-feature.

and marked as POP) and reusing vocabulary terms from the same vocabulary (using features $f_1 - f_4$ and marked as SAME) resemble the baseline. Our SLP-feature-based approach uses features $f_1 - f_5$. Each boxplot displays the different recommendations of an RDF type for resources in subject position (abbreviated as *sts*), a RDF type for resources in object position (abbreviated as *ots*), and a property (abbreviated as *ps*) for both the BTC 2014 and the DyLDO dataset, and comprises the measured values from each fold. The plot in bold font presents the best performing configuration.

In addition, Table 3 shows the average MAP and MRR@5 values (including the standard deviation) for the evaluations using the best performing algorithm *Random Forests* based on the BTC 2014 and the DyLDO datasets, respectively. They underpin the increase of the recommendation quality when adding the SLP-feature to the set of features, which is used by the ranking model. For the BTC 2014 dataset, using the SLP-feature provides on average a higher MAP and MRR@5 value by 29 % compared to the SAME baseline ($f_1 - f_4$), and by 36 % compared to the POP baseline ($f_1 - f_3$). For the DyLDO data, these differences are not as distinctive, but they are still 13 % compared to the baseline

Table 3. *MAP and MRR@5 values for BTC 2014 and DyLDO using the Random Forests algorithm.* Each row depicts the average MAP and MRR@5 values and their standard deviation for the Random Forests algorithm and the utilized set of features.

Data set	Features	sts		ps		ots		overall	
		MAP	MRR@5	MAP	MRR@5	MAP	MRR@5	MAP	MRR@5
BTC	POP	.32 (.20)	.40 (.21)	.26 (.12)	.28 (.12)	.45 (.17)	48 (.15)	.34 (16)	.39 (.16)
	SAME	52 (.16)	.56 (.15)	.37 (.14)	.39 (.14)	.49 (.16)	.50 (.17)	.46 (.15)	.48 (.15)
	SLP	.72 (.11)	.80 (.10)	.75 (.10)	.77 (.10)	.78 (.12)	.83 (.08)	.75 (.11)	.80 (.09)
DyLDO	POP	.44 (.29)	.55 (.31)	.35 (.28)	.36 (.28)	.43 (.25)	.49 (.26)	.41 (.27)	.47 (.28)
	SAME	.59 (.27)	.65 (.24)	.46 (.24)	.46 (.24)	.49 (.21)	.52 (.21)	.51 (.24)	.54 (.23)
	SLP	.65 (.26)	.70 (.24)	.63 (.25)	.63 (.24)	.64 (.17)	.68 (.15)	.64 (.23)	.67 (.21)

SAME and 23 % compared to the baseline POP. In summary, the overall best recommendation quality is provided by the point-wise L2R algorithm *Random Forests* using all features, including the SLP-feature.

(i) Differences between L2R algorithms. Comparing the three most competitive L2R algorithms and making use of all features, one can observe that with MAP ≈ 0.8 based on the BTC 2014 dataset, the algorithms *Coordinate Ascent* and *Random Forests* outperform the *LamdaMART* algorithm (MAP ≈ 0.6). Using the DyLDO data set, the differences between the L2R algorithms are not as noticeable, but are still significant, as a Friedman test shows ($\mathcal{X}^2 = 14,000, p = .001$). A subsequent pair-wise Wilcoxon signed-rank test with Bonferroni correction shows that the *Random Forests* algorithm provides significantly better recommendations than the *Coordinate Ascent* algorithm ($Z = -2.492, p = .013$) as well as than *LambdaMART* algorithm ($Z = -4.237, p < .001$).

The reason why *Random Forests* outperforms the other algorithms can be explained by the fact that we use only binary relevance, i.e., a recommendation candidate is either relevant or irrelevant, in our evaluation. Point-wise approaches perform better in such scenarios, especially if there are solely one or a few relevant recommendation candidates for most queries [26].

(ii) Impact of the SLP-feature. Most noticeable is the influence of the SLP-feature when using the BTC 2014 dataset as evaluation data. The median recommendation quality increases by about 30 % compared to the baseline of reusing solely popular vocabulary terms (POP) and by 20 % compared to the SAME baseline. Such differences are not as noticeable when performing the evaluation on the DyLDO dataset. However, they are still around 15–20 % compared to the baselines POP and SAME. In total, using the SLP-feature increases the average MAP value up to MAP ≈ 0.75. Applying a Friedman test ($\mathcal{X}^2 = 51,667, p < .001$) explains that the differences between using the SLP-feature and the baselines are significant. A post-hoc Wilcoxon signed-rank test shows that using the SLP-feature provides significantly better recommendations ($Z = -4.782, p < .001$ compared to the SAME baseline and $Z = -5.832, p < .001$ compare to the POP baseline).

Such a result shows to which extent the SLP-feature is relevant for providing valuable vocabulary term recommendations. It makes it very likely to recommend the engineer vocabulary terms that might lead to reducing heterogeneity in data representation.

(iii) Differences between recommendation types. Before modeling data as LOD, it is common to describe how the data objects are related.[13] One could first define a set of classes, which depict the objects, and then define relationships connecting these classes, or vice versa. Objects are often described using more than one RDF type[14], whereas there is usually one property to describe a relationship. For that reason, one can argue that TermPicker could provide better property recommendations, as there are many possible RDF types, of which only a few are considered relevant in our evaluation design. However, this was not an influencing factor in our experiment. Comparing the recommendations between RDF types for resources in subject position, RDF types in object position, or properties, only slight differences (between 5–10 %) in the recommendation quality can be perceived when utilizing all features. A Friedman test shows significant differences ($\mathcal{X}^2 = 14,000$, **n.s.**, $p = .449$).

7 Discussion

Our experiments show that using the SLP-feature significantly improves the recommendation quality of vocabulary terms. However, this improvement was not as noticeable when performing the evaluation with the DyLDO dataset. Further investigations showed that the training sets based on the BTC 2014 dataset and the DyLDO dataset had differences regarding their query-SLPs. In detail, the training sets based on the BTC 2014 dataset contained 37 % more relevant recommendation candidates with an SLP-feature value greater than zero ($f_5 > 0$) than the DyLDO data. This means the ranking model induced from the DyLDO data did not encounter the SLP-feature to be improving the recommendation quality. Using the BTC 2014 dataset, the number of SLPs simulating datasets on the LOD cloud is twice as high compared to the number of such SLPs using the DyLDO data. In general, the larger the data for calculating the feature values, the more representative are the generated results [26]. Therefore, we can argue that the results based on using the BTC 2014 dataset are more representative than the results of the evaluation based on the DyLDO data.

A potential threat to the validity of our experiments is the utilized evaluation design. It considers solely the recommendation candidates as relevant that have been selected and extracted from a query-SLP before providing this query-SLP as input for TermPicker (cf. Sect. 5.1). This leads to two major weaknesses. First, many recommendation candidates are identified as irrelevant, although they are appropriate considering the rdfs:domain and rdfs:range,

---

[13] https://www.w3.org/TR/ld-bp/#MODEL, last access 2/7/16.
[14] See the rdf:type information on Pelé: http://dbpedia.org/page/Pele, last access 12/12/15.

the owl:equivalentClass, or other information. For example, let us assume we selected and extracted the property foaf:made from an SLP, resulting in the query-SLP ({mo:MusicArtist}, $\varnothing$, $\varnothing$). The only relevant recommendation candidate for properties is this foaf:made property, as it was used in the original SLP. Properties, such as mo:produced or mo:remixed are considered as irrelevant in our evaluation, although it would make sense to reuse them as outgoing properties from mo:MusicArtist. Thus, an L2R algorithm may identify many *commonly used* vocabulary terms (with an SLP-feature value greater than zero) as irrelevant, which then can result in a ill-trained ranking model. Second, if SLPs from the training set contain terms that are used incorrectly, such as using the property foaf:made as data type property, and this term is randomly selected and extracted, then it is very likely that the SLP-feature value would be zero. This selected term is considered relevant though. Thus, the L2R algorithm learns to decrease the usefulness of the SLP-feature. However, addressing these limitation requires human judgment whether a recommendation candidate is relevant. We do not utilize human judgment, as the automatic evaluation enables us to use many queries and many recommendation candidates to establish a generalized ranking model using a lot of data. Otherwise, we would need a lot of domain experts from various domains judging whether a candidate is relevant or not, which is not feasible.

8   Conclusion

This paper presented TermPicker, a novel approach for recommending vocabulary terms for reuse based on the information how other data providers on the LOD cloud modeled their data. We introduced the notion of schema-level patterns (SLPs) that represent such information and presented a set of features (among them the SLP-feature) that is used by Learning To Rank algorithms to induce a ranking model. We demonstrated that using the SLP-feature increases the recommendation quality by about 35 % compared to the baselines of recommending vocabulary terms from popular vocabularies and recommending terms from the same vocabulary. In total numbers, the Mean Average Precision (MAP) is approximately 0.75 for our recommendation tasks used in the evaluation. Finally, based on the evaluation design that assesses the relevance of a recommendation candidate automatically, it seems that point-wise Learning To Rank (L2R) algorithms provide better results than pair-wise or list-wise L2R algorithms. As future work, we will perform a user study evaluating the recommendation quality by assessing the users' satisfaction with the recommendations.

References

1. Heath, T., Bizer, C.: Synthesis lectures on the semantic web. In: Linked Data: Evolving the Web into a Global Data Space. Morgan & Claypool Publishers, San Rafael (2011)

2. Vandenbussche, P.Y., Atemezing, G.A., Poveda-Villalón, M., Vatant, B.: Linked Open Vocabularies (LOV): a gateway to reusable semantic vocabularies on the Web. Semantic Web J. (Preprint) 1–16 (2015)

3. Auer, S., Demter, J., Martin, M., Lehmann, J.: LODStats – an extensible framework for high-performance dataset analytics. In: ten Teije, A., Völker, J., Handschuh, S., Stuckenschmidt, H., d'Acquin, M., Nikolov, A., Aussenac-Gilles, N., Hernandez, N. (eds.) EKAW 2012. LNCS, vol. 7603, pp. 353–362. Springer, Heidelberg (2012)

4. Fernandez, M., Cantador, I., Castells, P.: Core: a tool for collaborative ontology reuse and evaluation. In: 4th International Workshop on Evaluation of Ontologies for the Web (2006)

5. Schaible, J., Gottron, T., Scherp, A.: TermPicker: enabling the reuse of vocabulary terms by exploiting data from the linked open data cloud - an extended technical report. ArXiv e-prints, December 2015. http://arxiv.org/abs/1512.05685

6. d'Aquin, M., Baldassarre, C., Gridinoc, L., Sabou, M., Angeletou, S., Motta, E.: Watson: supporting next generation semantic web applications. In: Proceedings of the IADIS International Conference WWW/Internet 2007, pp. 363–371 (2007)

7. Cheng, G., Ge, W., Qu, Y.: Falcons: searching and browsing entities on the semantic web. In: Proceedings of the 17th International Conference on World Wide Web. ACM (2008)

8. García-Santa, N., Atemezing, G.A., Villazón-Terrazas, B.: The ProtégéLOV plugin: ontology access and reuse for everyone. In: Gandon, F., Guéret, C., Villata, S., Breslin, J., Faron-Zucker, C., Zimmermann, A. (eds.) ESWC 2015. LNCS, vol. 9341, pp. 41–45. Springer, Switzerland (2015)

9. Scharffe, F., Atemezing, G., Troncy, R., Gandon, F., et al.: Enabling linked-data publication with the datalift platform. In: AAAI 2012, 26th Conference on Artificial Intelligence (2012)

10. Cheng, G., Gong, S., Qu, Y.: An empirical study of vocabulary relatedness and its application to recommender systems. In: Aroyo, L., Welty, C., Alani, H., Taylor, J., Bernstein, A., Kagal, L., Noy, N., Blomqvist, E. (eds.) ISWC 2011, Part I. LNCS, vol. 7031, pp. 98–113. Springer, Heidelberg (2011)

11. Ramnandan, S.K., Mittal, A., Knoblock, C.A., Szekely, P.: Assigning semantic labels to data sources. In: Gandon, F., Sabou, M., Sack, H., dAmato, C., Cudre-Mauroux, P., Zimmermann, A. (eds.) ESWC 2015. LNCS, vol. 9088, pp. 403–417. Springer, Heidelberg (2015)

12. Taheriyan, M., Knoblock, C.A., Szekely, P., Ambite, J.L.: Learning the semantics of structured data sources. Web Semant. Sci. Serv. Agents World Wide Web (2016). ISSN: 1570-8268. doi:10.1016/j.websem.2015.12.003

13. Knoblock, C.A., et al.: Semi-automatically mapping structured sources into the semantic Web. In: Simperl, E., Cimiano, P., Polleres, A., Corcho, O., Presutti, V. (eds.) ESWC 2012. LNCS, vol. 7295, pp. 375–390. Springer, Heidelberg (2012)

14. Presutti, V., Aroyo, L.M., Gangemi, A., Adamou, A., Schopman, B., Schreiber, G.: A knowledge pattern-based method for linked data analysis. In: Proceedings of the Sixth International Conference on Knowledge Capture, pp. 173–174. ACM (2011)

15. Zhang, Z., Gentile, A.L., Blomqvist, E., Augenstein, I., Ciravegna, F.: Statistical knowledge patterns: identifying synonymous relations in large linked datasets. In: Alani, H., et al. (eds.) ISWC 2013, Part I. LNCS, vol. 8218, pp. 703–719. Springer, Heidelberg (2013)

16. Campinas, S., Perry, T.E., Ceccarelli, D., Delbru, R., Tummarello, G.: Introducing RDF graph summary with application to assisted SPARQL formulation. In: 23rd International Workshop on Database and Expert Systems Applications (DEXA), pp. 261–266. IEEE (2012)

17. Dudáš, M., Svátek, V., Mynarz, J.: Dataset summary visualization with LODSight. In: Gandon, F., Guéret, C., Villata, S., Breslin, J., Faron-Zucker, C., Zimmermann, A. (eds.) ESWC 2015. LNCS, vol. 9341, pp. 36–40. Springer, Heidelberg (2015). doi:10.1007/978-3-319-25639-9_7

18. Schaible, J., Gottron, T., Scherp, A.: Survey on common strategies of vocabulary reuse in linked open data modeling. In: Presutti, V., d'Amato, C., Gandon, F., d'Aquin, M., Staab, S., Tordai, A. (eds.) ESWC 2014. LNCS, vol. 8465, pp. 457–472. Springer, Heidelberg (2014)

19. Liu, T.Y.: Learning to rank for information retrieval. Found. Trends Inf. Retrieval **3**(3), 225–331 (2009)

20. Lodi, G., Maccioni, A., Scannapieco, M., Scanu, M., Tosco, L.: Publishing official classifications in linked open data. In: Proceedings of the 2nd International Workshop on Semantic Statistics (SemStats2014) in conjunction with the 13th International Semantic Web Conference (ISWC). Springer, Riva del Garda, Italy (2014)

21. Käfer, T., Abdelrahman, A., Umbrich, J., O'Byrne, P., Hogan, A.: Observing linked data dynamics. In: Cimiano, P., Corcho, O., Presutti, V., Hollink, L., Rudolph, S. (eds.) ESWC 2013. LNCS, vol. 7882, pp. 213–227. Springer, Heidelberg (2013)

22. Käfer, T., Harth, A.: Billion Triples Challenge data set (2014). http://km.aifb.kit.edu/projects/btc-2014/

23. Metzler, D., Croft, W.B.: Linear feature-based models for information retrieval. Inf. Retrieval **10**(3), 257–274 (2007)

24. Wu, Q., Burges, C.J., Svore, K.M., Gao, J.: Adapting boosting for information retrieval measures. Inf. Retrieval **13**(3), 254–270 (2010)

25. Breiman, L.: Random forests. Mach. Learn. **45**(1), 5–32 (2001)

26. Busa-Fekete, R., Szarvas, G., Elteto, T., Kégl, B., et al.: An apple-to-apple comparison of learning-to-rank algorithms in terms of normalized discounted cumulative gain. In: 20th European Conference on Artificial Intelligence, ECAI 2012, vol. 242 (2012)

Implicit Entity Linking in Tweets

Sujan Perera[1]([⊠]), Pablo N. Mendes[2], Adarsh Alex[1], Amit P. Sheth[1],
and Krishnaprasad Thirunarayan[1]

[1] Kno.e.sis Center, Wright State University, Dayton, OH, USA
{sujan,adarsh,amit,tkprasad}@knoesis.org
[2] IBM Research, San Jose, CA, USA
pnmendes@us.ibm.com

Abstract. Over the years, Twitter has become one of the largest communication platforms providing key data to various applications such as brand monitoring, trend detection, among others. Entity linking is one of the major tasks in natural language understanding from tweets and it associates entity mentions in text to corresponding entries in knowledge bases in order to provide unambiguous interpretation and additional context. State-of-the-art techniques have focused on linking explicitly mentioned entities in tweets with reasonable success. However, we argue that in addition to explicit mentions – i.e. *'The movie Gravity was more expensive than the mars orbiter mission'* – entities (movie Gravity) can also be mentioned implicitly – i.e. *'This new space movie is crazy. you must watch it!.'* This paper introduces the problem of implicit entity linking in tweets. We propose an approach that models the entities by exploiting their factual and contextual knowledge. We demonstrate how to use these models to perform implicit entity linking on a ground truth dataset with 397 tweets from two domains, namely, Movie and Book. Specifically, we show: (1) the importance of linking implicit entities and its value addition to the standard entity linking task, and (2) the importance of exploiting contextual knowledge associated with an entity for linking their implicit mentions. We also make the ground truth dataset publicly available to foster the research in this new research areacity.

Keywords: Implicit entities · Entity modeling · Entity linking · Contextual knowledge

1 Introduction

Data show that 350,000 tweets are generated per minute – 500 million per day.[1] These tweets have become a valuable source of information for trend detection, event monitoring, and opinion mining applications. Mining tweets poses unique challenges due to their short, noisy, context-dependent, and dynamic nature [4].

Entity linking in tweets has the potential to benefit all aforementioned applications. The term 'entity' in this paper refers to an unambiguous, terminal page in Wikipedia as in [7]. State-of-the-art entity linking solutions in tweets

---

[1] http://www.internetlivestats.com/twitter-statistics.

© Springer International Publishing Switzerland 2016
H. Sack et al. (Eds.): ESWC 2016, LNCS 9678, pp. 118–132, 2016.
DOI: 10.1007/978-3-319-34129-3_8

have mainly focused on explicitly mentioned entities [1,4,7,11]. However, we will show that entities may also be mentioned implicitly. For example, consider the two tweets: *'movie wasn't the story Veronica wrote. It was the story director hacksawed!!'* and *"whew' the movie was the WORST example of that book. Neil Burger completely rewrote that whole story!.'* State-of-the-art entity linking systems may link the entity mention *Veronica Roth* in the first tweet and *Neil Burger* in the second tweet to their corresponding entities in a knowledge base. However, they do not realize that both tweets have implicit mentions of the movie *Divergent.* We term entities that are being implicitly mentioned as 'implicit entities'.

Linking implicit entities in tweets is an important task that affects downstream analytics. If they were ignored, a sentiment analysis task wouldn't identify that the aforementioned tweets had negative sentiment towards the movie *Divergent.* An trend detection application wouldn't detect that *Oscar Pistorius* is trending if it was not able to identify the reference[2] to him in a tweet like *'Kinda sad to hear about that South African runner kill his girlfriend'* as he was referred to frequently with similar phrases in tweets.

We hypothesize that implicit entities are a common occurrence. In Sect. 2, we assess the prevalence of implicit versus explicit entities on a random sample of tweets. This experiment shows that 21 % of the entities are mentioned implicitly in the Movie domain while it is 40 % in the Book domain. Therefore, linking these entities will have significant impact on downstream applications.

The implicit entity linking problem (IEL) is notably different from explicit entity linking (EL) in several ways. While we elaborate our findings in Sect. 2, we summarize them here. Implicit entity mentions do not contain the entity name. The absence of the entity name is filled by leveraging different characteristics of the entity – the first tweet in the above examples uses the author of the book that the movie is based on while the second tweet uses its director to make a reference to the movie. Furthermore, the context that helps to resolve the implicit entity mentions changes overtime – the phrase 'space movie' may refer to distinct movies at different time intervals. These features of implicit entity mentions warrant a new approach to solve this problem.

In this paper, we propose an approach to the implicit entity linking problem in tweets that factors in the above features. Twitter users often rely on sources of context outside the current tweet, assuming that there is some shared understanding between them and their audience, or temporal context in the form of recent events or recently-mentioned entities [4]. This assumption allows them to constrain the message to 140 characters, yet make it understandable to the audience. Our approach models entities by encoding this shared understanding by harnessing factual and contextual knowledge of entities to complement the context expressed in the tweet text. The contextual knowledge captures temporally relevant topics and other entities associated with the entity of interest.

---

[2] We use the terms 'entity reference' or 'entity mention' interchangeably to signify the usage of a phrase that unambiguously evokes a given entity in a tweet.

Our evaluation shows that the proposed approach achieves 61 % of accuracy in linking implicit entities. Furthermore, we show that IEL helps boost the overall accuracy of entity linking when combined with the EL. Our contributions are:

- We introduce the IEL problem on tweets, assess the significance of implicit entity mentions in a sample of tweets and describe their characteristics,
- We propose and evaluate a model to capture and encode both factual and contextual knowledge of an entity to perform implicit entity linking, and
- We create ground truth data sets and make them publicly available to foster research on this important new problem.

2   Understanding Implicit Entity Mentions

In this section, we formally define the IEL problem w.r.t tweets and describe its main characteristics. We discuss the prevalence of implicit mentions on Twitter, the dynamicity of context associated with the entities, and the types of references-through-characteristics.

Definition 1. *Implicit Entity is an entity mentioned in a tweet where its name is not present nor it is a synonym/alias/abbreviation of an entity name or a co-reference of an explicitly mentioned entity in the tweet.*

The explicit entity linking (EL) task can be defined as "matching a textual entity mention, possibly identified by a named entity recognizer, to a KB entry, such as a Wikipedia page that is a canonical entry for that entity" [16]. Therefore, the input for EL is a tuple (s, c, t) where s is the mention string (i.e. surface form) and c is the entity type extracted by NER from text t. The output is an entity identifier e such that $\text{argmax}_e P(e|s, c, t)$.

We define the implicit entity linking for tweets as:

Definition 2. *Implicit Entity Linking (IEL): given a tweet with an implicit entity mention of a particular type (e.g. Movie, Book) output the entity mentioned by the tweet w.r.t a given knowledge base.*

Therefore, the input for IEL is a tuple (c, t) where c is the entity type and t is the text (e.g. tweet) where the implicit entity occurred. In contrast to EL the candidate set for an implicit entity mention is potentially much larger since it cannot be narrowed down based on the name of the entity.

2.1   Prevalence

To estimate the volume of implicit entity mentions on tweets, we performed a manual analysis on a sample set of tweets. We focused on the domains of Movie and Book, since they offered an agreeable level of difficulty for human annotators. We collected two random samples of tweets – one for the Movie domain using the keywords 'movie' and 'film', and another for the Book domain using the keywords 'book' and 'novel'. We subsequently annotated the tweets in the sample as 'explicit', 'implicit', and 'NIL' according to the following guidelines.

Consider three tweets in the Movie domain: (1) *'the movie trailer for 50 shades of grey looks really good,'* (2) *'ISRO sends probe to Mars for less money than it takes Hollywood to make a movie about it,'* and (3) *'How the hell is every movie the #1 movie in America?.'* The first tweet has a mention of the movie *Fifty Shades of Grey,* hence it is annotated as 'explicit.' The second tweet is annotated as 'implicit' since it has an implicit reference to the movie *Gravity.* The third tweet does not refer to any movie, hence it is annotated as 'NIL.' The tweets that have both explicit and implicit movie mentions are annotated with both labels.

This annotation exercise produced 416 and 114 tweets for the 'explicit' and 'implicit' categories respectively. This means 21 % of the tweets with mentions of movies are implicit references. In other words, this experiment showed that roughly for every four tweets with explicit mentions of entities, there is a tweet with an implicit mention in the Movie domain. A similarly constructed experiment in the Book domain found that roughly for every five tweets with explicit entity mentions there are two tweets with implicit mentions.

2.2   Characteristics

Dynamic Context. When a human annotator is trying to resolve an implicit entity mention, they often rely on domain knowledge outside the tweet text – for instance, they know which were the latest movies, or which actors starred in different movies, etc. The relevant domain knowledge may change dynamically.

Often, one phrase may refer to distinct entities at different time intervals. For instance, the phrase 'space movie' could refer to the movie *Gravity* in fall 2013 while the same phrase in fall 2015 would likely refer to the movie *The Martian.* On the flip side, the most salient characteristics of the movies may change over time, and so will the phrases used to refer to them. The movie *Furious 7* was frequently referred to with phrase 'Paul Walker's last movie' in November 2014. This was due to the actor's passing around that time. However, after the movie release in April 2015 the same entity was often mentioned through the phrase 'fastest film to reach the $1 billion.'

Types of References-Through-Characteristics. We observed that individuals resort to a diverse set of entity characteristics to make implicit references. For example, consider the following implicit references to the movie *'Boyhood'*[3]: (1) 'Richard Linklater movie,' (2) 'Ellar Coltrane on his 12-year movie role,' (3) '12-year long movie shoot,' (4) 'latest movie shot in my city Houston,' and (5) 'Mason Evan's childhood movie.' The first two tweet fragments refer to the movie through its director and actor, the third tweet fragment uses a distinctive feature (it was shot over a long period), the fourth example uses the shooting location of the movie, and last one refers to it with a character in the movie.

---

[3] We show only the fragments of the tweets that indicate a mention of an entity.

3   Related Work

Entity linking in tweets has recently gained attention in academia and industry alike. The literature on entity linking in tweets can be categorized as 'word-level entity linking' and 'whole-tweet entity linking' [4]. While the former task is focused on resolving the entity mentions in a tweet, the latter task is focused on deriving the topic of the tweet. The topic may be derived based on the explicit and implicit entity mentions in the tweets. For instance, the tweet *'Texas Town Pushes for Cannabis Legalization to Combat Cartel Traffic'* has an explicit mention of entity *Cannabis Legalization* and an implicit mention of entity *El Paso* city. The topic of the tweet would be 'Cannabis Legalization in El Paso.' Hence, it is worth noting that the work on deriving topics is neither comparable to explicit nor to implicit entity linking since they are extracting the topic of the tweet text, rather than actual mentions of an entity in the tweet [4]. We have not found literature that focuses on implicit entity linking in tweets. In this section, we will survey the literature on both word-level and whole-tweet entity linking and explain why techniques and features used by such solutions may not be applicable to implicit entity linking, hence it deserves special attention.

Meij, et al. [12] derives the topics of a tweet. They extract features from the tweet and the Wikipedia pages of entities, and apply machine learning algorithms to derive the topic. We found that this work has focused on deriving topic using explicit entities as their evaluation dataset contains only 16 tweets whose label of the manually annotated topic is not present in the tweet text (i.e. not a string match). Nevertheless, they are found to be either synonyms or related entities to the explicit entities in the tweets and not implicit entity mentions (e.g. New York and Big Apple, Almighty and God, stuffy nose and Rhinitis).

The word-level tweet linking has two main steps: (1) candidate selection, and (2) disambiguation. The word-level entity linking has been studied extensively for text like Wikipedia and News [2,5,9,13,14]; however, these approaches have proved to be ineffective on short and noisy text like tweets [4,6]. Here we will discuss the approaches taken to solve this problem in tweets. The first step was performed by matching the word sequences of the tweet to the page titles and the anchor texts in Wikipedia and consider all matching pages and pages redirected by matching anchor texts to be candidates [1,6,7]. The second step was performed by optimizing the relatedness calculated among the candidate entities [6,11] or based on the threshold defined over measures that calculate the similarity between the entity mention and entity representation [1], or by applying structural learning techniques [7]. An approach to solve the implicit entity linking in tweets has to take a fresh perspective since by definition it has neither anchor text nor the page title present in the tweet.

Our previous work on implicit entity linking has dealt with clinical entities in electronic medical records [15]. The main challenge in the medical setting resides in the heterogeneous usage of language by individuals mentioning entities implicitly including their negated mentions. Our approach focused on modeling the entities by using their static definitions and exploiting WordNet as the knowledge base to account for heterogeneity in language usage. The task of implicit entity linking in tweets is different from that of clinical text since, in addition to the differences in nature of the text, the heterogeneity arise with the usage of different characteristics of entities and their association with dynamic context as discussed in Sect. 2. Hence, the static descriptions/definitions of entities fall short in linking implicit entities as we will demonstrate in our evaluation.

4   Linking Implicit Entities in Tweets

Our approach models entities with factual knowledge and contextual knowledge by leveraging existing knowledge bases and relevant tweets. The entity models are integrated to create an entity model network (EMN) for each domain of interest to reflect the topical relationships among domain entities[4] at time t.

The first step of creating the EMN is to identify domain entities that are relevant at time t. This can be done, for instance, by running an off-the-shelf entity linking system over a corpus of recent tweets and identify the mentioned entities. The idea is that if an entity is relevant at time t, it will likely to be mentioned explicitly by tweets around that time. Even though an automatic annotation approach may not be perfect, it will identify at least one occurrence of explicit mention of entities within the corpus. This is sufficient as one appearance of an entity in the corpus qualifies it to be included in the EMN. We start building EMN by creating entity models for identified entities for time t.

4.1   Entity Model Creation

Consider two tweets about the movie *Gravity*: *'New Sandra Bullock astronaut lost in space movie looks absolutely terrifying,'* and *'ISRO sends probe to Mars for less money than it takes Hollywood to send a woman to space.'* The first tweet has a mention of its actress and that, along with other terms, helps to resolve its mention to the movie *Gravity*. This kind of **factual knowledge** (e.g. to relate actors and movies) can be extracted from a knowledge base. The second tweet does not have a mention of any entity associated with the movie *Gravity*; hence, the factual knowledge falls short in identifying its implicit mention. However, contemporary tweets with explicit mentions of movie *Gravity* often use phrases like 'ISRO', 'woman to space', 'less money' which helps to link the implicit mention of *Gravity* in the second tweet. We refer to these as **contextual knowledge**. The entity model also contains an estimate of **temporal salience** for each entity and it is computed by counting the corresponding Wikipedia page views within the last 30 days w.r.t time t. The entity model consists of phrases generated using the knowledge components and its temporal salience. Figure 1(a) shows the fragment of the model generated for the movie *Gravity*.

Acquiring Factual Knowledge: Factual knowledge of an entity can be acquired from existing knowledge bases and Linked Open Data. We have used

---

[4] The term 'domain entity' in this work refers to movies in the Movie domain and books in the Book domain.

DBpedia [10] as our knowledge base due to its wide coverage of domains and up-to-date knowledge. For a given entity e we retrieved triples where e appears as subject or object. However, for a given entity type not all relationships are important in modeling its entities. For example, a movie has relationships 'director' and 'starring' as well as 'billed' and 'license.' The former two relationships are more important when describing a movie than the latter. We capture this intuition by ranking the relationships based on their joint probability value with the given entity type as follows.

$$P(r, T) = \frac{number\ of\ triples\ of\ r\ with\ instances\ of\ T}{total\ number\ of\ triples\ of\ r}, \tag{1}$$

where T is the entity type (e.g. Movie) and r is the relationship. The instances of a given entity type can be obtained from DBpedia via 'rdf:type' relationship.

The triples of entity e of type T with one of the top m relationships are selected to build the entity model of e. We collect the 'rdfs:label' value of entities connected to e in these triples as the factual knowledge of entity e. In addition to the top m relationships, we also consider the value of 'rdfs:comment' relationship of e to build the entity model. 'rdfs:comment' gives a textual description of an entity that oftentimes complements the knowledge captured by the triples.

Acquiring Contextual Knowledge: The contextual knowledge can be extracted from contemporary tweets that explicitly mention the entity. We use 'rdfs:label' of the entity in DBpedia along with its type as the keyword to collect the 1,000 most recent tweets for that entity. For example, we used the phrases 'gravity movie' and 'gravity film' to collect tweets for movie *Gravity*; this will minimize the tweets with other meanings of the term 'gravity' in collected tweets.

Model Creation: The entity model consists of weighted phrases and unigrams generated using the acquired knowledge and the temporal salience of the entity. Firstly, the collected tweets needs to be cleaned before they can be used to create the entity model. We remove the punctuations, and emoticons, and normalizes the numbers to the pseudo-string 'NUMBER' in tweets. The hashtags and mentions that were written in camel case style were retained after decomposing (@VeronicaRoth → Veronica Roth, #MarkWahlberg → Mark Wahlberg) and others were retained by removing '#' and '@' symbols.

We use Wikipedia anchor texts and page titles to identify meaningful phrases in the acquired knowledge. We chunk the text in acquired knowledge (i.e. factual knowledge and cleaned tweets) into n-grams (n = 2, 3, 4) and the n-grams present as anchor text or as page titles are added to the entity model as phrases. However, Twitter users do not always use complete phrases; consider the reference to actress *Sandra Bullock* in the tweet: *'It's hard for me to imagine movie stars as astronauts, but the movie looks great! and who doesn't like **Sandra**.'* Therefore, the entity model should also contain the portions of the phrases, hence we include unigrams excluding stop words to fulfill this requirement.

An entity model consists of phrases as well as unigrams (collectively referred as clues), and we stored it as a graph as shown in the Fig. 1(a).

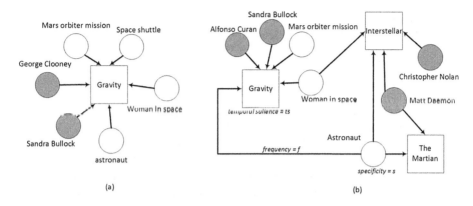

Fig. 1. Entity model network. The rectangles depict entities, shaded circles and plain circles show the clues generated with factual and contextual knowledge respectively. The properties of nodes and edges are shown for one connection in (b).

4.2 Entity Model Network (EMN) Creation

The entity models created for all the entities are integrated via common clues to generate the EMN as shown in Fig. 1(b). The specificity of the clue node c_j is inspired by the inverse document frequency measure and is calculated as $\log \frac{|N|}{|N_{c_j}|}$ where $|N|$ is the number of entity nodes in the EMN and $|N_{c_j}|$ is the number of adjacent nodes to c_j. The 'frequency' property value of the edge between clue node c_j and entity node e_i is calculated as the total number of times that the value of 'clue name' of node c_j is present in tweets collected for entity e_i.

Formally, an entity model network (EMN) is defined as a property graph $G_{EMN} = (V_e, V_c, E, \mu)$, where V_e and V_c represent the vertices of two types, E represents the edges, and μ represents the property map. The edges are directed (i.e. $E \subseteq (V_c \text{ X } V_e)$), and μ maps the properties of vertices and edges as keys to values (i.e. $\mu : (V_e \cup V_c \cup E)XR \rightarrow S$), where R is a set of keys and S denotes values. V_e represents the entities and has the properties 'name' and 'temporal salience' as keys and their values as key/value pairs. V_c represents the clues and has the properties 'clue name' and 'specificity' as keys and their values as key/value pairs. The edges in the graph has the property 'frequency' and its value as key/value pair.

4.3 Linking Implicit Entities with the EMN

To understand how to use the EMN to perform implicit entity linking for a given tweet, it is useful to divide the task into two steps: (1) candidate selection and ranking, and (2) disambiguation.

Candidate Selection and Ranking. The objective of the candidate selection and ranking step is to prune the search space so that disambiguation step does not have to evaluate all entities in EMN as candidates. In EL this is usually done by looking for candidates for a given surface form. In IEL, we took a different approach. The input tweet goes through the cleaning step as described in Sect. 4.1. Then we identify the phrases in the tweet using Wikipedia anchor text and page titles, and the terms that are not qualified as phrases are considered as unigrams. We refer to both phrases and unigrams extracted from tweets as 'tweet clues' and denotes it as C_t. The candidate selection step takes these tweet clues and match with clue nodes in EMN, the entities that have at least one edge from matching clues are selected as the initial set of candidates.

Formally, given a set of tweet clues C_t, the initial candidate entity set $\mathcal{E}_{\mathcal{IC}} = \{e_i | (c_j, e_i) \in E \text{ and } c_j \in C_t\}$.

The entities in the initial candidate set are scored based on the strength of evidences. The strength of evidences for entity $e_i \in \mathcal{E}_{\mathcal{IC}}$ (SC_{e_i}) is calculated as:

$$SC_{e_i} = \sum_{c_j \in C_t} specificity \ of \ c_j * frequency \ of \ edge \ (c_j, e_i)$$

The top k candidates based on these scores (denoted as E_c) are considered for disambiguation step.

Disambiguation. The objective of the disambiguation step is to sort the selected candidate entities such that the implicitly mentioned entity in a given tweet is at the top position of the ranked list. This is accomplished through a machine learned-ranking model based on the pairwise approach: all pairs of selected candidate entities (along with a feature set) are taken as input, and the model approximates the ranking as a classification problem that tells which of the entities in the pair is better than the other.

The feature set of a candidate entity consists of its similarity to the tweet and its temporal salience w.r.t temporal salience of other candidate entities.

The similarity between the candidate entity and the tweet is calculated via their vector representations. The vector representation of the candidate entity e_i is obtained via its incoming connections from other nodes. It is denoted as e_{i_v} and defined as $e_{i_v} =< v_1, v_2, ..., v_n >$ where $v_j = specificity \ of \ c_j * frequency \ of \ edge \ (c_j, e_i)$ for all $(c_j, e_i) \in E$. The vector representation of the tweet is created using tweet clues. The similarity between the candidate entity and the tweet is calculated by the cosine similarity of these vectors.

The temporal salience of the candidate entity e_i is normalized w.r.t the temporal salience of other candidate entities in E_c as:

$$\frac{temporal \ salience \ of \ e_i}{\sum_{e \in E_c} temporal \ salience \ of \ e} \tag{2}$$

We trained a SVM^{rank} model to solve the ranking problem. We used linear kernel, 0.01 as the trade-off between training error and margin, and the total number of swapped pairs summed over all queries as the loss function. SVM^{rank} is shown to perform well in similar ranking problems, specifically it is able to provide the best performance in ranking the top concept [12], which suits the characteristics of our problem.

5   Evaluation

We evaluated the implicit entity linking performance and its value to the explicit entity linking task in two domains, namely, Movie and Book. There is no standard dataset available to evaluate this task. Hence, we have created datasets for the two domains. We focus on answering three questions in our evaluation.

- How effective is the proposed approach in linking implicit entities?
- How important is the contextual knowledge in linking implicit entities?
- What is the value added by linking implicit entities?

5.1   Dataset Preparation

In order to prepare datasets for evaluation, we collected tweets using 'movie' and 'film' as keywords for the Movie domain and 'book' and 'novel' as keywords for the Book domain. This dataset was collected during August 2014 and it was independent of the dataset described in Sect. 2.1. The collected tweets were manually annotated by two individuals with 'explicit', 'implicit', and 'NIL' labels for movies and books following the guidelines described in Sect. 2.1. We included annotated tweets in the evaluation dataset that was agreed upon by both annotators. Table 1 shows the important characteristics of the annotated dataset and it is available in our project page at https://goo.gl/jrwpeo.

To perform IEL on the aforementioned evaluation dataset, we created the EMN for July 31st, 2014. We collected the most recent 15,000 tweets as of July 31st, 2014 for each domain using its type labels as the keywords (e.g. 'movie' and 'film' for the Movie domain) and applied a simple spotting mechanism to identify entity mentions. The spotting mechanism collects the labels of the domain entities from DBpedia (i.e. rdfs:label value of the instances of type Film) and then it checks for their presence in collected dataset for the domain. If the label is found within the tweets, we add that entity to the EMN. We collected the 1,000 most recent tweets that explicitly mention the identified entities to generate their contextual knowledge, extracted the factual knowledge from the

Table 1. Evaluation dataset statistics. Describes, per domain, the total number of tweets per annotation type (explicit, implicit, NIL), number of distinct entities annotated, and average tweet length.

Domain	Annotation	Tweets	Entities	Avg. length
Movie	Explicit	391	107	16.5 words
	Implicit	207	54	18 words
	NIL	117	0	16.4 words
Book	Explicit	200	24	18.5 words
	Implicit	190	53	18.5 words
	NIL	70	0	17.5 words

Table 2. Implicit entity linking performance

Domain	Candidate selection recall	Disambiguation accuracy
Movie	90.33	60.97
Book	94.73	61.05

DBpedia version created with May 2014 Wikipedia dumps, and obtained page hit counts of Wikipedia pages for the month of July 2014. We varied the number of top m relationships ranked according to Eq. 1 to extract the factual knowledge for modeling entities. The best results were obtained when $m = 10$, and we observed a dramatic decrease in the accuracy in linking movie entities when $m > 15$, potentially due to inclusion of relatively irrelevant knowledge. Hence, the factual knowledge component consists of knowledge extracted with the top 15 relationships. These collected resources are used to create entity models for each entity and the EMN as described in Sect. 4. The created EMNs for the Movie and Book domains had 617 entities and 102 entities respectively.

5.2   Implicit Entity Linking Evaluation

This section evaluates the implicit entity linking task in isolation. We show the results on both candidate selection and ranking, and disambiguation steps. The candidate selection and ranking is evaluated as the proportion of tweets that had the correct entity within the top k selected candidates for that tweet (denoted as Candidate Selection Recall). We experimented by varying k between 5 and 35 and found that results improves as we increased k and came to near plateau after k = 20. We demonstrate the results for k = 25 and interested readers can find detailed results on our project page. The disambiguation step is evaluated with 5-fold cross validation and report the results as proportion of the tweets in evaluation dataset that had correct annotation at the top position. Table 2 shows the results of this step on both domains.

Qualitative Error Analysis. The error analysis on implicit entity linking shows that errors are fourfold: (1) errors due to lack of contextual knowledge of the entity, (2) errors due to novel entities, (3) errors due to the cold start of entities and topics, and (4) errors due to multiple implicit entities in the same tweet. Table 3 shows an example tweet for each error type.

The first tweet in the table is annotated with the movie *White Bird in a Blizzard*. There were only 46 tweets for this movie. Hence, the contextual knowledge component of the entity model did not provide strong evidence in the disambiguation step. The second tweet is annotated with the movie *Deepwater Horizon*. The Wikipedia page for this movie was created on September 2014, hence it was not available to EMN. This is known as emerging entity discovery problem and requires separate attention as in the explicit entity linking [8]. A few entities and topics emerged among Twitter users only after July 31st, 2014.

Table 3. Example tweet for each error type.

Error	Tweet
1	*'That Movie Where Shailene Woodley Has Her First Nude Scene? The Trailer Is RIGHT HERE!: No one can say Shailene Woodley isn't brave!'*
2	*"'hey, what's wrawng widdis goose?" RT @TIME: Mark Wahlberg could be starring in a movie about the BP oil spill* http.//ti.me/1oZhb5V *'*
3	*'Video: George R.R. Martin's Children's Book Gets Re-release* http://bit.ly/1qNNH5r *'*
4	*'That moment when you realize that hazel grace and Augustus are brother and sister in one movie and in love battling cancer..'*

These entities and topics were not present in our EMN. One of them is the republication of George Martin's book *The Ice Dragon* which emerged in early August 2014 resulting tweets about the book. One such tweet is showed in the third row in Table 3. Both the entity and the topic were not known to the EMN, hence it couldn't link tweet to the book *The Ice Dragon*. This problem can be solved by implementing an evolution mechanism for EMN. Lastly, a couple of tweets in the dataset had two implicit entities. One such tweet is shown in the last row of the table which is annotated with the movies *The Fault in Our Stars* and *Divergent*. Since our method links only one entity per tweet, we had the choice to either remove tweets with two mentions or add it once for each mention. We did the latter to preserve the characteristics of the tweets. However, not surprisingly, both tweets were annotated with the same movie (*The Fault in Our Stars*) resulting in an incorrect annotation. Although this is a limitation of our approach, this phenomenon is not a frequent occurrence as the dataset had only 6 (=1.5 %) tweets with two implicit entity mentions.

5.3 Importance of Contextual Knowledge

One of the major components in our entity model is the contextual knowledge of the entity. This section evaluates its contribution to the proposed implicit entity linking solution by comparing the results obtained by EMN created with contextual knowledge and EMN created without contextual knowledge. As in Sect. 5.2, we evaluated this for both candidate selection and ranking, and disambiguation steps.

Table 4 shows the results of this experiment. As shown in Table 4, the contextual knowledge contributes to both candidate selection and ranking, and disambiguation steps on both domains. Quantitatively, the recall value of the first step increased by 14 % and 19 % while accuracy of second step increased by 15 % and 18 % for movies and books, respectively. This is due to the fact that people do not necessarily use associated entities when referring to entities implicitly, but rely on other clues. For example, consider tweets *"'My name was Salmon, like the fish; first name, Susie." Great book!'* and *"2 actors playing brother and*

Table 4. Contribution of the contextual knowledge component of entity model.

Step	Domain	Without ctx	With ctx
Candidate selection recall	Movie	77.29 %	90.33 %
	Book	76.84 %	94.73 %
Disambiguation	Movie	51.7 %	60.97 %
	Book	50.00 %	61.05 %

sister then plot twist new movie, but they have cancer and love each other." The first tweet has an implicit mention of the book *The Lovely Bones* and second tweet has implicit mentions of movies *The Fault in Our Stars* and *Divergent*. However, none of them contain any entities associated (e.g. author, publisher, actors, directors) with the book or the movie; hence, the factual knowledge component in the entity model fell short in these tweets. The contextual knowledge component fills in the gaps since it can build the association between the clues indicated in tweets and the respective entities.

5.4   Value Addition to Standard Entity Linking Task

The real-world datasets collected via keywords contain explicit and implicit entity mentions as well as tweets with no entity mentions. This experiment assess the impact of implicit entity linking in such datasets. To create a dataset for this experiment we followed the following steps:

- Select 40 % of tweets from the dataset with implicit entities as the test dataset. We use the rest to train the ranking model.
- Mix the selected test dataset with tweets containing explicit entity mentions by preserving the explicit:implicit ratio. This ratio is 4:1 in the Movie domain and 5:2 in Book domain.
- Add 25 % of tweets that have no mention of entity to account for NIL mentions.

The experiment setup used three well-known entity linking solutions with the configurations noted below: (1) DBpedia Spotlight [3] (confidence $= 0.5$, support $= 20$), (2) TagMe [6] ($\rho = 0.5$), and (3) Zemanta.[5] It first annotates the

Table 5. EL and IEL combined performance

	DBpedia spotlight		TagMe		Zemanta	
	Movie	Book	Movie	Book	Movie	Book
F1 (EL)	0.18	0.44	0.24	0.19	0.32	0.17
F1 (EL + IEL)	0.34	0.54	0.30	0.34	0.39	0.37

---

[5] http://www.zemanta.com.

prepared Movie dataset using DBpedia Spotlight. The tweets that are not annotated with movies of the output are sent to our proposed solution assuming they have implicit entity mentions. The same exercise is repeated for TagMe and Zemanta. Then we conducted this experiment for the Book dataset.

Table 5 shows the results of this experiment using an F1 measure. The precision (P) and recall (R) values for the F1 measure are calculated as follows:

$$P = \frac{c}{total\ tweets\ annotated\ with\ entity} \quad \text{and} \quad R = \frac{c}{total\ tweets\ with\ entity}$$

where c is the total number of correctly annotated tweets with an entity. These results demonstrate the value of adding IEL as a post-step to EL.

5.5 Discussion

While we collected tweets based on a limited set of keywords, we did not depend on these keywords to link the implicit entities. It is merely used as technique to collect tweets to create an evaluation dataset. Our approach can be applied to link the implicit entity mentions of a given type in the absence of these keywords.

We showed the value of contextual knowledge in implicit entity linking. We believe that it can play a similar role in the disambiguation step of the explicit entity linking task.

The requirement to evolve EMN is observed with the first experiment. We have identified the two events that can change the EMN over time: (1) a new entity becomes popular and people start to tweet about it or the popularity of an existing entity fades away, and (2) a new topic of interest emerges for an existing entity or with the introduction of a new entity, or the popularity of the existing topic fades away. In the future, we will implement the operators that will keep EMN up-to-date by continuously collecting tweets and injecting derived knowledge from them, and from DBpedia.

6 Conclusion and Future Work

We introduced the problem of implicit entity linking in tweets and studied its prevalence and characteristics. We proposed a solution that models the entities with their factual and contextual knowledge and demonstrated that these models are capable of linking implicit entities with higher accuracy than state-of-the-art entity linking approaches. In the future, we will extend our model to account for NIL mentions and expand our evaluation to more domains and larger datasets.

Acknowledgement. We thank US National Science Foundation for supporting this research with grant CNS 1513721 'Context-Aware Harassment Detection on Social Media.'

References

1. Chang, M.-W., Hsu, B.-J., Ma, H., Loynd, R., Wang, K.: E2e: an end-to-end entity linking system for short and noisy text. In: Making Sense of Microposts (2014)
2. Cucerzan, S.: Large-scale named entity disambiguation based on wikipedia data. In: EMNLP-CoNLL, vol. 7, pp. 708–716 (2007)
3. Daiber, J., Jakob, M., Hokamp, C., Mendes, P.N.: Improving efficiency and accuracy in multilingual entity extraction. In: Proceedings of the 9th International Conference on Semantic Systems (I-Semantics) (2013)
4. Derczynski, L., Maynard, D., Rizzo, G., van Erp, M., Gorrell, G., Troncy, R., Petrak, J., Bontcheva, K.: Analysis of named entity recognition and linking for tweets. Inf. Process. Manage. **51**(2), 32–49 (2015)
5. Dredze, M., McNamee, P., Rao, D., Gerber, A., Finin, T.: Entity disambiguation for knowledge base population. In: Proceedings of the 23rd International Conference on Computational Linguistics, pp. 277–285. Association for Computational Linguistics (2010)
6. Ferragina, P., Scaiella, U.: Tagme: on-the-fly annotation of short text fragments (by wikipedia entities). In: Proceedings of the 19th ACM International Conference on Information and Knowledge Management, pp. 1625–1628. ACM (2010)
7. Guo, S., Chang, M.-W., Kiciman, E.: To link or not to link? a study on end-to-end tweet entity linking. In: HLT-NAACL, pp. 1020–1030 (2013)
8. Hoffart, J., Altun, Y., Weikum, G.: Discovering emerging entities with ambiguous names. In: Proceedings of the 23rd International Conference on World Wide Web, pp. 385–396. International World Wide Web Conferences Steering Committee (2014)
9. Hoffart, J., Yosef, M.A., Bordino, I., Fürstenau, H., Pinkal, M., Spaniol, M., Taneva, B., Thater, S., Weikum, G.: Robust disambiguation of named entities in text. In: Proceedings of the Conference on Empirical Methods in Natural Language Processing, pp. 782–792. Association for Computational Linguistics (2011)
10. Lehmann, J., Isele, R., Jakob, M., Jentzsch, A., Kontokostas, D., Mendes, P.N., Hellmann, S., et al.: DDBpedia-a large-scale, multilingual knowledge base extracted from Wikipedia. Semant. Web **6**(2), 167–195 (2015)
11. Liu, X., Li, Y., Haocheng, W., Zhou, M., Wei, F., Yi, L.: Entity linking for tweets. In: ACL, vol. 1, pp. 1304–1311 (2013)
12. Meij, E., Weerkamp, W., de Rijke, M.: Adding semantics to microblog posts. In: Proceedings of the Fifth ACM International Conference on Web Search and Data Mining, pp. 563–572. ACM (2012)
13. Mendes, P.N., Jakob, M., García-Silva, A., Bizer, C.: DBpedia spotlight, shedding light on the web of documents. In: Proceedings of the 7th International Conference on Semantic Systems, pp. 1–8. ACM (2011)
14. Milne, D., Witten, I.H.: Learning to link with wikipedia. In: Proceedings of the 17th ACM Conference on Information and Knowledge Management, pp. 509–518. ACM (2008)
15. Perera, S., Mendes, P., Sheth, A., Thirunarayan, K., Alex, A., Heid, C., Mott, G.: Implicit entity recognition in clinical documents. In: Proceedings of the Fourth Joint Conference on Lexical and Computational Semantics (*SEM), pp. 228–238 (2015)
16. Rao, D., McNamee, P., Dredze, M.: Entity linking: finding extracted entities in a knowledge base. In: Poibeau, T., Saggion, H., Piskorski, J., Yangarber, R. (eds.) Multi-source, Multilingual Information Extraction and Summarization, pp. 93–115. Springer, Heidelberg (2013)

Machine Learning Track

Fast Approximate A-Box Consistency Checking Using Machine Learning

Heiko Paulheim$^{(\boxtimes)}$ and Heiner Stuckenschmidt

Data and Web Science Group, University of Mannheim, Mannheim, Germany
{heiko,heiner}@informatik.uni-mannheim.de

Abstract. Ontology reasoning is typically a computationally intensive operation. While soundness and completeness of results is required in some use cases, for many others, a sensible trade-off between computation efforts and correctness of results makes more sense. In this paper, we show that it is possible to approximate a central task in reasoning, i.e., A-box consistency checking, by training a machine learning model which approximates the behavior of that reasoner for a specific ontology. On four different datasets, we show that such learned models constantly achieve an accuracy above 95 % at less than 2 % of the runtime of a reasoner, using a decision tree with no more than 20 inner nodes. For example, this allows for validating 293M Microdata documents against the schema.org ontology in less than 90 min, compared to 18 days required by a state of the art ontology reasoner.

Keywords: Approximate ontology reasoning · Machine learning

1 Introduction

Ontologies have been a central ingredient of the Semantic Web vision from the beginning on [2,14]. With the invention of OWL, the focus has been on rather expressive ontologies that describe a rather limited set of instances. A number of advanced reasoning systems have been developed to exploit the logical semantics of OWL [11,13,42,47] to support tasks like consistency checking and the derivation of implicit information. While ontological reasoning has been shown to be useful for ensuring the quality of ontological models, it is also known that reasoning is very resource demanding both in theory and in practice. The tableaux algorithm for consistency checking in the Description Logic $\mathcal{SHOIN}$ that provides the formal underpinning of OWL-DL [16] is NExpTime-complete.[1] Further, for large ontologies, the size of the data structures created during the reasoning process easily exceeds the working memory of standard desktop computers.

Although there have been major improvements on the performance of Description Logic reasoners, using them for large real world data sets like DBpedia [20], YAGO [45], or schema.org [28] is not possible in real time. To overcome

---

[1] http://www.cs.man.ac.uk/~ezolin/dl/.

© Springer International Publishing Switzerland 2016
H. Sack et al. (Eds.): ESWC 2016, LNCS 9678, pp. 135–150, 2016.
DOI: 10.1007/978-3-319-34129-3_9

this problem, tractable fragments, i.e. OWL QL and OWL EL [27] have been proposed and corresponding reasoners have been developed that offer much better performance on models that adhere to these subsets [1,43]. However, real world ontology often do not adhere to the defined sublanguages, but rather use logical operators as needed for the purpose of capturing the intended semantics. Examples for popular ontologies outside the tractable fragments of OWL include FOAF, schema.org, and GoodRelations. This means that we need alternative solutions to be able to perform real time reasoning on large real world datasets.

On the other hand, there are good reasons to assume that in many practical settings, the reasoning process can be drastically simplified. For many ontologies, some constructs defined in the T-box are never or only scarcely used in actual A-boxes. For example, for the *schema.org* ontology, we have shown in [24] that a significant portion of the classes and properties defined in schema.org are never deployed on any web site. Similarly, for DBpedia, we have analyzed the reasoning explanations for A-box inconsistencies in [31]. We have observed that those inconsistencies are not equally distributed; on the contrary, no more than 40 different types of inconsistencies are responsible for 99 % of all inconsistencies in DBpedia. Such findings could help tailoring approximate reasoners for a specific ontology which focus on those parts of the ontology which are actually required to address the majority of all cases.

So far, the major part of research on ontology reasoning focuses on developing reasoning systems that are both sound and complete. However, we argue that reasoning results that are 100 % accurate are not required in many use cases for which ontology reasoning has been proposed and/or applied in the past, e.g., information retrieval [40], recommender systems [26], or activity recognition [4]. On the other hand, many of those use cases have very strict performance requirements, as they are usually applied in real time settings. For these settings, approximations of ontological reasoning are a promising approach. A common approach that has already been investigated in the early days of description logics is the idea of language weakening where a given model is translated into a weaker language that allows for more efficient reasoning [38]. While early ideas on approximating Description Logic reasoning have been shown to not be effective in practice [12], recent work on reducing OWL 2 to tractable fragments and using special data structures for capturing dependencies that cannot be represented in the weaker logic has been reported to provide good results [32].

In this paper, we show how to learn an approximate A-box consistency checking function automatically. To this end, we represent that task as a binary classification problem and apply machine learning method to construct a classifier that efficiently solves the problem. Such a classifier automatically learns to focus on those concepts and axioms in an ontology that are relevant in a particular setting, as discussed above. This idea can essentially be applied to any ontology and does not require any manual ground truth annotations, since we can use a sound and complete reasoner for generating the training data for the classifier. The contributions of this paper are the following:

- We present the general idea of imitating A-box consistency checking using machine learning.
- We apply the approach to the problem of consistency checking of a large number of A boxes adhering to the same T box.
- We evaluate the approach on four real world datasets, i.e. DBpedia, YAGO, schema.org Microdata, and GoodRelations RDFa.
- We show that the approach reaches more than 95 % accuracy, and performs the reasoning task at least 50 times faster than a state of the art Description Logic reasoner.

We see this work as a first step towards a more complete investigation of a principle of compiling complex logical reasoning into an efficient decision model that can be executed in the case of highly limited resources.

The paper is structured as follows. Section 2 discusses previous efforts on fast, approximate reasoning. We introduce our approach in Sect. 3, and an evaluation on four real-world Semantic Web datasets in Sect. 4. We conclude the paper with a summary and an outlook on future research directions.

2   Related Work

Some approaches have been proposed in the past to accelerate ontology reasoning. In essence, those encompass three families of techniques: approximating the deduction process itself (e.g. [3]), weakening the expressivity of the language, or compiling some intermediate results off-line (e.g. [32]).

Some reasoning approaches have been proposed that are specifically tailored to the needs of a particular model. For instance, the FacT reasoner, one of the first reasoners for OWL-DL, was originally designed for reasoning about knowledge in the GALEN Ontology [15]. More recently, the SnoRocket System was created to reason about the SnoMed Terminology [23], and in [46], Suda et al. present a special calculus and reasoning system for logical reasoning on the YAGO Ontology. Nevertheless, optimizing reasoning so far has focused on providing more efficient reasoning methods for a particular logic. In particular the EL fragment for OWL has been proposed as a language for very large models and highly optimized reasoners for OWL EL have been built (e.g. [18,43]) to solve the performance problem.

In contrast to those works, which always require some explicit assumptions on how to simplify a given reasoning problem (e.g., which deduction rules to weaken, which logic language concepts to drop, which intermediate results to materialize, etc.), the approach we follow is to train a machine learning model, where the learning algorithm automatically finds a good approximation for a given ontology, implicitly selecting an appropriate subset of strategies from all of the above.

The use of machine learning techniques to complement reasoning has been proposed for different purposes, e.g., as *ontology learning* for assisting ontology engineering and enrichment [19,48], or to introduce approximate class definitions, which are not part of standard ontology languages [33]. There have been attempts to apply machine learning techniques to reasoning problems in description logics ontologies. The corresponding approaches have addressed the problems query answering, classification, and instance retrieval, all of which can be reduced to deciding whether an instance belongs to a class expression in the ontology.

The connection between the classification and the reasoning problem is established through the definition of Kernel [8] or dissimilarity functions respectively that are defined based on joint membership of instances in ontological classes or their negation, where the membership is either determined using logical reasoning or approximated by a lookup of explicitly membership statements. These functions are then used as a basis for inductive methods, i.e. kNN [5] and Support Vector Machines [7].

In more recent work [36] from the same group, terminological extensions of decision tree [35] and random forest classifiers [34] have been proposed. The nodes of the decision trees may contain conjunctive concept expressions which are also represent the feature space. The induction of the trees is guided by an information gain criterion derived from positive, negative and unknown examples which correspond to instances known to belong to the class, to its negation, or that are not known to belong to any of the former. Similar extensions have been proposed for Bayesian classifiers and regression trees.

These methods were evaluated on a set of rather small ontologies with 50–100 classes (with one exception) and up to 1000 individuals.

The methods proposed in this paper share the motivation with the work above, the efficient approximation of deductive reasoning tasks by inductive methods. There are are some differences, however. While the work above focuses on the development of *new machine learning methods* for terminological data, our approach is based on *existing machine learning techniques*. Further, we address a different reasoning task, namely consistency checking instead of deciding class membership. We chose this task as it is the most basic operation in deductive reasoning. This allows us to later extend our approach to any kind of deductive reasoning by reduction to unsatisfiability checking. In particular, being able to approximately check inconsistency provides us with an approximate entailment operator using the principle of proof by refutation. Finally, to the best of our knowledge, our work is the first attempt to apply the idea to very large real world ontologies that are actually used in many applications. We consider models that are an order of magnitude larger than the ones considered in the works above, and it not clear whether the existing approaches scale to models of the size considered in this paper.

In another line of work, machine learning techniques have been used to predict the efficiency of terminological reasoning [17,37] – that work, however, is only weakly related as it uses different features and its goal is to predict the reasoning performance, not its result.

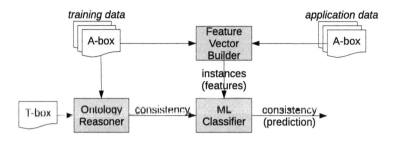

Fig. 1. Schematic view of our approach

3 Approach

With our approach, we try to approximate a reasoner that checks an A-box for consistency, given a T-box. Since the outcome of such a consistency is either true or false, we regard the problem as a binary classification problem when translating it to a machine learning problem.

Figure 1 depicts a schematic view of our approach. An ontology reasoner is run on a T-box and a small number of A-boxes, denoted as *training data*. At the same time, the A-boxes are translated to feature vectors for the machine learning classifier. The pairs of feature vector representations of the A-box and the consistency detection outcome of the reasoner are used as labeled training examples for the classifier, which then learns a model. This model can be applied to a (potentially larger) set of A-boxes (denoted as *application data*).

Formally, we train a classifier C for a T-box T that, given an A-box a in the set of all A-boxes A, determines whether that A-box is consistent or not:

$$C_T : A \rightarrow \{true, false\} \tag{1}$$

Standard machine learning classifiers do not work on A-boxes, but on *feature vectors*, i.e., vectors of nominals or numbers. Thus, in order to exploit standard classifiers, we require a translation from A-boxes to feature vectors, which is given by a function F (denoted as *feature vector builder* in Fig. 1). In this work, we use binary feature vectors, although other types of feature vectors would be possible:

$$F : A \rightarrow \{0, 1\}^n \tag{2}$$

With such a translation function, a binary machine learning classifier M, which operates on feature vectors, can be used to build C_T:

$$C_T := M(F(a)) \rightarrow \{true, false\}, where \ a \in A \tag{3}$$

In the work presented in this paper, we use *path kernels* (often referred to as *walk kernels* as well) as defined in [22] for the translation function F. We slightly altered the original definition such that for literals, we only use the datatype (xsd:string if none exists), but not the literal value as such. Those kernels generate an enumeration of all paths up to a certain lengths that exist

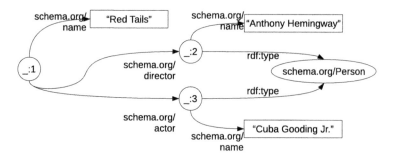

Fig. 2. Example excerpt of an RDF graph

in an RDF graph, and each path is used as a binary feature (i.e., the path is present in a given A-box or not).

Figure 2 depicts an example graph, which is a sub-part of an actual RDF graph in our schema.org Microdata dataset (see below). That graph would be represented by the following six features[2], which represent the number of different paths that exist in the graph:

```
schema.org/name_xsd:string
schema.org/director_schema.org/name_xsd:string
schema.org/director_rdf:type_schema.org/Person
schema.org/actor_schema.org/name_xsd:string
schema.org/actor_rdf:type_schema.org/Person
rdf:type_schema.org/Person
```

4   Evaluation

We evaluate our approach in four different settings:[3]

1. Validating relational assertions in DBpedia with DOLCE
2. Validating type assertions in YAGO with DOLCE
3. Validating entire Microdata documents against the schema.org ontology
4. Validating entire RDFa documents against the GoodRelations ontology

Furthermore, we compare four different learning algorithms for each setting: Decision tree, Naive Bayes, SVM with linear kernel[4], and Random Forest. *HermiT* [11] is used as a reasoner which produces the ground truth, and also as

---

[2] i.e., for those features, the feature value would be 1, for all others, it would be 0.

[3] All the datasets created in this paper are available online at http://dws.informatik. uni-mannheim.de/en/research/reasoning-approximation.

[4] In essence, since we use root paths, this is equivalent to using an SVM whose kernel function is the count of overlapping root paths of two instances.[6].

Table 1. Datasets used for the evaluation. The table shows the ontology complexity of the schema, the feature set size for the different sample sizes, as well as the percentage of consistent A-boxes in the samples.

Dataset	Complexity	# Features	% consistent
DBpedia-11	$\mathcal{SHIN}(\mathcal{D})$	61	72.73
DBpedia-111		216	84.68
DBpedia-1111		854	87.67
YAGO-11	$\mathcal{SHIN}(\mathcal{D})$	101	63.64
YAGO-111		621	73.87
YAGO-1111		4,344	71.38
schema.org-11	$\mathcal{ALCHI}(\mathcal{D})$	90	27.28
schema.org-111		250	35.14
schema.org-1111		716	35.10
GoodRelations-11	$\mathcal{SHI}(\mathcal{D})$	200	54.55
GoodRelations-111		508	60.36
GoodRelations-1111		862	63.10

a baseline for performance comparisons.[5] All experiments were conducted with *RapidMiner Studio*[6], for Random Forest, we use the implementation in the Weka extension. For the latter, the number of trees was set from 10 to 50 due to the large number of features in some datasets (see below), all other operators were used in their standard settings.

4.1 Datasets

While most type assertions in DBpedia are correct, the main sources of errors are relational and literal assertions [30,49]. Since the latter can hardly be detected by reasoners without literal range definitions (which do not exist in DBpedia), we concentrate on relational assertions. As a dataset, we use all mapping-based properties and types in DBpedia [20]. Since DBpedia as such comes with only few disjointness axioms, which are required for detecting inconsistencies, we add the top-level ontology DOLCE-Zero [9,10] to the reasoner. In this setting, we check whether a relational assertion is consistent with the subject's and object's type assertion. Hence, as proposed in [31], each A-box to be checked consists of a relation assertion and all the type assertions for the subject and object.

In YAGO, types are derived from Wikipedia categories and mapped to Word-Net [44]. Thus, in contrast to DBpedia, incompatible class assertions are possible

---

[5] A reasoner capable of handling the $\mathcal{SHIN}(\mathcal{D})$ complexity class is required for the experiments (cf. Table 1). In a set of preliminary experiments, we tried both Pellet [42] and HermiT, where only the latter was able to handle the DBpedia+DOLCE ontology.

[6] http://www.rapidminer.com.

in YAGO. For example, the instance *A Clockwork Orange* has both the types *1962 Novels* and *Obscenity Controversies*, among others, which are ontologically incompatible on a higher level (a controversy being subsumed under *Activity*, a novel being subsumed under *Information Entity*). Again, since the YAGO ontology as such does not contain class disjointness axioms, we use it in conjunction with the DOLCE top level ontology. Similar to the DBpedia case, the reasoner checks individual A-boxes, each consisting of all the type assertions of a *single entity*. Here, we use types which have already been materialized according to the full hierarchy.

For the schema.org case, we use a sample of documents from the WebDataCommons 2014 Microdata corpus [25], and the corresponding schema.org ontology version.[7] Here, a single document is the set of triples extracted from one Web page. Since schema.org does not come with disjointness axioms, we have defined a set of disjointness axioms between the high level classes as follows: Starting from the largest class, we inspect all the class definitions (both formal and textual), and insert disjointness axioms where appropriate as long as the T-box does not become inconsistent. The set of those inconsistencies is shown in Table 2.

Likewise, we proceed with the GoodRelations RDFa documents. From the WebDataCommons RDFa corpus, we use a subset of documents that use at least one concept from the GoodRelations namespace, and validate it against the GoodRelations 1.0 ontology[8].

In all cases, we use root paths of arbitrary lengths. This is possible since there are no cycles in the graphs, since we are only validating single relational statements for DBpedia, single instances for YAGO, and the documents in the schema.org Microdata are also cycle-free [29]. For the GoodRelations dataset, cycles would be possible in theory, however, we did not encounter any in our sample. If an A-box to test was not proper DL (e.g., an object property was

Table 2. Disjointness Axioms inserted for schema.org. An X denotes a disjointness axiom defined, an e denotes that the disjointness axiom was expected, but lead to a T-box inconsistency. Since *Local Business* in schema.org is both a subclass of *Organization* and *Place*, the latter two cannot be disjoint. Likewise, *Exercise Plan* is both a subclass of *Creative Work* and *Medical Entity*.

	Product	Place	Person	Organization	Medical Entity	Intangible	Event	Creative Work
Action	X	X	X	X	X		X	X
Creative Work		X	X	X	e		X	
Event	X	X	X	X	X			
Intangible		X						
Medical Entity		X	X	X				
Organization	X	e	X					
Person	X	X						
Place	X							
Product								

---

[7] We have followed the recommendations on this web page to create the OWL version: http://topbraid.org/schema/.

[8] http://purl.org/goodrelations/v1.

used with a literal object), the instance was directly marked as inconsistent, without passing it to the reasoner.

For each of the three use cases, we extracted three versions, i.e., one with 11, one with 111, and one with 1,111 A-box instances. The rationale for those numbers is that it is possible to evaluate the performance of our approach using 10, 100, and 1,000 training instances, respectively, in a ten-fold cross validation setting. Table 1 summarizes the characteristics of the datasets.

4.2 Results

For all four pairs of datasets, we compare the results for the four learning methods. All experiments were conducted in 10-fold cross validation. Table 3 depicts the accuracy of all approaches.

From the results, we can see that in general, 10 training instances are too few to train a classifier significantly outperforming the majority prediction baseline. For the other cases, the decision tree classifier usually provides the best results (on the DBpedia datasets, the difference to SVM is not significant). Unlike the other classifiers, the performance of Naive Bayes is worse when trained with 1,000 instances than when trained with 100. This can be explained with the increase in the number of features when increasing the number of training examples (cf. Table 1). At the same time, there are many interdependent features, which violate the independence assumption of Naive Bayes.

Figure 3 depicts ROC curve plots for the four classifiers and the six problems. Since the table shows that meaningful models are only learned for 100 and more instances, we only depict ROC curves for the 111 and 1,111 datasets. ROC curves

Table 3. Performance of the different classifiers on the three problems. The table depicts the accuracy for all classifiers, including the confidence intervals obtained across a 10-fold cross validation. The best result for each dataset is marked in bold. The baseline is formed by predicting the majority class.

Dataset	Baseline	Decision Tree	Naive Bayes	SVM	Random Forest
DBpedia-11	**.727**	.700 ± .458	600 ± .490	.600 ± .490	.600 ± .490
DBpedia-111	.847	.937 ± .058	.756 ± .101	**.946 ± .073**	.901 ± .049
DBpedia-1111	.877	.951 ± .016	.856 ± .043	**.961 ± .015**	.926 ± .017
YAGO-11	.636	.550 ± .472	**.850 ± .320**	.650 ± .450	.650 ± .450
YAGO-111	.739	**.955 ± .061**	.805 ± .153	.892 ± .066	.901 ± .064
YAGO-1111	.714	**.976 ± .015**	.672 ± .056	.929 ± .052	.913 ± .034
schema.org-11	**.727**	.700 ± .458	.600 ± 0.490	.700 ± .458	.600 ± .490
schema.org-111	.645	**.936 ± .071**	.730 ± .158	.695 ± .066	.749 ± .182
schema.org-1111	.649	**.977 ± .020**	.762 ± .066	.778 ± .060	.906 ± .038
GoodRelations-11	.545	**.700 ± .458**	**.700 ± .458**	**.700 ± .458**	**.700 ± .458**
GoodRelations-111	.604	**.901 ± .076**	.847 ± 0.129	**.901 ± .076**	.892 ± .078
GoodRelations-1111	.631	**.969 ± .016**	.874 ± .027	.945 ± 0.20	.945 ± .027

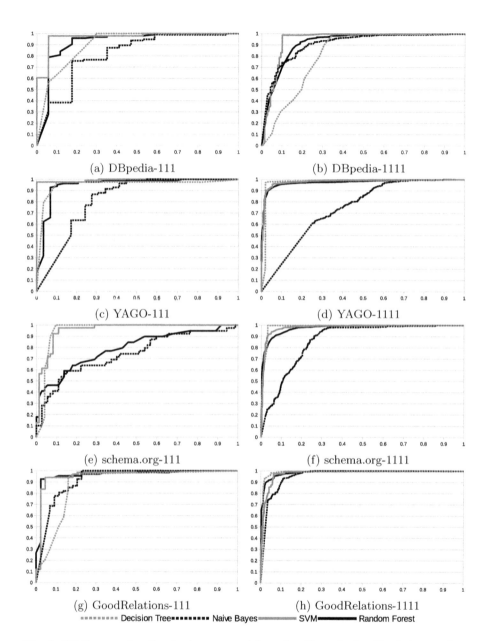

(a) DBpedia-111

(b) DBpedia-1111

(c) YAGO-111

(d) YAGO-1111

(e) schema.org-111

(f) schema.org-1111

(g) GoodRelations-111

(h) GoodRelations-1111

Decision Tree ▪▪▪▪▪▪▪▪ Naive Bayes ═══ SVM ▬▬▬ Random Forest

Fig. 3. ROC curves for the six datasets and four classifiers. A classifier is better than another if its curve runs above the other classifiers curve. A random prediction baseline would lead to to a diagonal.

Table 4. Computation times for 1,000 statements with HermiT vs. Decision Tree, and size of the models learned. For the tree size, the first number is the number of inner nodes, while the number in parantheses depicts the maximum path length to a leaf node.

Dataset	1000 k instances (ms)				All A-boxes (s)		
	HermiT	Features	Training	Classification	HermiT	Trained	DT Size
DBpedia	86,240	259	256	2	1,293,733	4,002	13 (8)
YAGO	20,624	179	814	6	186,022	556	9 (7)
schema.org	7,769	138	285	2	1,574,014	4,110	19 (19)
GoodRelations	15,577	236	229	2	5,545	101	18 (13)

do not take into account only correct and incorrect predictions, but also reflect how good the confidence scores computed by the classifiers are. We can see that the curves often run close to the upper left corner, i.e., the confidence scores are also usually quite accurate.

In addition to the qualitative results, we also look into computation times. Here, we compare the runtime for the reasoner to the runtime of a Decision Tree based classifier, which has been shown to perform best in terms of accuracy. We compare the runtime for classifying 1,000 A-boxes, as well as provide an estimate for classifying all A-boxes in the respective settings. For the latter, we consider all relational statements in DBpedia (ca. 15M in the `dbpedia-owl` namespace), all typed YAGO instances (ca. 2.9M), all URLs from which schema.org Microdata can be extracted (ca. 293M), and all URLs from which RDFa annotations with GoodRelations can be extracted (ca. 356k). For estimating the time for checking all A-boxes, we extrapolated from the time to check 1,000 A-boxes. The extrapolation of the time for the trained model includes the time for running the reasoner and training the Decision Tree on 1,000 instances, as well as the classification with that Decision Tree on the remaining instances. More formally, the extrapolated time for all A-boxes is

$$T_{ext} = |A| \cdot T_F + T_T + (|A| - 1000) \cdot T_C, \tag{4}$$

where T_F is the time to create features on a single A-box, T_T is the time to train the model on 1,000 instances, T_C is the time to check an A-box with the trained model, and $|A|$ is the total number of A-boxes.[9] In total, it can be observed that the computation time based on the trained model is lower than the reasoning based computation time by at least a factor of 50.

For the runtimes, only pure computation times for the consistency determination were measured, but no initialization time for the reasoner (i.e., loading the ontology for the first time) and no I/O times.[10] Furthermore, to make the

---

[9] Note that the features are generated on each A-box in isolation, so there is only a linear dependency on the number of A-boxes.

[10] All runtimes were measured on a Windows 7 laptop with an Intel i7 quadcore processor and 8 GB of RAM.

comparison fair for YAGO, the reasoner was also provided with fully materialized type statements, and the type hierarchy was not loaded by the reasoner.

Table 4 sums up the computation times. It can be observed that for the Decision Tree based approach, the larger amount of time is spent on the creation of feature vector representations of the instances, rather than the classification itself. For example, we see that checking all of DBpedia's relational assertions[11] with our approach would take less than 90 min, whereas HermiT would take around 15 days. Likewise, checking all the schema.org Microdata documents in the Web Data Commons Corpus would take more than 18 days with HermiT, while our approach would also finish that task in less than 90 min.

Another interesting observation is that the size of the models used for classification is relatively small. While the ontologies are fairly large (loading DBpedia and DOLCE into HermiT for validating a single statement consumes 1.5 GB of RAM), the corresponding Decision Trees are surprisingly small, with no more than 20 inner nodes. Furthermore, the longest path to a leaf node, which corresponds to the number of tests to be performed on the A-box for computing the results, is also strongly constrained, which explains the high computational performance of the approach.

5   Conclusion and Outlook

In this paper, we have shown that it is possible to approximate a reasoner for A-box consistency checking by training machine learning classifiers. Especially decision trees have turned out to be quite effective for that problem, reaching an accuracy of more than 95 % at computation times at least 50 times as fast as a standard ontology reasoner. The resulting trees are surprisingly small, with no more than 20 decision nodes and tests to perform. This makes the approach appealing to be applied in scenarios with very strongly limited computation resources where not only computation time, but also memory is constrained.

The fact that the learned decision trees use only a very small number of features would allow for even faster implementations of such reasoners, since the tests in the inner nodes could also be performed by pattern matching against the A-box instead of the explicit creation of the entire set of features.

So far, we have focused on consistency checking. However, we believe that the approach is versatile enough to be transferred to other reasoning tasks as well. For example, the task of instance classification could be modeled and solved as a hierarchical multi-label classification problem [41]. Similarly, the task of *explaining* an inconsistency could be modeled as a multi-label classification problem, with the target of predicting the axioms in the explanation.

The evaluation has shown that with decision tree classifiers, it is not only possible to make predictions at high accuracy. The ROC curves furthermore show that the classifier is aware of its quality, i.e., wrong predictions usually go together with low confidence values. This gives way to developing a hybrid

---

[11] Note that this is *not* equivalent to checking the consistency of the entire A-box of DBpedia as a whole.

system which could deliver high quality results by exploiting active learning [39], i.e., deciding by itself when to invoke the actual reasoner (and incorporate its result in the machine learning model).

For the learning task, a challenging problem is to use an appropriate feature vector representation and configuration of the learning algorithm. For the former, we assume that there is a correlation between the ontology's expressivity and the optimal feature vector representation – e.g., for ontologies incorporating numerical restrictions other than 0 and 1, numeric feature vector representations should be preferred over binary ones. For the latter, we assume that there is a correlation between certain characteristics of the ontology and the machine learning algorithm configuration – for example, the number of required decision trees in a Random Forest is likely to depend on the number of classes and properties defined in the ontology.

The models issued by the learning algorithms have, so far, not been taken into account for any other purpose than the prediction. However, looking at which features are actually used by the decision tree could be exploited as well, e.g., for ontology summarization [21], since they refer to concepts that are likely to influence a reasoner's outcome.

In summary, we have shown that it is possible to train rather accurate approximate reasoners using machine learning. The findings give way to fast implementations of reasoning in scenarios which do not require 100 % accuracy. On a theoretical level, the approach opens up a number of interesting research questions.

Acknowledgements. The authors would like to thank Aldo Gangemi for providing the DOLCE mappings for DBpedia and YAGO, and Robert Meusel for his assistance in providing suitable samples from the WebDataCommons corpora. This work has been supported by RapidMiner in the course of the RapidMiner Academia program.

References

1. Baader, F., Lutz, C., Suntisrivaraporn, B.: CEL — a polynomial-time reasoner for life science ontologies. In: Furbach, U., Shankar, N. (eds.) IJCAR 2006. LNCS (LNAI), vol. 4130, pp. 287–291. Springer, Heidelberg (2006)
2. Berners-Lee, T., Hendler, J., Lassila, O., et al.: The semantic web. Sci. Am. **284**(5), 28–37 (2001)
3. Cadoli, M., Schaerf, M.: Approximation in concept description languages. In: KR, pp. 330–341 (1992)
4. Chen, L., Nugent, C.: Ontology-based activity recognition in intelligent pervasive environments. Int. J. Web Inf. Syst. **5**(4), 410–430 (2009)
5. d'Amato, C., Fanizzi, N., Esposito, F.: Query answering and ontology population: an inductive approach. In: Bechhofer, S., Hauswirth, M., Hoffmann, J., Koubarakis, M. (eds.) ESWC 2008. LNCS, vol. 5021, pp. 288–302. Springer, Heidelberg (2008)
6. de Vries, G.K.D., de Rooij, S.: A fast and simple graph kernel for RDF. In: DMoLD, vol. 1082 (2013)

7. Fanizzi, N., d'Amato, C., Esposito, F.: Statistical learning for inductive query answering on OWL ontologies. In: Sheth, A.P., Staab, S., Dean, M., Paolucci, M., Maynard, D., Finin, T., Thirunarayan, K. (eds.) ISWC 2008. LNCS, vol. 5318, pp. 195–212. Springer, Heidelberg (2008)
8. Fanizzi, N., d'Amato, C., Esposito, F.: Induction of robust classifiers for web ontologies through kernel machines. J. Web Sem. **11**, 1–13 (2012)
9. Gangemi, A., Guarino, N., Masolo, C., Oltramari, A.: Sweetening WORDNET with DOLCE. AI Mag. **24**, 13–24 (2003)
10. Gangemi, A., Mika, P.: Understanding the semantic web through descriptions and situations. In: Meersman, R., Schmidt, D.C. (eds.) CoopIS 2003, DOA 2003, and ODBASE 2003. LNCS, vol. 2888, pp. 689–706. Springer, Heidelberg (2003)
11. Glimm, B., Horrocks, I., Motik, B., Stoilos, G., Wang, Z.: Hermit: an OWL 2 reasoner. J. Autom. Reasoning **53**(3), 245–269 (2014)
12. Groot, P., Stuckenschmidt, H., Wache, H.: Approximating description logic classification for semantic web reasoning. In: Gómez-Pérez, A., Euzenat, J. (eds.) ESWC 2005. LNCS, vol. 3532, pp. 318–332. Springer, Heidelberg (2005)
13. Haarslev, V., Möller, R.: Racer: a core inference engine for the semantic web. In: EON, vol. 87 (2003)
14. Hendler, J.: Agents and the semantic web. IEEE Intell. Syst. **2**, 30–37 (2001)
15. Horrocks, I., Rector, A.L., Goble, C.A.: A description logic based schema for the classification of medical data. In: KRDB, vol. 96, pp. 24–28. Citeseer (1996)
16. Horrocks, I., Sattler, U.: A tableau decision procedure for\ mathcal {SHOIQ}. J. Autom. Reasoning **39**(3), 249–276 (2007)
17. Kang, Y.-B., Li, Y.-F., Krishnaswamy, S.: Predicting reasoning performance using ontology metrics. In: Cudré-Mauroux, P., et al. (eds.) ISWC 2012, Part I. LNCS, vol. 7649, pp. 198–214. Springer, Heidelberg (2012)
18. Kazakov, Y., Krötzsch, M., Simančík, F.: The incredible ELK. J. Autom. Reasoning **53**(1), 1–61 (2014)
19. Lehmann, J., Auer, S., Bühmann, L., Tramp, S. (geb. Dietzold).: Class expression learning for ontology engineering. J. Web Seman. 9(1), 71–81 (2011)
20. Lehmann, J., Isele, R., Jakob, M., Jentzsch, A., Kontokostas, D., Mendes, P.N., Hellmann, S., Morsey, M., van Kleef, P., Auer, S., et al.: DBpedia-a large-scale, multilingual knowledge base extracted from wikipedia. Seman. Web J. **5**, 1–29 (2014)
21. Li, N., Motta, E., d'Aquin, M.: Ontology summarization: an analysis and an evaluation. In: International Workshop on Evaluation of Semantic Technologies (2010)
22. Lösch, U., Bloehdorn, S., Rettinger, A.: Graph kernels for RDF data. In: Simperl, E., Cimiano, P., Polleres, A., Corcho, O., Presutti, V. (eds.) ESWC 2012. LNCS, vol. 7295, pp. 134–148. Springer, Heidelberg (2012)
23. Metke-Jimenez, A., Lawley, M.: Snorocket 2.0: concrete domains and concurrent classification. In: ORE, pp. 32–38. Citeseer (2013)
24. Meusel, R., Bizer, C., Paulheim, H.: A web-scale study of the adoption and evolution of the schema. org vocabulary over time. In: 5th International Conference on Web Intelligence, Mining and Semantics (WIMS), pp. 15. ACM (2015)
25. Meusel, R., Petrovski, P., Bizer, C.: The WebDataCommons microdata, RDFa and microformat dataset series. In: Mika, P., et al. (eds.) ISWC 2014, Part I. LNCS, vol. 8796, pp. 277–292. Springer, Heidelberg (2014)
26. Middleton, S.E., De Roure, D., Shadbolt, N.R.: Ontology-based recommender systems. Handbook on Ontologies. International Handbooks on Information Systems, pp. 779–796. Springer, Heidelberg (2009)

27. Motik, B., Grau, B.C., Horrocks, I., Wu, Z., Fokoue, A., Lutz, C.: OWL 2 web ontology language: profiles. W3C recommendation, vol. 27, p. 61 (2009)
28. Patel-Schneider, P.F.: Analyzing schema.org. In: Mika, P., et al. (eds.) ISWC 2014, Part I. LNCS, vol. 8796, pp. 261–276. Springer, Heidelberg (2014)
29. Paulheim, H.: What the adoption of schema.org tells about linked open data. In: Dataset PROFIling & fEderated Search for Linked Data (2015)
30. Paulheim, H., Bizer, C.: Improving the quality of linked data using statistical distributions. Int. J. Seman. Web Inf. Syst. (IJSWIS) 10(2), 63–86 (2014)
31. Paulheim, H., Gangemi, A.: Serving DBpedia with DOLCE – more than just adding a cherry on top. In: Arenas, M., et al. (eds.) ISWC 2015. LNCS, vol. 9366, pp. 180–196. Springer, Heidelberg (2015). doi:10.1007/978-3-319-25007-6_11
32. Ren, Y., Pan, J.Z., Zhao, Y.: Soundness preserving approximation for tbox reasoning. In: AAAI, pp. 351–356 (2010)
33. Rizzo, G., dAmato, C., Fanizzi, N.: On the effectiveness of evidence-based terminological decision trees. In: Esposito, F., et al. (eds.) ISMIS 2015. LNCS, vol. 9384, pp. 139–149. Springer. Heidelberg (2015). doi:10.1007/978-3-319-25252-0_15
34. Rizzo, G., d'Amato, C., Fanizzi, N., Esposito, F.: Tackling the class-imbalance learning problem in semantic web knowledge bases. In: Janowicz, K., Schlobach, S., Lambrix, P., Hyvönen, E. (eds.) EKAW 2014. LNCS, vol. 8876, pp. 453–468. Springer, Heidelberg (2014)
35. Rizzo, G., d'Amato, C., Fanizzi, N., Esposito, F.: Towards evidence-based terminological decision trees. In: Laurent, A., Strauss, O., Bouchon-Meunier, B., Yager, R.R. (eds.) IPMU 2014, Part I. CCIS, vol. 442, pp. 36–45. Springer, Heidelberg (2014)
36. Rizzo, G., dAmato, C., Fanizzi, N., Esposito, F.: Inductive classification through evidence-based models and their ensembles. In: Gandon, F., Sabou, M., Sack, H., dAmato, C., Cudré-Mauroux, P., Zimmermann, A. (eds.) ESWC 2015. LNCS, vol. 9088, pp. 418–433. Springer, Heidelberg (2015)
37. Sazonau, V., Sattler, U., Brown, G.: Predicting performance of OWL reasoners: locally or globally? In: KR. Citeseer (2014)
38. Schaerf, M., Cadoli, M.: Tractable reasoning via approximation. Artif. Intell. 74(2), 249–310 (1995)
39. Settles, B.: Active learning literature survey. University of Wisconsin, Madison, vol. 52(55–66), p. 11 (2010)
40. Shah, U., Finin, T., Joshi, A., Cost, R.S., Matfield, J.: Information retrieval on the semantic web. In: Proceedings of the Eleventh International Conference on Information and Knowledge Management, pp. 461–468. ACM (2002)
41. Silla Jr., C.N., Freitas, A.A.: A survey of hierarchical classification across different application domains. Data Min. Knowl. Disc. 22(1–2), 31–72 (2011)
42. Sirin, E., Parsia, B., Grau, B.C., Kalyanpur, A., Katz, Y.: Pellet: A practical OWL-dl reasoner. Web Seman. Sci. Serv. Agents World Wide Web 5(2), 51–53 (2007)
43. Steigmiller, A., Liebig, T., Glimm, B.: Konclude: system description. Web Seman. Sci. Serv. Agents World Wide Web 27, 78–85 (2014)
44. Suchanek, F.M., Kasneci, G., Weikum, G.: Yago: a core of semantic knowledge. In: 16th International Conference on World Wide Web, pp. 697–706 (2007)
45. Suchanek, F.M., Kasneci, G., Weikum, G.: YAGO: a large ontology from wikipedia and wordnet. Web Seman. Sci. Serv. Agents World Wide Web 6(3), 203–217 (2008)
46. Suda, M., Weidenbach, C., Wischnewski, P.: On the saturation of YAGO. In: Giesl, J., Hähnle, R. (eds.) IJCAR 2010. LNCS, vol. 6173, pp. 441–456. Springer, Heidelberg (2010)

47. Tsarkov, D., Horrocks, I.: FaCT++ description logic reasoner: system description. In: Furbach, U., Shankar, N. (eds.) IJCAR 2006. LNCS (LNAI), vol. 4130, pp. 292–297. Springer, Heidelberg (2006)
48. Völker, J., Niepert, M.: Statistical schema induction. In: Antoniou, G., Grobelnik, M., Simperl, E., Parsia, B., Plexousakis, D., De Leenheer, P., Pan, J. (eds.) ESWC 2011, Part I. LNCS, vol. 6643, pp. 124–138. Springer, Heidelberg (2011)
49. Wienand, D., Paulheim, H.: Detecting incorrect numerical data in DBpedia. In: Presutti, V., dAmato, C., Gandon, F., dAquin, M., Staab, S., Tordai, A. (eds.) ESWC 2014. LNCS, vol. 8465, pp. 504–518. Springer, Heidelberg (2014)

Enriching Product Ads with Metadata from HTML Annotations

Petar Ristoski[1]([✉]) and Peter Mika[2]

[1] Data and Web Science Group, University of Mannheim, Mannheim, Germany
petar.ristoski@informatik.uni-mannheim.de
[2] Yahoo Labs, London, UK
pmika@yahoo-inc.com

Abstract. Product ads are a popular form of search advertizing offered by major search engines, including Yahoo, Google and Bing. Unlike traditional search ads, product ads include structured product specifications, which allow search engine providers to perform better keyword-based ad retrieval. However, the level of completeness of the product specifications varies and strongly influences the performance of ad retrieval.

On the other hand, online shops are increasing adopting semantic markup languages such as Microformats, RDFa and Microdata, to annotate their content, making large amounts of product description data publicly available. In this paper, we present an approach for enriching product ads with structured data extracted from thousands of online shops offering Microdata annotations. In our approach we use structured product ads as supervision for training feature extraction models able to extract attribute-value pairs from unstructured product descriptions. We use these features to identify matching products across different online shops and enrich product ads with the extracted data. Our evaluation on three product categories related to electronics show promising results in terms of enriching product ads with useful product data.

Keywords: Microdata · schema.org · Data integration · Product data

1 Introduction

Product ads are a popular form of search advertizing[1] that are increasingly used as a replacement for text-based search ads, and are currently offered as an option by Bing, Google and Yahoo under different trade names. Unlike traditional search ads that carry only a title, link and a description, product ads are more structured. They often include further details such as the product identifier, brand, model for electronics, or gender for clothing. These details are provided as part of data feeds that merchants transmit to the search engine, and they allow search engine providers to perform better keyword-based ad retrieval, and

---

[1] Search advertising is a method of placing online advertisements on web pages that show results from search engine queries.

© Springer International Publishing Switzerland 2016
H. Sack et al. (Eds.): ESWC 2016, LNCS 9678, pp. 151–167, 2016.
DOI: 10.1007/978-3-319-34129-3_10

to offer additional options such as faceted search over a set of product results. The level of completeness of the product specification, however, depends on the completeness of the advertisers' own data, their level of technical sophistication in creating data feeds and/or willingness to provide additional information to the search engine provider beyond the minimally required set of attributes. As a result of this, product ads are often very incomplete when it comes to the details of the product on offer.

In this paper, we address this problem by enriching product ads with structured data extracted from HTML pages that contain semantic annotations. Structured data in HTML pages is becoming more commonplace and it obviates the need for costly information extraction. In particular, annotations using vocabularies such as schema.org (an initiative sponsored by Bing, Google, Yahoo, and Yandex) and Facebook's OGP (Open Graph Protocol) are increasingly popular. To our knowledge, ours is the first work to investigate the potential of this data for enriching product ads, and to provide a targeted solution for matching product ads to product descriptions on the Web in order to exploit this data. In this work we focus on data annotated with the Microdata markup format using the schema.org vocabuiary. Recent works [10,11] have shown that the Microdata format is the most commonly used markup format, with highest domain and entity coverage. Also, schema.org is the most frequently used vocabulary to describe products.

Our method relies on a combination of highly accurate Information Extraction from unstructured text (titles and descriptions of products) and efficient and effective instance matching (also called reconciliation or duplicate detection) between product descriptions. More precisely, we use the structured product specifications in Yahoo's Gemini Product Ads as a supervision to build two feature extraction models, i.e., dictionary-based model and Conditional Random Field tagger, able to extract attribute-value pairs from unstructured text. Later, we use these features to build machine learning models able to identify matching products. An evaluation on three categories related to electronics shows that we are able to identify matching products across thousands of online shops with high precision and extract valuable structured data for enriching product ads.

The rest of this paper is structured as follows. In Sect. 2, we give an overview of related work. In Sect. 3, we formally define the problem of enriching product ads and we introduce our methodology. In Sect. 4, we present the results of matching unstructured product descriptions, followed by the results of the product ads enriching in Sect. 5. In Sect. 6, we adapt the proposed methodology for the task of product categorization. We conclude with a summary and an outlook on future work.

2   Related Work

While the task of enriching product ads with features from HTML annotations hasn't been studied so far, the problem of products matching and integration on the Web has been extensively studied in the recent years.

The approach presented by Ghani et al. [6] is the first effort for enriching product databases with attribute-value pairs extracted from product descriptions on the Web. The approach uses Naive Bayes in combination with semi-supervised co-EM algorithm to extract attribute-value pairs from text. An evaluation on apparel products shows promising results, however the system is able to extract attribute-value pairs only if both the attribute and the value appear in the text.

One of the closest works is the work by Kannan et al. [8]. The approach uses a database of structured product records to build a dictionary-based feature extraction model. Later, the features of the products are used to train Logistic Regression model for matching product offers. The approach has been used for matching offers received by Bing shopping data to the Bing product catalog.

The *XploreProducts.com* platform [16] is the first effort to integrate products from different online shops annotated using RDFa annotations. The approach is based on several string similarity functions for product matching. Once the matching products are identified, the system integrates the available ratings, offers and reviews into one system. The system is evaluated on an almost balanced set of 600 electronics product combinations. However, in real applications the problem of products matching is highly imbalanced. The approach is first extended in [1], using a hybrid similarity method. Later, the method is extended in [2], where hierarchical clustering is used for matching products from multiple web shops, using the same hybrid similarity method.

Similar to our CRF feature extraction approach, the authors in [9] propose an approach for annotating products descriptions based on a sequence BIO tagging model, following an NLP text chunking process. Specifically, the authors train a linear-chain conditional random field model on a manually annotated training dataset, to identify only 8 general classes of terms. However, the approach is not able to extract explicit attribute-value pairs.

The first approach to perform products matching on Microdata annotation is presented in [13]. The approach is based on the Silk rule learning framework [7], which is able to identify matching products based on their attributes. To do so, different combination of features from the product descriptions are used, e.g., bag of words, attribute-value pairs extracted using a dictionary, features extracted using manually written regular expressions, and combination of all. The work has been extended in [14], where the authors developed a genetic algorithm for learning regular expressions for extracting attribute-value pairs from products.

While there are several approaches concerned with products data categorization [8,12,13,15,16], the approach by Meusel et al. [11] is the most recent approach for exploiting Microdata annotations for categorization of products data. In this approach the authors exploit the already assigned *s:Category* property to develop distantly supervised approaches to map the products to set of target categories from an existing product catalog.

3 Approach

3.1 Problem Statement

We have a database A of structured product ads and a dataset of unstructured product descriptions P extracted from the Web. Every record $a \in A$ consist of title, description, URL, and a set of attribute-value pairs extracted from the title of the ad, where the attributes are numeric, categorical or free-text attributes. Every record $p \in P$ consist of title and description as unstructured textual fields. Our objective is to use the structured information from the product ads set A as supervision for identifying duplicate records in P, or matching products from P to one or more structured ads in A. More precisely, we use the structured information as a supervision for building a feature extraction model able to extract attribute-value pairs from the unstructured product descriptions in P. After the feature extraction model is applied, each product $p \in P$ is represented as a vector of attributes $F_p = \{f_1, f_2, ..., f_n\}$, where the attributes are numerical or categorical. Then we use the attribute vectors to build a machine learning model able to identify matching products. To train the model we manually label a small training set of matching and non-matching unstructured product offers.

3.2   Methodology

The approach we propose in this paper consist of three main steps: (i) *feature extraction*, (ii) *calculating similarity feature vectors* and (iii) *classification*. The overall design of our system is illustrated in Fig. 1. The products integration workflow runs in two phases: *training* and *application*. The training phase starts with preprocessing both the structured product ads and the unstructured product descriptions. Then, we use the structured product ads to build a feature extraction model. In this work we build two strategies for feature extraction (see Sect. 3.3): dictionary-based approach and Conditional Random Fields tagger. Next, we manually label a small training set of matching and non-matching unstructured pairs of product descriptions. We use the created feature extraction model to extract attribute-value pairs from the unstructured product descriptions. Then, we calculate the similarity feature vectors for the labeled training product pairs (see Sect. 3.5). In the final step, the similarity feature vectors are used to train a classification model (see Sect. 3.6). After the training phase is over, we have a trained feature extraction model and a classification model.

The application phase starts with preprocessing both of the datasets that are supposed to be matched[2]. Next, we generate a set M of all possible candidate matching pairs, which leads to a large number of candidates i.e., $|M| = n*(n-1)/2$, if we try to identify duplicates within a single dataset of n products, or $|M| = n*m$, if we try to match two datasets of products with size n and m, respectively. To reduce the search space we use the brand value for blocking, i.e., we apply the matcher only for pairs of product descriptions sharing the same brand. Then, we extract the attribute-value pairs using the feature extraction model and calculate the feature similarity vectors. In the final step we apply the previously built classification model to identify the matching pairs of products.

---

[2] We need to note that we apply the same approach for identifying duplicates within the dataset of unstructured product descriptions, and for identifying matches between the unstructured product descriptions and the structured product ads.

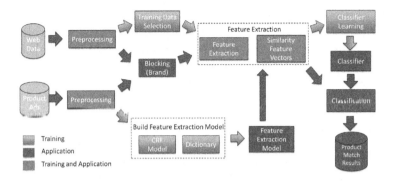

Fig. 1. System architecture overview (Color figure online)

3.3   Feature Extraction

In this section we describe two approaches for extracting attribute-value pairs from unstructured product title and description. In particular, both approaches take as an input unstructured text, and output a set of attribute-value pairs.

Dictionary-Based Approach: To implement the dictionary-based approach we were motivated by the approach described by Kannan et al. [8]. We use the set of product ads in A to generate a dictionary of attributes and values. Let F represent all the attributes present in the product ads A. The dictionary represents an inverted index D from A such that $D(v)$ returns the attribute name $f \in F$ associated with a string value v. Then, to extract features from a given product description p, we generate all possible n-grams $(n \leq 4)$ from the text, and try to match them against the dictionary values. In case of multiple matches, we choose the longest n-gram.

Conditional Random Fields: As the dictionary-based approach is able to extract only values that were seen, we need to use more advanced approach that is able to extract unseen attribute-value pairs. A commonly used approach for tagging textual descriptions in NLP are conditional random field (CRF) models. A CRF is a conditional sequence model which defines a conditional probability distribution over label sequences given a particular observation sequence. In this work we use the Stanford CRF implementation[3] in order to train product specific CRF models [5]. To train the CRF model the following features are used: current word, previous word, next word, current word character n-gram $(n \leq 6)$, current POS tag, surrounding POS tag sequence, current word shape, surrounding word shape sequence, presence of word in left window (size = 4) and presence of word in right window (size = 4).

To train the CRF model we use the structured product ads from database A. That means that the model is able to extract only attribute names that appear in the database A, but it can tag values that don't appear in the database.

---

[3] http://nlp.stanford.edu/software/CRF-NER.shtml.

Custom Feature Extraction: Beside the supervised extraction approaches, we extract several more features for all unstructured products. We use the product Web domain name, and the product URL (both are considered a long string in the following section). The rationale for using these two fields is that often important keywords can be found in the product URL, and the domain might be a good indicator about the type of the product.

Furthermore, we noticed that in some of the product title and/or description a so called *product code* is present, which in many cases uniquely identifies the product. For example, *UN55ES6500* is a unique product code for a *Samsung* TV. This attribute has a high relevance for the task of product matching. To extract the product code from the text we use a set of manually written regular expressions across all categories.

3.4   Attribute Value Normalization

Once all attribute-value pairs are extracted from the given dataset of offers, we continue with normalizing the values of the attributes. To do so, we first try to identify the data type of each of the attributes, using several manually defined regular expressions, which are able to detect the following data types: string, long string (string with more than 3 word tokens), number and number with unit of measurement. Additionally, the algorithm uses around 200 manually generated rules for converting units of measurements to the corresponding base unit (metric system), e.g. 5" will be converted to 0.127 m. In the end, the string values are lower cased, stop words and some special characters are removed.

Example Tagging: In Fig. 2 we give an example of feature extraction from a given product title. The extracted attribute-value pairs are shown in Table 1, as well as the normalized values, and the detected attribute data type.

Fig. 2. Example of attribute extraction from a product title

3.5   Calculating Similarity Feature Vectors

After the feature extraction is done, we can define an attribute space $F = \{f_1, f_2, ..., f_n\}$ that contains all of the extracted attributes. To measure the similarity between two products we calculate similarity feature vector $F(p_i, p_j)$ for each candidate product pair. For two products p_1 and p_2, represented with the attribute vectors $F_{p1} = \{f_1v, f_2v, ..., f_nv\}$ and $F_{p2} = \{f_1v, f_2v, ..., f_nv\}$, respectively, we calculate the similarity feature vector $F(p_1, p_2)$ by calculating the similarity value for each attribute f in the attribute space F. Let $p_1.\text{val}(f)$

Table 1. Attributes and values normalization

Attribute name	Attribute value	Normalized attribute value	Attribute data type
Brand	Samsung	samsung	string
Phone type	Galaxy S4	galaxy s4	string
Product code	GT-I9505	gt-i9505	string
Memory	16 GB	1.6e + 10 (B)	unit
Size	5.0 in.	0.127 (m)	unit
Compatible computer operating system	Android	android	string
Phone carrier	Sprint	sprint	string
Color	White Frost	white frost	string
Tagline	New Smartphone with 2-Year Contract	new smartphone with 2 year contract	long string

and $p_2.val(f)$ represent the value of an attribute f from p_1 and p_2, respectively. The similarity between p_1 and p_2 for the attribute f is calculated based on the attribute data type as shown in Eq. 1.

$$f(p_1, p2) = \begin{cases} 0, & if \quad p_1.val(f) = 0 \quad OR \quad p_2.val(f) = 0 \\ JaccardSimilarity(p_1.val(f), p_2.val(f)), & if \quad f \quad is \quad string \quad attribute \\ CosineSimilarity(p_1.val(f), p_2.val(f)), & if \quad f \quad is \quad long \quad string \quad attribute \\ p_1.val(f) == p_2.val(f) \quad ? \quad 1:0, & if \quad f \quad is \quad numeric \quad or \quad unit \quad attribute \end{cases} \quad (1)$$

The Jaccard similarity is calculated on character n-grams ($n \leq 4$), and the Cosine similarity is calculated on word tokens using TF-IDF weights.

3.6 Classification Approaches

Once the similarity feature vectors are calculated, we train four different classifiers that are commonly used for the given task: (i) Random Forest, (ii) Naive Bayes, (iii) Support Vector Machines (SVM) and (iv) Logistic Regression.

As the training dataset contains only a few matching pairs, and a lot of non-matching pairs, the dataset is highly imbalanced. To address the problem of classifying imbalanced datasets we use two sampling approaches [4]: (i) *Random Under Sampling (RUS):* removes samples from the majority class until the number of the samples of the minority class equals the number of samples of the majority class; (ii) *Random Over Sampling (ROS):* randomly samples instances from the minority class until the number of the samples of the minority class equals the number of samples of the majority class.

4   Evaluation

In this section, we evaluate the extent to which we can use the dataset of structured product ads for the task of matching unstructured product descriptions.

4.1   Datasets

For the evaluation we use Yahoo's Gemini Product Ads (GPA) for supervision, and we use a subset of the WebDataCommons (WDC) extraction[4].

Table 2. Datasets used in the evaluation

Dataset	#products	#matching pairs	#non-matching pairs
Televisions	344	236	58,760
Mobile phones	225	467	24,734
Laptops	209	146	25,521

Prouduct Ads - GPA Dataset: For our experiments, we are using a sample of three product categories from the Yahoo's Gemini Product Ads database. More precisely, we use a sample of 3,476 TVs, 3,372 mobile phones and 3,330 laptops. There are 35 different attributes in the TVs and mobile phones categories, and 27 attributes in the laptops category. We use this dataset to build the dictionary-based and the CRF feature extraction models.

Unstructured Product Offers - WDC Microdata Dataset: The latest extraction of WebDataCommons includes over 5 billion entities marked up by one of the three main HTML markup languages (i.e., Microdata, Microformats and RDFa) and has been retrieved from the CommonCrawl 2014 corpus[5]. From this dataset we focus on product entities annotated with Microdata using the schema.org vocabulary. To do so, we use a sub-set of entities annotated with http://schema.org/Product. The dataset contains *288,082,823* entities in total, or *2,829,523,589* RDF quads. *89,608* PLDs (10.9 %) annotate at least one entity as *s:Product* and *62,849* PLDs (7.6 %) annotate at least one entity as *s:Offer*. In our approach, we make use of the properties *s:name* and *s:description* for extracting attribute-value pairs.

   To evaluate the approach, we built a gold standard from the WDC dataset on three categories in the *Electronics domain*, i.e., TVs, mobile phones and laptops. We set some constraints on the entities we select: (i) the products must contain s:name and s:description property in English language, (ii) the s:name must contain between 3 and 50 words, (iii) the s:description must contain between 10 and 200 words, (iv) ignore entities from community advertisement websites (e.g.,

---

[4] http://webdatacommons.org/structureddata/index.html.
[5] http://blog.commoncrawl.org/2015/01/december-2014-crawl-archive-available/.

gumtree.com), (v) the product can be uniquely identified based on the title and description i.e., contains enough information to pinpoint the exact product.

The gold standard is generated by manually identifying matching products in the whole dataset. Two entities are labeled as matching products if both entities contain enough information to be uniquely identified, and both entities point to the same product. It is important to note that the entities do not necessarily contain the same set of product features. The number of entities, the number of matching and non-matching pairs for each of the datasets is shown in Table 2.

4.2 Experiment Setup

To evaluate the effectiveness of the approach we use the standard performance measures, i.e., Precision (P), Recall (R) and F-score (F1). The results are calculated using stratified 10-fold cross validation. For conducting the experiments, we used the RapidMiner machine learning platform and the RapidMiner development library.

We compare our approach with two baseline methods. First, we try to match the products based on TF-IDF cosine similarity. We report the best score on different levels of matching thresholds, i.e., we iterate the matching threshold starting from 0.0 to 1.0 (with step 0.01) and we assume that all pairs with similarity above the threshold are matching pairs[6].

As a second baseline we use the Silk Link Discovery Framework [7], an open-source tool for discovering links between data items within different data sources. The tool uses genetic algorithm to learn linkage rules based on the extracted attributes. For this experiment, we first extract the features from the product title and description using our CRF model, and then represent the gold standard in RDF format. The evaluation is performed using 10-fold cross validation.

4.3 Results

The results for both baseline approaches are shown in Table 3. We might conclude that both baseline approaches deliver rather poor results.

Table 3. Products matching baseline results using cosine similarity and Silk

	Cosine similarity TF-IDF (title)			Silk (CRF features)		
Dataset	P	R	F1	P	R	F1
Television	0.299	0.219	0.253	0.501	0.911	0.646
Mobile phones	0.375	0.383	0.379	0.406	0.840	0.547
Laptops	0.296	0.397	0.339	0.284	0.808	0.420

---

[6] We tried calculating the similarity based on different combination of title and description, but the best results were delivered when using only the product title.

Table 4 shows the results of our approach on the TVs dataset, using both CRF and Dictionary feature extraction approach. The best score is achieved using the CRF feature extraction approach, and Random Forest classifier without sampling. We can see that the same classifier performs a little bit worse when using the dictionary-based feature extraction approach.

Table 4. Products matching performance - televisions

Model	Sampling	CRF			Dictionary		
		P	R	F1	P	R	F1
Random forest	ROS	0.882	0.765	0.819	0.829	0.741	0.783
	RUS	0.697	0.826	0.756	0.663	0.779	0.716
	No sampling	0.921	0.739	**0.820***	0.805	0.741	**0.772**
Naive Bayes	ROS	0.069	0.911	0.128	0.074	0.941	0.137
	RUS	0.069	0.898	0.128	0.043	0.932	0.081
	No sampling	0.072	0.893	0.133	0.046	0.932	0.088
SVM	ROS	0.629	0.682	0.655	0.070	0.114	0.087
	RUS	0.679	0.660	0.669	0.622	0.630	0.626
	No sampling	0.849	0.639	0.729	0.708	0.431	0.536
Logistic regression	ROS	0.506	0.762	0.608	0.232	0.811	0.361
	RUS	0.486	0.742	0.587	0.134	0.821	0.231
	No sampling	0.519	0.769	0.619	0.219	0.806	0.345

Table 5 shows the results on the mobile phones dataset. As before, the best score is achieved using the CRF feature extraction approach, and Random Forest classifier using ROS sampling. We can note that the results using the dictionary-based approach are significantly worse than the CRF approach. The reason is that the GPA dataset contains a lot of trendy phones from 2015, while the WDC dataset contains phones that were popular in 2014, therefore the dictionary-based approach fails to extract many attribute-value pairs.

Table 6 shows the results of our approach on the Laptops dataset. Again, the best score is achieved using the CRF feature extraction approach, and Random Forest classifier without sampling. We can observe that for this dataset the results drop significantly compared to the other datasets. The reason is that the matching task for laptops is more challenging, because it needs more overlapping features to conclude that two products are matching[7].

The results show that our approach clearly outperforms both baseline approaches on all three categories. The Random Forest classifier delivers the best result for all three categories. We can observe that the other classifiers achieve high recall, i.e., they are able to detect the matching pairs in the dataset, but

---

[7] For example, two laptops might share the same brand, same CPU, and same HDD, but if the memory differs, then the laptops are not the same.

Table 5. Products matching performance - mobile phones

Model	Sampling	CRF			Dictionary		
		P	R	F1	P	R	F1
Random forest	ROS	0.885	0.756	**0.815***	0.374	0.659	0.477
	RUS	0.814	0.744	0.777	0.337	0.624	0.438
	No sampling	0.894	0.742	0.811	0.386	0.659	**0.487**
Naive Bayes	ROS	0.153	0.631	0.246	0.102	0.354	0.158
	RUS	0.125	0.575	0.205	0.102	0.325	0.155
	No sampling	0.128	0.535	0.206	0.102	0.310	0.153
SVM	ROS	0.002	0.007	0.003	0.416	0.163	0.234
	RUS	0.440	0.457	0.448	0.398	0.150	0.218
	No sampling	0.340	0.430	0.380	0.385	0.143	0.208
Logistic regression	ROS	0.413	0.489	0.448	0.258	0.323	0.287
	RUS	0.388	0.457	0.420	0.279	0.325	0.300
	No sampling	0.407	0.472	0.437	0.247	0.319	0.278

Table 6. Products matching performance - Laptops

Model	Sampling	CRF			Dictionary		
		P	R	F1	P	R	F1
Random forest	ROS	0.687	0.530	0.598	0.702	0.484	0.573
	RUS	0.588	0.619	0.603	0.535	0.560	0.547
	No sampling	0.815	0.513	**0.630***	0.741	0.498	**0.596**
Naive Bayes	ROS	0.050	0.778	0.094	0.095	0.567	0.163
	RUS	0.062	0.758	0.115	0.116	0.560	0.192
	No sampling	0.061	0.758	0.113	0.105	0.560	0.177
SVM	ROS	0.676	0.529	0.594	0.172	0.585	0.265
	RUS	0.690	0.531	0.600	0.725	0.503	0.594
	No sampling	0.694	0.565	0.623	0.858	0.434	0.576
Logistic regression	ROS	0.643	0.504	0.565	0.538	0.433	0.480
	RUS	0.438	0.587	0.502	0.561	0.475	0.514
	No sampling	0.651	0.510	0.572	0.301	0.543	0.387

they also misclassify a lot of non-matching pairs, leading to a low precision. It is also interesting to observe that the RUS sampling performs almost as good as the other sampling techniques, but it has considerably lower runtime.

CRF Evaluation: We also evaluate the Conditional Random Field model on the database of structured product ads. For each of the three product categories we select 70 % of the instances as a training set and the rest as a test

Table 7. CRF evaluation on structured product ads data

Dataset	#training	#test	#attributes	P	R	F1
Televisions	2,436	1,040	35	0.962	0.9431	0.9525
Mobile phones	2,220	1,010	35	0.9762	0.9613	0.9687
Laptops	2,330	1,000	27	0.9481	0.9335	0.9408

set. The results for each category, as well as the number of instances used for training and testing, and the number of attributes are shown in Table 7. The results show that the CRF model is able to identify the attributes in the text descriptions with high precision.

5 Data Fusion

As the evaluation of the approach showed that we are able to identify duplicate products with high precision, we apply the approach on the whole WDC and GPA products datasets. First, we try to identify duplicate products within the WDC dataset for top 10 TV brands. Then, we try to identify matching products in the WDC dataset for the product ads in the GPA dataset in the TV category.

Integrating Unstructured Product Descriptions: In the first experiment we apply the previously trained Random Forest model to identify matching products for the top 10 TV brands in the WDC dataset. To do so, we selected a sub-set of products from the WDC dataset that contain one of the TV brands in the *s:name* or *s:description* of the products. Furthermore, we apply the same constraints described in Sect. 4, which reduces the number of products. We use the brand name as a blocking approach, i.e., we generate candidate matching pairs only for products that share the same brand. We use the CRF feature extraction approach to extract the features and we tune the Random Forest model in a way that we increase the precision, on the cost of lower recall, i.e., a candidate products pair is considered to be positive matching pair if the classification confidence of the model is above *0.8*.

We report the number of discovered matches for each of the TV brands in Table 8. The second row of the table shows the number of candidate product descriptions after we apply the selection constraints on each brand. We manually evaluated the correctness of the matches and report the precision. The results show that we are able to find a large number of matching products with high precision. By relaxing the selection constraints of product candidates the number of discovered matches would increase, but it might also reduce the precision.

Furthermore, we try to identify how many matches and new attributes can be identified for single products. We randomly chose 5 different TVs and counted the number of discovered matches, *s:offers*, *s:reviews* and *s:aggregatedRating* properties from the WDC dataset, and how many new attribute-value pairs we discover from the *s:name* and *s:description* using the CRF model.

Table 8. Discovered matching products in the WDC dataset

Brand	Sony	LG	Samsung	RCA	Vizio	Panasonic	Philips	Magnavox	NEC	Proscan
#WDC products	3,673	14,764	4,864	3,961	563	1,696	1,466	141	23,845	30
#Matches	926	734	567	385	296	160	44	29	18	7
Precision	95	94	88.18	93.55	94.59	93.75	95.45	100	100	100

The results are shown in Table 9. The results show that we are able to identify a number of matches among products, and the aggregated descriptions have at least six new attribute-value pairs in each case.

Table 9. Extracted attributes for TV products

Product	#matches	#offers	#reviews	#ratings	#attributes
Vizio TV E241I-A1 24"	10	15	13	2	7
RCA TV 24G45RQ 24"	10	11	2	3	8
Samsung TV 55" UN55ES6500	8	12	2	1	14
LG TV 42LN5300 42"	6	8	4	0	6
Panasonic TV TH-32LRH30U 32"	4	6	0	1	6

Enriching Product Ads: In this experiment we try to identify matching products in the WDC dataset for the product ads in the GPA dataset. Similarly as before, we select WDC products based on the brand name and we apply the same filtering to reduce the sub-set of products for matching. To extract the features for the WDC products we use the CRF feature extraction model, and for the GPA products we use the already existing features provided by the merchants. To identify the matches we apply the same Random Forest model as before. The results are shown in Table 10. The second row reports the number of products of the given brand in the GPA dataset, and the third row in the WDC dataset.

The results show that we are able to identify small number of matching products with high precision. We have to note again that we are not able to identify any matches for the products in the GPA dataset that are released after 2014, because they do not appear in the WDC dataset.

Furthermore, we analyzed the number of new attributes we can discover for the GPA products from the matching WDC products. The distribution of matches, newly discovered attribute-value pairs, offers, ratings and reviews per GPA instance is shown in Fig. 3. The results show that for each of the product ads that we found a matching product description, at least 1 new attribute-value pair was discovered. And for some product ads even 8 new attribute-value pairs were discovered.

Table 10. Discovered matching products in the WDC dataset for product ads in the GPA dataset

Brand	Samsung	Vizio	LG	RCA	Sony	Proscan	NEC	Magnavox	Panasonic	Philips
#GPA products	560	253	288	10	102	18	22	28	41	11
#WDC products	4,864	563	14,764	3,961	3,673	30	23,845	141	1,696	1,466
#Matches	202	123	102	67	40	28	21	12	6	2
Precision	80.85	91.80	89.24	79.10	97.50	100.00	85.70	100.00	100.00	100.00

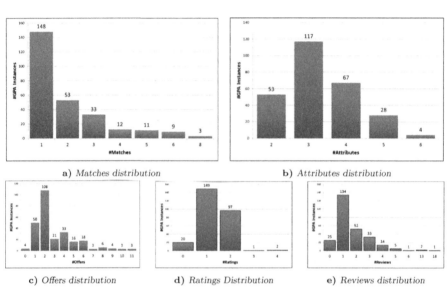

a) *Matches distribution*   b) *Attributes distribution*

c) *Offers distribution*   d) *Ratings Distribution*   e) *Reviews distribution*

Fig. 3. Distribution of newly discovered matches and attributes per product ad

6   Product Categorization

Categorization of products in a given product catalog is an important task. For example, online shops categorize products for easier navigation, and in the case of product ads, it allows easier ad retrieval and better user targeting. Also, identifying the category of the products before applying the matching approach might be used as a blocking approach. Here we examine to which extent we can use a structured database of product ads to perform categorization of unstructured products description. Again we use the database of structured product ads to extract features from unstructured product descriptions, which are then used to build a classification model.

Table 11. The best categorization results using the dictionary-based approach and baseline approach

GS1 Level	Features	Model	ACC	P	R	F1
1	Dictionary (BTO)	SVM	88.34	74.11	64.78	69.13
2	Dictionary (TF-IDF)	NB	79.87	46.31	40.08	42.97
3	Dictionary (TF-IDF)	NB	70.9	25.45	24.36	24.89

In this approach we use only the dictionary-based feature extraction approach as described in Sect. 3.3[8]. To build the dictionary, we use all product ads across all categories in the database of product ads. To generate the feature vectors for each instance, after the features from the text are extracted, the value of each feature is tokenized, lowercased, and removed tokens shorter than 3 characters. The terms of each feature are concatenated with the feature name e.g. for a value *blue* for the feature *color*, the final value will be *blue-color*.

Following Meusel et al. [11], in order to weigh the different features for each of the elements in the two input sets, we apply two different strategies, Binary Term Occurrence (BTO) and TF-IDF. In the end we use the feature vectors to build a classification model. We have experimented with 4 algorithms: Naive Bayes (NB), Support Vector Machines (SVM), Random Forest (RF) and k-Nearest Neighbors (KNN), where $k = 1$.

Gold Standard: For our experiments we use the GS1 Product Catalogue (GPC)[9] as a target hierarchy. The hierarchy is structured in six different levels, but in our experiments we try to categorize the products in the first three levels of the taxonomy: Segment, Family and Class. The first level contains 38 different categories, the second level 113 categories and the third level 783 categories.

To evaluate the proposed approach we use the Microdata products gold standard developed in [11]. We removed non-English instances from the dataset, resulting in 8,362 products. In our evaluation we use the *s:name* and the *s:description* properties for generating the features.

Evaluation: The evaluation is performed using 10-fold cross validation. We measure accuracy (Acc), Precision (P), Recall (R) and F-score (F1). We compare our approach to a TF-IDF and BTO baseline, where the text is preprocessed as before, but no dictionary is used.

---

[8] We were not able to build a sufficiently good CRF model that is able to annotate text with high precision because of the many possible attributes across all categories. A separate CRF model for each category in the structured database of product ads should be trained.

[9] http://www.gs1.org/gpc.

Due to space constraints we show only the best performing results for each of the three levels. The complete results can be found online[10]. The results show that the dictionary-based approach can be used for classification of products on different level of the hierarchy with high performance. Also, the results show that it outperforms the baseline approaches for all three levels of classification for both accuracy and F-score. Again, we have to note that the gold standard contains products that appear in 2014, while the GPA dataset contains relatively up to date products from 2015 (Table 11).

7   Conclusion

In this paper, we proposed an approach for enriching structured product ads with structured data extracted from HTML pages that contain semantic annotations. The approach is able to identify matching products in unstructured product descriptions using the database of structured product ads as supervision.

We identify the Microdata dataset as a valuable source for enriching existing structured product ads with new attributes. We showed that not only we could integrate some of the existing Microdata attributes, like *s:offers*, *s:aggregateRating* and *s:review*, but with our approach we can extract valuable attribute-value pairs from the textual information of the products that do not exist in the structured product ads. In future work, to validate the value of the new attributes we need to evaluate the influence of the new attributes on the ads ranking algorithm. We could also include other schema.org product properties in the approach, like *s:mpn*, *s:model* and *s:gtin*, which might be useful for identity resolution. Additionally, mining search engine query logs we could extract valuable features for identifying matching products.

Besides integrating products over different online shops and product categorization, our approach could be used for search query processing, which would undoubtedly improve the shopping experience for users [3].

Acknowledgements. We would like to acknowledge Roi Blanco (Yahoo Labs) and Christian Bizer (University of Mannheim) for their helpful comments to our work. We would also like to acknowledge the support, help and insights of the Yahoo Gemini Product Ads engineering and the Yahoo Labs Advertising Sciences teams, in particular Nagaraj Kota and Ben Shahshahani.

References

1. de Bakker, M., Frasincar, F., Vandic, D.: A hybrid model words-driven approach for web product duplicate detection. In: Salinesi, C., Norrie, M.C., Pastor, Ó. (eds.) CAiSE 2013. LNCS, vol. 7908, pp. 149–161. Springer, Heidelberg (2013)
2. van Bezu, R., Borst, S., Rijkse, R., Verhagen, J., Vandic, D., Frasincar, F.: Multi-component similarity method for web product duplicate detection (2015)

---

[10] http://data.dws.informatik.uni-mannheim.de/gpawdc/categorization/results/
FullResults.xlsx.

3. Bhattacharya, S., Gollapudi, S., Munagala, K.: Consideration set generation in commerce search. In: Proceedings of the 20th International Conference on World Wide Web, WWW 2011, pp. 317–326. ACM, New York, NY, USA (2011). http://doi.acm.org/10.1145/1963405.1963452

4. Chawla, N.V.: Data mining for imbalanced datasets: an overview. In: Maimon, O., Rokach, L. (eds.) Data Mining and Knowledge Discovery Handbook, pp. 853–867. Springer, Heidelberg (2005)

5. Finkel, J.R., Grenager, T., Manning, C.: Incorporating non-local information into information extraction systems by gibbs sampling. In: Proceedings of the 43rd Annual Meeting on Association for Computational Linguistics, pp. 363–370. Association for Computational Linguistics (2005)

6. Ghani, R., Probst, K., Liu, Y., Krema, M., Fano, A.: Text mining for product attribute extraction. ACM SIGKDD Explor. Newslett. **8**(1), 41–48 (2006)

7. Isele, R., Bizer, C.: Learning linkage rules using genetic programming. In: Proceedings of the International Workshop on Ontology Matching, pp. 13–24 (2011)

8. Kannan, A., Givoni, I.E., Agrawal, R., Fuxman, A.: Matching unstructured product offers to structured product specifications. In: 17th ACM SIGKDD International Conference on Knowledge Discovery and Data Mining, pp. 404–412 (2011)

9. Melli, G.: Shallow semantic parsing of product offering titles (for better automatic hyperlink insertion). In: Proceedings of the 20th ACM SIGKDD International Conference on Knowledge Discovery and Data Mining, pp. 1670–1678. ACM (2014)

10. Meusel, R., Petrovski, P., Bizer, C.: The webdatacommons microdata, RDFa and microformat dataset series. In: Mika, P., et al. (eds.) ISWC 2014, Part I. LNCS, vol. 8796, pp. 277–292. Springer, Heidelberg (2014)

11. Meusel, R., Primpeli, A., Meilicke, C., Paulheim, H., Bizer, C.: Exploiting microdata annotations to consistently categorize product offers at web scale. In: Stuckenschmidt, H., Jannach, D. (eds.) EC-Web 2015. LNBIP, vol. 239, pp. 83–93. Springer, Heidelberg (2015)

12. Nguyen, H., Fuxman, A., Paparizos, S., Freire, J., Agrawal, R.: Synthesizing products for online catalogs. Proc. VLDB Endowment **4**(7), 409–418 (2011)

13. Petrovski, P., Bryl, V., Bizer, C.: Integrating product data from websites offering microdata markup. In: Proceedings of the 23rd International Conference on World Wide Web Companion, pp. 1299–1304 (2014)

14. Petrovski, P., Bryl, V., Bizer, C.: Learning regular expressions for the extraction of product attributes from e-commerce microdata (2014)

15. Qiu, D., Barbosa, L., Dong, X.L., Shen, Y., Srivastava, D.: Dexter: large-scale discovery and extraction of product specifications on the web. Proc. VLDB Endowment **8**(13), 2194–2205 (2015)

16. Vandic, D., Van Dam, J.W., Frasincar, F.: Faceted product search powered by the semantic web. Decis. Support Syst. **53**(3), 425–437 (2012)

Iterative Entity Navigation via Co-clustering Semantic Links and Entity Classes

Liang Zheng, Jiang Xu, Jidong Jiang, Yuzhong Qu$^{(\boxtimes)}$, and Gong Cheng

National Key Laboratory for Novel Software Technology,
Nanjing University, Nanjing 210023, People's Republic of China
{zhengliang,jiangxu,jdjiang}@smail.nju.edu.cn, {yzqu,gcheng}@nju.edu.cn

Abstract. With the increasing volume of Linked Data, the diverse links and the large amount of linked entities make it difficult for users to traverse RDF data. As semantic links and classes of linked entities are two key aspects to help users navigate, clustering links and classes can offer effective ways of navigating over RDF data. In this paper, we propose a co-clustering approach to provide users with iterative entity navigation. It clusters both links and classes simultaneously utilizing both the relationship between link and class, and the intra-link relationship and intra-class relationship. We evaluate our approach on a real-world data set and the experimental results demonstrate the effectiveness of our approach. A user study is conducted on a prototype system to show that our approach provides useful support for iterative entity navigation.

Keywords: Entity navigation · Semantic link · Entity class · Co-clustering

1 Introduction

With the enrichment of available Linked Data on the Web, challenges in navigating the data space arise: large numbers of linked entities and high diversity of links, often make it hard for users to explore and find the entities of interest quickly. As semantic links and classes of linked entities are two key aspects to help users navigate, clustering links and classes may provide effective organizations about large numbers of links and classes.

As shown in Fig. 1, `Steven Spielberg` in DBpedia [2] is linked to 4 entities (`Falling Skies`, `Men in Black`, `A.I.` and `Medal of Honor`) through 3 semantic links (`executive producer`, `producer` and `writer`). Links can be clustered, such as {`producer`, `writer`} (based on the common linked entities they are linked to) and {`executive producer`, `producer`} (based on lexical similarity between their labels). 4 linked entities are associated with 3 entity classes (`Television Show`, `Film` and `Video Game`). Classes can also be clustered, such as {`Television Show`, `Film`} according to semantic similarity. Link clusters and class clusters offer effective organizations and provide a overview of overall information during users' navigation. However, link clusters and class clusters are utilized separately, and potential relationships between them are not taken into account.

© Springer International Publishing Switzerland 2016
H. Sack et al. (Eds.): ESWC 2016, LNCS 9678, pp. 168–181, 2016.
DOI: 10.1007/978-3-319-34129-3_11

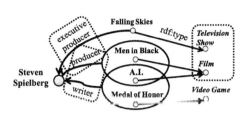

Fig. 1. The context of browsing an entity

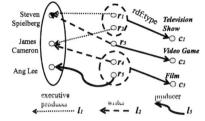

Fig. 2. The context of browsing a set of entities

In order to improve the efficiency of navigation, jointly utilizing link clusters and class clusters may provide users with an iterative refinement mode during entity navigation. For instance, through link cluster {producer, writer}, users find 3 linked entities (Men in Black, A.I. and Medal of Honor). With respect to the 3 linked entities, there are two class clusters ({Film} and {Video Game}). Then, users can locate Medal of Honor by using class cluster {Video Game}. This iterative navigation process can assist users to explore and understand the overall information space. Besides, there is a necessity for navigation paradigm to take into account not only single-entity-oriented transition, but also entity-set-oriented transition. Figure 2 shows the context of browsing the three best director academy award winners. We can capture the strength of the relation-ship between a link and a class. For instance, 2 directors (Steven Spielberg and James Cameron) are the executive producer of Television Show. These rich and meaningful inter/intra relationships among links and classes could be leveraged to improve entity navigation.

In this paper, we propose a co-clustering approach that organizes semantic links and entity classes to support iterative entity navigation. For a given context of entity browsing, we propose a notion of navigation graph to model the three aspects of navigation (links, linked entities and their classes) based on tripartite graph [12]. Then, links and classes are clustered simultaneously based on infor-mation theoretic co-clustering (ITCC) [4] over navigation graph. To improve the effect of co-clustering, we define a measure of intra-link similarity and intra-class similarity and incorporate them into ITCC.

The remainder of the paper is structured as follows. Section 2 defines the basic notion to be used and describes our co-clustering problem. Section 3 introduces similarity measuring scheme and our proposed approach. Section 4 provides our experimentation. The related work is reported in Sect. 5. Section 6 concludes this paper.

2   Problem Statement

In this section, we define the notion of navigation graph based on tripartite graph [12] to model the three aspects of navigation (links, linked entities and

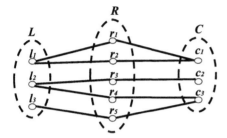

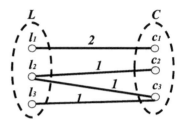

Fig. 3. An example of navigation graph.   **Fig. 4.** An example of link-class graph.

their classes) over an RDF graph and introduce the problem of co-clustering semantic links and entity classes.

In this paper, we do not consider literal and blank node in the RDF data model. Let U be the set of URI named entities and classes, L be a set of links including object properties and inverse of them and $T \subseteq U \times L \times U$ be a set of triples.

Definition 1 (Navigation Graph). *Given a set of entities $S \subseteq U$ being the focus, a navigation graph $G =< H, E >$ consists of a set of vertices $H = L' \cup R \cup C$ where $L' \subseteq L$ denotes a set of links, $R \subseteq U$ a set of linked entities and $C \subseteq U$ a set of classes, and a set of edges $E = \{(l, r) \mid \exists s \in S, (s, l, r) \in T\} \cup \{(r, c) \mid (r, rdf : type, c) \in T\}$.*

Suppose the user explores 3 best director academy award winners, Fig. 3 shows the navigation graph associated with Fig. 2. Furthermore, there are three bipartite graphs deduced from navigation graph G. These three graphs can model the associations between links and entities (graph LR), classes and entities (graph CR) and links and classes (graph LC).

Definition 2 (Link-Class Graph). *Given a navigation graph $G =< H, E >$, a weighted bipartite graph LC can be defined as follows: $LC =< V, E_{lc} >$ where $V = L' \cup C$ and $E_{lc} = \{(l, c) \mid \exists r \in R, (l, r) \in E \land (r, c) \in E\}$. $w : E_{lc} \rightarrow N^+$, $\forall e \in E_{lc}, w(e) = |\{s \mid \exists s \in S, (s, l, r) \in T \land (r, rdf : type, c) \in T\}|$.*

There is an example of link-class graph LC derived from navigation graph G, as shown in Fig. 4. $w(l_1, c_2) = 2$ represents that 2 directors are the executive producer of Television Show.

Problem 1 (Links-Classes Co-clustering). *Given a navigation graph $G =< H, E >$ and a similarity function sim, find k disjoint clusters of links $\{\hat{x}_1, \hat{x}_2, ..., \hat{x}_k\}$ and l disjoint clusters of classes $\{\hat{y}_1, \hat{y}_2, ..., \hat{y}_l\}$ such that $\sum_{i=1}^{k} \sum_{x,x' \in \hat{x}_i} sim(x, x')$ and $\sum_{j=1}^{l} \sum_{y,y' \in \hat{y}_j} sim(y, y')$ are maximized.*

Clearly the similarity function is a key facet in solving this problem and we define a measuring scheme in Sect. 3.1.

3   Co-clustering Links and Classes

In this section, we define a measuring scheme to compute the link similarity and class similarity in Sect. 3.1. Then we give a solution to the co-clustering problem based on information theoretic co-clustering (ITCC) [4] in Sect. 3.2. Finally we introduce a method for cluster labeling in Sect. 3.3.

3.1   Measuring Similarity

In our work, we focus on cosine similarity, lexical similarity and semantic similarity.

Cosine Similarity. As described in Sect. 2, we use the bipartite graph LC to model the link and class collections. First, link and class can be represented by each other based on vector space model [11]. Each class c can be modeled as a vector over the set of links. Likewise, each link l can be modeled as a vector over the set of classes. For example, as shown in Fig. 4, $c_1 = (2, 0, 0)$, $c_2 = (0, 1, 0)$ and $c_3 = (0, 1, 1)$. Then, the class similarity is defined as the cosine function of angle between two vectors of c_i and c_j:

$$sim_{cos}(c_i, c_j) = \frac{c_i \cdot c_j}{||c_i|| \cdot ||c_j||} \tag{1}$$

and the value of sim_{cos} is in the range $(0, 1]$. Likewise, the cosine similarity between links can be captured according to the above method.

Lexical Similarity. The label of a link or a class is useful for human understanding. When the labels of two links or classes share many common words, it may indicate some kind of similarity between them. Given two classes c_i and c_j, the lexical similarity is defined as the edit distance between two strings of c_i's label (s_{c_i}) and c_j's label (s_{c_j}):

$$sim_{edit}(c_i, c_j) = \frac{1}{1 + editDist(s_{c_i}, s_{c_j})} \tag{2}$$

where the value of sim_{edit} is in the range $(0, 1]$ and $editDist(s_{c_i}, s_{c_j})$ is the minimum number of character insertion and deletion operations needed to transform one string to the other [9]. Likewise, the lexical similarity between links can be captured according to the above method.

Semantic Similarity. In our work, semantic similarity refers to similarity between two concepts in a taxonomy. Due to a hierarchical structure in many taxonomies, we adopt a similarity measure based on path lengths between concepts [15]. The semantic similarity between class c_i and c_j is calculated as:

$$sim_{sem}(c_i, c_j) = \frac{2 \cdot depth(LCA)}{depth(c_i) + depth(c_j)} \tag{3}$$

where the value of sim_{sem} is in the range $(0, 1]$ and $depth(c_i)$ is the shortest path length from the root to c_i and LCA is the least common ancestor of c_i and c_j. The semantic similarity between semantic links can be also computed according to the above method.

Combination of Similarity. We discuss three similarity measures from different points of view. In order to get the total similarity, we combine these measures based on a natural way (linear combination). Thus, we define the similarity scoring function between two classes c_i and c_j as follows.

$$sim(c_i, c_j) = \alpha \cdot sim_{cos}(c_i, c_j) + \beta \cdot sim_{edit}(c_i, c_j) + \gamma \cdot sim_{sem}(c_i, c_j) \quad (4)$$

where $\alpha + \beta + \gamma = 1$ and $\alpha, \beta, \gamma \in [0, 1]$ indicate the weights for each similarity measure to be tuned empirically.

Also, the similarity between two links l_i and l_j can be captured according to the above method.

3.2 Information-Theoretic Co-clustering Links and Classes

In [4], Dhillon et al. define co-clustering as a pair of maps from rows to row clusters and from columns to column-clusters based on information theory. The optimal co-clustering is one that minimizes the difference ("loss") in mutual information between the original random variables and the mutual information between the clustered random variables.

Let $X = \{x_1, ..., x_m\}$ and $Y = \{y_1, ..., y_n\}$ be discrete random variables respectively. Let $p(X, Y)$ denote the joint probability distribution between X and Y. Let the k disjoint clusters of X be written as: $\hat{X} = \{\hat{x}_1, \hat{x}_2, ..., \hat{x}_k\}$, and the l disjoint clusters of Y be written as: $\hat{Y} = \{\hat{y}_1, \hat{y}_2, ..., \hat{y}_l\}$. An optimal co-clustering minimizes the loss of mutual information, defined as

$$I(X; Y) - I(\hat{X}; \hat{Y}) = D(p(X, Y) || q(X, Y)) \quad (5)$$

where $I(X; Y)$ is the mutual information between sets X and Y, $D(\cdot || \cdot)$ denotes the Kullback-Leibler(KL) divergence, and $q(X, Y)$ is a distribution of the form $q(x, y) = p(\hat{x}, \hat{y})p(x|\hat{x})p(y|\hat{y})$, $x \in \hat{x}$, $y \in \hat{y}$.

In our context, the link set L' can be considered as X and the classes C as Y. The joint probability distribution between links and classes can be captured based on entity set R over navigation graph $G = < L' \cup R \cup C, E >$. The joint probability of a link x and a class y is defined as follows:

$$p(x, y) = \frac{|\{r|\exists r \in R, (x, r) \in E\} \cap \{r|\exists r \in R, (r, y) \in E\}|}{|R|} \quad (6)$$

To improve co-clustering, many research studies utilize the intra-relationships (e.g., interdocument similarity, interword similarity) [14]. We take the similarity between links, and the similarity between classes as the intra-relationships, and incorporate them into the information theoretic co-clustering. We rewrite Eq. (5) as

$$I(X;Y) - I(\hat{X};\hat{Y}) - \lambda LCS - \mu CCS \tag{7}$$

$$LCS - \frac{1}{k}\sum_{i=1}^{k}\sum_{x,x' \in \hat{x}_i} \frac{sim(x,x')}{|\dot{x}_i| * (|x_i| - 1)} \tag{8}$$

$$CCS = \frac{1}{l}\sum_{j=1}^{l}\sum_{y,y' \in \hat{y}_j} \frac{sim(y,y')}{|\hat{y}_j| * (|\hat{y}_j| - 1)} \tag{9}$$

LCS and CCS are the total similarity within link clusters and class clusters respectively. $\lambda + \mu = 1$ and $\lambda, \mu \in [0,1]$ indicate the weights for the trade off among the loss of mutual information, LCC and CCS.

In our implementation, we use the ITCC algorithm with time complexity $O((nz(k + l) + km^2 + ln^2)\tau)$ provided by [14], where nz is the number of non-zeros in $p(X,Y)$ and τ is the number of iterations.

3.3 Making the Clusters Easy to Browse

To help user decide at a glance whether the contents of a cluster are of interest, we aim to provide concise and accurate cluster description. A heuristic method is to find the cluster's centroid which is most similar with other elements in the cluster as the label of cluster.

Given a class cluster $\hat{y}$ and a class $y \in \hat{y}$, we define the centricity of y as follows:

$$centricity(y) = \sum_{y' \in \hat{y}} sim(y, y'). \tag{10}$$

Thus, the centroid of cluster $\hat{y}$ can be defined as follows:

$$centroid(\hat{y}) = \arg\max_{y \in \hat{y}} centricity(y) \tag{11}$$

Also, the link cluster's centroid can be captured according to the above method.

4   Evaluation

In this section, we evaluate our approach compared with three baseline algorithms on real-world datasets. Our proposed approach is implemented in a prototype system and then compared with two Linked Data browsers via a user study.

Table 1. Statistics of experimental datasets

	Artist	City	Company	University
Number of entities	1233	2243	304	510
Number of links	139	280	174	163
Number of linked entities	59654	402580	88003	57487
Average num. of links per entity	25.8	30.1	27.2	28.9
Average num. of linked entities per entity	113.2	217.5	338.1	151.7
Average num. of classes per linked entity	6.9	8.7	5.8	8.9

4.1   Experimental Evaluation

Datasets. we used the DBpedia (version:2015-04)[1] *Mapping-based Properties* dataset, excluding RDF triples containing literals. We selected 4 common classes (i.e., Artist, City, Company, University). For each class, we collected those entities that each one has more than 15 semantic links. As to entity classes, we used the *Mapping-based Types* dataset. For the purpose of our task, we used the *DBpedia Ontology* dataset. The statistical results of these datasets are listed in Table 1.

Baselines. We compare our method (ITCC+) with three baselines (ITCC [4], co-clustering via bipartite spectral graph partition(BSGP) [3] and K-means [10]).

ITCC only focuses on the relationship between row and column in the co-occurrence matrix from mutual information aspect but neglects the intra-row and intra-column relationships.

BSGP considers the co-clustering problem in term of finding minimum cut vertex partitions in a weighted bipartite graph. In our context, the link-class bipartite graph $LC = < L' \cup C, E_{lc} >$ can be captured. We use the edge weighting method provided by [5] and the weight between a link l and a class c is computed with

$$w(l,c) = \max_{c_i \in l} sim(c_i, c) + \max_{l_j \in c} sim(l, l_j) \qquad (12)$$

where $l = \{c_i | \exists c_i \in C, (l, c_i) \in E_{lc}\}$ and $c = \{l_j | \exists l_j \in L', (l_j, c) \in E_{lc}\}$.

K-means measures the distance between two data points according to the similarity by Eq. (4). Since K-means is a one-sided clustering algorithm, the link and class collections are clustered separately.

Evaluation Metrics. We define three metrics (*cohesion, separation* and *overall*) to measure the quality of clustering. Given a cluster set $O = \{O_1, ..., O_k\}$ of N elememts, the three metrics of O are defined as follows.

$$cohesion(O) = \frac{1}{k} \sum_{i=1}^{k} coh(O_i), \ coh(O_i) = \frac{\sum_{o \in O_i, o' \in O_i} sim(o, o')}{|O_i| \cdot (|O_i| - 1)}. \qquad (13)$$

---

[1] http://wiki.dbpedia.org/Downloads2015-04.

Table 2. Comparsion of *overall* against different k and (α, β, γ)

		t_1	t_2	t_3	t_4	t_5	t_6
$k=3$	ITCC+	**0.874**	0.566	0.401	**0.467**	**0.911**	0.642
	ITCC	0.872	0.377	**0.486**	0.273	0.688	**0.729**
	BSGP	0.835	0.226	0.179	0.238	0.465	0.43
	K-means	0.852	**0.643**	0.463	0.356	0.673	0.577
$k=5$	ITCC+	**0.891**	0.566	**0.587**	**0.428**	**0.925**	0.535
	ITCC	0.887	0.384	0.427	0.372	0.749	0.471
	BSGP	0.797	0.273	0.209	0.193	0.522	0.383
	K-means	0.814	**0.762**	0.479	0.383	0.765	**0.61**
$k=8$	ITCC+	**0.857**	**0.672**	**0.52**	0.433	**0.887**	0.668
	ITCC	0.832	0.359	0.497	0.324	0.886	**0.729**
	BSGP	0.686	0.238	0.419	0.273	0.596	0.365
	K-means	0.823	0.663	0.478	**0.481**	0.829	0.649

$$separation(O) = \frac{1}{k}\sum_{i=1}^{k} sep(O_i), \quad sep(O_i) = \frac{\sum_{o \in O_i, o' \notin O_i} sim(o, o')}{|O_i| \cdot (N - |O_i|)}. \quad (14)$$

$$overall(O) = 1 - \frac{separation(O)}{cohesion(O)}. \quad (15)$$

Obviously, the higher the overall score the better the clustering quality is.

Results and Discussions. In the experiments, we randomly selected 200 entities from our experimental dataset. For each selected entity, we empirically conducted 10 runs using four algorithms (ITCC+, ITCC, BSGP and K-means) respectively and reported the average *overall*.

For parameter settings, we investigated the sensitivity with respect to the disjoint clusters size k (=3, 5, 8) and the balanced parameters $(\alpha, \beta, \gamma, \lambda, \mu)$. As to the parameter (α, β, γ) in similarity Eq. (4), we set $t_1 = (1, 0, 0)$, $t_2 = (0, 1, 0)$, $t_3 = (0, 0, 1)$, $t_4 = (0.33, 0.33, 0.33)$, $t_5 = (0.6, 0.1, 0.3)$ and $t_6 = (0.3, 0.1, 0.6)$ in computing link similarity and class similarity. As to the parameters (λ, μ) of loss in mutual information Eq. (7), we set $\lambda = \mu = 0.5$ based on equity. More parameters settings will be experimented in future work. Besides, we set the number of iterations $\tau = 20$ which is enough for convergence [4].

From Table 2, we have the following observations. In most cases, ITCC+ outperforms ITCC, BSGP and K-means. Since the core of ITCC+ and ITCC depends on the joint probability distribution between links and classes, ITCC+ and ITCC can bring better results when the cosine similarity has higher weight (e.g., t_1 and t_5).

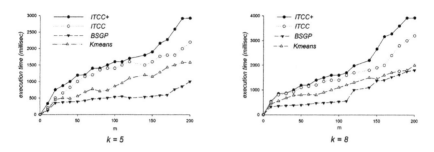

Fig. 5. Execution time of four algorithms with varying k and m.

Efficiency Evaluation. We evaluated the performance of ITCC+, ITCC, BSGP and K-means by measuring the average execution time for varying size of entities denoted by m. We randomly selected entities from our experimental dataset. The four algorithms were implemented in Java and carried out on an Intel Core2 Quad 2.66 GHz CPU, Windows 7 with 1.2 GB JVM.

As can be seen from Fig. 5, the four algorithms were reasonably fast in practice. BSGP was faster than ITCC+, ITCC and K-means. ITCC+ and ITCC need multiple iterations and recompute the distributions on every iteration. Besides, ITCC+ compute the link cluster similarity and class cluster similarity respectively on every iteration.

4.2   User Study

We conducted a user study to compare our approach with two Linked Data browsers (Rhizomer[2], SView[3]) and to evaluate the effectiveness of our approach.

Participant Systems. We implemented our proposed approach in a prototype system called CoClus[4], as shown in Fig. 6. Users can start browsing with an entity URI by entering into the input box (A). The left-hand side of the interface lists the link clusters (B). The right-hand side lists class clusters (C). There are connections between link clusters and class clusters (D). Users can click the button "browse all" to explore all the linked entities (E). Also, users can choose some link/class clusters to iteratively filter the target entities. Selected filters can be cancelled (F).

Rhizomer provides users with facet navigation to explore RDF data in DBpedia. Once users have zoomed by selecting the kind of entities from the navigation bar, facets are generated automatically and help users filter out those that are not interesting.

---

[2] http://rhizomik.net/html/rhizomer/.
[3] http://ws.nju.edu.cn/sview/.
[4] http://ws.nju.edu.cn/coclus/.

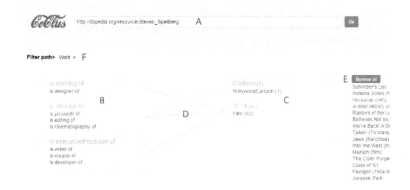

Fig. 6. A screenshot of CoClus.

Table 3. An example of navigation tasks about **John Lennon**

		Tasks
G1	*E1*	Explore the information related to **John Lennon**, and describe three main aspects of him
	F1	Find the albums written by **John Lennon**
G2	*E2*	Explore the information related to the band members of the Beatles, and describe three main aspects of them
	F2	Find the films starred by the band members of the Beatles

SView provides a navigation module (called "Link") which organizes the semantic links in the form of link patterns [16]. It supports a hierarchical and interactive interface to help users to find target linked entities.

Tasks. We selected 8 entities from experimental dataset in Sect. 4.1 at random as the starting points of user navigation. In a browsing scenario, navigation tasks can be divided into two types: *Explore* (a user has a fuzzy need) and *Find* (a user has a clear need) tasks. According to navigation paradigm, tasks can also be divided into two groups: single-entity-oriented (*G1*) and entity-set-oriented (*G2*) tasks. For each starting point, we established 4 navigation tasks. The navigation tasks about **John Lennon** are shown in Table 3.

Procedure. We invited 12 student subjects majoring in computer science who were familiar with the Web, but with no knowledge of our project. The evaluation was conducted in three phases. First, the subjects learned how to use the given systems through a 20 min tutorial, and had additional 10 min for free use and questions. Second, the subjects used each of the three systems arranged in random order. For each system, the subjects were randomly assigned to one starting point, and required to complete 4 navigation tasks. Meanwhile, the starting points of user navigation among the three systems were different. To enable

Table 4. Navigation questionnaire

	Questions
Q1:	The system helped me get an overview of all the information
Q2:	The number of navigation options was overwhelming
Q3:	The navigation options were well organized
Q4:	The navigation option titles were understood well
Q5:	It was easy to reorient myself in the navigation

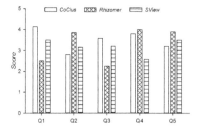

Fig. 7. Results of navigation questionnaire.

Fig. 8. Average time and number of operations for each task.

users to use the systems (e.g., Rhizomer is used to browse a whole dataset in a faceted manner) for carrying out such a task, prior to testing, we navigate to the desired starting point, from which users start their tasks. The subjects were asked to complete all the tasks in 10 min. We recorded their answers, their operations and the time they spent on each task.

With regard to each system, the subjects responded to the navigation questionnaire, as shown in Table 4. The questions in the questionnaires were responded by using a five-point Likert scale ranging from 1 (strongly disagree) to 5 (strongly agree). Finally, the subjects were asked to comment on the three systems.

Results and Discussions. Navigation questionnaire Q1–Q5 captured subjects' navigation experience with different systems in Fig. 7. Repeated measures ANOVA revealed that the differences in subjects' mean ratings were all statistically significant ($p < 0.01$). LSD post-hoc tests ($p < 0.05$) revealed that, according to Q1 ($F = 11.543$, $p = 0.0$), CoClus ($mean = 4.18$, $sd = 0.54$) provided a better overview of all the information than SView ($mean = 3.8$, $sd = 1.056$) and Rhizomer ($mean = 2.75$, $sd = 0.85$) due to the use of link and class clusters. According to Q2 ($F = 9.56$, $p = 0.0$), Rhizomer ($mean = 3.98$, $sd = 1.06$) provided too many faceted filtering views and links compared with CoClus ($mean = 2.85$, $sd = 0.56$) and SView ($mean = 3.42$, $sd = 1.04$). According to Q3 ($F = 13.65$, $p = 0.002$), CoClus ($mean = 3.78$, $sd = 0.96$) provided a better organization of links and classes than SView ($mean = 3.41$, $sd = 0.83$)

and Rhizomer ($mean = 2.4$, $sd = 0.61$). According to Q4 ($F = 10.36$, $p = 0.0$), Rhizomer ($mean = 4.12$, $sd = 1.03$) helped subjects more easily understand the label of links. Finally, according to Q5 ($F = 11.67$, $p = 0.0$), Rhizomer ($mean = 4.08$, $sd = 0.97$) helped subjects keep track of browsing and provided easy rollback.

Figure 8 shows the average time spent on each task and the number of operations which participants had conducted in one task. Since CoClus provided a overview using link clusters and class clusters, subjects required far less time and fewer interactions to complete these tasks.

We also summarized all the major comments of the subjects. As to Rhizomer, 10 subjects (83%) said that faceted navigation helped them filter out those entities that were not interesting but needed them to input manually and multiple interactions. As to SView, 8 subjects (67%) said that link patterns provided useful support to help users to locate target linked entities, but target entity collection was still large. As to CoClus, 11 subjects (92%) said that it was distinguished by its co-clustering of links and classes of linked entities, and it offered a different, more useful set of navigation functions. The link and class clusters provided a overview of information and enabled users to navigate iteratively. But 4 subjects (33%) said that it had some risks, such as misleading information because of some inappropriate cluster labels.

5   Related Work

Navigation as an important feature of Linked Data, has been supported by many Linked Data browsers. Tabulator [1] allows users to browse data by starting from a single resource and following links to other resources. It also allows users to select a resource for further exploration in a nest tree view. gFacet [8] is a tool that supports the exploration of the Web of data by combining graph-based visualization with faceted filtering functionalities. With gFacet it is possible to choose one class and then pivot to a related class keeping those filters for the instances of the second class connected to the filtered instances in the first class. VisiNav [7] is a system based on an interaction model designed to easily search and navigate large amounts of Web data. It provides four atomic operations over object structured datasets: keyword search, object focus, path traversal, and facet specification. Users incrementally assemble complex queries that yield sets of objects. Rhizomer [6] addresses the exploration of semantic data by applying the data analysis mantra of overview, zoom and filter. Users can interactively explore the data using facets. Moreover, facets also feature a pivoting operation. Visor [13] is a generic RDF data explorer that can work over SPARQL endpoints. In Visor, exploration starts by selecting a class of interest from the ontology. Then, users can pivot to related collections and continue browsing. Visor provides a hierarchical overview of the collections and also provides a spreadsheet requiring manual customization to filter the collection.

Whereas the above efforts mainly focus on providing the user with powerful interaction modes, we aim to cluster semantic links and entity classes simultaneously to support iterative entity navigation, which is complementary to all of them.

6   Conclusion

In this paper, we propose a co-clustering approach which clusters both links and classes simultaneously to provide users with an iterative entity navigation. To achieve this, we define a new notion of navigation graph based on tripartite graph to model the three aspects of navigation (links, linked entities and their classes). We also measure the link similarity and the class similarity, and incorporate them into the co-clustering algorithm. The proposed approach is implemented in a prototype system. The evaluation results demonstrate that it provides a better overview of all the information, and supports users' iterative entity navigation.

Acknowledgements. This work is supported in part by the 863 Program under Grant 2015AA015406, in part by the National Science Foundation of China under Grant Nos. 61223003 and 61572247, and in part by the Fundamental Research Funds for the Central Universities. We are also grateful to all the participants in the experiments of this work.

References

1. Berners-lee, T., Chen, Y., Chilton, L., Connolly, D., Dhanaraj, R., Hollenbach, J., Lerer, A., Sheets, D.: Tabulator: exploring and analyzing linked data on the semantic web. In: 3rd International Semantic Web User Interaction Workshop (2006)
2. Bizer, C., Lehmann, J., Kobilarov, G., Auer, S., Becker, C., Cyganiak, R., Hellmann, S.: DBpedia-a crystallization point for the web of data. J. Web Sem. **7**(3), 154–165 (2009)
3. Dhillon, I.S.: Co-clustering documents and words using bipartite spectral graph partitioning. In: 7th International Conference on Knowledge Discovery and Data mining, pp. 269–274 (2001)
4. Dhillon, I.S., Mallela, S., Modha, D.S.: Information-theoretic co-clustering. In: 9th International Conference on Knowledge Discovery and Data mining, pp. 89–98 (2003)
5. Giannakidou, E., Koutsonikola, V.A., Vakali, A., Kompatsiaris, Y.: Co-clustering tags and social data sources. In: 9th International Conference on Web-Age Information Management, pp. 317–324. IEEE (2008)
6. Garcia, R., Gimeno, J.M., Perdrix, F., et al.: Building a usable and accessible semantic web interaction platform. World Wide Web **13**(1–2), 143–167 (2010)
7. Harth, A.: VisiNav: a system for visual search and navigation on web data. J. Web Sem. **8**(4), 348–354 (2010)
8. Heim, P., Ertl, T., Ziegler, J.: Facet graphs: complex semantic querying made easy. In: Aroyo, L., Antoniou, G., Hyvönen, E., Teije, A., Stuckenschmidt, H., Cabral, L., Tudorache, T. (eds.) ESWC 2010, Part I. LNCS, vol. 6088, pp. 288–302. Springer, Heidelberg (2010)
9. Lin, D.: An information-theoretic definition of similarity. In: International Conference on Machine Learning, pp. 296–304 (1998)
10. MacQueen, J.: Some methods for classification and analysis of multivariate observations. In: Fifth Berkeley Symposium on Mathematical Statistics and Probability, pp. 281–297 (1967)

11. Manning, C.D., Raghavan, P., Schütze, H.: Introduction to information retrieval. Cambridge University Press, Cambridge (2008)
12. Mika, P.: Ontologies are us: a unified model of social networks and semantics. J. Web Sem. **5**(1), 5–15 (2007)
13. Popov, I.O., Schraefel, M.C., Hall, W., Shadbolt, N.: Connecting the dots: a multi-pivot approach to data exploration. In: Aroyo, L., Welty, C., Alani, H., Taylor, J., Bernstein, A., Kagal, L., Noy, N., Blomqvist, E. (eds.) ISWC 2011, Part 1. LNCS, vol. 7031, pp. 553–568. Springer, Heidelberg (2011)
14. Wu, J.S., Lai, J.H., Wang, C.D.: A novel co-clustering method with intra-similarities. In: 11th International Conference on Data Mining Workshops, pp. 300–306 (2011)
15. Wu, Z., Palmer, M.: Verbs semantics and lexical selection. In: 32nd Annual Meeting on Association for Computational Linguistics, pp. 133–138 (1994)
16. Zheng, L., Qu, Y., Jiang, J., Cheng, G.: Facilitating entity navigation through top-k link patterns. In: 14th International Semantic Web Conference, pp. 163–179 (2015)

DoSeR - A Knowledge-Base-Agnostic Framework for Entity Disambiguation Using Semantic Embeddings

Stefan Zwicklbauer[✉], Christin Seifert, and Michael Granitzer

University of Passau, 94032 Passau, Germany
{stefan.zwicklbauer,christin.seifert,michael.granitzer}@uni-passau.de

Abstract. Entity disambiguation is the task of mapping ambiguous terms in natural-language text to its entities in a knowledge base. It finds its application in the extraction of structured data in RDF (Resource Description Framework) from textual documents, but equally so in facilitating artificial intelligence applications, such as Semantic Search, Reasoning and Question & Answering. In this work, we propose DoSeR (Disambiguation of Semantic Resources), a (named) entity disambiguation framework that is knowledge-base-agnostic in terms of RDF (e.g. DBpedia) and entity-annotated document knowledge bases (e.g. Wikipedia). Initially, our framework automatically generates semantic entity embeddings given one or multiple knowledge bases. In the following, DoSeR accepts documents with a given set of surface forms as input and collectively links them to an entity in a knowledge base with a graph-based approach. We evaluate DoSeR on seven different data sets against publicly available, state-of-the-art (named) entity disambiguation frameworks. Our approach outperforms the state-of-the-art approaches that make use of RDF knowledge bases and/or entity-annotated document knowledge bases by up to 10 % F1 measure.

Keywords: Entity disambiguation · Neural networks · Linked Data · Semantic web

1 Introduction

Entity disambiguation refers to the task of linking phrases in a text, also called surface forms, to a set of candidate meanings, referred to as the knowledge base (KB), by resolving the correct semantic meaning of the surface form. It is an essential task in combining unstructured with structured or formal information; a prerequisite for artificial intelligence applications such as Semantic Search, Reasoning and Question &Answering. Regarding Example 1, an accurate (named) entity recognition system (focusing on persons, organizations and locations only) would return the surface forms **TS** and **New York**. A (named) entity disambiguation system, basing on the well-known DBpedia KB [12], maps the surface forms **TS** and **New York** to the DBpedia resources *Times_Square* and *New_York_City*.

© Springer International Publishing Switzerland 2016
H. Sack et al. (Eds.): ESWC 2016, LNCS 9678, pp. 182–198, 2016.
DOI: 10.1007/978-3-319-34129-3_12

Example 1. The **TS** has been a **New York** attraction for over a century.

While entity disambiguation systems have been well-researched so far, most approaches, such as DBpedia Spotlight [14], TagMe2 [4] or Wikifier [2], have been designed to work on a particular type of KB [22], namely either RDF-based KBs (RDF-KB) like DBpedia [12] or YAGO3 [13], or entity-annotated document KBs (EAD-KB) like Wikipedia. As a consequence these approaches might perform significantly worse after exchanging the underlying KB or do not work at all. Furthermore, RDF-KBs and EAD-KBs can complement each other in terms of entity coverage, i.e. the total number of entities available in a KB, and entity description, i.e. the completeness and quality of the description of one entity. In order to exploit the potential of different KBs, entity disambiguation approaches have to be knowledge-base-agnostic in terms of RDF-KBs and EAD-KBs, while at the same time maintaining simplicity, disambiguation accuracy and preprocessing/disambiguation performance.

In this work, we present DoSeR - (**D**isambiguation **o**f **Se**mantic **R**esources), a novel, knowledge-base-agnostic entity disambiguation framework. DoSeR achieves higher F-measures than the current, publicly available, state-of-the-art (named) entity disambiguation frameworks for DBpedia and Wikipedia entities. These results are achieved by first computing semantic entity embeddings using information from one or multiple underlying KBs and then applying the PageRank algorithm to an automatically created entity candidate graph.

Our **contributions** are the following:

– We present DoSeR, a (named) entity disambiguation framework that is knowledge-base-agnostic in terms of RDF and entity-annotated document KBs.
– We propose how to easily generate semantic entity embeddings to compute semantic similarities between entities regardless of the type of KBs available.
– We evaluate DoSeR on seven well-known and open source data sets showing that DoSeR outperforms current publicly available, state-of-the-art approaches by up to 10 % F1 measure.

The remainder of the paper is structured as follows: In Sect. 2, we review related work. Section 3 introduces the problem formally and outlines our approach. Sections 4 and 5 explain the implementation of DoSeR as well as the process of generating semantic entity embeddings. In Sect. 6, we present the results of DoSeR attained on seven data sets. Finally, we conclude our paper in Sect. 7.

2   Related Work

Entity disambiguation has been studied extensively in the past 10 years. One of the first frameworks to annotate and disambiguate Linked Data Resources was **DBpedia Spotlight** [14]. The framework uses DBpedia as underlying KB, is based on the vector space model and cosine similarity and is publicly available as web service. Further, it is able to disambiguate all classes of the DBpedia ontology. Another framework to disambiguate Linked Data resources is **AGDIS-TIS** [22], a knowledge-base-agnostic approach from 2014. It is based on string similarity measures, an expansion heuristic for labels to cope with co-referencing and the graph-based Hypertext-Induced Topic Search (HITS) algorithm. Focusing on named entities only, it is the current publicly available, state-of-the-art approach in terms of disambiguating named entities while exclusively using DBpedia knowledge. Hoffert et al. proposed **AIDA** [7], a named entity disambiguation framework which is based on YAGO2. It unifies prior approaches into a framework that combines the following three measures: prior probability of an entity, the similarity between the surface forms' context and entity candidates as well as the coherence among entity candidates for all surface forms together.

In contrast to the approaches mentioned before, *wikification* approaches link surface forms to Wikipedia pages. One of the first wikification approaches was proposed by Cucerzan et al. [3] who introduced topical coherence for entity disambiguation. The authors used the referent entity candidate and other entities within the same context to compute topical coherence by analyzing the overlap of categories and incoming links in Wikipedia. Milne and Witten [16] improved the exploitation of topical coherence using Normalized Google Distance and unambiguous entities in the context only. Further improvements on topical coherence were made by Kataria et al. [9] who proposed a new topic model (Wiki-based Pachinko Allocation Model). Their model uses all words of Wikipedia to learn entity-word associations and the Wikipedia category hierarchy to capture co-occurrence patterns among entities. Subsequent works are also based on topic models which perform well on data sets with sufficiently long textual contexts [5].

Another framework is Linden [21], an approach to link named entities in text with a knowledge base unifying Wikipedia and WordNet, by leveraging the rich semantic knowledge embedded in the Wikipedia and the taxonomy of the knowledge base. **Wikifier** [2], a well-known system from 2013, incorporates, along with statistical methods, richer relational analysis of the text. To our knowledge it is the current publicly available state-of-the-art system for linking surface forms to Wikipedia pages regarding the average accuracy across several data sets. Another publicly available framework that links to Wikipedia pages is **WAT** [17], which includes a redesign of TagMe [4] components and introduces two disambiguation families: graph-based algorithms for collective entity disambiguation and vote-based algorithms for local entity disambiguation [23].

The authors of [26] provide an evaluation of biomedical disambiguation systems with respect to three crucial properties: entity context, i.e. the way entities are described, user data, i.e. quantity and quality of externally disambiguated entities, and quantity and heterogeneity of entities to disambiguate.

Many state-of-the-art approaches rely on exhaustive data mining methods and compute, similar to us, semantic relatedness models for entity disambiguation [8]. However, in contrast to these models, DoSeR maintains simplicity, disambiguation accuracy on all data sets, preprocessing/disambiguation performance and extensibility (in terms of KBs). The bold highlighted frameworks are compared with DoSeR in Sect. 6.

3   Problem Statement and Approach

The goal of entity disambiguation is to find the correct semantic mapping between surface forms and entities in a KB. More formally, let $<m_1, ..., m_K>$ be a tuple of K surface forms in a document D, and $\Omega = \{e_1, ..., e_{|\Omega|}\}$ be a set of target entities in a KB. Let Γ be a possible entity configuration $<t_1, ..., t_K>$ with $t_i \in \Omega$, where t_i is the respective target entity for surface form m_i. In this definition, we assume that each entity in Ω is a possible candidate for a surface form m_i. The goal of entity linking can then be formalized as finding the optimal entity configuration Γ^*. Different to [18] we do not pose the optimization problem as maximizing the sum of the scores of a locality function ϕ and a coherence function ψ, which has been proven to be NP-hard [3]. Instead, we approximate the solution using the PageRank algorithm with priors [1,24] on a specially constructed graph G which encompasses both, the locality and the coherence function. In our work, the locality function reflects the likelihood that a target entity t_i is the correct disambiguation for m_i, whereas the coherence function computes a compatibility score describing the coherence between entities in Γ^*. The PageRank algorithm is a well-researched link-based ranking algorithm that simulates a random walk on the underlying graph and reflects the importance of each node. It has been shown to provide good performance for many applications [25], also in entity disambiguation tasks [6].

The graph we construct consists of one node for each entity candidate for all given surface forms. The graph is K-partite, with K being the number of surface forms, and only contains edges between entities of different surface forms. The edge weights are based on the similarity of the semantic embeddings of the entities. We show that semantic embeddings (vectors) can be estimated in a KB-agnostic way and thus, the graph can be constructed for any given KB.

The core steps of our solution are (i) the generation of entity candidates for the given surface forms, (ii) the estimation of semantic embeddings from EAD-KBs and RDF-KBs, (iii) the construction of the K-partite graph using priors and similarities of semantic embeddings, and (iv) the application of PageRank with priors on the graph.

The overall process is outlined in Sect. 4 and its subsections, whereas the estimation of semantic embeddings for different types of KB is presented in Sect. 5.

4   DoSeR - Disambiguation of Semantic Resources

In this section, we present the architecture of the DoSeR framework[1] (cf. Fig. 1). DoSeR consists of the following three main steps: (1) index creation (Sect. 4.1), (2) candidate generation (Sect. 4.2) and (3) the assignment of entities to surface forms (Sect. 4.3). The first step in the index creation process is to define a set of *core* KBs. The set of core KBs (depicted with a continuous line in Fig. 1) is used to specify the set of entities Ω which should be disambiguated by our

---

[1] https://github.com/quhfus/DoSeR/.

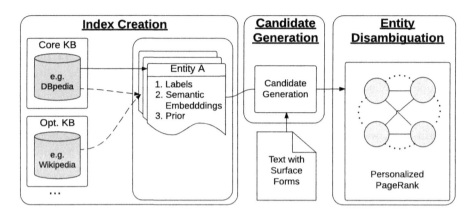

Fig. 1. Overview of DoSeR

framework. In the following, DoSeR processes the contents of all given (core and optional) KBs and stores available surface forms from entity-annotated documents, a semantic entity embedding as well as a prior probability for each entity (optional KBs are figured with a dashed line in Fig. 1). This KB preprocessing step is executed only the first time or if the data of a new KB should be integrated. After preprocessing, DoSeR accepts documents with surface forms (e.g. manually marked by users) that should be linked to entities.

In the candidate generation step, we identify a set of possible entity candidates for each surface form and, thus, significantly reduce the number of possible target entities. To this end we apply several heuristics proposed in [22] or make use of known surface forms. In the final disambiguation step we use this set of candidates to create an entity candidate graph. By applying a PageRank algorithm we attempt to find the best possible entity configuration. More specifically, each entity candidate of a surface form that provides the highest PageRank score denotes the disambiguated target entity for that surface form. In the following, we present each of the steps of DoSeR in more detail.

4.1   Index Creation

Before starting the index creation process we first choose one or multiple source KBs that contain entity describing data. Basically, DoSeR accepts RDF-KBs (e.g. DBpedia, YAGO3) and EAD-KBs (e.g. Wikipedia). Since EAD-KBs do not have a standardized format, DoSeR requires a unified format for annotated entities. Example 2 shows an annotation of the surface form "New York", with the *id* denoting a unique entity id:

Example 2. ...been a <e id="dbr:New_York_City">New York</e> attraction...

To have EAD-KBs in the denoted format we have to parse them in the first step.

Next, given a set of source KBs in the appropriate format, we select a set of *core* KBs. The set of all entities specified or annotated in these core KBs specifies our target entity set Ω. If the core KBs provide information about the entities' classes (e.g. *rdf:type*), Ω can be restricted to named entities only (i.e. persons, organizations and places). After specifying Ω, we use all core and optional KBs as data sources for the entities in Ω. Optional KBs complement the core KBs in terms of completeness and quality of entity descriptions. Overall, our approach is fully knowledge-base-agnostic in terms of RDF-KBs and EAD-KBs. In the next step, DoSeR creates an index comprising the following three entity describing information:

Labels: By default DoSeR extracts the *rdfs:label* attribute of all given RDF-KBs and stores them in a label field. Our approach can be configured to use any set of properties as label. Further, DoSeR searches for EAD-KBs in our specified KB set and, if available, extracts and stores surface forms which have been used to address a specific entity.

Semantic Embeddings: DoSeR automatically creates a semantic embedding (vector) for all entities regardless of the underlying KBs. The resulting embeddings are used to compute a semantic similarity between entities. The creation of these embeddings on the basis of different KBs is explained in Sect. 5 in detail.

Prior: Generally, some entities occur more frequently than others. Thus, these popular entities provide a higher probability to reoccur in other documents [14, 26]. The prior $p(e_i)$ describes the a-priori probability that entity e_i occurs. We use the core KBs to compute entity priors by analyzing the number of its annotations in the documents (EAD-KB) or the number of in and outgoing edges (RDF-KB). In the latter case, we regard the KB as a directed graph, where the nodes V are resources, the edges E are properties and $x, y \in V, (x, y) \in E \Leftrightarrow \exists p : (x, p, y)$ is an RDF triple. For those entities, which have been annotated in a core EAD-KB, we use the number of its annotations in these documents to compute the prior. For the other entities, we use the number of in- and outgoing edges in the RDF-KBs as quantity during the prior computation.

Given these information in an index, we disambiguate entities by selecting relevant candidates (Sect. 4.2) and computing the optimal entity assignments (Sect. 4.3).

4.2   Candidate Generation

Given a constructed index, our framework accepts documents that contain one or multiple surface forms that should be linked to entities. DoSeR disambiguates all surface forms within a document using a collective approach. In our entity disambiguation chain, entity candidate generation is the first crucial step (cf. Fig. 1). Our goal is to reduce the number of possible entity candidates for each input surface form by determining a set of relevant target entities. Hereby, we proceed as follows:

First, we compare the input surface form to those stored in the index. All entities in the index that provide an exact surface form matching serve as entity candidate.

Second, we use the candidate generation approach proposed by Usbeck et al. for AGDISTIS [22]. The authors apply a string normalization approach to the input text. It eliminates plural and genitive forms, removes common affixes such as postfixes for enterprise labels and ignores candidates with time information within their label. Similar to AGDISTIS [22], our system compares the normalized surface forms with the labels in our index by applying trigram similarity. The trigram similarity threshold ($\sigma = 0.82$) is constant in our system and experiments since it provides good results across all data sets and is the default setting in the AGDISTIS framework[2]. If an entity's label matches with the heuristically obtained label, while exceeding the trigram similarity threshold, and the entity is not yet a candidate for the surface form, the entity becomes a candidate.

4.3 Entity Disambiguation

After generating candidates for each surface form, DoSeR uses the set of candidates to create an entity candidate graph. On this graph we perform a random walk and determine the node relevance which can be seen as the average amount of its visits. The random walk is simulated by a PageRank algorithm that permits edge weights and non-uniformly-distributed random jumps [1,24].

First, DoSeR creates a **complete, directed** K-partite graph whose set of nodes V is divided in K disjoint subsets $V_1, ..., V_K$. K denotes the amount of surface forms and V_i is the set of generated entity candidates $\{e_1^i, ..., e_{|V_i|}^i\}$ for surface form m_i. Since our graph is K-partite, there are only directed, weighted edges between entity candidates that belong to different surface forms. Connecting the entities that belong to the same surface form would be wrong since the correct target entities of surface forms are determined by the other surface forms' entity candidates (coherence). Consequently, using a complete graph instead results in an accuracy decrease of $\approx 2\%$ points F1.

The edge weights in our graph represent entity transition probabilities (ETP) which describe the likelihood to walk from a node (entity) to the adjacent node. We compute these probabilities by normalizing our semantic similarity measurement (cf. Eq. 1). The semantic similarity between two entities is the cosine similarity (cos) of its semantic embeddings (vectors) $vec(e_u^i)$ and $vec(e_v^j)$ stored in the index.

$$ETP(e_u^i, e_v^j) = \frac{cos(vec(e_u^i), vec(e_v^j))}{\sum_{k \in (V \setminus V_i)} cos(vec(e_u^i), vec(k))} \tag{1}$$

Given the current graph, we additionally integrate a possibility to jump from any node to any other node in the graph during the random walk with probability $\alpha = 0.1$. Typical values for alpha (according to the original paper [24]) are in the range $[0.1, 0.2]$. We did not manually integrate jump edges in the graph (as in the transition case), but our PageRank algorithm simulates edges between all node pairs during PageRank computation. We compute a probability for

---

[2] http://aksw.org/Projects/AGDISTIS.html.

each entity candidate being the next jump target. For this purpose we use the already precomputed prior stored in our index. Figure 2 shows a possible entity candidate graph of our introducing example. The surface form "TS" has only one entity candidate and consequently has already been linked to the entity "Time Square". The second surface form "New York" is still ambiguous, providing two entity candidates. We omit the jump probability values in the figure to improve visualization

After constructing the disambiguation graph, we need to identify the correct entity candidate node. By applying the PageRank algorithm we compute a relevance score for each entity candidate. We empirically choose $it = 50$ PageRank iterations, which is the best trade-off between performance and accuracy in our experiments. Afterwards, the entity candidate e^i_j of a surface form candidate set V_i that provides the highest relevance score is our entity result for surface form m_i. To improve performance, we automatically thin out our

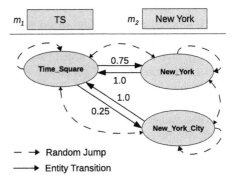

Fig. 2. Entity candidate graph with candidates for the surface forms "TS" and "New York".

disambiguation graph by removing 25 % of those edges, whose source and target entities have the lowest semantic similarity. Despite a loss of information, our disambiguation results nearly stay the same.

We also evaluated the graph-based HITS-algorithm [10]. In our experiments, this algorithm performs worse if the input documents contain a bulk of surface forms or the number of generated entity candidates is high (e.g. when using Wikipedia as KB).

5   Semantic Embedding

Embeddings are n-dimensional vectors of concepts which describe the similarities between these concepts via cosine similarity. This has already been well researched for words in literature [15]. In this work, we show how entity embeddings can be generated for the different types of source KBs. First, we briefly introduce Word2Vec, a set of models that are used to produce word embeddings, in Sect. 5.1. Second, we propose two algorithms to generate appropriate training corpora for Word2Vec in Sect. 5.2.

5.1   Learning Semantic Embeddings Using Word2Vec

Word2Vec is a group of state-of-the-art, unsupervised algorithms to create word embeddings from (textual) documents initially presented by Mikolov et al. [15]. To train these embeddings, Word2Vec uses a two-layer neural network to process non-labeled documents. The neuronal network architecture is based either on the continuous bag of words (CBOW) or the skip-gram architecture. Using CBOW, the input to the model could be $w_i - 2, w_i - 1, w_i + 1, w_i + 2$, the preceding and following words of the current word w_i. The output of the network is the probability of w_i being the correct word. The task can be described as predicting a word given its context. The skip-gram model works vice-versa: the input to the model is a word w_i and Word2Vec predicts the surrounding context words $w_i - 2, w_i - 1, w_i + 1, w_i + 2$. In contrast to other natural language neuronal network models, the training speed of Word2Vec is very fast and can be further significantly improved by using parallel training. The training time on the Wikipedia corpus (without tables) takes ≈ 90 min on our personal computer with a 4×3.4 GHz Intel Core i7 processor and 16 GB RAM.

An important property of Word2Vec is that it groups the vectors of similar words together in the vector space. If enough data is used for training, Word2Vec makes highly accurate guesses about a word's meaning based on its past appearances. Additionally, the resulting word embeddings capture linguistic regularities, for instance the vector operation $vec(\text{"President"}) - vec(\text{"Power"}) \approx vec(\text{"Prime Minister"})$. However, the semantic similarity between two words, which is important in the context of our work, denotes the cosine similarity between the words' Word2Vec vectors.

Since the current version of DoSeR considers the semantic similarity between two entities, we treat entities similar to words. Further, we build our semantic entity embeddings with the help of an entity corpus instead of a textual corpus containing sentences and paragraphs (cf. Sect. 5.2). To train our entity model, we apply the skip-gram architecture that performs better with infrequent words (less popular entities) [15].

5.2   Corpus Generation

Word2Vec typically accepts a set of corpora containing natural language text as input and trains its word vectors according to the words' order in the corpora. Since we want to learn entity representations only, we have to create an appropriate Word2Vec input corpus file that exclusively comprises entities. The entities' order in the corpus file reflects how entities occur in RDF-KBs or EAD-KBs. In the following, we present how DoSeR creates a suitable Word2Vec corpus basing on one or multiple KBs. Hereby, the outputs of both algorithms are concatenated to create a single corpus file.

EAD-KB: To create a Word2Vec corpus out of entity-annotated natural language documents, we assume to have the entity annotations in the format described in Sect. 4.1 (Example 2). Next, DoSeR iterates over all documents in the underlying corpus and replaces all available, linked surface forms with its respective target entity identifier. Further, all non-entity identifiers like words and punctuations are removed so that all documents consist of entity identifiers separated by whitespaces only. However, the collocation of entities is still maintained as given by the original document. The resulting documents are concatenated to create a single Word2Vec corpus file. The corpus creation approach for EAD-KBs is explicated in Algorithm 1.

Algorithm 1. Creating a Word2Vec corpus out of a EAD-KB

---

input : document corpus C, output file
output: word2vec corpus file
forall the $D \in C$ **do**
  $\quad D \longleftarrow replaceSFwithTargetID(D)$
  $\quad D \longleftarrow removeAllNonTargetIDs(D)$
  $\quad appendToOutputFile(D)$

RDF-KB: Originally, Word2Vec trains its word (entity) embeddings based on the order given in a document. Since an RDF-KB does not contain this kind of entity sequence, we model this sequence by random walks between resources in the KB. Hereby, we assume that resources that are directly connected via relations or connected via short relation paths provide cohesiveness. In our case, we create a sequence of those resources that are in our entity target set Ω. For this purpose, we regard an RDF-KB as an **undirected** graph $G_{KB} = (V, E)$ where the nodes V are resources of the KB, the edges E are properties of KB and $x, y \in V, (x, y) \in E \Leftrightarrow \exists p : (x, p, y) \vee \exists p : (y, p, x)$ is an RDF triple in the KB. After that, we perform a random walk on the graph G_{KB}. Whenever the random walk visits a node $x \in V$ we append the entity identifier of node x to the output corpus file, if $x \in \Omega$. The succeeding node $succ(x)$ of x is randomly selected by choosing an adjacent node of x, with probability $\frac{1}{edgesOf(x)}$. Hereby, the function $edgesOf$ counts the number of edges that contact node x. Additionally, we introduce a random variable X_x that provides probabilities to jump to a specific node if a random jump is performed:

$$X_x = P(X_x = x) = \frac{IEF(x)}{\sum_{k \in V} IEF(k)}, \text{ with } IEF(x) = log(\frac{|E|}{edgesOf(x)}) \quad (2)$$

We compute the jump probability from any node to a specific node x by normalizing the respective inverse edge frequency IEF of node x. In our experiments, we use the parameter $\alpha = 0.1$ to perform a random jump. However, values of $0.05 < \alpha < 0.25$ do not significantly affect the resulting Word2Vec model. Furthermore, the parameter θ specifies the number of random walks on the graph. We suggest to use $\theta = 5 * |E|$, which results in $\approx 50M$ random walks for DBpedia. Higher values of θ do not improve the entity embeddings but increase the training time. The corpus creation approach for RDF-KBs is explicated in Algorithm 2.

Algorithm 2. Creating a Word2Vec corpus out of a RDF knowledge base

---

input   : undirected graph G = (V, E), jump random variable X_x, output file
output   : word2vec corpus file
parameter: α node jump probability, θ number of walks
$x \longleftarrow drawRandomNode(X_x)$
$walks \longleftarrow 0$
while $walks < \theta$ do
    if $x \in \Omega$ then
      | $appendToOutputFile(x)$
    if $randomInt(100) > (\alpha * 100)$ then
      | $x \longleftarrow chooseNextNode(x)$;   // adjacent node with $p(succ(x)) = \frac{1}{edgesOf(x)}$
    else
      | $x \longleftarrow drawRandomNode(X_x)$
    $walks \longleftarrow walks + 1$;

---

6 Evaluation

Many disambiguation systems rely on RDF-KBs and additionally leverage knowledge from Wikipedia to significantly improve disambiguation accuracy (e.g. DBpedia Spotlight, AIDA). Thus, the aim in our evaluation is two-fold.

First, we compare DoSeR to the current state-of-the-art named entity disambiguation framework AGDISTIS [22] that exclusively makes use of RDF data by default. Therefore, we use the same KB in form of DBpedia and the same entity target set as in AGDISTIS, consisting of named entities only (cf. left column in Table 1).

Second, we compare DoSeR to publicly available entity disambiguation systems that rely on knowledge from Wikipedia. These are DBpedia Spotlight [14], AIDA [7], Wikifier [2] and WAT [17]. Wikifier and WAT use Wikipedia as underlying KB and link surface forms directly to Wikipedia pages (wikification). In contrast, DBpedia Spotlight and AIDA rely on the RDF-KBs DBpedia and YAGO2, while additionally making use of Wikipedia knowledge (e.g. entity priors). Since entities within these three KBs (DBpedia, Wikipedia and YAGO2) provide *sameAs* relations, we can easily compare the disambiguation accuracy while using the same data sets.

To evaluate DoSeR as well as the competitive disambiguation systems we use the GERBIL - General Entity Annotator Benchmark [23] which offers an easy-to-use platform for the agile comparison of annotators using multiple data sets. In GERBIL, we make use of the D2KB task, which evaluates entity disambiguation only. We report the F1, Precision and Recall, aggregated across surface forms (micro-averaged). Spotlight and WAT are integrated in GERBIL by default, whereas we manually downloaded Wikifier and AIDA and installed them on our server with its best settings. For AIDA we downloaded the default entity repository that is suggested as reference for comparison.

Within our experiment we train the Word2Vec model with Gensim [19], an open-source vector space modeling and topic modeling toolkit. On overview of our parameter settings, links to downloadable resources of other systems and

Table 1. Class constraints for named entities (persons, organizations and places) only in DBPedia and YAGO3. Prefix **dbo** stands for http://dbpedia.org/ontology/, **foaf** for http://xmlns.com/foaf/0.1/ and **yago** for http://yago-knowledge.org/resource/.

Class	DBpedia	YAGO3
Person	dbo:Person, foaf:Person	yago:yagoLegalActorGeo
Organization	dbo:Organization, dbo:WrittenWork	yago:yagoLegalActorGeo
Place	dbo:Place, yago:YagoGeoEntity	yago:yagoLegalActorGeo

some GERBIL result sheets of our experiments can be found here[3]. In the following Sect. 6.1, we briefly describe the data sets which are used in our experiments. In Sect. 6.2, we show how DoSeR performs against the other disambiguation systems.

6.1 Data Sets

In the following, we present seven well-known and publicly available data sets which are used in our evaluation. All data sets are integrated in GERBIL and strongly differ in document length and amount of entities per document. Table 2 shows the statistics of our test corpora.

Table 2. Statistics of our test corpora

Data set	Topic	#Doc	#Ent.	Ent./Doc.
ACE2004	News	57	253	4.44
AIDA/CO-NLL-TestB	News	231	4458	19.40
AQUAINT	News	50	727	14.50
MSNBC	News	20	658	32.90
N3-Reuters	News	128	621	4.85
IITB	Mixed	103	11245	109.01
Microposts-2014 Test	Tweets	1165	1440	1.24

1. **ACE2004.** This data set from Ratinov et al. [18] is a subset of the ACE2004 coreference documents and contains 57 news articles comprising 253 surface forms.
2. **AIDA/CO-NLL-TestB.** The AIDA data set [7] was derived from the CO-NLL 2003 task and contains 1.393 news articles. The corpora was split into a training and two test corpora. The second test set has 231 documents with 19.40 entities on average.
3. **AQUAINT.** Compiled by Milne and Witten [16], the data set contains 50 documents and 727 surface forms from a news corpus from the Xinhua News Service, the New York Times, and the Associated Press.
4. **MSNBC.** The corpus was presented in 2007 by Cucerzan et al. [3] and contains 20 news documents with 32.90 entities per document.
5. **N3-Reuters.** This corpus is based on the well-known Reuters-21578 corpus which contains economic news articles. Roeder et al. proposed this corpus in [20] which contains 128 short documents with 4.85 entities on average.

---

[3] https://github.com/quhfus/DoSeR/wiki/Configurations.

6. **IITB.** This manually created data set by Kulkarni et al. [11] with 123 documents displays the highest entity/document-density of all data sets ($\approx$109 entities/document).
7. **Microposts-2014 Test.** The tweet data set was introduced for the "Making Sense of Microposts" challenge and has very few entities per document on average [23].

6.2   Results

1. Experiment: We compare DoSeR against AGDISTIS, the current state-of-the-art named entity disambiguation framework from 2014, on DBpedia (i.e. DBpedia as core KB and no optional KBs). AGDISTIS is able to disambiguate all entity classes but achieves its best results on named entities [22]. To the best of our knowledge, AGDISTIS is the only available approach that is able to perform named entity disambiguation by using **only** DBpedia knowledge without implementation effort and significant accuracy drop. We also investigate whether DoSeR performs better on the up-to-date YAGO3 KB compared to the DBpedia KB. To provide a fair comparison in our experiment, we exclusively regard the same named entities (persons, organizations and places) as used in AGDISTIS (cf. Table 1). Further, we present the results after omitting the DBpedia category system during the training process of the semantic embeddings. (i.e. creating the Word2Vec corpus without including http://purl.org/dc/terms/subject).

DoSeR performs best on six out of seven data sets (cf. Table 3), with and without using the DBpedia category system (denoted as *DoSeR* and *DoSeR - No Categories*). Both variants attain similar results, but using the DBpedia categories further improves the F-measure by up to 3 % points. Despite applying the same candidate generation approach as proposed in AGDISTIS (because no external surface forms are available), DoSeR outperforms AGDISTIS by up to 10 % F1 measure (IITB data set). On the other data sets (except MSNBC) the advantage is $\approx$5–6 % F1 measure. We assume that the groundtruth entities in the MSNBC data set perfectly fit to available relations between entities in DBpedia. A look at the precision values shows that DoSeR links surface forms to entities more accurate (up to 18 % precision measure on Microposts-2014 Test). The bottle neck which prevents achieving higher F-measures is the absence of surface forms (resulting in a low recall) in the index. The results attained with the YAGO3 KB are sightly worse than those attained in DBpedia ($\approx$2–3 % F1 measure). However, we still outperform AGDISTIS on six data sets by $\approx$4–5 % F1 measure.

2. Experiment: We evaluate DoSeR against the entity disambiguation systems Wikifier, DBpedia Spotlight, AIDA and WAT which all leverage Wikipedia knowledge. Similar to the previous experiment we use DBpedia as core KB, but let DoSeR disambiguate all entities in DBpedia (all entities belonging to the *owl:thing* class) instead of named entities only. We present the results of DoSeR using DBpedia only (denoted as *DoSeR*), using DBpedia and surface forms extracted from Wikipedia (denoted as *DoSeR +WikiSF*) as well as using DBpedia as core KB and Wikipedia as optional KB (denoted as *DoSeR +Wiki*). On all variants, we make use of the DBpedia category system.

Table 3. Performance of DoSeR using relations only, DoSeR using relations and categories, and AGDISTIS on seven different data sets using micro F-measure (**F1**).

Data set	Corpus	Approach	F1-measure	Precision	Recall
ACE2004	DBpedia	DoSeR	0.702	0.795	0.629
		DoSeR - No Categories	**0.706**	**0.800**	**0.632**
		AGDISTIS	0.058	0.090	0.024
	YAGO3	DoSeR	0.679	0.778	0.602
AIDA/CONLL-TestB	DBpedia	DoSeR	**0.616**	**0.697**	**0.552**
		DoSeR - No Categories	0.602	0.684	0.537
		AGDISTIS	0.582	0.628	0.541
	YAGO3	DoSeR	0.608	0.662	0.550
AQUAINT	DBpedia	DoSeR	**0.646**	**0.820**	**0.533**
		DoSeR - No Categories	0.637	0.809	0.525
		AGDISTIS	0.596	0.739	0.499
	YAGO3	DoSeR	0.611	0.754	0.513
MSNBC	DBpedia	DoSeR	0.725	0.763	0.690
		DoSeR - No Categories	0.727	0.765	0.692
		AGDISTIS	**0.751**	**0.772**	**0.730**
	YAGO3	DoSeR	0.735	0.773	0.700
N3 Reuters	DBpedia	DoSeR	**0.731**	**0.817**	**0.661**
		DoSeR - No Categories	0.713	0.791	0.649
		AGDISTIS	0.658	0.721	0.605
	YAGO3	DoSeR	0.725	0.805	0.656
IITB	DBpedia	DoSeR	**0.515**	**0.773**	**0.386**
		DoSeR - No Categories	0.488	0.751	0.362
		AGDISTIS	0.412	0.637	0.304
	YAGO3	DoSeR	0.454	0.697	0.333
Microposts-2014 Test	DBpedia	DoSeR	**0.489**	**0.763**	**0.360**
		DoSeR - No Categories	0.478	0.750	0.351
		AGDISTIS	0.428	0.584	0.337
	YAGO3	DoSeR	0.465	0.703	0.347

Table 4 shows the F1 values of DoSeR using different data sources compared to other disambiguation approaches. We also provide the average F1 values across all data sets. Overall, the results of *DoSeR* are slightly worse than those of the previous experiment. This is because our index does not only contain named entities and thus, the entity target set Ω comprises more entities to be disambiguated [26]. Using surface forms from the Wikipedia corpus (*DoSeR +WikiSF*) improves the results from 61.4 % F1 to 67.4 % F1 on average, which is mainly caused by increased recall values. In this configuration, DoSeR outperforms Spotlight and AIDA by ≈7 % F1 measure on average.

Using Wikipedia as optional KB in DoSeR (*DoSeR +Wiki*) further increases the average F1 values by ≈10 % F1 percentage points and significantly out-

Table 4. F1 values of DoSeR, DBpedia Spotlight, AIDA, WAT and Wikifier on seven data sets.

Data set	DoSeR	DoSeR (+WikiSF)	DoSeR (+Wiki)	Wikifier	Spotlight	AIDA	WAT
ACE2004	0.681	0.768	**0.864**	0.824	0.713	0.741	0.800
AIDA/CONLL-TestB	0.597	0.735	0.722	0.776	0.593	0.806	**0.843**
AQUAINT	0.638	0.709	0.820	**0.862**	0.713	0.534	0.768
MSNBC	0.719	0.796	**0.881**	0.851	0.511	0.796	0.777
N3-Reuters	0.700	0.718	**0.727**	0.694	0.577	0.571	0.644
IITB	0.497	0.525	0.713	**0.755**	0.447	0.277	0.611
Microposts-2014 Test	0.469	0.464	**0.639**	0.586	0.623	0.412	0.595
Average	0.614	0.674	**0.767**	0.764	0.597	0.591	0.720

performs the other approaches. It also beats the current state-of-the-art approach Wikifier on four out of seven data sets (ACE2004, MSNBC, N3-Reuters Microposts2014-Test). Considering the AIDA-TestB data set, DoSeR performs comparatively poor with 72.2 % F1 compared to 84.3 % F1 by the WAT system. Analyzing the results on this data set shows that an analysis of the surface forms' textual context is necessary to perform better. In contrast, on the ACE2004 and MSNBC data sets our approach performs exceptionally well with 86.4 % F1 and 88.1 % F1 respectively. We also performed the same evaluation with disambiguating Wikipedia entities only (Wikipedia as core KB only). The results are very similar to *DoSeR +Wiki*, achieving an average F1 measure of 76.0 %.

Discussion: In our experiments, we show that the semantics of entity relations and categories in RDF-KBs as well as entity annotations in EAD-KBs can be optimally captured by Word2Vec's entity embeddings. In contrast to binary relations, the semantic embeddings allow to compute an accurate entity similarity even there is no direct relation in the KB between these entities. Additionally, these embeddings are robust against noisy information within these KBs. In contrast to other works, our approach does not regard the surrounding contextual words of the surface forms to disambiguate entities. This is a crucial issue especially on the AIDA test data set. For instance, given a set of location names as surface forms, our approach is not able to decide whether the surface forms refer to locations or football clubs. However, we are going to tackle this deficit in the near future. In terms of performance, DoSeR practically disambiguates as fast as AGDISTIS if only a moderate number of entity candidates is available (e.g. Experiment 1). The disambiguation performance decreases in our second experiment, but we will further optimize the PageRank computation by heuristic computations [25].

7   Conclusion

In this work, we present DoSeR a (named) entity disambiguation framework that is knowledge-base-agnostic in terms of RDF-KBs (e.g. DBpedia) and EAD-KBs (e.g. Wikipedia). In this context, we propose how to easily generate semantic entity embeddings with Word2Vec to compute semantic similarities between entities regardless of the type of KBs available (RDF-KBs or EAD-KBs). Our collective disambiguation algorithm relies on the PageRank algorithm, which is applied on an automatically created entity candidate graph. DoSeR outperforms the current state-of-the-art approach for named entity disambiguation on RDF-KBs on DBpedia by up to 10 % F1 measure. Further, our approach outperforms AIDA, DBpedia Spotlight, WAT and Wikifier (the current publicly available state-of-the-art disambiguation system for Wikipedia entities), when leveraging Wikipedia knowledge.

Acknowledgments. The presented work was developed within the EEXCESS project funded by the European Union Seventh Framework Programme FP7/2007–2013 under grant agreement number 600601.

References

1. Brin, S., Page, L.: The anatomy of a large-scale hypertextual web search engine. In: 7th WWW, pp. 107–117. Elsevier Science Publishers B.V., Amsterdam (1998)
2. Cheng, X., Roth, D.: Relational inference for wikification. In: Proceedings of the Conference on Empirical Methods in Natural Language Processing, EMNLP 2013 (2013)
3. Cucerzan, S.: Large-scale named entity disambiguation based on Wikipedia data. In: EMNLP-CoNLL, pp. 708–716. ACL, Prague, June 2007
4. Ferragina, P., Scaiella, U.: Fast and accurate annotation of short texts with wikipedia pages. IEEE Softw. **29**(1), 70–75 (2012)
5. Han, X., Sun, L.: An entity-topic model for entity linking. In: EMNLP-CoNLL, pp. 105–115. ACL, Stroudsburg (2012)
6. Han, X., Sun, L., Zhao, J.: Collective entity linking in web text: a graph-based method. In: SIGIR, pp. 765–774. ACM, New York (2011)
7. Hoffart, J., Yosef, M.A., Bordino, I., Fürstenau, H., Pinkal, M., Spaniol, M., Taneva, B., Thater, S., Weikum, G.: Robust disambiguation of named entities in text. In: EMNLP, pp. 782–792. ACL, Stroudsburg (2011)
8. Huang, H., Heck, L., Ji, H.: Leveraging deep neural networks and knowledge graphs for entity disambiguation. CoRR abs/1504.07678 (2015)
9. Kataria, S.S., Kumar, K.S., Rastogi, R.R., Sen, P., Sengamedu, S.H.: Entity disambiguation with hierarchical topic models. In: 17th SIGKDD, pp. 1037–1045. ACM, New York (2011)
10. Kleinberg, J.M.: Authoritative sources in a hyperlinked environment. J. ACM **46**(5), 604–632 (1999)
11. Kulkarni, S., Singh, A., Ramakrishnan, G., Chakrabarti, S.: Collective annotation of wiki-pedia entities in web text. In: 15th SIGKDD, pp. 457–466. ACM, New York (2009)

12. Lehmann, J., Isele, R., Jakob, M., Jentzsch, A., Kontokostas, D., Mendes, P., Hellmann, S., Morsey, M., van Kleef, P., Auer, S., Bizer, C.: DBpedia - a large-scale, multilingual knowledge base extracted from wikipedia. Semant. Web J. **6**, 167–195 (2014)
13. Mahdisoltani, F., Biega, J., Suchanek, F.M.: Yago3: a knowledge base from multi-lingual wikipedias (2015)
14. Mendes, P.N., Jakob, M., García-Silva, A., Bizer, C.: Dbpedia spotlight: shedding light on the web of documents. In: 7th I-Semantics, pp. 1–8. ACM, New York (2011)
15. Mikolov, T., Chen, K., Corrado, G., Dean, J.: Efficient estimation of word repre-sentations in vector space. CoRR abs/1301.3781 (2013)
16. Milne, D., Witten, I.H.: Learning to link with wikipedia. In: 17th CIKM, pp. 509–518. ACM, New York (2008)
17. Piccinno, F., Ferragina, P.: From TagMe to WAT: a new entity annotator. In: First International Workshop on Entity Recognition/Disambiguation, ERD 2014, pp. 55–62. ACM, New York (2014)
18. Ratinov, L., Roth, D., Downey, D., Anderson, M.: Local and global algorithms for disambiguation to wikipedia. In: ACL, pp. 1375–1384. ACL, Stroudsburg (2011)
19. Řehůřek, R., Sojka, P.: Software framework for topic modelling with large corpora. In: Proceedings of the LREC 2010 Workshop, pp. 45–50. ELRA, Valletta, May 2010
20. Röder, M., Usbeck, R., Hellmann, S., Gerber, D., Both, A.: N3 - a collection of datasets for named entity recognition and disambiguation in the NLP interchange format. In: 9th LREC, Reykjavik, Iceland, 26–31 May 2014 (2014)
21. Shen, W., Wang, J., Luo, P., Wang, M.: Linden: linking named entities with knowl-edge base via semantic knowledge. In: 21st WWW, pp. 449–458. ACM, New York (2012)
22. Usbeck, R., Ngonga Ngomo, A.-C., Röder, M., Gerber, D., Coelho, S.A., Auer, S., Both, A.: AGDISTIS - graph-based disambiguation of named entities using linked data. In: Mika, P., et al. (eds.) ISWC 2014, Part I. LNCS, vol. 8796, pp. 457–471. Springer, Heidelberg (2014)
23. Usbeck, R., Röder, M., Ngonga Ngomo, A.C., Baron, C., Both, A., Brümmer, M., Ceccarelli, D., Cornolti, M., Cherix, D., Eickmann, B., Ferragina, P., Lemke, C., Moro, A., Navigli, R., Piccinno, F., Rizzo, G., Sack, H., Speck, R., Troncy, R., Waitelonis, J., Wesemann, L.: GERBIL - general entity annotation benchmark framework. In: 24th WWW Conference (2015)
24. White, S., Smyth, P.: Algorithms for estimating relative importance in networks. In: 9th SIGKDD, pp. 266–275. ACM, New York (2003)
25. Xie, W., Bindel, D., Demers, A., Gehrke, J.: Edge-weighted personalized pager-ank: breaking a decade-old performance barrier. In: 21th SIGKDD, pp. 1325–1334. ACM, New York (2015)
26. Zwicklbauer, S., Seifert, C., Granitzer, M.: From general to specialized domain: analyzing three crucial problems of biomedical entity disambiguation. In: Chen, Q., Hameurlain, A., Toumani, F., Wagner, R., Decker, H. (eds.) DEXA 2015. LNCS, vol. 9261, pp. 76–93. Springer, Heidelberg (2015)

Embedding Mapping Approaches for Tensor Factorization and Knowledge Graph Modelling

Yinchong Yang[2]([⊠]), Cristóbal Esteban[1,2], and Volker Tresp[1,2]

[1] Siemens AG, Corporate Technology, Munich, Germany
[2] Ludwig-Maximilians-Universität München, Munich, Germany
yinchong.yang@hotmail.com

Abstract. Latent embedding models are the basis of state-of-the art statistical solutions for modelling Knowledge Graphs and Recommender Systems. However, to be able to perform predictions for new entities and relation types, such models have to be retrained completely to derive the new latent embeddings. This could be a potential limitation when fast predictions for new entities and relation types are required. In this paper we propose approaches that can map new entities and new relation types into the existing latent embedding space without the need for retraining. Our proposed models are based on the observable —even incomplete— features of a new entity, e.g. a subset of observed links to other known entities. We show that these mapping approaches are efficient and are applicable to a wide variety of existing factorization models, including nonlinear models. We report performance results on multiple real-world datasets and evaluate the performances from different aspects.

1 Introduction

Latent embedding models, aka *factorization models*, have proven to be powerful approaches for modelling Knowledge Graphs (KG) as described in [17,18]. A special case is Collaborative Filtering (CF) where latent embedding models have shown state-of-the-art performance [16]. The common key aspect of these models is that an observed link between multiple entities can be modelled as the interaction between their latent embedding vectors. Multi-linear models such as CP/PARAFAC [14] and Tucker [22] as well as RESCAL [19] are typical examples of models that use latent embeddings. Nonlinear Neural Network-based embedding models are derived in [8,20]. For a more detailed review of these works please see [18].

The latent embedding vectors can be used in several ways. For example, it has been shown that distances between entities in the latent space are more compact and meaningful than in the original observable feature space. Also, in entity resolution, entities close to each other in the latent space can sometimes be interpreted as duplicates [7]. Finally, it has been shown that unknown links between known entities can be predicted based on interactions of their latent embeddings [18].

© Springer International Publishing Switzerland 2016
H. Sack et al. (Eds.): ESWC 2016, LNCS 9678, pp. 199–213, 2016.
DOI: 10.1007/978-3-319-34129-3_13

A drawback of latent embedding models is that they need to be retrained when new entities are appended to the database. For large-scale databases and/or situations where the system is expected to perform immediate operations, such as entity resolution or link prediction on the new entities, this would be very costly and factorization models would find only limited applications.

In this paper, we propose a new class of approaches to handle new entities and new relation types by mapping them into the latent space learned by the factorization model. We emphasize that such mapping models can be learned in conjunction with the training of the factorization model. To map a new entity into the latent space we only require the observable features of the entity. In a KG, for instance, such observable features form a binary vector or matrix, representing the existence of links between this a entity and a subset of known entities in the database.

The rest of the paper is organized as follows: In Sect. 2 we give a brief review of selected embedding-based factorization models and illustrate the concept of an embedding mapping. We show that for certain specific factorization models there exist embedding mappings in closed form. In Sect. 3, we propose a general framework that describes a variety of factorization models on a more abstract level and derive a framework defining the mapping models and elaborate three options for training. In Sect. 4 we present experimental results on real-world datasets. Section 5 discusses related work and Sect. 6 contains conclusions and an outlook for further works.

Notations: A matrix A is represented as a bold capital letter and a multidimensional tensor $\mathcal{X}$ by a calligraphic bold capital letter. By default we assume a 3-dimensional tensor. In some applications the dimensions correspond to entities and relation types, which we sometimes treat as generalized entities. A matrix with indexing superscript as $A^{(l)}$ denotes the latent embedding matrix for entities of the l-th dimension of a matrix or tensor. The matrix derived by unfolding a tensor w.r.t. dimension l is noted using subscripts as $X_{(l)}$. Note that unfolding a matrix w.r.t. first and second dimension is equivalent to the matrix itself and its transpose, respectively. $X^{\dagger}$ stands for the Moore-Penrose pseudoinverse. A vector is denoted with bold small letters such as $x_{i,\bullet}$ and refers to the i-th row in a corresponding matrix X. We refer to a set using either a simple capital Greek letter such as Θ or —if we focus on the elements— using curly brackets as $\{A^{(l)}\}_{l=1}^{L}$. The concatenation operation is noted with squared brackets $[\bullet, ..., \bullet]_{+}$.

2   Factorization Models with Closed-Form Mappings

In this section we review a few well-studied factorization models that are based on latent embeddings and motivate our problem setting of mapping new entities into the latent embedding space.

Matrix Cases: First we review the Singular Value Decomposition (SVD) as a latent embedding model: For an SVD in form of $X = UDV^{T}$ we interpret the matrix U to consist of latent embedding vectors in rows, for each entity represented in the first dimension of X. The matrix X is constructed by a linear combination of the embeddings U with weights defined as rows of $(DV^T)^T$.

Then we consider $U = X(DV^T)^\dagger$, which is an inverse-relation, to be a mapping function from X to the latent embedding in U. It is generally assumed that this mapping relation also holds for a new observation which is not present in X, i.e.

$$u_{new}^T = x_{new}^T (DV^T)^\dagger \tag{1}$$

given that D and V are regarded as constant. We can generalize these relationships to Matrix Factorization(MF) $X = AB^T$ as used in [16]. The latent embeddings are now rows of A and the weights as rows of B. The mapping function now is

$$a_{new}^T = x_{new}^T (B^T)^\dagger. \tag{2}$$

In both cases (SVD and MF), instead of a complete recalculation of the factorization to derive the corresponding latent embedding vector, we simply need to apply a linear map to x_{new}, where the map is derived from the pseudo-inverse operation.

Tensor Cases: Following the notation in [15], we describe the CP/PARAFAC model [14] as well as its more general form, the Tucker decomposition [22], as $\mathcal{X} \approx \mathcal{G} \times_1 A \times_2 B \times_3 C$. A row in each of the three matrices, i.e., $a_{i,\bullet}, b_{j,\bullet}, c_{k,\bullet}$, stores the latent embedding of the i-, j- and k-th entity, respectively, in the corresponding dimensions of $\mathcal{X}$; and the core tensor $\mathcal{G}$ specifies the linear interaction between each triple of embedding vectors to derive the entry $x_{i,j,k}$. In the special case of CP, $\mathcal{G}$ takes the form of a hyper-diagonal tensor. By rewriting the model as $X_{(1)} = AG_{(1)}(C \otimes B)^T$ we could also interpret A as a latent embedding matrix and $G_{(1)}(C \otimes B)^T$ as the linear weights. Inverting this linear relation we can obtain a mapping of the form

$$a_{new}^T = x_{new}^T (G_{(1)}(C \otimes B)^T)^\dagger. \tag{3}$$

Such closed-form mappings cannot be derived in at least two cases: Firstly, for non-linear factorization models such as Multiway Neural Network (mwNN) [8], Neural Tensor Networks [20] and TransE [3]; secondly for models with shared embeddings such as RESCAL [19]. We derive solutions for those two cases in the next section. In the experimental part of this paper, we implement and test (logistic) MF to model data in matrix form and CP, RESCAL and mwNN for tensor data. A brief summary of the model architectures can be found in Table 1. To obtain proper probabilistic models, we introduce a natural parameter η for each entry x in the matrix or tensor.

3   General Models and Training Algorithms

In this section we introduce a generic framework describing factorization and **Em**bedding **Ma**pping (abbreviately termed 'Emma'). We also propose three

Table 1. A summary of selected factorization models within the scope of this paper. sig denotes the sigmoid or logistic function $sig(x) = \frac{1}{1+\exp(-x)}$; $\mathcal{N}$ and $\mathcal{B}$ denote Gaussian and Bernoulli distributions, respectively. We denote with MLP a standard three layered perceptron.

Model	Distr. assumption	Natural parameter
MF	$x_{i,j} \sim \mathcal{N}(\eta_{i,j}, \sigma)$	$\hat{\eta}_{i,j} = \sum_{r=1}^{R} a_{i,r}^{(1)} \cdot a_{j,r}^{(2)}$
Logistic MF	$x_{i,j} \sim \mathcal{B}(sig(\eta_{i,j}))$	
CP	$x_{i,j,k} \sim \mathcal{N}(\eta_{i,j,k}, \sigma)$	$\hat{\eta}_{i,j,k} = \sum_{r=1}^{R} a_{i,r}^{(1)} \cdot a_{j,r}^{(2)} \cdot a_{k,r}^{(3)}$
RESCAL	$x_{i,j,k} \sim \mathcal{N}(\eta_{i,j,k}, \sigma)$	$\hat{\eta}_{i,j,k} = \sum_{p}^{P} \sum_{p'}^{P} w_{p,p',k} \cdot a_{i,p}^{(1)} \cdot a_{j,p'}^{(1)}$
mwNN	$x_{i,j,k} \sim \mathcal{B}(sig(\eta_{i,j}))$	$\hat{\eta}_{i,j,k} = MLP([a_{i,\bullet}^{(1)}, a_{j,\bullet}^{(2)}, a_{k,\bullet}^{(3)}]+)$

novel approaches to train the mapping models. For the sake of generality we shall refer to matrices also as tensors.

3.1 General Models

The Factorization Model defines the interaction between latent embeddings to construct the tensor as

$$\widehat{\mathcal{H}} = f\left(\{\boldsymbol{A}^{(l)}\}_{l=1}^{L}; \, \Theta\right) \quad \text{with}$$

$$\widehat{\mathcal{H}} \in \mathbb{R}^{n_1 \times \ldots \times n_L}, \boldsymbol{A}^{(l)} \in \mathbb{R}^{n_l \times p_l}. \tag{4}$$

Here, the L-dimensional tensor $\mathcal{H}$ contains the natural parameters η. The matrices in set $\{\boldsymbol{A}^{(l)}\}_{l=1}^{L}$ store the latent embeddings in their rows. The tensor is reconstructed with operations defined by a parameterized function $f(\bullet; \, \Theta)$.

For instance, in case of CP factorization, the function f specifies the linear combination of rows in each embedding matrix without additional parameters ($\Theta = \emptyset$); and for mwNN, Θ consists of the weights in an NN model whose architecture is defined as part of f.

The Factorization Cost Function: We define the factorization cost function c_F and its regularized version c_F^p as

$$c_F = d_F(\boldsymbol{\mathcal{X}}, \widehat{\mathcal{H}}) = d_F(\boldsymbol{\mathcal{X}}, f(\{\boldsymbol{A}^{(l)}\}_{l=1}^{L}; \, \Theta)) \tag{5}$$

$$c_F^p = c_F + \sum_{l=1}^{L} \gamma(\boldsymbol{A}^{(l)}) + \rho(\Theta). \tag{6}$$

During training the cost function is optimized w.r.t. the parameters in Θ and embeddings in the $\boldsymbol{A}^{(l)}$'s. An example for the differentiable distance measure d_F is the Frobenius Norm. For binary tensors, the cross-entropy loss is more suitable. In Eq. (6) we included a regularization function $\gamma(\bullet)$ for the latent embeddings and a second one $\rho(\bullet)$ for the model parameters.

The Mapping Model defines a function $g(\bullet; \Psi_l)$ that maps each row in the tensor unfolding $\boldsymbol{X}_{(l)}$ to the corresponding row in the learned embedding matrix $\boldsymbol{A}^{(l)}$ as

$$\widehat{\boldsymbol{A}}^{(l)} = g\left(\boldsymbol{X}_{(l)}; \Psi_l\right) \quad \forall l \in [1, ..., L] \quad \text{with}$$
$$\boldsymbol{X}_{(l)} \in \mathbb{R}^{n_l \times \prod_{l' \neq l} n_{l'}}. \tag{7}$$

Note that in the input of the mapping function, each arbitrary row i in $\boldsymbol{X}_{(l)}$, is identical to the vectorized i-th hyper-slice of the tensor and consists of all available information about the i-th entity. For a KG, for instance, this could be the vector indicating the existence of relations between entity i and all other entities for all relation types. The function $g(\bullet)$ defines the architecture of the mapping model and Ψ_l consists of all parameters.

The Mapping Cost Function: We define the mapping cost function as

$$c_M = \sum_{l=1}^{L} d_M(\boldsymbol{A}^{(l)}, \widehat{\boldsymbol{A}}^{(l)}) = \sum_{l=1}^{L} d_M(\boldsymbol{A}^{(l)}, g(\boldsymbol{X}_{(l)}; \Psi_l)) \tag{8}$$

$$c_M^p = c_M + \sum_{l=1}^{L} \rho(\Psi_l). \tag{9}$$

Optimizing the mapping cost function involves adjusting Ψ_l for each l with a given $g(\bullet)$ so that the distance between the learned embedding $\boldsymbol{A}^{(l)}$ from factorization and mapped embedding $\widehat{\boldsymbol{A}}^{(l)}$ from the corresponding tensor unfoldings is minimized.

The Compact Model: Since the latent embeddings, e.g., the $\boldsymbol{A}^{(l)}$'s, are also adjustable parameters, we could write a more compact model by plugging Eq. (7) into Eq. (4) and obtain

$$\widehat{\mathcal{H}} = f\left(\{g\left(\boldsymbol{X}_{(l)}; \Psi_l\right)\}_{l=1}^{L}; \Theta\right). \tag{10}$$

Analogously, combining cost functions of the factorization model —Eqs. (5) and (6)— and those of the mapping model —Eqs. (8) and (9)— we obtain:

The Compact Cost Function as:

$$c = d_F(\boldsymbol{\mathcal{X}}, \widehat{\mathcal{H}}) = d_F\left(\boldsymbol{\mathcal{X}}, f\left(\{g\left(\boldsymbol{X}_{(l)}; \Psi_l\right)\}_{l=1}^{L}; \Theta\right)\right) \tag{11}$$

$$c^p = c + \rho(\Theta) + \sum_{l=1}^{L} \rho(\Psi_l), \tag{12}$$

where $\boldsymbol{A}^{(l)}$'s are not explicitly defined but could be derived from $g(\boldsymbol{X}_{(l)}; \Psi_l)$.

It should be noted that the tensor $\boldsymbol{\mathcal{X}}$ occurs both at the output and the input (as unfoldings) of the cost function. More specifically, for a tensor $\boldsymbol{\mathcal{X}}$ of three dimensions, the factorization and mapping models as a whole actually model

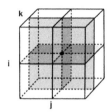

Fig. 1. Illustration of the Compact Model in case of a tensor: We model the each entry $x_{i,j,k}$ over the natural parameter $\eta_{i,j,k}$ based on the three slices of the tensor indexed by i, j and k, respectively.

each entry $x_{i,j,k}$ based on the i-th, j-th and k-th slices of 1st, 2nd and the 3rd dimension of the tensor, respectively, whereby the $x_{i,j,k}$ is the intersection of all these three slices. This aspect is illustrated in Fig. 1.

The factorization problem defined in Eq. (4) could be solved using a variety of well studied factorization models that optimize Eq. (5). In the following we focus on solving the mapping problem in Eq. (7) and the compact modelling problem in Eq. (10). Within the scope of this paper we specifically assume that the function $g(\bullet; \Psi_l)$ represents a linear relation between $\boldsymbol{A}^{(l)}$ and $\boldsymbol{X}_{(l)}$ that can be modelled by a matrix $\Psi_l = \{\boldsymbol{M}_l\}$, i.e.

$$\widehat{\boldsymbol{A}}^{(l)} = \boldsymbol{X}_{(l)} \boldsymbol{M}_l \ \forall l \in [1, ..., L]. \tag{13}$$

3.2 Training Approaches

Post Embedding Mapping: The most intuitive way to perform an Emma is to solve Eqs. (4) and (7) sequentially: Given that a certain factorization model has already derived the latent embeddings $\boldsymbol{A}^{(l)}$, we could consider the mapping from $\boldsymbol{X}_{(l)}$ to the embedding to be a linear system of n_l equations as suggested in Eq. (7). Since one is interested in small dimensions for the latent embeddings, i.e. $p_l < \prod_{l' \neq l} n_{l'}$, the system is overdetermined and can be approximately solved using Least-Square (LS) methods. Specifically: $\widehat{\boldsymbol{M}}_l = (\boldsymbol{X}_{(l)}^T \boldsymbol{X}_{(l)})^\dagger \boldsymbol{X}_{(l)}^T \boldsymbol{A}^{(l)}$.

It is easy to see that the inverting methods of Eqs. (1), (2) and (3) introduced in Sect. 2 are special cases of this LS estimation. For instance, with MF model $\boldsymbol{X} = \boldsymbol{A}\boldsymbol{B}^T$, the general LS estimation could be described as $\widehat{\boldsymbol{A}} = \boldsymbol{X}\boldsymbol{M}_A$ with $\boldsymbol{M}_A = (\boldsymbol{X}^T\boldsymbol{X})^\dagger \boldsymbol{X}^T \boldsymbol{A}$. Plugging in the information of the model definition, we obtain $\boldsymbol{M}_A = (\boldsymbol{B}\boldsymbol{A}^T\boldsymbol{A}\boldsymbol{B}^T)^\dagger \boldsymbol{A}\boldsymbol{B}^T \boldsymbol{A} = (\boldsymbol{B}^T)^\dagger$, which is the same as the inverting operation in Eq. (2). An apparently desirable feature of this Post Mapping approach is its simplicity. It is applicable for any known factorization model and does not affect the factorizing process, since this approach assumes the learned embeddings as fixed. In the following we shall refer to this approach as Emma-Post.

Embedding Mapping Learning with Hatting Algorithm: Alternatively, one could also integrate the Emma learning into the factorization learning

---

Algorithm 1. Hatting Algorithm Framework

---

for all $l \in [1, ..., L]$ **do:**
$$U^{(l)} \leftarrow (X_{(l)}^T X_{(l)} - \lambda I)^\dagger X_{(l)}^T$$
$$H^{(l)} \leftarrow X_{(l)} U^{(l)}$$
end for
for each epoch t in learning factorization **do:**
    $\{ A^{(l)} \}_{l=1}^L, \Theta$ ⟵ update w.r.t. c_F^n

Absolute Updating:	or	*Stochastic Updating with Late-Starting:*
		if $t > \tau$ **then:**
for $l \in [1, ..., L]$ **do:**		**for** $l \in [1, ..., L]$ **do:**
		$\pi^{(l)} = 1 - \max(0, R^2(A^{(l)}, H^{(l)}A^{(l)}))$
$A^{(l)} \leftarrow H^{(l)} A^{(l)}$		$A^{(l)} \leftarrow H^{(l)} A^{(l)}$ with probability $\pi^{(l)}$
end for		**end for**
		end if

end for
return:
    $\{A^{(l)}\}_{l=1}^L$ as latent embeddings;
    $\{M^{(l)} = U^{(l)} A^{(l)}\}_{l=1}^L$ as Emma matrices.

---

process: Instead of solving the LS problem after the factorization learning is completed, we suggest to fit the LS solution against the current latent embedding after each epoch of the factorization learning. Specifically, after each epoch of the factorization learning, we solve the LS problem based on the current embeddings and replace these with their LS estimates to satisfy the criterion of Eq. (8). In terms of notation, we replace $A^{(l)}$ with $\widehat{A}^{(l)} = H^{(l)} A^{(l)}$ i.e. we 'hat' the embedding matrix with the Hat-Matrix as in linear regression [13]. In the next epoch, the factorization algorithm proceeds from the LS estimates of the embeddings. In addition, to avoid collinearity and overfitting, it is also advisable to add a ridge regularization term to the Hat Matrix. We formulate this integrated Emma as an algorithmic framework in Algorithm 1 with the left-sided option termed *Absolute Updating.*

There are two major advantages of this algorithmic framework: Firstly, the LS updates are efficient to calculate since the Hat Matrix is only calculated once prior to the iterative factorization learning, during which one only needs to perform in each epoch one matrix multiplication for each dimension for the sake of LS-updating. Secondly, any factorization algorithm can be easily extended with this LS update as long as it is performed in a iterated manner, such as when using ALS or gradient based approaches.

Based on our experiments we also propose following practical adjustments of the algorithm:

i. Late Starting Strategy: Since the embeddings are usually initialized randomly by many algorithms, it is not necessary in such cases to perform LS updates during early epochs. It is advisable to start LS updates, for instance, when the embeddings are updated in smaller magnitudes i.e. where the cost function is locally flat. One could measure the gradient changes in each epoch or simply define a $\tau > 1$ to be the first iteration where the LS update commences.

ii. Stochastic Update: We suggest monitoring the quality of the linear mapping in terms of R^2, the Coefficient of Determination. As long as R^2 is small, it is always assumed to be necessary to further perform LS update. But as the training proceeds, R^2 often tends to get larger and converges to 1. In such cases it may again be unnecessary to perform an LS update in each iteration and one could save some runtime. Since the R^2 typically lies within $(0, 1)$ but could also be negative based on our definition, we define a coefficient $\pi = 1 - \max(0, R^2)$ to be interpreted as the probability or necessity to perform an LS update. Both improvements are taken into account in the right-sided option *Stochastic Updating with Late-Starting* of Algorithm 1. In the following we shall refer to this approach as Emma-Hatting.

Embedding Mapping Learning with Back-Propagation: It is easy to see that, in combination with Multi-way Neural Networks, the linear mapping between tensor unfoldings and the latent embeddings could also be considered as one more linear activated layer of the network. To this end the Hatting Algorithm could be replaced with the usual Error Back-Propagation. This aspect also applies to other factorization models as long as such a model can be formulated as an NN. In such cases the latent embeddings become the first hidden layer and the tensor unfoldings become the input layer. For such models the latent embeddings are not explicitly learned but are derived from the product of the input vector and the mapping matrices. This aspect corresponds to the compact model described in Eq. (10) and is illustrated in Fig. 2. Similar illustrations could also be found in [18] for RESCAL and the mwNN. Note that for MF and RESCAL there are no parameters other than the mapping matrices to be learned; while the mwNN also learns the weights following the embedding vectors. In the following we shall refer to this approach as Emma-BP.

In summary, the Emma-Post approach optimizes the cost functions in Eqs. (5) and (8) consecutively and separately. Therefore the two error terms of $d_F(\boldsymbol{\mathcal{X}}, \widehat{\boldsymbol{\mathcal{H}}})$ and $d_M(\boldsymbol{A}^{(l)}, \widehat{\boldsymbol{A}}^{(l)})$ are —though minimized to a certain extent— always present. The Emma-Hatting approach also considers these two cost functions. But the LS estimates are calculated more than once and the LS error term —in the long run— is expected to be smaller than that derived from Emma-Post. However, because of this LS update within the factorization algorithm, the gradient approach may become unpredictable with respect to whether the optima identified with LS correction are better than those without the LS update. The stochastic Hatting Algorithm, from this point of view, could be considered as a compromise between the post mapping and the absolute hatting approach. On the one hand it still regulates the factorization algorithm to satisfy the cost function Eq. (8); on the other hand it allows the factorization algorithm to minimize the cost function of Eq. (5) continuously as long as the error of Eq. (8) is relatively small. In other words, this approach enables better factorization quality by tolerating some acceptable mapping error. By omitting

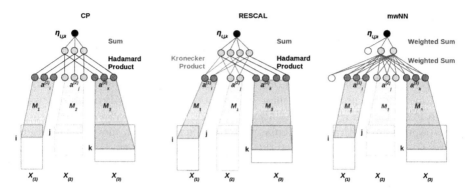

Fig. 2. Illustrating the Compact Model as NNs in 3-D case. Here the rows indexed by i, j and k in the tensor unfoldings $\boldsymbol{X}_{(1)}$, $\boldsymbol{X}_{(2)}$ and $\boldsymbol{X}_{(3)}$ correspond to the three coloured slices in Fig. 1, respectively. Note that for RESCAL there are only two mapping matrices instead of three since the entity embeddings are shared between subject and object. (Color figure online)

the explicit mapping error, the Emma-BP approach models the factorization and mapping as a whole only in terms of the factorization error of Eq. (11).

4 Experiments

In this section we present experiments on three real-world datasets and evaluate the results from two different aspects. First we evaluate our models on a user-item matrix and a KG dataset (both binary) in terms of prediction quality. Then, with another tensor dataset of real values, we focus on the interpretability of the mapped embeddings. The models were implemented in Keras [6] and its backend Theano [1].

4.1 Movielens Data

<u>Data</u>: The Movielens-100K [12] dataset is a user-item matrix containing 100000 ratings from 943 users on 1682 movies. The fact that a rating was performed on an item is encoded as a 1, otherwise we use a 0. Such binary-item matrix could be considered as a special case of a KG with two types of entities and one type of relation.

<u>Task</u>: The major task of a conventional recommender system is to predict the probability of a known user being interested in an known item, as long as event has not been observed in the past. (In terms of a KG this is equivalent to link prediction for one relation between two entities.) In contrast, we intend to predict the probabilities for a *new user* being interested in known items; or vice versa: for a new item to be of interest for known users.

<u>Settings</u>: First, we sample 20 % of users to hold out for test and train our models using the remaining 80 % with embedding sizes of 20 and 50. In the test phase,

we mask a sequence of proportions $[0, 0.1, 0.2, 0.3, 0.4, 0.5]$ of all watched movies of each test user and predict a distribution over all known movies, especially the masked ones. We measure the quality of each prediction in terms of NDCG@k [5]. Further we transpose the user-movie matrix and conduct the same experiment, i.e., we add new movies to the database and try to recommend each movie to the most likely user. Other than [11], we test a logistic MF combined with Emma-Post, Emma-Hatting and Emma-BP. As baseline model we perform a most-popular prediction: we calculate the frequencies of each movie for all users in the training set and interpret these as constant recommendation scores for a new user and a new movie, respectively. We repeat this process 5 times with different and mutually exclusive training and test samples in order to derive prediction stability in terms of mean and standard deviations.

Results: The results of the experiments are presented in Fig. 3. The mean and standard deviation of the NDCG@k scores of the three mapping techniques are visualized in corresponding colors. The horizontal axis represents the proportion of masked entries in the test set; the mean and standard deviation of baseline predictions are demonstrated as the horizontal line and the gray bars.

The performances of the models suggest different trends for new-user and new-movie cases. For new users, the predictions made by logistic MF with Emma-Hatting are suboptimal and may even drop below the baseline for larger masking proportions. Emma-Post and Emma-BP offer better and comparable prediction qualities, though the latter remains advantageous even as the masking proportion becomes quite large. Emma-Post, however, cannot even beat the baseline for new appended movies, while Emma-BP perform noticeably well for all possible masking proportions.

4.2   Knowledge Graph Data

Data: In following experiments we test our models on the Freebase dataset [2] as prepared in [17]. We sample entities that contain at least 500 known relations forming a binary tensor of shape $39 \times 115 \times 115$.

Task: Similar to recommender systems, the conventional link prediction in KGs is performed for each triple of known entities and relations [21]. With our Emma models we intend to predict the existence of links between known and new entities, given some observed but incomplete information about this new entity. More specifically, for an existing KG modelled with a binary tensor $\mathcal{X} \in \mathbb{R}^{I \times I \times K}$ we assume a further binary tensor $\mathcal{Z} \in \mathbb{R}^{I' \times I \times K}$ storing a subset of true links between I' new entities and the I known ones. Our task here shall be to predict the unobserved links in $\mathcal{Z}$ based on factorization and mapping models trained only on $\mathcal{X}$.

Settings: In order to also estimate the model's stability we perform a $20\,\%–80\,\%$ Cross-Validation on the data by splitting the entity set into 5 mutually exclusive groups. In each test set we mask and try to recover $20\,\%$ of known links for each entity with two approaches: (1) We map the entities with masked links into

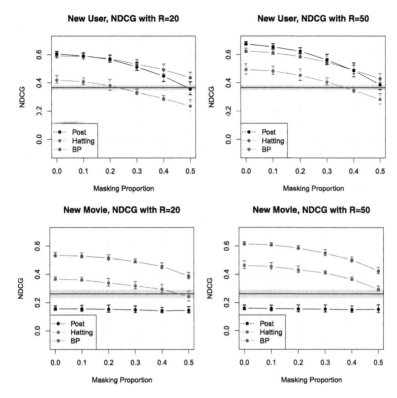

Fig. 3. Evaluation of prediction for new users and movies. The horizontal line and grey bar represent the baseline recommendation and its standard deviation. (Color figure online)

the latent space obtained by the Emma-Post, Emma-Hatting and Emma-BP models that have been trained with the corresponding training set and predict the masked links with the same models; (2) we train a RESCAL and a mwNN model on training *and* test sets, simulating a retraining scenario, and predict the masked links in the conventional way such as in [17]. In both cases we generate negative samples according to the Local-Closed-World-Assumption [17].

<u>Results:</u> In Table 2 we report the prediction quality of AUROC and AUPRC using models of RESCAL and mwNN in combination with all three Emma approaches as well as from retraining. In general, mwNN outperforms RESCAL in terms of larger means and smaller standard deviations in almost all cases, which could also be supported by the results reported in [17]. In predicting for new entities, mwNN combined with Emma-BP yields the best mean values. Especially in terms of AUPRC the advantage could be as large as 33 % compared to second best result produced by RESCAL+Emma-Hatting for $R = 10$ and 44 % compared with RESCAL+Emma-BP for $R = 30$. The minimal standard deviations are achieved in 5/8 cases by Emma-Post, though it almost always produces worst mean values in combination with any factorization models. As

Table 2. Prediction Qualities of RESCAL and mwNN in combination with all three mapping approaches on a FreeBase dataset.

Fac.	Mapping	R = 10		R = 30	
		AUROC	AUPRC	AUROC	AUPRC
RESCAL	Retraining	0.901 ± 0.039	0.820 ± 0.059	0.788 ± 0.054	0.616 ± 0.100
	Emma-Post	0.759 ± 0.096	0.600 ± 0.145	$0.693 \pm \mathbf{0.065}$	$0.432 \pm \mathbf{0.106}$
	Emma-Hatting	$\mathbf{0.778} \pm 0.112$	$\mathbf{0.605} \pm 0.152$	0.700 ± 0.091	0.481 ± 0.120
	Emma-BP	$0.700 \pm \mathbf{0.060}$	$0.485 \pm \mathbf{0.092}$	$\mathbf{0.740} \pm 0.090$	$\mathbf{0.509} \pm 0.134$
mwNN	Retraining	0.964 ± 0.008	0.886 ± 0.060	0.970 ± 0.010	0.923 ± 0.017
	Emma-Post	$0.844 \pm \mathbf{0.009}$	0.390 ± 0.101	$0.826 \pm \mathbf{0.035}$	$0.382 \pm \mathbf{0.063}$
	Emma-Hatting	0.847 ± 0.042	0.423 ± 0.118	0.843 ± 0.038	0.394 ± 0.081
	Emma-BP	$\mathbf{0.949} \pm 0.022$	$\mathbf{0.805} \pm 0.080$	$\mathbf{0.931} \pm 0.036$	$\mathbf{0.735} \pm 0.101$

expected, retraining always offers better predictions than Emma approaches. But do note that a prediction with an Emma model does not cost any run time; while a retraining process for one or multiple entities would demand a comparable amount of time as training an Emma model from scratch. Interpreting the retraining predictions as upper bound, it should also be noted that the combination of mwNN and Emma-BP achieves in most cases results relatively close to those of retraining. We speculate that such canonical model-algorithm combination might enjoy numerical advantage.

4.3   Amino Acid Data

With previous experiments we have shown that Emma models are able to predict links between every known entity and a newly appended entity based on incomplete information. With the following experiment we also show that Emma models can map a new entity into the latent space with high interpretability — here in terms of correlation coefficients.

<u>Data</u>: The Amino Acid Dataset [4] contains a three-way tensor $\mathcal{X} \in \mathbb{R}^{5 \times 51 \times 201}$ and a matrix $\boldsymbol{Y} \in \mathbb{R}^{5 \times 3}$. The latter one describes the proportion of 3 types of amino acid mixed according to 5 different recipes. The corresponding 5 samples are then measured by fluorescence with excitation 250–300 nm, emission 250–450 nm on a spectrofluorometer and the measurements are recorded in the tensor $\mathcal{X}$. With a CP factorization producing matrices of dimensions $\boldsymbol{A}^{(1)} \in \mathbb{R}^{5 \times 3}$, $\boldsymbol{A}^{(2)} \in \mathbb{R}^{201 \times 3}$ and $\boldsymbol{A}^{(3)} \in \mathbb{R}^{61 \times 3}$ it is expected that each column in $\boldsymbol{A}^{(1)}$ would strongly correlate with one column in the recipe matrix $\boldsymbol{Y}$. Note that the order of the columns in $\boldsymbol{A}^{(1)}$ is arbitrary and may not correspond to the column in $\boldsymbol{Y}$ at the same position. For more details please refer to [4].

<u>Task</u>: The latent embeddings learned from this data set are expected to be interpretable in terms of correlations with known recipes. If a new entity is correctly mapped into the latent space, its *mapped* embedding(s) along with other *learned* embeddings would also correlate with the corresponding column

in the recipe matrix. Here we assume the information observed on the new entity to be complete and do not perform any masking.

Settings: We remove each slice $X_{i,\bullet,\bullet}, i \in [1,5]$ (corresponding to a certain recipe) from the tensor and calculate CP factorizations with Emma-Post and Emma-Hatting based on the rest of the data. The slice held out is then mapped to the 3-D latent space with mapping matrix and appended to the learned embeddings of the other slices. We then calculate the mean of its column-wise Pearson correlation coefficients with Y. Further we conducted the same experiment with two slices removed at a time.

Results: With the first leaving-one-out experiment setting we derive accordingly 5 averaged correlations and for the second leaving-two-out setting we have $\frac{5!}{3! \cdot 2!} = 10$ values. We report that the means and standard deviations of the correlation coefficients derived from Emma-Post and Emma-Hatting to be both 0.999 ± 0.001. As for leaving-two-out experiments, Emma-Post achieves 0.991 ± 0.007 and Emma-Hatting yields the same mean value but a larger standard deviation of 0.015. To this end we conclude that both mapping approaches can map one or two new slices into the latent space in a way that its embedding(s) —along with other embeddings learned from factorization— correlates column-wise with the recipe matrix Y with a high correlation score. As expected, in leaving-two-out experiments the correlation coefficients decrease slightly due to the smaller training set. Note that once again Emma-Post seems to produce smaller standard deviations just as the KG experiments.

5   Related Works

[10] introduced a method to map user attributes (referred to as 'content information' in the context of content filtering) into the latent embedding space to solve cold-start problem for new entities. Despite the fact that we are not considering the cold-start problem and do not require content information, this model still shares a few aspects with ours: (1) In both approaches, the codomain of the mapping function is the latent space learned by factorization models with the purpose of finding latent embeddings for new entities; (2) both approaches use modular learning that can be combined with a variety of existing factorization models. An important difference is the domain of the mapping function. In our case it is the observable feature space of an entity, while in [10], it is the user attribute space. It would be interesting, though, to combine our models with such content information: In the first step, one could perform content filtering to produce some first recommendations. Secondly, one could interpret these recommendations as incomplete and enrich them using Emma models by mapping them into the latent space and perform link predictions.

Our proposed Emma-BP model shares the idea of learning the factorization jointly with an implicit latent embeddings with the Temporal Latent Embedding (TLEM) Model [9], where a concatenated NN is trained with observable features as inputs, which are mapped to some implicit latent features. In TLEM one intends to model the *consecutive* effect of a sequence of events on the next one; while we are interested in the *collaborative* effect among entities.

6   Conclusions

The major contribution of this paper is to propose three approaches for mapping new entities into the latent embedding space learned by a wide variety of factorization models.

Our approaches are not based on retraining while obtaining comparable quality to model retraining. Our framework describes factorization and mapping problems on an abstract level, which could inspire the development of further mapping approaches. During our experiments we also realized the model's restrictions in practice: Due to the unfolding operations in the mapping model, the dimensions of the model inputs increase quadratically with the dimensions of the tensor. For instance, in our KG experiment in Sect. 4 with a tensor of shape $39 \times 115 \times 115$, the mapping model requires inputs of dimensions 4485, 4485 and 13225, respectively. As part of future work, we will explore different approaches for dimensionality reduction. In addition, we plan to study non-linear extensions to Emma as well as the application to recurrent NNs.

References

1. Bergstra, J., Breuleux, O., Bastien, F., Lamblin, P., Pascanu, R., Desjardins, G., Turian, J., Warde-Farley, D., Bengio, Y.: Theano: a CPU and GPU math expression compiler. In: Proceedings of the SciPy (2010)
2. Bollacker, K., Evans, C., Paritosh, P., Sturge, T., Taylor, J.: Freebase: a collaboratively created graph database. In: ACM SIGMOD (2008)
3. Bordes, A., Usunier, N., Garcia-Duran, A., Weston, J., Yakhnenko, O.: Translating embeddings for modeling multi-relational data. In: NIPS (2013)
4. Bro, R.: Multi-way analysis in the food industry: models, algorithms, and applications. Ph.D. thesis, Københavns Universitet (1998)
5. Burges, C., Shaked, T., Renshaw, E., Lazier, A., Deeds, M., Hamilton, N., Hullender, G.: Learning to rank using gradient descent. In: Proceedings of the 22nd International Conference on Machine Learning, pp. 89–96. ACM (2005)
6. Chollet, F.: Keras: deep learning library for theano and tensorflow (2015). https://github.com/fchollet/keras
7. Culotta, A., McCallum, A.: Joint deduplication of multiple record types in relational data. In: ACM Information and Knowledge Management (2005)
8. Dong, X., Gabrilovich, E., Heitz, G., Horn, W., Lao, N., Murphy, K., Strohmann, T., Sun, S., Zhang, W.: Knowledge vault: a web-scale approach to probabilistic knowledge fusion. In: ACM SIGKDD (2014)
9. Esteban, C., Schmidt, D., Krompaß, D., Tresp, V.: Predicting sequences of clinical events by using a personalized temporal latent embedding model. In: ICHI (2015)
10. Gantner, Z., Drumond, L., Freudenthaler, C., Rendle, S., Schmidt-Thieme, L.: Learning attribute-to-feature mappings for cold-start recommendations. In: ICDM (2010)

11. Gantner, Z., Rendle, S., Freudenthaler, C., Schmidt-Thieme, L.: Mymedialite: a free recommender system library. In: ACM Conference on Recommender Systems (2011)
12. Herlocker, J.L., Konstan, J.A., Borchers, A., Riedl, J.: An algorithmic framework for performing collaborative filtering. In: ACM SIGIR (1999)
13. Hoaglin, D.C., Welsch, R.E.: The hat matrix in regression and anova. Am. Stat. **32**(1), 17–22 (1978)
14. Kiers, H.A.: Towards a standardized notation and terminology in multiway analysis. J. Chemometr. **14**(3), 105–122 (2000)
15. Kolda, T.G., Bader, B.W.: Tensor decompositions and applications. SIAM Rev. **51**, 455–500 (2009)
16. Koren, Y., Bell, R., Volinsky, C.: Matrix factorization techniques for recommender systems. Computer **42**(8), 30–37 (2009)
17. Krompaß, D., Baier, S., Tresp, V.: Type-constrained representation learning in knowledge graphs. In: ISWC (2015)
18. Nickel, M., Murphy, K., Tresp, V., Gabrilovich, E.: A review of relational machine learning for knowledge graphs. Proc. IEEE **104**(1), 11–33 (2016)
19. Nickel, M., Tresp, V., Kriegel, H.P.: A three-way model for collective learning on multi-relational data. In: ICML (2011)
20. Socher, R., Chen, D., Manning, C.D., Ng, A.: Reasoning with neural tensor networks for knowledge base completion. In: NIPS (2013)
21. Taskar, B., Wong, M.F., Abbeel, P., Koller, D.: Link prediction in relational data. In: NIPS (2003)
22. Tucker, L.R.: Some mathematical notes on three-mode factor analysis. Psychometrika **31**(3), 279–311 (1966)

Comparing Vocabulary Term Recommendations Using Association Rules and Learning to Rank: A User Study

Johann Schaible[1(✉)], Pedro Szekely[2], and Ansgar Scherp[3]

[1] GESIS – Leibniz Institute for the Social Sciences, Cologne, Germany
johann.schaible@gesis.org
[2] Information Sciences Institute, University of Southern California,
Los Angeles, CA, USA
pszekely@isi.edu
[3] ZBW – Leibniz Information Center for Economics, Kiel University, Kiel, Germany
asc@informatik.uni-kiel.de

Abstract. When modeling Linked Open Data (LOD), reusing appropriate vocabulary terms to represent the data is difficult, because there are many vocabularies to choose from. Vocabulary term recommendations could alleviate this situation. We present a user study evaluating a vocabulary term recommendation service that is based on how other data providers have used RDF classes and properties in the LOD cloud. Our study compares the machine learning technique Learning to Rank (L2R), the classical data mining approach Association Rule mining (AR), and a baseline that does not provide any recommendations. Results show that utilizing AR, participants needed less time and less effort to model the data, which in the end resulted in models of better quality.

1 Introduction

Linked Data engineers typically employ Resource Description Framework (RDF) vocabularies to represent data as Linked Open Data (LOD). Reusing vocabulary terms, i.e., RDF types and properties, from existing vocabularies is considered best practice and reduces heterogeneity in data representation [1]. However, this task has challenges: First, even when data engineers are focused on a specific domain, there are many vocabularies to choose from, and second, after choosing a vocabulary, there might be multiple terms in that vocabulary that seem to be correct (considering rdfs:domain and rdfs:range information) for modeling the data in the same way. In detail, data engineers need to select the vocabulary that adequately represents the semantics of the data and maximizes compatibility with existing LOD. For example, when modeling museum data as LOD, one should preferably reuse some vocabulary terms that are also used by other museums for modeling their data. However, this is not trivial. Figure 1(a) shows an example where several vocabularies contain terms that can represent publications and person data, i.e., there are various terms describing that a resource

© Springer International Publishing Switzerland 2016
H. Sack et al. (Eds.): ESWC 2016, LNCS 9678, pp. 214–230, 2016.
DOI: 10.1007/978-3-319-34129-3_14

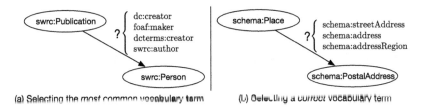

(a) Selecting the most common vocabulary term   (b) Selecting a correct vocabulary term

Fig. 1. *Problems when choosing vocabulary terms for reuse.* In (a) the engineer searches for a property that is most commonly used by other data providers to express the intended semantics of the relationship. In (b) the engineer searches for a property that connects two RDF types in a *correct* way considering the rdfs:domain and rdfs:range information.

of type swrc:Publication has a creator, maker, or author that is a resource of type swrc:Person. Figure 1(b) illustrates an example where multiple alternatives remain even after selecting one vocabulary, i.e., the schema.org vocabulary offers various properties which seem useful to describe that a place has a postal address. Without additional guidance, data engineers may not know which term to select. The examples in Fig. 1 illustrate the difficulties when selecting a property, but the full problem is also to select appropriate terms to represent the RDF types of the subject and object of triples in a dataset.

In this paper, we perform a user study comparing two RDF vocabulary term recommendation approaches and a baseline of no recommendations. The recommendation approaches are based on previous work [2], where the recommendations are based on the RDF types and properties used by other data providers in datasets contributed to the LOD cloud. The first approach applies the machine learning method *Learning To Rank* (L2R) to generate a list of recommended vocabulary terms. In addition, we implemented a second approach that calculates the recommendations via the data mining approach *Association Rule* (AR) mining (cf. Sect. 2). In the user study, we asked 20 participants to model three different datasets as LOD (cf. Sect. 3). For one dataset they received recommendations based on AR, for another they received recommendations based on L2R, and for the third dataset the participants received no recommendations. We measured task completion time, the recommendation-acceptance rate (number of times participants chose a recommended term vs. manual search), and the quality of the resulting LOD representation. To assess the quality, we compared the participants' LOD representation to representations generated by five different LOD experts independently, i.e., did the participants choose the same vocabulary terms as the experts. Additionally, we asked the participants to rate the two recommendation approaches and to report their level of satisfaction regarding the recommendations on a 5-point Likert-scale. The main contribution of this paper are the real-life results based on these measures, as there is no gold standard that can be used for an automatic evaluation.

The results show that the recommendations based on AR are effective, whereas the L2R-based recommendations do not seem to satisfy the participants. The task completion times for AR (4:13 min) were significantly faster than the completion times for L2R (5:41 min) and the baseline (5:26 min). Users accepted the term recommendations from AR most of the time (4.85 out of 7 times), but not from L2R (2.05 out of 7 times). AR also led to high quality results (75 % of the selected terms were also chosen by the experts), whereas the L2R-based recommendations did not perform as good (58 % of the selected terms were chosen by the experts). The user satisfaction survey resembles similar results and indicates a preference for the AR recommendations (cf. Sect. 4 and the discussion in Sect. 5).

2   Apparatus

To perform a user study evaluating the quality of the recommendations based on L2R and AR, we integrated the recommendation service into the data integration tool Karma [3].[1] This way, participants were able to use a graphical UI in order to explore the recommendations and to model data as LOD.

In the following, we describe Karma in more detail, especially the operations that the participants of the user study need to complete the modeling tasks (cf. Sect. 2.1). Furthermore, we provide insights on how the vocabulary term recommendations are calculated with Learning to Rank (L2R) and explain the Association Rule mining (AR) based approach in detail (cf. Sect. 2.2).

2.1   Modeling Data with Karma

Karma is an interactive tool that enables data engineers to model data sources as Linked Data. It provides a visual interface to help data engineers to incrementally select RDF vocabulary terms and to build a model for their data. Figure 2 shows a screenshot of Karma with a partially specified model. The bottom part shows a table with the data being modeled, and the top part shows a graphical representation of the model. The dark ovals represent classes, and the labeled edges represent properties. The edges connecting classes to the columns in the data source are called Semantic Types, as they specify the semantics of the data in a column. For example, in Fig. 2, the semantic type for the third column labeled imageDescription specifies that the contents of the column are notes of a resource of type ecrm:E38_Image, and is represented using the pair (ecrm:P3_has_Note, ecrm:E38_Image). The semantic type "uri" specifies that the corresponding resource is described via a URI, otherwise it is specified via a blank node. The edges connecting the dark ovals represent object properties specifying the relationships between the instances of the corresponding classes. For example, the edge labeled P138i_has_representation between E22_Man-Made_Object and E38_Image specifies that the image represents the object.

To build models and to refine partially specified models, users can click on the edge labels or on the classes to edit a model. When users click on an edge label, Karma presents commands to let users change the property depicted in

---

[1] http://usc-isi-i2.github.io/karma/, last access 12/12/15.

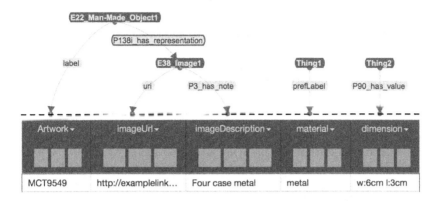

Fig. 2. *Karma User Interface.* The bottom part (below the dotted line) shows the table data being modeled and the top part shows the data's graphical representation including RDF classes (the dark ovals) and properties (the labeled edges). The edges connect classes either to table columns (Semantic Types specifying the semantics of the column data) or to other classes (object properties specifying the relationships between the instances of the corresponding classes)

the label, the source class of the edge, and the destination class of the edge (this last option is only available for object properties). Similarly, when users click on a class, Karma shows commands to change the class depicted in the oval and to add incoming or outgoing edges.

2.2 Vocabulary Term Recommendations

Recommending a vocabulary term means that the recommendation approach suggest the engineer to reuse a specific RDF class or property for describing a data entity or a relationship between data entities. To this end, our approach, which is described in [2] in detail, makes use of so-called *schema-level patterns* (SLPs) that represent the properties commonly used to connect instances of specific classes. The output is a ranked list of recommended vocabulary terms (ordered from most appropriate to least appropriate) which is calculated using the machine learning approach Learning To Rank (L2R) or the data mining approach Association Rule (AR) mining.

In the following, we explain the notion of SLPs as well as provide insights on the recommendations based on L2R and AR. For the interested reader, formalizations of SLPs as well as of the recommendation approach can be found in [2].

Schema-Level Patterns. Schema-level patterns contain three sets of vocabulary terms that describe the connection between two sets of RDF types via a set of properties. For example, the following schema-level pattern specifies that resources of type swrc:Publication are connected to resources of types foaf:Person and swrc:Person via the properties foaf:maker and dc:creator.

$$(\{swrc:Publication\}, \{foaf:maker, dc:creator\}, \{foaf:Person, swrc:Person\}) \qquad (1)$$

We make use of such SLPs in a two-fold way: First an input SLP (provided by the modeling tool such as Karma) is used to describe the input for the recommendation service based on L2R and AR, and second they simulate how other data providers use vocabulary terms to represent their data. The latter can be calculated from existing data sets on the LOD cloud using a straight-forward counting algorithm.

Learning to Rank. Learning to Rank (L2R) refers to a class of supervised machine learning algorithms for inducing a ranking model based on the utilized features [4,5]. We use an L2R algorithm to produce a generalized ranking model that orders vocabulary term recommendation from most appropriate to least appropriate. To this end, we use an SLP that represents a part of the current model as input for the L2R algorithm. For each recommendation candidate x (the candidate is a vocabulary term from the set of all vocabulary terms appearing in data sets on the LOD cloud) the algorithm calculates five features representing the popularity of x, whether the engineer already uses the vocabulary of x, and how many other data sets on the LOD cloud use x in a model that is similar to the engineer's model. In detail, these features are:

Feature 1. Number of datasets on the LOD cloud using the vocabulary term x
Feature 2. Number of datasets on the LOD cloud using the vocabulary of x
Feature 3. Total number of occurrences of vocabulary term x on the LOD cloud
Feature 4. Is term x from a vocabulary that the engineer already uses?
Feature 5. The SLP-feature: Number of SLPs calculated from the LOD cloud that represent the modeled dataset and additionally contain vocabulary term x.

During a training phase, the L2R algorithm uses these features to derive a generalized ranking model. This ranking model can then be used to rank vocabulary term recommendations in new and previously unknown situations. We chose to use the L2R algorithm *Random Forest* [6] for the user study following our prior evaluation (cf. [2]), in which the algorithm achieved the best results (Mean Average Precision (MAP) of vocabulary term recommendations with MAP ≈ 0.8).

Association Rule Mining. Association Rules (ARs) are *if/then* statements that help discover relationships between data in a data repository [7]. We use ARs to identify such relationships between vocabulary terms, i.e., find vocabulary terms that are frequently used together. For example, we can use ARs to find properties that are frequently used as outgoing datatype properties from the RDF class ecrm:E22-Man-Made-Object.

In detail, our data repository for calculating ARs is the set of SLPs calculated from data sets on the LOD cloud, and the quality of the association rules depends on two measures called *support* and *confidence*. These measures depict a threshold for generating an *if/then* statement, i.e., a rule, and for our use-case they specify how often a set of vocabulary terms appear in the set of SLPs calculated from the data on the LOD cloud. Let us assume, we have two sets of different vocabulary terms X and Y. The support specifies how often the union of X and Y occurs in the set of SLPs calculated from the data on the LOD cloud. The confidence denotes how often the union of X and Y occurs in the set of SLPs compared to the number of occurrences of X in the set of SLPs. If both the support and the confidence are higher than a self-defined threshold, then one can induce the rule $X \Rightarrow Y$ specifying "*if the vocabulary terms in X are used, then the vocabulary terms in Y* are used as well". For example, X comprises the vocabulary terms ecrm:E22_Man-Made_Object as well as ecrm:E38_Image and Y comprises the term ecrm:P138i_has_representation. If there are many SLPs containing the object property ecrm:P138i_has_representation together with ecrm:E22_Man-Made_Object and ecrm:E38_Image, then it is very likely that the AR algorithm generates the rule $X \Rightarrow Y$ specifying: If ecrm:E22_Man-Made_Object and ecrm:E38_Image are used together, *then* ecrm:P138i_has_representation is used as well. Formally, the support $Supp$ and the confidence $Conf$ for a rule $X \Rightarrow Y$ are defined as follows:

$$Supp(X,Y) = \frac{frq(X,Y)}{N} \qquad Conf(X,Y) = \frac{frq(X,Y)}{frq(X)} \qquad (2)$$

where frq denotes the frequency how often the arguments occur together and N is total number of SLPs calculated from the data on the LOD cloud. Generally, the higher the manually defined minimum values, the more precise are the rules. But with higher minimum values, various rules will not be considered relevant, and by setting the minimum values too high, it is likely to produce no rules at all [7]. Thus, in order to generate all possible rules, we specified that the support and confidence must solely be greater than zero ($Supp > 0$ and $Conf > 0$).

To generate the rules, we make use of the state of the art *Apriori* algorithm [8]. It is an iterative approach to find frequent sets of vocabulary terms that are used together based on the assumption that a subset of a frequent set of terms must also be a frequent set of terms. Based on this assumption, the algorithm is able to generate a set of frequent SLPs $\{slp_1, \ldots, slp_n\}$ to calculate the association rules. These rules are defined as

$$s_i \rightarrow (slp_i - s_i)$$

where s_i is a non-empty subset of slp_i with $1 \leq i \leq n$. This means that every non-empty subset of slp_i except s_i contains vocabulary terms that can be recommended, if the terms in s_i are provided as input SLP for the recommender.

3 Experiment Setup

3.1 Evaluation Procedure

To compare term recommendations based on L2R and AR, we integrated the recommendation algorithms in Karma. We configured Karma so that the experimenter can select the recommendation approach for a participant and for a modeling task. In our setup, Karma can be configured to offer no recommendation, or recommendations using L2R or AR in its menus for selecting vocabulary terms. We performed a within-subject [9] design user study with 20 participants (cf. Sect. 3.5). The subjects were invited to the lab to conduct some LOD modeling tasks. Each participant first performed a training modeling task to become familiar with the Karma user interface and the menus for selecting vocabulary terms. After completing the training (about 10 min), the participants worked on modeling three different data sets (6 min each) (cf. Sect. 3.2). We applied a Latin-square design to arrange the tasks and the recommendation approaches, i.e., we configured Karma so that we would be able to select one of the three recommendation conditions for each data set (no recommendation, L2R and AR). The no-recommendation condition is the baseline for comparison. We measured task completion time, the number of times participants chose a term from the recommendations vs. searching for the desired term manually, as well as the quality of the resulting LOD representation. Finally, the subjects filled in a user satisfaction questionnaire (Sect. 3.4).[2]

3.2 Modeling Assignments (Tasks)

In our study, we used three datasets from three different domains, namely music, museum objects, and product offers. We chose disparate domains to reduce the overlap in the classes and properties used to model them, thereby reducing learning effects across the modeling tasks. The music dataset includes information about musicians, their recordings, and their wikipedia page. The dataset on museum objects includes information about artworks, materials, dimensions, and images of them. The product offers dataset comprises information about the seller, location, and offered items.

One concern with the user evaluation is that modeling three full datasets takes a long time. To mitigate this problem, we gave participants partially-defined models, and asked them to complete these models. For example, Fig. 3(a) illustrates the modeling assignment for the music dataset. The dataset contains information about the musician's name, their band's name, albums, wikipedia page, as well as a column containing the URI for each musician. The elements above the table represent a partially-defined model of the data: for example, the Musician column is modeled as the name of a resource of type owl:Thing, whereas the desired model is the name of a resource of type mo:MusicArtist. The instructions to the participants explained that the models are incomplete, and asked them to complete the model:

– replace the owl:Thing classes and the rdfs:label properties, and
– define object properties specifying that a musician is a member of a band, that a musician recorded an album, and that a musician has a wikipedia page.

The instructions for the other datasets are similar. To complete the modeling assignments, participants needed to perform six or seven Karma operations for

---

[2] Accompanying material and the modeling results can be found at https://github.com/WanjaSchaible/termpicker_karmaeval_material, last access 3/6/16.

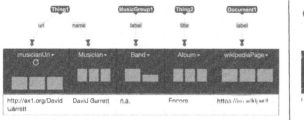

(a) Modeling Task from the Music Domain   (b) Menu with Recommendations

Fig. 3. *Example Modeling Task and Menu with Recommended Terms.* (a) depicts the modeling task from the Music Domain. In order to fulfill the modeling task, the engineers have to: (i) replace the owl:Thing classes and the rdfs:label properties, as well as (ii) define object properties specifying that a musician is a member of a band, that a musician recorded an album, and that a musician has a wikipedia page. (b) illustrates an example menu the participants can use to select a recommended term (bold) or search for another term manually using the search box.

each dataset. When users perform editing operations to change or add properties or classes, Karma uses the pre-configured recommendation approach (including no recommendations) to populate the menus that users see for selecting vocabulary terms (cf. Fig. 3(b)). These menus have two sections, one showing the recommended terms and one showing all the terms (all properties or all classes), if the user clicks on "more". In addition, Karma also provides a search box that filters terms containing the query string.

3.3 Dataset for Recommendations

To compute the vocabulary term recommendations, we calculated the L2R features and the association rules using the BTC 2014 dataset [10]. In order to reduce the memory requirements for SLP computation, we use approximately the first 34 million triples out of the 1.4 billion triples in BTC 2014. Our subset includes data from 3,493 pay-level domains[3], including dbpedia.org and bbc.co.uk. In total, it contains about 5.5 million distinct RDF classes and properties from 1,530 different vocabularies, which were sufficient to calculate 227,010 different SLPs. In order to adequately support the user evaluation, we verified that the 34 million triples covered all the types and properties needed to model the datasets in the evaluation.

---

[3] A pay-level domain (PLD) is a sub-domain of a top-level domain, such as *.org* or *.com*, or of a second-level country domain, such as *.de* or *.uk*. To calculate the PLD, we use the Google guava library: https://code.google.com/p/guava-libraries/, last access 12/12/15.

3.4   Measurements

To evaluate the recommendation approaches, we measured the participants' *effort* needed to complete an assignment, as well as the *quality* of the resulting LOD model.

We collected two measures of effort. One measure is the amount of time a participant spent on each modeling task. We limited participants to six minutes for each modeling assignment. The second measure is the recommendation-acceptance rate, i.e., the number of times the participants chose a recommended vocabulary term from the menu of recommendations, as opposed to searching for terms in the menu containing the full list of vocabulary terms. Our expectation is that participants would search the full list of terms when the recommended terms were inadequate, so that faster completion times would indicate adequacy of the recommendations. As each assignment comprises seven operations, choosing a recommended vocabulary term seven times in one assignment is the maximum score, whereas choosing a recommendation zero times is the minimum score.

To measure quality, we assembled a panel of five ontology engineering experts to construct a gold-standard model for each dataset. We sent each of the modeling assignments to each expert, together with a proposed model that we created. The experts provided feedback suggesting additional vocabulary terms to specify a data entity or a relation between data entities, and suggested replacing some of the vocabulary terms we proposed with better ones. Thus, the gold standard consists of a set of vocabulary terms for each modeling operation. A participant's selection of a vocabulary term for a modeling operation is correct, if this vocabulary term is in the set of vocabulary terms defined in the gold standard for that operation. The quality score for each dataset is the fraction of correct modeling operations.

3.5   Participants of the Study

Overall, $n = 20$ participants (5 female) took part in the user study, all working with LOD in academia and 2 also in industry. The participants' profession ranges from master students (2) over research associates (14) to post doctoral researchers (3) and professors (1) with an average age range between 30–35 years. On average the participants have worked for 3.05 years with LOD. However, they rated their own expertise consuming LOD as moderate ($M = 2.8, SD = 1.1$) and publishing LOD as little ($M = 2.1, SD = 1.6$) on a 5-point-Likert scale from 1 (*none at all experienced*) to 5 (*expert*). The same applies to the participants' knowledge about the domains of the data from the three modeling tasks ($M = 2.1, SD = 1.1$ across the three domains). In addition, 13 participants had never used Karma, whereas 7 specified they are experts or have a high knowledge about Karma. Summarizing, we observe that the participants of our study had a moderate knowledge of Linked Open Data, but they were quite inexperienced regarding LOD publishing and the domain of the modeling tasks' data.

We recruited the participants from the Information Integration group at the University of Southern California. These participants were familiar with Karma but not with the recommendation systems being evaluated in this work. We also recruited participants from GESIS - Leibniz Institute for the Social Sciences, who did not use Karma before. The participants were acquired in person or via personal email inviting them to participate within a private surrounding (laboratory or private office space), in which solely the principal investigator was present. Participants signed a consent form prior to their participation and received no compensation for their efforts.

4 Results

We illustrate and discuss the results based on the measurements considering the efficiency (Sect. 4.1), quality (Sect. 4.2), and level of satisfaction (Sect. 4.3).[4]

4.1 Effort

To calculate the average time the participants needed to complete a modeling task, we added up the times from participants who used the same recommendation approach for the modeling task and divided the sum by the number of these participants. For example, if seven out of the 20 participants modeled the music data using AR, we summed up the times these seven participants needed to complete modeling the music data, and divided it by the seven participants. Table 1 illustrates these calculated average times for each of the three data models and the combined total average time.

The results show that AR led to the fastest completion times (4:13 min), over a minute faster than the baseline of having no recommendations (5:28 min) and

Table 1. *Average Task Completion Time in Minutes.* The table shows the average time the participants needed to complete the modeling tasks followed by the standard deviation and the number of participants modeling the data with a recommendation approach

Recommendations	Music domain	Museum domain	Product Offer domain	Average in total
L2R	5:31 (0:41) - 6	5:48 (0:24) - 7	5:42 (0:25) - 7	5:41 (0:30) - 20
AR	4:50 (1:10) - 6	4:37 (1:02) - 7	3:16 (1:02) - 7	4:13 (1:03) - 20
No Rec	5:22 (0:59) - 8	5:28 (0:37) - 6	5:34 (0:42) - 6	5:28 (0:47) - 20

Table 2. *Recommendation-Acceptance Rate.* The table shows the average number of times the participants selected a recommended vocabulary term out of the number of operations and the standard deviation within the brackets.

Recommendations	Music domain	Museum domain	Product Offer domain	Average in total
L2R	2.00/7 (.89)	2.37/7 (.91)	1.67/6 (.81)	2.05/6.67 (.88)
AR	4.33/7 (.51)	4.85/7 (.69)	5.28/6 (.95)	4.85/6.67 (.95)

---

[4] Research data is available at http://dx.doi.org/10.7802/1206, last access 3/3/16.

about 90 s faster than using recommendations based on L2R (5:41 min). It is surprising that users took less time to complete tasks with the baseline system than with the L2R recommendations. The survey reveals participants were dissatisfied with the L2R recommendations (cf. Sect. 4.3). The results suggest that participants spent time evaluating the L2R recommendations (they used the L2R recommendations for about 2 out of 7 modeling operations), and proceeded to search for terms as they would have done in the baseline system, thus taking more time overall. The L2R recommendations use multiple features, including the popularity of a term and whether or not it is from an already used vocabulary. Thus, the recommendations sometimes include terms that are popular or from a commonly used vocabulary, but that do not fit the desired semantics. For example, instead of suggesting the datatype property foaf:name to specify the name of a person, it suggests the property foaf:knows. This occurred regardless of the domain of the data. Applying a Friedman test, we could show that the differences between the task completion time is significant ($\mathcal{X}^2 = 15.083$, $p = .001$) However, the Wilcoxon signed-rank test (with a Bonferroni correction applied) shows that recommendations based on L2R and having no recommendations at all is not significant ($Z = -1.256$, **n.s.**, $p = .209$), whereas having recommendations based on AR is significantly better (compared to L2R: $Z = -3.220$, $p = .001$; compared to the baseline of no recommendations: $Z - -3.018$, $p = .003$).

The average number of selected recommendations, which is displayed in Table 2, underpins the observations from Table 1. The participants needed more time to complete an assignment while given L2R recommendations, and did not use the recommended terms very often (on average 2.05 times out of 6.67). Compared to the number of selected recommendations based on AR (on average 4.85 selected recommendations out of 6.67 possibilities) it is significantly lower ($Z = -3.848, p < .001$).

4.2 Quality

The modeling tasks had either six or seven operations. Table 3 shows the average number of correctly selected vocabulary terms these six or seven operations. One can observe that the recommendations based on Association Rules (in total 4.97/6.67) helped the participants to select more correct vocabulary terms than the recommendations based on L2R (3.88/6.67) or not using any recommendations at all (3.57/6.67). This especially applies to modeling the data from the Product Offers domain (5.25/6), which uses mainly the schema.org vocabulary.

When searching for an appropriate vocabulary term, participants tried to find a term that matched the description in the assignment. For example, a task from the Product Offers domain states "Find a better data type property specifying that the table column contains the price of an item". The participants were instantly looking for a term that contained the string "price" in its specification, such as hasPrice. Even if the recommendations did not contain such a term, the participants easily found the property schema:price using Karma's string search and incorporated it into the model. However, not every assignment was as easy. Another example from the task with the Product Offers data states "Find a

Table 3. *Quality of the Resulting RDF Representation.* The table shows the average number of vocabulary terms that were also chosen by the LOD experts out of the number of operations and the standard deviation within the brackets.

Recommendations	Music domain	Museum domain	Product Offers domain	Average in total
L2R	4.02/7 (0.64)	3.75/7 (1.05)	3.87/6 (0.83)	3.88/6.67 (.14)
AR	4.87/7 (0.88)	4.75/7 (1.03)	5.25/6 (0.64)	4.95/6.67 (.26)
No Rec	3.84/7 (0.83)	3.37/7 (0.91)	3.52/6 (0.70)	3.57/6.67 (.24)

better RDF class specifying that the table column contains the address of a place". Only in the conditions with AR recommendations, participants were able to choose the correct vocabulary term (schema:PostalAddress). Recommendations based on L2R did not provide such good results. The quality of the resulting data is almost the same as modeling the data without any recommendations. A Friedman test shows that the differences are significant ($\mathcal{X}^2 = 13.125, p = .001$), but only the ARs seem to provide significantly better recommendations ($Z = -3.384, p = .001$).

4.3   Satisfaction

The satisfaction of the recommendation approaches was obtained using a questionnaire. We asked the participants various questions to identify their level of satisfaction considering the recommendations based on AR and L2R. First, we asked the participants to provide their general level of satisfaction on a 5-point Likert scale from "1 - Very Dissatisfied" to "5 - Very Satisfied", with a neutral position "3 - Unsure". They were rather unsure ($M = 3.00$, $SD = 1.1$) whether they were satisfied with the recommendations based on L2R. The recommendations based on AR were considered rather satisfying ($M = 4.23$, $SD = 0.7$). A Friedman test showed that this difference is significant ($\mathcal{X}^2 = 6.231, p = .013$).

In addition, we asked the participants to compare both recommendation approaches to each other. First, they were asked to compare the AR-based recommendation approach on a 5-point Likert scale from "1 - much worse than L2R" to "5 - much better than L2R". A total of 19 participant stated that the AR recommendation are either somewhat better or much better than the L2R recommendations (10 participants stated "much better than L2R" and only one stated "3 - About the same"). These results are supported by our second question to rank the two recommendation approaches (and the no recommendations baseline) from "best" to "worst" considering the help for reusing vocabulary terms. Recommendations based on AR were unanimously considered to be most helpful (each participants ranked AR at the first position). A Friedman showed that the difference in the ranking positions is significant ($\mathcal{X}^2 = 32.500, p < .001$). A post-hoc analysis using the Wilcoxon signed-rank test with Bonferroni correction illustrates that the difference between the recommendation approaches AR and L2R and between the AR-based approach and no recommendations are significant ($Z = -4.134, p < .001$ and $Z = -4.251, p < .001$, respectively).

However, the differences between getting recommendation based on L2R and getting no recommendations at all are not significant ($Z = -2.236$, **n.s.**, $p = .025$).

5   Discussion

Discussion of Results. The results indicate a quite clear picture: vocabulary term recommendations based on Association Rule (AR) mining are perceived to aid the participants more in reusing vocabulary terms than the recommendations based on the machine learning approach Learning To Rank (L2R). Both approaches use the SLPs calculated from datasets on the LOD cloud, in order to identify how other data providers model their data. However, L2R uses additional features describing the popularity of a recommendation candidate and whether it is from an already used vocabulary. Based on all these features, the list of recommendations can also contain very popular terms from a vocabulary that is already used, but that are not used by others in a similar model. Such recommendations are very likely irrelevant, and if the engineer is not sure whether the recommended terms are appropriate for reuse, she is more likely to overlook the relevant recommendations. This was also observed in the user study. In about 70 % of all assignments across the three modeling tasks, L2R recommended a correct vocabulary term at the fifth or tenth position of the list of all recommendations. The participants however either skipped it, or were not sure whether to select it, as the list of recommendations contained various terms at a better rank that seemed rather inappropriate. This uncertainty stopped most participants from selecting the correct vocabulary term from the recommendation list. On the contrary, AR does not rely on the popularity of a vocabulary term, or whether it is from an already used vocabulary. It solely exploits the information which terms other data sets on the LOD cloud used in conjunction with the terms in the modeling task. We observed that the recommendation lists generated by AR supported the participants select the correct terms. Even when participants were not aware of the existence of the correct term, they felt confident to select a term from the list. For example, the assignment to model an outgoing object property from mo:MusicArtist to mo:MusicGroup from the modeling task on music data, most participants were not aware of the existence of the property mo:member_of. Having no recommendations, they searched for "part_of" or "is_member", but with recommendation based on AR, they saw this property very quickly and selected it.

In summary, most participants stated that the L2R recommendations were useful in general, but they were unsure whether the recommended terms were appropriate for reuse. Recommendations based on AR alleviated this situation and supported the participants in selecting the correct terms. Therefore, Association Rules seem to provide the better vocabulary term recommendations, and they can be seen as the better option to reduce the heterogeneity in data representation on the LOD cloud for similar data.

Thread to Validity. Our user study was based on a within-subject design. The advantage is that it requires fewer participants. However, we needed to design three different modeling tasks that contained data from different domains, but were approximately equally difficult to model. One could argue that the modeling tasks we chose did not have the same complexity, and that the participants were able to model one task with more ease than the others. That is why we constantly changed which recommendation approach is used with which modeling task (Latin-square), and the results demonstrate that the AR recommendations performed best for all of the three tasks.

Another disadvantage of within-subject design is the problem of carry-over effects, where the first modeling task adversely influences the other. First, participants who were not familiar with Karma might have needed a longer time to complete the first modeling task, as they needed to get accustomed to Karma, but then the practice effect might make them more confident for the second and third modeling task. To address this issue, we let the participants first model a data set from the publications domain for practicing to use Karma and its menus containing the recommended vocabulary terms. The user study started only, as soon as the participants felt ready to begin with the actual modeling tasks. In addition, we changed the order of the modeling tasks and the recommendation approach for a task (based on Latin-square), such that no tasks or none of the two recommendation approaches would benefit from the carry-over effects. Second, the participants might get tired during the user study, such that it could decrease their performance on the last modeling task. To address this problem, we limited the time for each modeling task in the user study to six minutes. This way, each participant was able to complete all three modeling tasks within 18 min, such that it is less likely that their performance decreased during the last modeling task.

6   Related Work

To the best of our knowledge, no comparable user study has been conducted yet. Thus, we discuss the related word on other vocabulary term recommendation approaches that aid the engineer in reusing vocabulary terms.

Existing services that recommend RDF types and properties are generally based on syntactic and semantic similarity measures. Prominent examples are CORE (short for: Collaborative Ontology Reuse and Evaluation) [11] and the Watson [12] plug-in for the NeOn ontology engineering toolkit.[5] CORE determines a ranked list of domain-specific ontologies considerable for reuse whereas the Watson plug-in uses semantic information from a number of ontologies and other semantic documents published on the Web to recommend appropriate vocabulary terms. Whereas these services provide recommendations based on string analyses (as the input is a string value), they do not exploit any structural information on how vocabulary terms are connected to each other. In contrast,

---

[5] http://www.neon-project.org/, last access 12/18/15.

Falcons' Ontology Search [13] provides the engineer with such information. However, it is mainly designed to establish a general relatedness between vocabularies specifying that different vocabularies contain terms that describe similar data.

The "data2Ontology" module of the Datalift platform [14] provides suggestions to match data entities to a vocabulary term based linguistic proximity between the data entity and the vocabulary term as well as on the quality of the vocabulary using criteria from LOV [15]. LOV itself also offers a SPARQL endpoint that can be used to search for vocabulary terms that are equivalent or a sub-class-of another vocabulary term, or for properties that have a specific RDF type as rdfs:domain and/or rdfs:range. But such information needs to be encoded in the vocabulary specification, whereas using SLPs it can be directly derived from datasets on the LOD cloud. Karma itself also comprises two different types to recommend vocabulary terms. Recently, Karma comprises new algorithms that use previously created models as well as the axioms in the ontology to automate model construction [16,17]. A key difference between that work and the work presented in this paper is that the work presented in [16,17] learns from previous Karma models, while our recommendation approach learns from LOD data, irrespective of how it was created. The latest Karma work investigates algorithms to exploit patterns in the LOD cloud to automate model creation [18]. An important difference to our approach is that the data engineer first selects a set of ontologies for modeling the data. Karma then searches the LOD cloud for patterns of how the classes and properties in the selected ontologies are used in published LOD cloud, and then attempts to combine the patterns to construct a complete model for the source. The work presented in this paper does not require data engineers to select the ontologies, and its focus is on presenting recommendations interactively rather than on computing a model automatically.

7   Conclusion

This paper presented a user study comparing vocabulary term recommendations using Association Rules (AR) mining vs. using Learning To Rank (L2R). We illustrated the experiment set-up, which had the outcome that recommendations based on AR perform best. The participants were able to complete the modeling tasks with less effort, the quality of the resulting RDF representation was higher than using L2R recommendation or no recommendations, and the user satisfaction survey showed a clear favor towards the AR recommendations.

As future work, we plan to extend the implementation of the recommendation service into Karma and combine the Karma recommendation algorithms that reason about the axioms in the ontology together with the AR recommendations that don't require data engineers to pre-select a set of ontologies for modeling sources.

Acknowledgements. We would like to thank Laura Hollink, Benjamin Zapilko, Ruben Verborgh, Jérôme Euzenat, and Oscar Corcho for providing the essential contribution to generate the gold standard RDF representation of the assignment datasets.

We also thank Malte Knauf and Thomas Gottron for providing the basis of our Association Rule generation algorithm for calculating the vocabulary term recommendations. Naturally, we also thank the participants of the user study for helping us with our research.

References

1. Heath, T., Bizer, C.: Linked Data: Evolving the Web into a Global Data Space. Synthesis Lectures on the Semantic Web. Morgan & Claypool Publishers, San Rafael (2011)
2. Schaible, J., Gottron, T., Scherp, A.: TermPicker: enabling the reuse of vocabulary terms by exploiting data from the linked open data cloud - an extended technical report. ArXiv e-prints, December 2015
3. Knoblock, C.A., Szekely, P., Ambite, J.L., Goel, A., Gupta, S., Lerman, K., Muslea, M., Taheriyan, M., Mallick, P.: Semi-automatically mapping structured sources into the semantic web. In: Simperl, E., Cimiano, P., Polleres, A., Corcho, O., Presutti, V. (eds.) ESWC 2012. LNCS, vol. 7295, pp. 375–390. Springer, Heidelberg (2012)
4. Liu, T.Y.: Learning to rank for information retrieval. Found. Trends Inf. Retr. **3**(3), 225–331 (2009)
5. Hang, L.: A short introduction to learning to rank. IEICE Trans. Inf. Syst. **94**(10), 1854–1862 (2011)
6. Breiman, L.: Random forests. Mach. Learn. **45**(1), 5–32 (2001)
7. Zhang, C., Zhang, S. (eds.): Association Rule Mining: Models and Algorithms. LNCS (LNAI), vol. 2307. Springer, Heidelberg (2002)
8. Agrawal, R., Imieliński, T., Swami, A.: Mining association rules between sets of items in large databases. In: ACM SIGMOD Record, vol. 22, pp. 207–216. ACM (1993)
9. Charness, G., Gneezy, U., Kuhn, M.A.: Experimental methods: between-subject and within-subject design. J. Econ. Behav. Organ. **81**(1), 1–8 (2012)
10. Käfer, T., Harth, A.: Billion Triples Challenge data set (2014). http://km.aifb.kit.edu/projects/btc-2014/
11. Fernandez, M., Cantador, I., Castells, P.: Core: a tool for collaborative ontology reuse and evaluation. In: 4th International Workshop on Evaluation of Ontologies for the Web (2006)
12. d'Aquin, M., Baldassarre, C., Gridinoc, L., Sabou, M., Angeletou, S., Motta, E.: Watson: supporting next generation semantic web applications. In: Proceedings of the IADIS International Conference WWW/Internet 2007, pp. 363–371 (2007)
13. Cheng, G., Gong, S., Qu, Y.: An empirical study of vocabulary relatedness and its application to recommender systems. In: Aroyo, L., Welty, C., Alani, H., Taylor, J., Bernstein, A., Kagal, L., Noy, N., Blomqvist, E. (eds.) ISWC 2011, Part I. LNCS, vol. 7031, pp. 98–113. Springer, Heidelberg (2011)
14. Scharffe, F., Atemezing, G., Troncy, R., Gandon, F., et al.: Enabling linked-data publication with the datalift platform. In: AAAI 2012, 26th Conference on Artificial Intelligence - Semantic Cities. (2012)
15. Vandenbussche, P.Y., Atemezing, G.A., Poveda-Villalón, M., Vatant, B.: Linked Open Vocabularies (LOV): a gateway to reusable semantic vocabularies on the Web. Semant. Web J. (to appear). http://www.semantic-web-journal.net/

16. Taheriyan, M., Knoblock, C.A., Szekely, P., Ambite, J.L.: A graph-based approach to learn semantic descriptions of data sources. In: Alani, H., Kagal, L., Fokoue, A., Groth, P., Biemann, C., Parreira, J.X., Aroyo, L., Noy, N., Welty, C., Janowicz, K. (eds.) ISWC 2013, Part I. LNCS, vol. 8218, pp. 607–623. Springer, Heidelberg (2013)
17. Taheriyan, M., Knoblock, C., Szekely, P., Ambite, J.L., et al.: A scalable approach to learn semantic models of structured sources. In: 2014 IEEE International Conference on Semantic Computing (ICSC), pp. 183–190. IEEE (2014)
18. Taheriyan, M., Knoblock, C.A., Szekely, P., Ambite, J.L., Chen, Y.: Leveraging linked data to infer semantic relations within structured sources. In: Proceedings of the 6th International Workshop on Consuming Linked Data (COLD 2015) (2015)

Mobile Web, Sensors and Semantic Streams Track

Full-Text Support for Publish/Subscribe Ontology Systems

Lefteris Zervakis[1(✉)], Christos Tryfonopoulos[1(✉)], Spiros Skiadopoulos[1], and Manolis Koubarakis[2]

[1] Department of Informatics and Telecommunications,
University of Peloponnese, Tripolis, Greece
{zervakis,trifon,spiros}@uop.gr
[2] Department of Informatics and Telecommunications,
Univeristy of Athens, Athens, Greece
koubarak@di.uoa.gr

Abstract. In this work, we envision a publish/subscribe ontology system that is able to index large numbers of expressive continuous queries and filter them against RDF data that arrive in a streaming fashion. To this end, we propose a SPARQL extension that supports the creation of full-text continuous queries and propose a family of main-memory query indexing algorithms which perform matching at low complexity and minimal filtering time. We experimentally compare our approach against a state-of-the-art competitor (extended to handle indexing of full-text queries) both on structural and full-text tasks using real-world data. Our approach proves two orders of magnitude faster than the competitor in all types of filtering tasks.

1 Introduction

As the Web is growing continuously, a great amount of data is available to users, making it more difficult for them to discover interesting information by searching. For this reason, publish/subscribe (pub/sub) systems have emerged as a promising paradigm that enables the user to cope with the high rate of information production and avoid the cognitive overload of repeated searches. In a pub/sub system, users (or services that act on users' behalf) express their interests by submitting a continuous query and wait to be notified whenever a new event of interest occurs. The vast majority of modern pub/sub services and systems are typically content-based (contrary to previous decades, where they used to be topic/channel based); subscribers express their interest on the content of the publication (be it structure or data/text values) by appropriately specifying constraints in the submitted continuous queries.

In the early days of content-based pub/sub the structure of a publication was nothing more than a (usually static) collection of named attributes with

L. Zervakis and S. Skiadopoulos have been partly supported by Greek national funds through the EICOS project (Research Funding Program Thales, Operational Program "Education and Lifelong Learning", National Strategic Reference Framework.

© Springer International Publishing Switzerland 2016
H. Sack et al. (Eds.): ESWC 2016, LNCS 9678, pp. 233–249, 2016.
DOI: 10.1007/978-3-319-34129-3_15

values of different types (e.g., text) [19,21]. As XML gained popularity and started becoming the standard for data/information representation and exchange on the web, various XML-based pub/sub systems have, naturally, arised [3,6–8,14]. In those systems, publications were expressed in XML and extensions of XPath/XQuery were used to express continuous queries. All research in the field focused mainly on the structural/value matching between (indexed) continuous queries and incoming publications, but has largely ignored semantics. This gave rise to ontology-based pub/sub systems [12,15,16,20] that typically used RDF [18] for representing publications and SPARQL [17] extensions/modifications for expressing user interests through continuous queries.

Ontology-based pub/sub systems research [12,15,16,20] has naturally focused more on semantics and has delivered interesting results. What it currently lacks, though, compared to the technological arsenal of the traditional pub/sub research is the support of a complete full-text retrieval mechanism, beyond existing regular expression and equality support, with sophisticated algorithms and data structures to minimise processing and memory requirements.

In this work, we initially propose an *extension* of SPARQL with full-text operators, aiming at more expressive continuous queries that are able to support versatile user needs in applications like digital libraries or news filtering. To preserve the expressivity of SPARQL, we view the full-text operations as an additional filter of the query variables. In our setup, publications are ontology data that contain RDF literals in their property elements. A full-text expression is evaluated against a literal, and supported expressions involve the usual Boolean operators (i.e., conjunction, disjunction, negation), as well as word proximity and phrase matching. To efficiently filter the incoming publications against the stored queries, we present RTF (acronym for RDF Text Filtering), a family of trie-based, main-memory, (continuous) query indexing algorithms that support SPARQL queries with full-text constraints and are able to filter incoming publications in a few milliseconds. We propose indexing methods that exploit the commonalities between continuous queries at indexing time and leverage on the natural properties of RDF during the filtering procedure. To the best of our knowledge, our family of algorithms is the first in the literature that is able to support SPARQL queries with full-text constraints. To demonstrate the efficiency of our approach we extend iBROKER [12], a state-of-the-art query indexing and RDF publication filtering algorithm, with full-text capabilities and compare it against our approach both on structural and full-text filtering tasks; our approach proves more than two orders of magnitude faster for the structural and more than one order of magnitude faster for the full-text filtering tasks.

In the light of the above, our contributions are:

- We *extend* SPARQL with full-text operators and support Boolean, word proximity, and phrase matching operators.
- We *develop* a family of continuous query indexing algorithms that support full-text SPARQL queries and are able to filter the incoming RDF publications efficiently.

- We *extend* iBROKER [12], a third party algorithm for ontology pub/sub, to offer full-text support and use it as a state-of-the-art competitor.
- We *identify* algorithmic alternatives for query indexing and assess their performance with a real-word data set against the extended version of iBROKER.

The rest of the work is organised as follows. Section 2 presents an overview of our data and query model, while Sect. 3 introduces the RTF family of algorithms and outlines the competitor extensions. Subsequently, Sect. 4 presents the experimental evaluation of the developed algorithms with a real-world data set. Finally, Sect. 5 presents related research in ontology pub/sub systems, and Sect. 6 concludes the paper.

2 Query and Data Model

RDF constitutes a conceptual model and a formal language for representing resources in the Semantic Web; it is the building block of a metadata layer on top of the current structured information layer of the *World Wide Web*, which enables interoperability between different systems and facilitates the exchange of machine-understandable information. The SPARQL query language is currently the W3C recommendation for querying the Semantic Web; the graph model, over which it operates, naturally joins data together and supports several query forms for querying RDF datasets. However, it still lacks the support of a complete full-text mechanism for filtering purposes. Since we focus our attention on full-text filtering of ontology data we are interested only in property elements with a plain RDF literal as their content. In this context, the subject of an RDF triple is always a node element and the predicate denotes the relation to the literal. The object is the literal, which is expressed as a string.

In the spirit of [2], we propose an extension to the SPARQL syntax to support full-text continuous queries in RDF datasets. To preserve SPARQL expressibility we view the full-text operations as an additional filter of the (continuous) query variables. In this context, we define a new binary operator *ftcontains* (full-text contains), that takes as input a *variable* of the continuous SPARQL query and a *full-text expression* that operates on the values of this variable. The query signature of the operator is expressed as the function $xsd : boolean : ftcontains(var, ft_{expression})$. A full-text expression is evaluated only against a literal, so *var* is always the object of the SPARQL tuple pattern; the subject and/or predicate of the tuple pattern may be constants. The expressions supported involve the usual Boolean operators (denoted by ftAND, ftOR, etc.), as well as proximity (denoted by ftNEAR) and phrase matching as in [4]. To this end, we carefully designed a new set of full-text queries which currently can not be efficiently evaluated by existing pub/sub ontology systems.

The example SPARQL continuous query in Fig. 1 will match all publications that are of type *article* and have an attribute *title* with a string literal. The title of the publications must contain the terms "olympic" and "games". Additionally, the publications that match must have an attribute *body* that contains the terms

```
1 SELECT ?publication
2 WHERE {?publication type article.
3          ?publication title ?title.
4          ?publication body ?body.
5 FILTER ftcontains(?title, "olympic" ftAND "games")
6 FILTER ftcontains(?body, "olympic" ftAND "games" ftNEAR[0,2] "rio")
7 }
```

Fig. 1. An example SPARQL query with the proposed extended syntax.

"olympic", "games" and "rio", and the term "rio" is at least 0 and at most 2 words after the term "games" (due to the word proximity constraint).

In addition to the full-text extension of SPARQL we also support the *wildcard (*)* operator applied in RDF triples, i.e., queries where the subject, predicate and/or object of a triple may match any value of the publication. Such a combination of full-text and wildcard operations allows us to offer to users a rich set of tools that allow them to specify expressive continuous queries that will match their information needs. An example query of this type could be derived by substituting line 2 of Fig. 1 with "WHERE {? publication type *.".

A publication, in this context, is represented as a set of RDF triples containing additional fields, where needed, to store the text parts. Hence, the underlying model is a directed graph which contains a set of nodes that may serve as the subject or the object in a triple statement and are connected via properties that are expressed as the predicate.

3   Query Indexing Algorithms

In this section, we present RTF, a family of query indexing algorithms that utilise trie structures to exploit commonalities between continuous queries to achieve faster filtering times. Initially, we elaborate on the indexing algorithm RTFM[1], discuss its variation RTFS, and provide details for the common filtering procedure. Finally, we briefly discuss IBROKER, a state-of-the-art competitor that uses an inverted index to store submitted SPARQL queries.

3.1   Algorithm RTFM

Algorithm RTFM is indexing each continuous query by executing the following three steps:

1. Transforming the continuous query to conjunctions of tuples (quadruples or triples depending on the existence of a text constraint or not) and assigning a unique identifier to each tuple.
2. Registering all the discrete tuples produced from the previous step in a table that associates each continuous query with the tuple identifiers it contains.
3. Indexing of all the query tuples at the trie structure described below.

---

[1] No connection to the infamous initialism – https://en.wikipedia.org/wiki/RTFM.

In the following, we analyse each step and provide details on the data structures and algorithms utilised.

Step 1: Tuple Representation. Algorithm RTFM operates on tuples; in this section we define continuous queries as conjunctions of tuples, and, in the following sections, we illustrate how we exploit commonalities between those tuples to achieve better query indexing and thus faster filtering performance.

Definition 1. *We define a continuous query q as a series of $i, i \in 1 \ldots, n$, tuple conjuncts. Each tuple has three mandatory attributes, namely subject (S_i), predicate (P_i) and object (O_i). There is an additional, non-mandatory, attribute $\mathcal{F}_i$ that facilitates the representation of the full-text operators and their textual constraints. Thus, a continuous query may be represented as:*

$$q = t_1(S_1, P_1, O_1, \{\mathcal{F}_1\}) \wedge \cdots \wedge t_n(S_n, P_n, O_n, \{\mathcal{F}_n\})$$

Example 1. By applying Definition 1 to the continuous query q in the example of Fig. 1 we receive the following set of tuples:

```
(?publication, type, article) ∧
(?publication, title, ?title, ftcontains("olympic" ftAND "games")) ∧
(?publication, body, ?body,
                    ftcontains("olympic" ftAND "games" ftNEAR[0,2] "rio"))
```

Moving from a SPARQL query to a tuple-based representation is achieved by appropriate parsing of the continuous query with a tool like Sesame[2].

Step 2: Associating Queries with Tuple Identifiers. Following Step 1, RTFM receives a query q that consists of two fields, a unique *query identifier* and a set of tuples also associated with their unique tuple ids. RTFM proceeds by storing each continuous query along with the tuple identifiers into the Query Table (QT). QT is comprised of two fields: the unique identifier of each query q and a linked list that stores the unique identifiers of the continuous query tuples. For instance, for the continuous query of Fig. 1, RTFM will add three tuple identifiers into QT (as they are shown in the previous step). RTFM proceeds in a similar way to insert every new continuous query that is submitted in QT.

Step 3: Indexing Tuples in the Trie Forest. The *trie forest* is populated in order to store the tuples compactly by exploiting their common elements. Thus, every trie forest consists of a collection of tries, which in turn contain a number of trie nodes; in each node N the following information is stored:

- The node content, denoted by *content(N)*, that may represent either an RDF attribute/variable or a word contained in a text constraint of a query.
- The list of children nodes of N, denoted *children(N)*.
- The list of tuple identifiers, denoted by *tIDs(N)*, that are indexed under N.

---

[2] http://rdf4j.org.

When a new query q arrives RTFM iterates through the set of all query tuples and indexes every tuple in the trie forest. During the indexing phase RTFM searches the trie forest for a suitable place to index each tuple as follows.

The first tuple of the first continuous query that is submitted will naturally arrive in an empty trie forest and will create a number of nodes that depend on the form of the tuple. Specifically, for the structural constraints of the tuple, RTFM creates three new trie nodes one for each attribute specified. If the tuple contains also a full-text constraint with k distinct words, RTFM will create k more nodes (one for each distinct word). For illustration purposes, we use a pseudo-node "FT" to separate the structural from the word constraints and highlight the difference between the different RTF variants.

In general, when inserting a new tuple, RTFM considers storing it at an existing trie or creating a new trie. To insert a new tuple $t(\mathcal{S}, \mathcal{P}, \mathcal{O})$, RTFM examines the subject $\mathcal{S}$ of the tuple and utilises the trie structure to find if there is a candidate trie which has a root node R such that $content(R) = \mathcal{S}$. If such a trie is found, the indexing algorithm proceeds to examine $children(R)$ in order to determine if there is a child C such that $content(C) = \mathcal{P}$. The same applies for the object $\mathcal{O}$ of the tuple. Notice that variables (in subject/predicate/object) and wildcards in tuples are mapped onto the corresponding variable/wildcard nodes. If the new tuple contains full-text constraints, the trie is expanded with the distinct words contained in these tuple constraints in a similar manner.

If, during the indexing phase, RTFM fails to locate an appropriate trie position to store a new tuple, it proceeds in creating a new set of nodes that will index the remaining tuple fields. After locating (or creating) the appropriate trie that will store a tuple t, RTFM stores also the tuple id at $tIDs(N)$ of node N of this trie, so as to be able to identify the tuple at filtering time. Notice that different query insertion order will, naturally, give different tries, since query organisation is greedy, and depends on the already stored queries.

Indexing of proximity formulas and phrases in the trie forest of RTFM is performed in the same way as described above, since proximity is a more constrained case of conjunction. To accommodate the word distance in the proximity/phrase expression, we use an extra data structure that stores the proximity constraints in the spirit of [19]. Disjunctions are handled by creating separate queries (that have the same user as the notification recipient) for the different word operands.

Figure 2 shows the resulting trie after inserting three continuous queries, including the query q of Fig. 1. Additionally, the three tuples of q (shown in Example 1 above) are assigned ids $q.t_1, q.t_2, q.t_3$ respectively. From the indexing performed by RTFM in these queries notice that:

- Query q shares the same tuple ((?publication, type, article)) with query v, as two different tuple identifiers (namely $q.t_1$ and $v.t_1$) are stored in the same leaf node. Moreover, this tuple contains only structural constraints.
- Query q contains tuple $q.t_3$ that specifies both structural and full-text constraints.

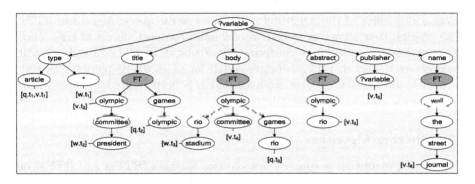

Fig. 2. Trie forest during the indexing phase of RTFM.

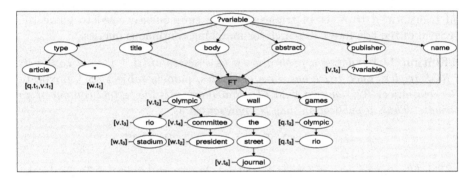

Fig. 3. Trie forest during the indexing phase of RTFS.

- Query q (with tuple $q.t_3$) shares the same structural constraints and also has the word `"olympic"` in common for the textual constraints part with query v (with tuple $v.t_4$).

Finally, note that Fig. 2 shows just one of the tries that would be created; typically, because of different query structure the resulting indexing structure is a *forest of tries*. Thus, a hash table (not shown in Fig. 2 to avoid cluttering) is used to provide fast access to trie roots.

Algorithm RTFS. Indexing of word constraints in the context of RTF may be performed in two different ways: (i) using *multiple* tries (hence the name RTFM) for indexing the word constraints depending on the structural part of the continuous queries as described in the previous section (and shown in Fig. 2), and (ii) using a *single* trie forest (hence the name RTFS) that is dedicated to all text components, regardless of the structural part of the continuous queries (shown in Fig. 3 – not described in detail due to space considerations).

In the former case (algorithm RTFM), the textual constraints are considered as a natural expansion of the structural ones, but there exist fewer clustering opportunities for words. Contrary, in the latter case (algorithm RTFS), the word constraints are considered as a different type of constraint and are clustered together regardless of the structural constraints of the query. Algorithm RTFs is a variation that allows us to construct a more compact[3] forest of tries since this organisation creates more clustering opportunities for the words. As we will demonstrate in Sect. 4, RTFs is better suited for cases where queries with text constraints are relatively sparse, whereas RTFM is better suited for cases where many queries contain text constraints.

3.2 Filtering Algorithm

The filtering algorithm is common for the two variants (RTFM and RTFs) of RTF. In this section, we present the filtering algorithm that allows RTF to filter incoming RDF publications and issue notifications to subscribed users.

The filtering process operates on triples; new RDF publications are parsed and transformed to a set of triples that are subsequently used to guide the traversal of the trie forest in search for matching continuous queries.

Definition 2. *We define a publication p as a series of $i, i \in 1 \ldots, n$ conjuncts of RDF triples. Each triple has three attributes, namely subject ($\mathcal{S}_i$), predicate ($\mathcal{P}_i$) and object ($\mathcal{O}_i$) or text field ($\mathcal{T}_i$) that represents the textual content of an attribute. Thus, a publication may be represented as:*

$$p = t_1(\mathcal{S}_1, \mathcal{P}_1, \mathcal{O}_1 | \mathcal{T}_1) \wedge \cdots \wedge t_n(\mathcal{S}_n, \mathcal{P}_n, \mathcal{O}_n | \mathcal{T}_n)$$

The filtering process proceeds as follows. For every triple $t(\mathcal{S}, \mathcal{P}, \mathcal{O})$, in the newly arrived publication p, the trie forest is examined and the root R for which $content(R) = \mathcal{S}$ is visited. Thereafter, RTF begins traversing the trie in a depth first manner and examines the nodes $children(R)$ in order to determine if there are matching tuples. In order to reach from the root node R to a leaf node, every node N in the path must fulfil the following requirements: $content(N) = \mathcal{P}$ and $content(N) = \mathcal{O}$, or $content(N) = \$variable$. If, at any point of the trie traversal a node N with a wildcard field ($content(N) = *$) is visited the traversal continues to $children(N)$, as this is considered a match for N. The traversal of the trie finishes when a leaf is reached.

For every triple $t(\mathcal{S}, \mathcal{P}, \mathcal{T})$, in the newly arrived publication p, the trie forest is examined as above. When a node N that represents a word constraint is visited, the traversal continues as follows. For every node $C, C \in children(N)$, for which $content(C)$ is contained in $\mathcal{T}$ the sub-trie that has C as a root is examined in a depth-first manner. The traversal of the trie continues recursively for as long as common words between the children of a visited node and $\mathcal{T}$ exist.

Notice that, independently of the structural or full-text constraints, the $tIDs(N)$ list at each node N gives implicitly all query tuples that match the incoming publication tuple. Thus, all $tIDs(N)$ of all traversed trie nodes are marked as matched in QT. Word distance constraints in phrase/proximity operations are checked for satisfaction after the trie traversal. In the end of the

---

[3] Notice that the trie of Fig. 3 has less nodes compared to that of Fig. 2 for the same queries and the same query insertion order.

Algorithm: FILTER

```
1  Function traverseTrie(node N, tuple t)
2      if content(N) is satisfied by t then
3          matchedTuples ← matchedTuples ∪ tIDs(N)
4          traverseTrie(C,t), where C ∈ children(N)

5  Function filterPublication(p)
6      matchedTuples ← Null
7      foreach tuple t in publication p do
8          foreach trie root R do
9              traverseTrie(R,t)

10     isSatisfied ← TRUE
11     foreach query q in QT do
12         foreach tuple t ∈ q do
13             if t is not marked as matched then
14                 isSatisfied ← FALSE
15                 break

16     if isSatisfied==TRUE then
17         notify subscriber
```

Fig. 4. Pseudocode for publication filtering.

processing of publication p (i.e., after processing all its tuples), a scan of QT allows us to determine the queries that have matched the incoming publication. The pseudocode of the filtering process for RTF variants is given in Fig. 4.

3.3   Algorithm IBROKER

To evaluate the efficiency of RTF we have also implemented IBROKER [12] as a *baseline competitor*. IBROKER is a continuous query indexing algorithm that supports SPARQL queries with structural and string matching constraints, and is currently *the only* state-of-the-art algorithm that is able to handle RDF queries with both structure and (some form of) text. In this section, we outline the basic idea behind IBROKER and the data structures upon which it operates and show how we extended its functionality to support full-text constraints.

Algorithm IBROKER utilises an inverted index to store the continuous queries. Its indexing structure consists of a hash table that is used to index the unique attributes of all triples that correspond to the submitted SPARQL queries. IBROKER uses the unique attribute names as hash keys to access the corresponding hash buckets, and each hash bucket contains references to lists of stored queries. These lists store: (i) the unique identifier of query q, (ii) a reference to a hash bucket, named *NextToMatch*, that contains the next attribute in q, (iii) the string that might be present in q named *Value*, and (iv) any possible variables in q.

This inverted index stores the queries in a chain-like manner. Every query may be recomposed by following the *NextToMatch* references to hash buckets until an empty *NextToMatch* field is visited. This procedure is applied by the algorithm IBROKER during the filtering of a publication event. As there is no defined hierarchy that outlines the filtering sequence, an incoming publication may need to examine many hash buckets looking for the beginning of a query. The result is that IBROKER must in this case examine all the continuous queries in the corresponding bucket list, and then proceed to examine their *NextToMatch* entries until there is none left to match.

ıBROKER implements string matching, but has no support for full-text operators. To evaluate it against RTF we have extended ıBROKER with full-text subscriptions by replacing the *Value* field with the list of words that appear in a full-text constraint. This modification enables ıBROKER to support both string and full-text constraints. Finally, for comparison purposes, we have also extended the functionality of ıBROKER to index and filter SPARQL queries that contain wildcard operators. For a more detailed description of ıBROKER and the specifics of its algorithms the interested reader is referred to [12].

4   Experimental Evaluation

In this section, we present a series of experiments that compare RTF against ıBROKER under a series of different scenarios.

Data and Query Set. For the experimental evaluation we utilised, the *DBpedia* corpus (http://dbpedia.org/Downloads2015-04) extracted from the *Wikipedia* domain that forms a structured knowledge database of more than 4 million items. A major part, namely 3.22 millions publications, of the *DBpedia* corpus, has been classified into an ontology resulting in 529 different classes which are described by 2.3 thousand properties. Additionally, publications bear textual information that originates from human generated content published at the *Wikipedia* domain. The vocabulary extracted from the *DBpedia* publications consists of 3.14 millions unique words. The maximum textual information present in a publication is $14,254$ words, while, the average is 53 words. The diversity in content of the *DBpedia* corpus accompanied with the information on structural and textual level, renders it as the perfect candidate for evaluating our algorithms indexing and filtering efficiency.

Query Set. The queries were constructed by utilising classes and properties extracted from the *DBpedia* corpus. Each query, contains at most 4 tuples. The query set, is formed by sets of tuples containing full-text operators with probabilities of $FT_{pr} = 0\%$, 50% or 100%. The full-text operators contain conjunctive terms that are selected equiprobably among the multi-set of words from the *DBpedia* vocabulary. Additionally, the full-text operators contain at most 3 terms. Queries with $FT_{pr} = 0\%$, examine the performance of the algorithms for structural matching only. For queries, with $FT_{pr} = 50\%$, we examine a mixed filtering scenario where half of the queries contain also textual constraints apart from the structural ones. Finally, queries with $FT_{pr} = 100\%$ demonstrate the scaling capabilities of the algorithms as they all contain full-text constraints.

Publication Set. In order to evaluate the query collections, described above, we selected $I_{pub} = 5$ K publications from the *DBpedia* corpus. The set selected, had structural and textual information as extracted and processed from the *Wikipedia* domain. The publications contain human-generated, real-life data, thus providing a realistic overview of the performance of the algorithm. We

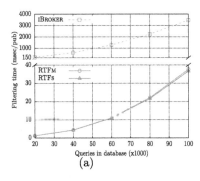 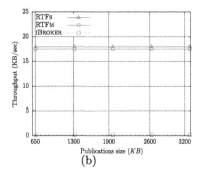

(a)   (b)

Fig. 5. Comparing (a) filtering time when varying DB and (b) filtering throughput when $DB = 100\,\mathrm{K}$, for queries of $FT_{pr} = 50\,\%$.

maintain the same publication set through the evaluation process against different query collections. Thus, it is asserted that the algorithms are evaluated based on their indexing capabilities and the nature of queries they index.

Metrics Employed. In our evaluation, we present and discuss the filtering time and throughput of each algorithm, i.e., the amount of time needed to locate all continuous queries satisfied by an incoming publication. We present and compare the memory requirements of the algorithm. As all algorithms index the same query databases, a lower memory requirement indicates a more compact clustering of data while a higher memory footprint a less compact database. Finally, we present the insertion time of each algorithm, i.e., the amount of time needed to index a set of queries $I_p = 20\,\mathrm{K}$ into the database.

Technical Configuration. All algorithms were implemented in C++, and an off-the-shelf PC (Core $i7$ 3.6 GHz, 8 GB RAM, Ubuntu Linux 14.04) was used. The time shown in the graphs is wall-clock time and the results of each experiment are averaged over 10 runs to eliminate fluctuations in time measurements.

4.1   Results When Varying the Query Database Size

In this section, we present the most significant findings for the proposed algorithms, when, varying the query database size DB for queries of $FT_{pr} = 50\,\%$.

Comparing Filtering Time. Figure 5(a) presents the time in milliseconds needed to filter an incoming publication for $I_{pub} = 5\,\mathrm{K}$ publications, when the DB size is increasing. Notice that the y-axis is split into two parts due to high differences in the performance of RTF variants and IBROKER. We observe that filtering time increases for all algorithms as the DB size grows. Algorithms RTFs and RTFM achieve the lowest filtering times, suggesting better performance. Algorithm IBROKER, is more sensitive to DB size changes compared to RTF due to its query indexing structures, i.e., an inverted index that does not implement any clustering techniques. More specifically, the results indicate that algorithm RTFM

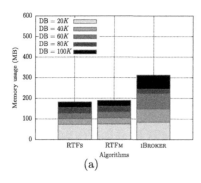
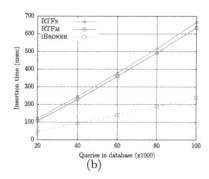

(a)   (b)

Fig. 6. Comparing (a) memory usage and (b) insertion time when varying DB for $FT_{pr} = 50\%$.

filters incoming publications 92 times faster compared to IBROKER. Finally, algorithm RTFS achieves the lowest filtering time (2.5 % faster than RTFM and 94 times faster than IBROKER), i.e., 36 ms/publication.

Comparing Filtering Throughput. We present the results concerning the algorithms' filtering throughput, when, indexing $DB = 100\,K$ queries for $I_{pub} = 5\,K$ incoming publications. Figure 5(b) presents the throughput all algorithms achieve during the filtering of $I_{pub} = 5\,K$ incoming publications. We observe that the throughput remains steady throughout the publication events. This is attributed to the nature of the algorithms, as their filtering capability is not affected by the publications size but from the indexing structures that store the queries. Algorithms RTFS and RTFM achieve the highest throughput, thus the best performance, compared to algorithm IBROKER. More specifically algorithms RTFS and RTFM achieve a throughput of more than 17 KB/s that corresponds to more than 27 publications/s. Contrary, IBROKER accomplishes a throughput of 0.18 KB/s that corresponds to 0.28 publications/s.

Comparing Memory Usage. In Fig. 6(a), we present the results for the memory requirements of each algorithm when increasing the query database DB by $I_p = 20\,K$ new continuous queries in each iteration. Algorithm RTFS has the lowest memory requirements using 183 MB for storing the whole query database $DB = 100\,K$. Algorithm's RTFM memory usage is 190 MB for the same $DB = 100\,K$, as it maintains multiple forests of tries for the indexing of textual constraints. We observe that RTF's variations reserve the majority of their memory when indexing the first $I_p = 20\,K$ to an empty database. This is due to the creation of many new tries at index structure initialisation. Namely, RTFS reserves 73 MB when indexing the first $I_p = 20\,K$ queries and RTFM reserves 75 MB. For every new $I_p = 20\,K$ inserted into the database RTFS and RTFM do not require more than 28 MB to facilitate the indexing of new queries due to the accommodation of new queries mostly in existing tries. Finally, algorithm IBROKER occupies 313 MB of memory to index a database of $DB = 100\,K$ queries. IBROKER reserves 84 MB for the first $I_p = 20\,K$ queries, while it requires more than 60 MB of memory to index every set of $I_p = 20\,K$ new queries.

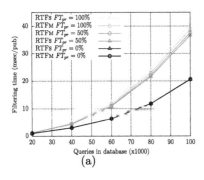 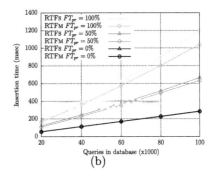

Fig. 7. Comparing RTF's (a) filtering and (b) insertion time when varying DB size and FT_{pr}.

Comparing Insertion Time. In this section, we discuss the query index-ing time of all algorithms. Figure 6(b) shows the insertion time in milliseconds required to insert $I_p = 20$ K queries when the DB size increases. We observe that the algorithms require more time to index new queries as the database size increases. Algorithms RTFs and RTFM need more time to index the same number of queries $I_p = 20$ K compared to IBROKER. This can be explained as follows. The variations of RTF utilise trie-based data structures to capture and index the common structural and textual constraints of the queries. Trie tra-versal results to high insertion time during the indexing phase. On the other hand, insertion in an inverted index (as done by IBROKER) is faster. Notice that insertion time is not critical in a pub/sub scenario; the most important dimension is filtering time/throughput that defines the processing rate of publications.

4.2 Results When Varying the Full-Text Percentage

This section presents the most interesting results concerning the RTF variants when varying the percentage of full-text constraints in the tuples. Notice that we do not show IBROKER (that has a significantly worse performance than the RTF variants as demonstrated in the previous sections) to avoid cluttering the graphs. We evaluate the structural matching performance of RTFM and RTFs when $FT_{pr} = 0\%$, and stress-test the algorithms when the query database contains the highest number of full-text constraints possible, i.e., when $FT_{pr} = 100\%$.

Comparing Filtering Time. Figure 7(a) shows the time needed to filter an incoming publication against full-text constraints with $FT_{pr} = 0\%, 50\%$ and 100%, when increasing the DB size. As expected, RTFs and RTFM achieve the lowest filtering times when $FT_{pr} = 0\%$, and exhibit the same performance as they utilise the same indexing structure for the structural constraints of the queries. Finally, RTFs and RTFM increase their filtering times when $FT_{pr} = 100\%$ with RTFM achieving better performance.

Comparing Insertion Time. Figure 7(b) shows the time required to insert $I_p = 20$ K queries when DB size is increasing and varying $FT_{pr} = 0\%, 50\%$

and 100 %. As expected, algorithms RTFs and RTFM increase their time needs, when more textual constraints ($FT_{pr} = 100\%$) are included in the continues queries, and reduce them when no textual constraints ($FT_{pr} = 0\%$) are present.

Comparing Memory Usage. We give an outline of RTFs's and RTFM's memory requirements, when $DB = 100\,K$, for queries of $FT_{pr} = 0\%$ and 100%. Both, RTFs and RTFM have the lowest memory requirements when $FT_{pr} = 0\%$, namely 168 MB. Finally, for $FT_{pr} = 100\%$, 196 MB and 203 MB are required for RTFs and RTFM respectively.

4.3   Summary of Results

Our experimental evaluation demonstrated the filtering effectiveness of algorithm RTFs for cases where queries with text constraints are not very often, whereas algorithm RTFM is better suited for cases where a high percentage or queries contain text constraints. Both algorithms RTFs and RTFM are over two orders of magnitude faster than IBROKER on average.

5   Related Work

In this section we discuss pub/sub approaches in centralised and distributed environments and contrast them to our approach.

Centralised Ontology Pub/Sub Systems. The S-ToPSS [15] system was among the first designs that supported pub/sub functionality in an ontological context. S-ToPSS was designed to enhance the matching process aiming at semantically similar but syntactically different information present in publications and user subscriptions. This was achieved by identifying synonyms and utilising concept taxonomies and hierarchies. Its successor, G-ToPSS [16], focused on information dissemination of RDF data on ontologies, emphasising on scalability and fast filtering of RDF data. G-ToPSS represented publications as directed labelled graphs, while a two-level hash table was used for the subscriptions; the matching algorithm traversed the publication and subscription graphs. In the same spirit, the Ontology-based Pub/Sub (OPS) system [20] supported events with complex data structures and aimed for a uniform representation. Subsequently, user subscriptions and publication events were processed into RDF graphs and thereafter indexed or filtered respectively by utilising graph matching algorithms. Finally, OPS examined the matching trees that emerged from the graph traversal to determine the matching subscriptions. The Sparkwave [10] system was built to perform continuous pattern matching over RDF streams by supporting expressive pattern definitions, sliding windows and schema-entailed knowledge. The C-SPARQL [1] extension enabled the registration of continuous SPARQL queries over RDF streams, thus, bridging data streams with knowledge bases and enabling stream reasoning.

Although all the aforementioned works focus on supporting pub/sub functionality in ontology systems, none has considered supporting any form of text extension. The work closest to ours is IBROKER [12], an OWL-based pub/sub mediator focused in filtering publications from OWL ontologies against a set of stored SPARQL queries. ɪBROKER matched incoming events generated from an ontology against user queries by resorting on an inverted index to represent the graph that indexed user subscriptions. Although there is no text support, ɪBROKER is able to perform string matching using the inverted index mentioned above. In this work, we extended ɪBROKER with full-text support and used it as a baseline competitor for our algorithms to showcase the performance gains.

Distributed Ontology Pub/Sub Systems. With the advent of distributed and P2P computing, decentralised ontology pub/sub systems naturally emerged. The first P2P pub/sub system based on RDF data was build by Chirita et al. [5]; the system utilised a super-peer architecture where super-peers were responsible for the routing of the content determined by the RDF schema, property or value, while peers were responsible for specific schemas and properties. At publication time, super peers routed the data to the responsible peers for the filtering process; the performance gain was achieved by utilising the content similarities present in subscriptions. Similarly, Liarou et al. [11] studied the problem of evaluating multi-predicate conjunctive queries in pub/sub systems; the aim of the system was to distribute the load of the matching process into a P2P network. In the same spirit, an RDF-based pub/sub P2P network was build by Pelegrino et al. [13] to study the messaging paradigm. The system supported the creation of queries by making use of SPARQL and publications by using RDF data. Users' subscriptions were indexed into a peer, determined by the CAN protocol. Data from a publication event that concerned a peer was stored while the event was forwarded to other peers. Finally, Kaoudi et al. [9] presented a study for distributed RDF reasoning and query answering. The work in [9] focused on implementing, optimising, and evaluating forward and backward chaining over a distributed hash-table for a subset of SPARQL.

None of the works mentioned above supports full-text in their data/query model. Finally notice that our solution can be extended in a decentralised environment by distributing the triple forest to different nodes and modifying the filtering process to visit only nodes that may contain matching queries.

Connection to Other Technologies. The vast majority of RDF stores (Apache Jena[4] Text module, Virtuoso[5], Allegrograph[6], OntoText GraphDB[7]) offers text indexing and full-text retrieval combined with SPARQL. Our solution shares ideas with these technologies, but pub/sub copes with different problems and challenges compared to traditional retrieval. Finally, pub/sub functionality on ontologies may complement many applications including LOD platforms such as Lotus[8] by enabling users to get notified for information of interest, or by providing a useful moderation/monitoring tool for curators/editors of such systems.

---

[4] http://jena.apache.org/.
[5] http://virtuoso.openlinksw.com/.
[6] http://franz.com/agraph/allegrograph/.
[7] http://ontotext.com/products/graphdb/.
[8] http://lotus.lodlaundromat.org/.

6   Conclusions and Outlook

In this work, we studied the problem of full-text support on ontology-based pub/sub systems. In this context, we proposed a full-text extension for SPARQL continuous queries and a family of query indexing algorithms that are two orders of magnitude faster at filtering tasks than a state-of-the-art competitor.

Currently, we are working on supporting VSM queries/text representation in SPARQL, and adapting our algorithms to multi-processor environments.

References

1. Barbieri, D.F., Braga, D., Ceri, S., Valle, E.D., Grossniklaus, M.: C-SPARQL: a continuous query language for RDF data streams. IJSC **4**, 3–25 (2010)
2. Case, P., Dyck, M., Holstege, M., Amer-Yahia, S., Botev, C., Buxton, S., Doerre, J., Melton, J., Rys, M., Shanmugasundaram, J.: XQuery and XPath Full Text. (2011). http://www.w3.org/TR/xpath-full-text-10/
3. Chan, C.Y., Felber, P., Garofalakis, M.N., Rastogi, R.: Efficient filtering of XML documents with XPath expressions. In: ICDE (2002)
4. Chang, C.C.K., Garcia-Molina, H., Paepcke, A.: Predicate rewriting for translating Boolean queries in a heterogeneous information system. ACM TOIS **17**, 1–39 (1999)
5. Chirita, P.-A., Idreos, S., Koubarakis, M., Nejdl, W.: Publish/subscribe for RDF-based P2P networks. In: Bussler, C.J., Davies, J., Fensel, D., Studer, R. (eds.) ESWS 2004. LNCS, vol. 3053, pp. 182–197. Springer, Heidelberg (2004)
6. Diao, Y., Altinel, M., Franklin, M.J., Zhang, H., Fischer, P.M.: Path sharing and predicate evaluation for high-performance XML filtering. ACM TODS **28**, 467–516 (2003)
7. Green, T.J., Gupta, A., Miklau, G., Onizuka, M., Suciu, D.: Processing XML streams with deterministic automata and stream indexes. ACM TODS **29**, 752–788 (2004)
8. Hou, S., Jacobsen, H.: Predicate-based filtering of XPath expressions. In: ICDE (2006)
9. Kaoudi, Z., Miliaraki, I., Koubarakis, M.: RDFS reasoning and query answering on top of DHTs. In: Sheth, A.P., Staab, S., Dean, M., Paolucci, M., Maynard, D., Finin, T., Thirunarayan, K. (eds.) ISWC 2008. LNCS, vol. 5318, pp. 499–516. Springer, Heidelberg (2008)
10. Komazec, S., Cerri, D., Fensel, D.: Sparkwave: continuous schema-enhanced pattern matching over RDF data streams. In: DEBS (2012)
11. Liarou, E., Idreos, S., Koubarakis, M.: Publish/subscribe with RDF data over large structured overlay networks. In: DBISP2P (2005)
12. Park, M.J., Chung, C.W.: iBroker: an intelligent broker for ontology based publish/subscribe systems. In: ICDE (2009)
13. Pellegrino, L., Huet, F., Baude, F., Alshabani, A.: A distributed publish/subscribe system for RDF data. In: GLOBE (2013)
14. Peng, F., Chawathe, S.S.: XPath queries on streaming data. In: SIGMOD (2003)
15. Petrovic, M., Burcea, I., Jacobsen, H.A.: S-ToPSS: semantic toronto publish/subscribe system. In: VLDB (2003)
16. Petrovic, M., Liu, H., Jacobsen, H.A.: G-ToPSS: fast filtering of graph-based metadata. In: WWW (2005)

17. Prud'hommeaux, E., Seaborne, A.: SPARQL query language for RDF (2008). http://www.w3.org/TR/rdf-sparql-query/

18. Schreiber, G., Raimond, Y.: RDF 1.1 Primer (2014). http://www.w3.org/TR/rdf11-primer/

19. Tryfonopoulos, C., Koubarakis, M., Drougas, Y.: Information filtering and query indexing for an information retrieval model. TOIS **27**, 1–47 (2009)

20. Wang, J., Jin, B., Li, J.: An ontology based publish/subscribe system. In: Jacobsen, H.-A. (ed.) Middleware 2004. LNCS, vol. 3231, pp. 232–253. Springer, Heidelberg (2004)

21. Yan, T.W., Garcia-Molina, H.: The SIFT information dissemination system. ACM TODS **24**, 529–565 (1999)

Heaven: A Framework for Systematic Comparative Research Approach for RSP Engines

Riccardo Tommasini, Emanuele Della Valle, Marco Balduini$^{(\boxtimes)}$,
and Daniele Dell'Aglio

DEIB, Politecnico of Milano, Milano, Italy
{riccardo.tommasini,emanuele.dellavalle,marco.balduini,
daniele.dellaglio}@polimi.it

Abstract. Benchmarks like LSBench, SRBench, CSRBench and, more recently, CityBench satisfy the growing need of shared datasets, ontologies and queries to evaluate window-based RDF Stream Processing (RSP) engines. However, no clear winner emerges out of the evaluation. In this paper, we claim that the RSP community needs to adopt a Systematic Comparative Research Approach (SCRA) if it wants to move a step forward. To this end, we propose a framework that enables SCRA for window based RSP engines. The contributions of this paper are: (i) the requirements to satisfy for tools that aim at enabling SCRA; (ii) the architecture of a facility to design and execute experiment guaranteeing repeatability, reproducibility and comparability; (iii) $\mathcal{H}$eaven – a proof of concept implementation of such architecture that we released as open source –; (iv) two RSP engine implementations, also open source, that we propose as baselines for the comparative research (i.e., they can serve as terms of comparison in future works). We prove $\mathcal{H}$eaven effectiveness using the baselines by: (i) showing that top-down hypothesis verification is not straight forward even in controlled conditions and (ii) providing examples of bottom-up comparative analysis.

1 Introduction

The Stream Reasoning (SR) [10] community agrees on the principle that Information Flow Processing approaches [9] and reasoning techniques can be coupled in order to reason upon rapidly changing information flows (for a recent survey see [18]). Currently, a W3C community group[1] is working towards the standardization of the following basic SR technologies:

- *RDF Streams*, which were introduced in [10], later on picked up in [2,16] and approached as *virtual* RDF stream in [8].
- *Continuous extension of SPARQL* such as, in chronological order, C-SPARQL [5], SPARQL$_{stream}$ [8], CQELS-QL [16] and EP-SPARQL [2,13].

---

[1] http://www.w3.org/community/rsp/.

H. Sack et al. (Eds.): ESWC 2016, LNCS 9678, pp. 250–265, 2016.
DOI: 10.1007/978-3-319-34129-3_16

- *Reasoning techniques* optimized for the streaming scenario such as, in chronological order, Streaming Knowledge Bases [27], IMaRS [6], a stream-oriented version of TrOWL [21], EP-SPARQL [2] and RDFox [20].

These concepts gave birth to a new class of systems, collectively called RDF Stream Processing (RSP) engines, which proved SR feasibility and drew the attention of the community to their evaluation.

The goal of a domain specific benchmark is to foster technological progress by guaranteeing a fair assessment. For window-based RSP engines - the class of systems in the scope of this paper - RDF Streams, ontologies, continuous queries and performance measurements were proposed in LSBench [17], SRBench [28], CSRBench [11] and CityBench [1]. However, existing benchmarks showed no absolute winner and it is even hard to claim when an engine outperforms another.

In this paper, we argue that SR needs to identify the situations under which an engine A works better than an engine B, either quantitatively, i.e. focusing on a very aspect of the performance, or qualitative, i.e. classifying the behavior by observing the dynamics. This approach is known as Systematic Comparative Research Approach (SCRA) [15] and it is adequate when the complexity of the study subject makes it hard to formulate top-down hypothesis. Consequently, we formulate our research question as:

How can we enable a SCRA for window based RSP engines?

Our answer to this research question comprises:

(1) A set of requirements to satisfy in order to enable a SCRA.
(2) A general architecture for an RSP engine Test Stand; a facility that: (i) makes possible to design and systematically execute experiments; (ii) provides a controlled environment to guarantee repeatability and reproducibility of experiments; (iii) collects performance measurements while the engines are running; and (iv) allows us to comparatively evaluate the engines post-hoc.
(3) *H*eaven, a proof-of-concept implementation that we released open source[2], which improves the one proposed in [25] as explained in Sect. 5.
(4) Two *baseline* RSP engine implementations that we propose as terms of comparison, since they are the kind of "simplified complex cases that combine known properties" advocated in SCRA [15] to highlight differences and similarities among real cases. They are released open source, too[2].

Demonstrating that the paper contributions positively answer our research question requires us to: (i) highlight that top-down hypothesis confirmation is not straight forward even in a controlled environment and when the observed RSP engines are as simple as the baselines are; (ii) prove *H*eaven effectiveness by showing the relevance of the bottom-up analysis that it enables. To this extent, we first executed 14 experiments that involve the baselines as subjects and we showed that we cannot confirm the following two hypotheses, which have been already investigated top-down by the SR community [12,20,21,26]:

---

[2] https://github.com/streamreasoning/heaven.

Hp.1 Materializing from scratch the ontological entailment of the window content, each time it slides, is faster than the incremental maintenance of the materialization, when changes are large enough (e.g. greater than 10 %).

Hp.2 Dually, if changes are small (e.g. less than 10 %), the incremental maintenance of the materialization of the window content ontological entailments, each time it slides, is faster than materializing it from scratch.

Then, we show some example of the bottom-up analysis that our approach enables. This part of our investigation is lead by four questions:

Q.a Qualitatively, is there a solution that always outperforms the others?

Q.b If no dominant solution can be found, when does a solution work better than another one?

Q.c Quantitatively, is there a solution that distinguishes itself from the others?

Q.d Why does a solution perform better than another solution under a certain experimental condition?

The remainder of the paper is organized as follows. Section 2 provides the minimum background knowledge required to understand the content of the paper. Section 3 presents the definition of RSP experiment and the requirements of a software framework to enable SCRA for RSP engines (i.e. an architecture and at least a baseline). Section 4 shows the architecture of the RSP engine Test Stand and its workflow. Section 5 reports the implementation experience of $\mathcal{H}$eaven. Section 6 describes two RSP engines baseline. Section 7 shows the evaluation of $\mathcal{H}$eaven. Section 8 positions this paper within the state-of-the-art of RSP benchmarking. Finally, in Sect. 9, we come to conclusions.

2   Background

Stream Reasoning (SR) [10] is a novel research trend which focuses on combining Information Flow Processing (IFP) [9] engines and reasoners to perform reasoning on rapidly changing information. IFP engines are systems capable to continuously process **Data Streams**, that are potentially infinite sequences of events. The data streams considered in SR are the **RDF Streams**, where events are described by timestamped RDF data. For example, if at time 1, an event e_1 states that a MicroPost : $tweet_1$ is posted by *:alice*, while at time 5, a data item e_2 states that *:instagram_post2* is posted by *:bob*. The data stream $((e_1, 1), (e_2, 5))$ can be represented by the RDF Stream:

$(:tweet_1 \; sioc : has\_creator : alice), 1$
$(:instagram\_post_2 \; sioc : has\_creator : bob), 5$

Most of the RSP query languages adopt a Triple-Based *RDF Stream Model*, i.e., events are represented by a single timestamped RDF statement. C-SPARQL [5], CQELS-QL [16] and SPARQL$_{stream}$ [8] are examples of this class of languages. However, a Graph-Based *RDF Stream Model* is possible and has been recently adopted by the RSP W3C group. An example of engine that adopts the Graph-Based RDF Stream model is SLD [4], that uses RDF graphs as a form of punctuation [23] to separate the events in the stream.

The IFP query languages and their processing methods can be grouped in two main membership classes:

- **Complex Event Processing (CEP) engines** that validate incoming primitive events and recognize patterns upon their sequences.
- **Data Stream Managements Systems (DSMSs)** [3] that exploit relational algebra to process portions of the data stream captured using special stream-to-relation (S2R) operators.

In this work, we target a class of Stream Reasoners known as window-based RDF Stream Processing (RSP) engines, inspired to DSMSs. The key operators of this class of engines are the S2R ones and, in particular, the sliding window operators. They allow to cope with the infinite nature of streams. Intuitively, a sliding window creates a view, named active window, over a portion of the stream that changes (slides) over time. **Time-based sliding windows** are sliding windows that create the active window and slide it accordingly to time constraints. They are defined by the width ω – i.e., the time range that has to be considered by the active window – and the slide β – i.e. how much time the active window moves ahead when it slides.

As advocated in the early works on Stream Reasoning [10,27], the most simple approach to create a window-based RSP engine is pipelining a DSMS with a reasoner. For example, the C-SPARQL engine[3] uses Esper as DSMS and Jena as framework to manage RDF and reasoning over the window content. The C-SPARQL engine processes C-SPARQL queries under RDFS entailment regime[4]. For instance, the following C-SPARQL query asks to report every day (see STEP 15 min) the people mentioned during the last week in the stream of those who have published a paper (see RANGE 1d):

SELECT ?micropost
FROM STREAM <http://www.ex.org/socialStream> *[RANGE 1d STEP* 15 min*]*
WHERE { ?micropost a :MicroPost}

The requested information is not explicitly stated in the RDF stream exemplified above, but being the range of the *sioc:has_creator* property a *:MicroPost*, an RDFS reasoner can deduce that :tweet$_1$ and :instagram_post$_2$ are of type :MicroPost and, thus, they belong to the answer of the query.

Other examples of window-based RSP engines are Morph$_{stream}$ [8] and CQELS [16], that adopt different approaches: the former is a native implementation of a window-based RSP engine, to achieve performance and adaptivity; the latter adopts an OBDA-like approach to rewrite the continuous query in a query to be DSMS and to be evaluated over a (non-RDF) data streams.

---

[3] https://github.com/streamreasoning/CSPARQL-engine.
[4] http://www.w3.org/TR/sparql11-entailment/#RDFSEntRegime.

3 Notion of RSP Experiment and SCRA Requirements

An experiment is a test under controlled conditions that is made to demonstrate a known truth, examine the validity of a hypothesis[5], and that guarantees results *reproducibility*, *repeatability*, and *comparability*. This notion is not new in the RSP benchmarking state-of-the-art [1], but a formal definition is still missing.

We define an RSP Experiment as a tuple $< \mathcal{E}, \mathcal{T}, \mathcal{D}, \mathcal{Q}, \mathcal{K} >$ where:

- $\mathcal{E}$ is the RSP engine used as subject in the experiment;
- $\mathcal{T}$ is an ontology and any data not subject to change during the experiment;
- $\mathcal{D}$ is the description of the input data streams;
- $\mathcal{Q}$ is the set of continuous queries registered into $\mathcal{E}$; and
- $\mathcal{K}$ is the set of key performance indicators (KPIs) to collect.

The result of an RSP Experiment is a report $\mathcal{R}$ that contains the trace of the RSP engine processing for the entire duration of the experiment.

On these definitions and on the notions of *Comparability, Reproducibility* and *Repeatability*[6], we elicit the requirements for the architecture of a test stand that enables SCRA starting from an RSP Experiment.

Comparability refers to the nature of the experimental results $\mathcal{R}$ and their relationship with the experimental conditions. It requires that the test stand is [R.1] RSP engine agnostic, i.e. as long as the experimental conditions do not change, the results must be comparable, independently from the tested RSP engine. To this end, the report $\mathcal{R}$ has to record at least: (i) the data stream (or Stimulus) sent to $\mathcal{R}$; (ii) the application time of the Stimulus; (iii) the system time at which the Stimulus is sent; (iv) any KPIs to be measured before $\mathcal{E}$ processes the Stimulus; (v) the Response, if any, of $\mathcal{E}$ to the Stimulus; (vi) the system time at which the Response, if any, is sent; (vii) any KPIs to be measured after $\mathcal{E}$.

Reproducibility refers to measurement variations on a subject under changing conditions. It requires that the test stand is: [R.2] *data independent*, which means allowing the usage of any data stream and any static data (e.g., those proposed in [1,11,17,28]); and [R.3] *query independent*, which means allowing the usage of any query from users' domains of interest (e.g., those proposed in [1,11,17,28])

Repeatability refers to variations on repeated measurements on a subject under identical conditions. It requires that the test stand [R.4] minimizes the experimental error, i.e., it has to affect the RSP engine evaluation as less as possible and in a predictable way.

All these experiment properties together require the test stand to be [R.5] *independent from the measured key performance indicators (KPIs)*, i.e., the KPIs set has to be extensible. According to [22] a set of meaningful KPIs comprises: (i) *query-execution-latency* – the delay between the system time at which a Stimulus is sent to the RSP engine and the system time at which a Response occurs, if any; (ii) *throughput* – the number of triples per unit of time processed

---

[5] http://bit.ly/experiment-definition.
[6] http://bit.ly/experiment-properties.

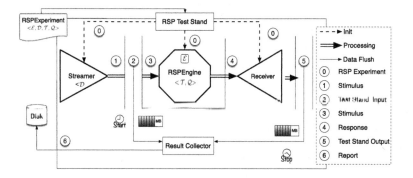

Fig. 1. RSP TEST STAND modules and workflow.

by the system; (iii) *memory usage* – the actual memory consumption of the RSP engine; (iv) *Completeness & Soundness* of query-answering results w.r.t the entailment regime that the RSP engine is exploiting.

Last, but not least, due to its case-oriented nature, SCRA exploits naive terms of comparison to draw research guidelines (Sect. 1). For this reason, it is necessary to identify at least one [R.6] RSP engine baseline, i.e., the minimal meaningful approaches to realize an RSP engine.

4   Architecture and Workflow of RSP Test Stand

In this section, we present the architecture of a TEST STAND for window based RSP engines (namely RSP TEST STAND) - a software facility to enable SCRA - it components and how they interact during the execution of an RSP Experiment.

The RSP TEST STAND consists of four components, whose position within the architecture is presented in Fig. 1:

- The STREAMER sets up the incoming streams w.r.t to $\mathcal{D}$;
- The RSP ENGINE represents the RSP engine $\mathcal{E}$, initialized with $\mathcal{T}$ and $\mathcal{Q}$;
- The RECEIVER continuously listens to $\mathcal{E}$ responses to $\mathcal{Q}$; and,
- The RESULT COLLECTOR compiles and persists the report $\mathcal{R}$.

During the execution, six kind of events are exchanged between the components: (i) RSP EXPERIMENT - the top-level input, defined in Sect. 3, used to initialize the test stand components; (ii) STIMULUS - a portion of the incoming data streams in which all the data have the same application time and whose complete specification is included in $\mathcal{D}$ (e.g., a real word data stream, a recorded data stream as those proposed in [1,11,17,28]), a synthetic data stream generated in order to stress the engine); (iii) RESPONSE - the answer of the queries specified in $\mathcal{Q}$; (iv) TEST STAND INPUT - the STIMULUS and the performance measurements collected as specified in $\mathcal{K}$ just before STIMULUS creation; (v) TEST STAND OUTPUT - the RESPONSE and the performance measurements collected as specified in $\mathcal{K}$ just after the RESPONSE creation; (vi) TEST STAND REPORT - the carrier for the data that the RESULT COLLECTOR has to persist.

Figure 1 also shows also how the TEST STAND components have to interact during the experiment execution. In step (0), the TEST STAND receives an RSP EXPERIMENT and it initializes its modules (dashed arrows): it loads the engine $\mathcal{E}$ and registers into it $\mathcal{T}$ and the query-set $\mathcal{Q}$; it sets up the STREAMER, according with the incoming streams definition $\mathcal{D}$; it connects the engine to the RECEIVER and it initializes the RESULT COLLECTOR to receives and saves the test stand outputs. The TEST STAND loops between steps (1) and (4), until the experiment ends. In step (1), the STREAMER creates and pushes a STIMULUS to the engine $\mathcal{E}$. In step (2), the TEST STAND intercepts the STIMULUS of step (1), collects the performance measurements specified in $\mathcal{K}$ (e.g., it starts a timer to measure latency and it calculates the memory consumption of the system) and it sends a TEST STAND INPUT to the RESULT COLLECTOR. In step (3), $\mathcal{E}$ receives a STIMULUS, it processes it according with the query-set $\mathcal{Q}$ and its own execution semantics [13] (i.e., we consider the RSP engine as a black box). Step (4) occurs when $\mathcal{E}$ outputs a RESPONSE and sends it to the RECEIVER. In step (5), the TEST STAND receives a RESPONSE from the RECEIVER, takes some measurements (e.g., it stops the timer and calculates the memory consumption of the system again), it creates and sends a TEST STAND OUTPUT to the RESULT COLLECTOR. Finally, in step (6), the RESULT COLLECTOR persists all the collected data as an TEST STAND REPORT.

5   Implementation Experience

In this section, we present $\mathcal{H}$eaven, an implementation of the RSP TEST STAND architecture described in Sect. 4. In the following, we show how $\mathcal{H}$eaven satisfies the requirements we posed in Sect. 3.

As a consequence to requirements [R.1, R.2, R.3], we implement $\mathcal{H}$eaven with an extensible design, i.e., each module can be replaced with another one with the same interface, but different behavior, without affecting the system stability.

To satisfy [R.2] we developed a STREAMER, namely RDF2RDF$_{Stream}$, that can define different types of workload. In the current implementation it allows to define (i) flows that remain stable over time; (ii) flows that remain stable for a while, then suddenly increase; and (iii) flows that change according to a distribution (e.g., Poisson or Gaussian).

The RDF2RDF$_{Stream}$ generates streams according with the $\mathcal{D}$ parameter, which comprises: DS the actual data to stream; M the data model used by the stream (e.g. RDF Streams encoded as timestamped RDF Triples or as timestamped RDF Graphs); and the function F that describes the STIMULUS content over time. The RDF2RDF$_{Stream}$ (1) retrieves and parses the data from the specified file; (2) it creates a STIMULUS w.r.t the specified data model M; it exploits the function F to determine the cardinality of the triple-set to stream; (3) it attaches a non-decreasing application timestamps to the STIMULUS which is immediately pushed to the tested RSP engine $\mathcal{E}$. The RDF2RDF$_{Stream}$ does not consider the semantic relation between triples, but it parse the incoming data in order.

A clear limitation of $\mathcal{H}$eaven regards the satisfaction of [R.4]. Due to our choice of using Java for prototyping, we cannot explicitly control the memory consumption. A fair workaround consist in implementing $\mathcal{H}$eaven as single thread and suspending the test stand while the RSP engine is under execution. We also reduced as much as possible the number of memory-intensive data structures used in the RSP TEST STAND and we limited the I/O operations.

According to [R.5], $\mathcal{H}$eaven must be independent to the KPIs set. Currently, it can measure query latency and memory consumption, but we developed it to be easily extensible with other KPIs, e.g. throughput or Quality of Service etc [22]. Actually, Completeness and Soundness of the query results can be evaluated post-hoc using the REPORT content persisted by the RESULT COLLECTOR as shown in CSRBench [11].

6   Baseline RSP Engines

The final contribution of this work consists in two baseline, that implement the minimal meaningful approaches to realize an RSP engine[7] (see Sect. 2) and, thus, they can be used as a initial terms of comparison to enable SCRA.

They currently support reasoning under the ρDF entailment regime [19] which is the RDF-S fragment that reduces complexity while preserving its normative semantics and core functionalities. This decision is motivated because we want to provide reasoning capabilities and ρDF is the minimal meaningful task for a Stream Reasoner [26].

The baselines cover two main design decisions: (i) data can flows from the DSMS to the reasoner via snapshots (i.e. Fig. 2A) or differences (Fig. 2B); and (ii) they exploit absolute time, i.e. their internal clock can be externally controlled [9].

Figure 2A shows the first approach, which is similar to what the C-SPARQL engine [7] does. The DSMS produces a snapshot of the active window content at each cycle and the reasoner materializes it from scratch (according to the ontology $\mathcal{T}$ and the entailment regime); then all the queries in $\mathcal{Q}$ are applied to the new materialization to generate the responses. Figure 2B shows the second approach, that is

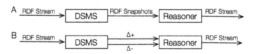

Fig. 2. The architecture of the two baselines inspired to the C-SPARQL engine (A) and of the two Incremental ones inspired to TrOWL and RDFox (B).

similar to what TrOWL and RDFox [20,24] do. The DSMS outputs the differences Δ^+ and Δ^- between the active window and the previous one. Δ^+ contains the triples that have just entered in the active window, while Δ^- contains the

---

[7] We implement the baselines pipelining Esper 5.3: http://esper.codehaus.org/, a mature open source DSMS, with the Jena 2.12.1 general purpose rule engine: http://jena.apache.org/documentation/inference/#rules, also open source.

triples that have just exited from the active window. The reasoner, using Δ^+ and Δ^-, incrementally maintains the materialization over time.

Exploiting absolute time allows us to ensure the Completeness and Soundness of baselines responses, since we can wait for the RSP engine to complete the evaluation before sending the next STIMULUS. In this way, the baselines cannot be overloaded, they can only violate the responsiveness constrain (A exhaustive reference is [9]). Moreover, we can observe the entire engine dynamics, because we fully control the engine behavior and, thus, we can relate to any given STIMULUS with the relative RESPONSE even in stressing condition.

7   Evaluation

In this section, we demonstrate what we argued in the Sect. 1, i.e. (i) top-down hypothesis confirmation is not straight-forward; and (ii) $\mathcal{H}$eaven is effective because it enables relevant bottom-up analysis. To this extent, we show how to design a set of experiments setting up a controlled environment. We provide evidences of the non-obvious evaluation of hypothesis Hp.1 and Hp.2 (presented in Sect. 1) and we answer questions Q.a, Q.b, Q.c, Q.d for the baselines.

7.1   Experiment Design

According with RSP Experiment definition presented in Sect. 4, to prepare an experiment we need to fill the tuple $<\mathcal{E}, \mathcal{D}, \mathcal{T}, \mathcal{Q}, \mathcal{K}>$.

As $\mathcal{E}$, the RSP engine to test, we consider our baseline implementations.

We used $RDF2RDF_{Stream}$ as STREAMER, that accepts in input any RDF file. We need to stress the reasoning capabilities of the baseline in a regular way and LUBM is a recognized benchmark for the reasoning domain that has been used before in the SR field [26][8]. Therefore, we configured $\mathcal{D}$ as follows: as dataset DS we opted for LUMB [14];

Table 1. The number of RDF triples in the active window as a function of the number of triples in each STIMULUS and the duration ω of the window. Note sliding of the window β is 100 ms and the application time unit is 100 ms, i.e., we send a Stimulus each 100 ms.

Stimulus	ω				
Size	.1 s	1 s	10 s	100 s	1000 s
1	1	10	10^2	10^3	10^4
10	10	10^2	10^3	10^4	
10^2	10^2	10^3	10^4		
10^3	10^3	10^4			

as data model M we opted for RDF Stream encoded as timestamped RDF Graphs; and as function F, since we want to keep the data stream regular, we chose one which keeps constant the number of triples in each STIMULUS forming the RDF Stream and we set the application time unit of 100 ms.

---

[8] LUBM is the easiest choice in this context. Indeed, among the existing SR benchmarks only SRbench includes queries that require reasoning, but the data streams do not flow regularly. On the other hand, LUBM can be streamed regularly through the $RDF2RDF_{Stream}$ without affecting the semantics. A research towards a better SR benchmark focused on reasoning [22] is worthy, but is out of the scope of this paper.

Table 2. Experiments comparison for average query latency (a) and average memory consumption (b) of the baselines. I:Incremental, N:Naive, $\simeq$: Even. Assumption: approach A dominates B when A/B is grater than 5 %. Highlighted cells indicate where Hp.1 or Hp.2 are not confirmed.

(a) Latency						(b) Memory					
Stimulus Size	ω					Stimulus Size	ω				
	.1s	1s	10s	100s	$10^3 s$		.1s	1s	10s	100s	$10^3 s$
1	I	I	I	I	I	1	N	N	I	I	I
10	I	I	I	I		10	N	N	I	I	
10^2	I	I	I			10^2	I	I	I		
10^3	N					10^3	I	$\simeq$			

Coherently with $\mathcal{E}$ and $\mathcal{D}$, we picked as $\mathcal{T}$ the ρDF subset of the LUBM TBox. As normally assumed in SR research, we consider this TBox static, therefore, the materialization of $\mathcal{T}$ is computed at configuration time.

As set of queries $\mathcal{Q}$, we choose the query that continuously asks for the entire materialization of the active window content w.r.t. $\mathcal{T}$ under ρDF entailment regime. This query is the most general query that can be registered in $\mathcal{E}$ and it is enough to support our claims. Indeed, adding more queries can only make the complex situation (without clear winners), which we discuss in Sect. 7.2, even more intricate. In different experiments, we used variants of the this query, fixing the sliding parameter to $\beta = 100$ ms and varying the total duration ω of the window, as summarized in Table 1.

Finally, as KPIs set $\mathcal{K}$, we measure the memory consumption and, since the baselines allows to exploit the absolute time for the computation, the query latency. To stress the baselines at maximum, we pushed the Stimuli in the baselines as soon as the baselines end the previous computation, reducing the probability of garbage collecting until the system finishes the free memory.

All experiment are 30000 STIMULI long and, each of them was executed 10 times on an 2009 iMac with 12 Gb of RAM and 3.06 GHZ with OSX 10.7.

7.2 Uncomfortable Truths in Hypothesis Verification

State-of-the-art results [1,11,17,28] show that top-down hypothesis confirmation is not straight-forward. This is still true even under the controlled experimental condition we just defined and for already investigated hypothesis like Hp.1 and Hp.2. In the following, we specify them more w.r.t the experimental setting and we show how they are not trivially verified.

Hp.1 When $\omega = \beta$ i.e., the window contains only one STIMULUS, the Naive approach is always faster than the Incremental one.

Hp.2 When the number of changes $\Delta+$ and $\Delta-$ (Sect. 6) is a small fraction of the content of the window an Incremental approach is faster than the Naive one.

Table 2(a) and (b) exploit the layout of Table 1 to compare respectively average query latency and average memory consumption values of the two approaches for

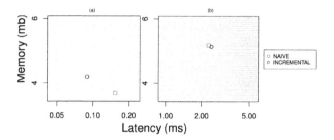

Fig. 3. Example of dashboard representation: average latency values on x-axis reports, average memory values on the y-axis in logarithmic scale. w.r.t Table 1: (a) Stimulus size = 1, $\omega = 1\,\mathrm{s}$ (b) Stimulus size = 1000, $\omega = 1\,\mathrm{s}$

all the 14 experiments. We assume that one approach dominates the other one when the differences in performance measurements are greater than the 5 %.

Table 2a seems to confirm Hp.2, but it only partially confirms Hp.1, because while the Incremental approach wins in all the settings where it was expected, the Naive one does not for two cases on four (first column).

The insights become more complex when we consider the memory consumption. Not only we have results where either Hp.1 or Hp.2 are not verified, but the results are not even coherent between memory consumption and the query latency.

Figure 3A and B abstract information of Table 2 in a dashboard representation: on the x-axis we have average latency values and on the y-axis we have average memory values both in logarithmic scale. Figure 3A shows a precise ordering between the two baselines: the Incremental is the best in term of latency while the memory consumption is quite similar. However, Fig. 3B presents an inverted situation, even though the Incremental solution has still less latency, their relative position is totally different.

We stated that any domain specific benchmark aim at fostering technological progress by guaranteeing a fair assessment, this means identify the best solution. However, this evaluation, as well as results in RSP benchmarking state-of-the-art, showed that this is not always possible: at this level of analysis, we can hardly identify in which situation a solution dominates another one and, thus, we should drill down the analysis to determine at least the situations where a solution dominates another one.

7.3   Systematic Comparative Research Approach

Answering the questions we posed in Sect. 1 represents a first step toward a SCRA, which does not refuse the top-down analysis, but extends it into a layered methodology that aims at catching bottom-up different aspects of the analysis.

Top down research tries to answer Q.a and Q.b applying the kind of analysis we did to verify Hp.1 and Hp.2. The top layer (dashboard) summarizes the performances of the benchmarked RSP engines into a solution space, where the engines are ordered by the experimental results. This allows to qualitatively identify the

Table 3. Pattern identification in memory for incremental baseline

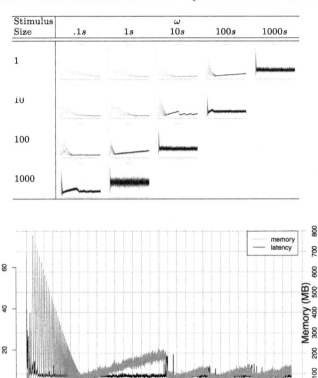

Fig. 4. Experiment 11 − STIMULUS Size = 10 − ω = 1 s − Incremental.

best solution (if such a dominance exists) and, thus, to answer Q.a. Intermediate layers (see Table 2) answer Q.b focusing on both *Inter Experiment Comparisons* – quantitatively contrasting the results of a single measurement in different experiments – and *Intra Experiment Comparisons* – pointing out the relation between multiple measurements within an experiment.

However, $\mathcal{H}$eaven allows also to answer Q.c and Q.d. With high level visual comparisons we can contrast many RSP engine dynamics at once and identify anomalies and/or patterns to answer Q.c. Table 3 shows how differently the memory evolves over time in different experiments just for one baseline (i.e., incremental). Finally, finer grain visual comparisons allows to contrast different variables within an experiment, and highlight the presence of internal trade-offs, that may explain the surprising results we obtained previously. Indeed, Fig. 4 is an example where we can observe that memory oscillations, probably due to the garbage collection, clearly influence the query latency, slowing down the system (see the latency spikes every time the memory is freed).

Enabling SCRA, by the means of $\mathcal{H}$eaven, makes possible to point out how memory and latency, as well as other KPIs, are related. For a deeper analysis of the results, we invite interested readers to check out our technical report[9].

8   Related Works in RSP Engines Benchmarking

The state-of-the-art for RSP engine benchmarking currently comprises four benchmarks [1,11,17,28], summarized in Table 4 and a first attempt towards a general architecture to systematically evaluate RSP engines [25].

LSBench [17] offers a social network data stream schema, the S2Gen generator, and 12 queries. It covers three classes of testing: i.e. (a) query language expressiveness; (b) maximum execution throughput and scalability in terms of query number and static data size; (c) result mismatch between different engines to assess the correctness of the answers.

SRBench [28] focuses on RSP engine query language coverage. It comprises 17 queries to cover three main use cases: (i) flow-data only (ii) flow and background data and (iii) flow and GeoNames and DBpedia datasets. It also targets the reasoning task testing, but not as a main use case. CSRBench [11] extends [28] by addressing the correctness verification. It adds to the query set aggregated queries, queries requiring to join triples with different timestamp and parametric sliding windows queries. Moreover, it provides an Oracle to automatically check correctness of query results.

CityBench [1] provides a real world data streams from the CityPulse project[10] (i.e., vehicle traffic data, weather data and parking spots data); a synthetic data streams about user location and air pollution. Real world static datasets about cultural events are also included. [1] evaluates query latency, memory consumption and result completeness as evaluation metrics. 13 continuous queries are available to test the RSP engine. It also provides a testbed infrastructure for experiment execution, that specifically targets smart city workloads and applications.

Table 4 shows the positioning of this work w.r.t [1,11,17,28]. Differently from the state-of-the-art, we do not propose new workloads, queries and ontologies, but we target the evaluation approach itself. Our aim is to enable SCRA for window based RSP engines rather than proposing yet another benchmark.

Moreover, it is worth to note that this work differs from the one we presented in [25], because in this paper we provides a broader set of requirements, we revised the architecture, we re-implement the proof of concept of the architecture; we also introduce two baselines to be used as terms of comparison and we evaluate the effectiveness of $\mathcal{H}$eaven using them.

---

[9] http://streamreasoning.org/TR/2015/Heaven/2015-results.pdf.
[10] http://citypulse.insight-centre.org/.

Table 4. RSP benchmarking state of the art. ✓: provided, X: not-provided, ≃: partially-provided

	Data stream	Ontologies	Queries	Facility	Baselines
LSBench [17]	✓	✓	✓	≃	X
SRBench [28]	✓	✓	✓ _reasoning_	X	X
CSRBench [11]	✓	✓	✓	X	~
CityBench [1]	✓	✓	✓	≃	X
_H_eaven	X	X	X	✓	✓

9   Conclusions

In this paper, we claimed that the RSP community needs to enable a Systematic Comparative Research Approach for window-based RSP engines. To support this claim and enable SCRA, we proposed: (i) set of requirements elicited on the experiment definition; (ii) a general architecture for an RSP engine Test Stand [25]; (iii) _H_eaven, a proof-of-concept of such architecture that is released as open source; and (iv) two _baselines_, i.e., the minimal meaningful approaches to realize an RSP engine, because the SR community still miss well defined terms of comparison that other communities have (e.g., machine learning algorithms).

We demonstrate the framework effectiveness running a set of experiments that involve _H_eaven and the baselines.

We successfully showed that top-down hypothesis verification is not straight forward by proving that, even when an RSP engines is extremely simple (i.e. the baselines), it is hard to formulate verifiable hypothesis. We also proved _H_eaven effectiveness by showing how it reduces the investigation biases, allowing to better understand ambiguous results, by enabling to catch cross-layered insights.

Finally, we positioned this work in the state of the art of RSP benchmarking, showing its novelty and the differences with similar solution [1,25].

This work is towards an RSP Engine lab, an web-based environment where a users can: (i) design an experiment, (ii) run it, (iii) visualize, and (v) compare the results. Therefore, we have to: (i) implement the adapting facade for the RSP engines to test, and (ii) include, in a experiments suite, existing workload, ontologies and queries (e.g. those of SRBench and LSBench and CityBench). Finally the baselines will shepherd future RSP comparative researches, due to the simplicity of their architectures, the availability of full modeled execution semantics [12,21] and their open source implementation that allows to fully observe their dynamics. Therefore, we want to pursuit the baselines development.

References

1. Ali, M.I., Gao, F., Mileo, A.: CityBench: a configurable benchmark to evaluate RSP engines using smart city datasets. In: Arenas, M., et al. (eds.) ISWC 2015 - Part II. LNCS, vol. 9367, pp. 374–389. Springer, Switzerland (2015)

2. Anicic, D., Fodor, P., Rudolph, S., Stojanovic, N.: EP-SPARQL: a unified language for event processing and stream reasoning. In: WWW, pp. 635–644 (2011)

3. Babcock, B., Babu, S., Datar, M., Motwani, R., Widom, J.: Models and issues in data stream systems. In: SIGMOD, pp. 1–16 (2002)

4. Balduini, M., Della Valle, E., Dell'Aglio, D., Tsytsarau, M., Palpanas, T., Confalonieri, C.: Social listening of city scale events using the streaming linked data framework. In: Alani, H., et al. (eds.) ISWC 2013, Part II. LNCS, vol. 8219, pp. 1–16. Springer, Heidelberg (2013)

5. Barbieri, D.F., Braga, D., Ceri, S., Della Valle, E., Grossniklaus, M.: C-SPARQL: a continuous query language for RDF data streams. IJSC 4(1), 3–25 (2010)

6. Barbieri, D.F., Braga, D., Ceri, S., Della Valle, E., Grossniklaus, M.: Incremental reasoning on streams and rich background knowledge. In: Aroyo, L., Antoniou, G., Hyvönen, E., Teije, A., Stuckenschmidt, H., Cabral, L., Tudorache, T. (eds.) ESWC 2010, Part I. LNCS, vol. 6088, pp. 1–15. Springer, Heidelberg (2010)

7. Barbieri, D.F., Braga, D., Ceri, S., Della Valle, E., Grossniklaus, M.: Querying RDF streams with C-SPARQL. SIGMOD Rec. 39(1), 20–26 (2010)

8. Calbimonte, J.-P., Corcho, O., Gray, A.J.G.: Enabling ontology-based access to streaming data sources. In: Patel-Schneider, P.F., et al. (eds.) ISWC 2010, Part I. LNCS, vol. 6496, pp. 96–111. Springer, Heidelberg (2010)

9. Cugola, G., Margara, A.: Processing flows of information: from data stream to complex event processing. ACM Comput. Surv. 44(3), 15:1–15:62 (2012)

10. Della Valle, E., Ceri, S., Barbieri, D.F., Braga, D., Campi, A.: A first step towards stream reasoning. In: Domingue, J., Fensel, D., Traverso, P. (eds.) FIS 2008. LNCS, vol. 5468, pp. 72–81. Springer, Heidelberg (2009)

11. Dell'Aglio, D., Calbimonte, J.-P., Balduini, M., Corcho, O., Della Valle, E.: On correctness in RDF stream processor benchmarking. In: Aroyo, L., et al. (eds.) ISWC 2013, Part II. LNCS, vol. 8219, pp. 326–342. Springer, Heidelberg (2013)

12. Dell'Aglio, D., Della Valle, E.: Incremental reasoning on RDF streams. In: Harth, A., Hose, K., Schenkel, R. (eds.) Linked Data Management, pp. 413–436. CRC Press, Boca Raton (2014). Chap. 16

13. Dell'Aglio, D., Della Valle, E., Calbimonte, J.P., Corcho, O.: RSP-QL semantics: a unifying query model to explain heterogeneity of RDF stream processing systems. Int. J. Semant. Web Inf. Syst. (IJSWIS) 10(4), 17–44 (2014)

14. Guo, Y., Pan, Z., Heflin, J.: LUBM: a benchmark for OWL knowledge base systems. J. Web Semant. 3(2–3), 158–182 (2005)

15. Kuehl, R.O.: Design of experiments stastistical principles of research design and analysis. No. Q182. K84 2000 (2000)

16. Le-Phuoc, D., Dao-Tran, M., Xavier Parreira, J., Hauswirth, M.: A native and adaptive approach for unified processing of linked streams and linked data. In: Aroyo, L., Welty, C., Alani, H., Taylor, J., Bernstein, A., Kagal, L., Noy, N., Blomqvist, E. (eds.) ISWC 2011, Part I. LNCS, vol. 7031, pp. 370–388. Springer, Heidelberg (2011)

17. Le-Phuoc, D., Dao-Tran, M., Pham, M.-D., Boncz, P., Eiter, T., Fink, M.: Linked stream data processing engines: facts and figures. In: Heflin, J., Sirin, E., Tudorache, T., Euzenat, J., Hauswirth, M., Parreira, J.X., Hendler, J., Schreiber, G., Bernstein, A., Blomqvist, E., Cudré-Mauroux, P. (eds.) ISWC 2012, Part II. LNCS, vol. 7650, pp. 300–312. Springer, Heidelberg (2012)

18. Margara, A., Urbani, J., van Harmelen, F., Bal, H.E.: Streaming the web: reasoning over dynamic data. J. Web Semant. 25, 24–44 (2014)

19. Muñoz, S., Pérez, J., Gutierrez, C.: Minimal deductive systems for RDF. In: Franconi, E., Kifer, M., May, W. (eds.) ESWC 2007. LNCS, vol. 4519, pp. 53–67. Springer, Heidelberg (2007)
20. Nenov, Y., Piro, R., Motik, B., Horrocks, I., Wu, Z., Banerjee, J.: RDFox: a highly-scalable RDF store. In: Arenas, M., et al. (eds.) ISWC 2015 - Part II. LNCS, vol. 9367, pp. 3–20. Springer, Switzerland (2015)
21. Ren, Y., Pan, J.Z.: Optimising ontology stream reasoning with truth maintenance system. In: CIKM, pp. 831–836 (2011)
22. Scharrenbach, T., Urbani, J., Margara, A., Della Valle, E., Bernstein, A.: Seven commandments for benchmarking semantic flow processing systems. In: Cimiano, P., Corcho, O., Presutti, V., Hollink, L., Rudolph, S. (eds.) ESWC 2013. LNCS, vol. 7882, pp. 305–319. Springer, Heidelberg (2013)
23. Tatbul, N., Cetintemel, U., Zdonik, S., Cherniak, M., Stonebraker, M.: Exploiting punctuation semantics in continuous data streams. IEEE Trans. Knowl. Data Eng. **15**(3), 555–568 (2003)
24. Thomas, E., Pan, J.Z., Ren, Y.: TrOWL: tractable OWL 2 reasoning infrastructure. In: Aroyo, L., Antoniou, G., Hyvönen, E., ten Teije, A., Stuckenschmidt, H., Cabral, L., Tudorache, T. (eds.) ESWC 2010, Part II. LNCS, vol. 6089, pp. 431–435. Springer, Heidelberg (2010)
25. Tommasini, R., Della Valle, E., Balduini, M., Dell'Aglio, D.: Heaven test stand: towards comparative research on RSP engines. In: OrdRing (2015)
26. Urbani, J., Margara, A., Jacobs, C., van Harmelen, F., Bal, H.: DynamiTE: parallel materialization of dynamic RDF data. In: Alani, H., et al. (eds.) ISWC 2013, Part I. LNCS, vol. 8218, pp. 657–672. Springer, Heidelberg (2013)
27. Walavalkar, O., Joshi, A., Finin, T., Yesha, Y.: Streaming knowledge bases. In: In International Workshop on Scalable Semantic Web Knowledge Base Systems (2008)
28. Zhang, Y., Duc, P.M., Corcho, O., Calbimonte, J.-P.: SRBench: a streaming RDF/SPARQL benchmark. In: Heflin, J., et al. (eds.) ISWC 2012, Part I. LNCS, vol. 7649, pp. 641–657. Springer, Heidelberg (2012)

Natural Language Processing
and Information Retrieval Track

Bridging the Gap Between Formal Languages and Natural Languages with Zippers

Sébastien Ferré[(✉)]

IRISA, Université de Rennes 1, Campus de Beaulieu, 35042 Rennes, France
ferre@irisa.fr

Abstract. The Semantic Web is founded on a number of Formal Languages (FL) whose benefits are precision, lack of ambiguity, and ability to automate reasoning tasks such as inference or query answering. This however poses the challenge of mediation between machines and users because the latter generally prefer Natural Languages (NL) for accessing and authoring knowledge. In this paper, we introduce the N<A>F design pattern based on Abstract Syntax Trees (AST), Huet's zippers and Montague grammars to zip together a natural language and a formal language. Unlike question answering, translation does not go from NL to FL, but as symbol N<A>F suggests, from ASTs (A) of an intermediate language to both NL (N<A) and FL (A>F). ASTs are built interactively and incrementally through a user-machine dialog where the user only sees NL, and the machine only sees FL.

1 Introduction

The Semantic Web is founded on a number of formal languages to represent and reason on facts (RDF), logical axioms (OWL), or queries (SPARQL). Those languages make data processable by machines but they also constitute a language barrier to the production and consumption of semantic data by end users. An important issue in the Semantic Web is to bridge the gap between formal languages (FL) designed for the machines, and natural languages (NL) understood by humans. A crucial aspect of this issue is the *adequacy* between the expressivity of FL and NL. Users should be guided to NL expressions that have a FL counterpart to avoid the *habitability* problem [11]; and all or most of the *expressivity* of the FL should be expressible through NL if we are to avoid two-class citizenship among users.

We propose the N<A>F design pattern to bridge the gap between NL and FL, along with techniques to ease its implementations. As a design pattern, it is not a finished design but a reusable template. We have successfully used it to deal with the NL-FL gap in different tasks: querying SPARQL endpoints [5], authoring RDF descriptions [8], and completing OWL ontologies [7]. However, in those previous works, the design pattern was mixed with other aspects, and only presented through its concrete applications. We here give a detailed and

This research is supported by ANR project IDFRAud (ANR-14-CE28-0012-02).

H. Sack et al. (Eds.): ESWC 2016, LNCS 9678, pp. 269–284, 2016.
DOI: 10.1007/978-3-319-34129-3_17

independent account of the N<A>F design pattern in order to facilitate its reuse in other contexts.

The N<A>F design pattern drastically differs from Question Answering (QA) approaches [13] in two ways. First, translation does not go from NL to FL *through* intermediate representations, but as symbol N<A>F suggests, *from* an intermediate language (A) to *both* NL (N<A) *and* FL (A>F). Second, questions are not written but *interactively and incrementally built* through a user-machine dialogue. That approach provides a number of added values compared to existing approaches (see Sect. 6 for details). Compared to QA, it ensures a strong reliability because it has no habitability problem, and it allows for higher expressivity. Compared to Controlled Natural Languages (CNL) with auto-completion [10], it offers similar expressivity with more flexibility in query construction, and more semantic and more fine-grained guidance. Compared to graphical query builders [11], it hides FL behind NL, and also offers more semantic and more fine-grained guidance. In truth, those added values do not guarantee user preference, a *subjective* value, but they are *objective* values along which different approaches can be characterized and compared precisely. Still, user preference is our ultimate goal, and this is why we develop and maintain a number of tools, one of which (Sparklis [5]) has already attracted a large number of users[1]. The design pattern is computationally lightweight, and can adapt to all sorts of FL and tasks. Its main limitation is that it cannot be used to process NL expressions that were not edited through it, typically existing texts.

The paper is structured as follows. Section 2 presents the theoretical foundations: Abstract Syntax Trees (AST), Huet's zippers, and Montague grammars. Section 3 presents an overview of our design pattern, and Sect. 4 illustrates it in detail to SPARQL-based querying. Section 5 demonstrates the generality of our design pattern by reporting on three different applications. Finally, Sect. 6 discusses related work, and Sect. 7 concludes.

2   Theoretical Foundations

2.1   Abstract Syntax Trees

Abstract syntax is a way to describe the structure of the sentences of a language, while abstracting over their concrete representation. Abstract syntax is generally not represented as a text but as a tree data structure called Abstract Syntax Tree (AST). In the compilation of programming languages, ASTs are commonly used as an intermediate representation between the source code (a text) and the target code (a binary). They here play the same role between NL and FL.

ASTs are best defined with algebraic datatypes that allow to recursively compose base types[2]. For example, assuming base type *term* for SPARQL variables and RDF nodes, the following datatypes define a simplified subset of SPARQL

---

[1] http://www.semantic-web-journal.net/system/files/swj1236.pdf.

[2] In OO programming, algebraic datatypes are modelled with the Composite pattern.

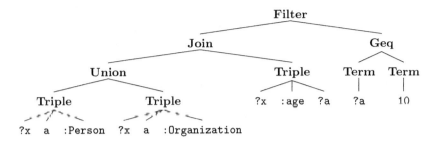

Fig. 1. AST graphical representation of the SPARQL graph pattern { ?x a :Person } UNION { ?x a :Organization } ?x :age ?a FILTER (?a >= 10)

graph patterns (*gp*), and expressions (*expr*):

$$gp \quad ::= \textbf{Triple}(\textit{term}, \textit{term}, \textit{term}) \mid \textbf{Filter}(\textit{gp}, \textit{expr})$$
$$\mid \textbf{Join}(\textit{gp}, \textit{gp}) \mid \textbf{Union}(\textit{gp}, \textit{gp}) \mid \textbf{Optional}(\textit{gp}, \textit{gp})$$
$$expr ::= \textbf{Term}(\textit{term}) \mid \textbf{Geq}(\textit{expr}, \textit{expr}) \mid \textbf{Regex}(\textit{expr}, \textit{expr})$$

Each italic name is a datatype, and each boldface name is a *constructor* (AST node label). Every datatype may have any number of alternative constructors, and every constructor may have any number of arguments (including 0 for constant constructors). Figure 1 shows a graphical representation of the AST of a SPARQL graph pattern.

2.2 Huet's Zippers

In his "Functional Pearl" [9], Huet introduced the *zipper* as a technique for traversing and updating a data structure in a purely functional way, and yet in an efficient way. Purely functional programming completely avoids modification in place of data structures, and makes it much easier to reason on program behaviour, and hence to ensure their correctness [17]. We here use zippers for the incremental construction of ASTs. A simple and illustrative example is on simply chained lists, their traversal, and the insertion of elements. Given a base type *elt* for list elements, the *list* datatype is defined with two constructors: one for the empty list, one for adding an element at the head of another list.

$$list ::= \textbf{Nil} \mid \textbf{Cons}(\textit{elt}, \textit{list})$$

The AST of list $[1, 2, 3]$ is $\textbf{Cons}(1, \textbf{Cons}(2, \textbf{Cons}(3, \textbf{Nil})))$. The zipper idea is to keep a location in the list such that it is easy and efficient to insert an additional element at that location, and also to move that location to the left or to the right. A location (e.g. at element 2 in the above list) splits the data structure in two parts: the *sub-structure* at the location ($[2, 3]$), and the surrounding *context* ($[1, \_]$). It has been shown that the context datatype corresponds to a data structure with one hole, and can be seen as the *derivative* of the structure

$$
\begin{array}{rl}
s & \to np\ vp \\
vp & \to vt\ np \\
np & \to e \\
& |\ det\ nn \\
det & \to \mathbf{a(n)} \\
& |\ \mathbf{every} \\
e & \to \mathbf{John}\ |\ \dots\ |\ \mathbf{Mary} \\
nn & \to \mathbf{man}\ |\ \dots\ |\ \mathbf{woman} \\
vt & \to \dots\ |\ \mathbf{loves}
\end{array}
\qquad
\begin{array}{r}
(np\ vp) \\
\lambda x.(np\ \lambda y.(vt\ x\ y)) \\
\lambda d.(d\ e) \\
\lambda d.(det\ nn\ d) \\
\lambda d_1.\lambda d_2.(\exists X.((d_1\ X) \wedge (d_2\ X))) \\
\lambda d_1.\lambda d_2.(\forall X.((d_1\ X) \Rightarrow (d_2\ X))) \\
\mathbf{Mary} \\
\lambda x.(\mathbf{woman}(x)) \\
\lambda x.\lambda y.(\mathbf{loves}(x,y))
\end{array}
$$

Fig. 2. Montague grammar of a small fragment of English

datatype [1]. We therefore name *list'* the context datatype for lists, and define it as follows.

$$list' ::= \mathbf{Root}\ |\ \mathbf{Cons}'(elt, list')$$

That definition says that a list occurs either as a root list or as the right-argument of constructor **Cons**, which has in turn its own context. For example, the context at location 3 of list $[1, 2, 3]$ is $\mathbf{Cons}'(2, \mathbf{Cons}'(1, \mathbf{Root}))$. In fact, a list context is the reverse list of elements before the location. Finally, a zipper data structure combines a structure and a context: *zipper* ::=$\mathbf{List}(list, list')$. A zipper contains all the information of a data structure plus a location in that structure. That location is also called "focus".

A zipper makes it easy to move the location to neighbour locations, and to apply local transformations such as insertions or deletions. For example, to insert an element x in a list zipper $\mathbf{List}(l, l')$ is as simple as returning the zipper $\mathbf{List}(\mathbf{Cons}(x, l), l')$. Given a zipper $\mathbf{List}(l, \mathbf{Cons}'(e, l'))$, the location can be moved to the left by returning the zipper $\mathbf{List}(\mathbf{Cons}(e, l), l')$.

2.3   Montague Grammars

Montague grammars [4] are an approach to translation from NL to FL that is based on context-free grammars and simply typed λ-calculus. In a Montague grammar, each rule is decorated by a λ-expression that denotes the FL semantics of the syntactic construct defined by the rule. We can use them for translation from ASTs to FL because our ASTs have a syntactic structure close to NL.

Figure 2 shows an example of Montague grammar for a small fragment of English, and its semantics in first-order logic. That fragment covers sentences like 'John loves Mary', and 'every man loves a woman'. The semantics is defined in a fully compositional style, i.e., the semantics of a construct is always a composition of the semantics of sub-constructs. The first rule in Fig. 2 says that a sentence (s) whose syntax is made of a noun phrase (np) followed by a verb phrase (vp) has its semantics defined by the λ-application ($np\ vp$) of the semantics of the noun phrase to the semantics of the verb phrase. The semantics of a sentence is a logical formula, e.g. $\forall X.(\mathbf{man}(X) \Rightarrow \exists Y.(\mathbf{woman}(Y) \wedge \mathbf{loves}(X, Y)))$ for the sentence 'every man loves a woman'[3]. The semantics of a verb phrase is a formula λ-abstracted over an entity x (λx.) because a verb phrase misses an entity e to form a complete sentence. If we call such an abstraction a *description*, then the semantics of a noun phrase is a formula λ-abstracted over a description d because a noun phrase misses a verb phrase vp to form a complete sentence. A noun nn has the same semantic type as a verb phrase, i.e. a formula missing an entity. The semantics of a transitive verb vt is a formula missing two entities, a subject x and an object y. The semantics of a determiner det is a quantifier introducing a fresh logical variable X, i.e. a formula abstracted over two descriptions.

Given a sentence, its semantics is obtained by the bottom-up composition of λ-terms from the grammar through the parse tree. For example, from the sentence 'every man loves a woman' whose parse tree is '$[_s[_{np}[_{det}\text{every}]$ $[_{nn}\text{man}]]$ $[_{vp}[_{vt}\text{loves}]$ $[_{np}[_{det}\text{a}]$ $[_{nn}\text{woman}]]]]$', we obtain the λ-term

$$((\lambda d_1.\lambda d_2.(\forall X.((d_1\ X) \Rightarrow (d_2\ X)))\ \lambda x.(\mathbf{man}(x)))$$
$$\lambda x.((\lambda d_1.\lambda d_2.(\exists Y.((d_1\ Y) \wedge (d_2\ Y)))\ \lambda x.(\mathbf{woman}(x)))$$
$$\lambda y.(\lambda x.\lambda y.(\mathbf{loves}(x,y))\ x\ y)))$$

which simplifies by β-reduction to the expected formula $\forall X.(\mathbf{man}(X) \Rightarrow \exists Y.(\mathbf{woman}(Y) \wedge \mathbf{loves}(X,Y)))$.

Although initially designed for translating NL to formal logic, Montague grammars as a technique can be used with other kinds of FLs as a target: e.g., SPARQL [6]. Their main benefits are to bridge the gap from natural to formal languages, and to offer a fully-compositional semantics. Fully-compositional semantics, like purely-functional data structures, makes it easier to ensure the correctness of the translation because there are no side-effects.

3   Overview of the Design Pattern

The principles of the N<A>F design pattern are schematized as a "suspended bridge over the NL-FL gap" in Fig. 3. The central pillar is made of *AST zippers*, i.e. AST tree structures with a focus on one AST node. The nature of the intermediate language abstracted by the ASTs depends on the application: e.g., queries, descriptions, logical axioms. The AST zipper is initialized by the system, and modified by users applying structural *transformations*, never by direct textual input. For that reason, it is important to design a *complete* set of transformations, so that every AST is reachable through a finite sequence of transformations. Conversely, only *safe* transformations should be suggested to users, so as to avoid syntactic and semantic errors. In the case of querying, a semantic error could be applying a transformation that leads to an empty result. In the case of ontology construction, it could be constructing an axiom that leads to inconsistency.

---

[3] We use capital letters for logical variables to distinguish them from λ-variables.

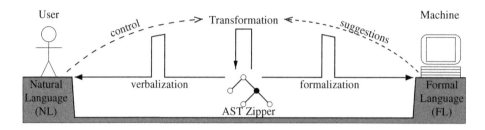

Fig. 3. Principle of zipper-based edition for bridging the gap between NL and FL

In order for the AST structure to be understood by both the user and the machine, *verbalization* translates ASTs to NL, and *formalization* translates ASTs to FL. In addition to those translations, verbalization supports user control by showing suggested transformations in NL, and formalization supports the computation of suggestions by taking into account the semantics of the AST. A key issue in the design of ASTs is to make the two translations semantically transparent, and simple enough. First, the AST structure should reproduce the syntactic structure of NL (e.g., sentences, noun phrases, verb phrases), while abstracting over as many details as possible. Indeed, starting with a flat representation like SPARQL, it is possible to produce a NL version [15], but it is difficult to make it stable across transformations. Second, the AST structure should semantically align with the target FL. Indeed, every AST that can be obtained by a sequence of transformations must have a semantics that is expressible in FL.

4   Application to a Core RDF Query Language

In this section, we explain in detail the application of the N<A>F design pattern to a core RDF query language (CRQL for short)[4]. CRQL covers a significant fragment of SPARQL 1.0 SELECT queries (i.e., triple patterns, UNION, simple filters) but restricted to tree patterns, and extended with negation (NOT EXISTS). Section 5 discusses variations around CRQL to deal with more expressive querying and other tasks.

4.1   AST Zippers

The following datatype definitions describe the structure of CRQL ASTs. Given base types for RDF nodes (*node*), RDFS classes (*class*), RDFS properties (*prop*), and literals (*lit*), we define abstract sentences (s), abstract noun phrases (np), and abstract verb phrases (vp).

s ::= **Select**(np)
np ::= **Something** | **Some**(*class*) | **Node**(*node*) | **That**(np, vp)
vp ::= **IsA**(*class*) | **Has**(*prop, np*) | **IsOf**(*prop, np*) | **Geq**(*lit*)
        | **True** | **And**(vp, vp) | **Or**(vp, vp) | **Not**(vp)

---

[4] We also provide online the source code in two programming languages: Java and OCaml. Visit http://www.irisa.fr/LIS/ferre/pub/CRQL/.

Node np_{ex} in Fig. 4 points to the tree representation of an example CRQL AST. It has type np, and specifies *"any film that has director Steven Spielberg and that has release date something that is after January 1st, 2010"*. The AST definitions reflect NL syntax with noun phrases and verb phrases, but is indeed abstract because all sorts of syntactic distinctions are ignored. The two constructors **Has** and **IsOf** account for the traversal direction of a property, but not whether the property is rendered by a verb, a noun, or a transitive adjective. Similarly, constructor **Geq** has different NL renderings depending on the datatype of the literal.

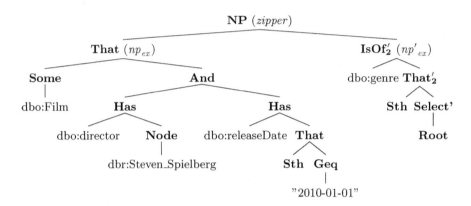

Fig. 4. Example CRQL zipper made of a sub-structure (np_{ex}), and a context (np'_{ex})

The following datatype definitions describe the structure of CRQL contexts. They are automatically obtained as the derivatives of the AST datatypes (see Sect. 2.2). When a constructor has several arguments (e.g., **And**), the derived constructors are indexed by the position of the focus (e.g., **And$'_2$** for a focus on the second argument).

s' $::=$ **Root**
np' $::=$ **Select$'$**(s') | **That$'_1$**(np', vp) | **Has$'_2$**$(prop, vp')$ | **IsOf$'_2$**$(prop, vp')$
vp' $::=$ **That$'_2$**(np, np') | **Not$'$**(vp')
     | **And$'_1$**(vp', vp) | **And$'_2$**(vp, vp') | **Or$'_1$**(vp', vp) | **Or$'_2$**(vp, vp')

Node np'_{ex} in Fig. 4 points to the tree representation of an example CRQL context. It has type np', and specifies the one-hole AST *"select something that is the genre of _"*, where the underscore (hole) gives the location of the zipper substructure. Finally, the following datatype definition describes a CRQL zipper, combining an AST datatype and its derivative.

$zipper$ $::=$ **S**(s, s') | **NP**(np, np') | **VP**(vp, vp')

Therefore, when editing a CRQL query, the focus can be put on the whole sentence, on any noun phrase (i.e. on any entity involved in the query), or on any particular verb phrase (i.e. on any description of any entity in the query). Figure 4 displays the tree representation of an example CRQL zipper that represents the NL question *"Give me the genre of films directed by Steven Spielberg and whose release date is after January 1st, 2010"*, where the focus is on films.

4.2 A Complete Set of Zipper Transformations

Because ASTs are only built by the successive and interactive application of transformations, it is important to define an initial zipper and a set of zipper transformations that makes the building process complete.

Definition 1 (Completeness). *An initial zipper and a set of zipper transformations is* complete *w.r.t. AST datatypes iff every AST zipper can be reached by applying a finite sequence of transformations starting with the initial zipper.*

We start by defining an initial AST x_0 for each AST datatype x: $s_0 := \mathbf{Select}(np_0)$, $np_0 := \mathbf{Something}$, $vp_0 := \mathbf{True}$. The initial zipper is then defined as $zipper_0 := \mathbf{S}(s_0, \mathbf{Root})$. It corresponds to the totally unconstrained query that returns the list of everything.

We continue by defining a number of zipper transformations. A zipper transformation is formally defined as a partial mapping from zipper to zipper. Transformations are denoted by all-uppercase names, and are defined by unions of mappings from a zipper pattern to a zipper expression. For example, transformations that introduce constructors **That**, **Not**, **And**, **Or** are defined as follows. Transformation NOT toggles the application of negation; other transformations coordinate a sub-structure x with an initial AST x_0, and move the focus to x_0.

$$
\begin{aligned}
THAT &:= \mathbf{NP}(np, np') \to \mathbf{VP}(vp_0, \mathbf{That}'_2(np, np')) \\
NOT &:= \mathbf{VP}(\mathbf{Not}(vp), vp') \to \mathbf{VP}(vp, vp') \\
&\ \mid\ \mathbf{VP}(vp, vp') \to \mathbf{VP}(\mathbf{Not}(vp), vp') \\
AND &:= \mathbf{VP}(vp, vp') \to \mathbf{VP}(vp_0, \mathbf{And}'_2(vp, vp')) \\
OR &:= \mathbf{VP}(vp, vp') \to \mathbf{VP}(vp_0, \mathbf{Or}'_2(vp, vp'))
\end{aligned}
$$

In order to allow transformations at an arbitrary focus, it is important to allow moving the focus location through the AST. To this purpose, we define four transformations UP, $DOWN$, $LEFT$, $RIGHT$ to move the focus respectively up to the parent AST node, down to the leftmost child, to the left sibling, and to the right sibling. We only provide the definitions for constructor **That** as other constructors work in a similar way.

$$
\begin{aligned}
DOWN &:= \mathbf{NP}(\mathbf{That}(np, vp), np') \to \mathbf{NP}(np, \mathbf{That}'_1(np', vp)) \\
UP &:= \mathbf{NP}(np, \mathbf{That}'_1(np', vp)) \to \mathbf{NP}(\mathbf{That}(np, vp), np') \\
&\ \mid\ \mathbf{VP}(vp, \mathbf{That}'_2(np, np')) \to \mathbf{NP}(\mathbf{That}(np, vp), np') \\
RIGHT &:= \mathbf{NP}(np, \mathbf{That}'_1(np', vp)) \to \mathbf{VP}(vp, \mathbf{That}'_2(np, np')) \\
LEFT &:= \mathbf{VP}(vp, \mathbf{That}'_2(np, np')) \to \mathbf{NP}(np, \mathbf{That}'_1(np', vp))
\end{aligned}
$$

Most transformations are performed by inserting or deleting a sub-structure at the focus. We define a generic transformation $INSERT$ that works on any zipper with focus on an initial AST, $\mathbf{X}(x_0, x')$, where x stands for any AST datatype.

$$INSERT(x) := \mathbf{X}(x_0, x') \to \mathbf{X}(x, x')$$

Note that $INSERT$ takes an AST of type x as an argument, the sub-structure to insert. Because ASTs cannot be textually edited, the insertable sub-structures must be picked from a finite collection, and have to be suggested by the system. A generic $DELETE$ transformation can also be defined to undo insertions.

The above set of transformations can be proved complete provided that a proper collection of insertable elements is defined.

Theorem 1 (Completeness). *Assume the following insertable elements for each datatype, where 'node', 'class', and 'prop' hold for any value present in the RDF dataset, and where 'lit' holds for any filtering value input by users:*

$np :$ **Some**$(class)$, **Node**$(node)$,
$vp :$ **IsA**$(class)$, **Has**$(prop, np_0)$, **IsOf**$(prop, np_0)$, *and* **Geq**(lit).

Then, initial zipper$_0$ and the set of transformations defined above ($INSERT(x)$ applying to insertable elements) is complete.

$T(\mathbf{Select}(np))$	$:= DOWN; T(np); UP$
$T(\mathbf{Something})$	$:= ID$
$T(\mathbf{Some}(class))$	$:= INSERT(\mathbf{Some}(class))$
$T(\mathbf{Node}(node))$	$:= INSERT(\mathbf{Node}(node))$
$T(\mathbf{That}(np, vp))$	$:= T(np); THAT; T(vp); UP$
$T(\mathbf{IsA}(class))$	$:= INSERT(\mathbf{IsA}(class))$
$T(\mathbf{Has}(prop, np))$	$:= INSERT(\mathbf{Has}(prop, np_0)); DOWN; T(np); UP$
$T(\mathbf{IsOf}(prop, np))$	$:= INSERT(\mathbf{IsOf}(prop, np_0)); DOWN; T(np); UP$
$T(\mathbf{Geq}(lit))$	$:= INSERT(\mathbf{Geq}(lit))$
$T(\mathbf{True})$	$:= ID$
$T(\mathbf{And}(vp_1, vp_2))$	$:= T(vp_1); AND; T(vp_2); UP$
$T(\mathbf{Or}(vp_1, vp_2))$	$:= T(vp_1); OR; T(vp_2); UP$
$T(\mathbf{Not}(vp))$	$:= T(vp); NOT$

Fig. 5. Recursive definition of the transformation sequence $T(x)$ from x_0 to AST x

Proof. First, it is easy to show that the moving transformations give access to all zippers of an AST. Therefore, to prove completeness, it is enough to prove that every zipper in the form $\mathbf{S}(s, \mathbf{Root})$ is reachable. Figure 5 defines a recursive function $T(x)$ that returns a transformation sequence from an initial AST x_0 to any AST x, where x stands for any AST datatype. For each case, it can be proved that the transformation sequence $T(x)$ indeed leads from x_0 to x, and that every recursive call is well defined (sub-structure x_0 at focus). ∎

The above proof provides an algorithm T for computing the transformation sequence leading to any AST x. The example zipper given in Fig. 4 can be reached with the following sequence:

$DOWN$; $THAT$; $INSERT(\mathbf{IsOf}(\texttt{dbo:genre}, np_0))$; $DOWN$;
$INSERT(\mathbf{Some}(\texttt{dbo:Film}))$;
$THAT$; $INSERT(\mathbf{Has}(\texttt{dbo:director}, np_0))$; $DOWN$;
$INSERT(\mathbf{Node}(\texttt{dbr:Steven\_Spielberg}))$; UP;
AND; $INSERT(\mathbf{Has}(\texttt{dbo:releaseDate}, np_0))$; $DOWN$;
$THAT$; $INSERT(\mathbf{Geq}(\texttt{"2010-01-01"}))$; UP; UP; UP; UP.

In practice, it appears useful to tune transformations so as to minimize the number of interaction steps for users. For example, moving down after inserting a property can be made automatic given its frequency.

4.3   Formalization to SPARQL

AST zippers are given a semantics by translating them to a formal language. Here, we translate CRQL to SPARQL. Translating queries to SPARQL makes it easy to evaluate them with SPARQL engines and through SPARQL endpoints. Before the translation itself, we apply to zippers a transformation $NORM$ to normalize them into sentence zippers $\mathbf{S}(s, \mathbf{Root})$. This has the advantage to simplify the translation by restricting it to ASTs, and making it unnecessary for zipper contexts.

$$
\begin{aligned}
NORM := \; & \mathbf{S}(s, \mathbf{Root}) \rightarrow \mathbf{S}(s, \mathbf{Root}) \\
| \; & \mathbf{VP}(vp_1, \mathbf{Or}'_1(vp', vp_2)) \rightarrow NORM(\mathbf{VP}(vp_1, vp')) \\
| \; & \mathbf{VP}(vp_2, \mathbf{Or}'_2(vp_1, vp')) \rightarrow NORM(\mathbf{VP}(vp_2, vp')) \\
| \; & \mathbf{VP}(vp, \mathbf{Not}'(vp')) \rightarrow NORM(\mathbf{VP}(vp, vp')) \\
| \; & \mathbf{X}(x, x') \rightarrow NORM(UP(\mathbf{X}(x, x')))
\end{aligned}
$$

By default, normalizing a zipper is simply moving the focus up repeatedly (last line), until the sentence level is reached (first line). However, constructors **Or** (union), and **Not** (negation) are removed when the focus is in their scope (middle lines). The reason is to make sure that the query results are defined and relevant to the focus. For example, if the focus is in one branch of an union, then the other branch can be temporarily ignored in the semantics. In SPARQL, a variable that is introduced inside a negation cannot be returned in results. Moving the focus in the scope of a negation is then a way to access the values of such a variable, by temporarily ignoring negation.

Once a sentence AST s is obtained by normalization, its translation to a SPARQL SELECT query can be defined by the Montague grammar in Fig. 6. SPARQL strings are delimited with quotes, and concatenated with $+$. One SPARQL variable '$?x_i$' is introduced for each indefinite head (constructors **Something** and **Some**). For example, the translation of the zipper given in Fig. 4 produces the following SPARQL query. The three variables correspond respectively to the film's genre, the film, and the film's release date.

```
SELECT ?x1 ?x2 ?x3 WHERE
```

$$
\begin{array}{lll}
s & ::= \mathbf{Select}(np) & \text{'SELECT ?x}_1 \ \ldots \ \text{?x}_n \ \text{WHERE } \{ \text{'} + (np \ \lambda x.(\text{''})) + \text{'}\}\text{'} \\
np & ::= \mathbf{Something}_i & \lambda d.((d \ \text{'?x}_i\text{'})) \\
& | \ \ \mathbf{Some}(class)_i & \lambda d.(\text{'?x}_i \ \text{a'} + class + \text{'.'} + (d \ \text{'?x}_i\text{'})) \\
& | \ \ \mathbf{Node}(node) & \lambda d.((d \ node)) \\
& | \ \ \mathbf{That}(np, vp) & \lambda d.(np \ \lambda x.((d \ x) + \text{'.'} + (vp \ x))) \\
vp & ::= \mathbf{IsA}(class) & \lambda x.(x + \text{'a'} + class) \\
& | \ \ \mathbf{Has}(prop, np) & \lambda x.((np \ \lambda y.(x + prop + y))) \\
& | \ \ \mathbf{IsOf}(prop, np) & \lambda x.((np \ \lambda y.(y + prop + x))) \\
& | \ \ \mathbf{Geq}(lit) & \lambda x.(\text{'FILTER('} + x + \text{'>='} + lit + \text{')'}) \\
& | \ \ \mathbf{True} & \lambda x.(\text{''}) \\
& | \ \ \mathbf{And}(vp_1, vp_2) & \lambda x.((vp_1 \ x) + \text{'.'} + (vp_2 \ x)) \\
& | \ \ \mathbf{Or}(vp_1, vp_2) & \lambda x.(\text{'}\{\text{'} + (vp_1 \ x) + \text{'}\}\ \text{UNION } \{\text{'} + (vp_2 \ x) + \text{'}\}\text{'}) \\
& | \ \ \mathbf{Not}(vp) & \lambda x.(\text{'FILTER NOT EXISTS } \{\text{'} + (vp \ x) + \text{'}\}\text{'})
\end{array}
$$

Fig. 6. Formalization to SPARQL of CRQL ASTs

```
{ ?x2 a dbo:Film . ?x2 dbo:genre ?x1 .
  ?x2 dbo:director dbr:Steven_Spielberg .
  ?x2 dbo:releaseDate ?x3 . FILTER (?x3 >= "2010-01-01") }
```

4.4   Verbalization to English

ASTs are given a concrete and user-friendly syntax by verbalizing them to NL. It is simpler than direct verbalizations of FL expressions, e.g. SPARQL queries [15], because ASTs follow the phrase structure of NL. A good quality verbalization requires linguistic resources about lexical and syntactic aspects (e.g., Lemon lexicons [14], WordNet, Grammatical Framework [16]). However, because humans are more robust than machines at language understanding, a naive verbalization can be good enough in practice for simple languages like CRQL. We here sketch such a verbalization, showing that the technique of Montague grammars is also useful for AST to NL translation.

$$
\begin{array}{lll}
s & ::= \mathbf{Select}(np) & \text{'Give me'} + np \\
np & ::= \mathbf{That}(np, vp) & np + (vp \ 0) \\
vp & ::= \mathbf{IsA}(class) & \lambda n.(\text{'that'} + (is \ n) + \text{'a(n)'} + class) \\
& | \ \ \mathbf{Has}(prop, np) & \lambda n.(\text{'whose'} + prop + (is \ n) + np) \\
& | \ \ \mathbf{And}(vp_1, vp_2) & \lambda n.((vp_1 \ n) + (and \ n) + (vp_2 \ n)) \\
& | \ \ \mathbf{Not}(vp) & \lambda n.(vp \ \overline{n})
\end{array}
$$

Fig. 7. Verbalization to English of CRQL ASTs (partial)

Before verbalization, an AST zipper is normalized like for SPARQL translation (Sect. 4.3), except that constructors **Or/Not** in the context are not removed but marked for dimmed display. Similarly, the zipper sub-structure is highlighted to show the focus. For a naive verbalization, we assume that nodes, classes and properties are verbalized to their labels, which we assume to be common nouns: e.g., 'film', 'director', 'release date'. The Montague grammar translates each AST to a string, and translations of verb phrases are relative clauses abstracted by a Boolean flag, n, in order to propagate negation as a modifier to the relevant verbs. Figure 7 shows a subset of the translation rules. Function *is* verbalizes the copula depending on negation: $(is\ 0) := $ 'is', $(is\ 1) := $ 'is not'. Function *and* encodes De Morgan laws: $(and\ 0) := $ 'and', $(and\ 1) := $ 'or'. The translation of an AST can be post-processed to further simplify it. For example, the sequence 'something that is *NP*' can be replaced by *NP*, and the sequence 'is something after' by 'is after'. Then, the verbalization of the example in Fig. 4 results in 'Give me the genre of a film whose director is Steven Spielberg and whose release date is after January 1st, 2010'.

5   Validation of the Design Pattern

As the purpose of a design pattern is to provide a same solution to different problems, we validate the N<A>F design pattern by showing how it has been used to address the NL-FL gap in three quite different tasks related to the Semantic Web. Each task brought its own contribution relative to the task, and was properly evaluated w.r.t. expressivity, usability, and/or performance. They constitute a demonstration of the effectiveness and genericity of our design pattern. In this section, we shortly describe each tool, and how it can be seen as a variation of querying with CRQL (detailed in Sect. 4).

5.1   SPARKLIS: Querying SPARQL Endpoints

SPARKLIS [5] is a tool for querying SPARQL endpoints. Its target FL is hence SPARQL. The covered subset of SPARQL includes CRQL, extended with arbitrary basic graph patterns (including cycles), additional filters, OPTIONAL, aggregation and grouping, and solution modifiers (ordering, HAVING). Cyclic graph patterns are verbalized with anaphoras: e.g., 'a film that is starring the director of <u>the film</u>'. Aggregations are verbalized with head modifiers: e.g., 'Give me <u>the number of</u> film whose director is Tim Burton'. SPARQL endpoints are used not only to retrieve query answers, but also to compute the insertable elements that do not lead to empty answers, i.e. that provide information relevant to the current answers. For example, considering flights arriving in Heraklion, only showing departure cities, not all cities.

SPARKLIS is available online[5], had about 1000 unique users over 18 months, and was used on more than 100 different endpoints. It scales to datasets as large as DBpedia (several billions triples). In an experiment on QALD-3 DBpedia questions, the median query construction time was 30 s, the maximum time was 109 s, and only one question led to a timeout.

---

[5] Sparklis online at http://www.irisa.fr/LIS/ferre/sparklis/.

5.2   UTILIS: **Authoring RDF Descriptions**

UTILIS [8] is a functionality integrated to SEWELIS[6] for authoring RDF descriptions. Its target FL is hence RDF. A subset of CRQL can be used to cover all of RDF. Constructor **Select**(np) is replaced by constructor **Describe**($node, vp$), where $node$ acts as the RDF node being described, and vp acts as its description. Constructors **Or**, **Not**, and filters become irrelevant. Constructors **Something** and **Some** correspond to the introduction of blank nodes. The key difference with SPARKLIS lies in the semantics of AST zippers. Their semantics is a ranking of the RDF nodes of a dataset, according to their similarity with the entity at the zipper focus. Given the example zipper of Fig. 4, one gets first *films by Spielberg since 2010*, then *films by Spielberg until 2010*, then *films by other directors*, etc. The system uses the description of most similar nodes to suggest insertable elements.

A user study comparing UTILIS to PROTÉGÉ has shown that users prefered the fine-grained suggestions of UTILIS to the static entity lists of PROTÉGÉ. We have also observed that those suggestions improve consistency across RDF descriptions without the rigidity of a prescriptive schema.

5.3   PEW: **Completing OWL Ontologies**

PEW[7] is a tool for completing OWL ontologies. Its target FL is hence OWL. The covered subset of OWL is made of expression classes with existential restrictions, nominals, conjunctions, disjunctions, atomic negations, and inverse roles. Like for UTILIS, a subset of CRQL can be used, excluding filters, and restricting **Not** to atomic verb phrases. Constructor **Select**(np) is replaced by **Sat**(np) to express the fact that the semantics of an AST is the satisfiability of the corresponding class expression. The suggested insertable elements are those that preserve satisfiability. Negated elements are also suggested, and are the only way to introduce negation in the AST. Because only satisfiable class expressions can be built, PEW allows the exploration of the "possible worlds" of an ontology (its models). When undesirable "possible worlds" are found, the tool allows to automatically produce an OWL axiom so as to exclude them.

We have found the tool useful, and even playful, for completing an ontology with negative axioms (e.g., class disjointness), which are often missing. An experiment has been performed on the well-known Pizza ontology, and has revealed many missing negative axioms through undesirable situations: e.g., *"a country that is also some food"* (missing disjointness), *"a country of origin that is not a country"* (missing range), and even *"a vegetarian pizza that can have meat as ingredient"* (the definition uses `hasTopping` instead of `hasIngredient`).

6   Related Work

The FL-NL gap is not a new issue, and many solutions have been proposed to reduce it. Question Answering (QA) consists in translating NL expressions to

---

[6] Download and screencasts at http://www.irisa.fr/LIS/softwares/sewelis/.

[7] Download and screencast at http://www.irisa.fr/LIS/softwares/pew/.

FL, generally going through one or several intermediate representations (e.g., parse trees). In the Semantic Web, many systems translate English questions to SPARQL queries (see [13] for a survey), and the QALD[8] challenge is devoted to that task. NL interfaces are attractive for their ease-of-use, and definetely have a role to play, but they suffer from a weak adequacy: (habitability) spontaneous NL expressions often have no FL counterpart or are ambiguous, (expressivity) only a small FL fragment is covered in general. This makes it difficult to interpret an empty answer: Does it reflect actual data? an out-of-scope question? a lack of expressivity or simply a misunderstanding by the system? Habitability can be addressed in part by suggesting reformulations of user questions, e.g. based on templates [2], but the number of suggestions tends to grow exponentially with expressivity (query size and number of features). An alternative to spontaneous NL is Controlled NL (CNL) [12]. A CNL typically defines a NL fragment that is adequate to the target FL, and that eliminates or reduces ambiguity. However, while much easier to read than FL, a CNL remains difficult to write because of the constrained syntax. That is why CNL-based editors often come with auto-completion to suggest the next possible words during edition: e.g., ACE Wiki for facts and rules [10], Ginseng for queries [11], Atomate it! for reactive rules [18], Halo for problem solving [3]. Auto-completion offers a limited flexibility because suggestions are only at the end of a partial sentence, and because translation to FL, and hence semantics, is only available when the sentence is complete. The latter limitation also implies that suggestions are not based on semantics, but on syntax and schema only. Note that, by *semantics*, we here mean not only the FL expression, but also any interpretation that may come with it (e.g., query results, satisfiability checks of OWL class expressions).

A noticeable approach that is not based on NL is structural edition, e.g. *query builders* like SemanticCrystal [11], where a point-and-click interface is used to build FL expressions incrementally. They avoid the habitability problem but the presentation of the FL expression is generally based either on its syntax tree, and hence very close to the FL, or on a graphical view that is more intuitive but often limits expressivity. Grammatical Framework (GF) [16] improves structural edition by distinguishing abstract syntax and concrete syntax, only showing the later to users. In fact, the tools and linguistic resources of GF can be used to implement the verbalization part of our N<A>F design pattern. However, the GF equivalent of AST transformations are not customizable to the target task and semantics; and ASTs are not guaranteed to have a fully-defined semantics at every edition step. This is because GF is about syntax, not about task-specific semantics.

7   Conclusion

The gap between formal languages and natural languages is deep, and we do not pretend to fill it. However, we believe that our design pattern based on AST zippers provides a powerful strategy to build bridges over the gap (see Fig. 3).

---

[8] http://greententacle.techfak.uni-bielefeld.de/~cunger/qald/.

For people on the natural language side, those bridges offer a safe and large access to the benefits of formal languages. Once they engage on such a bridge, they cannot fall in the gap (no *habitability* problem), and their access to the formal language side is open to a large area (high *expressivity*). We think that it provides an interesting alternative to question answering by avoiding the major difficulties of NL understanding thanks to NL-based interaction. We have shown the versatility of our design pattern with applications to three different formal languages and tasks. The perspectives consists in creating new applications, and improving them on three axes: formal language coverage (*expressivity*), quality of the verbalization (*readability*), and intelligence of system suggestions (*guidance*).

References

1. Abbott, M., Altenkirch, T., McBride, C., Ghani, N.: ∂ for data: differentiating data structures. Fundamenta Informaticae **65**(1), 1–28 (2005)
2. Chaudhri, V.K., Clark, P.E., Overholtzer, A., Spaulding, A.: Question generation from a knowledge base. In: Janowicz, K., Schlobach, S., Lambrix, P., Hyvönen, E. (eds.) EKAW 2014. LNCS, vol. 8876, pp. 54–65. Springer, Heidelberg (2014)
3. Clark, P., Chaw, S.Y., Barker, K., Chaudhri, V., Harrison, P., Fan, J., John, B., Porter, B., Spaulding, A., Thompson, J., Yeh, P.: Capturing and answering questions posed to a knowledge-based system. In: International Conference Knowledge Capture, pp. 63–70. ACM (2007)
4. Dowty, D.R., Wall, R.E., Peters, S.: Introduction to Montague Semantics. Reidel Publishing Company, Dordrecht (1981)
5. Ferré, S.: Expressive and scalable query-based faceted search over SPARQL endpoints. In: Mika, P., Tudorache, T., Bernstein, A., Welty, C., Knoblock, C., Vrandečić, D., Groth, P., Noy, N., Janowicz, K., Goble, C. (eds.) ISWC 2014, Part II. LNCS, vol. 8797, pp. 438–453. Springer, Heidelberg (2014)
6. Ferré, S.: SQUALL: The expressiveness of SPARQL 1.1 made available as a controlled natural language. Data Knowl. Eng. **94**, 163–188 (2014)
7. Ferré, S., Rudolph, S.: Advocatus diaboli – exploratory enrichment of ontologies with negative constraints. In: ten Teije, A., Völker, J., Handschuh, S., Stuckenschmidt, H., d'Acquin, M., Nikolov, A., Aussenac-Gilles, N., Hernandez, N. (eds.) EKAW 2012. LNCS, vol. 7603, pp. 42–56. Springer, Heidelberg (2012)
8. Hermann, A., Ferré, S., Ducassé, M.: An interactive guidance process supporting consistent updates of RDFS graphs. In: ten Teije, A., Völker, J., Handschuh, S., Stuckenschmidt, H., d'Acquin, M., Nikolov, A., Aussenac-Gilles, N., Hernandez, N. (eds.) EKAW 2012. LNCS, vol. 7603, pp. 185–199. Springer, Heidelberg (2012)
9. Huet, G.: Functional pearl - the zipper. J. Funct. Program. **7**(5), 549–554 (1997)
10. Kaljurand, K., Kuhn, T.: A multilingual semantic wiki based on attempto controlled english and grammatical framework. In: Presutti, V., Hollink, L., Rudolph, S., Cimiano, P., Corcho, O. (eds.) ESWC 2013. LNCS, vol. 7882, pp. 427–441. Springer, Heidelberg (2013)
11. Kaufmann, E., Bernstein, A.: Evaluating the usability of natural language query languages and interfaces to semantic web knowledge bases. J. Web Semant. **8**(4), 377–393 (2010)
12. Kuhn, T.: A survey and classification of controlled natural languages. Comput. Linguist. **40**(1), 121–170 (2013)

13. Lopez, V., Uren, V.S., Sabou, M., Motta, E.: Is question answering fit for the semantic web? a survey. Seman. Web **2**(2), 125–155 (2011)

14. McCrae, J., Spohr, D., Cimiano, P.: Linking lexical resources and ontologies on the semantic web with lemon. In: Antoniou, G., Grobelnik, M., Simperl, E., Parsia, B., Plexousakis, D., De Leenheer, P., Pan, J. (eds.) ESWC 2011, Part I. LNCS, vol. 6643, pp. 245–259. Springer, Heidelberg (2011)

15. Ngomo, A.C.N., Bühmann, L., Unger, C., Lehmann, J., Gerber, D.: Sorry, I don't speak SPARQL: translating SPARQL queries into natural language. In: WWW, pp. 977–988 (2013)

16. Ranta, A.: Grammatical framework. J. Funct. Program. **14**(02), 145–189 (2004)

17. Turner, D.: Functional programming and proofs of program correctness. In: Néel, D. (ed.) Tools and Notions for Program Construction: An Advanced Course, pp. 187–209. Cambridge University Press, Cambridge (1982)

18. Van Kleek, M., Moore, B., Karger, D., André, P., Schraefel, M.: Atomate it! end-user context-sensitive automation using heterogeneous information sources on the web. In: International Conference World Wide Web, pp. 951–960. ACM (2010)

Towards Monitoring of Novel Statements in the News

Michael Färber$^{(\boxtimes)}$, Achim Rettinger, and Andreas Harth

Karlsruhe Institute of Technology (KIT), Karlsruhe, Germany
{michael.faerber,rettinger,harth}@kit.edu

Abstract. In media monitoring users have a clearly defined information need to find so far unknown statements regarding certain entities or relations mentioned in natural-language text. However, commonly used keyword-based search technologies are focused on finding relevant documents and cannot judge the novelty of statements contained in the text. In this work, we propose a new semantic novelty measure that allows to retrieve statements, which are both novel and relevant, from natural-language sentences in news articles. Relevance is defined by a semantic query of the user, while novelty is ensured by checking whether the extracted statements are related, but non-existing in a knowledge base containing the currently known facts. Our evaluation performed on English news texts and on CrunchBase as the knowledge base demonstrates the effectiveness, unique capabilities and future challenges of this novel approach to novelty.

Keywords: Semantic novelty measures · Novelty detection · Statement extraction

1 Motivation

End users – both in a private or professional setting – increasingly face the challenge to screen and analyze large amounts of natural language text in order to find novel statements which they were not aware of previously. Consider for example a Web user who is interested in the latest technical achievements in the smart phone domain. Also, a stock broker might look for acquisitions of certain companies mentioned in the news. When using current technology, all potentially relevant text documents are first roughly selected by keyword search, before being checked manually for statements which are novel to the user.

In this work, we propose an approach to support the task of novel statement detection by automatically extracting so far unknown statements from natural language sentences. There exist numerous techniques for information extraction (IE) on text, i.e. systems which convert sentences into formal representations such as triples or other n-ary relational data. However, the detection of genuinely new facts in text differs from traditional web search or monitoring systems, since typically only relevance is taken into account as a selection

© Springer International Publishing Switzerland 2016
H. Sack et al. (Eds.): ESWC 2016, LNCS 9678, pp. 285–299, 2016.
DOI: 10.1007/978-3-319-34129-3_18

criterion. All existing novelty detection systems are based on syntactical and statistical techniques and are not able to assess written statements w.r.t. novelty on a semantic level. In contrast, our semantic novelty measure allows to satisfy the user's information need to an extent which could not be achieved before: Our novelty detection system (i) determines the semantic novelty by checking against a background Knowledge Base (KB) containing the current knowledge on this domain, (ii) presents only those statements to the user which are relevant to the user's current individual information need expressed by a semantic query and (iii) intuitively shows the user what the novel aspect in a statement is by assigning it to certain novelty classes.

Our proposed novelty search system is domain independent and can be applied in different settings ranging from news monitoring and news summarization to evaluating human generated summaries of documents for completeness. Our main contributions are:

- providing a semantic measure for the novelty of statements based on a background KB,
- proposing a semantic novelty detection system for statements in text documents,
- performing an evaluation of our approach on real-world data to demonstrate its unique capabilities in media monitoring tasks, like forecasting of facts, KB population and impact quantification.

The remainder of this paper is organized as follows: First we present our definition of a semantic novelty measure for facts in Sect. 2, before proposing our semantic novelty detection system in Sect. 3. After discussing our evaluation in Sect. 4 and giving an overview of related work in Sect. 5, we conclude in Sect. 6.

2   Measuring Semantic Novelty of Statements

In this work we assume that each statement[1] we are interested in, i.e. we want to extract from the text, can be represented as a Resource Description Framework (RDF) triple (s, p, o) where $p \in U$ is the binary relation between a subject $s \in U$ and an object $o \in U$, where U is the set of unique identifiers.[2] We further assume that there is a KB represented in RDF which contains all knowledge in the domain of interest of the user. This KB acts as a reference point to assess novelty. There are several tools for personalized knowledge management that produce RDF, like semantic wikis. Also, the constantly growing knowledge graphs like Linked Data sources such as DBpedia as well as company-internal knowledge graphs (Google's Knowledge Graph, Microsoft's Satori, etc.) can be used as an initial collection of knowledge.

In our scenario, the user wants to retrieve triples which are both relevant and novel (see Fig. 1) and which can be added automatically to his KB.

---

[1] We use the words "statements", "facts", and "triples" interchangeably in this paper.
[2] We do not consider triples where the object is a literal.

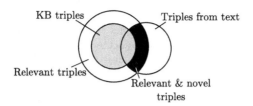

Fig. 1. Triples relevant to the user, KB triples, and triples extracted from text visualized as different sets in a Venn diagram. The aim of our novel triple extraction system is to retrieve triples which are novel (not yet in the KB), but relevant (black area).

Our three-step filtering process facilitates this: Firstly, all extracted triples must be relevant according to the given user query. Secondly, all triples have to be novel in the sense that they do not occur in the KB yet. Thirdly, the novel, but relevant triples need to complement the existing knowledge seamlessly. This is ensured by requiring the novel statements to partially overlap with the KB:

1. Triples extracted from text where an identical triple exists in the KB are not novel.
2. Triples which consist of two parts – (s, p), (p, o), or (s, o) – existing in the KB and one part (entity or relation) which is not yet in the KB, i.e. unknown, are regarded as novel and contextually fitting, since the extracted facts are related to known elements, but also contain new knowledge.
3. Triples which consist of two or three unknown parts are novel, but are assumed unrelated w.r.t. the KB, since the gap between these extracted triples and triples in the KB is too wide resulting in mostly irrelevant statements.

In Table 1 we formally define the different cases of novelty.

By using this semantic definition of novel facts, we do not limit ourselves to a notion of novelty relying on the creation date of each document to assess whether the information is novel or not [1,2]. Instead, we express the fact as a binary relationship. A change over time is intrinsically expressed via a different structure and/or semantics of the novel triple.

3 The Novel Statement Extraction System

Figure 2 presents an overview of our novel triple extraction system. In principle, we have implemented our system as a three step process with the steps *Textual Triple Extraction*, *KB Linking*, and *Novelty Detection*. In the following, we give a description of each of these steps.

Textual Triple Extraction. In this step each sentence of each input document is transformed into propositions, i.e., statements consisting of a subject, a relation, and none, one, or more arguments (e.g., grammatical direct object). For each proposition found in a sentence and apparently compatible with our RDF knowledge representation, we retrieve so called textual triples (see Fig. 3).

Table 1. The different classes of novel triples with textual descriptions, formal representations, and examples. Dotted lines indicate the novel items in the triple. s_t, p_t, and o_t represent the textual mentions of a subject $s \in U$, predicate $p \in U$, and object $o \in U$, respectively, in a sentence. U is the set of unique identifiers. f is a function which maps a textual resource (s_t or o_t) to its corresponding RDF resource: $f(s_t) = s$ and $f(o_t) = o$. The function g maps a textual predicate p_t to an RDF predicate: $g(p_t) = p$. In cases of unsuccessful matches, we write $\emptyset$.

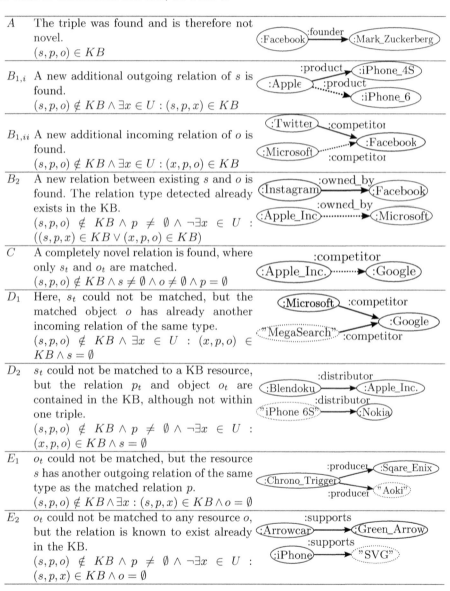

A	The triple was found and is therefore not novel. $(s, p, o) \in KB$	
$B_{1,i}$	A new additional outgoing relation of s is found. $(s, p, o) \notin KB \wedge \exists x \in U : (s, p, x) \in KB$	
$B_{1,ii}$	A new additional incoming relation of o is found. $(s, p, o) \notin KB \wedge \exists x \in U : (x, p, o) \in KB$	
B_2	A new relation between existing s and o is found. The relation type detected already exists in the KB. $(s, p, o) \notin KB \wedge p \neq \emptyset \wedge \neg\exists x \in U : ((s, p, x) \in KB \vee (x, p, o) \in KB)$	
C	A completely novel relation is found, where only s_t and o_t are matched. $(s, p, o) \notin KB \wedge s \neq \emptyset \wedge o \neq \emptyset \wedge p = \emptyset$	
D_1	Here, s_t could not be matched, but the matched object o has already another incoming relation of the same type. $(s, p, o) \notin KB \wedge \exists x \in U : (x, p, o) \in KB \wedge s = \emptyset$	
D_2	s_t could not be matched to a KB resource, but the relation p_t and object o_t are contained in the KB, although not within one triple. $(s, p, o) \notin KB \wedge p \neq \emptyset \wedge \neg\exists x \in U : (x, p, o) \in KB \wedge s = \emptyset$	
E_1	o_t could not be matched, but the resource s has another outgoing relation of the same type as the matched relation p. $(s, p, o) \notin KB \wedge \exists x : (s, p, x) \in KB \wedge o = \emptyset$	
E_2	o_t could not be matched to any resource o, but the relation is known to exist already in the KB. $(s, p, o) \notin KB \wedge p \neq \emptyset \wedge \neg\exists x \in U : (s, p, x) \in KB \wedge o = \emptyset$	

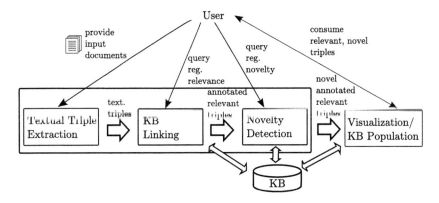

Fig. 2. Overview of our novel triple extraction system with the three steps of (i) *Textual Triple Extraction*, (ii) *KB Linking*, and (iii) *Novelty Detection*.

Textual triples are of the form (s_t, p_t, o_t), where s_t and o_t are natural-language mentions of entities and p_t is a relation found in the sentence. For the example sentence "The sixth generation iPhone was developed by Apple and features a touchscreen display.", we retrieve the textual triples (i) ("Apple", "develop","sixth generation iPhone") and (ii) ("sixth generation iPhone", "feature","touchscreen display").

"The sixth generation iPhone was developed by Apple and features a touchscreen display."

1. Dependency tree generation

2. Clause detection and clause type determination

Clause (i) of type subject-verb-object,   Clause (ii) of type subject-verb-object,
passive voice   active voice

3. Textual triple generation

Textual triple I: Textual triple II:
"Apple" "develop" "sixth generation iPhone"   "sixth generation iPhone" "feature" "touchscreen display"

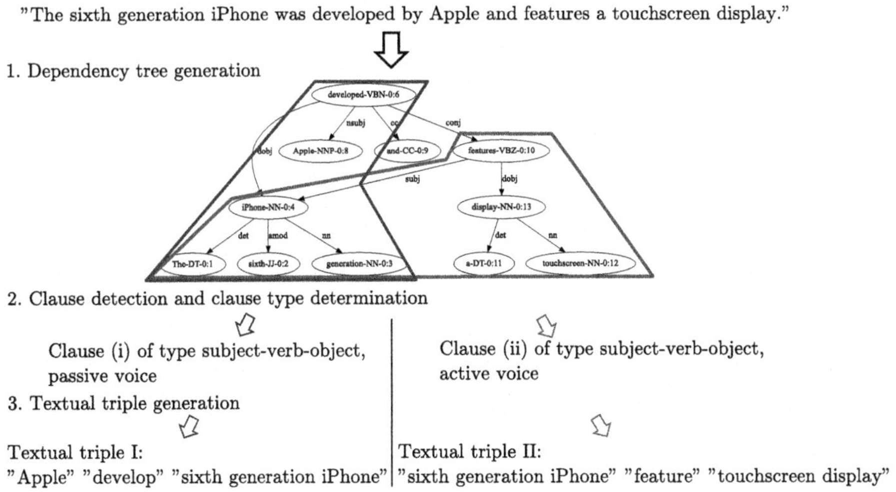

Fig. 3. Different steps of our Textual Triple Extraction module.

Technically, the Textual Triple Extraction step is based on the tool ClausIE [3]. However, we need to modify ClausIE in terms of linguistic processing to our requirements:

– Adjectives cannot be encoded into RDF triples easily and even prevent that the entity mention or predicate can be mapped to the correct KB resource or relation, respectively. We therefore remove all adjectives from the dependency tree.
– Temporal aspects expressed in the input sentence (point in time or period of time) such as "yesterday" are difficult to attach to a triple, but can be represented more adequately as separate information units. Hence, we extract temporal phrases with the help of the inherent Stanford PoS tagger and store them separately.
– In sentences which contain linguistic complements, as they often occur in case of indirect speech ("X said that Y has Z"), we focus on the extraction of the proposed fact itself (which is the complement), not the subject of the sentence (e.g. the person X who said something).
– If a clause in a sentence is written in passive voice, we transform it into active voice.

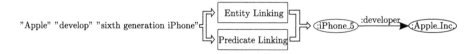

Fig. 4. KB Linking illustrated by an example. Here it is assumed that all parts of the textual triple can be mapped.

KB Linking. In this second step we map the textual phrases s_t, p_t, and o_t to the corresponding KB entities and, respectively, KB properties as far as possible (see Fig. 4). For entity linking we exploit *xLiD lexica* [4] by which we can link textual mentions of an entity to its corresponding resource in the KB DBpedia.[3] Due to the ambiguity problem (e.g., the mention "Apple" can represent the DBpedia entity :Apple_Inc.,[4] the fruit :Apple and several other entities), we integrate a disambiguation step based on [5] in order to find the most likely entity for each mention. For a sufficient disambiguation, we take all mentions in the current sentence into account.[5]

---

[3] See http://dbpedia.org, requested on Mar 7, 2016. DBpedia is widely used for entity linking in general domain settings. However, also other KBs can be used as far as a suitable entity linking component is available.
[4] We avoid the DBpedia namespaces for better readability.
[5] For our example sentence (see Fig. 3), this would be "sixth generation iPhone", "Apple", and "touchscreen display".

Furthermore, relations expressed in the text need to be linked to the corresponding KB properties. For instance, the textual property "bought" needs to be linked to the KB property :acquired. However, linking the textual predicate p_t to the corresponding KB property p is a hard task [6], since KB properties are typically not equipped with the information how they occur in natural-language texts. Trying to match the label of the KB property (e.g., "acquired") with the textual predicate p_t (e.g., "bought") is generally inefficient. One reason for that is that the KB property can be expressed by a variety of expressions. In our work, we therefore set up a two-step process for predicate linking, which is similar to the approach of [7]: In a *training phase* which is performed before the actual novel triple extraction phase, our system performs the Textual Triple Extraction and the KB Linking. If both the textual subject s_t and object o_t of an extracted textual triple could be mapped to KB entities and if the classes of these KB entities match with the pre-defined domain and range of the target KB property, the corresponding p_t might express the target property p. In a semi-automatic fashion a user then confirms or rejects the found mappings $p_t \rightarrow p$ (e.g., "buy" $\rightarrow$:acquired). In case the mapping is confirmed, it is added to the mappings $p_t \rightarrow p$. At the end we have for all considered KB properties mappings $p_t \rightarrow p$ learned from text. In the application phase, i.e. the actual novel triple extraction phase, the Textual Triple Extraction, the KB Linking, and the Novelty Detection step is run. Here, all novel triples except from novelty class C (see Table 1) are extracted, since novelty class C was used in the training to find mappings from textual predicates to KB properties.

At the end of the KB Linking step, we have textual triples which are mapped to KB triples either partly or completely. Given a semantic user query regarding the relevance of the extracted triples (consisting of basic graph patterns and implemented as SPARQL query; a query expressed in natural language might be:"Retrieve all acquisitions of companies in the smartphone domain."), all triples are filtered out which are irrelevant.

Novelty Detection. We determine which of the remaining triples are novel w.r.t. our KB and classify these triples into the different novelty classes defined in Sect. 2. Our system is designed for allowing the user to choose which novelty classes should be considered by the system. The user query is, hence, extended by the information need regarding the novelty classes (e.g., only triples of the novelty classes $B_{1,i}$, $B_{1,ii}$, and B_2 should be retrieved).

4   Evaluation

In the following, we first present the data sets used for evaluating our approach.[6] We then describe the evaluation settings and finally show the results of our evaluation.

---

[6] The data sets and evaluation results are available at http://people.aifb.kit.edu/mfa/novel-triple-extraction/.

4.1   Data Used

KB: As KB we used CrunchBase[7] in combination with DBpedia.[8] CrunchBase consists of structured information about organizations (including companies), people, products, investments, and several other items, and is edited by a Web community. We built an RDF KB using the CrunchBase API and integrated `owl:sameAs`-relations to the corresponding entities in DBpedia. By bridging to DBpedia, we can use the existing entity linking module [4,5] for mapping the textual subjects and objects. This approach is more robust than using simple string matching based on the CrunchBase entity labels.

In our evaluation, we focus on relations between organizations and on relations between persons and organizations (see Fig. 5). Our CrunchBase RDF KB contains 16,706 entities of type organization and 26,468 entities of type person – in both cases with corresponding `owl:sameAs`-links to DBpedia. There are 16,509, 60,936, 151,722, and 83,470 distinct facts regarding the KB properties `cb:acquired`, `cb:competesWith`, `cb:founded`, and `cb:isBoardMemberOrAdvisor`, respectively.

Documents: We used English news articles from the IJS newsfeed[9] [8] as input text documents for our novel triple extraction system. For learning the textual predicate mappings in the training phase (cf. Sect. 3), we used all English news from May 1 until May 15, 2015 (607,289 articles) and ignored those paragraphs which contained no known label of a person or organization in our KB (leading to 136,907 paragraphs). For the actual triple extraction phase, we chose the time range of May 16 until August 31, 2015, (3,642,771 articles) and applied the same filter, resulting in 797,224 paragraphs of text.

4.2   Evaluation Setting

Our evaluation addresses the following claims:

1. **Fact Forecast:** We claim that we can detect facts, such as acquisitions, which are sometimes leaked, rumored or discussed publicly before they are officially announced. This is of great interest in media monitoring.
2. **Improved KB Population:** Given true new facts, our system provides a comfortable way to insert these facts to the KB: (i) Our method can detect and extract novel and known facts mentioned in the news faster than if they were added to the KB manually. (ii) Our method already provides new facts in a semantically-structured format, ready for inserting it to a KB. (iii) Our method can provide links to the news articles (together with other meta-data) in which relevant novel facts are mentioned. This provenance information can be added to the KB and provides evidence for the facts (fact verification).

---

[7] See http://crunchbase.com, requested on Mar 7, 2016.
[8] See http://dbpedia.org, requested on Mar 7, 2016.
[9] See http://newsfeed.ijs.si, requested on Mar 7, 2016.

3. **Impact Quantification:** For all facts stored in the KB, our system can show when and how often these facts have been mentioned in the news. This feature can be used for tracking facts, e.g., in the context of beat reporting.

Hence, our novel triple extraction system was evaluated in two parts:

1. We evaluated whether our system can achieve the above mentioned goals 1, 2, and 3 regarding acquisition facts. The query can be formulated in natural language as: "Extract all novel triples (considering all novelty classes) with the relation cb:acquired."
2. In a second evaluation, we expanded the query to retrieve all novel facts with the KB properties cb:acquired, cb:competesWith, cb:founded, and cb:isBoardMemberOrAdvisor (see Fig. 5) and evaluated how many facts were correctly extracted from the news, thereby considering goal 2.[10]

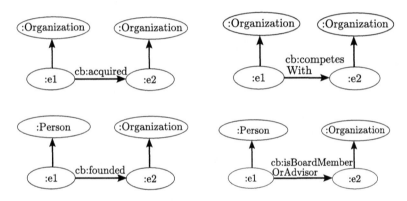

Fig. 5. KB properties with their domains and ranges as used in the second evaluation. owl:sameAs links to DBpedia entities and their rdf:type relations are not shown for convenience.

4.3   Evaluation Results

Evaluation Part 1

Regarding Fact Forecast: Given the news of the specified time range (May 16 until August 31, 2015; 797,224 paragraphs), our novelty detection tool was able to extract 32 distinct acquisition facts (89 in total) which were also in CrunchBase given the state of October 1, 2015. Out of these 32, two acquisitions (i) were announced within the specified time range according to CrunchBase and

---

[10] Goal 1 and 3 are not considered here since facts with the chosen KB properties do not occur often.

(ii) were detected by our tool back then before they were inserted into Crunch-Base (see Fig. 6). This shows that our system is able to detect facts before they might actually become true. Our manual evaluation on all 1,333 retrieved `cb:acquired` facts (see Fig. 2) revealed that the extracted hypothetically formulated facts about acquisitions can be found across the novelty classes (3 6, 3, 2, 3, 0, 8, and 3 occurrences for the novelty classes A, $B_{1,i}$, $B_{1,ii}$, B_2, D_1, D_2, E_1, and E_2, respectively).

Regarding Improved KB Population: Out of the 32 distinct extracted acquisitions, 4 acquisitions (i) had been announced – according to CrunchBase – and inserted to CrunchBase in the specified time frame of the news, i.e., those triples are really novel (Class B) considering the CrunchBase state from May 15, 2015 and (ii) were not written as hypotheses. In total, 69 acquisitions were announced according to CrunchBase within the specified time range. However, 21 of the 63 not-detected acquisitions were not mentioned at all in the selected news, 34 were mentioned at most five times.[11] The approximative *recall* is therefore at least 6/42=0.143.

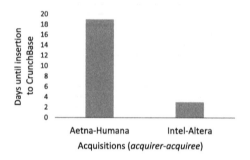

Fig. 6. Days between the retrieval dates of the news where the facts were extracted from and the dates of manual insertion into CrunchBase. Shown are only acquisitions with positive value.

Regarding Impact Quantification: The most repeated acquisitions were `:Facebook :acquired :Oculus VR` (18 times), `:Verizon_Communications :acquired :AOL` (7 times), and `:Apple_Inc. :acquired :Beats_Electronics` (6 times). Figure 7 shows how such repeatedly mentioned facts can be visualized w.r.t. the fact `:Facebook :acquired :Oculus_VR` extracted from the news of March 25–28, 2014.

---

[11] This analysis was performed by evaluating all sentences containing two labels of the entities of acquisitions which were missed and containing the phrase "acquire"/"acquisition"/"buy"/"purchase" etc.

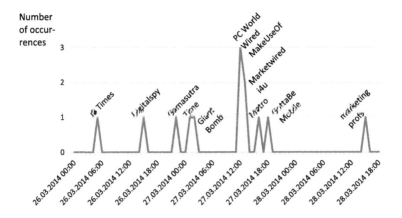

Fig. 7. Publishing dates of the news of March 25–28, 2014, where the fact :Facebook :acquired :Oculus_VR was extracted, together with the source such as "IB Times".

Table 2. Number of correct novel triples (as far as evaluated) and number of extracted novel triples per novelty class.

KB property	A	$B_{1,i}$	$B_{1,ii}$	B_2	D_1	D_2	E_1	E_2
:acquired	89/89	33/36	4/4	13/13	24/86	14/71	63/636	48/398
:competesWith	7/8	35/35	11/13	19/20	72	57	171	240
:founded	33/33	0/0	1/2	3/3	63	73	145	311
:isBoardMemberOrAdvisor	1/2	7/9	18/18	7/8	58	70	70	450

Evaluation Part 2 – Regarding Improved KB Population: Using the CrunchBase version of October 1, 2015, and the news of the specified time range and not considering the announcement dates, our system was able to retrieve known and novel facts as presented in Table 2. Regarding the 89 extracted facts about acquisitions which were classified as known (Class A), we found out that four of them would have been detected by our tool before they were added to CrunchBase.

The high *precision* of the novel facts of the novelty classes $B_{1,i}$, $B_{1,ii}$, and B_2 (on average $281/293 = 0.959$) shows that especially in cases where the entities are already known (as it is usually the case), our system is able to retrieve high-quality novel facts. The numbers also indicate that CrunchBase misses a significant number of facts.

The manual assessment on the correctness of the extracted novel triples with the relation :acquired and with the novelty classes D_1, D_2, E_1 and E_2 revealed a *precision* of $149/1192 = 0.125$. Note, however, that regarding all incorrect triples, about every second triple is almost completely correct (520 occurrences, i.e., 49.8 %). In those cases, all mappings of the annotated triple are correct and the non-mapped part does not contain any ballast. In those cases, the non-mapped part of the triple contains some additional information mentioned in the text (mostly nominal phrases, e.g., "Chicago-based DTZ") which prevents a mapping. A proposition often resulted in an incorrect triple of Class D or E and simultaneously in a correct triple of Class B. Hence, triples of class D and E may be ignored. Further common failures are that coreferences cannot be resolved (161 cases, i.e., 15.4 %). In 324 cases (31.0 %), although the extracted statement cannot be represented in the triple format or the non-mapped entity was too abstract for being relevant for the KG, the statement was output by the system and, hence, judged as incorrect by the assessor. Determining those cases automatically is non-trivial and left for future work.

5   Related Work

There are several areas of work which are worthwhile to mention as related work (see Table 3 for an overview). First of all, there are approaches to information extraction where the relations and/or entities have a grounding in a knowledge base. They are either based on shallow parsing or deep natural language processing. FRED [9] is one approach of deep NLP where text is transformed into a complex semantic model via Discourse Representation Theory (DRT). In this way, complex statements can be expressed, not only simple binary relations, which we focus on and which makes a comparison with a KB easier. The system of Carvalho [10] is similar to FRED in terms of constructing a structured data graph (SDG). LODifier [11] embraces several existing tools for deep semantic analysis for transforming text into RDF. However, in contrast to our system, the focus of the LODifier is high recall instead of high precision and the tool is designed for scenarios with no a-priori schema information.

Table 3. Summary of related work.

	Extraction of and queries on statements (graph/tupel)	Grounding of extracted entities in a KB	Grounding of extracted relations in a KB	Novelty detection task implemented	Implemented and evaluated on a Web scale
Presutti et al. [9]	✓	✓	✓		✓
Carvalho et al. [10]	✓	✓			
Augenstein et al. [11]	✓	✓	✓		
Fader et al. [12]	✓				✓
Del Corro et al. [3]	✓				✓
Mausam et al. [13]	✓				✓
Zhang et al. [14]				✓	✓
Gabrilovich et al. [1]				✓	✓
Karkali et al. [2]				✓	✓
Li et al. [15,16]				✓	✓
Systems for TREC Novelty Track 2002-2004 [17]				✓	✓
Systems for TREC KBA (2013–2014)	✓			✓	✓
Clarke et al. [18]				✓	✓

Secondly, there are approaches [3, 13] based on the idea of Open Information Extraction (OIE). OIE has become prominent as a method for extracting relations in web documents on a huge scale. The aim of OIE is to build a database for textually expressed relations plus associated textually grounded instances without any schema information. RDF – as we need it in our case as final output of our text processing pipeline – is not supported by OIE tools. Converting OIE triples to RDF is non-trivial, however explored by Dutta et al. [19]. As we saw in Sect. 3, we use the tool ClausIE [3] as part of our processing pipeline, but we need to modify it to our requirements.

Our semantic novelty detection approach is, to the best of our knowledge, the first system which introduces a semantic novelty measure by relying on an RDF KB. By formalizing novelty on a triple level, we go beyond pure statistical approaches (often in combination with named entity recognition) [1, 2, 14–16] for novelty detection.

For the TREC novelty track [17] of the years 2002–2004, systems had to solve different tasks regarding novelty retrieval. The most similar task to our scenario is stated as follows: Given all documents of a broad topic, identify all relevant and novel sentences. Events and opinions form the two types of topics which were provided. In 2004, the number of topics was limited to 50. Each of the 50 topics was defined by a short description and a task narrative.[12] Contrary to that, we generate a formal representation of new statements in terms of an RDF graph and compare novelty on a triple basis. Instead of broad topics, we focus on single relations between entities.

For the subtask *Vital Filtering* of the TREC Knowledge Base Acceleration (KBA) call[13], systems judge the utility of documents mentioning an entity. However, the used tags such as *vital* are not appendant to specific properties of entities. The subtask *Streaming Slot Filling* is about gathering attribute values of specific entities from the text. The set of possible slots and entities is fixed. However, (i) the ground truth for that task in TREC KBA 2014 does not provide information about where in the corpus the slot values were found; (ii) there is no grounding of the slot values, only textual phrases from the text are provided. Regarding the data sets of 2012/2013 and 2015, the TREC Dynamic Domain Track, we face similar differences to our approach. Clarke et al. [18] present an evaluation framework which rewards novelty. Regarding novelty, ranked lists are considered where the relevance of each element is dependent on the proceeding ones.

6   Conclusion and Future Work

Targeted search for novel, formal and grounded facts in unstructured text is an open issue, since existing novelty detection systems primarily regard novelty as

---

[12] For instance, for the topic "Diana Car Accident", the task was to find novel information about where the accident happened, who was killed, the extent of injuries, how it happened, and who else was involved.

[13] See http://trec-kba.org, requested on Mar 7, 2016.

a statistical filtering step of sentences or documents. In this paper we presented a conceptually new approach that can satisfy the user's information need in a fine grained manner by extracting novel statements in the form of RDF triples. Novelty is hereby measured w.r.t. a background KB and semantic novelty classes.

Our experiments demonstrated that our prototypical system can facilitate (i) *fact forecast*, i.e., detecting hypothetically formulated statements before they are officially announced; (ii) *improved KB population*, i.e., retrieving both novel and relevant facts in a semantically-structured format, with references to the news, potentially even before the fact is inserted manually to the KB by the community; (iii) *impact quantification*, i.e., monitoring the frequency of certain statements over time and thus the impact of this statement.

While this paper constitutes a step towards a more precise and meaningful monitoring of novelty in news, the biggest challenge towards establishing it in a professional setting is to improve recall. A promising next step to achieve this is to improve the Textual Triple Extraction step. This includes a more elaborated textual subject/object extraction (eliminating noise). A further increase of recall could be archived by implementing coreference resolution, so that more subjects and objects can be linked to the corresponding KB entities. Last but not least, recall might be significantly improved by considering relations which are expressed by other means than verbs such as nominalized verbs (e.g., "the acquisition of X by Y"). We believe that the qualitative improvements of our approach justify future research efforts to close the quantitative performance gap to traditional novelty approaches.

Acknowledgement. This work was carried out with the support of the German Federal Ministry of Education and Research (BMBF) within the Software Campus project *SUITE* (Grant 01IS12051).

References

1. Gabrilovich, E., Dumais, S., Horvitz, E.: Newsjunkie: providing personalized news-feeds via analysis of information novelty. In: Proceedings of the 13th International Conference on World Wide Web, WWW 2004, pp. 482–490. ACM, New York (2004)
2. Karkali, M., Rousseau, F., Ntoulas, A., Vazirgiannis, M.: Efficient online novelty detection in news streams. In: Lin, X., Manolopoulos, Y., Srivastava, D., Huang, G. (eds.) WISE 2013, Part I. LNCS, vol. 8180, pp. 57–71. Springer, Heidelberg (2013)
3. Del Corro, L., Gemulla, R.: ClausIE: clause-based open information extraction. In: Proceedings of the 22nd International Conference on World Wide Web, WWW 2013, Republic and Canton of Geneva, Switzerland, pp. 355–366. ACM (2013)
4. Zhang, L., Färber, M., Rettinger, A.: xLiD-Lexica: cross-lingual linked data lexica. In: Proceedings of the Ninth International Conference on Language Resources and Evaluation (LREC 2014), pp. 2101–2105. European Language Resources Association (2014)
5. Zhang, L., Rettinger, A.: X-LiSA: cross-lingual semantic annotation. PVLDB **7**(13), 1693–1696 (2014)

6. Welty, C., Fan, J., Gondek, D., Schlaikjer, A.: Large scale relation detection. In: Proceedings of the NAACL HLT 2010 First International Workshop on Formalisms and Methodology for Learning by Reading. FAM-LbR 2010, Stroudsburg, PA, USA, pp. 24–33. Association for Computational Linguistics (2010)

7. Gerber, D., Ngonga Ngomo, A.C.: Bootstrapping the linked data web. In: 1st Workshop on Web Scale Knowledge Extraction @ ISWC 2011 (2011)

8. Trampuš, M., Novak, B.: Internals of an aggregated web news feed. In: Proceedings of the Fifteenth International Information Science Conference IS SiKDD 2012, pp. 431–434 (2012)

9. Presutti, V., Draicchio, F., Gangemi, A.: Knowledge extraction based on discourse representation theory and linguistic frames. In: ten Teije, A., Völker, J., Handschuh, S., Stuckenschmidt, H., d'Acquin, M., Nikolov, A., Aussenac-Gilles, N., Hernandez, N. (eds.) EKAW 2012. LNCS, vol. 7603, pp. 114–129. Springer, Heidelberg (2012)

10. Carvalho, D.S., Freitas, A., da Silva, J.C.P.: Graphia: extracting contextual relation graphs from text. In: Cimiano, P., Fernández, M., Lopez, V., Schlobach, S., Völker, J. (eds.) ESWC 2013. LNCS, vol. 7955, pp. 236–241. Springer, Heidelberg (2013)

11. Augenstein, I., Padó, S., Rudolph, S.: LODifier: generating linked data from unstructured text. In: Simperl, E., Cimiano, P., Polleres, A., Corcho, O., Presutti, V. (eds.) ESWC 2012. LNCS, vol. 7295, pp. 210–224. Springer, Heidelberg (2012)

12. Fader, A., Soderland, S., Etzioni, O.: Identifying relations for open information extraction. In: Proceedings of the Conference on Empirical Methods in Natural Language Processing. EMNLP 2011, Stroudsburg, PA, USA, pp. 1535–1545. Association for Computational Linguistics (2011)

13. Mausam, S., M., Bart, R., Soderland, S., Etzioni, O.: Open language learning for information extraction. In: Proceedings of the 2012 Joint Conference on Empirical Methods in NLP and Computational Natural Language Learning. EMNLP-CoNLL 2012, Stroudsburg, PA, USA, pp. 523–534. ACL (2012)

14. Zhang, Y., Callan, J., Minka, T.: Novelty and redundancy detection in adaptive filtering. In: Proceedings of the 25th annual international ACM SIGIR conference on Research and development in information retrieval. SIGIR 2002, pp. 81–88. ACM, New York (2002)

15. Li, X., Croft, W.B.: An information-pattern-based approach to novelty detection. Inf. Process. Manag. 44(3), 1159–1188 (2008)

16. Li, X., Croft, W.B.: Novelty detection based on sentence level patterns. In: Proceedings of the 14th ACM International Conference on Information and Knowledge Management. CIKM 2005, pp. 744–751. ACM, New York (2005)

17. Soboroff, I., Harman, D.: Novelty detection: the trec experience. In: Proceedings of Human Language Technology Conference and Conference on Empirical Methods in Natural Language Processing, Vancouver, British Columbia, Canada, pp. 105–112. Association for Computational Linguistics (2005)

18. Clarke, C.L., Kolla, M., Cormack, G.V., Vechtomova, O., Ashkan, A., Büttcher, S., MacKinnon, I.: Novelty and diversity in information retrieval evaluation. In: Proceedings of the 31st Annual International ACM SIGIR Conference on R&D in Information Retrieval. SIGIR 2008, pp. 659–666. ACM, New York (2008)

19. Dutta, A., Meilicke, C., Stuckenschmidt, H.: Semantifying triples from open information extraction systems. In: STAIRS 2014 : Proceedings of the 7th European Starting AI Researcher Symposium, IOS Press, pp. 111–120, Clifton, VA (2014)

AskNow: A Framework for Natural Language Query Formalization in SPARQL

Mohnish Dubey[1]([⊠]), Sourish Dasgupta[2], Ankit Sharma[3], Konrad Höffner[4], and Jens Lehmann[1,5]

[1] Computer Science Institute, University of Bonn, Bonn, Germany
dubey@cs.uni-bonn.de
[2] DA-IICT, Gandhinagar, India
sourish@rygbee.com
[3] State University of New York, Buffalo, USA
ankitkai@buffalo.edu
[4] AKSW Group, University of Leipzig, Leipzig, Germany
konrad.hoeffner@uni-leipzig.de
[5] Fraunhofer IAIS, Sankt Augustin, Germany
jens.lehmann@iais.fraunhofer.de

Abstract. Natural Language Query Formalization involves semantically parsing queries in natural language and translating them into their corresponding formal representations. It is a key component for developing question-answering (QA) systems on RDF data. The chosen formal representation language in this case is often SPARQL. In this paper, we propose a framework, called *AskNow*, where users can pose queries in English to a target RDF knowledge base (e.g. DBpedia), which are first normalized into an intermediary canonical syntactic form, called Normalized Query Structure (*NQS*), and then translated into SPARQL queries. NQS facilitates the identification of the desire (or expected output information) and the user-provided input information, and establishing their mutual semantic relationship. At the same time, it is sufficiently adaptive to query paraphrasing. We have empirically evaluated the framework with respect to the syntactic robustness of NQS and semantic accuracy of the SPARQL translator on standard benchmark datasets.

1 Introduction

With the advent of massive scale knowledge bases (such as DBpedia [1], YAGO [2], Freebase [3], Google Knowledge Vault [4], Microsoft Satori, etc.), the need to have a user-friendly interface for querying them became relevant. However, users usually are not deft in (and in most cases lack the knowledge of) writing formal queries. *Natural language query formalization (NLQF)* is a formal and systematic procedure of translating a user query in natural language (NL) into a query expression in a target formal query language. In this paper, we scope the problem of NLQF to RDF/RDF-S knowledge bases only. Within this context, the target formal query language chosen is SPARQL [5] – the W3C recommended and widely adopted query language for RDF data stores.

© Springer International Publishing Switzerland 2016
H. Sack et al. (Eds.): ESWC 2016, LNCS 9678, pp. 300–316, 2016.
DOI: 10.1007/978-3-319-34129-3_19

NLQF into SPARQL for question-answering on RDF data stores is non-trivial. This can be attributed to several reasons: (a) a semantic interpretation of natural language query is intrinsically complex and error-prone, (b) the schema of the target dataset is not fixed, (c) a partial lack of rich schema structures of RDF datasets leading to syntactic mismatches, (d) lexical mismatches of query tokens, and (e) mismatches due to lack of explicit entailed relations in an RDF store. One of the key linguistic challenges is the accurate identification of the *query desire* (also known as *query intent* or *answer type* in the literature) of a user-query. Another major challenge is that a query can be paraphrased into multiple forms, thereby triggering the potential of lexico-syntactic mismatches. Also, there is no unique way to create schemata in RDF, i.e. the same fact can be written in different triple forms and could also be expressed using multiple triples.

In this paper, we propose an NLQF framework, called *AskNow*, for posing queries in English to target RDF data stores. AskNow uses an intermediate canonical syntactic structure, called *Normalized Query Structure (NQS)*, into which an *NQS fitting* algorithm normalizes English queries. One of the primary objectives of the algorithm is to normalize paraphrased queries into a common structure. Another objective is to help a SPARQL translator to easily identify the query desire, query input (i.e. additional information provided by the user), and their mutual semantic relation. As an example, given the query: *"What is the capital of India?"*, the algorithm will be able to differentiate the query desire (i.e. instances of class *capital*) and relate it to the input *India* via the relation *of*. Here, the input plays an important role in automatically constructing the declarative formal description of the desire. After a query is normalized into an NQS instance, the SPARQL translator then maps the query tokens in the NQS instance so as to entities defined in the RDF data store. This is done to solve the potential problem of lexical and schematic mismatch mentioned earlier. We show empirically, using QALD [6] benchmark datasets, that the devised NQS Fitting algorithm is accurate in correctly characterizing (both in terms of syntax and semantics) most NL queries. Our contributions in this paper is as follows:

- A novel paraphrase resilient query characterization structure (and algorithm), called *NQS*, is proposed. NQS is less sensitive to structural variation. It supports complex queries, hence, serves as a robust intermediary formal query representation.
- An NQS to SPARQL translation algorithm (and tool) is proposed that supports user queries to be agnostic of the target RDF store structure and vocabulary.
- An evaluation of AskNow in terms of: (i) assessing robustness using the Microsoft Encarta data set, and (ii) evaluating accuracy using a community query dataset built on the OWL-S TC v.4.0 dataset and the QALD-4/5 datasets.

The paper is organized into the following sections: Sect. 3 *Approach*, where the formal notions of *NQS* query is defined, Sect. 4 in which the *Architectural*

Pipeline of *AskNow* is elaborated, Sect. 5 *Evaluation*, where various evaluation criteria are discussed, and Sect. 6 *Related Work* outlining some of the major contributions in NL query processing.

2 Preliminaries

Definition 1 (Simple Query): A simple query consists of a single and unconstrained query-desire (explicit or implicit) and a single, unconstrained, and explicit query-input. For example, in *"What is the capital of USA?"* the query-desire (i.e. *Capital*) is explicit, single, and not constrained by any clausal phrase. The query-input (i.e. *USA*) is also explicit, single, and unconstrained. The query-desire can be implicit: For example, in the query *"What is a tomb?"*, the implicit query-desire is the *definition* of *tomb*, while the query-input *Tomb* is single and unconstrained.

Definition 2 (Complex Query): A complex query consists of a single query-desire (explicit or implicit, constrained or unconstrained) and multiple, explicit query-inputs (constrained or unconstrained). For example, the query *"What is the capital of USA during World War II?"* is a complex query where the implicit, unconstrained query-desire is single (*Capital*) while the query-input is multiple (*USA, World War II*).

Definition 3 (Compound Query): A compound query consists of conjunction/disjunction operator connectives between one or more simple or complex queries. An example of a compound query is *"What are the capitals of USA and Germany?"*.

3 Approach

3.1 Motivation

Our chosen query representation language is SPARQL, in which basic graph patterns consist of subject, predicate and object. So, the primary objective of query parsing should be to identify the query desire/s and describe it in terms of the query predicate/s and query input/s. The identification of such a constraint relation is called *query desire-input dependency*. One of the key tasks for solving the problem of NLQF is to do *syntactic normalization* of NL queries. Syntactic normalization is the process that re-structures queries having different syntactic structural variations into a common structure so that subsequent formalization can be executed using a standard translation algorithm on this structure. Such normalization is difficult to achieve through a *query desire-input dependency* identification process alone. It is in this direction that we propose a chunker-styled *pseudo-grammar*, called *Normalized Query Structure (NQS)*.

3.2 Normalized Query Structure (NQS)

NQS is the proposed basis structure of natural language queries. It acts like a surface level syntactic template for queries, defining the universal linguistic dependencies (i.e. query desire-input dependency) between the various generic sub-structures (i.e. chunks) of any query. The primitive sub-structures of any query is the query token (which determines the query type), query desire, the query input, and the dependency relation that connects them. Each of the primitive sub-structures can be assigned a linguistic characterization (i.e. type). An example characterization are POS (part-of-speech) tag based chunks (such as noun phrase, verb phrase, etc.). For instance, desire and input can be hypothesized to assume a noun form, while the dependency relation can be assumed to be a verb form. An example of such characterization can be seen in the query: *"In which country is New York located?"*. Here, the query desire is the noun phrase *country*, the query input is the noun phrase *New York*, and the dependency relation is the verb phrase *located in*. Since the number of query tokens is finite and there are only three query forms (simple, complex, compound), natural language queries can be categorized into a (finite) set of generic NQS templates.

We now introduce the NQS syntax definitions as follows:

Simple Query NQS: A *simple query* can be characterized according to the following NQS structure:

$$[Wh]\ [R_1]\ [D]\ [R_2]\ [I]$$

here,

$$[D] = Q_D^? M_D^* D \text{ and } [I] = Q_I^? M_D^* I$$

where the notation is defined as follows[1]:

$[D]$: Query desire class/instance-value is restricted to the following POS tags: *NN, NNP, JJ, RB, VBG*. When and where queries have $[D = NULL]$, and NQS automatically annotates D as *TIME* and *LOCATION* respectively.

$[I]$: Query input class/instance-value restricted to the POS tags: *NN, NNP, JJ, RB, VBG*.

$[R_1]$: Auxiliary relation - includes lexical variations of the set: {*is, is kind of, much, might be, does*}.

$[R_2]$: Relation that acts as (i) predicate having D as the subject and I as the object or (ii) action role having I as the actor - value restricted to the POS tags: *VB, PP, VB-PP*.

$Q_D^?$ or $Q_I^?$: Quantifier of D or I - values restricted to the POS tag: DT. The ? indicates that Q can occur zero or one time before D or I.

M_D^* or M_I^*: Modifier of D or I - value restricted to the POS tags: *NN, JJ, RB, VBG*. The * indicates that M can occur zero or multiple time before D or I.

---

[1] All POS-tag notations follow Penn Treebank.

Characteristics of Relation Tokens: R_1 serves as a good indicator for resolving several linguistic ambiguities. For example, in a *how*-query, if R_1 is *much* (or its lexical variations) then it is a quantitative query. However, in a *who*-query if R_1 is *does* (or its lexical variations) then the associated verb is an activity (i.e. Gerund; ex: *"Who does the everyday singing in the church?"* - *everyday singing* is an activity in this case). R_2 is a relation that can either be associated with D as the subject or I as the subject but not both. If R_2 is positioned after D in the original NL query then R_2's subject is D. For example, in the simple NL query *"What is the capital of USA?"* the subject of R_2 (*of*) is D (*Capital*) and the object is I (*USA*). However, if R_2 is positioned after I in the original query then its subject is I. For example, in the query *"Which country is California located in?"* the subject of R_2 (*located in*) is I (*California*) and object is D (*Country*).

NQS of Complex & Compound Wh-Queries: A complex *Wh*-query can be characterized according to:

$[Wh]\ [R]\ [D]\ [Cl_D^?]\ [R_2]\ [I_1^1]\ [((CC]\ [I_2^1])^*]^*... \ [Cl_2^?]\ [R_3]\ [I_1^2]\ [((CC]\ [I_2^2])]^*...$
$...[Cl_N^?]]\ [R_{N+1}]\ [I_1^N]\ [((CC]\ [I_2^N])]^*$

where:

Cl_D: clausal lexeme (constraining D). Example of clausal lexemes: *wh*-tokens, *that, as, during/while/before/after*, etc. It is to be noted that clausal lexemes generates nested sub-queries which themselves may (or may not) be processed independently to the parent query. An example where the sub-query (in bold) has a dependency is: *"Which artists where born on the same date as **Rachel Stevens**?"*

Cl_2: second clausal lexeme (constraining I_1)

Cl_k: clausal lexeme associated with k-th sub-structure

$[CC]$: conjunctive/disjunctive lexeme for I

$[D]$: query desire - value restricted to POS tags: {*NN, NNP, JJ, RB, VBG*}

$[I_l^k]$: l-th query input for k-th structure - value restricted to POS tags: {NN, NNP, JJ, RB, VBG}

$[R_{k+1}]$: relation associated with the k-th clause that acts as (i) predicate of D as the subject and I as the object or (ii) action role of I as the actor - value restricted to POS tags: {VB, PP, VB-PP}.

Notation with ? may occur zero or one time.

Notation with $*$ may occur zero or multiple time.

In the given complex NQS, we see the possible repetition of the structure: $[I_1^k][([CC][I_2^k])]$. Within this structure, there is an optional substructure $[([CC][I_2^k])]$ that may add to the number of inputs within each of such structures. A clausal lexeme in a complex clausal *wh*-query is always associated with such a structure. The number of clausal lexemes is the same as the number of such structures in a given query. It should be noted that there must be at least two such structures for a query to qualify as complex. Clausal lexemes are optional and hence, the NQS also works for complex non-clausal wh-queries. We name the following structure as *clausal structure* (*CS*):

$$[Cl_D^?] \ [R_2^?] \ [I_1^1] \ [([CC][I_2^1])^*]^? ... [Cl_2^?] \ [R_3^?] \ [I_1^2] \ ([CC] \ [I_2^2])^? ...$$
$$...([Cl_N^?][R_{N+1}^?][I_1^N]([CC][I_2^N])^*[?].$$

A compound *Wh-query* can then be characterized according to:

$$[Wh^1] \ [R_1^{1?}] \ [D_1^{1?}] \ [Cl_D^?] \ [R_2^?] \ [I_1^1] \ ([CC] \ [I_2^1])^* \ [Cl_2^?][R_3^?][I_1^2]([CC][I_2^2])^* ....$$
$$([Cl_N^?][R_{N+1}^?][I_1^N]([CC][I_2^N])^*)^*[?].$$

4 AskNow Architectural Pipeline

We have outlined the architectural pipeline of AskNow, in Fig. 1 with two basic components: the NQS Instance Generator and the NQS to SPARQL converter.

4.1 NQS Instance Generation

As mentioned in the previous section, the objective of NQS is not to propose yet another grammar but rather to provide a modular format to the internal sub-structures of a query. Therefore, an efficient template-fitting algorithm that can parse the natural language query, identify the sub-structures (using a standard POS tagger), and then fit them into their corresponding *cells* within the larger generic NQS template is required. Our proposed template-fitting algorithm is called *NQS Instance Generator*. Through the fitting process the query-desire, query-input, and other relevant information can be extracted. A fitted NQS is called an *NQS instance*. The fitting process automatically leads to normalization. Also, it is resilient to paraphrasing of queries since the sub-structures in a paraphrase typically remain unaffected[2]. The change is only in the inter sub-structure positioning (e.g.: "*New York* is *located in* which *country*?" vs. "*Which is the country where New York is located?*")[3]. Note that the original NL query may lose its syntactic structure during the NQS instance generation process and also, it does not guarantee the grammatical correctness of the normalized query.

In summary, the flexibility of NQS modeling is to be attributed to the NQS Instance Generator algorithm. All internal components of the NQS Instance Generator are described as follows:

Query Processor: This module initiates the NQS query processing system by initializing other modules. It calls the POS-tagger (in our case we used Stanford coreNLP[4]) so as to tag every query token. Then it breaks the query text into

---

[2] In certain cases minor splitting has to be handled in the dependency relation, where the query starts with a preposition, e.g.: "*In* which *country* is *New York* **located** ?".

[3] Paraphrasing may include lexical substitution of synonymous query tokens and morphological changes of the tokens.

[4] http://nlp.stanford.edu/software/tagger.shtml.

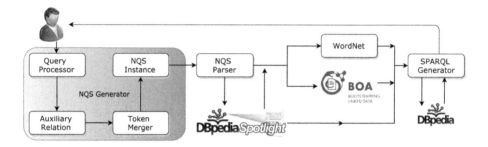

Fig. 1. Architectural pipeline of AskNow

individual POS-tagged query tokens. Subsequently, the *Syntactic Normalizer* transforms original queries to have a common syntactic structure. For example, it normalizes each query to start with *wh*-token, handling apostrophe, etc.

Auxiliary Relation Handler: The module to extract $R1$. More details regarding the utility of the auxiliary relation is given in previous section.

Token Merger: This module merges (or *chunks*) tokens that together form a single meaningful lexeme. Based on the POS tags of the tokens in the original NL query, the token merging module can guess the possible tokens to be combined so that they can fit the NQS. For example, when the query, *"Who is the Prime Minister of India?"* is passed to the POS-Tagger we get the resulting answer: *"Who$_{WP}$ is$_{VBZ}$ the$_{DT}$ Prime$_{NNP}$ Minister$_{NNP}$ of$_{IN}$ India$_{NNP}$?"*. Then two *consecutive tokens* are taken at a time and checked, using a token-merging map, whether they can be combined or not. We have manually bootstrapped the token-merging map on different types of token-pair lexico-syntactic patterns, using the M.S. Encarta 98 query dataset. The map keeps getting updated as and when other valid token-pairs are identified in future.

NQS Instance Generator: After the individual chunks have been identified, the query then goes through the NQS Instance Generator (see Fig. 1). It uses the following two hypothesis about the generic structure of a query (which was observed to be empirically true when tested on Microsoft Encarta 98, which is a large-scale query dataset).

Hypothesis 1: The query desire is always a noun phrase.

Hypothesis 2: The query desire always precedes the query input in the normalized NL query.

The algorithm also utilizes the characteristics of desire-input dependency relations, as discussed in the previous section. Every time it encounters a noun phrase chunk it treats it as a candidate desire. Depending upon the availability of conjunctive connectives, it then does a conflict resolution among all candidate desires by verifying the positioning of the verb phrase. As an example, the query *"Desserts from which country contain fish?"* has three candidate desires: *dessert*, *country*, *fish* (based on Hypothesis 1). The main relation *contain* is positioned after *country*. Therefore, the *potential* subject of *contain* is identified to

Table 1. QALD-5 example on AskNow

NL Query	List down all the Swedish holidays
NQS values	[WH = What], [R1 = is], [D = list], [R2 = of], [M = Swedish], [I = holiday]
Type	List
SPARQL	SELECT DISTINCT ?uri WHERE { ?uri rdf:type dbo:Holiday. ?uri dbo:country res:Sweden }
NL Query	In which country is Mecca located?
NQS values	[WH = which], [R1 = is], [D = country], [R2 = located In], [I = Mecca]
Type	Property Value
SPARQL	SELECT ?num WHERE { res:Mecca dbo:country ?num . }
NL Query	How many ethnic groups live in Slovenia
NQS values	[WH = How many], [R1 = *null*], [D = count(ethnic group)], [R2 = live in], [I = Slovenia]
Type	Count
SPARQL	SELECT COUNT(DISTINCT ?uri) WHERE { res:Slovenia dbo:ethnicGroup ?uri . }
NL Query	Who is the heaviest player of the Chicago Bulls?
NQS values	[WH = Who], [R1 = is], [M = heaviest], [D = player], [R2 = of], [I = the Chicago Bulls]
Type	Ranking
SPARQL	SELECT DISTINCT ?uri WHERE { ?uri rdf:type dbo:Person . ?uri dbo:weight ?num . ?uri dbp:team res:ChicagoBulls} ORDER BY DESC(?num) OFFSET 0 LIMIT 1

be *country*[5]. Now according to Hypothesis 2, the desire must precede the input in the NQS instance. So *fish* is resolved not to be a candidate desire any more, but rather an input. Now, the query has another main relation *from*, the subject being *dessert* and the object being a query token *which*. Thus, the algorithm resolves that *country* is the desire while *dessert* is another input. Finally, the algorithm analysizes that *country* being the desire, and also having the inverse relation *from* to the input *dessert*, cannot have the relation *contain* to the second input *fish*. Therefore, it is the input *dessert* which is the true subject of the relation *contain* to the object (i.e. the second input) *fish*. The final NQS will be: $[wh = which][R_1 = null][D = country][R_2 = from][I_1 = dessert][R_3 = contain][I_2 = fish]$. It is to be noted that there is an implicit nested dependent sub-query: "*Which desserts contain fish?*" because of the clausal connective *whose* (*Which country **whose** desserts . . .*) that is an inverse of the relation *from*. This example illustrates that the previously outlined NQS syntax definitions are not static templates, but rather dynamically fitted.

---

[5] A standard dependency parser could also be used to understand the subject of the relation *contain*.

input : NQS instance ℵ, knowledge base KB
output: SPARQL query results
```
// Step 1: NQS analysis
```
1 $D \longleftarrow$ queryDesire(ℵ);
2 $wh \longleftarrow$ getWhQuestionType(ℵ);
3 $t \longleftarrow$ determineQueryType(D, wh);
4 $I \longleftarrow$ queryInput(ℵ);
```
// Step 2: SPARQL preparation
```
5 $i \longleftarrow$ mapInput(I);
6 $S = \{(p,v)|(i,p,v) \in KB\};$ `// construct predicate object map`
7 init p_{match};
8 **foreach** $(p,v) \in S$ **do**
    ```
    // label matching
    ```
9 | **if** $lm(p)==D$ **then** $p_{match} = p$; break;
    ```
    // WordNet synonyms of desire
    ```
10 | **if** $wns(D) ==p$ **then** $p_{match} = p$; break;
    ```
    // BOA library
    ```
11 | **if** $BOA(D) ==p$ **then** $p_{match} = p$; break;

```
// Step 3: SPARQL generation and retrieval
```
12 $q \longleftarrow$ generateQuery(i, p_{match}, KB);
13 $R \longleftarrow$ executeQuery(q, KB);
14 **return** R

Algorithm 1. NQS to SPARQL Algorithm.

4.2   NQS to SPARQL Conversion

Given an NQS instance, the NQS2SPARQL module translates it to a SPARQL query and returns the result from the SPARQL endpoint. There are four main steps in this module as shown in Algorithm 1.

NQS Analysis: Once we have an NQS instance for a query, the system treats it as per its category. The categories are the expected query types, specifically: (i) Boolean (ii) Ranking (iii) Count (iv) Set (List) and (v) Property Value. In a Boolean query a user asks whether a specific statement is True of False. For instance "Is Barack Obama a democrat?" A Ranking query requires ranking the answers based on some entity dimensions, e.g. "Which is the highest mountain in Asia?". In a Count query the user intent is to get the number of times a certain condition is repeated. A Set query will generate a list of items which satisfy a required condition. In a Property value query the user intent is to ask for the value of a property of the given input. As an example, in the query "What is the capital of India?" the user intents to extract the value of the property "capital" given the input "India". Query-types are chosen based on desire (D) and wh-type (wh) of the NQS instance. Each category is processed by a different SPARQL query syntax converter.

Entity Mapping: The basic operation here is to retrieve the knowledge base entity matching the spotted query desire, query input and their relation. For the QALD experiments described later, we annotated the query using DBpedia

Spotlight [7]. As a result of the mapping, we get the knowledge base entity equivalent of the query input I which has been identified in the NQS instance. We denote this entity as i. The mapping approach then collects properties related to i (where i is a resource) and their values in set (denoted S).

Subsequently, each element (a pair of property and value) of S is observed. The next goal is to identify the entity which matches the desire (itself denoted as D) and denote it as d. This is done using three mapping functions as follows: The first test is made by a simple label matching function(lm). If this fails, then the second test for mapping($D \rightarrow d$) is through the WordNet synonym (wns)function. It finds the synonym of user desire using WordNet [8] within set S. If this test fails we move to next test. Here we use BOA pattern library [9] for the same purpose. When this is unsuccessful, then we declare that the query is unprocurable by the system.

SPARQL Generation: This component creates the final SPARQL query using information provided by above two steps. NQS analysis basically gives the SPARQL pattern possible. Where as i, d provide the key DBpedia information (vocabulary) required for SPARQL. Examples are given in Table 1. Currently NQS2SPARQL is functional for DBpedia only. However, we can plugin any other RDF store using suitable corresponding entity mapping module.

5   Evaluation

5.1   Evaluation Goal and Metric

Goal I. Syntactic Robustness: *Syntactic robustness* of *NQS* measures its structuring capacity after normalization. Ideally, the *NQS* algorithm should be correct By *correct structuring* we mean that there should not be any mismatch between the POS tag of a linguistic constituent and its corresponding *NQS* cell. At the same time, the algorithm should be complete (i.e. there should not be any valid English query that is not accepted by the algorithm, either fully or partially). To evaluate robustness we decided on a simple measure called *Structuring Coverage* (*SC*). We measure SC in the following three different perspectives:

(i) **SC-Precision:** Given a test set of NL queries, *SC-Precision* is calculated as the ratio of the number of correct *NQS-structured* queries (N_{CI}) and the total number of *NQS-structured* queries in the test set (N_I). It largely depends upon the accuracy of the POS tagger used.

(ii) **SC-Recall:** Given a test set of NL queries, *SC-Recall* is calculated as the ratio of the number of correct *NQS-structured* queries (N_{CI}) and the total number of queries in the test set (N).

(iii) **SC-F1:** The Simple Harmonic Mean of *SC-Precision* and *SC-Recall.*

Goal II. Sensitivity to Structural Variation: *Sensitivity to structural variation* of *NQS* measures the degree to which *NQS* can *correctly fit* queries having same *desire* (and its relationship with *input*) yet different syntactic structures. To evaluate *sensitivity to structural variation* we introduce following two measures:

(i) **Variational-Precision (*VP*):** Given a test set of NL queries, the *VP* is calculated as the ratio of the number of correct *NQS-structured* queries (i.e. without any of their variations getting incorrectly fitted) (N_{VI}) and the total number of identified queries in the test set (N_I).

(ii) **Variational-Recall (*VR*):** Given a test set of NL queries, the *VR* is calculated as the ratio of N_{VI} and total number of queries in the test set (N).

Goal III. Semantic Accuracy: *Semantic accuracy* of *NQS* measures the degree to which the query *desire* and its relation with query *inputs* has been properly identified. To evaluate this we use the following measures:

(i) **Semantic-Precision (*SP*):** Given a test set of NL queries, the *SP* is calculated as the ratio of the number of correctly identified queries (i.e. in terms of *desire-identification*, *input-identification*, and *desire-input relation identification*) (N_{SI}) with respect to a human-judgement benchmark, and the total number of identified queries in the test dataset (N_I).

(ii) **Semantic-Recall (*SR*):** Given a test set of NL queries, the *SR* is calculated as the ratio of N_{SI} with respect to a human-judgement benchmark, and the total number of queries in the test dataset (N).

Here are examples to give a better understanding of purpose of each measure:
Failed NQS (i.e. no instance): [Wh = NULL] [R1 = is] [D = Berlin] [R2 = NULL] [I = country][?]
Incorrectly structured NQS instance: [Wh = In which country] [R1 = is] [D = Berlin] [R2 = located] [I = NULL]. This will be considered as identified query (i.e. one in N_I).
Correctly structured NQS instance (i.e. in $N_C I$): [Wh = Which] [R1 = is] [D = Berlin] [R2 = located in] [I = country]. We use SC (and also VP, VR) to test $N_C I$ with respect to N_I and total queries (N).
Correctly "identified" NQS instance (i.e. in $N_S I$): [Wh = Which] [R1 = is] [D = country] [R2 = located in] [I = Berlin][?]. We use SP and SR to test this.

Goal IV Accuracy of the AskNow System: The final goal of the evaluation is to test the system on the QALD-5 [10] benchmark (Multilingual question answering over DBpedia). Here, we have queries in English language which are answered with NQS translated SPARQL.

5.2   Datasets

In order to evaluate *syntactic robustness* (for goal-I), we have used the Microsoft Encarta 98[6] query test set. The test set contains 1365 usable English *wh*-queries. There are total 522 queries of procedural *how* and *why* that have been excluded. We also created an extensive query set based on OWLS-TC v4[7] for evaluation

---

[6] http://research.microsoft.com/en-us/downloads/88c0021c-328a-4148-a158-a42d73 31c6cf/.
[7] http://projects.semwebcentral.org/projects/owls-tc/.

Table 2. *SC* evaluation on different datasets

	QALD 5			M.S. Encarta			OWL-S TC			Total			Result		
	N	N_I	N_{CI}	N	N_I	N_{CI}	N	N_I	N_{CI}	N	N_I	N_{CI}	SC_R	SC_P	SC_{F1}
How	31	31	31	165	158	158	4	4	2	200	193	191	95.50	98.96	97.20
What	37	37	37	406	392	392	1711	1709	1608	2154	2138	2037	94.57	95.28	94.92
When	12	12	12	39	35	35	0	0	0	51	47	47	92.16	100	95.92
Where	5	5	5	85	82	82	20	20	19	110	107	106	96.36	99.07	97.70
Which	81	81	81	5	5	5	316	316	308	402	402	394	98.01	98.01	98.01
Who	48	48	48	143	143	143	166	166	166	357	357	357	100	100	100
Total	214	214	214	843	815	815	2217	2215	2215	3274	3244	3226	98.53	99.45	98.99

of both *sensitivity to structural variation* (goal-II) and *semantic accuracy* (goal-III). Three research assistants independently formulated wh-queries for every web service of OWLS-TC v4 dataset, such that the query desire matches the given service output, and the query input matches the required service input. We had 1083 services to make three different query versions for each service. Similar syntactic structure queries were excluded resulting in a total of 2217 queries It is to be noted that the goal of the experiment (cf.: Goal II) was to test the robustness of an NQS Instance Generator, in terms of POS-tag pattern fitting (i.e. syntactic accuracy), over different syntactic variations of the same query. 90 % of the queries were complex or compound queries. Ideally, the extracted query desire by NQS should be semantically equivalent to the output parameter of the corresponding web service specification. Based on this notion, we have calculated *SC-accuracy*, *VP/VR*, and *SP/SR* for each of the three versions of query dataset. We also used the QALD-5 [10] datasets for Goal-IV and QALD-4 [11] for evaluating Goal-II.

5.3   Results

Result I. Syntactic Robustness: We first performed the evaluation of *structural robustness* in terms of SC-Accuracy over different *query*-types on *Microsoft Encarta 98* dataset. We observe 100 % SC-Precision for all types of wh-queries, which shows that the NQS is theoretically sound. The *SC-Recall* came out to be 96.68 %. We then performed the same experiment over different *wh*-types on 2 more datasets: Training set of QALD-5's Multilingual tract (only *english* queries) and *OWLS-TC*. We observed a high overall SC-F1 of 98.99 %. The evaluation results are given in Table 2.

Result II. Sensitivity to Structural Variation: We performed evaluation of *sensitivity to structural variation* of *NQS* over the OWL-S TC query dataset (three versions) and the QALD-4 dataset (three versions). *NQS* was able to correctly fit 919 out of the 1083 OWLS-TC queries (along with all their syntactic variation), giving high *VP* of 96.43 %. All 24 out of 24 QALD-4 queries, with all there syntactic variations, were correctly fitted in *NQS*, giving a high sensitivity to structural variation.

Table 3. Evaluation of sensitivity to structural variation and semantic accuracy

Dataset	N_{Wh}	N_I	N_{VI}	N_{SI}	$VR\%$	$VP\%$	$SR\%$	$SP\%$
OWL S TC	1083	953	919	876	84.85	96.43	80.88	91.92
QALD-4	24	24	24	21	100	100	87.50	87.50
Total	1107	977	943	897	85.18	96.51	81.03	91.81

Result III. Semantic Accuracy: We observed an *SP* of 91.92 % for the OWL-S TC query dataset. For QALD-4 dataset, it was observed that 21 out of 24 queries (with their variations) were correctly fitted in *NQS*. Analysis of the fail case clearly indicates that NQS failure is dependent upon syntactic and POS Tag failures (Tables 3 and 4).

Results IV. Accuracy of AskNow: We used the benchmark data set of the 5th Workshop on Question Answering over Linked Data (QALD), which defines 50 questions to DBpedia and their answers. Here we compare the our results with the result published by QALD-5 [10]. Out of 50 questions provided by the benchmark we have successfully answered 16 correct and 1 partially correct. There were 5 questions where NQS algorithm fails to correctly identify the Inputs and Desire hence they could not be answered by translating them into SPARQL. The failure analysis of Result IV are as follows:

> **NQS failure:** Queries where NQS failed were not further processed successfully. NQS failed only 5 times, which was due to incorrect dependency analysis.
>
> **Entity Mapping:** There are 13 questions where AskNow could not map the DBpedia equivalent of correctly identified input and desire. In some cases, the correct mapping was presented but insufficient to answer the query. As an example, the query *"Who killed John Lennon?"* is correctly processed by NQS and forwarded to DBpedia Spotlight for annotation. It maps *JohnLennon* to http://dbpedia.org/resource/John_Lennon which is a correct mapping in general terms. But we can not answer the question based on this resource. For that we would require http://dbpedia.org/resource/Death_of_John_Lennon.
>
> **Relation Mapping:** In some cases, system could not resolve the *R2* (relation between input and desire) to the correct DBpedia property. Relations such as *study* and *graduated* were not mapped to the required DBpedia property *almaMater*.

6   Related Work

Over the last decade, several *NLQF* approaches have been proposed. Several of them attempt to translate NL queries into SPARQL-like formalisms. Early works in this direction includes GiNSENG [12]. It is a guided input NL search engine, that does not understand NL queries, but uses menus to formulate NL queries in

Table 4. Results on the QALD 5 benchmark.

	Processed	Right	Partial	Recall	Precision	F_1	F_1 Global
Xser	42	26	7	0.72	0.74	0.73	0.63
AskNow	27	16	1	0.63	0.60	0.61	0.33
QAnswer	37	9	4	0.35	0.46	0.40	0.30
APEQ	26	8	5	0.48	0.40	0.44	0.23
SemGraphQA	31	7	3	0.32	0.31	0.31	0.20
YodaQA	33	8	2	0.25	0.28	0.26	0.18

small and specific domains and allow users to query OWL knowledge bases in a controlled language akin to English. Subsequently, Semantic Crystal [13] was proposed, which is also a guided and controlled graphical query language. Systems such as AquaLog [14] and its advancement, PowerAqua [15], are based on mapping linguistic structures to ontology-compliant semantic triples. PowerAqua is the first system to perform QA over structured data, providing a single NL query interface for integrating information from heterogeneous resources. The limitation of PowerAqua is the lack of support for query aggregation functions. Along the same lines, FREyA [16] allows users to enter queries in any form, and uses ontology reasoning to learn more generic rules. It also provides a better handling of ambiguities over heterogeneous domains. But FREyA requires some level of effort in KB structure understanding to efficiently clarify disambiguation. Also, it highly depends on modeling and vocabulary of the data at the user-end, making it inadequate for a naive user. Other works such as NLPReduce [17] allow users to pose questions in full or slightly controlled English. NLPReduce is a domain-independent system, which leverages the lexico-syntactic pattern structures of query input to find better matches in the KB. It maps query tokens with synonym enhanced triple stores in the target corpus, based on which it generates SPARQL statements for those matches. QTL [18] is a feedback mechanism for question answering using supervised machine learning on SPARQL. TBSL [19] uses so called BOA patterns as well as string similarities to fill the missing URIs in query templates and bridge the lexical gap.

Recently, Question Answering over Linked Data (QALD) has become a popular benchmark. In QALD-3 [20], SQUALL2SPARQL [21] achieved the highest precision in the QA track. SQUALL2SPARQL takes an inputs query in SQUALL, which is a special English based language, and translates it to SPARQL. Since no linguistic resource is required, it results in a high performance. But on the other hand, it makes the SQUALL query unnatural to the end-user and requires manual annotations of the URIs. In QALD-4 [11], GFMed [22] achieved the highest precision in the biomedical track. It is based on Grammatical Framework (GF) [23] and a Description Logics based methodology and proposes an algorithm to translate a query in natural language into SPARQL queries using GF resources. It can support complex queries, but only works with controlled languages and biomedical datasets. In POMELO [24], predicates of the RDF triples are mapped to frame predicates while the subjects and objects are mapped to core frame elements. Then after a question abstraction step, the final SPARQL query is generated. POMELO is based on closed environment (biomedical) and fails to relate the disconnected semantic entities. gAnswer [25] proposes a graph mining algorithm to map natural language phrases to top-k possible predicates to form a paraphrase dictionary. It also proposes a novel approach to perform disambiguation in query evaluation phase, which improves the precision and speed up query processing time greatly.

Xser [26], the most successful system in QALD-4 and QALD-5, uses a two-step architecture. It first understands the NL query by extracting phrases and labeling them as *resource, relation, type* or *variable* to produce a Directed Acyclic Graph (DAG). This semantic parser works independent of any Knowledge Base (KB). Then these semantic entities are instantiated with the given KB. However, it requires too much human involvement in manually annotating the questions with phrase dependency DAG to train the system. In comparison to Xser, AskNow requires no training data. The NQS instance generation step is independent of both the query dataset and the target knowledge base (KB). Xser uses semantic parser with DAG as linguistic analyzer. AskNow use POS-tag and NER to find the main entity of the query. As linguistic analyzer an NQS instance has ability to further distinguish between query desire (D), query input (I) and their relation (R) apart from spotting the main entity only. Xser uses wikipedia miner tool to generate the candidate set of DBpedia entities, AskNow uses DBpedia Spotlight for annotating the query Input to DBpedia entities Xser uses PATTY to map phrases to predicates and categories of DBpedia whereas, AskNow do this by mapping query relation or query desire to DBpedia equivalent using WordNet and BOA pattern library.

APEQ [10], from QALD-5 [10], uses a graph traversal based approach, where it first extracts the main entity from the query and then tries to find its relations with the other entities using the given KB. APEQ uses Graph traversal technique to determine the main entity by graph exploration. Finally, the graph with the best scoring entities is returned as the answer.

7   Conclusion

In this paper, we propose *AskNow*, a NLQF framework, based on a novel syntactic structure *Normalized Query Structure* (NQS). We empirically show, using benchmark datasets, that NQS is robust in terms of syntactic variation, and also highly accurate in identifying the query desire (along with its relationship to the query input). Hence, we show that NQS serves as a strong intermediary model for translating NL queries into formal queries. We empirically demonstrated this by converting NQS to SPARQL.

Acknowledgements. This work was supported by a grant from the EU H2020 Framework Programme provided for the project Big Data Europe (GA no. 644564).

References

1. Auer, S., Bizer, C., Kobilarov, G., Lehmann, J., Cyganiak, R., Ives, Z.G.: DBpedia: a nucleus for a web of open data. In: Aberer, K., et al. (eds.) ASWC 2007 and ISWC 2007. LNCS, vol. 4825, pp. 722–735. Springer, Heidelberg (2007)
2. Suchanek, F.M., Kasneci, G., Weikum, G.: Yago: a core of semantic knowledge. In: Proceedings of the 16th International Conference on World Wide Web, pp. 697–706. ACM (2007)
3. Bollacker, K., Evans, C., Paritosh, P., Sturge, T., Taylor, J.: Freebase: a collaboratively created graph database for structuring human knowledge. In: Proceedings of the ACM SIGMOD International Conference on Management of Data, pp. 1247–1250. ACM (2008)
4. Dong, X., Gabrilovich, E., Heitz, G., Horn, W., Lao, N., Murphy, K., Strohmann, T., Sun, S., Zhang, W.: Knowledge vault: a web-scale approach to probabilistic knowledge fusion. In: Proceedings of the 20th ACM SIGKDD International Conference on Knowledge Discovery and Data Mining, pp. 601–610. ACM (2014)
5. Pérez, J., Arenas, M., Gutierrez, C.: Semantics and complexity of SPARQL. In: Cruz, I., Decker, S., Allemang, D., Preist, C., Schwabe, D., Mika, P., Uschold, M., Aroyo, L.M. (eds.) ISWC 2006. LNCS, vol. 4273, pp. 30–43. Springer, Heidelberg (2006)
6. Lopez, V., Unger, C., Cimiano, P., Motta, E.: Evaluating question answering over linked data. Web Semant. Sci. Serv. Agents World Wide Web **21**, 3–13 (2013)
7. Mendes, P.N., Jakob, M., García-Silva, A., Bizer, C.: Dbpedia spotlight: shedding light on the web of documents. In: Proceedings of the 7th International Conference on Semantic Systems, pp. 1–8. ACM (2011)
8. Miller, G.A.: Wordnet: a lexical database for English. Commun. ACM **38**(11), 39–41 (1995)
9. Gerber, D., Ngonga Ngomo, A.-C.: Bootstrapping the linked data web. In: 1st Workshop on Web Scale Knowledge Extraction @ ISWC (2011)
10. Unger, C., Forascu, C., Lopez, V., Ngomo, A.-C.N., Cabrio, E., Cimiano, P., Walter, S.: Question answering over linked data (QALD-5). In: Working Notes for CLEF Conference (2015)
11. Unger, C., Forascu, C., Lopez, V., Ngomo, A.-C.N., Cabrio, E., Cimiano, P., Walter, S.: Question answering over linked data (QALD-4). In: Working Notes for CLEF Conference (2014)
12. Bernstein, A., Kaufmann, E., Kaiser, C.: Querying the semantic web with ginseng: a guided input natural language search engine. In: 15th Workshop on Information Technologies and Systems, Las Vegas, NV, pp. 112–126. Citeseer (2005)
13. Kaufmann, E., Bernstein, A.: How useful are natural language interfaces to the semantic web for casual end-users? In: Aberer, K., et al. (eds.) ASWC 2007 and ISWC 2007. LNCS, vol. 4825, pp. 281–294. Springer, Heidelberg (2007)
14. Lopez, V., Uren, V., Motta, E., Pasin, M.: Aqualog: an ontology-driven question answering system for organizational semantic intranets. Web Semant. Sci. Serv. Agents World Wide Web **5**(2), 72–105 (2007)
15. Lopez, V., Fernández, M., Motta, E., Stieler, N.: Poweraqua: supporting users in querying and exploring the semantic web. Semant. Web **3**(3), 249–265 (2012)
16. Damljanovic, D., Agatonovic, M., Cunningham, H.: FREyA: an interactive way of querying linked data using natural language. In: García-Castro, R., Fensel, D., Antoniou, G. (eds.) ESWC 2011. LNCS, vol. 7117, pp. 125–138. Springer, Heidelberg (2012)

17. Kaufmann, E., Bernstein, A., Fischer, L.: NLP-reduce: a naïve but domain-independent natural language interface for querying ontologies. In: ESWC, Zurich (2007)
18. Lehmann, J., Bühmann, L.: AutoSPARQL: let users query your knowledge base. In: Antoniou, G., Grobelnik, M., Simperl, E., Parsia, B., Plexousakis, D., De Leenheer, P., Pan, J. (eds.) ESWC 2011, Part I. LNCS, vol. 6643, pp. 63–79. Springer, Heidelberg (2011)
19. Unger, C., Bühmann, L., Lehmann, J., Ngonga Ngomo, A.-C, Gerber, D., Cimiano, P.: Template-based question answering over RDF data. In: Proceedings of the 21st International Conference on World Wide Web, pp. 639–648. ACM (2012)
20. Cimiano, P., Lopez, V., Unger, C., Cabrio, E., Ngonga Ngomo, A.-C., Walter, S.: Multilingual question answering over linked data (QALD-3): lab overview. In: Forner, P., Müller, H., Paredes, R., Rosso, P., Stein, B. (eds.) CLEF 2013. LNCS, vol. 8138, pp. 321–332. Springer, Heidelberg (2013)
21. Ferré, S.: squall2sparql: a translator from controlled English to full SPARQL 1.1. In: Working Notes of Multilingual Question Answering over Linked Data (QALD-3) (2013)
22. Marginean, A.: GFMed: question answering over biomedical linked data with grammatical framework. In: CLEF (2014)
23. Ranta, A.: Grammatical Framework: Programming with Multilingual Grammars. CSLI Publications, Stanford (2011)
24. Hamon, T., Grabar, N., Mougin, F., Thiessard, F.: Description of the pomelo system for the task 2 of QALD-2014. In: CLEF (2014)
25. Zou, L., Huang, R., Wang, H., Yu, J.X., He, W., Zhao, D.: Natural language question answering over RDF: a graph data driven approach. In: Proceedings of the ACM SIGMOD International Conference on Management of Data, pp. 313–324. ACM (2014)
26. Xu, K., Zhang, S., Feng, Y., Zhao, D.: Answering natural language questions via phrasal semantic parsing. In: Zong, C., Nie, J.-Y., Zhao, D., Feng, Y. (eds.) NLPCC 2014. CCIS, vol. 496, pp. 333–344. Springer, Heidelberg (2014)

Knowledge Extraction for Information Retrieval

Francesco Corcoglioniti, Mauro Dragoni[(✉)], Marco Rospocher,
and Alessio Palmero Aprosio

Fondazione Bruno Kessler, Trento, Italy
{corcoglio,dragoni,rospocher,aprosio}@fbk.eu

Abstract. Document retrieval is the task of returning relevant textual resources for a given user query. In this paper, we investigate whether the semantic analysis of the query and the documents, obtained exploiting state-of-the-art Natural Language Processing techniques (e.g., Entity Linking, Frame Detection) and Semantic Web resources (e.g., YAGO, DBpedia), can improve the performances of the traditional term-based similarity approach. Our experiments, conducted on a recently released document collection, show that Mean Average Precision (MAP) increases of 3.5 % points when combining textual and semantic analysis, thus suggesting that semantic content can effectively improve the performances of Information Retrieval systems.

1 Introduction

Recent years have seen the growing maturity of Knowledge Extraction (KE) from natural language text. State-of-the-art KE approaches, such as FRED [1], News-Reader [2], and PIKES [3], exploit Natural Language Processing (NLP) techniques as well as Semantic Web (SW) and Linked Open Data (LOD) resources (e.g., DBpedia [4]) to extract semantic content from textual resources, linking it to well-known ontologies and to a growing body of LOD background knowledge.

In this paper we investigate the benefits of using the semantic content automatically extracted from text for Information Retrieval (IR). The goal in IR is to determine, for a given user query, the relevant documents in a text collection, ranking them according to their relevance degree for the query. Our approach aims to overcome known limitations of traditional IR approaches. Let us consider the following query example: "astronomers influenced by Gauss". Traditional IR approaches match the terms or possible term-based expansions (e.g., synonyms, related terms) of the query and the documents, but relevant documents may not necessarily contain all the query terms (e.g., the term "influenced" or "astronomers" may not be used at all in a relevant document); similarly, some relevant documents may be ranked lower than other ones containing all three terms, but in an unrelated way (e.g., documents about some astronomers born centuries before Gauss, influenced by Leonardo Da Vinci).

In our approach, both queries and documents are processed to extract semantic content pertaining to the following *semantic layers*: (i) entities, e.g., dbpedia:Carl_Friedrich_Gauss from "Gauss"; (ii) types of entities, either

© Springer International Publishing Switzerland 2016
H. Sack et al. (Eds.): ESWC 2016, LNCS 9678, pp. 317–333, 2016.
DOI: 10.1007/978-3-319-34192-3_20

explicitly mentioned, such as yago:Astronomer109818343 from "astronomers", or indirectly obtained from external resources for mentioned entities, such as yago:GermanMathematicians from "Gauss"; (iii) semantic frames and frame roles, such as framebase:Subjective_influence from "influenced"; and, (iv) temporal information, either explicitly mentioned in the text or indirectly obtained from external resources for mentioned entities, e.g., via DBpedia properties such as dbo:dateOfBirth (1777) and dbo:dateOfDeath (1855) for entity dbpedia: Carl_Friedrich_Gauss. We then match queries and documents considering both their textual and semantic content, according to a simple retrieval model based on the Vector Space Model (VSM) [5]. This way, we can match documents mentioning someone who is an astronomer (i.e., entities having type yago:Astronomer109818343) even if "astronomers", or one of its term-based variants, is not explicitly written in the document text. Similarly, we can exploit the entities and the temporal content to better weigh the different relevance of documents mentioning dbpedia:Carl_Friedrich_Gauss and dbpedia:GAUSS_(software), as well as to differently rank documents about Middle Age and 17th/18th centuries astronomers.

While other ontology-based IR approaches typically builds only on terminological knowledge (e.g., classes, subclasses), to the best of our knowledge our work is the first in exploiting such a variety of automatically extracted semantic content (i.e., entities, types, frames, temporal information) for IR.

We developed a first implementation of our approach, named KE4IR (read: *kee-fer*), using PIKES for KE and Apache Lucene[1] for indexing documents and evaluating IR queries. We performed a first assessment of the approach on a recently released dataset [6], showing that enriching textual information with semantic content outperforms retrieval performances over using textual data only.

The paper is structured as follows. In Sect. 2, we briefly review the state of the art in IR and KE. Section 3 presents the KE4IR approach, detailing the semantic layers and the retrieval model used for combining semantic and textual information. In Sect. 4, we describe the actual implementation of our approach, while in Sect. 5, we report a first assessment of the effectiveness of adding semantic content for IR, discussing in details some outcomes and findings. Section 6 concludes with some final remarks and future work directions.

2   State of the Art

Previous works have exploited some semantic information for IR. An early tentative in injecting domain knowledge information for improving the effectiveness of IR systems is presented in [7]. In this work, authors manually built a thesaurus supporting the expansion of terms contained in both documents and queries. Such a thesaurus modeled a set of relations between concepts including synonymy, hyponymy and instantiation, meronymy and similarity. An approach based on the same philosophy was presented in [8], where the authors propose an

---

[1] http://lucene.apache.org/.

indexing technique where WordNet [9] synsets, extracted from each document word, are used in place of textual terms in the indexing task.

In the last decade, semantic IR systems started to embed ontologies for addressing the task of retrieving domain-specific documents. An interesting review on IR techniques based on ontologies is presented in [10], while in [11] the author studies the application of ontologies to a large-scale IR system for Web usage. Two models for the exploitation of ontology base knowledge bases are presented in [12,13]. The aim of these models is to improve search over large document repositories. Both models include an ontology-based scheme for the annotation of documents, and a retrieval model based on an adaptation of the classic Vector Space Model (VSM) [5]. Finally, in [14] an analysis of the usefulness on using ontologies for the retrieval task is discussed. More recently, approaches combining many different semantic resources for retrieving documents have been proposed. In [15], the authors describe an ontology-enhanced IR platform where a repository of domain-specific ontologies is exploited for addressing the challenges of IR in the massive and heterogeneous Web environment.

A further problem in IR is the ranking of retrieved results. Users typically make short queries and tend to consider only the first ten to twenty results [16]. In [17], a novel approach for determining relevance in ontology-based IR is presented, different from VSM. When IR approaches are applied in a real-world environment, the computational time needed to evaluate the match between documents and the submitted query has to be considered too. Systems using the well-known VSM have typically higher efficiency with respect to systems adopting more complex models to account for semantic information. For instance, the work presented in [18] implements a non-vectorial data structure that exhibits high computational times for both indexing and retrieving documents.

In the last few years, several approaches and tools performing comprehensive analyses to extract quality knowledge from text were presented. FRED [1] extracts Discourse Representation Structures (DRSs), mapping them to linguistic frames that in turn are transformed in RDF/OWL via Ontology Design Patterns.[2] In NewsReader [2], a comprehensive processing pipeline extracts and corefer events and entities from large (cross-lingual) news corpora. PIKES[3] [3] is an open-source frame-based knowledge extraction framework that combines the processing of various NLP tools to distill knowledge from text, aligning it to Linked Data resources such as DBpedia and FrameBase[4] [19], a recently released broad-coverage SW-ready inventory of frames based on FrameNet.[5] We exploit PIKES for the implementation of our approach, as described in Sect. 4.

3   Approach

Standard IR systems look at documents and queries as bags of terms (e.g., stemmed tokens). In KE4IR we augment textual terms with additional terms

---

[2] http://ontologydesignpatterns.org/.

[3] http://pikes.fbk.eu/.

[4] http://framebase.org/.

[5] http://framenet.icsi.berkeley.edu/.

coming from different semantic annotation layers produced using NLP-based KE techniques as well as Linked Open Data background knowledge (Sect. 3.1), and then propose a simple retrieval model that makes use of this additional semantic information to find and rank the documents matching a query (Sect. 3.2).

3.1   Semantic Layers

We consider four semantic layers—URI, TYPE, FRAME, TIME—that complement the base TEXTUAL layer with 'semantic terms'. These terms can be obtained using KE techniques that identify *mentions* (i.e., snippets of text) denoting entities, events and relations. From each mention, a set of semantic terms is extracted, and the number of mentions a term derives from can be used to quantify its relevance for a document. Table 1 (first three columns) reports an example of terms and associated mentions that can be extracted from the simple query of Sect. 1: "astronomers influenced by Gauss".

URI Layer. The semantic terms of this layer are the URIs of entities mentioned in the text, disambiguated against external knowledge resources such as DBpedia. Disambiguated URIs are the result of two NLP/KE tasks:[6] Named Entity Recognition and Classification (NERC), which identifies proper names of certain entity classes (e.g., persons, organizations, locations) in a text, and Entity Linking (EL), which disambiguates those names against the individuals of a knowledge base. The Coreference Resolution NLP task can be also exploited to 'propagate' the URI associated to a disambiguated named entity to its coreferring mentions in the text, so to proper count the number of entity mentions.

TYPE Layer. This layer contains as terms the URIs of the ontological types (and super-types) associated to noun phrases in the text. For disambiguated named entities (resulting from NERC and EL), associated types can be retrieved from external background knowledge resources describing those entities (e.g., DBpedia). For common nouns, disambiguation against WordNet via Word Sense Disambiguation (WSD) provides synsets, which can be mapped to types of known ontologies using existing mappings. Given these two extraction techniques, an ontology particularly suited to this layer is the YAGO taxonomy [20], due both to its WordNet origins and the availability of YAGO types for DBpedia entities.

TIME Layer. The terms of this layer are temporal values mentioned in the text, either because explicitly expressed in a time expression (e.g., "the eighteenth century") recognized via the Temporal Expression Recognition and Normalization (TERN) NLP task, or because associated via some property to a disambiguated entity in the background knowledge (e.g., the birth date associated to dbpedia:Carl_Friedrich_Gauss). We propose to represent time at different granularities—day, month, year, decade, and century—in order to support both

---

[6] We report in this section the main NLP/KE tasks for the extraction of semantic terms. Some of them typically build on additional NLP analyses, such as Tokenization, Part-of-Speech tagging, Dependency Parsing and Constituency Parsing.

precise and fuzzy temporal matching. Therefore, each mentioned date time value (e.g., 2015-12-18) is mapped to (max) five TIME terms, one for each granularity level (e.g., day:2015-12-18, month:2015-12, year:2015, decade:201, century:20).

FRAME Layer. A semantic frame is a star-shaped structure that represents an event or n-ary relation, having a certain frame type (e.g., framebase:frame-Subjective_influence, from "influenced") and zero or more participants (e.g., dbpedia:Carl_Friedrich_Gauss) playing a specific semantic role in the context of the frame. Semantic frames are typically extracted using NLP tools for Semantic Role Labeling (SRL) based on certain predicate models, such as FrameNet, and then mapped to an ontological representation using an RDF/OWL frame-based ontology aligned to the predicate model, such as FrameBase [19]. Semantic frames provide relational information that helps matching queries and documents more precisely. Different approaches can be used to transform a star-shaped semantic frame into a set of terms of the FRAME layer. In this work, we propose to map each ⟨frame type, participant⟩ pair whose participant is a disambiguated entity (e.g., the pair ⟨framebase:frame-Subjective_influence, dbpedia:Carl_Friedrich_Gauss⟩) to a term, including also the terms obtainable by replacing the frame type URI with the URIs of its super-classes in the ontology. We investigated also using non-disambiguated participant entities, obtaining however worse results.

3.2   Retrieval Model

We adopt a retrieval model inspired to the Vector Space Model (VSM). Given a document collection D, we represent each document $d \in D$ (resp. query q) with a vector $\mathbf{d} = (d_1 \ldots d_n)$ ($\mathbf{q} = (q_1 \ldots q_n)$) where each element d_i (q_i) is the weight corresponding to a term t_i and n is the number of distinct terms in collection D. Differently from text-only approaches, terms of our model come from multiple layers [21], both textual and semantic, and each document (query) vector can be thought as the concatenation of smaller, layer-specific vectors. Given a term t, we denote its layer with $l(t) \in L = \{\text{TEXTUAL, URI, TYPE, FRAME, TIME}\}$.

To compute the similarity between a document $d \in D$ and a query q, we use a similarity function $sim(d, q)$. The documents matching q are the ones with $sim(d, q) > 0$, and they are ranked according to decreasing similarity values. To derive $sim(d, q)$ we start from the cosine similarity of VSM:

$$sim_{\text{VSM}}(d, q) = \frac{\mathbf{d} \cdot \mathbf{q}}{|\mathbf{d}| \cdot |\mathbf{q}|} = \frac{\sum_{i=1}^{n} d_i \cdot q_i}{\sqrt{\sum_{i=1}^{n} d_i^2} \cdot \sqrt{\sum_{i=1}^{n} q_i^2}} \tag{1}$$

and we remove the normalization by $|\mathbf{d}|$ and $|\mathbf{q}|$, thus obtaining:

$$sim(d, q) = \mathbf{d} \cdot \mathbf{q} = \sum_{i=1}^{n} d_i \cdot q_i \tag{2}$$

Normalizing by $|\mathbf{q}|$ does not affect the ranking and only serves to compare scores of *different* queries, thus we drop it for simplicity. Normalizing by $|\mathbf{d}|$ has the

Table 1. Terms extracted from the query "astronomers influenced by Gauss", with mentions m_1 = "astronomers", m_2 = "influenced", m_3 = "Gauss"; the TEXTUAL layer is weighted 0.5; the four semantic layers are weighted 0.125 each.

Layer l	Term t_i	$M(t_i, q)$	$\text{tf}_q(t_i, q)$	$\text{idf}(t_i, q)$	$w(l)$	q_i
TEXTUAL	astronom	m_1	1.0	2.018	0.5	1.009
TEXTUAL	influenc	m_2	1.0	3.404	0.5	1.702
TEXTUAL	gauss	m_3	1.0	1.568	0.5	0.784
URI	dbpedia:Carl_Friedrich_Gauss	m_3	1.0	3.404	0.125	0.426
TYPE	yago:GermanMathematicians	m_3	0.030	2.624	0.125	0.010
TYPE	yago:NumberTheorists	m_3	0.030	2.583	0.125	0.010
TYPE	yago:FellowsOfTheRoyalSociety	m_3	0.030	1.057	0.125	0.004
TYPE	...other 18 terms ...	m_3	0.030	...	0.125	...
TYPE	yago:Astronomer109818343	m_1, m_3	0.114	1.432	0.125	0.020
TYPE	yago:Physicist110428004	m_1, m_3	0.114	0.958	0.125	0.014
TYPE	yago:Person100007846	m_1, m_3	0.114	0.003	0.125	~0
TYPE	...other 9 terms ...	m_1, m_3	0.114	...	0.125	...
FRAME	⟨Subjective_influence-influence.v, dbpedia:Carl_Friedrich_Gauss⟩	m_2	0.333	5.802	0.125	0.242
FRAME	⟨Subjective_influence, dbpedia:Carl_Friedrich_Gauss⟩	m_2	0.333	5.802	0.125	0.242
FRAME	⟨Frame, Carl_Friedrich_Gauss⟩	m_2	0.333	3.499	0.125	0.146
TIME	day:1777-04-30	m_3	0.1	3.404	0.125	0.043
TIME	day:1855-02-23	m_3	0.1	3.404	0.125	0.043
TIME	century:1700	m_3	0.1	0.196	0.125	0.002
TIME	...other 7 terms	m_3	0.1	...	0.125	...

effect of making the similarity score obtained by matching m terms in a small document higher than the score obtained by matching the same m terms in a longer document. This normalization is known to be problematic in some document collections, is implemented differently and optionally disabled in real systems (e.g., Lucene and derivatives), and we deem it inappropriate in our scenario, where the document vector is expanded with a large amount of semantic terms whose number is generally not proportional to the document length.

We assign the weights of document and query vectors starting from the usual product of Term Frequency (tf) and Inverse Document Frequency (idf):

$$d_i = \text{tf}_d(t_i, d) \cdot \text{idf}(t_i, D) \tag{3}$$
$$q_i = \text{tf}_q(t_i, q) \cdot \text{idf}(t_i, D) \cdot w(l(t_i)) \tag{4}$$

The values of tf are computed differently for documents (tf_d) and queries (tf_q), while weights $w(l(t_i))$ quantify the importance of each layer. Given the form of Eq. 2, it suffices to apply $w(l(t_i))$ only to one of $\mathbf{d}$ and $\mathbf{q}$: we chose $\mathbf{q}$ so to allow selecting weights on a per-query basis. Table 1 reports the tf_q, idf, w, and q_i values for the terms of the example query "astronomers influenced by Gauss".

Several schemes for computing tf and idf have been proposed in the literature. Among them, we adopt the following scheme,[7] where $f(t, o)$ and $f'(t, o)$ are two measures of the frequency of a term t in a text (document or query) x:

$$\text{tf}_d(t, d) = 1 + \log(f(t, d)) \tag{5}$$

$$\text{tf}_q(t, q) = f'(t, q) \tag{6}$$

$$\text{idf}(t, D) = \log \frac{|D|}{|\{d \in D | f(t, d) > 0\}|} \tag{7}$$

The *raw frequency* $f(t, x)$ is defined as usual as the number of occurrences of term t in x. To account also for semantic terms, we denote with $M(t, x)$ the set of mentions in text x from where term t has been extracted, valid also for textual terms whose mentions are simply their occurrences in the text, and let $f(t, x) = |M(t, x)|$. The *normalized frequency* $f'(t, x)$ is newly introduced to account for the fact that multiple terms can be extracted from a single mention for the same semantic layer, differently from the textual case. It is defined as:

$$f'(t, x) = \sum_{m \in M(t, x)} \frac{1}{|T(m, l(t))|} \tag{8}$$

where $T(m, l)$ is the set of terms of layer l extracted from mention m. Since $|T(m, \text{TEXTUAL}))|$ is always 1, $f(t, x) = f'(t, x)$ for TEXTUAL terms. Note that Eq. 7 can indifferently use $f(t, x)$ or $f'(t, x)$.

The formulation of $f'(t, x)$ and its use in Eq. 6 aim at giving each mention the same importance when matching the query against the document collection. To explain, let's consider a query containing two mentions m_1 and m_2, with respectively n_1 and n_2 disjoint terms of a certain semantic layer (e.g., TYPE) extracted from each mention, $n_1 > n_2$; also assume that these terms have equal idf and tf_d values in the document collection. If we give these terms equal tf_q values, then a document matching the n_1 terms of m_1 (and nothing else) will be scored and ranked higher than a document matching the n_2 terms of m_2 (and nothing else). However, the fact that $n_1 > n_2$ does not reflect a preference of m_1 by the user; rather, it may merely reflect the fact that m_1 is described more richly than m_2 in the background knowledge. Our definition of normalized frequency corrects for this bias by assigning each mention a total weight of 1, which is equally distributed among the terms extracted from it for each semantic layer (e.g., weight $1/n_1$ assigned to terms of m_1, $1/n_2$ to terms of m_2).

For similar reasons, the use of $f'(t, x)$ in place of $f(t, x)$ in Eq. 5 would be inappropriate. Consider a query whose vector has a single TYPE term t (similar considerations apply to other semantic layers). Everything else being equal (e.g., idf values), two documents mentioning two entities of type t the same number

---

[7] Given the lack of normalization in $sim(d, q)$, our scheme can be roughly classified as ltn.ntn using the SMART notation; see http://bit.ly/weighting_schemes [22].

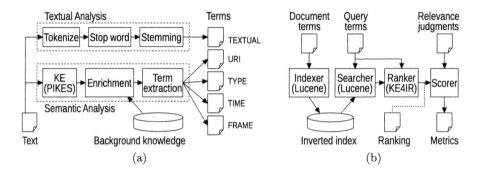

Fig. 1. Implementation: (a) term extraction; (b) indexing, searching and scoring.

of times should receive the same score. While this happen with $f(t, x)$, using $f'(t, x)$ the document mentioning the entity with fewest TYPE terms (beyond t) will be scored higher, although this clearly does not reflect a user preference.

4 Implementation

We built an evaluation infrastructure that implements the KE4IR approach presented in Sect. 3 and allows applying it on arbitrary documents and queries, measuring retrieval performances against gold relevance judgments. All the code is available for download on KE4IR website.[8]

Figure 1a shows the pipeline used to extract terms (with raw frequencies) from documents and queries, combining both textual and semantic analysis of input texts. Textual analysis generates the TEXTUAL layer employing standard components for text tokenization, stop word filtering, and stemming from Apache Lucene. Semantic analysis makes use of a KE tool for transforming the input text into an RDF knowledge graph where each instance is grounded to one or more mentions. This graph is then enriched with triples about selected URIs (DBpedia entities, YAGO types) retrieved from a persistent key-value store previously populated with the required LOD background knowledge;[9] the enrichment is done recursively and RDFS reasoning is done at the end to materialize inferences. The enriched graph is finally queried to extract semantic terms.

As KE extraction tool we use PIKES [3], a frame-based KE framework adopting a 2-phase approach. First—phase 1: *linguistic feature extraction*—an RDF graph of mentions is built by distilling the output of several state-of-the-art NLP tools, including Stanford CoreNLP[10] (tokenization, POS-tagging, lemmatization, NERC, TERN, parsing and coreference resolution), UKB[11] (WSD),

---

[8] http://pikes.fbk.eu/ke4ir.html.
[9] Subset of FrameBase ontology used in PIKES. Mapping-based properties with xsd:date, xsd:dateTime, xsd:gYear, and xsd:gYearMonth objects, YAGO types and type hierarchy from DBpedia 2015-04. All data available on KE4IR website.
[10] http://nlp.stanford.edu/software/corenlp.shtml.
[11] http://ixa2.si.ehu.es/ukb/.

DBpedia Spotlight[12] (EL), Mate-tools[13] and Semafor[14] (SRL). Then—phase 2: *knowledge distillation*—the mention graph is processed to distill the knowledge graph using SPARQL-like mapping rules, which are evaluated using RDFpro[15] [23], an RDF manipulation tool used also for RDFS reasoning. KE and the NLP tasks it relies on are computationally expensive. Using PIKES on a server with 24 cores (12 physical) and 192 GB RAM we obtained a throughput of ~700K tokens/h (~30K tokens/h core), corresponding to ~1200 documents/h for the document collection of Sect. 5 (570 tokens/document). While potentially inappropriate for a Web-scale deployment, this throughput is however adequate for small to medium-sized document collections (e.g., as encountered in corporate environments). Furthermore, larger collections can also be processed, with some loss in retrieval performances, by disabling the extraction of some of the layers.[16]

Figure 1b shows the workflow implemented for indexing extracted terms, executing queries and computing evaluation metrics. Document terms are directly indexed in a Lucene inverted index with their raw frequencies. At search time, query terms are OR-ed in a Lucene query that locates the documents containing at least one term (for which $sim(d, q) > 0$). Matched documents are scored and ranked externally to Lucene (for ease of testing) according to the KE4IR retrieval model of Sect. 3.2, starting from the term vectors of the query and the matched documents, and computing the necessary tf_d, tf_q, and idf values based on raw and normalized term frequencies and some statistics produced by Lucene (number of documents and document frequencies). The resulting ranking is compared with the gold relevance judgments to compute a comprehensive set of evaluation metrics, which are averaged along different queries.

5   Evaluation

In this section, we present an evaluation of KE4IR and discuss some insights emerged from it. All the evaluation materials are available on KE4IR website.

5.1   Evaluation Setup

KE4IR has been validated on the *ad-hoc* IR task, consisting in performing a set of queries over a document collection for which the list of relevance judgments is available. For the presented evaluation, we adopted the document collection created in [6], composed by a set of 331 documents and 35 queries. The relevance

---

[12] http://spotlight.dbpedia.org/.

[13] http://code.google.com/p/mate-tools/.

[14] http://www.cs.cmu.edu/~ark/SEMAFOR/.

[15] http://rdfpro.fbk.eu/.

[16] To give an idea, the impact of each semantic layer on the whole processing time for the document collection of Sect. 5 is: URI (3.5 %), TYPE (16.3 %), TIME (2.9 %), FRAME (77.3 %). Note also that substantial improvements of KE4IR indexing throughput can be achieved with further engineering and optimization, out-of-scope here.

of each document is expressed in a multi-value scale with scores going from 5 (the document contains exact information with respect to what the user is looking for) to 1 (the document is of no interest for the query). The peculiarity of this collection is the underlying semantic purpose with which it has been built. Indeed, the set of queries has been selected by varying from queries very close to keyword-based search (i.e., the query "Romanticism") to queries requiring semantic capabilities for retrieving relevant documents (i.e. "Aviation pioneers' publications"). In that work, the authors discuss some techniques exploiting manual annotations for semantic IR purposes. Unfortunately, their results and our results described next are not directly comparable, as the semantic techniques described in [6] are evaluated over annotations manually validated by experts, whereas we rely on totally automatic (and thus inevitably noisy) annotations.

Thus, we compared our approach against the two baselines introduced below:

- *Google baseline*: we exploited the Google custom search API for indexing pages containing our documents. The rationale behind this choice is to assess the performances of a commercial search engine, having as main challenge the "scalability" of indexing and retrieving documents, when a more custom document analysis is required. Google can be considered the same way as a black box, and we were not able to customize the way it analyzes text and computes document scores with respect to performed queries;
- *Textual baseline*: we indexed the raw text by adopting the standard Lucene library customized with the scoring formula described in Sect. 4. In our experiments, this customization provides the same (actually, slightly better) performances of a standard Lucene configuration,[17] and it also allows properly assessing the impact of semantic layers by excluding any interference related to slight differences in the definition of the scoring formula.

The protocol we used has been inspired by TREC [24]; however, due to the small size of the collection, we had to carry out some changes. Instead of drawing the precision/recall curve, we computed the precision values after the first (Prec@1), fifth (Prec@5), and tenth (Prec@10) document, respectively. The rationale behind this decision is the fact that the majority of search result click activity (89.8 %) happens on the first page of search results [16] corresponding to a set varying from 10 to 20 documents. Then, we provided two further metrics: (i) the Mean Average Precision (MAP) and (ii) the Normalized Discounted Cumulated Gain (NDCG) [25], computed both on the entire rank and after the first ten documents retrieved (resp., MAP@10 and NDCG@10). Validation on the NDCG metric is necessary in scenarios where multi-value relevance is used.

---

[17] For comparison, on **KE4IR** website we make available for download an instance of SOLR (a popular search engine based on Lucene) indexing the same document collection used in our evaluation, and we report on its performances on the test queries.

Table 2. Comparison of KE4IR against the two baselines.

Approach/System	Prec@1	Prec@5	Prec@10	NDCG	NDCG@10	MAP	MAP@10
Google	0.543	0.411	0.343	0.434	0.405	0.255	0.219
Textual	0.943	0.669	0.453	0.832	0.782	0.733	0.681
KE4IR	**0.971**	**0.680**	**0.474**	**0.854**	**0.806**	**0.758**	**0.713**
KE4IR vs. Textual	3.03 %	1.71 %	4.55 %	2.64 %	3.00 %	3.50 %	4.74 %
p-value (paired t-test)	0.324	0.160	0.070	**0.003**	**0.015**	**0.024**	**0.029**
p-value (approx. random.)	1.000	0.496	0.111	**0.003**	**0.020**	**0.020**	**0.030**

Table 3. KE4IR results using different layer combinations.

Layers	Prec@1	Prec@5	Prec@10	NDCG	NDCG@10	MAP	MAP@10
TEXTUAL,URI,TYPE, FRAME,TIME	**0.971**	**0.680**	**0.474**	**0.854**	**0.806**	**0.758**	**0.713**
TEXTUAL,URI,TYPE,FRAME	**0.971**	**0.680**	**0.474**	0.853	0.804	0.757	0.712
TEXTUAL,URI,TYPE,TIME	**0.971**	**0.680**	**0.474**	0.851	0.802	0.757	0.712
TEXTUAL,URI,TYPE	**0.971**	**0.680**	**0.474**	0.849	0.801	0.755	0.710
TEXTUAL,URI,FRAME,TIME	**0.971**	0.674	0.465	0.844	0.796	0.750	0.702
TEXTUAL,URI,FRAME	**0.971**	0.674	0.465	0.842	0.795	0.749	0.702
TEXTUAL,URI,TIME	**0.971**	0.674	0.465	0.840	0.791	0.747	0.700
TEXTUAL,URI	**0.971**	0.674	0.465	0.837	0.791	0.747	0.700
TEXTUAL,TYPE,FRAME,TIME	0.943	0.674	0.471	0.848	0.799	0.745	0.700
TEXTUAL,TYPE,TIME	0.943	0.674	0.471	0.843	0.794	0.743	0.697
TEXTUAL,TYPE,FRAME	0.943	0.674	0.468	0.847	0.797	0.743	0.695
TEXTUAL,FRAME,TIME	0.943	0.674	0.462	0.842	0.793	0.741	0.693
TEXTUAL,TYPE	0.943	0.674	0.468	0.842	0.792	0.740	0.693
TEXTUAL,TIME	0.943	0.669	0.462	0.836	0.786	0.737	0.689
TEXTUAL,FRAME	0.943	0.674	0.453	0.839	0.789	0.737	0.686

5.2 Overall Evaluation Results

We report here an overview of the results obtained, using equal weights for textual and semantic information in KE4IR, i.e., w(TEXTUAL) = w(SEMANTICS) = 0.5, with w(SEMANTICS) divided equally among semantic layers. We also provide a first analysis of KE4IR behavior using different layer combinations.

Comparison with the Baselines. Table 2 shows the comparison between the results achieved by KE4IR exploiting all the semantic layers, and the ones obtained by the proposed baselines. It is possible to see that KE4IR matches or outperforms the baselines for all the considered metrics. With respect to the textual baseline, the higher improvements are registered on the MAP, MAP@10, and Prec@10 values that quantify the capability of the proposed approach of producing an effective documents ranking. While the gains on the MAP and MAP@10 metrics assess only the retrieval of relevant documents without considering their relevance scores, the improvements obtained on the NDCG and NDCG@10 metrics highlight that produced rankings are effective also from a quality point of view. The improvements over the textual baseline are statistically significant for MAP, MAP@10, NDCG, and NDCG@10 (significance threshold 0.05), based on the p-values computed with the paired t-test (claimed as one of the best tests for IR in [26]) and the approximate randomization test [27]. With respect to the Google baseline, the marked difference of performances derives from Google returning far less results than KE4IR for the evaluation queries. Indeed, large-scale (web-scale) IR systems such as Google are heavily tuned for precision, as any query usually matches a large number of documents and the problem is to discard the irrelevant ones. In our context, small-scale IR, systems as our tool deal with fewer documents and hence they must be tuned also for recall.

Impact of Various Layer Combinations. A detailed analysis of the results obtained using different layer combinations in KE4IR is shown in Table 3. Combining all the semantic layers produces the best performances for all the considered metrics. In particular, the URI layer seems to be the most effective, as it is always included in the top settings for MAP. These results show that the integration of different semantic information leads to a general improvement of the effectiveness of the IR task, in line with the purpose of the proposed approach.

5.3   Query-by-Query Analysis

To complete the analysis of the overall results, we investigate the performances of the system query-by-query, discussing four representative queries more in-depth.

Impact of Each Single Layer. Table 4 shows, query by query, the impact of each semantic layer on system effectiveness. The first column contains the query identifier; from the second to the fifth columns, we show the comparison between the NDCG@10 computed using textual and single-layer semantic information, and the NDCG@10 computed by using only the textual information. From the sixth to the ninth columns, the same values are shown for the MAP metric. We selected only the MAP and the NDCG@10 metrics because those are the most indicative metrics for evaluating the performances of IR systems in general (MAP), and for deployment in a real-world environment (NDCG@10).

The TYPE layer affects the highest number of queries, but for some of them (e.g., "q28") its contribution is negative. This issue is likely a consequence of the large quantity of information inserted when the query is expanded with TYPE terms, especially the ones corresponding to super-types of entities and concepts mentioned in the query, which may lead documents scarcely related to the query to be matched by the system. Indeed, by injecting too much information in queries, it is possible to obtain a detrimental effect as shown in [28], where the authors discuss query expansion trade-offs and impact on IR effectiveness.

The URI layer also impacts on many queries with both positive and negative effects. Differences on NDCG@10 and MAP scores are larger than the ones resulting from the TYPE layer (see, e.g., queries "q16" and "q28"), reflecting the

Table 4. Differences on NDCG@10 and MAP obtained using TEXTUAL and a single semantic layer with respect to TEXTUAL only. Queries with no performance change are omitted ("q01", "q04", "q06", "q12", "q26", "q32", "q34" and "q46"). Note that semantic information may be available even if no difference is observed.

Query	Δ NDCG@10				Δ MAP			
	URI	TYPE	FRAME	TIME	URI	TYPE	FRAME	TIME
q02						0.001		
q03	0.002							
q07						0.005		
q08		0.049				0.012		
q09	0.029							
q10		0.093				0.061		
q13	0.066	0.066						
q14		0.015				0.005		
q16	0.018				0.283			
q17						0.026		0.015
q18		0.002				0.011		
q19						0.014		0.007
q22						0.021		−0.001
q23						0.012		0.003
q24					−0.012	−0.009		
q25	0.007	0.002			0.004	0.004		
q27	−0.007	0.011	0.218		0.011	0.020	0.135	
q28	−0.117	−0.016			−0.090	−0.042		
q29	0.002							
q36		−0.016						
q37						0.013		
q38		0.028				−0.003		
q40	0.054	0.007	0.017	0.005	0.030	0.021	0.021	0.021
q41	0.104	−0.004			0.140	0.010		0.032
q42		0.011				0.023		
q44	0.149	0.091		0.141	0.131	0.049		0.088
q45	−0.002	−0.002		−0.002		−0.007		−0.007

fact that URI terms impact more on document scores (with respect to the textual layer) since they are generally more selective (high idf) and often correspond to entities mentioned multiple times in a document (high tf).

The FRAME and TIME layers, where available, have almost always a positive impact on the performances (esp. for "q27" and "q44"). The FRAME layer affects the smallest number of queries. As described in Sect. 3, this layer describes relations between entities detected in the text, and thus requires to have a query structure that is more complex with respect to a simple keyword-based one.

Analysis of Selected Queries. While a comprehensive analysis of the performances of each query is not doable due to lack of space, we select four queries giving hints about pros and cons of using semantic information in IR. Table 5 shows these queries and their performances when using all the semantic layers.

Table 5. Results for selected queries using all the semantic layers.

Query	Query text	Δ NDCG@10	Δ MAP
q27	Nazis confiscate or destroy art and literature	0.154	0.099
q28	Modern age in English literature	−0.117	−0.095
q44	Napoleon's Russian campaign	0.151	0.147
q46	First woman who won a nobel prize	0	0

Query "q46" is an example where semantic information has no effects. This is because entities at different granularities are injected in the URI layers of query and documents. Specifically, the query is annotated with dbpedia:Nobel_Prize, while relevant documents have annotations like dbpedia:Nobel_Prize_in_X, where X is one of the disciplines for which Nobel Prizes are assigned. Unfortunately, these entities are not related in DBpedia (also in terms of types), thus it is not possible to expand the query in order to find matches with relevant documents.

Query "q28" is an example where worse performances are achieved by using semantic information, due to Entity Linking errors. From the query, two URI terms (and related TYPE terms) are correctly extracted: dbpedia:Modern_history, with no matches in the document collection, and dbpedia:English_literature, with 12 matches. Of these matches, 11 are incorrect and refer to irrelevant documents where dbpedia:English_literature is wrongly linked to mentions of other "English" things (e.g. "English scholar", "English society", "English medical herbs").

Queries "q27" and "q44" are examples where semantic information significantly boost performances. In "q44", the correct link to dbpedia:Napoleon and the type and time information associated to that entity in DBpedia allow extracting URI, TYPE and TIME terms that greatly help ranking relevant documents higher. In "q27", the major improvement derives from the extraction and matching of FRAME term ⟨framebase:frame-Destroying, dbpedia:Nazism⟩; while TIME information is also available (as dbpedia:Nazism is linked to category dbc:20th_century in DBpedia), our KE4IR implementation is not sophisticated enough to exploit it.

5.4   Balancing Semantic and Textual Content

In our work, we combine both textual and semantic content to improve the performances of IR. While in previous analyses we assigned equal weights to

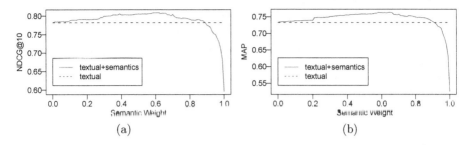

Fig. 2. Graphs showing the trend of (a) NDCG@10 and (b) MAP based on the quantity of semantic information considered with respect to the textual one.

semantic and textual information, here we experiment with different balances. Figure 2 shows how the NDCG@10 and MAP metrics change when the importance given to the semantic information changes as well. On the y-axis, we have the NDCG@10 (Fig. 2a) and MAP (Fig. 2b) values, while on the x-axis we have the weight w(SEMANTICS) assigned to all the semantic information and divided equally among semantic layers, with w(TEXTUAL) = 1 − w(SEMANTICS); a w(SEMANTICS) value of 0.0 means that only textual information is used (and no semantic content), while a value of 1.0 means that only semantic information is used (and no textual content). Results show that semantic information impacts positively on system performances up to w(SEMANTICS) ≤ 0.89 for NDGC@10 and w(SEMANTICS) ≤ 0.92 for MAP, reaching the highest scores around 0.61 and 0.65, respectively. Similar behaviors can be observed for NDCG and MAP@10. The highest scores obtained (NDCG@10 = 0.809, MAP = 0.763) are better than the scores reported where equal textual and semantic weights were intuitively used, suggesting the importance of a methodology for optimal weight tuning.

6   Concluding Remarks and Future Work

In this paper we investigated the benefits of using semantic content automatically extracted from text for Information Retrieval. Building on the Vector Space Model, we designed and implemented an approach, KE4IR, where both queries and documents are processed to extract semantic content such as entities, types, semantic frames, and temporal information. By evaluating our approach on a state-of-the-art document collection, we showed that complementing the textual information of queries and documents with the content resulting by processing them with typical knowledge extraction tools, enables to outperform document retrieval performances when only textual information is exploited.

Performance measured with different layer combinations shows that the aggregation of different semantic layers leads to effective rankings of relevant documents even in a multi-value relevance setting. The analysis of the NDCG and MAP values, representing the most meaningful metrics for evaluating an IR system, both in general and with respect to common user behaviors, validated the possibility of deploying KE4IR in a real-world environment.

Starting from the results obtained in this first experience, future work on the platform will touch different aspects. The evaluation here reported may be considered a first step for observing the behavior of the approach under different configurations and for enabling an analysis of the impact of each semantic layer. Extending the evaluation campaign to additional, larger document collections (e.g., TREC WT10g, ClueWeb) will be the next step for comparing the presented platform in different environments where further issues have also to be addressed as, for example, the scalability of the entire pipeline.

Results gave interesting insights about the components that should be improved for augmenting the effectiveness of the retrieval system. As shown in Table 5, concerning queries "q28" and "q46", single issues in the linking phase may lead to poor results. Thus, instead of trying to enrich as much text as possible with linked information coming from different knowledge bases, the use of approaches favoring precision of suggested links instead of recall may be the best strategy for obtaining a better average improvement of system effectiveness. Similar considerations can be done about all the other semantic layers.

Finally, in the presented version of KE4IR, we considered only general-purpose knowledge bases for enriching documents. However, the deployment in more domain-specific contexts would require the use of domain-specific resources able to provide more effective annotations. For instance, domain-specific KBs can be used with KB-agnostic entity linking tools to extract domain-specific URI and TYPE terms. For FRAME terms, domain-specific frames can be defined and annotated in a corpus to retrain the SRL tools used. We plan to validate these strategies to assess the usability of KE4IR with domain-specific document collection.

References

1. Gangemi, A., Draicchio, F., Presutti, V., Nuzzolese, A.G., Recupero, D.R.: A machine reader for the semantic web. In: Demos of ISWC, pp. 149–152 (2013)
2. Rospocher, M., van Erp, M., Vossen, P., Fokkens, A., Aldabe, I., Rigau, G., Soroa, A., Ploeger, T., Bogaard, T.: Building event-centric knowledge graphs from news. J. Web Semant. (to appear)
3. Corcoglioniti, F., Rospocher, M., Palmero Aprosio, A.: A 2-phase frame-based knowledge extraction framework. In: Proceedings of ACM Symposium on Applied Computing (SAC 2016) (2016, to appear)
4. Lehmann, J., Isele, R., Jakob, M., Jentzsch, A., Kontokostas, D., Mendes, P.N., Hellmann, S., Morsey, M., van Kleef, P., Auer, S., Bizer, C.: DBpedia - a large-scale, multilingual knowledge base extracted from Wikipedia. Semantic Web 6(2), 167–195 (2015)
5. Salton, G., Wong, A., Yang, C.S.: A vector space model for automatic indexing. Commun. ACM 18(11), 613–620 (1975)
6. Waitelonis, J., Exeler, C., Sack, H.: Linked data enabled generalized vector space model to improve document retrieval. In: Proceedings of NLP & DBpedia 2015 Workshop in Conjunction with 14th International Semantic Web Conference (ISWC 2015). CEUR Workshop Proceedings (2015)
7. Croft, W.B.: User-specified domain knowledge for document retrieval. In: Bernardi, L.R., Rabitti, F. (eds.) SIGIR, pp. 201–206. ACM (1986)

8. Gonzalo, J., Verdejo, F., Chugur, I., Cigarrán, J.: Indexing with WordNet synsets can improve text retrieval. CoRR (1998)
9. Fellbaum, C. (ed.): WordNet: An Electonic Lexical Database. MIT Press, Cambridge (1998)
10. Dridi, O.: Ontology-based information retrieval: overview and new proposition. In: RCIS, pp. 421–426 (2008)
11. Tomassen, S.L.: Research on ontology-driven information retrieval. In: Meersman, R., Tari, Z., Herrero, P. (eds.) OTM 2006 Workshops. LNCS, vol. 4278, pp. 1460–1468. Springer, Heidelberg (2006)
12. Castells, P., Fernández, M., Vallet, D.: An adaptation of the vector-space model for ontology-based information retrieval. IEEE Trans. Knowl. Data Eng. **19**(2), 261–272 (2007)
13. Vallet, D., Fernández, M., Castells, P.: An ontology-based information retrieval model. In: Gómez-Pérez, A., Euzenat, J. (eds.) ESWC 2005. LNCS, vol. 3532, pp. 455–470. Springer, Heidelberg (2005)
14. Jimeno-Yepes, A., Llavori, R.B., Rebholz-Schuhmann, D.: Ontology refinement for improved information retrieval. Inf. Process. Manage. **46**(4), 426–435 (2010)
15. Fernández, M., Cantador, I., Lopez, V., Vallet, D., Castells, P., Motta, E.: Semantically enhanced information retrieval: an ontology-based approach. J. Web Sem. **9**(4), 434–452 (2011)
16. Spink, A., Jansen, B., Blakely, C., Koshman, S.: A study of results overlap and uniqueness among major web search engines. Inf. Process. Manage. **42**(5), 1379–1391 (2006)
17. Stojanovic, N.: An approach for defining relevance in the ontology-based information retrieval. In: Web Intelligence, pp. 359–365 (2005)
18. Baziz, M., Boughanem, M., Pasi, G., Prade, H.: An information retrieval driven by ontology: from query to document expansion. In: RIAO (2007)
19. Rouces, J., de Melo, G., Hose, K.: FrameBase: representing n-ary relations using semantic frames. In: Gandon, F., Sabou, M., Sack, H., d'Amato, C., Cudré-Mauroux, P., Zimmermann, A. (eds.) ESWC 2015. LNCS, vol. 9088, pp. 505–521. Springer, Heidelberg (2015)
20. Hoffart, J., Suchanek, F.M., Berberich, K., Weikum, G.: YAGO2: a spatially and temporally enhanced knowledge base from Wikipedia. Artif. Intell. **194**, 28–61 (2013)
21. da Costa Pereira, C., Dragoni, M., Pasi, G.: Multidimensional relevance: prioritized aggregation in a personalized information retrieval setting. Inf. Process. Manage. **48**(2), 340–357 (2012)
22. Manning, C.D., Raghavan, P., Schütze, H., et al.: Introduction to Information Retrieval, vol. 1. Cambridge University Press, Cambridge (2008)
23. Corcoglioniti, F., Rospocher, M., Mostarda, M., Amadori, M.: Processing billions of RDF triples on a single machine using streaming and sorting. In: ACM SAC, pp. 368–375 (2015)
24. Voorhees, E., Harman, D.: Overview of the sixth text retrieval conference (trec-6). In: TREC, pp. 1–24 (1997)
25. Järvelin, K., Kekäläinen, J.: Cumulated gain-based evaluation of IR techniques. ACM Trans. Inf. Syst. **20**(4), 422–446 (2002)
26. Sanderson, M., Zobel, J.: Information retrieval system evaluation: effort, sensitivity, and reliability. In: SIGIR, pp. 162–169. ACM (2005)
27. Noreen, E.W.: Computer-Intensive Methods for Testing Hypotheses: An Introduction. Wiley, New York (1989)
28. Abdelali, A., Cowie, J., Soliman, H.: Improving query precision using semantic expansion. Inf. Process. Manage. **43**(3), 705–716 (2007)

Efficient Graph-Based Document Similarity

Christian Paul[1], Achim Rettinger[1(✉)], Aditya Mogadala[1(✉)],
Craig A. Knoblock[2(✉)], and Pedro Szekely[2(✉)]

[1] Institute of Applied Informatics and Formal Description Methods (AIFB),
Karlsruhe Institute for Technology, 76131 Karlsruhe, Germany
{rettinger,aditya.mogadala}@kit.edu
[2] Information Sciences Institute, University of Southern California,
Marina Del Rey, CA 90292, USA
{knoblock,pszekely}@isi.edu

Abstract. Assessing the relatedness of documents is at the core of many
applications such as document retrieval and recommendation. Most similarity approaches operate on word-distribution-based document representations - fast to compute, but problematic when documents differ in language, vocabulary or type, and neglecting the rich relational knowledge
available in Knowledge Graphs. In contrast, graph-based document models can leverage valuable knowledge about relations between entities - however, due to expensive graph operations, similarity assessments tend to
become infeasible in many applications. This paper presents an efficient
semantic similarity approach exploiting explicit hierarchical and transversal relations. We show in our experiments that (i) our similarity measure
provides a significantly higher correlation with human notions of document
similarity than comparable measures, (ii) this also holds for short documents with few annotations, (iii) document similarity can be calculated
efficiently compared to other graph-traversal based approaches.

Keywords: Semantic document similarity · Knowledge graph based
document models · Efficient similarity calculation

1 Introduction

Searching for related documents given a query document is a common task
for applications in many domains. For example, a news website might want to
recommend content with regards to the article a user is reading. Implementing
such functionality requires (i) an efficient method to locate relevant documents
out of a possibly large corpus and (ii) a notion of document similarity.

Established approaches measure text similarity statistically based on the
distributional hypothesis, which states that words occurring in the same context tend to be similar in meaning. By inferring semantics from text without
using explicit knowledge, word-level approaches become susceptible to problems caused by polysemy (ambiguous terms) and synonymy (words with similar meaning)[23]. Another problem arises when using distributional measures

© Springer International Publishing Switzerland 2016
H. Sack et al. (Eds.): ESWC 2016, LNCS 9678, pp. 334–349, 2016.
DOI: 10.1007/978-3-319-34129-3_21

across heterogeneous documents: due to different vocabularies and text length (e.g. news articles, Tweets) or languages, each type may underlie a different word distribution, making them hard to compare. Also, documents of different modalities (images, video, audio) may provide metadata, but no continuous text at all.

Semantic technologies help to address both these shortcomings. Knowledge bases like DBpedia [14] or Wikidata[1] unambiguously describe millions of entities and their relationship as a semantic graph. Using tools such as the cross-lingual text annotator xLisa [25], documents of different natures can be represented in the common format of knowledge graph entities. By using entities instead of text, heterogeneous content can be handled in an integrated manner and some disadvantages of statistical similarity approaches can be avoided.

In this paper, we present a scalable approach for related-document search using entity-based document similarity. In a pre-processing step called *Semantic Document Expansion*, we enrich annotated documents with hierarchical and transversal relational knowledge from a knowledge graph (Sect. 2). At search time, we retrieve a candidate set of semantically expanded documents using an inverted index (Sect. 3). Based on the added semantic knowledge, we find paths between annotations and compute path-based semantic similarity (Sects. 3.1, 3.2). By performing graph traversal steps only during pre-processing, we overcome previous graph-based approaches' performance limitations.

We evaluate the performance of our document similarity measure on two different types of data sets. First, we show on the standard benchmark for document-level semantic similarity that our knowledge-based similarity method significantly outperforms all related approaches (Sect. 5.2). Second, we demonstrate that we even achieve superior performance on sentence-level semantic similarity, as long as we find at least one entity to represent the sentence (Sect. 5.3). This suggest, that with growing knowledge graphs and improving entity linking tools, document models based on explicit semantics become competitive compared to the predominant vector-space document models based on implicit semantics.

2   Knowledge Graph Based Document Model

Given a document annotated with knowledge graph entities, Semantic Document Expansion enriches the annotations with relational knowledge. Following Damljanovic et al. [7], we distinguish between two types of exploited knowledge depending on the type of edge that is traversed to obtain it: an edge is classified as **hierarchical** if it represents a child-parent-relationship and denotes membership of an entity in a class or category. A **transversal** edge expresses a semantic, non-hierarchical predicate. Both groups of edge types have the potential to add value to our semantic measures: whereas connectivity via hierarchical edges indicates common characteristics on some categorical level, transversal paths express a relationship between entities independent of their intrinsic or type-based relatedness.

---

[1] https://www.wikidata.org.

2.1   Hierarchical Expansion

Hierarchically expanding an entity means enriching it with all information required for hierarchical similarity computation, so that it can be performed between any two expanded entities without accessing the knowledge graph. For each of a document's annotations, we locate its position within the hierarchical subgraph of our knowledge base and add all its *parent* and *ancestor* elements to it. Figure 1 shows a hierarchically expanded DBpedia entity using the Wikipedia Category System, with its parents and ancestors signaled by rectangular nodes.

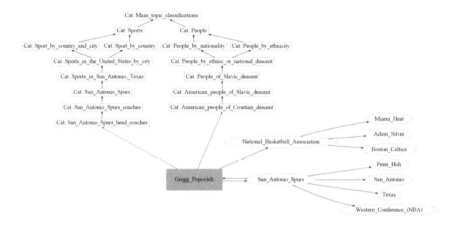

Fig. 1. Excerpt of hierarchically (top left) and transversally (bottom right, expansion radius = 2) expanded DBpedia entity `Gregg_Popovich`. (Cat.= Category)

2.2   Transversal Expansion

Transversal expansion resembles the Spreading Activation method: starting from a knowledge graph entity, it traverses semantic, non-hierarchical edges for a fixed number L of steps, while weighting and adding encountered entities to the document. We call L the entity's **expansion radius**. Formally, for each document annotation a, for each entity e encountered in the process, a weight is assigned according to the formula $w_a(e) = \sum_{l=1}^{L} \beta^l * |paths_{a,e}^{(l)}|$. Paths of length l are penalized by a factor of β^l, expressing the intuition that more distant entities are less relevant to annotation a than closer ones. Also, the more paths connect a with e, the higher a weight is assigned. We consider only outgoing edges - although this risks missing connections, we argue it also significantly reduces noise since the indegree for knowledge graph nodes can be very high (e.g. DBpedia entity `United_States`: $\approx$220 k). This weighting notion is closely related to the graph measure *Katz centrality* [11]. Nunes et al. used the same principle to determine connectivity between two nodes [16]. We set $\beta = 0.5$, as this value yielded the best results in our experiments.

3   Semantic Document Similarity

Our main contribution is an approach for efficient assessment of document relatedness based on entity-level semantic similarity. In order to provide a scalable solution, it is essential that at search time, only light-weight tasks are performed. Assuming we have a corpus of semantically annotated documents, our approach operates in four steps, where only steps 3 and 4 are performed online, during similarity calculation of the query document:

1. **Expand query document:** Enrich query document with hierarchically and transversally related entities from a knowledge graph.
2. **Store expanded document:** Add expanded query document to existing corpus of documents so that, in future search queries, it can also be found as a result.
3. **Pre-search:** Use an inverted index to locate and rank documents that appear most related to the query based on entity overlap. Return a candidate set consisting of the top n ranking entries.
4. **Full search:** Determine pairwise similarity between query document and candidate documents on the annotation level. Rank candidates accordingly and return top k.

In terms of similarity notions, the pre-search step performs a rough preselection from the full body of entities in a document, independent of the specifics of how they tie into the document.

Reducing the number of documents to be processed to a significantly smaller number n, allows the application of the more granular, yet more expensive *full search*. Pairwise similarity scores between the query document and each candidate document are computed on the entity level to better capture sub-document and entity-level affiliations than by considering a document as a whole.

3.1   Entity Similarity

In this section, we describe the entity similarity measures that underlie document similarity (Sect. 3.2). Using the enriched annotations in semantically expanded documents, we are able to compute entity similarity metrics efficiently and without performing graph traversal. Analogous to document expansion, similarity computation is divided into hierarchical and transversal parts. To benefit of both, we combine transversal and hierarchical scores into one entity similarity score, using normalized (by mean and variance) versions of *hierSim* and *transSim* :

$$sim_{ent}(e_1, e_2) = transSim^{norm}(e_1, e_2) + hierSim^{norm}(e_1, e_2) \qquad (1)$$

Hierarchical Entity Similarity: In hierarchical document expansion, each annotation gets enriched with the name and depth of its ancestor categories. We define **hierarchical entity similarity**:

$$hierSim(e_1, e_2) = 1 - d(e_1, e_2) \qquad (2)$$

For d, we use one of the two taxonomical distance measures $\mathbf{d_{ps}}$ [18] and $\mathbf{d_{tax}}$ [4], as inspired by Palma et al. [17]. Both d_{ps} and d_{tax} utilize the graph-theoretic concept of Lowest Common Ancestor (LCA). For nodes x and y, we define the LCA as a random representative from the subset of deepest nodes of x's and y's ancestors overlap.

Let $d(a, b) = |depth(a) - depth(b)|$, if a is an ancestor of b or vice versa, and $d(a, b) = 0$ otherwise. d_{tax}, as shown in Eq. 3 follows the notion that the closer two nodes are to their LCA compared to their overall depth, the higher the score. d_{ps} (Eq. 4) expresses distance in terms of the distance of two entities to their LCA compared to the LCA's depth:

$$d_{tax}(x, y) = \frac{d(lca(x, y), x) + d(lca(x, y), y)}{d(root, x) + d(root, y)} \tag{3}$$

$$d_{ps}(x, y) = 1 - \frac{d(root, lca(x, y))}{d(root, lca(x, y)) + d(lca(x, y), x) + d(lca(x, y), y)} \tag{4}$$

Transversal Entity Similarity: Given two annotations a_1, a_2 and expansion radius L, we find paths of length up to $2 * L$ that connect them, then compute a score depending on the length and number of those paths. We first compute $trans(a_1, a_2)$ in Eq. 5, with $paths^{(l)}_{(a_1,a_2)}$ the set of paths of length l based on outgoing edges from the annotations.

$$trans(a_1, a_2) = \sum_{l=0}^{L*2} \beta^l * |paths^{(l)}_{(a_1,a_2)}| \tag{5}$$

The formula is inspired by Nunes et al.'s *Semantic Connectivity Score* in [16]. However, instead of finding paths through graph traversal, we use the weights assigned to all entities in a_1's and a_2's respective L-step transversal neighborhood during document expansion. Let $paths_{(a_1,e,a_2)}$ denote the concatenation of $paths_{(a_1,e)}$ and $paths_{(a_2,e)}$, i.e. all paths from either annotation that connect it to e. With a_i's neighborhood $N(a_i)$, it is $\sum_{l=0}^{L*2} \beta^l * |paths^{(l)}_{(a_1,a_2)}|$

$$= \sum_{e \in N(a_1) \cap N(a_2)} \left(\sum_{l=0}^{L*2} \beta^l * |paths^{(l)}_{(a_1,e,a_2)}| \right) \tag{6}$$

$$= \sum_{e \in N(a_1) \cap N(a_2)} \left(\left(\sum_{i=0}^{L} \beta^i * |paths^{(i)}_{(a_1,e)}| \right) * \left(\sum_{j=0}^{L} \beta^j * |paths^{(j)}_{(a_1,e)}| \right) \right)$$

$$= \sum_{e \in N(a_1) \cap N(a_2)} w_{a_1}(e) * w_{a_2}(e)$$

This makes $trans_{a_1,a_2}$ easy to compute. Also, it is easily extendable to work with bidirectional expansion and paths: when considering edges of both directionalities, paths of length $> 2L$ will overlap on multiple nodes. This effect can be counteracted by penalizing path contribution depending on its length.

Finally, we receive our **transversal similarity** measure by normalizing the score:

$$transSim(a_1, a_2) = \frac{trans(a_1, a_2)}{trans(a_1, a_1)} \tag{7}$$

3.2 Document Similarity

Analyzing pairwise relationships between two documents' annotations makes it possible to explicitly assess how each single annotation corresponds to another document. We regard two documents as similar if many of a documents' annotations are related to at least one annotation in the respective other document. In other words, given two documents, we want to connect each entity of both annotation sets with its most related counterpart. Unlike some other approaches [5,17], we do not aim for a 1-1 matchings between annotations - we argue that equal or similar concepts in two documents can be represented by varying numbers of annotations, in particular when using automatic entity extraction. Our approach works in three steps:

1. **Full bipartite graph**: for each annotation pair (a_1, a_2) $(a_i$: annotation of document i), compute entity similarity score.
2. **Reduce graph**: start with empty *maxGraph*. For each annotation, add adjacent edge with maximum weight to the *maxGraph*.
3. **Compute document score**: with $matched(a_1), a_1 \in A_1$ denoting the annotation $a_2 \in A_2$ that a_1 has an edge to in *maxGraph*,

$$sim_{doc}(d_1, d_2) = \frac{\sum\limits_{a_{1i} \in A_1} (sim_{ent}(a_{1i}, matched(a_{1i})))}{|A_1| + |A_2|} \tag{8}$$

Fig. 2 illustrates an example of our approach. While edges are displayed as undirected, each edge $e = (v, w)$ carries $sim_{ent}(v, w)$ close to e's end towards v, and vice versa at its end towards w.

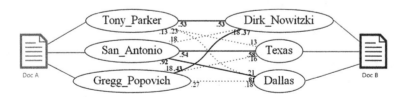

Fig. 2. Bipartite graph between sample documents 1 and 2. Bold lines constitute *maxGraph*

3.3   Computational Complexity

Our indexing of expanded entities and documents as well as the resulting possibility of a pre-search step does no less than *enable* large-scale search applications to use graph-based similarity. When faced with millions of documents, the number of computations between entities of a query and all documents would soon become overwhelming.

To compute pairwise entity similarity, any shortest-path-based algorithm ought to traverse all edges in the entities' neighborhoods in order to find connecting paths between them. For any subgraph G that is explored in the process, it holds that $|E| \leq \frac{|V|(|V|-1)}{2}$, i.e. that the maximum number of edges in E grows quadratically with the number of vertices in V. Another way of looking at this is that the number of edges that need to be traversed grows exponentially with the intended path length. In comparison, by traversing the graph and computing node scores at indexing time, we reduce this search-time complexity to be linear in $|V|$: the nodes of subgraph G can simply be retrieved and its node scores then be used in pairwise document similarity.

4   Related Work

We divide our related work into two categories based on the nature of their similarity measure.

4.1   Word-Distribution-based Document Similarity

Document search requires a method for efficient retrieval of relevant documents along with a scoring metric to rank candidates. Traditional text search approaches rely on the bag-of-words model and the distributional hypothesis. More sophisticated statistical approaches involve other sources of information in order to create more meaningful features in a document: Explicit Semantic Analysis(ESA) [8] represents text as a vector of relevant concepts. Each concept corresponds to a Wikipedia article mapped into a vector space using the TF-IDF measure on the article's text. Similarly, Salient Semantic Analysis (SSA) [9] use hyperlinks within Wikipedia articles to other articles as vector features, instead of using the full body of text.

While quick to compute, distributional metrics can perform poorly due to a lack of explicit information. Figure 3 demonstrates this for query "Gregg Popovich", coach of the San Antonio Spurs basketball team: while ESA ranks NBA players Bryant, Leonard, Nowitzki and Parker with no intuitive order, our knowledge-based method correctly recognizes a closer relationship between Gregg Popovich and his own (Spurs) players Kawhi Leonard and Tony Parker.

4.2   Graph-Based Document Similarity

Several approaches for text similarity were proposed based on the lexical knowledge graph WordNet[2]. These measures identify similarity on a lexicographic

---

[2] https://wordnet.princeton.edu/.

Rank	Entity	Score
1	San Antonio	0.026
2	Kobe Bryant	0.013
3	Kawhi Leonard	0.006
4	Dirk Nowitzki	0.006
5	Tony Parker	0.006
6	Phil Jackson	0.004

Rank	Entity	Score
1	Tony Parker	0.921
2	Kawhi Leonard	0.827
3	San Antonio	0.644
4	Kobe Bryant	0.604
5	Phil Jackson	0.533
6	Dirk Nowitzki	0.506

Fig. 3. Entity similarity scores using ESA (left) and our knowledge-based similarity using DBpedia (right) for "Gregg Popovich".

level, whereas we are interested in conceptual semantic knowledge, as can be found in DBpedia. Metrics such as PathSim [21] and HeteSim [20] assess the similarity of entities in heterogeneous graphs based on paths between them. Bhagwani et al. [5], Leal et al. [13] and Lam et al. [12] suggest methods for measuring relatedness of DBpedia entities. In order to accommodate DBpedia's heterogeneity, Leal et al.'s approach accepts a domain configuration to restrict DBpedia to a subgraph; Lam et al. apply a TF-IDF-inspired edge weighting scheme and Markov Centrality to rank entities by similarities with respect to a query entity.

Nunes et al. [16] present a DBpedia-based document similarity approach, in which they compute a document connectivity score based on document annotations. In a follow-up paper [15] they point out that for the "Semantic Search" use case, they use traditional TF-IDF because their pairwise document similarity measure is too complex.

Thiagarajan et al. [23] present a general framework how spreading activation can be used on semantic networks to determine similarity of groups of entities. They experiment with Wordnet and a Wikipedia Ontology as knowledge bases and determine similarity of generated user profiles based on a 1-1 annotation matching. In the Wikipedia Ontology, they restrict investigated concepts to parent categories. We do use spreading activation on transversal edges, but also apply specialized taxonomical measures for hierarchical similarity.

Palma et al. [17] describe an annotation similarity measure *AnnSim* with which they evaluate similarity of interventions/drugs through biomedical annotations using a 1-1 matching of annotations. We incorporated ideas from *AnnSim* into our hierarchical similarity measure.

Schuhmacher and Ponzetto's work [19] features entity and document similarity measures based on DBpedia entity linking and analysis of entity neighborhoods, making it particularly similar to our transversal similarity. However, they lack a notion for hierarchical similarity and their similarity metric differs in that it is based on *Graph Edit Distance*, and limits the maximum length of paths explored between entities to two, while we have successfully experimented with lengths of up to six (see $GBSS_3$ in Sect. 5.2).

A major difference to all graph-based approaches mentioned above relates to the computational complexity of graph traversal at "query time", as discussed in Sect. 3.3.

5   Evaluation

Our evaluation aims at showing (i) how different settings influences the performance of our approach, (ii) that it can be computed quickly, (iii) that our graph-based document similarity outperforms all related approaches for multiple-sentence documents (Sect. 5.2) and (iv) even single sentences as soon as they have at least one annotated entity (Sect. 5.3).

5.1   Experimental Setup

We implemented our related-document search using the following resources:

- **DBpedia:**[3] While the methods we present in this paper can be applied on any suitable semantic knowledge base, we choose DBpedia for our implementation because of its general-purpose, multilingual nature and comprehensiveness. Following Damljanovic et al. [7], we define the predicate types `skos:broader`, `rdf:type`, `rdfs:subclassOf` and `dcterms:subject` as hierarchical, while we consider semantic, non-hierarchical links from the *DBpedia Ontology* as transversal.
  Wikipedia Category Hierarchy: DBpedia contains multiple classification systems, namely YAGO, Wikipedia Categories and the hierarchical subgraph of the DBpedia Ontology. According to Lam et al. [12], the Wikipedia Category system has the highest coverage of entities among all three options. However, it does not have tree structure, plus various sources (e.g. [10,12]) confirm that it contains cycles. To overcome these issues, we use the *Wikipedia Category Hierarchy* by Kapanipathi et al. [10].
- **Lucene:** For the candidate document retrieval needed in the pre-search step, we leverage the built-in indexing capabilities of Lucene through *MoreLikeThis* queries. We store semantically expanded documents by adding two fields *transversal* and *hierarchical* to each document, for which we store term vectors: in each entry, the term represents the entity, while the term frequency captures the weight.
- **Jena:**[4] We use Jena TDB triplestores to operate DBpedia locally. We also store semantically expanded documents in a dedicated TDB store from where they can be efficiently retrieved at search time.
- **xLisa semantic annotator:** [25] We use xLisa to annotate text documents with DBpedia entities.

---

[3] http://wiki.dbpedia.org/data-set-2014.
[4] https://jena.apache.org/documentation/tdb.

5.2   Comparing Mulitple-Sentence Documents

We use the standard benchmark for multiple sentence document similarity to evaluate how well our metrics approximate the human notion of similarity. This corpus was compiled by Lee et al. of 50 short news articles of between 51 to 126 words each with pairwise ratings of document similarities.[5]

Semantic Similarity Evaluation: We assess Pearson and Spearman correlation plus their harmonic mean, as well as ranking quality using *Normalized Discounted Cumulative Gain (nDCG)*. The stated correlation metrics are also used in related work and thus allow us to compare our results to other approaches. With nDCG, we aim at measuring how well *relevant* documents are discovered, which is an important criterion for the related-document search use case. To capture relevant documents only, we confine the quality evaluation to the top $m(q)$ elements. For query document q, it is defined as twice the number of documents that humans scored greater than or equal to 3.0.

nDCG scores reported in this section represent the average nDCG score obtained by using each Lee50 document as a query document once. Figure 4 lists document similarity correlation and ranking quality based on the different measures we developed in our work:

– **Transversal Semantic Similarity** (TSS): Similarity score is solely based on transversal edges, as described in 3.1. The applied expansion radius is indicated in the subscript (e.g. $TSS_{r=2}$).
– **Hierarchical Semantic Similarity** (HSS): Similarity score is solely based on hierarchical edges, using one of the metrics d_{ps} or d_{tax}.
– **Graph-based Semantic Similarity** ($GBSS$): Combination of TSS and HSS, as described at the top of Sect. 3.1. The subscript denotes the expansion radius used in transversal similarity assessment.

Figure 4 shows that using d_{ps} in hierarchical similarity yields better results than d_{tax}. Transversal similarity achieves peak performance for expansion radius set to two - interestingly, it fares very well for ranking quality, while falling behind on correlation. Upon closer examination, many transversal document similarities turned out to be zero: while hierarchically, even strongly unrelated entities tend to share features on some abstract level and thus yield a score greater than zero, there is often no (short) transversal path between them. Moreover, results suggest that transversal paths longer than four or five (2∗ expansion radius) contain little value but add noise to the calculation.

By combining transversal and hierarchical (d_{ps}) scores for each entity in the $GBSS$ method, we achieved the best results across correlation and ranking quality. This demonstrates the different notions behind transversal and hierarchical similarity and that they can both add value to semantic measures.

---

[5] Lee50 dataset available at https://webfiles.uci.edu/mdlee/LeePincombeWelsh.zip.

	Correlation			Ranking
	r	ρ	μ	$nDCG$
$TSS_{r=0}$	0.59	0.46	0.517	0.811
$TSS_{r=1}$	0.641	0.424	0.510	0.846
$TSS_{r=2}$	0.663	0.437	0.527	0.851
$TSS_{r=3}$	0.62	0.442	0.516	0.802
HSS_{dtax}	0.652	0.51	0.572	0.827
HSS_{dps}	0.692	0.511	0.588	0.843
$GBSS_{r=1}$	0.7	0.507	0.588	0.863
$GBSS_{r=2}$	**0.714**	0.511	0.596	**0.870**
$GBSS_{r=3}$	0.704	**0.519**	**0.598**	0.863

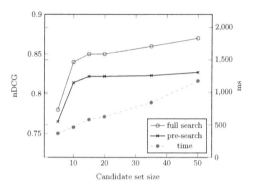

Fig. 4. Left: Correlation (Pearson (r), Spearman (ρ) and their harmonic mean μ) and $nDCG$ ranking quality for different measures in our approach. Right: $nDCG$ for pre-search, full search, and execution time.

Related-Document Search: The plot in Fig. 4 illustrates ranking quality after pre-search and full search as well as the average processing time per related-document search. We chose $GBSS_{r=2}$ as similarity metric in the full search step because it performed best on ranking quality. The plot shows that generally, the larger the candidate set, the better the quality. The fact that full search achieves higher nDCG scores than pre-search confirms the successful re-ordering that takes place in full search based on pairwise entity-based similarity computation.

Except for pre-search, which is performed offline, our approach's speed is independent of corpus size and only depends on candidate set size: the gray line shows that processing time grows linearly with candidate set size and confirms the efficiency of our search approach. Figure 5 breaks down the total execution time into its elements: given semantically expanded documents, the pairwise similarity computations in full search prove to be very fast. The bottleneck of our implementation turns out to be the retrieval of semantically expanded documents from a Jena TDB; using different means of storage, this should be easy to improve.

Comparison to Related Work: Figure 5 lists the performance for our two best-performing similarity measures $GBSS_{r=2}$ and $GBSS_{r=3}$, as well as for the following related approaches:

- **TF-IDF:** Distributional baseline algorithm.
- **AnnOv:** Similarity score based on annotation overlap. Corresponds to $TSS_{r=0}$ in Figure 4.
- **Explicit Semantic Analysis (ESA):** Via the public ESA REST endpoint,[6] we computed pairwise similarities for all document pairs.
- **Graph Edit Distance (GED):** correlation value for GED was taken from Schuhmacher [19]

---

[6] http://vmdeb20.deri.ie:8890/esaservice.

- **Salient Semantic Analysis (SSA), Latent Semantic Analysis (LSA):** correlation values for SSA and LSA were taken from Hassan and Mihalcea [9].

Figure 5 clearly shows that our approach significantly outperforms the to our knowledge most competitive related approaches, including Wikipedia-based SSA and ESA. While ESA achieves a rather low Pearson correlation and SSA comparably low Spearman correlation, our approach beats them in both categories. To compare ranking quality, we also computed nDCG for the best-scoring related approach ESA, where it reaches 0.845: as Figure 4 shows, our approach scores also beats that number significantly.

		Correlation		
		r	ρ	μ
Baseline	$TF-IDF$	0.398	0.224	0.286
	$AnnOv$	0.59	0.46	0.517
Related	LSA	0.696	0.463	0.556
	SSA	0.684	0.488	0.569
	GED	0.63	-	-
	ESA	0.656	0.510	0.574
Ours	$\mathbf{GBSS}_{r=2}$	**0.712**	0.513	0.596
	$\mathbf{GBSS}_{r=3}$	0.704	**0.519**	**0.598**

Operation	Time(ms)
1 document expansion	121.48
Generate candidate set (size 50)	51.0
Retrieve 50 expanded documents	794.42
50 sim_{doc} computations	209.18
Total time	1176.08

Fig. 5. Left: Correlation (Pearson (r), Spearman (ρ) and their harmonic mean μ) in comparison with related work. Right: Processing times for elements of our related-document search (candidate set size: 50)

5.3   Comparing Sentences

Since we base similarity on annotations, our approach requires documents for which high-quality annotations can be produced. Entity linking tools like xLisa [25] tend to produce better results for longer document (several sentences or more). To investigate if our approach is also competitive on very short documents, e.g. single sentences, we performed experiments on the SemEval task about Semantic Textual Similarity (henceforth STS)[7]. Each document only contains around 8–10 words and around 1–3 entities.

Datasets: We picked two datasets in the STS task from different years that have a good representation of entities and considered those sentences with at least one linkable entity mention.

- **2012-MSRvid-Test:** Sentence descriptions from the MSR Video Paraphrase Corpus[8], each summarizing the action of a video. For the SemEval-2012[9] task,

---

[7] http://ixa2.si.ehu.es/stswiki/index.php/Main_Page.

[8] http://research.microsoft.com/en-us/downloads/38cf15fd-b8df-477e-a4e4-a4680caa 75af/.

[9] https://www.cs.york.ac.uk/semeval-2012/task6/index.html.

750 pairs of sentences for training and 750 for testing were used. We used the dedicated testing part of the dataset on which baseline scores were reported.

- **2015-Images:** A subset of the larger Flickr dataset containing image descriptions. The dataset consists of around 750 sentence pairs.

Competing Approaches: To have a fair comparison, we only report those related approaches that perform unsupervised semantic similarity assessment, meaning they do not exploit the given manual similarity assessment to train or optimize the model, but only use these target values for evaluation of the approach.

- **STS-12:** Baseline approach reported in SemEval-2012 [2] for MSRvid-Test dataset.
- **STS-15:** Baseline approach reported in SemEval-2015 [1] for Images dataset.
- **Polyglot:** Word embeddings obtained from Al-Rfou et al. [3] is used to calculate the sentence embeddings by averaging over word embeddings. Sentence words that are not observed in the word embedding database are ignored.
- **Tiantianzhu7:** Uses a graph-based word similarity based on word-sense disambiguation.
- **IRIT:** Uses a n-gram comparison method combined with WordNet to calculate the semantic similarity between a pair of concepts.
- **WSL:** Uses an edit distance to include word order by considering word context.

Results: Table 1 shows Pearson correlation (r) scores for both datasets. Again, we clearly outperform related approaches with both graph-based and word-distribution-based document representations. This indicates that as long as we obtain at least one correct entity to represent a document, our sophisticated hierarchical and transversal semantic similarity measure can compete with the state-of-the-art even for very short text.

Table 1. Correlation (Pearson (r) in comparison with related work.

		Sentence Semantic Similarity	
		2012-MSRvid-Test	2015-Images
Baseline	STS-12	0.299	-
	STS-15	-	0.603
Related	Polyglot [3]	0.052	0.194
	Tiantianzhu7 [24]	0.594	-
	IRIT [6]	0.672	-
	WSL [22]	-	0.640
Ours	**GBSS**$_{r=2}$	0.666	**0.707**
	GBSS$_{r=3}$	**0.673**	0.665

6   Conclusion

In this paper, we have presented a new approach for efficient knowledge-graph-based semantic similarity. Our experiments on the well-established Lee50 document corpus demonstrate that our approach outperforms competing approaches in terms of ranking quality and correlation measures. We even achieve superior performance for very short documents (6–8 words in the SemEval task) as long as we can link to at least one entity. By performing all knowledge graph-related work in the *Semantic Document Expansion* preprocessing step, we also achieve a highly scalable solution. The strong performance of our similarity measure demonstrates that semantic graphs, including automatically generated ones like DBpedia contain valuable information about the relationship of entities. Moreover, similarity measures can be developed that compete with traditional word-distribution based approaches in every aspect. For future work, testing on diverse corpora with documents differing in language, vocabulary and modality seems promising.

Acknowledgments. This material is based on research supported by the European Union Seventh Framework Programme (FP7/2007-2013) under grant agreement no. 611346, and in part by the National Science Foundation under Grant No. 1117913.

References

1. Agirre, E., Carmen, B.: Semeval-2015 task 2: semantic textual similarity, English, Spanish and pilot on interpretability. In: Proceedings of the 9th International Workshop on Semantic Evaluation. Association for Computational Linguistics (2015)
2. Agirre, E., Mona, D., Daniel, C., Gonzalez-Agirre., A.: Semeval-2012 task 6: a pilot on semantic textual similarity. In: Proceedings of the Sixth International Workshop on Semantic Evaluation, Sofia, pp. 385–393. Association for Computational Linguistics (2012)
3. Al-Rfou, R., Perozzi, B., Skiena, S.: Polyglot: distributed word representations for multilingual NLP. In: Proceedings of the Seventeenth Conference on Computational Natural Language Learning, Sofia, pp. 183–192. Association for Computational Linguistics, August 2013
4. Benik, J., Chang, C., Raschid, L., Vidal, M.-E., Palma, G., Thor, A.: Finding cross genome patterns in annotation graphs. In: Bodenreider, O., Rance, B. (eds.) DILS 2012. LNCS, vol. 7348, pp. 21–36. Springer, Heidelberg (2012)
5. Bhagwani, S., Satapathy, S., Karnick, H.: Semantic textual similarity using maximal weighted bipartite graph matching. In: Proceedings of the First Joint Conference on Lexical and Computational Semantics, vol. 1: Proceedings of the Main Conference and the Shared Task, vol. 2: Proceedings of the Sixth International Workshop on Semantic Evaluation, SemEval 2012, pp. 579–585. Association for Computational Linguistics, Stroudsburg (2012)
6. Buscaldi, D., Tournier, R., Aussenac-Gilles, N., Mothe, J.: Irit: textual similarity combining conceptual similarity with an n-gram comparison method. In: Proceedings of the Sixth International Workshop on Semantic Evaluation, pp. 552–556. Association for Computational Linguistics (2012)

7. Damljanovic, D., Stankovic, M., Laublet, P.: Linked data-based concept recommen- dation: comparison of different methods in open innovation scenario. In: Simperl, E., Cimiano, P., Polleres, A., Corcho, O., Presutti, V. (eds.) ESWC 2012. LNCS, vol. 7295, pp. 24–38. Springer, Heidelberg (2012)
8. Gabrilovich, E., Markovitch, S.: Computing semantic relatedness using wikipedia- based explicit semantic analysis. In: IJCAI. vol. 7, pp. 1606–1611 (2007)
9. Hassan, S., Mihalcea, R.: Semantic relatedness using salient semantic analysis. In: AAAI (2011)
10. Kapanipathi, P., Jain, P., Venkataramani, C., Sheth, A.: Hierarchical interest graph, 21 January 2015. http://wiki.knoesis.org/index.php/Hierarchical_Interest_ Graph
11. Katz, L.: A new status index derived from sociometric analysis. Psychometrika 18(1), 39–43 (1953)
12. Lam, S., Hayes, C., Deri, N.U., Park, I.B.: Using the structure of dbpedia for exploratory search. In: Proceedings of the 19th ACM SIGKDD International Con- ference on Knowledge Discovery and Data Mining, KDD 2013. ACM, New York (2013)
13. Leal, J.P., Rodrigues, V., Queirós, R.: Computing semantic relatedness using dbpedia. In: OASIcs-OpenAccess Series in Informatics. vol. 21. Schloss Dagstuhl- Leibniz-Zentrum fuer Informatik (2012)
14. Lehmann, J., Isele, R., Jakob, M., Jentzsch, A., Kontokostas, D., Mendes, P.N., Hellmann, S., Morsey, M., van Kleef, P., Auer, S., et al.: Dbpedia-a large-scale, multilingual knowledge base extracted from wikipedia. Semant. Web 6(2), 167– 195 (2015)
15. Nunes, B.P., Fetahu, B., Dietze, S., Casanova, M.A.: Cite4me: a semantic search and retrieval web application for scientific publications. In: Proceedings of the 2013th International Conference on Posters & Demonstrations Track, vol. 1035, pp. 25–28. CEUR-WS.org (2013)
16. Nunes, B.P., Kawase, R., Fetahu, B., Dietze, S., Casanova, M.A., Maynard, D.: Interlinking documents based on semantic graphs. Procedia Comput. Sci. 22, 231– 240 (2013)
17. Palma, G., Vidal, M.E., Haag, E., Raschid, L., Thor, A.: Measuring relatedness between scientific entities in annotation datasets. In: Proceedings of the Inter- national Conference on Bioinformatics, Computational Biology and Biomedical Informatics, BCB 2013, pp. 367–376. ACM, New York (2013)
18. Pekar, V., Staab, S.: Taxonomy learning: factoring the structure of a taxonomy into a semantic classification decision. In: Proceedings of the 19th International Conference on Computational Linguistics, vol. 1, pp. 1–7 (2002)
19. Schuhmacher, M., Ponzetto, S.P.: Knowledge-based graph document modeling. In: WSDM, pp. 543–552. ACM (2014)
20. Shi, C., Kong, X., Huang, Y., Philip, S.Y., Wu, B.: Hetesim: a general framework for relevance measure in heterogeneous networks. IEEE Trans. Knowl. Data Eng. 10, 2479–2492 (2014)
21. Sun, Y., Han, J., Yan, X., Yu, P.S., Wu, T.: Pathsim: meta path-based top-k similarity search in heterogeneous information networks. In: VLDB (2011)
22. Takagi, N., Tomohiro., M.: Wsl: Sentence similarity using semantic distance between words. In: SemEval. Association for Computational Linguistics (2015)
23. Thiagarajan, R., Manjunath, G., Stumptner, M.: Computing semantic similarity using ontologies. In: The International Semantic Web Conference (ISWC 2008) (2008)

24. Tiantian, Z., Man, L.: System description of semantic textual similarity (STS) in the semeval-2012 (task 6). In: Proceedings of the Sixth International Workshop on Semantic Evaluation. Association for Computational Linguistics (2012)
25. Zhang, L., Rettinger, A.: X-LiSA: Cross-lingual Semantic Annotation. Proc. VLDB Endowment (PVLDB) **7**(13), 1693–1696 (2014). The 40th International Conference on Very Large Data Bases (VLDB)

Semantic Topic Compass – Classification Based on Unsupervised Feature Ambiguity Gradation

Amparo Elizabeth Cano$^{(\boxtimes)}$, Hassan Saif, Harith Alani, and Enrico Motta

Knowledge Media Institute, The Open University, Milton Keynes, UK
{amparo.cano,h.saif,h.alani,e.motta}@open.ac.uk

Abstract. Characterising social media topics often requires new features to be continuously taken into account, and thus increasing the need for classifier retraining. One challenging aspect is the emergence of ambiguous features, which can affect classification performance. In this paper we investigate the impact of the use of ambiguous features in a topic classification task, and introduce the Semantic Topic Compass (STC) framework, which characterises ambiguity in a topics feature space. STC makes use of topic priors derived from structured knowledge sources to facilitate the semantic feature grading of a topic. Our findings demonstrate the proposed framework offers competitive boosts in performance across all datasets.

Keywords: Topic classification · Feature engineering · Semantics

1 Introduction

Much research focused on understanding what is being discussed on Social Media. From opinion and sentiment mining [16] to event detection [11], one persistent challenge in making sense of this data is the task of assigning topic labels to microposts, which is a necessary step in supervised classification tasks.

Topic characterisation in Social Media poses various challenges due to the event-dependent nature of topics discussed on this outlet. Changes on a topic's representation involve the introduction of event-dependent features, which bring along ambiguous semantic relevance to the topic. For example the word Bataclan, referring to the Bataclan Theatre in Paris is commonly related to Entertainment, however during the November 2015 terrorist attacks in France it became relevant to the Topic Violence. The constant change of a topic's feature space makes apparent the need to be able to characterise the most discriminative features, while identifying ambiguous ones.

Existing feature selection methods such as Information Gain [3] and Odds Ratio [13], assess the problem of feature relevance but perform poorly when a dataset present ambiguous features. More recently, the problem of characterising ambiguous features has been approached using the Ambiguity Measure [12], which enables the selection of the most unambiguous features from a feature set. However such an approach relies on labelled data, and thus renders it less

© Springer International Publishing Switzerland 2016
H. Sack et al. (Eds.): ESWC 2016, LNCS 9678, pp. 350–367, 2016.
DOI: 10.1007/978-3-319-34129-3_22

adequate when modelling topics for social media, where labelling data is costly and becomes rapidly outdated.

In this paper we introduce the Semantic Topic Compass (STC) Framework, which is an unsupervised method that facilitates the semantic feature grading of a topic. This approach relies on the incorporation of feature priors derived from an external corpus to reweigh a Twitter corpus features in an unsupervised manner. Such feature representation partitions a Topic's feature space into four quadrants each representing the level of relevance and ambiguity of a feature to the Topic. To the best of our knowledge none of the existing approaches characterise ambiguity of a topic feature space on unlabelled corpora. The main contributions of this paper can be summarised as follows:

(1) We propose a novel unsupervised approach for topic feature representation based on polar coordinates;
(2) such representation enhances existing ones in characterising features based on both topic relevance and ambiguity;
(3) We propose a weighting strategy that proxies penalties to four feature types characterised by our framework: strongly related, weakly-related, weakly-unrelated and strongly unrelated features.
(4) We evaluate the effectiveness of the proposed framework on a classification task applied over three datasets using both lexical and semantic features.
(5) Our findings demonstrate that the proposed framework offers competitive boosts in performance across all datasets.

2   Related Work

Topic classification on Twitter consist of labelling tweets messages as being either topic-related or topic-unrelated [2]. Most existing works approach this task by training binary machine learning classifiers (e.g., Naive Bayes, SVM) on lexical features extracted either from tweets (i.e., Lexical Features) [10,15,20] and/or from external knowledge sources (i.e., semantic features) [2,6,19]. As such, these works can be divided as lexical approaches and semantic approaches. As for lexical features, Genc et al. [6], proposed the use of unigrams features to map a tweet to the most similar Wikipedia[1] articles which denote the tweets' topic. Sriram et al. [20] classified tweets to a predefined set of topics based on Twitter-specific features such as abbreviations, slangs, user mentions (i.e., @username) and opinionated words.

Rather than relying on lexical features in tweets for topic classification, other approaches proposed enriching the tweets' content with features extracted from external knowledge sources (KS) [2,6,14,19]. For example, [19] mapped a tweet's terms to the most likely resources in the Probbase KS. These resources were used as additional features in a clustering algorithm which outperformed the simple bag of words approach. Muñoz-García et al. [14] proposed an unsupervised vector space model for assigning DBpedia URIs to tweets in Spanish.

---

[1] http://wikipedia.org.

Cano et al. [2] performed cross-epoch topic classification based on four types of semantic features extracted from DBpedia knowledge graph, including the DBpedia resources, class types, categories and properties of named entities extracted from the tweets.

A persistent issue of both semantic and lexical approaches is the high dimensionality of the feature spaced used for training classifiers, which can reach the order of millions on large Twitter corpora. A large feature space usually affects both, the runtime complexity and the performance of classifiers [9]. To reduce the dimensionality of a feature space, feature selection techniques for topic classification are often used. Feature selection concerns about finding the most discriminative features in a given feature set, aiming at reducing the dimensionality of the classifier's feature space by excluding features of low discrimination power and maintaining high classification performance [5]. Wide range methods have been proposed for automatic selection of features, such as, Information Gain [3], Chi-Squared [5], term frequency and inverse term frequency (TF-IDF) [8], and Odds Ratio [13], etc. Most of these methods function by estimating the probability that a feature belongs to a specific class (topic) and the probability that the feature does not belong to that class. A Common limitation of these methods is that they are not tolerant to imbalanced class distributions in datasets. In other words, they tend to assign high discrimination scores to features that belong to the dominant class in the data (i.e., the class with the highest number of training samples). Also, these methods often perform poorly in the case of ambiguous features, where the presence and absence of a feature with a given class is almost identical. Instead of identifying and filtering out these type of features, they are still assigned a high discrimination score by methods like Odds Ratio, TF-IDF and Chi-Squared.

To address the above limitations Mengle and Goharian [12] proposed the Ambiguity Measure (AM) feature selection method. AM identifies ambiguous features by assigning a higher discrimination score to features pointing to one class than those pointing to more than one class. Their results show that feature selection based on AM outperforms Odds Ratio, Information Gain and Chi-Squared methods. However, similar to these methods AM functions in a supervised fashion, i.e., it requires tweets labelled with their topical orientation. In contrast in this paper we introduce the semantic topic compass framework, which is an unsupervised approach that enables the partition of a topic's feature space characterising ambiguous features. As opposed to previous work, which rely on labelled data for disambiguating features, our proposed approach only relies on topic feature priors extracted from knowledge sources.

3   Ambiguity in Topic Representation

In topic classification, the most discriminative features of a topic are generally those that are semantically-related and semantically-unrelated to the topic. On the other hand, the least discriminative features are those that are weakly-related or weakly-unrelated to the topic, and thus considered ambiguous due to their low discriminative power.

Spectrum of the semantic level of relatedness/ambiguity of words for the topic War_Conflict

For example, for the topic "War_Conflict", depicted in Fig. 1, words such as "ISIS"and "Assad" are semantically-related to the topic since their under-lying semantics (i.e., "Jihadist_Group", "Syria_President") denote a higher association with the topic "War_Conflict". In contrast, the words "Jay-Z" and "Mashable" are semantically-unrelated to "War_Conflict" as their seman-tics (i.e., "American_Rapper" and "Digital_Media_Website") are irrelevant to the topic. In between this spectrum lay ambiguous terms, which are not com-pletely relevant nor irrelevant to the topic. For example, the words "Greece" and "Al-Jazeera" are considered ambiguous as their underlying semantics (i.e., "Country", "News_Agency") are considered to be weakly associated with the topic "War_Conflict" in this example.

Identifying the level of ambiguity of a feature can aid in providing a better representation of a topic. It can also aid in filtering out features keeping only the most discriminatory ones, reducing in this way the dimensionality of the feature space. However to the best of our knowledge there are only few approaches to address the identification of ambiguous features. Moreover existing approaches rely on labelled data (i.e., are supervised) and are only able to discriminate features as being ambiguous or non-ambiguous but they do not differentiate the tendency of the ambiguous word towards the topic (weakly-related or weakly-unrelated).

This paper proposes a novel unsupervised approach to topic feature represen-tation which enables both the relevance/irrelevance feature weighting while giv-ing an ambiguity orientation to a feature. In this paper we propose the use of such topic feature representation to characterise a topic on a topic classification task.

4   Semantic Topic Compass Framework

Since the discriminative power of features used for topic classification relies on the relative use of those features within the topic, we propose to make use of distributional and conceptual semantics to characterise ambiguity in a topic's feature representation. We aim at using such topic characterisation to learn a representation of the topic for classification purposes.

The proposed Semantic Topic Compass (STC) Framework breaks down into three main phases: (1) Semantic Topic Representation: Given a collection of

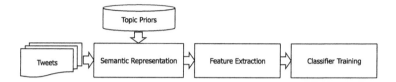

Fig. 2. Pipeline of the proposed topic representation and feature extraction approach.

tweets a semantic representation of a topic is constructed based on the features' semantic relatedness to the topic; (2) Feature Weighting: the semantic representation of the topic is then used to grade and extract features for topic classification.; and (3) Training of a topic classifier: Lastly, the extracted features are used to train the topic classifier.

In the following subsections we describe each of the three steps in the pipeline of our framework in more detail.

4.1 Semantic Representation of Topics' Feature Space

As mentioned before, our semantic topic compass framework relies on incorporating the semantics of words into the feature space of the studied topic, aiming at characterising the relevance and ambiguity of the these features. Hence, this step extracts first the latent semantics of words under a topic, and then incorporates these semantics into the topic's feature space.

Two main approaches have been extensively used in the literature for extracting the semantics of words, namely: the *Distributional Semantic Approach* [4,7] and the *Conceptual Semantic Approach* [18]. The distributional semantic approach (a.k.a statistical semantic approach) relies on the co-occurrence patterns of words in the text for words' semantic extraction, while the conceptual semantic approach makes use of external knowledge sources (e.g., DBpedia) for mapping words with their explicit semantic concepts.

In this paper we investigate the use of both semantic extraction approaches in our topic compass framework. First, describe in this section how to extract and use the distributional semantics of words for space representation and feature grading of a topic. After that, we explain in Sect. 5 how to enrich the space representation of the topic with the conceptual semantics of words.

1. Extracting Words' Distributional Semantics: To extract the distributional semantics of a word, we follow the distributional semantic hypothesis that *words that are used and occur in the same contexts tend to purport similar meanings.*[2] For example, the semantics of the word "ISIS" when it occurs with words like "Kill" and "Behead" denotes that "ISIS" refers to the terrorist militia organisation in the Middle East.

Given a tweet collection T of a topic P (e.g., "War_Conflict"), let's represent each term m in T (e.g., "ISIS") as a vector $c = (c_1, c_2, \ldots, c_n)$ of terms co-occurring with term m in any tweet in T (e.g., "Kill" , "Behead" , "Blood").

---

[2] Also known as Statistical Semantics.

We define the degree of correlation between each context word $c_i \in \boldsymbol{c}$ and m based on the *TF-IDF* weighting scheme as follows:

$$corr(m, c_i) = f(c_i, m) \times \log N/N_{c_i} \qquad (1)$$

where $f(c_i, m)$ is the number of times c_i occurs with m in tweets, N is the total number of terms, and N_{c_i} is the total number of terms that occur with c_i. Since our main task here is to measure the level of relatedness and ambiguity of a word to the studied topic, we also assign to each context term c_i a topic prior $p(c_i) \in [-1, +1]$, a numerical value representing the initial degree of relatedness of the context term to the topic. Section 4.3 describes how the topic priors of words in the Tweet collection are extracted.

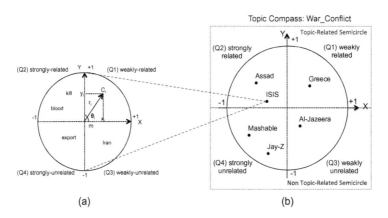

Fig. 3. Semantic representation of: (a) of a term m, and (b) the feature space of the term's related topic $\mathcal{P}$

To extract the collective semantics of the term m we resort to representing the vector $\boldsymbol{c}$ using the polar coordinate system inspired by [16]. In particular, the context vector $\boldsymbol{c}$ is transformed into a 2d circle representation as depicted in Fig. 3, which we will call S_m. The center of this circle represents the target term m and points within the circle denote the context terms of m. The position of c_i is defined jointly by a radius $r_i = corr(m, c_i)$ and an angle measured based on its topic prior $\theta_i = p(c_i)$ as:

$$x_i = r_i \cos \theta_i \qquad\qquad y_i = r_i \sin \theta_i \qquad (2)$$

The above representation partitions the context terms of the target term m (e.g., "ISIS") into four independent quadrants (Q1, Q2, Q3, Q4) as shown in Fig. 3a. Terms lying on the upper left quadrant (Q2) (e.g., "kill", "blood") are strongly related to the topic "War_Conflict", while terms lying on the upper right quadrant (Q1) (e.g., "Oil") are weakly-related to the topic. Also, terms residing in the lower left quadrant (Q4) (e.g., "export") are strongly-unrelated to the topic while terms residing on the lower right quadrant (Q3) (e.g., "Iran") are weakly-unrelated to the topic.

2. Constructing the Topic's Feature Space: Now we have the semantics of each term m in the tweet collection is represented by a circle S_m. The next step is to derive a global feature space representation for the topic $\mathcal{P}$. To this end, we also use the polar coordinate system, where we represent the topic's feature space as circle S_P, centred at the origin as depicted in Fig. 3b. Each point in the topic's circle S_P denote a term's circle S_m, and is positioned based on the geometric median point of S_m which can be calculated as:

$$g_m = \arg \min_{g_m \in \mathbb{R}^2} \sum_{i=1}^{n} ||c_i - g_m||_2 \qquad (3)$$

where the geometric median is a point $g_m = (x_k, y_k)$ in which its Euclidean distances to all the points c_i (context terms) in S_m the is minimum. We can notice that the feature space of the topic (i.e., the topic's Circle S_P) can be also partitioned, in similar way to the circle representation of the terms (i.e., term's circle S_m), into four different quadrants denoting the level of relatedness and ambiguity of the terms under the topic[3].

In the following subsection we show how to use the circle representation of the topic's feature space to weight a tweet for topic classification.

4.2   Feature Weighting for Classifier Training

Representing the feature space of a topic with the proposed framework in the polar coordinate system enhances the standard Euclidean vector space representation in two main aspects: (1) by providing a strength of the relative semantic relevance of a feature to a topic; (2) by augmenting the possible orientations of such relevance to the topic. In this section we propose a method to make use of this information by encoding it into a feature weighting strategy that can be used to weight features in a tweet collection to address a topic classification task.

Let $\mathcal{T}$ be a corpus of tweets denoted as $\mathcal{T} = \{t_1, t_2, .., t_T\}$; where each tweet consists of a sequence of N_t terms denoted by $t = (m_1, m_2, .., m_{Nt})$. Algorithm 1 presents the proposed steps for weighting a tweet t's features based on the topic representation S_P.

The proposed weighting strategy generates a metric that assigns weights to features based on their relevance to the topic[4]. Such relevance is considered based on the position of a feature within the two semicircles (i.e., topic related and non-topic related) described in Fig. 3b. Steps described in Algorithm 1 can be outlined as follows: (i) Given the term frequency vector of a document, iterate over these feature; (ii) For each feature obtain its coordinates in the topic circle

---

[3] We provide samples of the generated topic circles for the three topics at the following link http://tweenator.com/stc.php.

[4] Features appearing within an axis are considered ambiguous and are smoothed down to a low weight.

Algorithm 1. Feature Weighting based on a Topic's Circle Representation

Input: Term frequency features of t, topic circle S_P, penalties for quadrants $(pQ1,pQ2,pQ3,pQ4)$
Output: Weighted features for tweet t
1: **for** each term $m_i \in t$ **do**
2:    Extract m_i representation on S_P.
3:    Compute angle of m_i as $\theta = arctan(y, x)$ in degrees, where x, y are the coordinate representation of m_i in the circle.
4:    Compute the Euclidean distance of m_i from the circle's origin $(0, 0)$ as $l(m_i) = (x^2 + y^2)^{1/2}$
5:    **if** $x > 0 \wedge y > 0$ (first quadrant) **then**
6:        weight of m_i, $w(m_i) = tf(m_i) * pQ1 * l(m_i)/(180 - angle)/360$
7:    **end if**
8:    **if** $x < 0 \wedge y > 0$ (second quadrant) **then**
9:        weight of m_i, $w(m_i) = tf(m_i) * pQ2 * l(m_i)/(angle - 90)/360$
10:    **end if**
11:    **if** $x > 0 \wedge y < 0$ (third quadrant) **then**
12:        weight of m_i, $w(m_i) = tf(m_i) * pQ3 * l(m_i)/(angle)/360$
13:    **end if**
14:    **if** $x < 0 \wedge y < 0$ (fourth quadrant) **then**
15:        weight of m_i, $w(m_i) = tf(m_i) * pQ4 * l(m_i)/(angle)/360$
16:    **end if**
17: **end for**

representation; (iii) Based on the coordinates of the feature, weight it considering its magnitude, orientation in the circle and term frequency within the document. The proposed strategy generates a metric which assigns weights from highest to lowest in the following order Q2-Q1-Q3-Q4. Where the highest weight is provided to the strongly-related features (quadrant Q2) and the lowest weight to the strongly unrelated features (quadrant Q4). Both weakly-related and weakly-unrelated features fall close midway within the metric. The proposed penalties for each quadrants enables to emphasize a quadrant's feature or to bring down a quadrant's features relevance. This enables for example to filter out ambiguous features $(pQ1 = 0, pQ2 = 0)$, or to highlight the relevance of strongly related features $(pQ2 > 1.0)$. This weighting strategy provides a weighted representation of a document that can be used for training a topic classifier.

4.3   Extracting Topic Feature Priors from Semantic Knowledge Sources

In this paper we refer to a topic feature prior as the probability distribution that would express one's beliefs about this feature relevance/irrelevance to a topic before any other evidence is taken into account. Topic feature priors enable us to have a preliminary model of the language related to a topic when no other information about the topic is provided. For example for the topic "War_Conflict" such prior information maps violence polarity into violence words such as looting, war, drugs and non-violent polarity to background words such as today, afternoon, happy.

Word prior lexicon generation relies on the use of a positive and a negative samples of a topic. The feature prior representation of a Topic consists on getting all features (e.g., words) of a topic dataset and assigning to each of them a weight representing how well the feature is relevant to the topic. Social knowledge sources provide a rich textual information covering a large number of topics. In this work we use as a positive sample of a topic the set of articles' abstracts

belonging to categories and subcategories derived for a topic in DBpedia. As a
negative sample we use a set of tweets which are not related to this topic[5]. So
feature priors for the topic war for example would look like (feature:explosion
war: 0.8 non-war: 0.2; feature: sandwich war: 0.02 non-war: 0.98; and so on for
each feature in the War corpus).

To derive lexical features we use bag of words over the dataset. To derive
semantic features we extract and disambiguate entities appearing on these
abstracts, using AlchemyAPI. We then SPARQL queried DBpedia to obtain
specific semantic features about each entity e.g., categories, class type. Based on
these datasets to derive topic feature priors we employ the widely-used information gain method to select highly discriminative words under each class.

5   Conceptual Semantic Enrichment

In the previous sections we showed our proposed framework to facilitate feature grading of topics using the words' distributional semantics. However, using
the distributional patterns of a word (i.e., word's context) to detect its semantics in tweets is sometimes insufficient. For example, the word "ISIS" in "ISIS
continues spreading like a malignant tumor!" lack enough context to determine its semantics. Nonetheless, existing knowledge sources provide a wealth
of structure data that can be used to address this issue. For example the
word "ISIS" is a resource in DBpedia associated with the semantic category
"Jihadist_Group". Such association denotes a stronger relatedness with the topic
"War_Conflict". To account for semantic relatedness we propose to enrich our
topic compass framework, with the explicit or conceptual semantics of words in
tweets. To this end, we follow two main steps:

1. Entity Extraction and Semantic Mapping: This step extracts named
entities appearing in a tweet collection (e.g., "ISIS", "Bashar_Al_Assad",
"Barack_Obama") using the semantic extraction tool, AlchemyAPI.[6] Then, each
entity (e.g., "Bashar_Al_Assad") is mapped to a (i) semantic concept provided
by AlchemyAPI (Alc:Person); (ii) DBpedia Category (dbc:Presidents_of_Syria);
and (iii) DBpedia Class (dbo:Arab---Politician), shown in Table 1.

2. Conceptual Semantic Enrichment: This step incorporates the conceptual
semantics extracted from the previous step into the semantic representation of
the topics' feature space. As mentioned in Sect. 4.1, the context of a term m is
represented as a vector $c = (c_1, c_2, \ldots, c_n)$ of terms that occur with m in a given
tweet collection. Our semantic enrichment is done on this vector as follows:

– For AlchemyAPI Concepts, we extend the contextual vector c with the semantics $s = (s_1, s_2, \ldots, s_m)$ of named entities $e = (e_1, e_2, \ldots, e_m)$ that occur with
  m in the tweet collection as:

$$c_s = c + s = (c_1, c_2, \ldots, c_n, s_1, s_2, \ldots, s_m) \tag{4}$$

---

[5] Notice that the tweet sample used for deriving priors is independent of the corpus
used for topic classification in the experiments section.
[6] www.alchemyapi.com.

Table 1. Example of named entities extracted from tweets and mapped to their associated AlchemyAPI Concept, DBpedia Category, and DBpedia Class

Entity	Alchemy concept	Dbpedia category	Dbpedia class
ISIS	Organization	Jihadist_Groups	Populated_Place
Barack_Obama	Person	Presidents_of_the_US	Politician
Syria	Country	Middle_Eastern_Countries	Location
Bashar_Al_Assad	Person	Presidents_of_Syria	Arab_Politician

– For DBpedia Categories and Classes, we replace the entire contextual vector c with the semantic categories $o = (o_1, o_2, \ldots, o_m)$ or the semantic classes $l = (l_1, l_2, \ldots, l_m)$ of the entities in e as:

$$c_o = o = (o_1, o_2, \ldots, o_m) \tag{5}$$

$$c_l = l = (l_1, l_2, \ldots, l_m) \tag{6}$$

where c_s, c_o and c_l are the new semantically-enriched contextual vectors of m, which will be subsequently used instead of c to extract the semantic circle representation of m as described in Sect. 4.1. It is worth noting that the semantic enrichment done through Eqs. 5 and 6 results in topics' feature spaces completely represented by the entities' semantic categories or classes. Conversely, the feature spaces inferred from Eq. 4 are mix of words, named entities and AlchemyAPI concepts. The reason behind this representation variation is twofold. First, investigate the impact of words' semantics when solely used in our framework for feature grading. Secondly, unlike the large variety of DBpedia categories and types, the number of distinct concepts retrieved by AlchemyAPI from our datasets is limited to 41 concepts only. Relying on these concepts in our framework leads to sparse feature space representation of topics, which often results in low topic classification performance [17].

6   Experimental Setup

Here we present the experimental set up used to assess our proposed topic compass framework. We evaluate the effectiveness of our STC framework in a topic classification task. Specifically, we apply our framework on different Twitter datasets for feature extraction and grading. Then, the extracted features are used to train supervised classifiers for topic classification. Thus, our evaluation setup requires the selection of (i) Twitter datasets for feature extraction, (ii) baselines methods for cross-comparison, and (iii) the knowledge source from which the topic's prior are extracted. All these elements will be explained in the following subsections.

6.1   Datasets

To assess the performance of the classification task we require the use of datasets annotated with a topic label. For this work we selected three evaluation datasets, previously used in the literature of topic classification on Twitter [2]. These datasets consist of a collection tweets of Violence-related topics: *Disaster_Accident*, *Law_Crime* and *War_Conflict*. Tweets in each dataset are manually labelled with negative and positive scores denoting their relatedness to the topic.[7] Size and number of word unigrams, within each dataset are summarised in Table 2.

Table 2. Statistics of the three datatsets used for evaluation

Dataset	Tweets	Unigrams	Categories	Classes
Disaster_Accident	2,528	6,341	4,522	124
Law_Crime	1,967	4,540	3,582	113
War_Conflict	1,939	4,502	3,533	110

6.2   Baselines

As mentioned in the Sect. 2 different types of lexical and semantic features have been used in multiple works on topic classification. In this paper we choose to compare the features extracted by the STC framework against the following state-of-the-art lexical and semantic feature types.

Lexical Feature Baselines

TF Features: denoting word unigrams weighted by their term frequency in the tweets.

TF-IDF Features: denoting word unigrams weighted by using term frequency inverse document frequency.

LDA Features: referring to word unigrams weighted by the latent topic extracted from tweets using the probabilistic generative model, LDA [1]. To extract these latent topics from our datasets we use an implementation of LDA provided by Mallet.[8] LDA requires defining the number of topics to extract before applying it on the data. We experimented with different numbers of topics. Among all choices, 10 topics was the optimal number giving the highest classification performance for this baseline.

AM Features: referring to features weighted based on the Ambiguity Measure feature selection method [12].

---

[7] Details about the construction and the annotation of these datasets are provided in [2].
[8] http://mallet.cs.umass.edu/.

Semantic Feature Baselines

DBpedia Features: refer to two different types of semantic features obtained from DBpedia: (i) Semantic Categories (*Cat*) and (ii) Semantic Classes (*Cls*). To extract these features, we first extract the named entities in the Twitter datasets and map them after that to their classes and categories in DBpedia. Table 2 shows the number of semantic categories and classes extracted from each dataset.

AlchemyAPI Concepts: this type of features refers to the semantic concepts appearing in tweets. We extract these features using the Alchemy semantic extraction service. The number of the unique concepts extracted from our datasets is 41.

Examples of the above three types of semantic features are provided in Table 1.

7   Evaluation and Results

In this section we report the evaluation results obtained from using the features extracted by our STC framework in topic classification task. To this end, we use Naive Bayes classifiers. Our baselines of comparison are classifiers trained from the 7 types of lexical and semantic features described in Sect. 6.2. Results in all experiments are computed using 2-fold cross validation over 5 runs of different random splits of the data to test their significance. Statistical significance is done using the T-Test.

Evaluation in the subsequent sections consists of 4 main steps:

1. Investigate the impact of the feature weighting in our STC framework on the classification performance (Sect. 7.1).
2. Measure and compare performance of the STC framework against other supervised and unsupervised feature representation and selection models (Sect. 7.2).
3. Study the effect of enriching the STC framework with conceptual semantics on the topic classification performance (Sect. 7.3).

7.1   Feature Weighting with the STC Framework

The first task in our evaluation is to assess the performance of our semantic feature grading framework. As described in Sect. 4, STC provides an unsupervised approach for weighting a topic feature space. In this section we investigate how such topic representation along with the weighting strategy presented in Algorithm 1 performs in a classification task. Table 3 shows the results of binary topic classification performance for the three datasets following the weighting approach of the proposed STC framework. The table reports four sets of precision (P), recall (R), and F1-measure (F1), one for each dataset, and the fourth one shows the averages of the three.

The first column of the Table presents the penalties assigned to the four quadrants in Algorithm 1. When these penalties are higher or lower than 1.0 they enable to highlight or lessen respectively the weights assigned to features on a particular quadrant. We first analysed a base setting in which all penalties are set to 1.0. In this setting all weights derived directly from the topic circle are kept except for those situated on the axis which are smoothed down. This setting yields a consistent significant boost in P-measure on the three datasets with an increment in P of 7.36 % over the TF baseline (t-test with $\alpha < 0.01$). High precision in this setting shows the effectiveness of the topic circle approach to distribute topic independent features within the axis, which aids in improving the topic classification task.

Table 3. Performance of the TF and STC based classifiers. Tuples on the left side of the second section of the table represent the weights assigned to penalties $pQ1, pQ2, pQ3, pQ4$ respectively. The values highlighted in bold correspond to the best results obtained for each topic. A $\star$ denotes that the F-measure of a given weighted feature significantly outperforms the corresponding TF baseline. Significance levels: $p\text{-value} < 0.01$.

	Diss_Acc			Law_Crime			War_Conflict			Average		
	P	R	$F1$	P	R	$F1$	P	R	$F1$	P	R	$F1$
TF	0.8634	0.8886	0.8758	0.8245	0.882	0.8523	0.8205	0.8892	0.8535	0.8361	0.8866	0.8605
1.0,1.0,1.0,1.0	0.9383	0.7613	0.8405	0.8966	0.8380	0.8661	0.8889	0.8259	0.8558	0.9079$\star$	0.8084	0.8541
0.0,1.0,0.0,1.0	0.9024	0.6783	0.7744	0.8413	0.7573	0.7965	0.8141	0.6387	0.7156	0.8526	0.6914	0.7621
0.0,1.0,1.0,0.0	0.8826	0.8843	0.8835	0.8468	0.8699	0.8581	0.8466	0.8865	0.8660	0.8586	0.8802	0.8692$\star$
0.1,1.0,0.1,1.0	0.9024	0.7130	0.7965	0.8624	0.7636	0.8098	0.8276	0.7005	0.7585	0.8641	0.7257	0.7882
2.0,1.0,2.0,1.0	0.9358	0.8075	0.8669	0.8620	0.8689	0.8651	0.8457	0.8728	0.8588	0.8811	0.8497	0.8636$\star$
0.1,2.0,0.1,2.0	0.9043	0.7048	0.7921	0.8530	0.7527	0.7996	0.8183	0.6832	0.7446	0.8585	0.7135	0.7787
2.0,2.0,1.0,1.0	0.9400	0.7879	0.8572	0.8353	0.8819	0.8579	0.8372	0.8723	0.8542	0.8708	0.8473	0.8564
2.0,2.0,2.0,2.0	0.9317	0.8118	0.8676	0.8679	0.8711	0.8693	0.8709	0.8696	0.8698	0.8901$\star$	0.8508	0.8689$\star$

The last section of Table 3 presents results for different penalty settings. In particular we find that keeping weights on quadrant 2 (strongly-related) and 4 (strongly unrelated) while lessening Q1, Q3 (ambiguous quadrants) (i.e., setting (0.0,1.0,0.0,1.0)) boost performance but decreases recall. This result indicates the importance of ambiguous terms for a classifier to learn how to discriminate relevant from irrelevant topics.

We also find that highlighting the irrelevance of weakly-unrelated features (Q3) while keeping the strongly-related (Q2) weights and smoothing those features which are strongly unrelated (Q4) and weakly-related (Q1) (i.e., (0.0,1.0,1.0,0.0)) provides the best boost in F measure when compared against the TF baseline. This setting improves P in 2.25 % in average over the TF baseline (t-test with $\alpha < 0.01$). This result stresses the importance of balancing weights between the two types of ambiguous features identified by this framework. In particular we find that stressing the irrelevance of the weakly-unrelated features by smoothing down the weakly-related aids in improving performance. Given that this latter setting provides the best boost in F-measure we keep this setting to perform a cross comparison of the different type of features and weighting baselines in the following section.

7.2   Cross Comparison Results

Here we evaluate the performance of STC against both, the lexical and semantic baselines described in Sect. 6.2. The first section of Table 4 shows the performance obtained with the lexical baselines: (1) Term frequency (TF); (2) TF-IDF; (3) LDA; and (4) Ambiguity metric (AM). All TF-IDF, LDA and AM offer competitive results to the standard TF metric for all datasets. Specifically, TF-IDF offers an overall boost in P; however it's the LDA baseline the one that consistently outperforms the TF baseline on both P and F1 in all datasets. The second section of Table 4 shows the performance of the semantic baselines: (5) Semantic Categories (Cat), (6) Semantic Classes (Cls), and (7) AlchemyAPI Concepts (Alc). In particular, the Cat baseline consistently outperforms in P all lexical baselines in the datasets.

Average results for the lexical baselines (Avg_{Lex}) and the semantic baselines (Avg_{Sem}) in Table 4 show that the semantic baselines slightly outperform the lexical ones, but give lower performance in R and F1. Unlike feature weighting in the lexical baselines which considers all terms in the datasets, the feature weighting in the semantic baselines considers the named-entities only (see Sect. 6.2). This might explain the low recall and F1 of the semantic baselines.

From the set of baselines, Cat offers the best performance, while LDA offers the best boost in F1 across datasets. The third and fourth sections of Table 4 present results for the STC framework with distributional and conceptual semantics respectively. For the first case we present two weighting settings: (i) The default setting STC_{Def}, where all penalties are set to 1.0 (1.0, 1.0, 1.0, 1.0), and the STC_{Bal} setting (0.0, 1.0, 1.0, 0.0) which has shown to yield a balanced performance in P, R, and F1 among other settings, as described in Sect. 7.1.

Here STC_{Def} consistently outperform all 7 baselines in P across all datasets, with an average boost of 7 % (significant at $p < 0.01$) when compared to TF and 4 % (significant at $p < 0.01$) when compared to the highest baseline in P (Cat). In particular STC_{Def} provides best results for Law_Crime with a boost in P of 7.4 % when compared with its TF baseline and of 5.8 % when compared to its highest lexical baseline (LDA). STC_{Def} setting, leaves out features lying on the axes, which can also be considered as ambiguous features, this might explain the boost in precision. STC_{Bal} offers competitive results for P outperforming all lexical baselines, however the semantic baseline Cat outperforms it. Nonetheless, STC_{Bal} shows a consistent but slight boost in F1 for all datasets. The STC_{Bal} removes strongly unrelated (Q4) and weakly-related (Q1) features, this might explain the boost in P and F, however it also shows that removing these features can impact performance in R.

We believe that the above results show the effectiveness of our STC framework for feature grading over the baselines reported in this paper. While STC_{Def} and STC_{Bal} consistently boost P and F1 respectively across all dataset, each

Table 4. Cross-comparison results of the STC framework against the lexical and semantic baselines. The values highlighted in bold correspond to the best results obtained for each topic. A $\star$ denotes that the F-measure of a given weighted feature significantly outperforms the baselines. Significance levels: p-value < 0.01. The values highlighted in bold correspond to the best results obtained for each case.

	Diss_Acc			Law_Crime			War_Conflict			Average		
	P	R	$F1$	P	R	$F1$	P	R	$F1$	P	R	$F1$
TF	0.8634	0.8886	0.8758	0.8245	0.8820	0.8523	0.8205	0.8892	0.8535	0.8361	0.8866	0.8605
TFIDF	0.8702	0.8693	0.8697	0.8449	0.8467	0.8457	0.8397	0.8588	0.849	0.8449	0.8467	0.8457
LDA	0.8797	0.8876	0.8836	0.8368	0.8790	0.8573	0.8380	0.8910	0.8637	0.8515	0.8858	0.8682
AM	0.8587	0.8891	0.8736	0.816	0.8726	0.8433	0.8145	0.8866	0.849	0.8297	0.8827	0.8553
Avg_{Lex}	0.868	0.8836	0.8756	0.8305	0.8700	0.8496	0.8281	0.8814	0.8538	0.8422	0.8783	0.8597
Cat	0.9015	0.6238	0.7372	0.8500	0.5656	0.6791	0.8519	0.6692	0.7493	0.8678	0.6195	0.7219
Cls	0.8627	0.8211	0.8414	0.8164	0.8281	0.8221	0.8091	0.8324	0.8206	0.8294	0.8272	0.8280
Alc	0.8763	0.8789	0.8776	0.8372	0.8730	0.8547	0.8408	0.8797	0.8598	0.8514	0.8772	0.8640
Avg_{Sem}	0.8802	0.7746	0.8187	0.8345	0.7556	0.7853	0.8339	0.7938	0.8099	0.8486	0.7746	0.8046
STC with Distributional Semantics												
1.0,1.0,1.0,1.0	0.9383	0.7613	0.8405	0.8966	0.8380	0.8661	0.8889	0.8259	0.8558	0.9079$\star$	0.8084	0.8541
.0,0,1.0,0,0.0	0.8826	0.8843	0.8835	0.8468	0.8699	0.8581	0.8466	0.8865	0.8660	0.8586	0.8802	0.8692$\star$
STC with Conceptual Semantics												
STC_Cat	0.8456	0.8986	0.8713	0.7815	0.9145	0.8427	0.7899	0.8827	0.8336	0.8056	0.8986$\star$	0.8492
STC_Cls	0.86	0.8672	0.8635	0.7967	0.8669	0.8302	0.79	0.8817	0.8332	0.8152	0.8720	0.8422
STC_Alc	0.9244	0.7814	0.8469	0.8299	0.8866	0.8571	0.8202	0.8741	0.8459	0.8581$\star$	0.8473	0.8499

dataset has a different boost for each setting, this is expected since ambiguity and specificity are topic-dependent features. The following section presents results for STC with conceptual semantics.

7.3   Evaluation of the STC Framework with Semantic Enrichment

Here we evaluate the impact of the semantic enrichment when applying the STC framework. The last section of Table 4 presents results for STC with conceptual semantics for all datasets. In average STC_Cat boosts R with 19.1 % over the Cat baseline offering on average a slight boost of 1.2 % over the highest R baseline (TF). Considering the independent results in Table 4 we see that STC_Cat offers a boost in R for all datasets. In particular it provides the highest boost for Law_Crime with 3.35 % over TF, which provides the highest baseline.

The STC_Cls and STC_Alc also offer competitive baselines for P however they don't outperform the best results obtained with STC with distributional semantics. While the semantic enrichment improves upon the baselines in recall, it is the STC with distributional semantics the feature that provides the overall best performance in F1, outperforming also the baselines in P and offering a competitive R.

8   Discussion

In this paper we introduced a novel approach for topic feature representation and weighting which enhances the state-of-the-art feature weighting approaches in providing an ambiguity orientation to each feature. This approach facilitates the identification and filtering of relevant, weakly-related, weakly-unrelated, and unrelated features of a topic's feature space. The geometric nature of the proposed approach facilitates the partition of the feature space, allocating ambiguous feature over the axes and over quadrants 1 and 3.

In order to discuss the effect of ambiguous features in the classification performance task we performed a correlation analysis over gain in performance. For this analysis we focus on the gain provided by the (0.0, 1.0, 0.0, 1.0) setting which lessens the relevance of ambiguous features while highlighting the strongly relevant and strongly irrelevant features. We computed Pearson's correlation between the gain in P, R and F-measure of this setting for lexical and semantic feature versus their corresponding following ratios: (1) ratio of number of weakly-related (WR) to strongly-related (SR) features (WR/SR); (2) ratio of number of weakly-unrelated (WU) to strongly-related (SR) features (WU/SR); (3) ratio of number of the sum of WR and WU to SR ($WR + WU/SR$).

The computed correlations for these ratios are presented in Fig. 4 (statistically significant at $p < 0.05$). These results show the impact of filtering/keeping weakly-related or weakly-unrelated features in boosting performance on a classification task. In particular they reveal the compromise of the use of ambiguous features on this task. Lowering the weight of weakly related features has a slightly positive impact on Precision, while having a positive moderate effect in increasing Recall. Moreover the correlation analysis show that lessening the effect of both WR and WU has a high positive effect in increasing F-measure on the classification task (significant at $p < 0.05$).

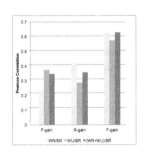

Fig. 4. Pearson correlation for P, R, F-measure gain versus the following ratios: weakly-related (WR) to strongly-related (SR) (WR/SR); weakly-unrelated (WU) to strongly-related (SR) (WU/SR) and ($WR + WU/SR$). Correlation windows: Negligible $(0 - 0.19)$; Weak $(0.2 - 0.39)$; Moderate $(0.4 - 0.69)$; High (>0.69). Statistically significant at $p < 0.05$.

In the previous section we demonstrated the positive effect of the use of STC framework in improving performance upon the TF weighting scheme. We also show that the use of distributional semantics improve performance over the lexical baselines. While Category features weighted with the STC improved upon both STC with distributional semantics and baselines in Recall. This is an expected results since the use of semantic categories provides a generalisation over the type of entities contained on a tweet. However, in our results semantic features did not outperformed in P and F-measure. When computing the density of quadrants for these features we observed that for both *Cat* and

Cls, the percentage of features appearing on Q1 and Q3 is less than 3 %. In this case lessening the ambiguity of those features does not have an apparent effect on the classification performance.

9   Conclusions

In this paper we introduced the Semantic Topic Compass Framework (STC) which enables to characterise the orientation of features towards the relevancy/irrelevancy of a topic. STC is an unsupervised approach relying only on the use of topic feature priors derived from semantic knowledge sources. It is based on the use of distributional and conceptual semantics to characterise feature ambiguity using a polar representation of a topic's feature space. Based on such feature representation we proposed a weighting strategy which encodes both ambiguity orientation and topic relevance. The proposed strategy proved useful in the topic classification task. To the best of our knowledge this is the first approach to address feature ambiguity characterising a feature's ambiguity-orientation towards being relevant/irrelevant to a topic. In particular our results show that there is a compromise between the use and filtering of weakly related and weakly unrelated features. Future work includes to iterate the process of characterising ambiguity within an active learning setting.

Acknowledgment. This work was supported by the EU-FP7 project SENSE4US (grant no. 611242) and the UK HEFCE project MK:Smart.

References

1. Blei, D.M., Ng, A.Y., Jordan, M.I.: Latent dirichlet allocation. J. Mach. Learn. Res. **3**, 993–1022 (2003)
2. Cano, A.E., He, Y., Alani, H.: Stretching the life of twitter classifiers with time-stamped semantic graphs. In: Mika, P., et al. (eds.) ISWC 2014, Part II. LNCS, vol. 8797, pp. 341–357. Springer, Heidelberg (2014)
3. Cover, T.M., Thomas, J.A.: Elements of Information Theory. Wiley, New York (2012)
4. Firth, J.R.: A Synopsis of Linguistic Theory. Studies in Linguistic Analysis (1930–1955) (1957)
5. Galavotti, L., Sebastiani, F., Simi, M.: Experiments on the use of feature selection and negative evidence in automated text categorization. In: Borbinha, J.L., Baker, T. (eds.) ECDL 2000. LNCS, vol. 1923, pp. 59–68. Springer, Heidelberg (2000)
6. Genc, Y., Sakamoto, Y., Nickerson, J.V.: Discovering context: classifying tweets through a semantic transform based on wikipedia. In: Schmorrow, D.D., Fidopiastis, C.M. (eds.) FAC 2011. LNCS, vol. 6780, pp. 484–492. Springer, Heidelberg (2011)
7. Harris, Z.S.: Distributional structure. Word **10**(2–3), 146–162 (1954)
8. How, B.C., Narayanan, K.: An empirical study of feature selection for text categorization based on term weightage. In: Proceedings of the 2004 IEEE/WIC/ACM International Conference on Web Intelligence, pp. 599–602. IEEE Computer Society (2004)

9. Janecek, A., Gansterer, W.N., Demel, M., Ecker, G.: On the relationship between feature selection and classification accuracy. J. Mach. Learn. Res. **4**, 90–105 (2008)

10. Lee, K., Palsetia, D., Narayanan, R., Patwary, M.M.A., Agrawal, A., Choudhary, A.: Twitter trending topic classification. In: 2011 IEEE 11th International Conference on Data Mining Workshops (ICDMW), pp. 251–258. IEEE (2011)

11. McCreadie, R., Macdonald, C., Ounis, I., Osborne, M., Petrovic, S.: Scalable distributed event detection for twitter. In: Proceedings of the 2013 IEEE International Conference on Big Data, Santa Clara, pp. 543–549, 6–9 October 2013

12. Mengle, S.S., Goharian, N.: Ambiguity measure feature-selection algorithm. J. Am. Soc. Inf. Sci. Technol. **60**(5), 1037–1050 (2009)

13. Mladeni'c, D., Grobelnik, M.: Feature selection for classification based on text hierarchy. In: Text and the Web, Conference on Automated Learning and Discovery CONALD-98. Citeseer (1998)

14. Muñoz-García, O., García-Silva, A., Corcho, O., Higuera Hernández, M., Navarro, C.: Identifying topics in social media posts using dbpedia (2011)

15. Phan, X.H., Nguyen, L.M., Horiguchi, S.: Learning to classify short and sparse text & web with hidden topics from large-scale data collections. In: Proceedings of the 17th International Conference on World Wide Web, pp. 91–100. ACM (2008)

16. Saif, H., Fernandez, M., He, Y., Alani, H.: SentiCircles for contextual and conceptual semantic sentiment analysis of twitter. In: Presutti, V., d'Amato, C., Gandon, F., d'Aquin, M., Staab, S., Tordai, A. (eds.) ESWC 2014. LNCS, vol. 8465, pp. 83–98. Springer, Heidelberg (2014)

17. Saif, H., He, Y., Fernandez, M., Alani, H.: Semantic patterns for sentiment analysis of twitter. In: Mika, P., et al. (eds.) ISWC 2014, Part II. LNCS, vol. 8797, pp. 324–340. Springer, Heidelberg (2014)

18. Sheth, A., Ramakrishnan, C., Thomas, C.: Semantics for the Semantic Web. Idea Group Publishing, p. 1 (2005)

19. Song, Y., Wang, H., Wang, Z., Li, H., Chen, W.: Short text conceptualization using a probabilistic knowledgebase. IJCAI **3**, 2330–2336 (2011)

20. Sriram, B., Fuhry, D., Demir, E., Ferhatosmanoglu, H., Demirbas, M.: Short text classification in twitter to improve information filtering. In: Proceedings of the 33rd International ACM SIGIR Conference on Research and Development in Information Retrieval, pp. 841–842. ACM (2010)

Reasoning Track

Supporting Arbitrary Custom Datatypes in RDF and SPARQL

Maxime Lefrançois[(✉)] and Antoine Zimmermann

École Nationale Supérieure des Mines, FAYOL ENSMSE,
Laboratoire Hubert Curien, 42023 Saint-Étienne, France
{maxime.lefrancois,antoine.zimmermann}@emse.fr

Abstract. In the Resource Description Framework, literals are composed of a UNICODE string (the lexical form), a datatype IRI, and optionally, when the datatype IRI is `rdf:langString`, a language tag. Any IRI can take the place of a datatype IRI, but the specification only defines the precise meaning of a literal when the datatype IRI is among a predefined subset. Custom datatypes have reported use on the Web of Data, and show some advantages in representing some classical structures. Yet, their support by RDF processors is rare and implementation specific. In this paper, we first present the minimal set of functions that should be defined in order to make a custom datatype usable in query answering and reasoning. Based on this, we discuss solutions that would enable: (i) data publishers to publish the definition of arbitrary custom datatypes on the Web, and (ii) generic RDF processor or SPARQL query engine to discover custom datatypes on-the-fly, and to perform operations on them accordingly. Finally, we detail a concrete solution that targets arbitrarily complex custom datatypes, we overview its implementation in Jena and ARQ, and we report the results of an experiment on a real world DBpedia use case.

Keywords: Literals · Datatypes · RDF · Linked data

1 Introduction

The Resource Description Framework empowers the Web of Data with three kinds of entities: IRIs, blank nodes, and literals [3]. IRIs are obviously central as they allow the interlinking of datasets and serendipitous discovery of more data. Blank nodes have been the subject of several papers (a comprehensive review is found in [9]). Literals are extremely important since they are, after all, the carriers of the data that is eventually processed. In fact, we argue that IRIs are only crucial insofar as they offer a way of traversing linked data towards the discovery of literal values.

RDF defines literals as being composed of a UNICODE string and a datatype IRI[1], the latter being an *arbitrary* IRI that may refer to any datatype conforming

This work has been supported by ITEA2 project SEAS 12004.

[1] And optionally a language tag when the datatype IRI is `rdf:langString`, but for the purpose of this paper, we will simply consider literals as pairs.

© Springer International Publishing Switzerland 2016
H. Sack et al. (Eds.): ESWC 2016, LNCS 9678, pp. 371–386, 2016.
DOI: 10.1007/978-3-319-34129-3_23

to the definition in [3, Sect. 5]. The datatype that the IRI refers to gives meaning to the literals having that type. Indeed, by definition, a datatype defines what value the UNICODE string represents in that type.

An RDF processor that is able to distinguish the values of literals for a given datatype IRI is said to *recognise* the IRI. It is possible to program an RDF processor such that it recognises a fixed set of IRIs by implementing the associated set of specifications. Usually, the set of recognised datatypes is the set of XSD datatypes. However, even some RDF processors don't process them in a uniform way [5]. And even then, processors cannot compare literals with datatype IRIs they do not recognise. In this paper, we want to address the case of a processor that does not necessarily recognise a fixed set of IRIs but is able to determine the datatype associated with an IRI on the fly. We provide motivating use cases for this in Sect. 2.

To achieve this, we first show that an RDF processor does not necessarily need to "know" the actual datatype (which is a mathematical structure that cannot always be represented in a computer format). Instead, for some reasoning or query answering purposes, recognising a datatype amounts to using a small set of functions that can usually be provided in a computer language. We describe these functions in Sect. 3. In Sect. 4, we show several options for implementing an RDF processor that can take advantage of a computerised description of these functions such that it can recognise some new datatypes on the fly. We present our own implementation in Sect. 5. Our evaluation in Sect. 6 demonstrates that the approach does not introduce significant overhead while it makes both publishing data easier, and writing more concise queries when compared to an approach purely based on standard datatypes. Section 7 provides a critical discussion of the overall approach and our specific implementation, with an overview of future work.

2 Use Cases for on-the Fly Support of a New Datatype

This section introduces several motivating use cases for enabling on-the-fly support of custom datatypes in RDF processors and SPARQL query engines.

Sharing Energy Related Data. In the ITEA2 SEAS project that partly funds this research, industrial partners want to share energy-related data such as energy consumption and production, capacities, temperatures, and they use various custom datatypes for representing these data. Sometimes, they use different datatypes to represent similar information, such as `ex1:wattHour`, `ex2:barrelOfOilEquiv`, and `ex3:GJ` for energy quantities. RDF processors and SPARQL query engines cannot be updated for the support of each individual datatype in use. Also, it is impossible to write a SPARQL query that selects consumptions or productions that are within a given range. Instead, processors could rely on a generic mechanism for automatically retrieving sufficient information for the data to be processed, queried, and compared.

Distributed Computation. In distributed and collaborative computing, it is necessary to transfer the state of a program execution in a serialised form. A state can be shared as a combination of metadata and serialised OOP objects that can be adequately represented in RDF, with serialised objects being written as literals with a type indicating the class membership. In this situation, it could be desirable to know which executions reach the same state. This would be possible with a SPARQL query, had there been a mechanism for associating the datatype IRI to the appropriate datatype definition.

Well Known Text Literals. In the OGC standard GeoSPARQL [11], a datatype is defined for serialising geolocated region of space, such as `"LINESTRING(0 0, 1 1, 1 2, 2 2)"^^geo:wktLiteral`.[2] However, wktLiteral can also specify a coordinate reference system that differs from the default CRS84 by adding a URI at the beginning of the literal, e.g., `"<http://www.opengis.net/def/crs/EPSG/0/4326> Point(33.95 -83.38)"^^geo:wktLiteral`. There is no restriction on the URI being used at this position in the standard, so some wktLiterals may not be understood even by processors that implement the GeoSPARQL standard. Had the coordinate system been given as a datatype IRI, and assuming a mechanism as we propose to dynamically obtain the specification of the datatype, new coordinate systems could be supported as soon as they appear.

3 Requirements for on-the Fly Support of a New Datatype

In this section, we describe required functionalities to effectively recognise a datatype IRI, but first we provide preliminary definitions. For clarity, we will then restrain to the case when it is assumed that value spaces are pairwise disjoint before addressing the more general case.

3.1 Preliminaries

As mentioned in footnote 1, we only focus here on literals that do not have a language tag. Therefore, from now on, a literal will be a pair comprising a UNICODE string called the *lexical form* and an IRI called the *datatype IRI.* When we need to refer to an arbitrary IRI, we use names of the form a, b, etc. with letters from the beginning of the alphabet, while for arbitrary UNICODE string, we use names like s, t, etc. with letters from the end of the alphabet. We first recall necessary definitions from the RDF 1.1 specifications.

Definition 1 (Datatype). *A datatype D is a structure comprising the following components:*

---

[2] Subsequently, we will use `geo:` for http://www.opengis.net/ont/geosparql. Similarly, we will use usual prefixes `rdf:`, `rdfs:`, `xsd:`, and `owl:` in all examples.

– a set $L(D)$ of UNICODE strings, called the lexical space;
– a set $V(D)$, called the value space of D;
– a mapping $L2V(D) : L(D) \rightarrow V(D)$, called the lexical-to-value mapping, *that maps all strings in the lexical space to a value in the value space.*

To avoid paraphrasing RDF 1.1 Semantics, we only refer the most relevant definitions in [7]. In this paper, we rely heavily on the notion of *recognised IRI*, *simple D-interpretation*, and *D-entailment* (or *simple entailment recognising D*) defined in [7, Sect. 7]. We also utilise the extensions to *RDF* and *RDFS-entailment recognising D* from [7, Sects. 7 and 8]. When an RDF processor recognises an IRI identifying a datatype D_a, we say that it *supports* D_a.

3.2   Pairwise Disjoint Value Spaces

An RDF processor that supports a datatype D_a identified by an IRI a must be able to check two things: whether a UNICODE string belongs to the lexical space of D_a or not, and whether two literals with datatype D_a share the same value.

Well-formedness. Given a UNICODE string s, is the lexical form s well formed in D_a, i.e., Does it belong to the lexical space of D_u? Or equivalently, is literal "s"^^a well typed? i.e., $s \in L(D_a)$.
For example, "12.5" is well formed in xsd:decimal, while "abc" is not (that is, "12.5"^^xsd:decimal is well typed, and "abc"^^xsd:decimal is ill-typed).
Equality. Given two UNICODE strings s,t, do "s"^^a and "t"^^a share the same value? i.e., $L2V(D_a)(s) = L2V(D_a)(t)$.
For example, "0.50"^^xsd:decimal and ".5"^^xsd:decimal share the same value.

Note that if a UNICODE string is not in the lexical space of a datatype, then it does not have a value. Hence, it would never be equal to any other literal value.

Concerning SPARQL query engines, basic graph matching only requires being able to join values, and therefore nothing more than what precedes is needed. Now, SPARQL offers an extension point related to filtering and ordering literals: SPARQL implementations may extend the XPath and SPARQL Tests operators $\{=, ! =, <, >, <=, >=\}$ [6]. Apart from testing equality, SPARQL engines may need to test the ordering of literals.

Value comparison. Given two UNICODE strings s,t, is the value of "s"^^a lower (resp., greater) than the value of "t"^^a? i.e., $L2V(D_a)(s) < L2V(D_a)(t)$ (resp., $L2V(D_a)(s) > L2V(D_a)(t)$).

These functionalities are sufficient to check for simple D-entailment between RDF graphs, and even RDFS entailment recognising D, as long as the value spaces are infinite (and we assume here they are disjoint), as shown in [4]. The case of RDFS reasoning recognising datatypes of finite size is tricky and discussed in Sect. 7.

3.3   Overlapping Value Spaces

Now, as justified by the use cases, datatypes value spaces may overlap. Practically, we need to extend the equality and value comparison checking to different datatypes.

Cross-datatype equality. Given two datatypes D_a and D_b respectively identified by IRIs a and b, given two UNICODE strings $(s, t) \in L(D_a) \times L(D_b)$, do "s"^^a and "t"^^b share the same value? i.e., $L2V(D_a)(s) = L2V(D_b)(t)$. For example, "1"^^ex2:barrelOfOilEquiv and "6.1178632e9"^^ex3:GJ share the same value.

Cross-datatype value comparison. Given two datatypes D_a and D_b respectively identified by IRIs a and b, given two UNICODE strings $(s,t) \in L(D_a) \times L(D_b)$, is the value of "s"^^a lower (resp., greater) than the value of "t"^^b? i.e., $L2V(D_a)(s) < L2V(D_b)(t)$ (resp., $L2V(D_a)(s) > L2V(D_b)(t)$). For example, "1"^^ex2:barrelOfOilEquiv is bigger than "1"^^ex3:GJ.

These functionalities are again sufficient to check for simple D-entailment between RDF graphs. However, they may not be sufficient for RDFS entailment, even with infinite value spaces. [4] proved that if any intersection of value spaces is infinite or empty, then these functions would be sufficient to do correct and complete RDFS reasoning recognising D (see Sect. 7 for more details). Note that if these constraints are not met, it is still possible to perform sound reasoning that is complete on graphs that only use the datatype IRIs in literals rather than as subject, predicate, or object. For example, graphs can contain this type of triples: :s :p "1"^^xsd:int, but not this type of triples: :p rdfs:range xsd:integer.

4   Implementation Options

RDF processors that have to deal with a datatype IRI for which they do not have hard-coded implementation should be able to retrieve a processable version of the functions described in Sect. 3. This assumes that these functions can be computed. In general, it is not the case. For instance, a datatype could encode a FOL formula, with the value space being the set of equivalent class wrt FOL entailment. In this paper, we want to address the most general case, namely when equality, well-formedness, and comparison are all computable functions (i.e., that the associated decision problems are decidable).

For cross-datatype comparisons, our requirements suggest that it should be possible to compare literals from any datatype to literals from any other. It is not practically doable, so any solution would be partial. However, we want to provide a mechanism that makes it possible to extend to an arbitrary large finite set of supported datatypes.

Clearly, any solution must involve an agreement between both the publisher[3] and the consumer on a common mechanism for presenting and exploiting the required functionalities. In an ideal situation, a standard would exist that would

---

[3] Here, the publisher is the one that specifies the datatype associated with an IRI.

reduce the need for coordinating between publishers and consumers. These functions could be provided by a centralised datatype registration service, where publishers submit their datatype specification. However, such a solution is unpractical and at odd with basic web principles.

Therefore, in what follows, we focus on solutions that work on the principle that the requested functions are accessible by way of dereferencing the datatype IRI. As a matter of fact, this is precisely what RDF 1.1 Semantics suggests in Sect. 7. Therefore, in this section, all the solutions that we describe require that datatype IRIs are HTTP IRIs.

Using Processor-Specific Modules. ARQ and SESAME offer ways to register classes that implement custom SPARQL filter functions. The support of custom datatypes could be done in a similar way. Hence, the information these implementations would need to get from the datatype IRI would be a jar with the necessary class, for instance. This solution is reasonably simple, but it is implementation-specific, and the custom datatype publisher would require to write one class for each RDF processor. It also presents serious security issues, unless the RDF engine implements complicated control measures to avoid executing harmful unknown compiled code.

Using Functions Defined in a Script. Instead of using a compiled class for each implementation, this solution consists in providing the code of the required functions. The burden of interpreting the code would then fall on the designers of RDF processors. Nonetheless, a pivot language such as JavaScript, for which engine integration exists in many programming languages, would make this solution viable. Moreover, it uses the full expressivity of a programming language and hence enables the specification of arbitrary custom datatype. We chose to follow this approach for our implementation described in Sect. 5.

Using a Web Service. An alternative to provide the code directly would be to offer the same functionalities encapsulated in a web service. The drawback of this approach compared to a script is that the service needs high availability, the code cannot be cached, compiled, and optimised. Otherwise, this approach is worth investigating and we expect to do so in future work.

Declarative Vocabulary-Based Description. Using the full expressivity of a programming language to describe a custom datatype is excessive in many cases. It would hence be interesting to describe a datatype using a vocabulary, possibly inspired from the OWL 2 datatype restrictions [16]. We are currently undergoing research to define and use such a vocabulary, and this solution will be further discussed in Sect. 7.

In the context of this paper, we will focus on the script-based solution.

5   Script-Based Support of Arbitrary Custom Datatypes

We focus on a solution where the custom datatype specification is defined using a scripting language. This section defines a specific solution for this, where we use the JavaScript language. Javascript has already been proposed as a language to implement custom funtions in SPARQL [17]. The solution we propose consists of: (i) guidelines for custom datatype publishers, including the definition of an API that the code in the JavaScript document must implement (Sect. 5.1); (ii) guidelines for RDF processors and SPARQL engines (Sect. 5.2). In a realistic setting, not all publishers will be following our guidelines, so we provide more functionalities than strictly needed for datatype support in order to make the approach robust to errors, corner cases, and missing information. This can serve as a model for other implementations, whether they are script-based, service-based, or declarative.

5.1 Guidelines for Datatype Publishers

The proposed solution requires to use an HTTP IRI **a** to identify datatype D_a, and to enable RDF processors and SPARQL engines to retrieve a JavaScript document from the datatype IRI when they look up **a** with a HTTP Accept header field that contains `application/javascript` (i.e., use content negotiation). Multiple datatypes may be defined in the same document, such as `xsd:string` and `xsd:int` that are defined in the same document at location http://www. w3.org/2001/XMLSchema. Hence the RDF processor would not know what part of the code it should execute for each datatype. Let D_a be a datatype identified by IRI **a**. We propose that the code implements a simple interface `CustomDatatypeFactory`, with a unique function `getDatatype(iri)`. When called with the string **a**, this function returns an object that holds the specification of datatype D_a, i.e., an instance of an interface `CustomDatatype`. We describe the methods in interfaces `CustomDatatypeFactory` and `CustomDatatype`, and sketch the expected behaviour of their implementations. This API and a formal set of constraints is described at http://w3id.org/lindt. All these methods take string parameters, and can generate errors as specified below.

Interface `CustomDatatype` defines a single method, `getDatatype`.

```
CustomDatatype getDatatype (iri)
```

A retrieved document contains a specification of custom datatype D_a identified by an IRI **a** if and only if `getDatatype(a)` returns an object **da** that implements interface `CustomDatatype`, and that complies with the set of constraints defined below. Such an object is called the *specification object* of datatype D_a.

Interface `CustomDatatype` defines the following set of methods.

```
String getIri()
Boolean isWellFormed (lexicalForm)
Boolean recognisesDatatype (datatypeIri)
String[] getRecognisedDatatypes ()
Boolean isEqual (lexForm1, lexForm2[, datatypeIri2])
Integer compare (lexForm1, lexForm2[, datatypeIri2])
String getNormalForm (lexicalForm1)
String importLiteral (lexicalForm, datatypeIri)
String exportLiteral (lexicalForm, datatypeIri)
```

Let da be the implementation of CustomDatatype returned by a call to getDatatype(a), i.e., da is the specification object of D_a. First suppose that the value space of the defined datatype is disjoint with that of every other datatypes.

isWellFormed. A string s is in the lexical space of D_a if and only if a call to da.isWellFormed(s) returns boolean true.

isEqual. Two literals "s"^^a and "t"^^a have equal values if and only if a call to da.isEqual(s, t) returns boolean true. This method must generate an error if either s or t is not in the lexical space of D_a. Finally, this method must be reflexive, symmetric, and transitive.

getNormalForm. It is of great interest for RDF processors to be able to normalize lexical forms. For instance in the context of datatype xsd:float, the normal form of lexical form 42.0 is lexical form 4.2E1. Method getNormalForm must return a string if the lexical form given as parameter is in the lexical space, or generate an error otherwise. Finally, this method is coherent with da.isEqual. Among other constraints, it is idempotent.

compare. This method must return a negative integer, zero, or a positive integer, depending on if the value of the first parameter is lower, equal, or greater than the value of the second parameter, respectively. It must generate an error if one of the parameters is not well formed, or if the literals are not comparable. Finally this method must be such that for any three well formed lexical forms s, t, and u,

- da.isEqual(s,t) ⟺ da.compare(s,t) = 0;
- da.compare(s,t) × da.compare(t,s) ≤ 0;
- (da.compare(s,t) ≥ 0) ∧ (da.compare(t,u) ≥ 0) ⟹ (da.compare(s,u) ≥ 0).

For datatypes whose value space is considered to be disjoint with that of any other datatype, the set of methods described above is sufficient to enable effective querying and RDFS reasoning recognising D, as justified in Sect. 3.2. As the value space of a datatype may intersect with that of other datatypes, interface CustomDatatype is completed as follows:

recognisesDatatype. Suppose the publisher of datatype ex1:wattHour is aware of the existence of datatype ex3:GJ, and knows how to compare values of ex1:wattHour literals with ex3:GJ literals, while the inverse is not true. In this case, the methods of the object that represents ex1:wattHour should be used to compare ex1:wattHour literals with ex3:GJ literals, and not the opposite. A datatype must recognise itself, but it does not need to recognise a datatype whose value space is disjoint with its own. Datatype D_a recognises datatype D_b identified by IRI b if and only if da.recognisesDatatype(b) returns boolean true.

isEqual. This method has an optional parameter which is the datatype IRI of the second literal. It must generate an error if the given IRI is not recognised. Given datatypes D_a, D_b, D_c identified by IRIs a, b, c, such that D_a and D_b are custom datatypes with specification objects da and db, D_a recognising D_b and D_c, D_b recognising D_c, lexical forms s and t well formed in D_a, u well formed in D_b, and v well formed in D_c, all of the following must be true:

- da.isEqual(s,t,a) = da.isEqual(s,t)
- a=c ⇒ da.isEqual(s,u,b) = db.isEqual(u,s,a)
- da.isEqual(s,u,b) and db.isEqual(u,v,c) ⇒ da.isEqual(s,v,c)

compare. This methods has an optional parameter which is the datatype IRI of the second literal. It must generate an error if the given IRI is not recognised. Given the same conditions as for isEqual, all of the following must be true:
- da.compare(s,t,a) − da.compare(s,t);
- da.isEqual(s,v,c) ⇔ da.compare(s,v,c) = 0;
- da.compare(s,u,b) × db.compare(u,s,a) ≤ 0;
- (da.compare(s,u,b) ≥ 0)∧(db.compare(u,v,c) ≥ 0) ⇒ (da.compare(s,v,c) ≥ 0).

importLiteral. Given a datatype D_a identified by IRI a and with specification object da, this method takes as input a lexical form t and a datatype D_b identified IRI b, and returns a well formed lexical form s such that $L2V(D_a)(s) = L2V(D_b)(t)$. If D_a does not recognise b or if there exists no such well formed lexical form, then the method must generate an error. Else, the following must be true:
- da.isEqual(da.importLiteral(t,b),t,b)

exportLiteral. Given a datatype D_a identified by IRI a and with specification object da, this method takes as input a lexical form s and another datatype D_b identified by IRI b, and returns a well formed lexical form t such that $L2V(D_a)(s) = L2V(D_b)(t)$. If D_a does not recognise b, if s is not well formed, or if there exists no such well formed lexical form, then the method must generate an error. Else, the following must be true:
- da.isEqual(s,da.exportLiteral(s,b),b)

5.2   Guidelines for RDF/SPARQL Engines

When an RDF processor or a SPARQL query engine encounters a literal with an unknown datatype D_a identified by IRI a, it may attempt to retrieve the JavaScript document located at URL a, using an HTTP GET request with an Accept header field that contains application/javascript.

If it retrieves such a document, it may then call method getDatatype(a) to get a specification object da of datatype D_a. Lexical form validation or value comparisons between literals must then be equivalent to calling methods of da, as specified in the previous section. Finally, SPARQL query engines implement the following addition to SPARQL 1.1 Sect. 15.1 recommendation [6]: given datatypes D_a and D_b identified by IRIs a and b, D_a being a custom datatype with specification object da and recognising D_b, then when a SPARQL query engine compares two literals "s"^^a and "t"^^b, the ordering of these two literals must match the one given by function da.compare(s,t,b).

To avoid security issues, the code may be executed in a sandbox environment without further precaution; it may undergo some static formal verifications; or it may be submitted to a trusted web service for approval. If the RDF processor or the SPARQL query engine decides not to use the datatype specification object, the datatype must be treated as an unrecognised datatype.

6   Implementation and Experiment

This section reports the implementation of these guidelines in Jena and ARQ, and the results of an experiment on a real world DBpedia use case.

6.1   Publication of a Simple Custom Datatype for Length

For illustration purposes, we introduce a custom datatype to represent lengths. This datatype is identified by IRI http://w3id.org/lindt/v1/custom_datatypes# length, abbreviated as `cdt:length`. Its lexical space is the concatenation of the lexical form of an `xsd:double`, an optional space, and a unit that can be either a metric length unit, or an imperial length unit, in abbreviated form or as full words, in singular or plural form. The value space corresponds to the set of lengths, as defined by the International Systems of Quantities, i.e., any quantity with dimension distance. The lexical-to-value mapping maps lexical forms with units in the metric system to their corresponding length according to the International Systems of Quantities, while the forms with an imperial unit are mapped to their equivalent length according to the International yard and pound agreement. For example, all literals below are well typed and share the same value.

```
"1 mile"^^cdt:length            "1.609344 km"^^cdt:length
"5280 ft"^^cdt:length           "1609.344 metre"^^cdt:length
"63360 in."^^cdt:length         "1.609344E+6 mm"^^cdt:length
```

We published a JavaScript implementation of the specification of `cdt:length`, following the guidelines of Sect. 5.1. We further followed best practices for data on the Web, and serve the most appropriate document using content negotiation:

- if the HTTP header option Accept contains `text/html` or `application/ xhtml+xml`, then a HTML document containing a human readable description of datatype Length is served;
- if it contains `text/turtle`, then a short RDF description of datatype Length is served;
- if it contains `application/javascript`, then a JavaScript document that contains the actual specification of custom datatype Length is served. This is equivalent to calling http://w3id.org/lindt/v1/custom_datatypes.js#length.

6.2   Implementation in Jena and ARQ

We implemented the support for on-the-fly custom datatype recognition in both the Jena RDF processor, and the ARQ SPARQL engine.[4] It follows guidelines from Sect. 5.2, but for now it only supports custom datatypes whose value space is disjoint from that of any other datatype.

---

[4] https://github.com/maximelefrancois86/jena.

A new attribute `enableDiscoveryOfCustomDatatypes` has been added to class `JenaParameters`, and package `com.hp.hpl.jena.datatypes` has been slightly modified as follows: If Jena parameter
`JenaParameters.enableDiscoveryOfCustomDatatypes` is set to true, then:

1. When method `getSafeTypeByName` from class `TypeMapper` is called with an unknown datatype IRI `a`, it calls static method `getCustomDatatype` of a new Jena class `CustomDatatype` for a new instance of `RDFDatatype`.
2. This method first makes an HTTP call to `a`, with an `Accept: application/javascript` HTTP header field, and follows redirects. If a JavaScript document is retrieved, its code is evaluated in the default JavaScript script engine (Oracle Nashorn in Java 1.8). An instance of interface `CustomDatatypeFactory` (see Sect. 5.1) is then compiled. Its method `getDatatype` is called and an instance of interface `CustomDatatype` is compiled. This instance is wrapped in an instance of Jena class `CustomDatatype`, and sent back to the `TypeMapper`.
3. Most methods of Jena class `CustomDatatype` wrap calls to the compiled instance of interface `CustomDatatype`.

In ARQ, the main modification concerns class `NodeValue` in package `com.hp.hpl.jena.sparql.expr`, which is used for the SPARQL operators equal and less-or-equal, and for the ORDER BY clause. When comparing node values, if their datatype is an instance of `CustomDatatype`, then calls to the compiled instance of interface `CustomDatatype` are made. A few other minor modifications have also been required:

- a new instance of `ValueSpaceClassification` has been added;
- in package `com.hp.hpl.jena.sparql.expr.nodevalue`, class `NodeValueCustom` has been added, and class `NodeValueVisitor` has been modified.

6.3   Experiment

In this section we present the results of evaluating the proposed protocol, and report on the performances on loading and querying three datasets based on DBpedia but with different approaches for representing lengths. All the details, resources, and instructions that enable the reproduction of this experiment can be found at URL http://w3id.org/lindt.

Datasets. We base our datasets on the DBpedia 2014 English specific mapping-based properties dataset, which contains 819,764 triples with 21 custom datatypes among those DBpedia defines. From this, we extracted the 223,768 triples that describe lengths,[5] i.e., those with the following datatypes:

- `http://dbpedia.org/datatype/millimetre`
- `http://dbpedia.org/datatype/centimetre`

---

[5] This dataset is available at http://wiki.dbpedia.org/Downloads#3.

– http://dbpedia.org/datatype/metre
– http://dbpedia.org/datatype/kilometre

For instance, the following triple represents the length of the Bathyscaphe Trieste submarine.

```
dbpedia:Bathyscaphe_Trieste
  <http://dbpedia.org/ontology/MeanOfTransportation/length>
    "17983.2"^^dbpdt:millimetre .
```

We call this dataset DBPEDIA. From this dataset, we generated the dataset CUSTOM by making all literals use the same datatype Length. For example, the same fact is represented as follows:

```
dbpedia:Bathyscaphe_Trieste
  <http://dbpedia.org/ontology/MeanOfTransportation/length>
    "17983.2 mm"^^cdt:length .
```

Finally, we generated a third dataset, QUDT, which used the QUDT [8] ontology to model the same facts. This is among the alternative choices for representing physical measures that only relies on standard datatypes and encode the relationship between the value and the unit in a graph, using an ontology of quantities.[6] As an example, the length of the Bathyscaphe Trieste submarine may be modelled as follows with the QUDT ontology:

```
dbpedia:Bathyscaphe_Trieste
  <http://dbpedia.org/ontology/MeanOfTransportation/length>
  [ qudt:quantityValue
    [ qudt:numericValue "17983.2"^^xsd:double ;
      qudt:unit qudt-unit:millimetre ] ] .
```

Finally, each dataset has been derived in four datasets, that contain the first 100 %, 50 %, 25 %, and 12.5 % of the original dataset.

Queries. Besides evaluating the loading time of each dataset, we evaluated the querying time of the following simple query: *Return the 100 triples that concern the biggest lengths that are lower than 5 m, order the results according to the descending order of the length.* Depending on the dataset, this query writes differently. Let us just note the conciseness of the query for dataset CUSTOM:

```
PREFIX cdt: <http://w3id.org/lindt/v1/custom_datatypes#>
SELECT ?x ?prop ?length WHERE {
  ?x ?prop ?length .
  FILTER( datatype(?length) = cdt:length
        && ?length < "5m"^^cdt:length )
}
ORDER BY DESC( ?length )
LIMIT 100
```

---

[6] Note that using complex graph structures for representing physical quantities would solve the problem of datatype support, but it displaces the problem to the level of ontologies, as there exists many for describing measurements (in chronological order, UCUM in OWL [2], MUO [13], QUDV [1], OM [14], QUDT [8]).

Table 1. Average and standard deviation of loading or querying time of datasets (in ms).

	DBPEDIA	CUSTOM, cold start	CUSTOM, hot start	QUDT
12.5%	155(±10)	915(±53)	469(±85)	381(±82)
25%	350(±12)	1434(±46)	980(±36)	892(±13)
50%	731(±16)	2498(±57)	2013(±46)	1829(±33)
100%	1640(±57)	4659(±155)	4173(±128)	3796(±68)

(a) Average and standard deviation of loading time of datasets (in ms).

	DBPEDIA	CUSTOM	QUDT
12.5%	416(±38)	195(±12)	267(±66)
25%	833(±16)	391(±13)	532(±26)
50%	2371(±42)	784(±26)	1079(±118)
100%	4158(±256)	1593(±98)	1143(±73)

	CUSTOM	QUDT
12.5%	35(±1)	12(±47)
25%	78(±4)	213(±13)
50%	154(±3)	411(±26)
100%	402(±35)	329(±84)

(b) Average and standard deviation of query-ing time of datasets (in ms).

(c) Average and standard deviation of evalua-tion of query HEIGHT (in ms).

Experiment Protocol and Results. For a given dataset, the experiment consists in repeating 100 times: (i) resetting the `TypeMapper` instance, (ii) loading the dataset, and (iii) querying the dataset and iterating through all the results. Duration of steps ii and iii were measured, and we report below the average duration and the standard deviation of these durations. We led twice the experiment for datasets CUSTOM: once with "cold start", where the custom datatype is discovered and loaded during step (ii), and once with "hot start", where the custom datatype is manually loaded before step (ii). This difference only affects loading times. The experiments were run on a server with a 64 bits Intel Xeon® CPU E5-1603 v3 processor with 4 cores at 2.80 GHz, it has 32 GB DDR3 RAM and is running Ubuntu 14.04 LTS.

Table 1a and b report loading and querying times, respectively. Loading times of datasets CUSTOM are very close to those of datasets QUDT, with on average 468 ms penalty for discovering and loading the custom datatype in the case of cold start. On the other hand, datasets CUSTOM have the best performance regarding querying time.

- Querying time of datasets CUSTOM is between 33 % and 47 % that of datasets DBPEDIA. This can be explained by the fact that the query for dataset DBPEDIA hides actually 4 queries: one for each datatype that represents a length. We believe this difference would grow if dbpedia was using more custom datatypes to represent lengths.
- Querying datasets CUSTOM is also slightly faster than querying datasets QUDT, except for 100 % of triples. Yet, the query for datasets QUDT actually has an anchor IRI to start with, whereas the base of the query for dataset CUSTOM has none.

To evaluate the impact of having an IRI anchor, we derived a second SPARQL query, HEIGHT, by fixing the predicate URI to http://dbpedia.org/ontology/Person/height. Table 1c reports querying times of this query on datasets CUSTOM and QUDT. Fixing this IRI has a greater impact on querying time of datasets CUSTOM than on datasets QUDT, because this query already had anchor IRIs. These results show that custom datatypes have low impact on loading and querying time, while increasing the genericity of this solution.

7   Discussion

Our proposal is only partially addressing the problem of dealing with custom datatypes in a generic way. It also has shortcomings that we discuss here, with possible ways to avoid them. We first discuss the drawbacks of our implementation. We then describe possible extensions of our work to more completely support custom datatypes processing, and emphasise the relationship with OWL 2 custom datatype definitions. We then examine how well our proposal enables D-entailment reasoning.

Drawbacks. Executing code found online presents a potential security threat. However, the `CustomDatatypeFactory` indirection already represents a kind of protection. Actual custom datatype specification objects may be stored as private members of function `getDatatype(iri)`. Then, the next loaded executable code cannot modify its definition. The RDF processor must be sure that the next loaded executable code could not modify previously loaded executable codes, and that it calls the right `getDatatype(iri)` method to get the definition of D: the one that has been retrieved at its IRI. The RDF processor could also execute the code in a sandbox environment, or perhaps apply static analysis to identify harmful code.

Besides, the use of a full-fledged programming language is a bit of an overkill for simple cases such as restricting existing datatypes. We discuss the case of a declarative description of the datatype.

Extending Datatype Description. In several cases, a simple declarative description of a datatype is sufficient. As a matter of fact, OWL 2 already provides means to define datatypes that restrict some of the W3C-standard datatypes using constraining facets `xsd:length`, `xsd:minLength`, `xsd:maxLength`, and `xsd:pattern`. Similarly, we could provide a declarative description of custom datatypes based on other existing ones. Examples of what this vocabulary could represent include:

– A datatype for lengths could be derived from a datatype for measured quantities in all units, as proposed by the Unified Code for Units of Measure [15];
– Describing composite datatypes, formed from the combination of lexical separators and multiple standard literals (e.g., vectors of `xsd:integer`). An RDF processor could then use its support of the derived datatypes to support the composite datatype;

- XSD type definition components and facets [12] could be provided declaratively;
- Direct relationships between datatypes could be used, such as disjointness or subtyping. From an operational point of view, such relations could speed up decisions but would have complicated consequences on reasoning.

Such a vocabulary would favour the interlinking of datatypes.

Reasoning with Custom Datatypes. The expressiveness of RDF with custom datatypes is unlimited. To make this clear, consider a datatype where the lexical space is the set of Turtle documents, and the value space contains the equivalent classes of RDF graphs according to the OWL 2 RDF-based semantics entailment regime (a.k.a OWL 2 Full). The lexical-to-value mapping is the obvious mapping from the documents to their class of equivalent OWL Full ontologies. Equivalence in OWL 2 Full is known to be undecidable [10] and therefore, D-entailment when D contains such a datatype is undecidable. Therefore, reasoning with datatypes is generally undecidable. In fact, even simple datatypes can impact D-entailment reasoning deeply, as witnessed by the following example:

```
rdfs:Resource rdfs:subClassOf xsd:nonNegativeInteger,
    xsd:nonPositiveInteger .
```

These two triples are inconsistent in RDFS recognising {xsd:nonNegative Integer, xsd:nonPositiveInteger} but reasoners implemented in Jena, Corese, and Sesame are unable to detect it. However, as noted in Sect. 3, under certain constraints on datatype value spaces or input graphs, our solution allow correct and complete reasoning for RDFS recognising D. Even in a more general case, reasoning is at least sound.

8   Conclusions

Custom datatypes are currently frown upon because they do not facilitate interoperability. If custom datatypes could be more easily supported generically, it would ease the publication of some domain-specific datasets which otherwise are difficult to represent with standard datatypes. We defined requirements for supporting arbitrary datatypes in reasoning and querying and proposed a concrete solution that requires that the designers of new datatypes follow guidelines that are in line with Linked Data principles. Assuming these guidelines are followed, RDF processors and SPARQL engines can effectively take advantage of custom datatypes on-the-fly, modulo a little overhead in implementing support for our proposal. We empirically demonstrated that performance is not much impacted, compared to a standard implementation. In some cases, relying on custom datatypes leads to better results than restructuring the data to only use standard ones. Arguably, in the use cases we identified, custom datatypes make data publishing more flexible, intuitive, and efficient. Nonetheless, we are conscious of some of the shortcomings of our approach and are investigating other directions for concretely implementing the requirements, based on a linked datatype vocabulary and web services. Finally, we want to investigate more deeply real needs from data publishers in exposing their own datatypes to the open Web.

References

1. Quantities, Units, Dimensions, Values (QUDV). SysML 1.2 Revision Task Force Working draft, Object Management Group, 30 October 2009
2. Bermudez, L.: The unified code for units of measure in OWL. OWL Ontology (2006). https://marinemetadata.org/files/mmi/ontologies/ucum,accessed12/04/2016
3. Cyganiak, R., Wood, D., Lanthaler, M.: RDF 1.1 Concepts and Abstract Syntax, W3C Recommendation, 25 February 2014
4. de Bruijn, J., Heymans, S.: Logical foundations of RDF(S) with datatypes. J. Artif. Intell. Res. **38**, 535–568 (2010)
5. Emmons, I., Collier, S., Garlapati, M., Dean, M.: RDF literal data types in practice. In: Proceedings of the 7th International Workshop on Scalable Semantic Web Knowledge Base Systems, vol. 1 (2011)
6. Harris, S., Seaborne, A.: SPARQL 1.1 Query Language - W3C Working Draft 5. W3C Working Draft, W3C, 5 January 2012
7. Hayes, P., Patel-Schneider, P.F.: RDF 1.1 Semantics, W3C Recommendation 25. W3C Recommendation, W3C, 25 February 2014
8. Hodgson, R., Keller, P.J., Hodges, J., Spivak, J.: QUDT - Quantities, Units. Dimensions and Data Types Ontologies. Technical report, NASA (2014)
9. Hogan, A., Arenas, M., Mallea, A., Polleres, A.: Everything you always wanted to know about blank nodes. J. Web Semant. **27**, 42–69 (2014)
10. Motik, B., Grau, B.C., Horrocks, I., Wu, Z., Fokoue, A., Lutz, C.: OWL 2 Web Ontology Language Profiles (2nd edn.). W3C Recommendation, W3C, 11 December 2012
11. Perry, M., Herring, J.: OGC GeoSPARQL - A Geographic Query Language for RDF Data. Ogc implementation standard, Open Geospatial Consortium, 10 September 2012
12. Peterson, D., Gao, S., Malhotra, A., Sperberg-McQueen, C.M., Thompson, H.S.: W3C XML Schema Definition Language (XSD) 1.1 Part 2: Datatypes, W3C Recommendation, W3C, 5 April 2012
13. Polo, L., Berrueta, D.: MUO - Measurement Units Ontology, Working Draft DD April 2008. Working draft, Fundación CTIC (2008)
14. Rijgersberg, H., van Assem, M., Top, J.L.: Ontology of units of measure and related concepts. Semant. Web J. **4**(1), 3–13 (2013)
15. Shadow, G., McDonald, C.J.: The Unified Code for Units of Measure. Technical report, Regenstrief Institute Inc., 22 October 2013
16. W3C OWLWorking Group: OWL 2 Web Ontology Language Document Overview (Second Edition), W3C Recommendation 11 December 2012. Technical report, W3C (2012)
17. Williams, G.: Extensible SPARQL functions with embedded javascript. In: Proceedings of the Workshop on Scripting for the Semantic Web (2007)

Handling Inconsistencies Due to Class Disjointness in SPARQL Updates

Albin Ahmeti[1,3(✉)], Diego Calvanese[2], Axel Polleres[3], and Vadim Savenkov[3]

[1] Vienna University of Technology, Favoritenstraße 9, 1040 Vienna, Austria
albin.ahmeti@gmail.com
[2] Faculty of Computer Science, Free University of Bozen-Bolzano, Bolzano, Italy
[3] Vienna University of Economics and Business,
Welthandelsplatz 1, 1020 Vienna, Austria

Abstract. The problem of updating ontologies has received increased attention in recent years. In the approaches proposed so far, either the update language is restricted to sets of ground atoms or, where the full SPARQL update language is allowed, the TBox language is restricted so that no inconsistencies can arise. In this paper we discuss directions to overcome these limitations. Starting from a DL-Lite fragment covering RDFS and concept disjointness axioms, we define three semantics for SPARQL instance-level (ABox) update: under cautious semantics, inconsistencies are resolved by rejecting updates potentially introducing conflicts; under brave semantics, instead, conflicts are overridden in favor of new information where possible; finally, the fainthearted semantics is a compromise between the former two approaches, designed to accommodate as much of the new information as possible, as long as consistency with the prior knowledge is not violated. We show how these semantics can be implemented in SPARQL via rewritings of polynomial size and draw first conclusions from their practical evaluation.

1 Introduction

RDF has become one of the most important data formats for interoperability, knowledge representation and querying. SPARQL, the W3C standardized language for managing RDF data [11], has grown to offer great power and flexibility of querying, including support for efficient reasoning, rooted in more than a decade of intensive research in description logics. With respect to updates however, SPARQL is currently far less mature. In particular, the interplay between updates and reasoning remains completely open.

In [1], we discussed semantics of SPARQL updates for RDFS ontologies, for the cases in which the knowledge base ABox is fully materialized or to the contrary, is reduced to its minimal core that cannot be derived using TBox axioms. The present paper continues this study of SPARQL updates focusing on the role of *inconsistency* in supporting SPARQL ABox updates over materialized stores. As a minimalistic ontology language allowing for inconsistencies, we consider RDFS_, an extension of RDFS [12] with *class disjointness axioms* of the form $\{P \text{ disjointWith } Q\}$ from OWL [16].

© Springer International Publishing Switzerland 2016
H. Sack et al. (Eds.): ESWC 2016, LNCS 9678, pp. 387–404, 2016.
DOI: 10.1007/978-3-319-34129-3_24

As a running example, we assume a triple store G with an RDFS$_\neg$ ontology (TBox) T encoding an educational domain, asserting a range restriction plus mutual disjointness of the concepts like professor and student (we use Turtle syntax [2], in which dw abbreviates OWL's disjointWith keyword, and dom and rng respectively stand for the domain and range keywords of RDFS).

$T = \{$:studentOf dom :Student. :studentOf rng :Professor.
    :Professor dw :Student. $\}$

Consider the following SPARQL update [8] request u in the context of the TBox T:

INSERT {?X :studentOf ?Y} **WHERE** {?X :attendsClassOf ?Y}

Consider an ABox with data on student tutors that happen to attend each other's classes: $\mathcal{A}_1 = \{$:jim :attendsClassOf :ann. :ann :attendsClassOf :jim$\}$. Here, u would create two assertions :jim :studentOf :ann and :ann :studentOf :jim. Due to the range and domain constraints in T, these assertions result in clashes both for Jim and for Ann. Note that all inconsistencies are in the new data, and thus we say that u is *intrinsically inconsistent* for the particular ABox $\mathcal{A}_1$. We discuss how such updates can be fixed using SPARQL rewritings.

Now, let $\mathcal{A}_2$ be the ABox {:jim :attendsClassOf :ann. :jim a :Professor}. It is clear that after the update u, the ABox will become inconsistent with respect to T due to the property assertion :jim :studentOf :ann, implying that Jim is both a professor and a student which contradicts the disjointness axiom. In contrast to the previous case, the clash here is between the prior knowledge and the new data. Based on [1] we propose *three update semantics* for this case, and provide *efficient SPARQL rewriting algorithms* for implementing them in the RDFS$_\neg$ setting.

The topic of knowledge base updates is extremely broad. Our aim in this paper is to adapt the basic belief revision operators for efficient implementation of ABox updates expressed in SPARQL 1.1, in the presence of RDFS$_\neg$ TBox axioms. In contrast to our setting, most of existing works on knowledge base evolution consider updates based on sets of ground facts to be inserted or deleted. Restricting negation to class disjointness allowed us to keep the presentation clear. It is not difficult to lift our rewritings to theories with role disjointness, functionality and inequality (owl:differentFrom). We discuss related work in more detail in Sect. 6.

In the remainder of the paper, after some short preliminaries (Sect. 2) we discuss checking for intrinsic inconsistencies in Sect. 3. Then in Sect. 4 we present three semantics for dealing with general inconsistencies in the context of materialized triple stores. Sect. 5 describes our practical evaluation of the semantics. Finally, Sect. 6 puts our work in the context of existing research and provides concluding remarks.

2   Preliminaries

We introduce basic notions about RDF graphs, RDFS$_\neg$ ontologies, and SPARQL queries. We will use RDF and DL notation interchangeably, treating RDF graphs without non-standard RDFS$_\neg$ vocabulary use [19] as a sets of TBox and ABox assertions.

Table 1. *DL-Lite*$_{\text{RDFS}_\neg}$ *assertions vs. RDF(S), where A, A′ denote concept (or, class) names, P, P′ denote role (or, property) names, Γ is the set of IRI constants (excl. the OWL/RDF(S) vocabulary) and x, y ∈ Γ. For RDF(S), we use abbreviations (rsc, sp, dom, rng, a) as introduced in [17].*

TBox	RDFS$_\neg$	TBox	RDFS$_\neg$	TBox	RDFS$_\neg$	ABox	RDFS$_\neg$
1. $A' \sqsubseteq A$	A' sc A.	3. $\exists P \sqsubseteq A$	P dom A.	5. $A' \sqsubseteq \neg A$	A' dw A.	6. $A(x)$	x a A.
2. $P' \sqsubseteq P$	P' sp P.	4. $\exists P^- \sqsubseteq A$	P rng A.			7. $P(x,y)$	x P y.

Definition 1 (RDFS$_\neg$ ABox, TBox, Triple Store). *We call a set $\mathcal{T}$ of inclusion assertions of the forms 1–5 in Table 1 an (RDFS$_\neg$) TBox, a set $\mathcal{A}$ of assertions of the forms 6–7 in Table 1 an (RDF) ABox, and the union $G = \mathcal{T} \cup \mathcal{A}$ an (RDFS$_\neg$) triple store.*

Definition 2 (Interpretation, Satisfaction, Model, Consistency). *An interpretation $\langle \Delta^{\mathcal{I}}, \cdot^{\mathcal{I}} \rangle$ consists of a non-empty set $\Delta^{\mathcal{I}}$ and an interpretation function $\cdot^{\mathcal{I}}$, which maps*

- *each atomic concept A to a subset $A^{\mathcal{I}}$ of $\Delta^{\mathcal{I}}$,*
- *each negation of atomic concept to $(\neg A^{\mathcal{I}}) = \Delta^{\mathcal{I}} \setminus A^{\mathcal{I}}$,*
- *each atomic role P to a binary relation $P^{\mathcal{I}}$ over $\Delta^{\mathcal{I}}$, and*
- *each element of Γ to an element of $\Delta^{\mathcal{I}}$.*

For expressions $\exists P$ and $\exists P^-$, the interpretation function is defined as $(\exists P)^{\mathcal{I}} = \{x \in \Delta^{\mathcal{I}} \mid \exists y.(x,y) \in P^{\mathcal{I}}\}$ and $(\exists P^-)^{\mathcal{I}} = \{y \in \Delta^{\mathcal{I}} \mid \exists x.(x,y) \in P^{\mathcal{I}}\}$, resp. An interpretation $\mathcal{I}$ satisfies an inclusion assertion $E_1 \sqsubseteq E_2$ (of one of the forms 1– 5 in Table 1), if $E_1^{\mathcal{I}} \subseteq E_2^{\mathcal{I}}$. Analogously, $\mathcal{I}$ satisfies ABox assertions of the form $A(x)$, if $x^{\mathcal{I}} \in A^{\mathcal{I}}$, and of the form $P(x,y)$, if $(x^{\mathcal{I}}, y^{\mathcal{I}}) \in P^{\mathcal{I}}$. An interpretation $\mathcal{I}$ is called a model of a triple store G (resp., a TBox $\mathcal{T}$, an ABox $\mathcal{A}$), denoted $\mathcal{I} \models G$ (resp., $\mathcal{I} \models \mathcal{T}$, $\mathcal{I} \models \mathcal{A}$), if $\mathcal{I}$ satisfies all assertions in G (resp., $\mathcal{T}$, $\mathcal{A}$). Finally, G is called consistent, if it does not entail both $C(x)$ and $\neg C(x)$ for any concept C and constant $x \in \Gamma$, where entailment is defined as usual.

As in [1], we treat only ABox updates with WHERE clauses restricted to unions of conjunctive queries (without projection) over DL ontologies:

Definition 3 (BGP, CQ, UCQ, Query Answer). *A conjunctive query (CQ) q, or basic graph pattern (BGP), is a set of atoms of the form 6–7 from Table 1,* *where now $x, y \in \Gamma \cup V$, V a countably infinite set of variables (written as '?'-prefixed alphanumeric strings). A* union of conjunctive queries *(UCQ) Q, or* UNION *pattern, is a set of CQs. We denote with $V(q)$ (or $V(Q)$) the set of variables from V occurring in q (resp., Q). An* answer *(under RDFS¬ Entailment) to a CQ q over a triple store G is a substitution θ of the variables in $V(q)$ with constants in Γ such that every model of G satisfies all facts in qθ. We denote the set of all such answers with $ans_{rdfs}(q, G)$ (or simply $ans(q, G)$). The set of answers to a UCQ Q is $\bigcup_{q \in Q} ans(q, G)$.*

Query answering in the presence of ontologies is done either by rule-based pre-materialization of the ABox or by query rewriting. In the RDFS¬ case, materialization in polynomial time is feasible. Let $mat(G)$ be the triple store obtained from exhaustive application of the inference rules in Fig. 1 on a consistent triple store G. We also define a special notation chase$(q, \mathcal{T})$ to denote the "materialization" (also known as chase) of an ABox resp. a BGP q w.r.t. the TBox $\mathcal{T}$. We call all triples occurring in chase$(q, \mathcal{T})$ but not in q the *effects of q w.r.t. $\mathcal{T}$.*

We now adapt the semantics for SPARQL update operations from [1].

Definition 4 (SPARQL Update Operation, Simple Update of a Triple Store). *Let P_d and P_i be BGPs, and P_w a BGP or UNION pattern. Then an update operation $u(P_d, P_i, P_w)$ has the form*

DELETE P_d **INSERT** P_i **WHERE** P_w

Let $G = \mathcal{T} \cup \mathcal{A}$ be a triple store then the *simple update* of G w.r.t. $u(P_d, P_i, P_w)$ is defined as $G_{u(P_d, P_i, P_w)} = (G \setminus \mathcal{A}_d) \cup \mathcal{A}_i$, where $\mathcal{A}_d = \bigcup_{\theta \in ans(P_w, G)} gr(P_d \theta)$, $\mathcal{A}_i = \bigcup_{\theta \in ans(P_w, G)} gr(P_i \theta)$, and $gr(P)$ denotes the set of ground triples in pattern P.

We call a triple store G (resp. the ABox of G) *materialized* if the equality $G \setminus \mathcal{T} = mat(G) \setminus \mathcal{T}$ holds. In this paper, we will always consider G to be materialized and focus on "materialization preserving" semantics for SPARQL update operations, which we dubbed $\mathbf{Sem}_2^{mat}$ in [1] and which preserves a materialized triple store. We recall the intuition behind $\mathbf{Sem}_2^{mat}$, given an update $u = (P_d, P_i, P_w)$: *(i)* delete the instantiations of P_d *along with all their causes*; *(ii)* insert the instantiations of P_i *plus all their effects*.

The notion of "causes" is made precise as follows. Given an ABox assertion A, $A^{\mathrm{caus}} = \{B \mid A \in \text{chase}(\{B\}, \mathcal{T})\}$. In the definition of A^{caus}, if A is a class

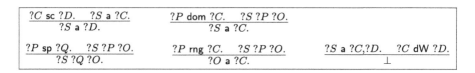

Fig. 1. Minimal RDFS rules from [17]plus class disjointness "clash" rule from OWL2 RL [16].

membership (x a C) where $x \in \Gamma \cup \mathcal{V}$, then B is one of (x a C'), (x P ?Y), (?Y P x) for some fresh variable $?Y$, class C' and role P. If A is a role participation assertion (x R z), B is of the form (x P z), for some role P. For a SPARQL triple (possibly with variables) C we use C^{caus} to denote a BGP computed in the same way as for the ABox assertion A above.

Definition 5 (Sem$_2^{mat}$[1]). *Let $u(P_d, P_i, P_w)$ be an update operation. Then*

$$G_{u(P_d, P_i, P_w)}^{\textbf{Sem}_2^{mat}} = G_{u(P_d^{\text{caus}}, P_i^{\text{eff}}, \{P_w\}\{P_d^{fvars}\})}$$

Here, $P_d^{\text{caus}} = \bigcup_{A \in atoms(P_d)} A^{\text{caus}}$; $P^{\text{eff}} = \text{chase}(P, \mathcal{T})$ and P_d^{fvars} is a pattern that binds variables occurring in P_d^{caus} but not in P_d to the constants from Γ occurring in G.

We refer to [1] for further details, but stress that as such, **Sem**$_2^{mat}$ is not able to detect or deal with inconsistencies arising from extending G with instantiations of P_i. In what follows, we will discuss how this can be remedied.

Remark 1. Note that although the DELETE clause P_d is syntactically a BGP, its semantics is different. Namely, triples occurring in P_d are mutually independent (cf. Definition 4), so that for every $\theta \in ans(P_w, G)$, each atom in $P_d\theta \cap G$ is deleted from G no matter which other atoms of $P_d\theta$ occur in G. Therefore, P_d^{caus} is computed atom-wise, unlike CQ rewriting [4]. Note that $|A^{\text{caus}}| = O(\|\mathcal{T}\|)$ where $\|\mathcal{T}\|$ denotes the vocabulary size of $\mathcal{T}$: in each RDFS$_\neg$ derivation, a class membership assertion can occur at most once for each class in $\mathcal{T}$, and a role membership assertion can occur at most twice for every role in $\mathcal{T}$. Thus, $|P_d^{\text{caus}}| \leq 2|P_d| \cdot \|\mathcal{T}\|$ and $|P_i^{\text{eff}}| \leq |P_i| \cdot \|\mathcal{T}\|$, so both can be computed in poly-time. This underpins the polynomial complexity of our rewritings.

3 Checking Consistency of a SPARQL Update

In the literature on the evolution of DL-Lite knowledge bases [5,7], updates represented by pairs of ABoxes $\mathcal{A}_d, \mathcal{A}_i$ have been studied. However, whereas such update might be viewed to fit straightforwardly to the corresponding $\mathcal{A}_d, \mathcal{A}_i$ in Definition 4, it is typically assumed that $\mathcal{A}_i$ is consistent with the TBox, and thus one only needs to consider how to deal with inconsistencies between the update and the old state of the knowledge base. However, this a priori assumption may be insufficient for SPARQL updates, where concrete values for inserted triples are obtained from variable bindings in the WHERE clause, and depending on the bindings, the update can be either consistent or not. This is demonstrated by the update u from Sect. 1 which, when applied to the ABox $\mathcal{A}_1$, results in an inconsistent set $\mathcal{A}_i$ of insertions. We call this *intrinsic inconsistency* of an update *relative to a triple store* $G = \mathcal{T} \cup \mathcal{A}$.

Definition 6. *Let G be a triple store. The update u is said to be* intrinsically consistent *w.r.t. G if the set of new assertions $\mathcal{A}_i$ from Definition 4 generated by applying u to G, taken in isolation from the ABox of G, does not contradict the TBox of G. Otherwise, the update is said to be* intrinsically inconsistent *w.r.t. G.*

Algorithm 1. constructing a SPARQL ASK query to check intrinsic inconsistency (for the definition of P_i^{eff}, cf. Definition 5)

Input: RDFS$_\neg$ TBox $\mathcal{T}$, SPARQL update $u(P_d, P_i, P_w)$
Output: A SPARQL ASK query returning $True$ if u is intrinsically inconsistent
1 **if** $\bot \in P_i^{\text{eff}}$ **then**
2 $\quad$ **return** ASK $\{\}$ //u *contains clashes in itself, i.e., is inconsistent for any* *triple store*
3 **else**
4 $\quad$ $W := \{$ FILTER$(False)\}$;   //*neutral element w.r.t. union*
5 $\quad$ **foreach** *pair of triple patterns* $(?X$ a $P)$, $(?Y$ a $R)$ *in* P_i^{eff} **do**
6 $\quad\quad$ **if** $P \sqsubseteq \neg R \in \mathcal{T}$ **then**
7 $\quad\quad\quad$ $W := W$ UNION $\{\{P_w \theta_1 [?X \mapsto ?Z]\} . \{P_w \theta_2 [?Y \mapsto ?Z]\}\}$ for a fresh $?Z$
8 $\quad$ **return** ASK WHERE $\{W\}$

Intrinsic inconsistency of the update differs crucially from the inconsistency w.r.t. the old state of the knowledge base, illustrated by the ABox $\mathcal{A}_2$ from Sect. 1. This latter case can be addressed by adopting an update policy that prefers newer assertions in case of conflicts, as studied in the context of DL-Lite KB evolutions [5], which we will discuss in Sect. 4 below. Intrinsic inconsistencies however are harder to deal with, since there is no cue which assertion should be discarded in order to avoid the inconsistency. Our proposal here is thus to discard *all* mutually inconsistent pairs of insertions.

We first present an algorithm for checking intrinsic inconsistency by means of SPARQL ASK queries and then a safe rewriting algorithm. This rewriting is based on an observation that clashing triples can be introduced by a combination of two bindings of variables in the WHERE clause, as the example in the Sect. 1 (the ABox $\mathcal{A}_1$) illustrates. To handle such cases, two copies of the WHERE clause P_w are created by the rewriting in Algorithms 1 and 2, for each pair of disjoint concepts according to the TBox of the triple store. These algorithms use notation described in Remark 2 below.

Remark 2. Our rewriting algorithms rely on producing fresh copies of the WHERE clause. Assume θ, θ_1, θ_2, ... to be substitutions replacing each variable in a given formula with a distinct fresh one. For a substitution σ, we also define $\theta[\sigma]$ resp. $\theta_i[\sigma]$ to be an extension of σ, renaming each variable at positions not affected by σ with a distinct fresh one. For instance, let F be a triple $(?Z$: studentOf $?Y)$. Now, $F\theta$ makes a variable disjoint copy of F: $?Z_1$: studentOf $?Y_1$ for fresh $?Z_1, ?Y_1$. $F[?Z \mapsto ?X]$ is just a substitution of $?Z$ by $?X$ in F. Finally, $F\theta[?Z \mapsto ?X]$ results in $?X$: studentOf $?Y_2$ for fresh $?Y_2$. We assume that all occurrences of $F\theta[\sigma]$ stand for syntactically the same query, but that $F\theta[\sigma_1]$ and $F\theta[\sigma_2]$, for distinct σ_1 and σ_2, can only have variables in $range(\sigma_1) \cap range(\sigma_2)$ in common. That is, the choice of fresh variables is defined by the parameterizing substitution σ. ∎

Algorithm 2. Safe rewriting safe(u)

Input: RDFS$_\neg$ TBox $\mathcal{T}$, SPARQL update $u(P_d, P_i, P_w)$
Output: SPARQL update safe(u)

1 **if** $\bot \in P_i^{\text{eff}}$ **then**
2 $\quad$ **return** $u(P_d, P_i, \text{FILTER}(\textit{False}))$
3 $W := \{\text{FILTER}(\textit{False})\};$ $\quad$ //neutral element w.r.t. union
4 **foreach** pair of triple patterns $(?X\ \mathsf{a}\ P)$, $(?Y\ \mathsf{a}\ R)$ in P_i^{eff} **do**
5 $\quad$ **if** $P \sqsubseteq \neg R \in \mathcal{T}$ **then**
6 $\quad\quad$ //cf. Remark 2 for notation $\theta[\dots]$
7 $\quad\quad$ $W := W$ UNION $\{P_w\theta_1[?X \mapsto ?Y]\}$ UNION $\{P_w\theta_2[?Y \mapsto ?X]\}$
8 **return** $u(P_d, P_i, P_w$ MINUS $\{W\})$

Using this notation, the possibility of unifying two variables $?X$ and $?Y$ in P_w on a given triple store can be tested with the query $\{P_w\theta_1[?X \mapsto ?Z]\}\{P_w\theta_2[?Y \mapsto ?Z]\}$ where θ_1 and θ_2 are variable renamings as in Remark 2 and $?Z$ is a fresh variable.

In order to check the intrinsic consistency of an update, this condition should be evaluated for every pair of variables of P_w, the unification of which leads to a clash. A SPARQL ASK query based on this idea is produced by Algorithm 1. Note that it suffices to check only triples of the form $\{?X\ \mathsf{a}\ ?C\}$ at line 5 of Algorithm 1, since disjointness conditions can only be formulated for concepts, according to the syntax in Table 1. Furthermore, since we are taking the facts in P_i^{eff} extended by all facts implied by $\mathcal{T}$, at line 6 of Algorithm 1 it suffices to check the disjointness conditions explicitly mentioned in $\mathcal{T}$ and not all those which are implied by $\mathcal{T}$. Note also that the DELETE clause P_d plays no role in this case, since we only consider clashes within inserted facts.

Example 1. Consider the update u from Sect. 1, in which the INSERT clause P_i can create clashing triples. To identify potential clashes, Sect. 1 first applies the inference rule for the range constraint, and computes $P_i^{\text{eff}} = \{?X\ \mathsf{a}\ :\mathsf{Student}\ .\ ?Y\ \mathsf{a}\ :\mathsf{Professor}\}$. Now both variables $?X, ?Y$ occur in the triples of type (6) from Sect. 1 with clashing concept names. The following ASK query is produced by Sect. 1.

$\quad$ **ASK WHERE {** ?X :attendsClassOf ?Y . ?Y :attendsClassOf ?X1 **}**
(In this and subsequent examples we omit the trivial FILTER(*False*) union branch used in rewritings to initialize variables with disjunctive conditions, such as W in Algorithm 1) ∎

Suppose that an insert is not intrinsically consistent for a given triple store. One solution would be to discard it completely, should the above ASK query return *True*. Another option which we consider here is to only discard those variable bindings from the WHERE clause, which make the INSERT clause P_i inconsistent. This is the task of the *safe rewriting* safe($\cdot$) in Algorithm 2, removing all variable bindings that participate in a clash between different triples of P_i. Let P_w be a WHERE clause, in which the variables $?X$ and $?Y$ should not be unified to avoid clashes. With θ_1, θ_2 being "fresh" variable renamings as in Remark 2, Algorithm 2 uses the union of $P_w\theta_1[?X \mapsto ?Y]$ and $P_w\theta_2[?Y \mapsto ?X]$ to eliminate unsafe bindings that send $?X$ and $?Y$ to the same value.

Example 2. Algorithm 2 extends the WHERE clause of the update u from Sect. 1 as follows:

INSERT{?X :studentOf ?Y**} WHERE{**?X :attendsClassOf ?Y
 MINUS{{?X1 :attendsClassOf ?X**} UNION {**?Y :attendsClassOf ?Y2**}}}**

Note that the safe rewriting can make the update void. For instance, safe(u) has no effect on the ABox $\mathcal{A}_1$ from Sect. 1, since there is no cue, which of :jim :attendsClassOf :ann, :ann :attendsClassOf :jim needs to be dismissed to avoid the clash. However, if we extend this ABox with assertions both satisfying the WHERE clause of u and not causing undesirable variable unifications, safe(u) would make insertions based on such bindings. For instance, adding the fact :bob :attendsClassOf :alice to $\mathcal{A}_1$ would assert :bob :studentOf :alice as a result of safe(u). ∎

A rationale for using MINUS rather than FILTER NOT EXISTS in Algorithm 2 (and also in a rewriting in forthcoming Sect. 4) can be illustrated by an update in which variables in the INSERT and DELETE clauses are bound in different branches of a UNION:

DELETE {?V a :Professor**} INSERT {**?X :studentOf ?Y**}**
WHERE {{?X :attendsClassOf ?Y**} UNION {**?V :attendsClassOf ?W**}}**

A safe rewriting of this update (abbreviating :attendsClassOf as :aCo) is

  DELETE {?V a :Professor**} INSERT {**?X :studentOf ?Y**}**
 WHERE { {{?X :aCo ?Y**} UNION {**?V :aCo ?W**}}**
       MINUS{ {{?X1 :aCo ?X**} UNION {**?V1 :aCo ?W1**}}**
         UNION {{?Y :aCo ?Y2**} UNION {**?V2 :aCo ?W2**}} } }**

It can be verified that with FILTER NOT EXISTS in place of MINUS this update makes no insertions on all triple stores: the branches {?V1 :aCo ?W1} and {?V2 :aCo ?W2} are satisfied whenever {?X :aCo ?Y} is, making FILTER NOT EXISTS evaluate to *False* whenever {?X :aCo ?Y} holds.

We conclude this section by formalizing the intuition of update safety. For a triple store G and an update $u = (P_d, P_i, P_w)$, let $[\![P_w]\!]_G^u$ denote the set of variable bindings computed by the query "SELECT?$X_1, \ldots, ?X_k$ WHERE P_w" over G, where $?X_1, \ldots, ?X_k$ are the variables occurring in P_i or in P_d.

Theorem 1. *Let $\mathcal{T}$ be a TBox, let u be a SPARQL update (P_i, P_d, P_w), and let query q_u and update safe(u) = (P_d, P_i, P'_w) result from applying Algorithm 1 resp. Algorithm 2 to u and $\mathcal{T}$. Then, the following properties hold for an arbitrary RDFS$_\neg$ triple store $G = \mathcal{T} \cup \mathcal{A}$:*

(1) $q_u(G) = True$ iff $\exists \mu, \mu' \in [\![P_w]\!]_G^u$ s.t. $\mu(P_i) \wedge \mu'(P_i) \wedge \mathcal{T} \models \bot$;
(2) $[\![P_w]\!]_G^u \setminus [\![P'_w]\!]_G^u = \{\mu \in [\![P_w]\!]_G^u \mid \exists \mu' \in [\![P_w]\!]_G^u$ s.t. $\mu(P_i) \wedge \mu'(P_i) \wedge \mathcal{T} \models \bot\}$.

4 Materialization Preserving Update Semantics

In this section we discuss resolution of inconsistencies between triples already in the triple store and newly inserted triples. Our baseline requirement for each update semantics is formulated as the following property.

Definition 7 (Consistency-preserving). *Let* G *be a triple store and* $u(P_d, P_i, P_w)$ *an update. A materialization preserving update semantics Sem is called* consistency preserving *in RDFS$_\neg$ if the evaluation of update* u, *i.e.,* $G^{Sem}_{u(P_d,P_i,P_w)}$, *results in a consistent triple store.*

Our consistency preserving semantics are respectively called *brave, cautious* and *fainthearted.* The brave semantics always gives priority to newly inserted triples by discarding all pre-existing information that contradicts the update. The cautious semantics is exactly the opposite, discarding inserts that are inconsistent with facts already present in the triple store; i.e., the cautious semantics never deletes facts unless explicitly required by the DELETE clause of the SPARQL update. Finally, the fainthearted semantics executes the update partially, only performing insertions for those variable bindings which do not contradict existing knowledge (again, taking into account deletions).

All semantics rely upon incremental update semantics $\mathbf{Sem}_2^{mat}$, introduced in Sect. 2, which we aim to extend to take into account class disjointness. Note that for the present section we assume updates to be intrinsically consistent, which can be checked or enforced beforehand in a preprocessing step by the safe rewriting discussed in Sect. 3. In this section, we lift our definition of update operation to include also updates (P_d, P_i, P_w) with P_w produced by the safe rewriting Algorithm 2 from some update satisfying Definition 4. What remains to be defined is the handling of clashes between newly inserted triples and triples already present in the triple store.

The intuitions of our semantics for a SPARQL update $u(P_d, P_i, P_w)$ in the context of an RDFS$_\neg$ TBox are as follows:

- *brave semantics* $\mathbf{Sem}_{brave}^{mat}$: *(i)* delete all instantiations of P_d and their causes, *plus all the non-deleted triples in* G *clashing with instantiations of triples in* P_i *to be inserted*, again also including the causes of these triples; *(ii)* insert the instantiations of P_i plus all their effects.
- *cautious semantics* $\mathbf{Sem}_{caut}^{mat}$: *(i)* delete all instantiations of P_d and their causes; *(ii)* insert all instantiations of P_i plus all their effects, *unless they clash with some non-deleted triples in* G: in this latter case, do not perform the update.
- *fainthearted semantics* $\mathbf{Sem}_{faint}^{mat}$: *(i)* delete all instantiations of P_d and their causes; *(ii)* insert those instantiations of P_i (plus all their effects) which *do not clash with non-deleted triples in* G.

Remark 3. Note that $\mathbf{Sem}_2^{mat}$ is not able to cope with so called "dangling" effects – that is, triples inserted at some point for the sake of materialization, whose causes have been subsequently deleted. As pointed out in [1], one way to

---

Algorithm 3. *Brave semantics* $\mathbf{Sem}_{brave}^{mat}$

Input: Materialized triple store $G = \mathcal{T} \cup \mathcal{A}$, SPARQL update $u(P_d, P_i, P_w)$

Output: $G_{u(P_d, P_i, P_w)}^{\mathbf{Sem}_{brave}^{mat}}$

1   $P_d' := P_d^{caus}$;

2   **foreach** *triple pattern* $(?X \; \mathsf{a} \; C)$ *in* P_i^{eff} **do**

3 | **foreach** C' *s.t.* $C \sqsubseteq \neg C' \in \mathcal{T}$ *or* $C' \sqsubseteq \neg C \in \mathcal{T}$ **do**

4 | | **if** $(?X \; \mathsf{a} \; C') \notin P_d'$ **then**

5 | | | $P_d' := P_d' \, . \, \{?X \; \mathsf{a} \; C'\}^{caus}$

6   **return** $G_{u(P_d', P_i^{\mathrm{eff}}, \{P_w\} P_d^{fvars})}$

---

deal with this issue is to combine $\mathbf{Sem}_2^{mat}$ with marking of explicitly inserted triples. This approach was implemented as a semantics $\mathbf{Sem}_{1b}^{mat}$ in [1], splitting the ABox $\mathcal{A}$ into the explicit part $\mathcal{A}_{ex}$ and the implicit part $\mathcal{A}_{im} = \mathcal{A} \setminus \mathcal{A}_{ex}$. $\mathcal{A}_{ex}$ can be maintained, e.g., in a separate RDF graph using a straightforward update rewriting. Now, deleting P_d would not only retract P_d^{caus} from $\mathcal{A}$, but also the triples in chase$(P_d^{caus}, \mathcal{T}) \setminus$ chase$(\mathcal{A}_{ex} \setminus P_d^{caus}, \mathcal{T})$. That is, the effects of P_d^{caus} are removed unless they can be derived from facts remaining in $\mathcal{A}$ after enforcing the deletion P_d. Such an aggressive removal of dangling triples can lead to counterintuitive behavior (cf. Example 9 in [1]), and requires maintaining the explicit ABox $\mathcal{A}_{ex}$, which is why we opted to preserve dangling effects in our rewritings.

We will now describe implementations of the three semantics above via SPARQL rewritings, which can be shown to be materialization preserving and consistency preserving.

4.1   Brave Semantics

The rewriting in Algorithm 3 implements the brave update semantics $\mathbf{Sem}_{brave}^{mat}$; it can be viewed as combining the idea of *FastEvol*[5] with $\mathbf{Sem}_2^{mat}$ to handle inconsistencies by giving priority to triples that ought to be inserted, and deleting all those triples from the store that clash with the new ones.

Example 3. Example 2 in Sect. 3 provided a safe rewriting safe(u) of the update u from Sect. 1. According to Algorithm 3, this safe update is rewritten to:

```
DELETE {?X a :Professor . ?X1 :studentOf ?X .
         ?Y a :Student . ?Y :studentOf ?Y1}
INSERT {?X :studentOf ?Y . ?X a :Student . ?Y a :Professor}
WHERE {{?X :attendsClassOf ?Y
MINUS{{?X2 :attendsClassOf ?X} UNION {?Y :attendsClassOf ?Y2}}}
OPTIONAL {?X1 :studentOf ?X} OPTIONAL {?Y :studentOf ?Y1} }
```

The DELETE clause removes potential clashes for the inserted triples. Note that also property assertions implying clashes need to be deleted, which introduces fresh variables $?X1$ and $?Y1$. These variables have to be bound in the WHERE clause, and therefore P_d^{fvars} adds two optional clauses to the WHERE clause, which is a computationally reasonable implementation of the concept P^{fvars} from Definition 5. ∎

The DELETE clause P_d' of the rewritten update is initialized in Algorithm 3 with the set P_d of triples from the input update. Rewriting ensures that also all "causes" of deleted facts are removed from the store, since otherwise the materialization will re-insert deleted triples. To this end, line 1 of Algorithm 3 adds to P_d' all facts from which P_d can be derived. Then, for each triple implied by P_i (that is, for each triple in P_i^{eff}) the algorithm computes the patterns of clashing triples and adds them to the DELETE clause P_d', along with their causes. Note that it suffices to only consider disjointness assertions that are syntactically contained in $\mathcal{T}$ (and not those implied by $\mathcal{T}$), since we assume that the store G is materialized. Finally, the WHERE clause of the rewritten update is extended to satisfy the syntactic restriction that all variables in P_d' must be bound: bindings of "fresh" variables introduced to P_d' due to the domain or range constraints in $\mathcal{T}$ are provided by the part P_d^{fvars}, cf. Definition 5 and Example 3. The rewritten update is evaluated over the triple store, computing its new materialized and consistent state.

In the RDFS$_\neg$ ontology language and under the restriction that only ABox updates are allowed, the brave semantics is a belief revision operator [10,20], performing a minimal change of the RDF graph (which due to materialization can be seen both as a deductive closure of the formula representing the ABox as well as the minimal model of this formula). There is a unique way of resolving inconsistencies since the only deduction rule with more than one ABox assertion in the premise, is the clash due to class disjointness (Fig. 1): assuming intrinsic consistency, the choice of which class membership assertion to remove in order to avoid clash is univocal (new knowledge is always preferred).

Theorem 2. *Algorithm 3, given a SPARQL update u and a consistent materialized triple store $G = \mathcal{T} \cup \mathcal{A}$, computes a new consistent and materialized state w.r.t. brave semantics. The rewriting in lines 1–6 takes time polynomial in the size of u and $\mathcal{T}$.*

4.2 Cautious Semantics

Unlike $\mathbf{Sem}_{brave}^{mat}$, its *cautious* version $\mathbf{Sem}_{caut}^{mat}$ always gives priority to triples that are already present in the triple store, and dismisses any inserts that are inconsistent with it. We implement this semantics as follows: *(i)* the DELETE command does not generate inconsistencies and thus is assumed to be always possible; *(ii)* the update is actually executed only if the triples introduced by the INSERT clause do not clash with state of the triple graph *after all deletions have been applied.*

Cautious semantics thus treats insertions and deletions asymmetrically: the former depend on the latter but not the other way round. The rationale is that deletions never cause inconsistencies and can remove clashes between the old and the new data.

Algorithm 4. *Cautious semantics* $\mathbf{Sem}_{caut}^{mat}$

Input: Materialized triple store $G = \mathcal{T} \cup \mathcal{A}$, SPARQL update $u(P_d, P_i, P_w)$

Output: $G_{u(P_d,P_i,P_w)}^{\mathbf{Sem}_{caut}^{mat}}$

1   $W := \{\, \mathsf{FILTER}(\textit{False}) \,\}$ // *neutral element w.r.t. union*

2   **foreach** $(?X \mathsf{\ a\ } C) \in P_i^{\text{eff}}$ **do**

3 | **foreach** C' *s.t.* $C \sqsubseteq \neg C' \in \mathcal{T}$ *or* $C' \sqsubseteq \neg C \in \mathcal{T}$ **do**

4 | | $\Theta_{C'}^- := \{\, \mathsf{FILTER}(\textit{False}) \,\}$

5 | | **foreach** $(?Y \mathsf{\ a\ } C') \in P_d^{\text{caus}}$ **do**

6 | | | $\Theta_{C'}^- := \Theta_{C'}^- \mathsf{\ UNION\ } \{P_w \theta[?Y \mapsto ?X]\}$

7 | | $W := W \mathsf{\ UNION\ } \{\{?X \mathsf{\ a\ } C'\} \mathsf{\ MINUS\ } \{\Theta_{C'}^-\}\}$

8   $Q := \mathsf{ASK\ WHERE}\ \{\{P_w\}.\{W\}\}$;

9   **if** $Q(G)$ **then**

10 | **return** G

11   **else**

12 | **return** $G_{u(P_d,P_i,P_w)}^{\mathbf{Sem}_{brave}^{mat}}$

As in the case of brave semantics, cautious semantics is implemented using rewriting, presented in Algorithm 4. First, the algorithm issues an ASK query to check that no clashes will be generated by the INSERT clause, provided that the DELETE part of the update is executed. If no clashes are expected, in which case the ASK query returns *False*, the brave update from the previous section is applied.

For a safe update $u = (P_d, P_i, P_w)$, the ASK query is generated as follows. For each triple pattern $\{?X \mathsf{\ a\ } C\}$ among the effects of P_i, at line 3 Algorithm 4 enumerates all concepts C' that are explicitly mentioned as disjoint with C in $\mathcal{T}$. As in the case of brave semantics, this syntactic check is sufficient due to the assumption that the update is applied to a materialized store; by the same reason also no property assertions need to be taken into account.

For each concept C' disjoint with C, we need to check that a triple matching the pattern $\{?X \mathsf{\ a\ } C'\}$ is in the store G and will not be deleted by u. Deletion happens if there is a pattern $\{?Y \mathsf{\ a\ } C'\} \in P_d^{\text{caus}}$ such that the variable $?Y$ can be bound to the same value as $?X$ in the WHERE clause P_w. Line 6 of Algorithm 4 produces such a check, using a copy of P_w, in which the variable $?Y$ is replaced by $?X$ and all other variables are replaced with distinct fresh ones. Since there can be several such triple patterns in P_d^{caus}, testing for clash elimination via the DELETE clause requires a disjunctive graph pattern $\Theta_{C'}^-$ constructed at line 6 and combined with $\{?X \mathsf{\ a\ } C'\}$ using MINUS at line 7.

Finally, the resulting pattern is appended to the list W of clash checks using UNION. As a result, $\{P_w\}.\{W\}$ queries for triples that are not deleted by u and clash with an instantiation of some class membership assertion $\{?X \mathsf{\ a\ } C\} \in P_i^{\text{eff}}$.

Theorem 3. *Algorithm 4, given a SPARQL update u and a consistent materialized triple store $G = \mathcal{T} \cup \mathcal{A}$, computes a new consistent and materialized state*

Algorithm 5. *Fainthearted semantics* $\mathbf{Sem}^{mat}_{faint}$

Input: Materialized triple store $G = \mathcal{T} \cup \mathcal{A}$, SPARQL update $u(P_d, P_i, P_w)$

Output: $G^{\mathbf{Sem}^{mat}_{faint}}_{u(P_d, P_i, P_w)}$

1 $W := P_w$

2 **foreach** *triple pattern* $(x \; a \; C)$ *in* P_i^{eff} **do**

3   **foreach** C' *s.t.* $C \sqsubseteq \neg C' \in \mathcal{T}$ *or* $C' \sqsubseteq \neg C \in \mathcal{T}$ **do**

4        $\Theta^-_{C'} := \{\,\mathsf{FILTER}(False)\,\}$;

5   **foreach** $(z \; a \; C') \in P_d^{\text{caus}}$ **do**

6           $\Theta^-_{C'} := \Theta^-_{C'} \; \mathsf{UNION} \; \{P_w \theta[z \mapsto x]\}$;

7        $W := \{W\} \; \mathsf{MINUS} \; \{x \; a \; C' \; \mathsf{MINUS} \; \{\Theta^-_{C'}\}\}$;

8 $W := \{W\} \; \mathsf{UNION} \; \{P_w \theta_1 \; . \; P_d^{fvars} \theta_1\}$;

9 **return** $G_{u(P_d^{\text{caus}} \theta_1, \; P_i^{\text{eff}}, \; W)}$

---

w.r.t. cautious semantics. *The rewriting in lines 1–8 takes time polynomial in the size of u and $\mathcal{T}$.*

Example 4. Algorithm 4 rewrites the safe update safe(u) from Example 2 as follows:

```
ASK WHERE{{?X :attendsClassOf ?Y
 MINUS{{?X1 :attendsClassOf ?X} UNION {?Y :attendsClassOf ?Y2}}}
 .{{?Y a :Student} UNION {?X a :Professor}}}
```

Now, consider an update u' having both INSERT and DELETE clauses:

```
DELETE {?Y a :Professor} INSERT{?X a :Student}
WHERE {?X :attendsClassOf ?Y}
```

The update u' inserts a single class membership fact and thus is always intrinsically consistent. The ASK query in Algorithm 4 takes the DELETE clause of u' into account:

```
ASK WHERE {{?X :attendsClassOf ?Y}
.{{?X a :Professor} MINUS {?Z :attendsClassOf ?X }}}
```
■

4.3 Fainthearted Semantics

Our third, *fainthearted* semantics is meant to take an intermediate position between the cautious semantics and the brave one. A shortcoming of the cautious semantics is that massive update can be retracted because of only a few clashing triples. Not to discard an update completely in such a case, the user can decide either to override the existing knowledge — that is, opt for the brave semantics — or to apply insertions only for those variable bindings which are not clashing with the existing state, which is what the fainthearted semantics does.

Our realization of the idea of accommodating non-clashing inserts is based on *decoupling the insert and the delete* part of an update: whereas the delete is executed for *all* variable bindings satisfying the WHERE clause, one dismisses the inserts for variable bindings that yield clashes with the state of the store *after the delete*. That is, we deviate from the notion of update as an atomic operation in a different way than in the safe rewriting where *both* deletions and insertions are dismissed for variable bindings leading to clashes. Our motivation for such a design decision is explained next.

Assume that for each variable binding μ returned by the WHERE pattern, we want to either insert $\mathrm{gr}(P_i\mu)$ along with deleting $\mathrm{gr}(P_d\mu)$, or dismiss μ altogether. As an example, consider the update u' from Example 4 and the ABox {:jim :attendsClassOf :ann. :jim a :Professor. :bob :attendsClassOf :jim}. With the variable binding $\mu_1 = [?X \mapsto$:jim, $?Y \mapsto$:ann] we insert :jim a :Student knowing that the clashing fact :jim a :Professor will be deleted by the binding $\mu_2 = [?X \mapsto$:bob, $?Y \mapsto$:jim]. However, if the update is atomic, this anticipated deletion will only happen if $\mathrm{gr}(P_i\mu_2)$ does not introduce clashes. Assume this is the case (i.e. also {:bob a :Professor} is in the ABox): we have to look one more step ahead and check if this triple will be deleted by some variable binding μ_3, and so on. This behaviour could be realized with SPARQL path expressions, which would however stipulate severe syntactic restrictions on the WHERE clause P_w of the original update.

As mentioned above, our interpretation of fainthearted semantics assumes independence between the INSERT and DELETE parts of the update. To implement this, we rely on SPARQL's flexible handling of variable bindings. Namely, we rename the variables in the DELETE clause apart from the rest of the update, and put this renamed apart copy of the WHERE clause in a new UNION branch. The original WHERE clause is then rewritten (using MINUS operator, similarly to the case of cautious semantics) to ensure that insertions are only done for variable bindings where clashes are removed by the DELETE clause with some variable binding. The implementation can be found in Algorithm 5.

Example 5. The update u' from Example 4 is rewritten as follows by Algorithm 5:

DELETE {?Y1 a :Professor } **INSERT** {?X a :Student}
WHERE {{?X2 :attendsClassOf ?Y1} **UNION** {?X :attendsClassOf ?Y.
      {MINUS {?X a :Professor **MINUS** {?X3 :attendsClassOf ?X}}}}}
The first union branch binds the variables in the DELETE clause (both using fresh variables). The second branch binds the variable ?X in the INSERT clause, using MINUS to remove variable bindings for which a non-deleted clash exists. The test that a clash will not be deleted is expressed using the inner MINUS operator. ∎

We conclude with a claim of correctness and polynomial complexity of rewriting, similar to those made for the brave and cautious semantics.

Theorem 4. *Algorithm 5, given a SPARQL update u and a materialized triple store $G = \mathcal{T} \cup \mathcal{A}$ w.r.t. fainthearted semantics, computes a new consistent and materialized state. The rewriting in lines 1–9 takes time polynomial in the size of u and $\mathcal{T}$.*

5   Experimental Evaluation

For each of the three semantics discussed in the previous section, we provided a preliminary implementation using the Jena API (http://jena.apache.org) and evaluated them against Jena TDB triple store which implements the latest SPARQL 1.1 specification. As before, for computing the initial materialization of a triple store $mat(G)$ we rely on-board, forward-chaining materialization in Jena TDB using the minimal RDFS rules as in Fig. 1.

For our experiments, we used the data generated by the EUGen generator [15] of for the size range of 5 to 50 Universities. We opted for using this generator as it extends the LUBM ontology [9] with chains of subclasses, making the rewritings more challenging. In our case we have used the default of $i = 20$ subclasses for each LUBM concept (e.g., Subj$i$Students) and made such subclasses pairwise disjoint. Moreover, we have added more disjointness axioms where appropriate, e.g., :AssociateProfessor dw :FullProfessor. All these TBox axioms are merged with our previous reduced RDFS version of LUBM used in our previous work [1]. To compare the experimental results with the previous work, for our experiments we adapted the seven updates from [1]. Our prototype, as well as files containing the data, ontology, and the updates used for experiments, are made available on a dedicated Web page[1].

The results summarized in Table 2 show that the LUBM 50 dataset (507 MB uncompressed, 8.7 M triples after materialization) can be handled in seconds on a quad-core Intel i7 3.20 GHz machine with 16 GB RAM. For each of the three semantics, we have compared the time elapsed for rewriting and for the evaluation of the resulting update. The last line in Table 2 is the evaluation time for the original, non-rewritten update. One can notice that brave semantics $\mathbf{Sem}_{brave}^{mat}$ is often the most expensive one, since it performs most modifications. When the number of inconsistent inserts is low though, the situation is different, and the brave semantics slightly outperforms the fainthearted semantics $\mathbf{Sem}_{faint}^{mat}$ (Update #6 and #7), due to the more complex checks in the WHERE clause produced by Algorithm 5. For the cautious semantics $\mathbf{Sem}_{caut}^{mat}$, the numbers in the table are construction and evaluation time of the ASK query checking for the feasibility of update (cf. Algorithm 4). In case this ASK query returns $False$, the runtime of brave semantics should be added in order to obtain the total runtime of the update. Update #4 demonstrates that $\mathbf{Sem}_{caut}^{mat}$ can perform significantly worse than $\mathbf{Sem}_{faint}^{mat}$ when the number of instantiations in the original WHERE clause is high. This is because the ASK query in $\mathbf{Sem}_{caut}^{mat}$ looks for instantiations of the WHERE clause which can lead to clashes with the existing tuples (using a conjunctive condition), whereas $\mathbf{Sem}_{faint}^{mat}$ reduces the set of solutions of the original WHERE clause using MINUS, which is apparently more efficient in the Apache TDB.

---

[1] http://dbai.tuwien.ac.at/user/ahmeti/sparqlupdate-inconsistency-resolver/.

Table 2. Evaluation results in seconds for LUBM 50

Update #	1	2	3	4	5	6	7
$\mathbf{Sem}_{brave}^{mat}$	12,4	14,8	0,1	22,1	46,0	15,3	13,6
$\mathbf{Sem}_{caut}^{mat}$	0,3	0,2	0,2	44,0	0,2	3,9	2,3
$\mathbf{Sem}_{faint}^{mat}$	2,2	2,8	0,01	17,4	3,3	16,7	15,3
Original	0,2	0,2	0,2	10,2	0,2	6,6	5,4

6   Related Work and Conclusions

In this paper we have taken a step further from our previous work, in combining SPARQL Update and RDFS entailment by adding concept disjoints as a first step towards dealing with inconsistencies in the context of SPARQL Updates. We distinguish the case of intrinsic inconsistency, localized within instantiations of the INSERT clause of a SPARQL update, and the usual case when the new information is inconsistent with the old knowledge. In the former case, our solution was to discard all solutions of the WHERE query that participate in an inconsistency. For the latter case, we discussed several reconciliation strategies, well suited for efficient implementation in SPARQL. Our preliminary implementation shows the feasibility of all proposed approaches on top of an off-the-shelf triple store supporting SPARQL and SPARQL update (Apache TDB).

The problem of knowledge based update and belief revision has been extensively studied in the literature, although not in the context of SPARQL updates where facts to be deleted or inserted come from a query. As argued in Sect. 4.1, brave semantics implements the most established approach of adapting the new information fully via a minimal change, which is feasible in the setting of fixed RDFS¬ TBoxes. Also semantics deliberating between accepting and discarding change are known (see [10] for a survey). In [18] an approach involving user interaction to decide whether to accept or reject an individual axiom is considered, with some part of the update being computed automatically in order to ensure its consistency. We do not consider interactive procedures here (although they clearly make sense in the case of more complex TBoxes or for TBox updates). Instead, we rely on the resolution strategies which are simple for the user to understand and can be efficiently encoded in SPARQL. In a practical KB editing system, one should probably combine the two approaches, e.g. for resolving the intrinsic inconsistency. Likewise, the approaches [3,7,13] consider grounded updates only, whereas our focus is on implementation of updates in SPARQL. The approach in [7] captures RDFS and several additional types of constraints and is close in spirit to our brave semantics.

Intrinsic consistency of an update is a common assumption in knowledge base update (e.g. [5–7,14]), which can be easily violated in the case of SPARQL updates. It is worth noting that our resolution strategy for intrinsic inconsistency called safe rewriting can be combined with all three update semantics using just the basic SPARQL operators.

Much interesting work remains to be done in order to optimize rewritten updates. Moreover, we plan to further extend our work towards increasing coverage of more expressive logics and OWL profiles, namely additional axioms from OWL2 RL or OWL 2 QL [16].

Acknowledgements. This work was supported by the Vienna Science and Technology Fund (WWTF), project ICT12-SEE, and EU IP project Optique (*Scalable End-user Access to Big Data*), grant agreement n. FP7-318338.

References

1. Ahmeti, A., Calvanese, D., Polleres, A.: Updating RDFS ABoxes and TBoxes in SPARQL. In: Mika, P., et al. (eds.) ISWC 2014, Part I. LNCS, vol. 8796, pp. 441–456. Springer, Heidelberg (2014)
2. Beckett, D., Berners-Lee, T., Prud'hommeaux, E., Carothers, G.: RDF 1.1 Turtle - Terse RDF Triple Language. W3C Recommendation, World Wide Web Consortium, February 2014
3. Benferhat, S., Bouraoui, Z., Papini, O., Würbel, E.: A prioritized assertional-based revision for DL-Lite knowledge bases. In: Fermé, E., Leite, J. (eds.) JELIA 2014. LNCS, vol. 8761, pp. 442–456. Springer, Heidelberg (2014)
4. Calvanese, D., De Giacomo, G., Lembo, D., Lenzerini, M., Rosati, R.: Tractable reasoning and efficient query answering in description logics: the DL-Lite family. J. Autom. Reasoning **39**(3), 385–429 (2007)
5. Calvanese, D., Kharlamov, E., Nutt, W., Zheleznyakov, D.: Evolution of $DL - Lite$ knowledge bases. In: Patel-Schneider, P.F., Pan, Y., Hitzler, P., Mika, P., Zhang, L., Pan, J.Z., Horrocks, I., Glimm, B. (eds.) ISWC 2010, Part I. LNCS, vol. 6496, pp. 112–128. Springer, Heidelberg (2010)
6. De Giacomo, G., Lenzerini, M., Poggi, A., Rosati, R.: On instance-level update and erasure in description logic ontologies. J. Log. Comput. **19**(5), 745–770 (2009)
7. Flouris, G., Konstantinidis, G., Antoniou, G., Christophides, V.: Formal foundations for RDF/S kb evolution. Knowl. Inf. Syst. **35**(1), 153–191 (2013)
8. Gearon, P., Passant, A., Polleres, A.: SPARQL 1.1 update. W3C Recommendation, World Wide Web Consortium, March 2013
9. Guo, Y., Pan, Z., Heflin, J.: LUBM: a benchmark for OWL knowledge base systems. J. Web Seman. **3**(2–3), 158–182 (2005)
10. Hansson, S.: A survey of non-prioritized belief revision. Erkenntnis **50**(2–3), 413–427 (1999)
11. Harris, S., Seaborne, A.: SPARQL 1.1 query language. W3C Recommendation, World Wide Web Consortium, March 2013
12. Hayes, P., Patel-Schneider, P.: RDF 1.1 semantics. W3C Recommendation, World Wide Web Consortium, February 2014
13. Kharlamov, E., Zheleznyakov, D., Calvanese, D.: Capturing model-based ontology evolution at the instance level: the case of DL-Lite. J. Comput. Syst. Sci. **79**(6), 835–872 (2013)
14. Liu, H., Lutz, C., Milicic, M., Wolter, F.: Updating description logic aboxes. In: Doherty, P., Mylopoulos, J., Welty, C.A. (eds.) KR, pp. 46–56. AAAI Press (2006)
15. Lutz, C., Seylan, I., Toman, D., Wolter, F.: The combined approach to OBDA: taming role hierarchies using filters. In: Kagal, L., et al. (eds.) ISWC 2013, Part I. LNCS, vol. 8218, pp. 314–330. Springer, Heidelberg (2013)

16. Motik, B., Grau, B.C., Horrocks, I., Wu, Z., Fokoue, A., Lutz, C.: Owl 2 web ontology language profiles, 2nd edn. W3C Recommendation, World Wide Web Consortium, December 2012
17. Muñoz, S., Pérez, J., Gutierrez, C.: Minimal deductive systems for RDF. In: Franconi, E., Kifer, M., May, W. (eds.) ESWC 2007. LNCS, vol. 4519, pp. 53–67. Springer, Heidelberg (2007)
18. Nikitina, N., Rudolph, S., Glimm, B.: Interactive ontology revision. Web Seman. Sci. Serv. Agents World Wide Web **12–13**, 118–130 (2012). reasoning with context in the Semantic Web
19. Polleres, A., Hogan, A., Delbru, R., Umbrich, J.: RDFS and OWL reasoning for linked data. In: Rudolph, S., Gottlob, G., Horrocks, I., van Harmelen, F. (eds.) Reasoning Weg 2013. LNCS, vol. 8067, pp. 91–149. Springer, Heidelberg (2013)
20. Winslett, M.: Updating Logical Databases. Cambridge University Press, Cambridge (2005)

A Contextualised Semantics for `owl:sameAs`

Wouter Beek[(✉)], Stefan Schlobach, and Frank van Harmelen

Department of Computer Science, VU University Amsterdam,
Amsterdam, Netherlands
{w.g.j.beek,stefan.schlobach,frank}@vu.nl

Abstract. Identity relations are at the foundation of the Semantic Web and the Linked Data Cloud. In many instances the classical interpretation of identity is too strong for practical purposes. This is particularly the case when two entities are considered the same in some but not all contexts. Unfortunately, modeling the specific contexts in which an identity relation holds is cumbersome and, due to arbitrary reuse and the Open World Assumption, it is impossible to anticipate all contexts in which an entity will be used. We propose an alternative semantics for `owl:sameAs` that partitions the original relation into a hierarchy of subrelations. The subrelation to which an identity statement belongs depends on the dataset in which the statement occurs. Adding future assertions may change the subrelation to which an identity statement belongs, resulting in a context-dependent and non-monotonic semantics. We show that this more fine-grained semantics is better able to characterize the actual use of `owl:sameAs` as observed in Linked Open Datasets.

1 Introduction

Identity relations are at the foundation of the Semantic Web and the Linked Data initiative. They allow to state and relate properties of an object using multiple names for that object, and conversely, they allow to infer that different names actually refer to the same object. The Semantic Web consists of sets of assertions that are published on the Web by different authors operating in different contexts, often using different names for the same object. Identity relations allow the interlinking of these multiple descriptions of the same thing. However, the traditional notion of identity expressed by `owl:sameAs` [17] is problematic when objects are considered the same in some contexts but not in others. According to the standard semantics, identical terms can be replaced for one another in all (non-modal) contexts *salva veritate*. Practical uses of `owl:sameAs` are known to violate this strict condition [10,11]. The standing practice in such cases is to use weaker relations of relatedness such as `skos:related` [16]. Unfortunately, these relations suffer from the opposite problem of having almost no formal semantics, thereby limiting reasoners in drawing inferences. In this paper we introduce an alternative semantics for `owl:sameAs` that is parameterized over the particular properties that are taken into account when deciding on identity. This allows formally specified context-specific adaptations of the identity relation. We give

© Springer International Publishing Switzerland 2016
H. Sack et al. (Eds.): ESWC 2016, LNCS 9678, pp. 405–419, 2016.
DOI: 10.1007/978-3-319-34129-3_25

the formal definition, provide working examples and present a small-scale implementation.

The rest of the paper is structured as follows. In the next section we analyze problems caused by the traditional notion of identity. After surveying existing work in Sect. 2 we present our approach in Sect. 5 and enumerate some of the applications of this new semantics in Sect. 6. We illustrate the results of applying our formalism to Linked Datasets in Sect. 7 based on a working implementation. Section 8 concludes.

2   Related Work

Existing research suggests the following six solutions for the problem of identity.

Introduce weaker versions of `owl:sameAs` [10,15] Candidates for replacement are the SKOS concepts `skos:related` and `skos:exactMatch` [16]. The former is not transitive, thereby limiting the possibilities for reasoning. The latter is transitive but is said to only be used in certain contexts without stating what those contexts of use are [16]. As example we will quote the intended use of property `skos:exactMatch` according to the SKOS specification: "[exactMatch] is used to link two concepts, indicating a high degree of confidence that the concepts can be used interchangeably across a wide range of information retrieval applications." From this it follows that the meaning of some SKOS relations changes over time, as IR applications become more advanced. Another problem with using weaker notions such as relatedness, is that everything is related to everything in *some* way.

Restrict the applicability of identity relations to specific contexts. In terms of Semantic Web technology, identities are expected to hold within a named graph or within a namespace but not necessarily outside of it [11]. de Melo [4] has successfully used the Unique Name Assumption within namespaces in order to identify many (arguably) spurious identity statements.

Introduce additional vocabulary that does not weaken but extend the existing identity relation. Halpin et al. [10] mentions an explicit distinction that could be made between mentioning a term and using a term, thereby distinguishing an object and a Web document describing that object. Other possible extensions of `owl:sameAs` may take the fuzziness and/or uncertainty of identity statements into account [14].

Use domain-specific identity relations [15] For instance "x and y have the same medical use" for identity in the domain of medicine and "x and y are the same molecule" for identity in the domain of chemistry. The downside to this solution is that domain-specific links are only locally valid, thereby limiting knowledge reuse.

Change modeling practice Possibly in a (semi-)automated way, by adapting visualization and modeling toolkits to produce notifications upon reading SW data or by posing additional restrictions on the creation and alteration of data. For example, adding an RDF link could require reciprocal confirmation from the maintainers of the authorities of the respective relata [5,11]. The problem with introducing checks on editing operations is that it violates one of the fundamental underpinnings of the SW according to which anybody is allowed to say anything about anything (AAA) [2].

Extract network properties of owl:sameAs datasets Ding et al. [6] shows that network analysis can provide insights into the ways in which identity is used on the Semantic Web. However, results from network analytics research have not yet been related to the semantics of the identity relation. We believe that utilizing network theoretic aspects in order to determine the meaning of identity statements may be interesting for future research.

What the existing approaches have in common is that many adaptations have to be made – introducing terminology, instructing modelers, converting datasets – in order to resolve only some of the problems of identity. Our approach provides a way of dealing with the heterogeneous real-world usage of identity in the Semantic Web that can be automated and that does not require changes to modeling practices or existing datasets.

Our work bears some resemblance to existing work on key discovery: the practice of finding sets of properties that allow subject terms to be distinguished [20]. In particular, our notion of indiscernibility properties (Definition 2) is identical to the notion of a key in key discovery.

3   Motivation

Entities that are the same share the same properties. This 'indiscernibility of identicals' (Principle 1) is attributed to Leibniz [7] and its converse, the 'identity of indiscernibles' (Principle 2), states that entities that share the same properties are the same. Ψ denotes the set of all properties.

Principle 1 (Indiscernibility of identicals). $a = b \rightarrow (\forall \phi \in \Psi)(\phi(a) = \phi(b))$

Principle 2 (Identity of indiscernibles). $(\forall \phi \in \Psi)(\phi(a) = \phi(b)) \rightarrow a = b$

Although Principles 1 and 2 provide necessary and sufficient conditions for identity, they do not point towards an effective procedure for enumerating the extension of the identity relation. Moreover, the principle is circular since $a = b$ implies that a and b share the properties "$= a$" and "$= b$". Even though this principle does not allow a positive identification of identity pairs, it does provide an exclusion criterion; namely objects that are known to not share some property are also known to not be identical.

Identity poses several problems that are not specific to the SW. Firstly, identity does not hold across (all) modal contexts, allowing Lois Lane to believe that Superman saved her without requiring her to believe that Clark Kent saved her. Secondly, identity is context-dependent [9]. For instance, two medicines may be considered the same in terms of their chemical substance while not being considered the same commercial drug (e.g., because they are produced by different companies). Thirdly, identity over time poses problems since a ship may still be considered the same ship, even though all its original components have been replaced by new ones [13]. Lastly, there is the problem of identity under counterfactual assertions, that allow *any* property of an individual to be negated [12]. E.g., "If my parents would not have met then I would not have been born." These four problems indicate that a real-world semantics of identity should be context-dependent and non-monotonic.

Besides the generic problems of identity there are problems that are specific to the Semantic Web and its particular semantics and pragmatics. The OWL semantics for identity is given in Definition 1, where $\mathcal{I}$ is the interpretation function mapping terms to resources and EXT is the extension function mapping properties to pairs of resources.

Definition 1 (Semantics of owl:sameAs).

$$\langle \mathcal{I}(a), \mathcal{I}(b) \rangle \in EXT(\mathcal{I}(owl\!:\!sameAs)) \iff \mathcal{I}(a) = \mathcal{I}(b)$$

Notice that Definition 1 defines owl:sameAs in terms of the identity relation '=' that we have previously argued to be highly problematic. Identity assertions are extra strong on the Semantic Web because of the Open World Assumption. Stating that two entities are the same implies that from now on no new property can be stated about only one of those entities. This follows from Definition 1 in combination with the principle of substitutivity *salva veritate*. For instance, if one source asserts that medicines b and c are the same based on them having the same chemical composition, this prohibits a future source from stating that b and c are produced by different companies, without resulting in an inconsistent state. In other words: every identity assertion makes a very strong claim that quantifies over the entire set Ψ (see Principles 1 and 2). Moreover, on the Semantic Web the set of properties Ψ is constantly increasing. In fact, since an RDF property has both in- and extension, the number of properties is not even limited by the size of the universe of discourse, as different properties may have the same extension. Finally, whether or not two objects share the absence of a property, i.e., a property of the form "does not have the property ϕ", cannot be concluded based on the absence of a property assertion. Such 'negative knowledge' must be provided explicitly using, e.g., class restrictions. All this amounts to saying that *there can in principle not be an effective procedure for establishing the truth of owl:sameAs assertions.* (Establishing the falsehood of such assertions is of course possible, see our comments above.)

When we take the social component of the Semantic Web into account as well, we observe that modelers sometimes have different opinions about whether two objects are identical or not. While in some cases this may be due to a difference in modeling competence, there is also the more fundamental problem that two modelers may be constructing (parts of) the same knowledge base from within different contexts. Since Semantic Web knowledge is intended to be re-used in unanticipated contexts, the presence of knowledge from different perspectives is one of its inherent characteristics. In addition, the term owl:sameAs is overloaded to not only denote the *semantics* of identity but also the *practice* of linking datasets together. The fifth star of Linked Open Data (LOD) publishing [3] states that you should "Link your data to other people's data to provide context," and data is almost exclusively linked using the `owl:sameAs` property [1]. Concluding, from the social point of view today's requirements on Semantic Web modelers are unreasonably high when they are required to anticipate future additions by others while asserting identity links in accordance with the strict semantics. At the same time Linked Data best practices state that modelers should make those links in order to contextualize their knowledge.

Based on the above analysis, we can state the following desiderata for a semantics of identity that does not suffer from the identified problems:

1. The uniform identity relation should be reinterpreted in multiple subrelations that should be characterized in terms of the contexts in which those subrelations appear.
2. An alternative semantics for identity should be able to derive entailment results with respect to a given context.
3. Based on an existing identity relation, semantically motivated feedback should be given to the modeler about the different context-dependent subrelations that are currently expressed.
4. The quality of an identity relation should be quantified in terms of the consistency with which its context-dependent subrelations are applied to the data. Specifically, suggestions for extending or limiting the identity subrelations should be derived by automated means.

4   Preliminaries

Here we introduce the terminology and symbolism that is used throughout the rest of this paper.

RDF syntax. RDF terms (RDF_T) come in three flavors: blank nodes (RDF_B), IRIs (RDF_I) and literals (RDF_L). Statements in RDF are triples $\langle s, p, o \rangle$ that are members of $(RDF_B \cup RDF_I) \times RDF_I \times RDF_T$. In a triple, s is called the subject term, p the predicate term and o the object term of that particular triple. A set of triples forms a graph G. Based on the positionality of terms appearing in the triples of G we distinguish between the subjects (S_G), predicates (P_G) and objects (O_G) of a graph. The nodes of a graph are defined as $N_G = S_G \cup O_G$.

Equivalence. An equivalence relation $\equiv$ is a binary relation that is reflexive, symmetric and transitive. The identity relation is the smallest equivalence relation. The equivalence class of an RDF node $x \in N_G$ under $\equiv$ is $[x]_\equiv = \{y \in N_G \mid x \equiv y\}$.

Set theory. We use the phrase "universe of discourse" to denote the instances that are formally described in a given dataset. Specifically, the universe of discourse for an RDF graph G is N_G. We use the capital letters X and Y to denote arbitrary sets. Elements of these sets are denoted by $x_1, \ldots, x_n$ and $y_1, \ldots, y_n$ respectively.

Modeling identity. It is common modeling practice to denote identity on the instance level with `owl:sameAs` and equivalence on the schema level with `owl:equivalentProperty` for properties and `owl:equivalentClass` for classes. We use ~ to indicate a set of pairs that are explicitly specified to be the same, using either of these three properties. In addition, `owl:differentFrom` is used by modelers to indicate that two terms do not denote the same resource. We use ≁ for a set of pairs that are explicitly indicated to *not* be the same.

Rough Set Theory. Relations are called 'attributes' in Rough Set Theory. They are functions that map to an arbitrary set of value labels. We only consider functions that map from binary input into the set of Boolean truth values, and therefore use the term 'predicates' to denote these functions. We recognize that extensions to multi-valued logics would require a richer set of value labels. Rough Set Theory has been related to Formal Concept Analysis, e.g. in [8].

Formal Concept Analysis. Formal Concept Analysis (FCA) takes a context $\langle O, A, M \rangle$ consisting of a set of objects O, a set of attributes A and a mapping M from the former to the latter. For a given set of objects $X \subseteq O$ one can calculate the attributes that are shared by those objects as $X' = \{y \in A \mid (\forall x \in X)(M(x,y))\}$. For a given set of attributes $Y \subseteq A$ one can calculate the objects that have (at least) those attributes as $Y' = \{x \in O \mid (\forall y \in Y)(M(x,y))\}$. A formal concept is a pair $\langle X, Y \rangle \in \mathcal{P}(O) \times \mathcal{P}(A)$ such that $X' = Y$ and $Y' = X$. The two functions $(\cdot)'$ are called the *polars* of M. For a given context, the set of concepts is denoted $\mathcal{B}(O, A, M)$. The concepts form a lattice $\langle \mathcal{B}(O, A, M), \{\langle \langle X_1, Y_1 \rangle, \langle X_2, Y_2 \rangle \rangle \in (\mathcal{P}(O) \times \mathcal{P}(A))^2 \mid X_1 \subseteq X_2\}\rangle$.

5 Approach

We start with a given identity relation ~ that partitions the universe of discourse N_G into equivalence classes. Since the identity relation is the smallest equivalence relation, it is also the most fine-grained partition of N_G. As we saw in Principles 1 and 2, identity is indiscernibility with respect to all (possible) properties Ψ. Besides identity, there are many other instances of indiscernibility: one corresponding to each set of properties $\Phi \subseteq \Psi$. According to this generalization, x and y are indiscernible with respect to a set of properties Φ iff $(\forall \phi \in \Phi)(\phi(x) = \phi(y))$. Every indiscernibility relation is also an equivalence relation, although not necessarily the smallest one. Every indiscernibility relation defined over domain N_G is also an identity relation, just over a different domain [19]. For instance, the set of properties $\Phi = \{$"has an income of 1,000 euro's"$\}$ does not uniquely identify people (since two people may have the same income), but does uniquely identify income groups.

Let us consider two medicines Baspirin (`abox:baspirin`) and Caspirin (`abox:caspirin`) that both contain acetylsalicylic acid as their chemical compound (`tbox:chemComp`). A chemist observes that they have the same substance and asserts that they are identical (`owl:sameAs` or ~), resulting in the graph in Fig. 1. However, Basperin and Casperin are produced (`tbox:prod`) by different companies: B Inc. (`abox:binc`) and C Inc. (`abox:cinc`). Basperin and Casperin cannot be told apart in a language that only contains the properties "is a" and "has chemical compound". However, if the language also includes the property "is produced by" then these medicines can be told apart. In other words: *we can look at the set of properties as a parameter that can be adjusted in order to obtain an equivalence relation that is more or less fine-grained, as required in different contexts* (in our example: contexts where the commercial supplier does or does not play a role in distinguishing two drugs).

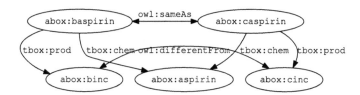

Fig. 1. Graph showing some of the assertions that we use as examples.

We now reinterpret the identity relation $\sim$ as if it were an indiscernibility relation $\approx_\Phi$ whose set of properties Φ is implicit in the data. Based on the extensional specification of the identity relation we can explicate the set of properties to which it is indiscernible with Definition 2, where $\{x_1, \ldots, x_n\}$ is one of the equivalence classes closed under $\sim$.

Definition 2 (Indiscernibility Properties).

$$P^+(\{x_1, \ldots, x_n\}) = \{p \in P_G \mid (\exists p_1, \ldots, p_n \in [p]_\equiv)($$
$$[\{o \in O_G \mid \langle x_1, p_1, o\rangle\}]_\equiv = \ldots = [\{o \in O_G \mid \langle x_n, p_n, o\rangle\}]_\equiv)\}$$

For instance, by using Definition 2 we can deduce that the indiscernibility properties of Basperin and Casperin include `rdf:type`, `tbox:chemComp` and, by definition `owl:sameAs`. Notice that both the predicate and object terms are closed under identity. Performing these closures is important in order to identify the relevant indiscernibility properties. For instance, chemical compound, or `tbox:chemComp`, in one dataset may be the same property as chemical substance, or `ex:chemSubst`, in another. Besides the indiscernibility properties, there may also be discernibility properties (Definition 3), i.e., properties that indicate that two terms should *not* be considered to denote the same resource. As with the identity relation $\sim$, we assume that we are given a 'different-from' relation $\nsim$ of pairs $\langle x_1, x_2\rangle$.

Definition 3 (Discernibility Properties).

$$P^-(\{x_1, x_2\}) = \{p \in P_G \mid (\exists p_1, p_2 \in [p]_\equiv)($$
$$\langle x_1, p_1, o_1\rangle, \langle x_2, p_2, o_2\rangle \in G \wedge (\exists \langle y_1, y_2\rangle \in \nsim)(o_1 \in [y_1]_\equiv \wedge o_2 \in [y_2]_\equiv))\}$$

In our example, the discernibility properties for Basperin and Casperin includes tbox:prod. Using the indiscernibility and discernibility properties we can define the indiscernibility relation (Definition 4).

Definition 4 (Indiscernibility relation).

$$x \approx_\Phi y \iff P^+(\{x, y\}) = \Phi \wedge P^-(\{x, y\}) \cap \Phi = \emptyset$$

For our example we derive that Baspirin and Caspirin are the same with respect to the type and chemical compound properties (Example ex1) and that they are not the same with respect to the producer property (Example ex2). Another way of phrasing this is: Basperin and Caspirin are the same drug in terms of their chemical compound, but they are different medical products.

$$\text{abox:baspirin} \approx_{\{\text{owl:sameAs,rdf:type,tbox:chemComp}\}} \text{abox:caspirin} \qquad \text{(ex1)}$$

$$\text{abox:baspirin} \not\approx_{\{\text{tbox:prod}\}} \text{abox:caspirin} \qquad \text{(ex2)}$$

Now that we have defined the indiscernibility properties for a given set of resources, we go on to say that two pairs of resources are *semi-discernible* iff their indiscernibility properties are the same. When we look at the pairs that constitute (the extension of) a given identity relation $\sim$, all identity assertions look the same. But when we redefine identity in terms of indiscernibility and semi-discernibility, we see that within a given identity relation there are pairs that are indiscernible with respect to different properties. Stating this formally, semi-discernibility is an equivalence relation on pairs of resources that induces a partition of the Cartesian product of the universe of discource. Definition 5 makes this concrete in terms of the earlier definitions.

Definition 5 (Semi-discernibility).

$$\langle x_1, y_1 \rangle \equiv_\Phi \langle x_2, y_2 \rangle \iff P^+(\{x_1, y_1\}) = P^+(\{x_2, y_2\}) = \Phi$$
$$\wedge P^-(\{x_1, y_1\}) \cap P^-(\{x_2, y_2\}) = \emptyset$$

For example, Baspirin and Caspirin are semi-discernible to Bicotine and Nicotine, two stimulant drugs (indiscernibility property rdf:type) whose chemical compound (indiscernibility property tbox:chemComp) is nicotine. An example of semi-discernible pairs from another application domain are $\langle \text{dbr:Amsterdam}, \text{dbr:Rotterdam} \rangle$ and $\langle \text{dbr:Netherlands}, \text{dbr:Germany} \rangle$, since the former are both cities and the latter are both countries (discernibility property rdf:type and each pair is part of the same geographic region (Amsterdam and Rotterdam are part of the Netherlands; the Netherlands and Germany are part of Europe).

Notice that the partitions obtained by $\equiv_\Phi$ contain but are not limited to the original identity pairs. Therefore, for sets of pairs closed under semi-discernibility we can distinguish between the following three categories:

1. All pairs in the set are identity pairs. This characterizes a consistent subrelation of the identity relation, since no semi-discernible pair is left out.

2. Only some pairs in the set are identity pairs. This characterizes a subrelation of the identity relation that is not applied consistently with respect to the semi-discernibility relation that can be observed in the data.
3. No pairs in the set are identity pairs. This characterizes a subrelation of the collection of pairs that is consistently kept out of the identity relation.

Each member of the semi-discernibility partition that is not of the third kind, i.e., every set of pairs that contains at least some identity pair, can be thought of as an identity subrelation. Not only is the uniform set of `owl:sameAs` assertions partitioned into subrelations, but each subrelation is described in meaningful terms that are drawn from the dataset vocabulary.

Now that we have determined the subrelations of identity we go on to define how these subrelations are related. Borrowing insights from Formal Concept Analysis we take N_G^2 as our set of FCA objects and P_G as our set of FCA attributes. The mapping from the former to the latter is $M(\langle x, y \rangle) = \Phi(\{x, y\})$. Because the number of FCA objects is quadratic in the size of the universe of discourse it is not practical to calculate the full concept lattice. However, we are only interested in the identity subrelations and how they are related to one another. Indeed, for every pair $\langle x, y \rangle \in \sim$ we can calculate the formal concept $\langle \{\langle x, y \rangle\}'', \{\langle x, y \rangle\}' \rangle$ by using the polars $(\cdot)'$. What FCA adds to the picture is a partial order $\leq$ between the identity subrelations (Definition 6).

Definition 6 (Indiscernibility Lattice). *For a given identity relation $\sim$, the poset of indiscernibility subrelations is $\langle B, \leq \rangle$ with $B = \{\langle \{\langle x, y \rangle\}'', \{\langle x, y \rangle\}' \rangle \mid \langle x, y \rangle \in \sim\}$ and $\langle \{\langle x_1, y_1 \rangle\}'', \{\langle x_1, y_1 \rangle\}' \rangle \leq \langle \{\langle x_2, y_2 \rangle\}'', \{\langle x_2, y_2 \rangle\}' \rangle$ iff $\Phi(\{x_1, y_1\}) \subseteq \Phi(\{x_2, y_2\})$.*

Every node in the lattice corresponds to a different set of indiscernibility properties, i.e., to a different subrelation of the identity relation. Each subrelation corresponds to an identity assertion context. Specifically, the indiscernibility properties Φ denote the aspects that are important in that context. Results derived/entailed in one context may not be derived in another. Asserting/retracting statements changes the indiscernibility lattice (even if the identity relation is kept the same). The indiscernibility lattice for the graph in Fig. 1 is given in Fig. 2.

Now that we have defined the indiscernibility lattice that comes with a given identity relation $\sim$, we can define the possible identity contexts (Definition 7) in which only part of an identity relation can be used, namely the part that is relevant relative to that identity context.

Definition 7 (Identity Context). *For a given identity relation $\sim$ and its indiscernibility lattice $\langle B, \leq \rangle$ an identity context is a subset of formal concepts $B' \subseteq B$ such that for all $\langle o_1, a_1 \rangle, \langle o_2, a_2 \rangle \in B'$ we have that (i) $a_1 \cap a_2 = \emptyset$ and (ii) $(\forall x, y \in N_G)(P^+(\{x, y\}) \not\subseteq a_1 \vee P^-(\{x, y\}) \not\subseteq a_2)$.*

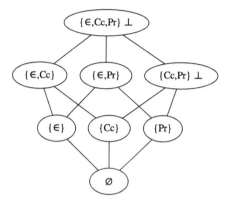

Fig. 2. The indiscernibility lattice for the graph in Fig. 1. For readability we abbreviate tbox:chemComp as Cc, tbox:prod as Pr, rdf:type as $\in$, and owl:sameAs as $\sim$. Two of the indiscernibility relations cannot be chosen without resulting in an inconsistent state.

6 Applications

Modeling. In Fig. 2 every node denotes an indiscernibility relation $\approx_\Phi$ based on a different set of indiscernibility properties Φ. We can define the *precision* of each node by quantifying how many of the pairs that are indiscernible with respect to Φ are also in the original identity relation: $|\sim \cap \approx_\Phi| \, / \, |\approx_\Phi|$. We can also define the *recall* of each node by quantifying how much of the original identity relation is characterized by Φ: $|\sim \cap \approx_\Phi| \, / \, |\sim|$.

The identity lattice annotated with precision and recall numbers can be used to deliver feedback to the modeler. For instance, low precision nodes indicate the absence of identity criteria that are explicit in the data. In practice, many identity links depend on special knowledge the modeler had at the time of assertion. If such special knowledge is not encoded in the data then another data user can no longer validate whether these links are correct. Automatic calculation of the precision of nodes in the identity lattice may prompt a modeler to either (i) make the identity criteria explicit or (ii) remove the identity assertion altogether. The latter may be the case for very low precision nodes, possibly indicating accidental or erroneous identity assertions. Another way in which the identity lattice can support the modeler is by using high-precision nodes in order to give automated suggestions for identity assertion. Specifically, pairs that are indiscernible according to the same criteria as many of the identity pairs may be considered good candidates for identity assertion.

Reasoning with Inconsistencies. As we saw in Sect. 3, one of the main problems of the current use of identity is that terms are considered the same in some but not all contexts. As we saw in Sect. 2 this either results in too many entailments and contradictions, or it results in the use of syntactic alternatives like `skos:related` that do away with entailment altogether. An example of the former can be given with respect to the example shown in Fig. 1, where the identity assertion of the two medicines based on their shared chemical compound results in the substitution of the two medicines in other contexts as well. Specifically, following the OWL2 rule in ent1 we derive that both medicines are produced by companies B Inc. and C Inc., which is unlikely to be the case.

$$\langle s, p, o \rangle \wedge s \sim s' \Rightarrow \langle s', p, o \rangle \tag{ent1}$$

Now that we have the indiscernibility lattice from Fig. 2 we can choose an identity context that can be used to calculate some, but not all entailments. This is supported by condition (ii) in Definition 7 that excludes contexts that result in an inconsist state. The OWL2 rule in ent1 is adapted to take into account an identity context Con, resulting in rule ent2. Other entailment rules require similar adaptations.

$$(\exists \Phi \in Con)(\langle s, p, o \rangle \wedge s \approx_\Phi s' \wedge p \in \Phi \Rightarrow \langle s', p, o \rangle \tag{ent2}$$

Quality Assessment. Borrowing insights from Rough Set Theory we can determine the quality of a given identity relation. The lower approximation of identity is the union of the indiscernibility relations that only contain identical pairs (Definition 8a). The higher approximation of identity is the union of indiscernibility relations that contain some identical pair (Definition 8b).

Definition 8 (Lower and Higher Approximation).

$$x_1 \underset{\sim}{\sim} y_1 \iff \forall_{\langle x_2, y_2 \rangle \in N_G^2} (\langle x_1, y_1 \rangle \equiv_\Phi \langle x_2, y_2 \rangle \rightarrow x_2 \sim y_2) \tag{8a}$$

$$x_1 \overset{\sim}{\sim} y_1 \iff \exists_{\langle x_2, y_2 \rangle \in N_G^2} (\langle x_1, y_1 \rangle \equiv_\Phi \langle x_2, y_2 \rangle \wedge x_2 \sim y_2) \tag{8b}$$

Based on these two approximations we can give the rough set representation $\langle \underset{\sim}{\sim}, \overset{\sim}{\sim} \rangle$ of the identity relation $\sim$ [18]. The quality of a rough set representation is given in Definition 9 and is always a number in $[0, 1]$.

Definition 9 (Quality). $\alpha(\sim) = |\underset{\sim}{\sim}| / |\overset{\sim}{\sim}|$

The quality of the identity relation is higher if the two approximations are closer to each other, and quality is highest if the two approximations are the same. The intuition behind this is that in a high-quality dataset the identity relation should be based on indiscernibility criteria that are explicit in the data. Formally this means that the semi-discernibility partition should consist of partition members that contain either no identity pairs (small value for $\overset{\sim}{\sim}$) or only identity pairs (large value for $\underset{\sim}{\sim}$). If a member of the semi-discernibility partition contains only some identity pairs then this means that the difference between identical and non-identical pairs cannot be based on the properties that are asserted in the data. As with the per-node precision and recall calculations (Sect. 6), the use of data-external identity criteria makes it more difficult to validate identity statements. The quality of a dataset can be improved by making explicit the properties two entities must share in order for them to be considered the same. Adding such indiscernibility properties results in a higher quality metric.

7   Implementation

The approach outlined in Sect. 5 was implemented and tested on datasets published in the instance matching track of the Ontology Alignment Evaluation Initiative. Figure 3 shows an indicative example of an indiscernibility lattice that is calculated for such datasets. Each rectangular box represents an indiscernibility relation. The set notation shows the indiscernibility properties Φ for each indiscernibility relation. For each box the precision quantifies how many pairs that are indiscernible with respect to Φ are in the original identity relation, i.e., $|\sim \cap \approx_\Phi| \ / \ |\approx_\Phi|$. For each box the recall quantifies how much of the original identity relation is characterized by Φ, i.e., $|\sim \cap \approx_\Phi| \ / \ |\sim|$.

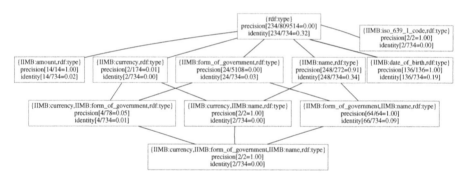

Fig. 3. Example of the identity subrelations for a dataset in the instance matching track of the Ontology Alignment Evaluation Initiative. This is the 16*th* variant of the IIMB datasets in the 2012 challenge (Color figure online).

Since in this figure a partition is only drawn when there is at least one identity pair that is indiscernible with respect to some set of predicates, the higher approximation amounts to the entire figure. The lower approximation only consists of those partition sets that contain at least one identity pair, and that contain no non-identity pair; these are distinguished by green borders. For each box the precision number indicates the ratio of identity pairs for each subrelation. By definition, subrelations in the lower approximation have precision 1.0 and that subrelations in the higher approximation have a non-zero precision.

Figure 3 shows that the uniform identity relation consists of conceptually different indiscernibility subrelations. For instance, some entities are considered the same based on their {IIMB:amount, rdf:type} properties (movies with the same budget are indiscernible in this dataset) and some entities are considered the same based on their {`IIMB:date_of_birth`, `rdf:type`} properties (people with the same birth data are indiscernible in this dataset). Notice that in both cases strict identity would indeed be too strong, since two movies might have the same budget and two people might have the same birth data. The figure also shows that approximately 30 % of the given identity relation extension is applied consistently with respect to the calculated indiscernibility lattice, i.e., the green boxes. The red boxes with high precision are able to isolate a limited number of pairs that are indiscernible in the same way as identity pairs but that are not in the given identity relation. An example of this is {`IIMB:name`, `rdf:type`}. These may either be candidates for identity assertions under the same condition, or some additional facts may be asserted about them in order to distinguish them from identity pairs. Finally, the figure shows that approximately one third of the original identity relation's extension are only indiscernible with respect to their `rdf:type` property. This is insufficient to set them apart from many non-identity pairs and results in a lower quality metric.

Calculation of the indiscernibility lattice is implemented in SWI-Prolog and its ClioPatria triple store [22]. Identity statements are either loaded from VoID linksets or are loaded from EDOAL (Expressive and Declarative Ontology Alignment Language) alignment files. The code is available at http://github.com/wouterbeek/IOTW/. For the 60 IIMB datasets in the OAEI 2012 Instance Matching track this naive implementation on average takes 15 s to calculate the identity lattice.

8   Conclusion

The identity relation `owl:sameAs` is a crucial element of the Semantic Web. It is therefore alarming that its semantics is both computationally ineffective and epistemilogically inadequate. Computationally, it is in principle impossible to define an effective procedure for establishing the truth of `owl:sameAs` assertions, because the open world assumption implies that the set of properties to be checked for indiscernibility is unknown; and epistemilogically it is impossible to model the situation that two given objects may be regarded as equal in one context, but not equal in another.

In this paper we presented a new approach for defining the identity relation. Instead of checking indiscernability with respect to all properties we explicity parameterise the identity relation over the set of properties that are taken into account for establishing identity. This gives both a computationally effective procedure and allows us to define different identity relations in different contexts.

Section 3 enumerates four desiderata for a semantics of identity. (i) The semi-discernibility partition allows the uniform identity relation to be characterized in terms of discernibility subrelations based on different sets of properties Φ. (ii) Since entailment can be defined with respect to a context, or collection of discernibility properties, it can be scoped to contexts in which entities are considered identical, preserving some of the benefits of entailment without resulting in an inconsistent state. (iii) Since the criteria for the discernibility subrelations are explicit in the data, the new semantics opens up possibilities for providing feedback to the modeler. (iv) A quality metric can be calculated for the identity relation of a dataset, indicating the consistency with which identity can be described in terms of the properties that occur in the data, rather than being based on knowledge left implicit by the original modeler.

The implications for OWL2 entailment under the here proposed semantics must be further investigated. Existing entailment languages such as RIF must be extended so that an identity context can be expressed. The current implementation is only a naive proof of concept and needs to be improved by using recent advances in calculating FCA's, e.g. [21], in order to be applicable to larger datasets. Quality metrics for identity could extend existing data quality metrics. Finally, the feedback mechanisms that are supported by the here presented semantics may be implemented as a plugin for an often used modeling editor such as Protégé in order to allow the utility of such features to be measured in practice.

References

1. Alexander, K., Cyganiak, R., Hausenbals, M., Zhao, J.: Describing linked datasets with the VoID vocabulary, March 2011
2. Antoniou, G., Groth, P., van Harmelen, F., Hoekstra, R.: A Semantic Web Primer, 3rd edn. The MIT Press, Cambridge (2012)
3. Berners-Lee, T.: Linked Data (2010). http://www.w3.org/DesignIssues/LinkedData.html
4. de Melo, G.: Not quite the same: identity constraints for the web of linked data. In: Proceedings of the American Association for Artificial Intelligence (2013)
5. Ding, L., Shinavier, J., Finin, T., McGuinness, D.L.: OWL: sameAs and linked data: an empirical study. In: Proceedings of the Web Science (2010)
6. Ding, L., Shinavier, J., Shangguan, Z., McGuinness, D.L.: SameAs networks and beyond: analyzing deployment status and implications of owl:sameAs in linked data. In: Patel-Schneider, P.F., Pan, Y., Hitzler, P., Mika, P., Zhang, L., Pan, J.Z., Horrocks, I., Glimm, B. (eds.) ISWC 2010, Part I. LNCS, vol. 6496, pp. 145–160. Springer, Heidelberg (2010)
7. Forrest, P.: The identity of indiscernibles. In: Zalta, E.N. (ed.) The Stanford Encyclopedia of Philosophy (2008)
8. Ganter, B.: Non-symmetric indiscernibility. In: Wolff, K.E., Palchunov, D.E., Zagoruiko, N.G., Andelfinger, U. (eds.) KONT 2007 and KPP 2007. LNCS, vol. 6581, pp. 26–34. Springer, Heidelberg (2011)
9. Geach, P.T.: Identity. Rev. Metaphysics **21**, 3–12 (1967). Reprinted in Geach, pp. 238–247 (1972)
10. Halpin, H., Hayes, P.J., McCusker, J.P., McGuinness, D.L., Thompson, H.S.: When owl: sameAs isn't the same: an analysis of identity in linked data. In: Patel-Schneider, P.F., Pan, Y., Hitzler, P., Mika, P., Zhang, L., Pan, J.Z., Horrocks, I., Glimm, B. (eds.) ISWC 2010, Part I. LNCS, vol. 6496, pp. 305–320. Springer, Heidelberg (2010)

11. Halpin, H., Hayes, P.J., Thompson, H.S.: When owl: sameAs isn't the same redux: towards a theory of identity, context, and inference on the semantic web. In: Christiansen, H., Stojanovic, I., Papadopoulos, G.A. (eds.) CONTEXT 2015. LNCS, vol. 9045, pp. 47–60. Springer, Heidelberg (2015)
12. Kripke, S.: Naming and Necessity. Harvard University Press, Cambridge (1980)
13. Lewis, D.: On the plurality of worlds. Basil Blackwell, Oxford (1986)
14. Liu, C., Qi, G., Wang, H., Yong, Y.: Fuzzy reasoning over RDF data using OWL vocabulary. In: Proceedings of the International Conferences on Web Intelligence and Intelligent Agent Technology, pp. 162–169 (2011)
15. McCusker, J., McGuinness, D.: Towards identity in linked data. In: Proceedings of OWL Experiences and Directions Seventh Annual Workshop (2010)
16. Miles, A., Bechhofer, S.: SKOS simple knowledge organization system reference, August 2009
17. Motik, B., Grau, B.C., Patel-Schneider, P.: OWL 2 web ontology language direct semantics, 2nd edn., December 2012
18. Pawlak, Z.: Rough Sets: Theoretical Aspects of Reasoning About Data. Kluwer Academic Publishing, Dordrecht (1991)
19. Van Orman Quine, W.: Identity, ostension, and hypostasis. J. Philos. **47**(22), 621–633 (1950)
20. Soru, T., Marx, E., Ngomo Ngonga, A.-C.: Rocker: a refinement operator for key discovery. In: Proceedings of the 24th International Conference on World Wide Web Conferences Steering Committee, pp. 1025–1033 (2015)
21. Vilem, V.: A new algorithm for computing formal concepts. In: Cybernetics and Systems, pp. 15–21 (2008)
22. Wielemaker, J., Beek, W., Hildebrand, M., van Ossenbruggen, J.: Cliopatria: A logical programming infrastructure for the semantic web. Semant. Web J. (2015)

Semantic Data Management, Big Data, Scalability Track

The Lazy Traveling Salesman – Memory Management for Large-Scale Link Discovery

Axel-Cyrille Ngonga Ngomo[✉] and Mofeed M. Hassan

AKSW Research Group, University of Leipzig,
Augustusplatz 10, 04103 Leipzig, Germany
{ngonga,mounir}@informatik.uni-leipzig.de
http://limes.sf.net

Abstract. Links between knowledge bases build the backbone of the Linked Data Web. In previous works, several time-efficient algorithms have been developed for computing links between knowledge bases. Most of these approaches rely on comparing resource properties based on similarity or distance functions as well as combinations thereof. However, these approaches pay little attention to the fact that very large datasets cannot be held in the main memory of most computing devices. In this paper, we present a generic memory management for Link Discovery. We show that the problem at hand is a variation of the traveling salesman problem and is thus NP-complete. We thus provide efficient graph-based algorithms that allow scheduling link discovery tasks efficiently. Our evaluation on real data shows that our approach allows computing links between large amounts of resources efficiently.

1 Introduction

A wide variety of data publishers are now making Linked Data available.[1] With this variety come millions of resources that need to be linked together to generate real five-star Linked Data. While one can rely on hardware architectures such as cloud-based processing to link large knowledge bases, previous works have shown that the bottleneck of having to transfer local data to the Cloud makes cloud-based solutions rather unattractive [13]. Local solutions on the other hand only provide limited amounts of memory that must commonly be shared with other processes. This in turn means that efficient memory management approaches for Link Discovery (LD) are needed to enable LD on large datasets while relying on local hardware. While the time complexity of LD has been studied considerably over the last years (see [10] for a survey), LD's space complexity has not been paid much attention to. In this paper, we address exactly this research gap and investigate a *generic memory management approach* for LD.

The rationale behind our approach, GNOME, is to allow deploying the paradigms commonly used for efficient LD (e.g., PPJoin+ [22], MultiBlock [9] and $\mathcal{HR}^3$ [11]) on very large datasets that do not fit in the main memory of the

---

[1] See, e.g., http://lod-cloud.net/ and http://stats.lod2.eu.

© Springer International Publishing Switzerland 2016
H. Sack et al. (Eds.): ESWC 2016, LNCS 9678, pp. 423–438, 2016.
DOI: 10.1007/978-3-319-34129-3_26

computer at hand. We implement this vision through the following research contributions: (1) We introduce best-effort and greedy graph-based algorithms for determining how a LD problem should be addressed to ensure good memory usage. (2) We justify the use of best-effort and greedy algorithms by mapping the problem of loading data into the memory in the right order to the traveling salesman problem (TSP), which is known to be NP-complete. (3) We show that while our algorithms are not optimal, they can deal with large datasets efficiently. (4) We provide empirical evidence to substantiate the use of greedy and best-effort approaches. To this end, we evaluate our approach on large real datasets derived from DBpedia and LinkedGeoData.

The rest of this paper is structured as follows: we present a formal specification of the problem at hand in Sect. 2. Then, we present the formal model behind GNOME in Sect. 3. In Sect. 4, we present GNOME in detail. We first give an overview of the intuitions behind the approach. In particular, we map the memory management problem to an edge-partitioning problem in combination with the TSP. The evaluation of GNOME and its components on datasets of different sizes is presented in Sect. 5. Here, we show that we can perform large-scale link discovery on commodity hardware. Finally, we give a brief overview of related work (Sect. 6) and conclude.

2   Preliminaries

The LD problem can be formalized as follows: Given a two sets of RDF resources S and T as well as a relation R, compute the set $M = \{(s, t) \in S \times T : R(s, t)\}$. Declarative LD frameworks usually compute an approximation $M' = \{(s, t) \in S \times T : \sigma(s, t) \geq \theta\}$ of M, where σ is a (complex) similarity function and θ is a similarity threshold. Naïve approaches to computing M' have a quadratic time complexity and at least a linear space complexity. The quadratic time complexity is due to the need to execute σ on all elements of $S \times T$. The linear space complexity comes from the need to load both S and T into the main memory of the computing device at hand. Over the last years, time-efficient approaches have been developed with the aim of reducing the runtime of declarative link discovery frameworks. One key insights that underlies GNOME is that while these approaches are based on varied models such as sequences of filters (e.g., PPJoin+ [22]), blocking (e.g., MultiBlock [9]) and space tiling (e.g., $\mathcal{HR}^3$ [11]) they all reduce the overall time complexity of LD by reducing the number of comparisons that need to be carried out to determine M' completely. This reduction is commonly achieved by determining automatically which (possibly overlapping) subsets of S must be compared with which (possibly overlapping) subsets of T to compute M' fully. We call approaches that abide by this model *divide-and-merge* LD approaches.

3   A Task Model for Efficient Link Discovery

Formally, our insight pertaining to divide-and-merge approaches translates into these approaches operating as follows: Given S, T, σ and θ (as defined above), they determine

1. the set $\mathcal{S}$ of subsets $S_1, \ldots, S_n$ of S, the set $\mathcal{T}$ of subsets $T_1, \ldots, T_m$ of T and
2. the mapping function $\mu : \mathcal{S} \to 2^{\mathcal{T}}$ such that
3. the elements of each S_i must only be compared with the elements of all sets in $\mu(S_i)$ and
4. the union of the results over all $S_i \in \mathcal{S}$ is exactly M', i.e., $M' = \{(s,t) : s \in S_i \wedge t \in T_j \wedge T_j \in \mu(S_i) \wedge \sigma(s,t) \geq \theta\} = \{(s,t) \in S \times T : \sigma(s,t) \geq \theta\}$.

We can thus model the computation of M' as the execution of a sequence of tasks E_{ij}, where each task E_{ij} consists of comparing all elements of S_i with all elements of T_j. We denote the set of all tasks with $\mathcal{E}$. We will regard the main memory of a computing device as a storage solution C with a limited capacity $|C|$ and three simple operators: load (D), which loads the data item D into C, evict(D) which deletes D from C and get(D) which (1) returns D if C contains D or (2) evicts as many elements from C as necessary to be able to load D into C and to return it. If $|S| + |T| \leq |C|$, then the computation of M' can be carried in the main memory of the device at hand. In this work, we will be concerned with cases where S and T cannot be held in C, i.e., $|S| + |T| > |C|$. In this case, if S_i or T_j is not available in C, then E_{ij} can only be computed once the missing items are loaded into C. In many cases, this will require deleting some of the sets S'_i and T'_j that are already in C. Determining which sets are to be evicted from a storage solution is commonly solved by relying on *caching strategies*.

The main goal of this paper is correspondingly to *determine the right order for executing tasks against C so as to maximize the locality of data*, i.e., to minimize the amount of data transferred between the hard drive and the main memory. Our first intuition pertaining to this goal is that μ *provides hints towards which tasks should form a subsequence* of the sequence of tasks. We implement this insight by clustering tasks and carrying the tasks within a cluster after each other. Our second intuition is that *certain clusters rely on overlapping data*. Clusters that share a large amount of data should be executed after one another to improve the overall locality. We implement this insight by providing a best-effort solution to the corresponding TSP.

Throughout this paper, we assume that $\forall S_i \; \forall T_j \in \mu(S_i) : |S_i| + |T_j| \leq |C|$. We say that C is *sufficient* to compare S with T if it fulfills this condition. The sufficiency of C is necessary to ensure that all elements of S_i can be compared with all elements of all $T_j \in \mu(S_i)$. Note that mappings μ that do not abide by this condition can be extended to mappings that do abide by this restriction simply by partitioning S_i and T_j into smaller datasets and updating μ correspondingly.

4   Approach

This section presents our approach to memory management formally. We assume that we are given the sets $\mathcal{S}$ and $\mathcal{T}$, the mapping function μ and a fixed and sufficient amount of main memory C with size $|C|$. We begin by defining the formal concepts necessary to understand our approach. Then, we present graph-clustering-based approaches that allow improving the usage of C while also improving the data locality of the problem at hand. Finally, we show that determining the right sequence for processing clusters is NP-complete and present best-effort approaches to approximate this sequence.

4.1   From Tasks to Task Graphs

Our approach begins by modeling the tasks to a given LD problem as an undirected weighted graph $G = (V, E, w_v, w_e)$, which we call a *task graph*. The set V of vertices of G is the set $\mathcal{S} \cup \mathcal{T}$, which we call the set of *data items*. Two data items $S_i, T_j \in V$ are related by an edge $e = \{S_i, T_j\}$ iff $T_j \in \mu(S_i)$. The weight function $w_v : V \to \mathbb{N}$ maps each $v \in V$ to the total amount of main memory required to store it, i.e., $w_v(v) = |v|$. The weight function $w_e : V \to \mathbb{E}$ maps each $e = \{S_i, T_j\}$ to $w_v(e) = |S_i||T_j|$, i.e., to the total number of comparisons needed to process $S_i \times T_j$. Note that the edge $\{S_i, T_j\}$ corresponds exactly to the task E_{ij}. As a running example, we will consider a case where (1) $|C| = 7$, (2) S is subdivided into S_1, S_2 and S_3, (3) T is divided into T_1, T_2 and T_3. The corresponding task graph with sizes for the S_i and T_j is shown in Fig. 1a. For the sake of simplicity, we will use $V(X)$ resp. $E(X)$ in this paper to denote the set of vertices resp. of edges of a graph X.

Within this model, the insights behind our approach can be translated as follows: Maximizing the locality of data in C corresponds to finding the sequence of tasks that minimizes the total volume of the `load` operations by C. We address this problem in two steps. First, we devise several approaches to clustering the graph G so as to find clusters of tasks that should form a subsequence of the sequence of tasks to execute. Then, we approximate the order in which these clusters of tasks should be executed.

4.2   Clustering Tasks

The aim of clustering tasks is to detect portions of the graphs that describe groups tasks which should be executed after each other with the aim of ensuring high data locality. Formally, clustering is equivalent to detecting subsequences of the sequence of all tasks. We developed two types of approaches to achieve this goal: a naïve approach, which makes use of the locality of subsets of S and a greedy approach which aims to maximize an objective function efficiently.

Naïve Approach. A naïve yet time-efficient approach towards clustering lies in making use of G being a bi-partite graph by virtue of subsets of S being

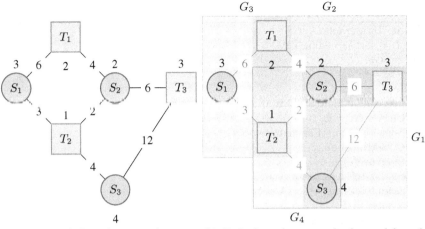

(a) Task dependency graph.

(b) Task dependency graph clustered by edge weight.

Fig. 1. Example task graph. The nodes from S are circles while the nodes from T are rectangles. The weights of nodes are displayed next to the nodes. The weights of the edges are displayed above edges.

linked to subsets T exclusively. Hence, we can cluster G by simply creating a cluster $G_i = G(S_i)$ for each S_i with (1) $V(G(S_i)) = \{S_i\} \cup \mu(S_i)$ and (2) $E(G(S_i)) = \{e \in E(G) : e \subseteq V(G(S_i))\}$. For our example, we get a.o. $E(G_1) = E(G(S_1)) = \{\{S_1,T_1\},\{S_1,T_2\}\}$ while $V(G_1) = \{S_1,T_1,T_2\}$. Note that the result of this approach is complete as it is guaranteed to cover all edges in G. The main advantage of the naïve approach is that is very time-efficient with a worst-case time complexity of $O(|\mathcal{E}|)$. However, it is clearly suboptimal. We thus also propose a clustering approach which generates solutions which have a higher locality.

Greedy Approach. Our more intricate approach towards clustering tasks takes $|C|$ into consideration by aiming to discover the portions of G that can fit in memory simultaneously while maximizing the number of comparisons that can be carried out while having this data in memory. Based on the graph model above, this aim translates to aiming to find a *division* of G into connected graphs $G_k = (V_k, E_k)$ that

1. are *node-maximal* w.r.t $|C|$, i.e., such that $\sum_{v \in V_k} |v| \leq |C|$ but building a connected graph $G'_k = (V'_k, E'_k)$ with $V_k \subset V'_k$ would lead to a graph with $\sum_{v \in V'_k} |v| > |C|$
2. are *complete*, i.e., $\bigcup_k V_k = V$ and $\bigcup_k E_k = E$
3. implement an *edge partitioning* of G, i.e., $\forall e \in E(G) \; \exists E_k : e \in E_k$ and $k \neq k' \rightarrow E_k \cap E_{k'} = \emptyset$.

Algorithm 1. Greedy Task Clustering Algorithm

Input: Task graph G, $|C|$

1 List L = sortAscending($E(G)$);
2 Cluster G_k;
3 clusters = $\emptyset$;
4 **while** $|L| > 0$ **do**
5   List candidateEdges = $\emptyset$;
6   Cluster $G_k = \emptyset$;
7   Edge e = L.firstElement();
8   G_k.addEdge(e);
9   L.remove(e);
10   candidateEdges.add(e.getRelatedEdges());
11   candidateEdges = sortAscending(candidateEdges);
12   counter = 0;
13   **while** $++counter < candidateEdges.size()$ **do**
14   e = candidateEdges.get(counter);
15   **if** $(canBeAdded(G_k, e))$ **then**
16   candidateEdges.remove(e);
17   L.remove(e);
18   G_k.addEdge(e);
19   candidateEdges.add(e.getRelatedEdges() $\cap L$);
20   candidateEdges = sortAscending(candidateEdges);
21   counter = 0;

22   clusters.add(G_k);
23 **return** clusters;

The nodes of each G_k are the data items that should be in memory at the same time while the edges of G_k allow deriving which tasks should be carried out when G_k is in memory. Hence, determining an appropriate division of G can be carried out as shown in Algorithm 1. We begin by sorting the edges in E by weight (see Line 11). Then, we create a new element G_k of the graph division. We select the unassigned edge e with the largest weight and add it to E_k while the corresponding nodes are added to V_k (see Line 7). e is then removed from the set of edges to assign (see Line 9). All unassigned edges that are reachable from V_k are then added to the set of candidate edges. We then pick the edge with the highest weight from this set of candidates, add it to E_k and add the corresponding nodes to V_k if does not break any of the conditions above (Lines 10–11). This procedure is repeated (see Line 13) until no edge can be added to G_k without breaking the node-maximality condition. We create new connected graphs G_k until all edges are covered, thus ensuring the completeness of the results generated by our approach.

For our running example, the approach begins with E_{33} as it has a size of 12. The edges E_{23} and E_{32} are then added to the list of candidates. Moreover, E_{33} is removed from the list of edges to process. As E_{23} nor E_{32} can be added without going against the maximality condition ($|S_3| + |T_3| = 7 = |C|$), the first cluster G_1 is considered completed. The next cluster is started with E_{11}, which is one of the edges with the highest weights. Upon completion, our approach returns the clusters shown in Fig. 1b.

4.3   Scheduling Clusters of Tasks

While the approaches above allows deriving how to make good use of C locally (i.e., at the level of subsequences), they do not allow determining the order in which the clusters should be loaded into C to minimize the traffic between the hard drive and main memory. Our main insight towards addressing this problem is that all tasks can be interpreted as a weighted undirected graph H with nodes G_k and edge weights $w(\{G_k, G_{k'}\}) = \sum\limits_{v \in V_k \cap V_{k'}} |v|$. Thus, finding a good sequence of nodes G_k is equivalent to finding a path in H that covers all nodes of H and bears a large weight, as a large overall weight signifies that several data items can be reused. Finding this path is equivalent solving a TSP where the total distance is to be maximized. We dub this variation of TSP the lazy TSP, as our salesman wants to stay on the road as much as possible to rest. The TSP being known to be NP-complete, we will not attempt to derive an optimal solution for the scheduling of tasks. Instead, we choose to use a best-effort approach.

Best-Effort Solution. We introduce the cluster overlap function, which measures the total weights of the nodes that two clusters have in common, i.e.,

$$o(G_i, G_j) = \sum_{v \in V(G_i) \cap V(G_j)} |v|. \tag{1}$$

For our example shown in Fig. 1b, the overlap between G_1 and G_2 is 5 as they share the nodes S_2 and T_3. We extend this overlap function to sequences as follows:

$$o(G_1, \ldots, G_N) = \sum_{i=1}^{N-1} o(G_i, G_{i+1}). \tag{2}$$

For example, the overlap of the sequence of clusters shown in Fig. 2a is 7. The basic idea behind our best-effort solution is to begin by the clusters set in a random order $G_1, \ldots, G_N$ and to find new permutations of the graphs G_i which improve the overlap function. The scalability of this approach is based on the following observation: After the permutation of the clusters with indexes i and j from a given sequence $G_1, \ldots, G_N$, the difference Δ in overlap is given by

$$\Delta = (o(G_{i-1}, G_i) + o(G_i, G_{i+1}) + o(G_{j-1}, G_j) + o(G_j, G_{j+1})) - \\ (o(G_{i-1}, G_j) + o(G_j, G_{i+1}) + o(G_{j-1}, G_i) + o(G_i, G_{j+1})). \tag{3}$$

If $\Delta > 0$, then a sequence that leads to a higher locality was found. Note that this approach only requires computing 8 overlap scores for each new permutation and thus scales well.

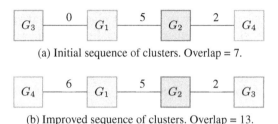

(a) Initial sequence of clusters. Overlap = 7.

(b) Improved sequence of clusters. Overlap = 13.

Fig. 2. Examples of sequences of clusters. The overlap of pairs of clusters are displayed above the corresponding edges.

Algorithm 2. Overview of GNOME

    Input: Cache C, set S, set T, function μ
1   Graph G = computeTaskGraph(S, T, μ);
2   List L = cluster(G);
3   L = schedule (L);
4   Mapping $M = \emptyset$;
5   **for** $G_k \in L$ **do**
6   **for** $\{S_i, T_j\} \in E_k$ **do**
7   source − C.get(S_i);
8   target = C.get(T_j);
9          $M = M \cup compare(source, target)$;
10 **return** M;

Our best-effort approach makes use of this insight by beginning with a random sequence. Then, it selects two indexes i and j at random and computes Δ for the sequence that would occur if G_i and G_j were swapped in the sequence. If $\Delta > 0$, then i and j are swapped. We iterate this procedure until a condition (e.g., a maximal runtime) is achieved and return the current best sequence found by the approach. In the example derived from Fig. 1b shown in Fig. 2a , let $i = 1$ and $j = 4$. This lead to G_3 (which has the index $i = 1$ in the sequence) being swapped with G_4 (index $j = 4$). The permutation (see Fig. 2b) leads to an increase of the overlap score to 13. Hence, it is kept by the approach. The main advantage of using such an approach is scalability as the overall overlap function does not need to be computed given that we can limit ourselves to finding permutations which improve the overlap. Hence, the approach can be set to run for a predefined amount of time or iterations (we hence dub it a best-effort approach) on any input dataset. For a fixed number of iterations I, the complexity of the approach is $O(I)$ as it always computes exactly $8I$ overlap scores per iteration.

Greedy Approach. For the sake of comparison, we also implemented a greedy approach. It starts with the first node within the input and finds a path that covers all clusters G_i by always iteratively choosing the next node with the highest overlap to the current node that does not yet belong to the path. For our example, the approach would return the path (G_3, G_2, G_1, G_4) with an overlap score of 13. Note that while this approach is global and can return results with a higher overlap score than the best-effort approach, it has a worst-case complexity of $O(N^2)$, where N is the number of clusters.

Algorithm 2 summarizes our approach and shows how all the components interact. Given the sets $\mathcal{S}$ and $\mathcal{T}$ as well as the function μ, we begin by computing the corresponding task graph. The graph is then clustered. The generated clusters are forwarded to the scheduling approach, which determine the sequence in which the clusters are to be executed. Finally, the clusters are executed in the order given by the scheduler.

5   Evaluation

5.1   Experimental Setup

Within our experiments, we aimed to answer the following questions:

Q1 *How do the alternative configurations of our approach perform?*
Q2 *How does our approaches perform against existing caching strategies?*
Q3 *How well does our approach scale?*

To answer these questions, we evaluated our approach against baseline approaches on two real datasets derived from DBpedia and LinkedGeoData.[2] We only considered deduplication experiments, where $S = T$.

Datasets. We selected the first 1 million labels from DBpedia version 04-2015[3] as our first dataset DBP. To compare these labels, we used the trigram similarity as similarity measure [22]. We computed the sets S_i by first computing all trigrams found in labels in S. Then, each S_i was set to be the set of all resources (1) whose labels had a particular length and (2) whose labels contained a particular trigram. Note that this means that $\exists i, j : i \neq j \wedge S_i \cap S_j \neq \emptyset$. Let $n(S_i)$ resp. $tri(S_i)$ be length of the strings in S_i resp. the trigram they all contain. To compute $\mu(S_i)$, we selected all S_j which (1) contained strings of length between $n(S_i) \times \theta$ and $n(S_i)/\theta$ (where θ is the similarity threshold) and (2) contained $tri(S_i)$. These equations were derived from [22] and guarantee the completeness of our results.

Our second dataset (LGD) contained 800,000 places from LinkedGeoData as well as their latitude and longitude. The computation of the S_i and $\mu(S_i)$ was based entirely on $\mathcal{HR}^3$ [11]. Let τ be the distance threshold corresponding to the similarity threshold θ (i.e., $\frac{1}{1+\theta}$) expressed in degrees. Each S_i was the region

---

[2] The datasets and the corresponding result files are available at https://github.com/ AKSW/LIMES/tree/master/LazyTravelingSalesMan.
[3] http://dbpedia.org/.

which contained resources with latitudes in $\left[\frac{k\tau}{4}, \frac{(k+1)\tau}{4}\right[$ and with longitudes in $\left[\frac{k'\tau}{4}, \frac{(k'+1)\tau}{4}\right[$ with $(k, k') \in \mathbb{Z}^2$. The set $\mu(S_i)$ contained all S_j that surrounded S_i and could contain points such that the euclidean distance between any point of S_j and of S_i could be less than τ. Note that in contrast to the approach used on DBpedia, the S_i here did not overlap.

Hardware. All experiments were carried out on a Linux Server running *Open-JDK* 64-Bit Server 1.8.0.66 on Ubuntu 14.04.3 LTS on Intel(R) Xeon(R) E5-2650 v3 processors clocked at 2.30 GHz. Each experiment was repeated three times and allocated 10 G of memory and 1 core to emulate good commodity hardware.

Measures. We measured the *runtime* of each approach by subtracting the current system time at the start of each approach from the current system time at the end of the same execution. The *number of hits* and misses was measured as follows: Each time that a data item D was required for computing a task E_{ij}, the required data was requested from the cache. If the data was found in the cache (cache hit), we added $|D|$ to the hit count *hits*, which was initialized with 0. If the data was not found in the cache, we proceeded accordingly with the number of misses *misses*. We report the *hit ratio* given by *hit ratio* = *hits*/(*hits*+*misses*).

Baseline. Different existing caching strategies can be used to improve the memory management by reducing the time necessary to transfer data from the hard drive. We implemented five caching strategies with different characteristics and used them as baseline as well as in combination with our approach. These strategies are: (1) **FIFO** (First-In First-Out): Once a cache following this strategy is full, it evicts the element that stayed the longest in the cache. (2) **FIFO2ndChance** (First-In First-Out Second Chance): This cache modifies the FIFO strategy by evicting the recording elements that have let to hits and giving them a second chance if they are to be evicted by reinserting them into the cache as new entries. The oldest element in the cache that has not led to a hit is then removed. (3) **LRU** (Least Recently Used): Here, the entry that has been referenced the furthest in the past is evicted. (4) **LFU** (Least Frequently Used): The evicted element is the one with the smallest number of references (hits). (5) **SLRU** (Segmented Least Recently Used): an extension of LRU where the cache has two segments, a protected and unprotected segment. First, new elements are cached in the unprotected segment. Once the cache recorded a hit for such element, it is transfered to the protected segment. Eviction occurs in both segments when they are full. While eviction out of whole cache occurs to the elements in the unprotected segment, the elements in protected segment are transfered to the unprotected one.

5.2   Evaluation of Clustering

We compared the two clustering approaches developed herein w.r.t. their runtime and their hit ratio. In each experiment, we used the first 1,000 resp.

Table 1. Average clustering results on LinkedGeoData and DBpedia. The results for 1,000 resp. 10,000 resources are shown in the top resp. bottom section of the table.

$|C|$	Runtimes (ms)				Hit ratio			
	LGD		DBP		LGD		DBP	
	Naive	*Greedy*	*Naive*	*Greedy*	*Naive*	*Greedy*	*Naive*	*Greedy*
100	568.0	646.3	888.0	31,973.7	0.57	0.77	0.50	0.54
200	518.3	594.0	937.0	32,563.0	0.66	0.80	0.50	0.54
400	532.0	593.3	9,014.0	32,180.7	0.67	0.80	0.50	0.54
1,000	5,974.0	118,454.7	9,014.0	1,991,841.7	0.51	0.64	0.49	0.57
2,000	6,168.0	115,450.0	10,018.0	1,848,703.0	0.51	0.63	0.50	0.57
4,000	7,118.3	121,901.7	11,001.7	1,947,228.7	0.50	0.63	0.50	0.57

10,000 resources described in the datasets at hand. As caching approach, we used the FIFO strategy. Our results on LGD and DBP are shown in Table 1. The table shows clearly that while the greedy and naïve approach achieve similar runtimes on the LinkedGeoData fragment with 1,000 resources, the greedy clustering approach is orders of magnitude slower than the naïve approach in all other cases. Still, the results also show that a better clustering of tasks as performed by greedy clustering leads to higher hit ratios, thus suggesting that clustering alone can already be beneficial for improving the scheduling of link discovery tasks. Overall, we opted to use the naïve approach in all subsequent experiments.

5.3 Evaluation of Scheduling

We compared the scheduling approaches using the same setting as for the clustering. The runtime for the naïve scheduler was set to 250 ms. The results shown in Table 2 suggest similar insights as with clustering (the evaluation of LinkedGeoData led to similar results). While the more complex approach followed by the greedy scheduler leads to more hits (e.g., 68.6 % more hits on 10,000 resources on LGD), the total runtime that it requires makes it unusable for large datasets. Hence, we chose to use the best-effort approach with a threshold set to 250 ms for the rest of our experiments.

5.4 Combination of GNOME with Existing Caching Approaches

After completing the evaluation of the components of GNOME, we aimed to determine the caching approach with which GNOME performed best. To achieve this goal, we compared the run times and the number of hits achieved by the five caching approaches presented at the beginning of this section on the same data as in the precedent section. The results of our evaluation are shown in Table 3. While the number of hits achieved with the different approaches varies only

Table 2. Average scheduling results on LGD and DBP. The results for 1,000 resp. 10,000 resources are shown in the top resp. bottom section. BE stands for best-effort.

$|C|$	Runtimes (ms)				Hit ratio			
	LGD		DBP		LGD		DBP	
	BE	*Greedy*	*BE*	*Greedy*	*BE*	*Greedy*	*BE*	*Greedy*
100	571.3	1,599.3	887.0	64288.7	0.56	0.68	0.50	0.65
200	565.7	1,448.3	860.3	62305	0.66	0.85	0.50	0.65
400	581.0	1,379.3	918.7	60458.7	0.67	0.88	0.50	0.65
1,000	5,666.0	814,271.7	8825.7	3851388.3	0.51	0.86	0.49	0.73
2,000	6,268.0	810,855.0	9347	3,385,014.3	0.51	0.86	0.50	0.73
4,000	6,675.7	814,041.7	10,666.7	3364521.3	0.50	0.86	0.50	0.73

slightly, the run times achieved by our approach when combined with FIFO are clearly less than those achieved with any other method. These results confirm that our combination of clustering and path ensures a high locality of the data independently of the caching approach used. The difference in runtime is due to the partly complex operations that have to be performed by the cache to detect the data item(s) that is (are) to be evicted when a cache miss comes about and the cache is full. The FIFO strategy being simple means that the cache itself leads to a small overhead, leading to a smaller computation time. Given that these observations hold on both datasets, we opted to combine our method with a FIFO as default setting for carrying out Link Discovery.

Table 3. Average results of our approach using different caches on LGD (top of the table) and DBP (bottom of the table). F2 stands for FIFO second chance. NA means that the approach timed out (time out = 6 hours)

$|C|$	Runtimes (ms)					Hit ratio				
	FIFO	*F2*	*LFU*	*LRU*	*SLRU*	*FIFO*	*F2*	*LFU*	*LRU*	*SLRU*
1,000	4,996.7	7,064.3	8,457.3	14,922.7	17,466.0	0.51	0.50	0.51	0.51	0.51
2,000	5,139.3	7,368.3	9,192.3	15,497.3	17,464.7	0.51	0.51	0.51	0.51	0.50
4,000	5,789.3	7,738.0	9,612.0	15,778.3	18,240.7	0.51	0.51	0.51	0.51	0.51
1,000	8,331.3	11,879.0	NA	8,881.3	12,483.3	0.50	0.50	NA	0.50	0.50
2,000	8,919.0	13,023.7	NA	9,411.0	13,473.3	0.50	0.50	NA	0.50	0.50
4,000	9,866.7	13,684.0	NA	10,385.7	14,431.3	0.50	0.50	NA	0.50	0.50

5.5   Comparison with Existing Approaches

We compared the performance our approach combined with FIFO with that of existing caching approaches, which we used as baselines. To achieve this goal, we measured the overall runtime and hit ratio achieved by our approach with

Table 4. Comparison of GNOME + FIFO with baselines on LGD

Runtimes (ms)

$|C|$	GNOME +FIFO	FIFO	F2	LFU	LRU	SLRU
1,000	5,974.0	37,161.0	42,090.3	45,906.7	54,194.3	56,904.3
2,000	6,168.0	31,977.0	39,071.3	39,872.0	45,473.0	46,795.0
4,000	7,118.3	21,337.0	40,860.0	28,028.3	26,816.7	27,200.0

Hit ratio

$|C|$						
1,000	0.51	0.17	0.16	0.19	0.17	0.17
2,000	0.51	0.29	0.30	0.32	0.30	0.30
4,000	0.51	0.54	0.55	0.59	0.55	0.56

Table 5. Comparison of GNOME + FIFO with baselines on BDP

Runtimes (ms)

$|C|$	GNOME +FIFO	FIFO	F2	LFU	LRU	SLRU
1000	9014.3	10276.3	17620	25870.7	40374.7	47713.7
2000	10018.3	11798.7	19588.3	32793	41693	49243
4000	11001.7	13296	21069.7	44744.7	43189	51317.7

Hit ratio

$|C|$						
1000	0.5	0.23	0.23	0.23	0.23	0.23
2000	0.5	0.23	0.23	0.23	0.23	0.23
4000	0.5	0.24	0.24	0.24	0.24	0.24

that of the baselines approaches on the same data as in the precedent section. Our results are shown in Tables 4 and 5. W.r.t. runtime, we are up to an order of magnitude faster. This result is especially significant when one considers the small number of resources in S and T. We achieve a two-fold improvement of the hit ratio, except for high values of $|C|$ on LGD, where the hit ratio of our approach is approximately 10 % worse than the other caching approaches but where GNOME is still 3 to 5 times faster than standard caching approaches.

5.6 Scalability

We were also interested in how well our approach scales with a growing amount of data. To measure the approach scalability, we used growing dataset sizes and measured the runtimes achieved by GNOME. The results, presented in Table 6, show our performance with a cache size of 10 % of the dataset size, an optimization time of 2.5 s and a threshold of 0.95. Over all experiments, GNOME achieves a hit ratio of 0.5. Moreover, our results show that the runtime of our approach grows linearly with the number of mappings generated by our approach. For example, on LGD, we generate approx. 360 mappings/ms. Interestingly, this

Table 6. Scalability results of GNOME. k stands for 10^3.

	100k	200k	400k	800k
LGD	362,141.3	1,452,922.0	5,934,038.7	20,001,965.7
DBP	434,630.7	1,790,350.7	6,677,923.0	12,653,403.3

number grows slightly with the size of the dataset and reaches 370 mappings/ms for $|S| = |T| = 8 \times 10^5$ resources.

Overall, the evaluation results presented above allow answering all three of the questions which guided our experimental design: pertaining to Q1, our experiments show clearly that the naïve resp. the best-effort approach are to be preferred over the more complex greedy approaches as they lead to a good tradeoff between runtime and hit ratio. The answer to Q2 is that our approach clearly outperforms all the baseline approaches in all settings. This is a positive result as it means that the overhead generated by clustering the data and determining the right sequence for the clusters pays off. Finally, concerning Q3, the results suggest that GNOME scales linearly with the number of mappings generated and thus scales well.

6   Related Work

This work is a contribution to the research area of LD. Several frameworks have been developed to this goal. The LIMES framework [12], in which GNOME is embedded, provides time-efficient algorithms for atomic measures (e.g., PPJoin+ [22] and $\mathcal{HR}^3$ [11]) and combines them by using set operators and filters. Most other systems rely on blocking. For example, SILK [9] relies on MultiBlock to execute LS efficiently. A similar approach is followed by the KnoFuss system [15]. Other time-efficient systems include [21] which present a lossy but time-efficient approach for the efficient processing of LS. Zhishi.links on the other hand relies on a pre-indexing of the resources to improve its runtime [16]. The idea of pre-indexing is also used by ScSLINT [14], who however give up the completeness of results to this end, a concessing we were not willing to make. CODI uses a sampling-based approach to compute anchor alignments to reduce its runtime [8]. Other systems descriptions can be found in the results of the Ontology Alignment Evaluation Initiative [4].[4] The idea of optimizing the runtime of schema matching has also been considered in literature [20]. For example, [17] presents an approach based on rewriting. Still, to the best of our knowledge, GNOME is the first generic approach for link discovery that allows scaling up divide-and-merge algorithms.

Our approach towards improving the scalability of LD is related to clustering, caching, scheduling and heuristics for addressing NP-complete problems. Each of these areas of research has a long history and a correspondingly large body of literature attached to it. Modern graph clustering approaches (see [19] for a survey) will play a central role in our endeavors to parallelize GNOME. The caching

---

[4] http://ontologymatching.org.

problem was studied in the context of hardware and software development. Surveys such as [1,18] point to the large number of solutions that have seen the day of light over the years. We limited ourselves to caching approaches that are used commonly in this work. Similarly to caching, a large number of approaches have been developed to scale up to the TSP, of which fall into the categories deterministic or non-deterministic [6]. While some rely on using more efficient hardware (see,e.g., [5]), most approaches are heuristics that can be run on any hardware and include approaches ranging from genetic programming (see,e.g., [7]) to collective [3] reinforcement learning [2]. Within this work, we wanted to show that the paradigm underlying GNOME outperforms the state of the art. Hence, we limited ourselves to exploring simple approaches towards addressing the TSP. An inclusion of high-performance TSP solvers remains future work.

7 Conclusion and Future Work

We presented an approach for the efficient computation of large divide-and-merge tasks on commodity hardware. We showed that our approach performs well even on large datasets. Moreover, we showed that the scheduling performed by our approach leads to improved runtimes when compared with state-of-art caching approaches. GNOME can be extended in a number of ways. First, we will consider the parallel processing of this approach. To this end, GNOME will be extended with a partitioning algorithm applied to the task graph. Moreover, we will address the determination of the best possible configuration for our approach (e.g., w.r.t. the optimization time used by the best-effort scheduling algorithm). This will also be the subject of future works.

Acknowledgement. This project has received funding from the European Union's Horizon 2020 research and innovation programme under grant agreement No 688227, the DFG project LinkingLOD and the BMWI project SAKE.

References

1. Ali, W., Shamsuddin, S.M., Ismail, A.S.: A survey of web caching and prefetching. Int. J. Adv. Soft Comput. Appl. **3**(1), 18–44 (2011)
2. Dorigo, M., Gambardella, L.M.: Ant-q: a reinforcement learning approach to the traveling salesman problem. In: Proceedings of ML-1995, Twelfth International Conference on Machine Learning, pp. 252–260 (2014)
3. Dorigo, M., Gambardella, L.M.: Ant colony system: a cooperative learning approach to the traveling salesman problem. IEEE Trans. Evol. Comput. **1**(1), 53–66 (1997)
4. Euzenat, J., Ferrara, A., Robert, W., van Hage, L., Hollink, C.M., Nikolov, A., Ritze, D., Scharffe, F., Shvaiko, P., Stuckenschmidt, H., Sváb-Zamazal, O., dos Santos, C.T.: Results of the ontology alignment evaluation initiative. In: OM, 2011 (2011)

5. Fujimoto, N., Tsutsui, S.: A highly-parallel TSP solver for a GPU computing platform. In: Dimov, I., Dimova, S., Kolkovska, N. (eds.) NMA 2010. LNCS, vol. 6046, pp. 264–271. Springer, Heidelberg (2011)
6. Goyal, S.: A survey on travelling salesman problem. In: Proceedings of 43rd Midwest Instruction and Computing Symposium (MICS), 2010 (2010)
7. Grefenstette, J., Gopal, R., Rosmaita, B., Van Gucht, D.: Genetic algorithms for the traveling salesman problem. In Proceedings of the first International Conference on Genetic Algorithms and their Applications, pp. 160–168. Lawrence Erlbaum, New Jersey (1985)
8. Huber, J., Sztyler, T., Nößner, J., Meilicke, C.: Codi: combinatorial optimization for data integration: results for OAEI. In: OM, 2011 (2011)
9. Isele, R., Jentzsch, A., Bizer, C.: Efficient multidimensional blocking for link discovery without losing recall. In: WebDB (2011)
10. Nentwig, M., Hartung, M., Ngonga Ngomo, A.C., Rahm, E.: A survey of current Link Discovery frameworks. Semant. Web, 1–18 (2015) (Preprint)
11. Ngonga Ngomo, A.-C.: Link discovery with guaranteed reduction ratio in affine spaces with Minkowski measures. In: Cudré-Mauroux, P., et al. (eds.) ISWC 2012, Part I. LNCS, vol. 7649, pp. 378–393. Springer, Heidelberg (2012)
12. Ngomo, A.-C.N.: On link discovery using a hybrid approach. J. Data Semant. **1**, 203–217 (2012)
13. Ngomo, A.-C.N., Kolb, L., Heino, N., Hartung, M., Auer, S., Rahm, E.: When to reach for the cloud: using parallel hardware for link discovery. In. Cimiano, P., Corcho, O., Presutti, V., Hollink, L., Rudolph, S. (eds.) ESWC 2013. LNCS, vol. 7882, pp. 275–289. Springer, Heidelberg (2013)
14. Nguyen, K., Ichise, R.: ScSLINT: time and memory efficient interlinking framework for linked data. In: Proceedings of the 14th Internation Semantic Web Conference Posters and Demonstrations Track (2015)
15. Nikolov, A., D'Aquin, M., Motta, E.: Unsupervised learning of data linking configuration. In: Proceedings of ESWC (2012)
16. Niu, X., Rong, S., Zhang, Y., Wang, H.: Zhishi links results for OAEI. In: OM, 2011 (2011)
17. Peukert, E., Berthold, H., Rahm, E.: Rewrite techniques for performance optimization of schema matching processes. In: EDBT, pp. 453–464 (2010)
18. Podlipnig, S., Böszörmenyi, L.: A survey of web cache replacement strategies. ACM Comput. Surv. (CSUR) **35**(4), 374–398 (2003)
19. Schaeffer, S.E.: Graph clustering. Comput. Sci. Rev. **1**(1), 27–64 (2007)
20. Shvaiko, P., Euzenat, J.: Ontology matching: state of the art and future challenges. IEEE Trans. Knowl. Data Eng. **25**(1), 158–176 (2013)
21. Song, D., Heflin, J.: Automatically generating data linkages using a domain-independent candidate selection approach. In: Aroyo, L., Welty, C., Alani, H., Taylor, J., Bernstein, A., Kagal, L., Noy, N., Blomqvist, E. (eds.) ISWC 2011, Part I. LNCS, vol. 7031, pp. 649–664. Springer, Heidelberg (2011)
22. Xiao, C., Wang, W., Lin, X., Jeffrey, X.: Efficient similarity joins for near duplicate detection. In WWW, pp. 131–140 (2008)

RDF Query Relaxation Strategies Based on Failure Causes

Géraud Fokou, Stéphane Jean[⊠], Allel Hadjali, and Mickaël Baron

LIAS/ISAE-ENSMA, University of Poitiers,
1, Avenue Clement Ader, 86960 Futuroscope Cedex, France
{fokou,jean,hadjali,baron}@ensma.fr

Abstract. Recent advances in Web-information extraction have led to the creation of several large Knowledge Bases (KBs). Querying these KBs often results in empty answers that do not serve the users' needs. Relaxation of the failing queries is one of the cooperative techniques used to retrieve alternative results. Most of the previous work on RDF query relaxation compute a set of relaxed queries and execute them in a similarity-based ranking order. Thus, these approaches relax an RDF query without knowing its *failure causes* (*FCs*). In this paper, we study the idea of identifying these FCs to speed up the query relaxation process. We propose three relaxation strategies based on various information levels about the FCs of the user query and of its relaxed queries as well. A set of experiments conducted on the LUBM benchmark show the impact of our proposal in comparison with a state-of-the-art algorithm.

1 Introduction

Recent projects like DBpedia [1] or Knowledge Vault [2] have created Knowledge Bases (KBs) with millions of facts represented in the RDF format. Despite their large size, KBs face a significant amount of incomplete factual knowledge, which makes query answering over them often unsuccessful. For instance, a recent study on SPARQL endpoints [3] shows that ten percent of the submitted queries between May and July 2010 over DBpedia returned empty answers.

Relaxation of the failing queries is one of the cooperative techniques used to retrieve alternative results in order to serve the users' needs. In the context of RDF, current approaches generate multiple relaxed queries using different techniques such as logical relaxation based on RDFS entailment and RDFS ontologies [4–7], query rewriting rules [8], statistical language models [9] or matching functions [10]. Most of these approaches compute the similarities between the obtained relaxed queries and the user failing query and then proceed to the execution of the relaxed queries in a similarity-based ranking order. A major drawback of the above approaches is the fact they relax the user query without knowing its *Failure Causes* (*FCs*).

In our previous work [11], we have addressed the issue of finding the FCs of an RDF query by computing a set of *Minimal Failing Subqueries* (*MFSs*) and argued that they provide the user with a clear explanation about the reasons of

© Springer International Publishing Switzerland 2016
H. Sack et al. (Eds.): ESWC 2016, LNCS 9678, pp. 439–454, 2016.
DOI: 10.1007/978-3-319-34129-3_27

the empty answer retrieved. In this paper, we investigate the idea of using MFSs to perform the query relaxation process. The main idea is that MFSs can speed up this relaxation process by avoiding executing relaxed queries that still contain one or several FCs. This approach applies both for the user query, as well as for the failing relaxed queries. However, as enumerating the MFSs of a query is an NP-hard problem [12], identifying them could be sometimes disadvantageous since the MFSs computation time may be greater than the execution time of the relaxed queries avoided thanks to them. Thus, we show that there is a tradeoff between not knowing any MFSs and identifying the MFSs of each relaxed query. To do so, we propose three strategies that leverage different levels of information about the MFSs of the user query and of its relaxed queries as well. The main contributions made in this paper are the following.

1. Based on previous work [4,5], we define the necessary data structures for relaxing the triple patterns of an RDF query and the query itself.
2. We review the state-of-the-art relaxation strategy and propose three new approaches. By doing so, we cover the full spectrum of information about the MFSs as they are respectively not, partially and fully taken into account in these strategies.
3. We provide a set of experiments on several datasets of the LUBM benchmark that were run on top of Jena TDB and Virtuoso. The analysis of the results shows that to guarantee a relaxation process with an acceptable computation time, a balancing between the information pertaining to the MFSs and the relaxed queries is needed.

This paper is organized as follows. In Sect. 2, we introduce the basic notions used in this paper. The data structures needed for our query relaxation strategies are then defined in Sect. 3. We detail these strategies in Sect. 4 and present our experiments to evaluate them in Sect. 5. Finally, we discuss related work in Sect. 6 and conclude in Sect. 7.

2   Preliminaries and Problem Statement

This section formally describes the parts of RDF and SPARQL that are necessary to our proposal using the definitions given in [13]. We also recall an RDF query relaxation model borrowed from [5].

2.1   Notion of Minimal Failing Subquery (MFS)

An *RDF triple* is a triple (subject, predicate, object) $\in (U \cup B) \times U \times (U \cup B \cup L)$ where U is a set of URIs, B is a set of blank nodes and L is a set of literals. We denote by T the union $U \cup B \cup L$. An *RDF database* (or triplestore) stores a set of RDF triples in a triples table or one of its variants.

An *RDF triple pattern* t is a triple (subject, predicate, object) $\in (U \cup V) \times (U \cup V) \times (U \cup V \cup L)$, where V is a set of variables disjoint from the sets U, B and L. We denote by $var(t)$ the set of variables occurring in t. We consider *RDF queries* defined as a conjunction of triple patterns: $Q = t_1 \wedge \cdots \wedge t_n$. Let D be an RDF database, t a triple pattern and Q an RDF query, the evaluation of t and Q over D are respectively denoted by $[[t]]_D$ and $[[Q]]_D$. This evaluation can be done under different entailment regimes as defined in the SPARQL specification. In this paper, the examples as well as our experiments are based on the RDFS entailment regime.

Given a query $Q = t_1 \wedge \cdots \wedge t_n$, a query $Q' = t_i \wedge \cdots \wedge t_j$ is a *subquery* of Q, $Q' \subseteq Q$, iff $\{t_i, \cdots, t_j\} \subseteq \{t_1, \cdots, t_n\}$. If $\{t_i, \cdots, t_j\} \subset \{t_1, \cdots, t_n\}$, we say that Q' is a *proper subquery* of Q ($Q' \subset Q$). A *Minimal Failing Subquery MFS* of a query Q is defined as follows: $[[MFS]]_D = \emptyset \wedge \nexists\, Q' \subset MFS$ such that $[[Q']]_D = \emptyset$. The set of all MFSs of a query Q is denoted by *mfs(Q)*. Examples of MFSs are given in the next section.

2.2   Query Relaxation Model

Given a triple pattern t, t' is a relaxed triple pattern obtained from t, denoted by $t \prec t'$, if $t' \neq t$ and over every RDF databases D for the given schema, $[[t]]_D \subseteq [[t']]_D$. A *relaxation rule* is a rewrite rule such that its application to a triple pattern results in a relaxed triple pattern. In this paper, we consider the following relaxation rules (*sc* and *sp* are respectively the shorter names for *subClassOf* and *subPropertyOf*):

- *Class relaxation (R1)*. $(s,\ type,\ c_1) \Rightarrow (s,\ type,\ c_2)$ if $(c_1,\ sc,\ c_2)$.
- *Property relaxation (R2)*. $(s,\ p_1,\ o) \Rightarrow (s,\ p_2,\ o)$ if $(p_1,\ sp,\ p_2)$.
- *Constant to variable relaxation (R3)*. If c is a constant occurring in t, then $t \Rightarrow t'$ where t' is the triple obtained by replacing c by a variable $v \notin var(t)$.

Given a triple pattern $t = (s,\ p,\ o)$ and its relaxed triple pattern $t' = (s',\ p',\ o')$, the similarity between t and t' can be defined as follows [5]:

$$Sim(t,t') = \frac{1}{3} * Sim(s,s') + \frac{1}{3} * Sim(p,p') + \frac{1}{3} * Sim(o,o')$$

We use the following similarity measures for the considered relaxation rules:

- If c is a subclass of c', $Sim(c,c') = \frac{IC(c')}{IC(c)}$ where $IC(c) = -logPr(c)$ and $Pr(c) = \frac{|Instances(c)|}{|Instances|}$ (*Instances(c)* is the set of instances of c and *Instances* the set of all instances of the RDF database).
- If p is a subproperty of p', $Sim(p,p') = \frac{IC(p')}{IC(p)}$ where $IC(p) = -logPr(p)$ and $Pr(p) = \frac{|Triples(p)|}{|Triples|}$ (*Triples(p)* is the set of triples concerning p and *Triples* the set of all triples of the RDF database).
- If c is a constant and v a variable, $Sim(c,v) = 0$.

Given a user query $Q = t_1 \wedge \cdots \wedge t_n$ and a query $Q' = t'_1 \wedge \cdots \wedge t'_n$, Q' is a relaxed query of Q, denoted by $Q \prec Q'$, if (i) for each triple pattern t_i, $t_i \preceq t'_i$ (either $t_i = t'_i$ or $t_i \prec t'_i$) and (ii) for at least one triple pattern t_j, $t_j \prec t'_j$.

Triples		
subject	predicate	object
s_1	type	Lecturer
s_1	teacherOf	SW
s_1	age	45
s_2	type	Lecturer
s_2	nationality	US
s_2	age	46
s_3	type	FullProfessor
s_3	teacherOf	DB
s_3	age	46

(a) RDF triples

```
SELECT ?p ?n WHERE {
  ?p type Lecturer      (t₁)
  ?p nationality ?n     (t₂)
  ?p teacherOf SW       (t₃)
  ?p age 46 }           (t₄)
```

(c) The query Q

$mfs(Q) = \{t_2 \wedge t_3, t_3 \wedge t_4\}$
$mfs(Q') = \{t_2 \wedge t_3^{(1)}, \ t_1 \wedge t_3^{(1)} \wedge t_4\}$

(b) The MFSs of Q and Q'

```
SELECT ?p ?n WHERE {
  ?p type Lecturer      (t₁)
  ?p nationality ?n     (t₂)
  ?p teacherOf ?c       (t₃⁽¹⁾)
  ?p age 46 }           (t₄)
```

(d) A relaxed query Q'

Fig. 1. Example of a relaxed query of Q

Figure 1 presents an example of a relaxed query Q' of $Q = t_1 \wedge t_2 \wedge t_3 \wedge t_4$ with their MFSs. In this example $t_3^{(1)}$ is a relaxed triple pattern of t_3.

As for the similarity between a query Q and its relaxed query Q', we use $Sim(Q, Q') = \prod_{i=1}^{n} Sim(t_i, t_i')$. Let D be an RDF database, if $\mu \in [[Q']]_D$ and $\mu \notin [[Q]]_D$, then μ is an *approximate answer* of Q. The approximate answers are ranked thanks to a score defined by: $Score(\mu, Q) = \{ \ max(Sim(Q, Q')) \mid Q \prec Q' \wedge \mu \in [[Q']]_D \ \}$.

Problem Statement. Knowing the set of MFSs of a failing RDF query Q, we are concerned with finding the top-k approximate answers of Q efficiently.

3 Query Relaxation Data Structures

We first define the data structures needed for the proposed relaxation strategies.

3.1 Triple Pattern Relaxation

Let t be a triple pattern. One or several relaxation rules may be applied to t. The same relaxation rules may also be applied several times to the same triple pattern. We denote by $t^{(0)}$ the original triple pattern, by $t^{(i)}$ the i-th best relaxation of t in terms of similarity with t and by $nbRel(t)$ the number of relaxed triple patterns of t. By definition, if $i < j$, then $Sim(t^{(i)}, t) \geq Sim(t^{(j)}, t)$. However, the following relationship does not necessarily hold $t^{(i)} \preceq t^{(j)}$.

Example. Let us assume that *FullProfessor* is a subclass of *Professor*. By applying the previous relaxation rules to the triple pattern $(?X, type, FullProfessor)$, we find the following relaxed triple patterns of t: $t^{(1)} = (?X, type, Professor)$

where $Sim(t^{(1)}, t) = 0.9$ and $t^{(2)} = (?X, ?Y, FullProfessor)$ where $Sim(t^{(2)}, t) = \frac{2}{3}$. Yet, $t^{(1)} \npreceq t^{(2)}$.

Let t be a triple pattern and *ApplyRules(t)* be a function that returns all the relaxed triple patterns of t resulting from the application of the three considered relaxation rules. Algorithm 1 computes the relaxed triple patterns of t ordered by similarity. An example of relaxation of the triple pattern $(?X, type, FullProfessor)$ using our three relaxation rules ($R1$, $R2$ and $R3$) is presented in Fig. 2.

Algorithm 1. Computation of the relaxation of t ordered by similarity

---

Relax(t)

    input : A triple pattern t;

    output: the list of relaxed triple patterns of t : $t^{(0)} \cdots t^{(n)}$;

1    $T \leftarrow \emptyset$; $Res \leftarrow \emptyset$; // Res: resulting list of $t^{(i)}$ sorted by sim

2    $T.enqueue(t)$; // T: priority queue of $t^{(i)}$ sorted by sim

3   **while** $T \neq \emptyset$ **do**

4        $t_i = T.dequeue()$;

5        $Res.enqueue(t_i)$;

6   **foreach** *triple pattern* $t_j \in ApplyRules(t_i)$ **do**

7   **if** $t_j \notin T$ **then**

8                $T.enqueue(t_j)$;

9   **return** Res;

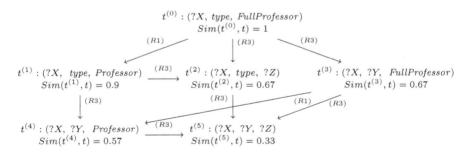

Fig. 2. Relaxation of a triple pattern

3.2   Query Relaxation Graph

Let $Q = t_1^{(0)} \wedge \cdots \wedge t_n^{(0)}$ be the original failing RDF query. The set of relaxed queries of Q is $\{ Q' = t_1^{(i_1)} \wedge \cdots \wedge t_n^{(i_n)} \mid \exists k \in [1, n] : i_k > 0 \}$. Inspired by [5][1],

---

[1] The proposed relaxation graph is not equivalent to the one proposed in [5]. Indeed, an edge between two queries Q_1 and Q_2 does not necessarily mean $Q_1 \preceq Q_2$. This property simplifies the computation of the children of a node in the graph.

we organize this set of relaxed queries in a graph structure. The initial query is at the top of this graph and each relaxed query is a node of this graph. An edge from node $Q_i = t_1^{(i_1)} \wedge \cdots \wedge t_n^{(i_n)}$ to $Q_j = t_1^{(j_1)} \wedge \cdots \wedge t_n^{(j_n)}$ exists if and only if (i) for one triple pattern t_l, $i_l = j_l - 1$ and (ii) for each other triple patterns t_k, $i_k = j_k$. Thus, by construction, $Sim(Q, Q_i) \geq Sim(Q, Q_j)$. This graph has different levels according to the lengths of the paths from the root to relaxed queries. At level h we find all relaxed queries $Q' = t_1^{(i_1)} \wedge \cdots \wedge t_n^{(i_n)}$ such as $\sum_{k=1}^{n} i_k = h$. The number of relaxed queries in this relaxation graph is $\prod_{i=1}^{n}(nbRel(t_i) + 1)$.

Figure 3 gives an example of a relaxation graph for our sample query $Q = t_1 \wedge t_2 \wedge t_3 \wedge t_4$. For simplification, this example assumes that each triple pattern can only be relaxed a single time. We do not give the algorithm to compute this complete query relaxation graph as it is incrementally built in the relaxation strategies proposed in the next section.

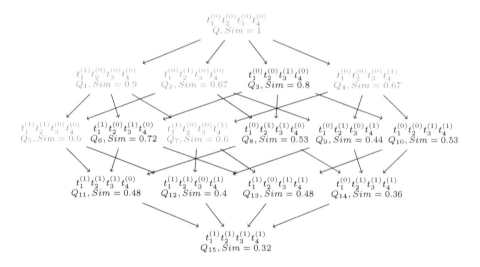

Fig. 3. Query relaxation graph

4   Query Relaxation Strategies

In this section, we first review a state-of-the-art strategy for exploring the query relaxation graph introduced in the previous section. Then we propose three MFS-based strategies.

4.1   Best-First Search (BFS)

It can be easily shown that the h-th best relaxed query Q' of Q is at level h or less of the query relaxation graph. Thanks to this property, the top-k approximate answers could be found with an algorithm that executes the relaxed queries of the graph in the ranking order such as the one proposed in [5]. For example, this algorithm explores the query relaxation graph depicted in Fig. 3 in the following order : $Q_1, Q_3, Q_6, Q_2, Q_4, Q_5, Q_7, Q_8, Q_{10}, Q_{11}, Q_{13}, Q_9, Q_{12}, Q_{14}, Q_{15}$.

In the best-case scenario, this algorithm will only execute one relaxed query to find the top-k approximate answers. In the worst-case scenario, it has to execute all the queries of the graph. As there is an exponential number of relaxed queries (in terms of query size), this algorithm may require an exponential time.

As the causes of the query's failure are unknown in this algorithm, it may execute queries that cannot have any answers and/or relax triple patterns that do not need to be modified. As it will be seen later, the MFSs provide important clues to avoid these pitfalls.

4.2   MFS-Based Search (MBS)

As stated in the following propositon, the MFSs of the failing query identify some relaxed queries that will necessarily fail.

Proposition 1. *Let Q' be a relaxed query of Q. If Q' does not relax at least one triple pattern of each MFS of Q, then Q' is failing.*

Proof. If there is one MFS of Q, denoted Q^*, such as none of its triple patterns has been relaxed in Q', then $Q^* \subseteq Q'$. A query that contains a failing query, also fails. By definition of an MFS, Q^* is a failing query. Thus Q' is also failing.

Based on Proposition 1, we have devised the Algorithm 2 named *MFS-Based Search (MBS)*. This algorithm uses a priority queue RQ of relaxed queries ordered by their similarities with Q. Initially the query Q is added to this queue. It explores each query enqueued in RQ and stops when RQ is empty or when the number of expected answers is obtained (line 3). Each query of RQ is explored as follows. If this query is not labelled as failing, it is executed and its answers are added to the result Res (lines 5-6). Then, all the children of this query that have not already been proceeded (labelled as marked) are added to RQ (lines 7–10). If the added child contains an MFS of Q, this query is labelled as failing (lines 12–13). This way, MBS prunes the search space of the query relaxation graph with failing RDF queries identified with MFSs. In this process, Algorithm 1 is used to find the relaxed triple patterns $t^{(i)}$ of each triple pattern t as well as its number of relaxed triple patterns $nbRel(t)$.

If the MFSs of our query $Q = t_1 \wedge t_2 \wedge t_3 \wedge t_4$ are $t_2 \wedge t_3$ and $t_3 \wedge t_4$, then all the queries in red in Fig. 3 (Q_1, Q_2, Q_4, Q_5, Q_7) can be pruned from the relaxation graph thanks to Proposition 1. Thus, MBS executes the queries in the following order : $Q_3, Q_6, Q_8, Q_{10}, Q_{11}, Q_{13}, Q_9, Q_{12}, Q_{14}, Q_{15}$.

4.3   Optimized MFS-Based Search (O-MBS)

In the previous approach, we only use the MFSs of the initial query to prune the search space. The idea behind the *Optimized MFS-Based Search (O-MBS)*

Algorithm 2. MFS-Based Query Relaxation

Relax(Q, $mfs(Q)$, D, k)

    inputs : A failing query Q ; the set of MFSs of Q : $mfs(Q)$;
            an RDF database D ; the number of expected answers k
    output: a set of top-k approximate results of Q denoted by Res;

1    $Res \leftarrow \emptyset$; $RQ \leftarrow \emptyset$; // the relaxed queries ordered by similarities
2    $RQ.enqueue(Q)$; label Q as failing;
3   **while** $RQ \neq \emptyset \wedge |Res| < k$ **do**
4        $Q' = RQ.dequeue()$;
5   **if** Q' is not labelled as failing **then**
6            $Res \leftarrow Res \cup [[Q']]_D$;

7   **foreach** triple pattern $t_k^{(i_k)} \in Q'$ such that $i_k < nbRel(t_k)$ **do**
8            $Q_c \leftarrow t_1^{(i_1)} \wedge \cdots \wedge t_k^{(i_k+1)} \wedge \cdots \wedge t_n^{(i_n)}$; // a child of Q'
9   **if** Q_c is not labelled as marked **then** // not explored
10               $RQ.enqueue(Q_c)$;
11   label Q_c as marked;
12   **if** $\exists Q^* \in mfs(Q)$ such that $Q^* \subseteq Q_c$ **then**
13   label Q_c as failing;

14   **return** Res;

---

is that the MFSs of the initial query Q give some clues on the MFSs of a relaxed query Q' of Q. Intuitively, a relaxed query Q' of Q fails if and only if at least one MFS of Q has not been repaired in Q' or if there is a failing query in Q' that was not minimal in Q. More formally, let M_Q be an MFS of Q, we denote by $M_Q^{\uparrow Q'}$ the query that corresponds to M_Q in Q'. By extension, we denote by $mfs^{\uparrow Q'}(Q)$ the queries corresponding to the MFSs of Q in Q'. For instance, in the example given in Fig. 1, $mfs^{\uparrow Q'}(Q) = \{t_2 \wedge t_3^{(1)}, t_3^{(1)} \wedge t_4\}$. This example also shows that the following relationship does not necessarily hold: $mfs(Q') \subseteq mfs^{\uparrow Q'}(Q)$. However, as we now prove, each MFS of Q' includes a query of $mfs^{\uparrow Q'}(Q)$.

Proposition 2. *For any MFS $M_{Q'}$ of Q' there is an MFS M_Q of Q such that $M_Q^{\uparrow Q'} \subseteq M_{Q'}$.*

Proof. Let $M_{Q'}$ be an MFS of Q'. By definition $M_{Q'}$ is failing. As $[[M_{Q'}^{\uparrow Q}]] \subseteq [[M_{Q'}]]$, $M_{Q'}^{\uparrow Q}$ is also failing. So, $M_{Q'}^{\uparrow Q}$ contains an MFS M_Q of Q and thus $M_Q^{\uparrow Q'} \subseteq M_{Q'}$.

Thus, if an MFS has been repaired, there can still be some queries that include this MFS and fail. Identifying these new MFSs is not easy. Indeed, the number of queries that include the repaired MFS is exponential in terms of the number of query triple patterns. Thus, the O-MBS strategy is only based on the MFSs that are not repaired. It extends the MBS algorithm (Algorithm 2) as follows. For each relaxed query Q' explored in the query relaxation graph, each query $M_{Q'} \in mfs^{\uparrow Q'}(Q)$ is executed. If the query $M_{Q'}$ is failing, $M_{Q'}$ is an MFS of Q' and thus, all queries that contains $M_{Q'}$ can be pruned from the query relaxation graph (thanks to Proposition 1). To optimize this process, the discovered MFSs of each query Q' explored are recorded. They are denoted $dmfs(Q')$. When a query Q_1 is explored, the O-MBS strategy only executes the MFSs in $dmfs(Q_0)$, where Q_0 is the last explored query such that $Q_0 \prec Q_1$. Indeed, it is unnecessary to execute the MFSs that was already repaired previously by Q_0.

Coming back to our example depicted in Fig. 3, let us assume that the query Q_3 does not repair the MFS $t_2 \wedge t_3$ (i.e., $t_2 \wedge t_3^{(1)}$ is failing). Then, the queries Q_6, Q_{10} and Q_{13} are pruned from the relaxation graph. If none of the MFSs of the following explored queries are discovered, then O-MBS executes the queries in the following order : $Q_3, Q_8, Q_{11}, Q_9, Q_{12}, Q_{14}, Q_{15}$.

4.4   Full MFS-Based Search (F-MBS)

In the previous strategy, all the MFSs of an explored relaxed query are not necessarily discovered. In this section, we propose an approach to compute this complete set of MFSs. By proposing this approach, we want to investigate if it is worth computing the set of MFSs of each explored node of the query relaxation graph, i.e., if this computation time is acceptable in comparison with the number of relaxed queries that are pruned thanks to the discovered MFSs. This strategy called *Full MFS-Based Search* (*F-MBS*) is based on the two following corollaries that are directly derived from Proposition 2.

Corollary 1. *If all the queries $M_Q \in mfs^{\uparrow Q'}(Q)$ are failing, then: $mfs(Q') = mfs^{\uparrow Q'}(Q)$.*

Corollary 2. *Each MFS $M_{Q'}$ of Q' contains the triple patterns that are shared by the queries of $mfs^{\uparrow Q'}(Q)$.*

Thanks to these corollaries, F-MBS extends the O-MBS strategy as follows. For each relaxed query Q', we execute all the MFSs of $mfs^{\uparrow Q'}(Q_0)$, where Q_0 is the last explored query such that $Q_0 \prec Q'$. If all these queries are failing, then $mfs(Q') = mfs^{\uparrow Q'}(Q_0)$ (thanks to Corollary 1). Otherwise, we execute an optimized version of the LBA algorithm [11] to find the MFSs of Q'. As in the previous strategies, the queries that include at least one of the identified MFSs of Q' are pruned from the query relaxation graph.

Because of space limitation, we only describe the main principle of the optimized version of LBA. Let us first describe the main steps of the original version of this algorithm. The LBA algorithm explores the lattice of subqueries of a query Q' built by removing some triple patterns of Q'. It follows a three-steps procedure: (1) find an MFS of Q', (2) compute the maximal queries that do not include the MFS previously found and (3) apply this process recursively on the failing queries previously computed.

Thanks to the discovered MFSs of Q', $dmfs(Q')$, this algorithm is optimized as follows. Instead of executing the first two steps, it directly computes the maximal queries that do not include the MFSs of $dmfs(Q')$. Moreover, using Corollary 2, the search for the next MFS is simplified as we know that it contains the triple patterns shared by the MFSs of $mfs^{\uparrow Q'}(Q_0)$. In the worst case scenario when none of the MFSs were discovered and no triple pattern is shared by the MFSs of $mfs^{\uparrow Q'}(Q_0)$, LBA is executed in its original version (it may cost exponential time in the worst case). In the best case scenario where only one MFS is missing and most of its triple patterns are included in the discovered MFSs, LBA will only execute one query for each missing triple pattern in this MFS.

Consider again the example depicted in Fig. 3 and let us assume that $mfs(Q_3) = \{t_2 \wedge t_3^{(1)},\ t_1 \wedge t_3^{(1)} \wedge t_4\}$. Then, the queries Q_6, Q_{10}, Q_{13} and Q_8 are pruned from the query relaxation graph. If the MFSs of the following explored queries do not help in pruning further the graph, then F-MBS executes the queries in the following order : $Q_3, Q_{11}, Q_9, Q_{12}, Q_{14}, Q_{15}$.

5   Experimental Evaluation

Experimental Setup. We have implemented the MBS, O-MBS and F-MBS algorithms in JAVA 1.8 64 bits. These algorithms take as inputs a failing SPARQL query and a number of expected answers k. They return a maximum of k approximate answers of this query. These algorithms are based on the MFSs of the failing query, which are computed with the LBA algorithm [11]. This implementation can be run on top of any triplestore that supports the SPARQL language. In our experiments, they were run on top of Jena TDB (version 3.0.0) and Virtuoso (version 7.2.1). Our implementation is available at http://www.lias-lab.fr/forge/projects/qars.

Our experiments were conducted on a Ubuntu Server 14.04.02 LTS system with Intel XEON CPU E5-2630 v3 @2.4 Ghz CPU and 32 GB RAM. All times presented are the average of five consecutive runs of the algorithms. Before the actual measured run starts, we run the algorithm once.

Dataset and Queries. As in previous work on RDF query relaxation [5], we used datasets generated with the LUBM benchmark. The used datasets range from LUBM100 (17M triples) to LUBM1K (167M triples). These datasets include both the initial triples generated with the LUBM benchmark and the implicit triples entailed by the RDFS semantics. Statistics on these datasets are precomputed and used later by our algorithms. They are composed of the classes and properties hierarchies, the number of instances by class, the number of triples by property and the total number of instances and triples.

As the workload used in [5] only involves queries with a maximum of 5 triple patterns and 1 MFS, we have modified these 7 queries. The resulting queries given in Table 1[2] cover the main query patterns (star, chain and composite), range between 1 and 15 triple patterns and include 1 up to 4 MFSs.

---

[2] For readability, we shorten URIs and omit namespaces.

Table 1. Workload queries

Q1 (1 MFS)	SELECT * WHERE { ?X type FullProfessor . ?X title 'Dr' }
Q2 (3 MFSs)	SELECT * WHERE { UndergraduateStudent33 advisor ?Y1 . ?Y1 doctoralDegreeFrom ?Y2 . ?Y2 hasAlumnus ?Y3 . ?Y3 title ?Y4 }
Q3 (4 MFSs)	SELECT * WHERE { ?X type FullProfessor . ?X publicationAuthor ?Y1 . ?X worksFor ?Y2 . ?Y3 advisor ?X . ?X title ?Y4 }
Q4 (3 MFSs)	SELECT * WHERE { ?X type UndergraduateStudent . ?X memberOf ?Y1 . ?X mastersDegreeFrom University822 . ?X emailAddress ?Y2 . ?X advisor FullProfessor0 . ?X takesCourse ?Y3 . ?X name ?Y4 }
Q5 (3 MFSs)	SELECT * WHERE { ?X type FullProfessor . ?X doctoralDegreeFrom ?Y1 . ?X memberOf ?Y2 . ?X headOf ?Y1 . ?X title ?Y3 . ?X officeNumber ?Y4 . ?X researchInterest ?Y5 . ?Y6 advisor ?X . ?Y6 name ?Y7 }
Q6 (4 MFSs)	SELECT * WHERE { ?X type Faculty . ?X doctoralDegreeFrom ?Y1 . ?X memberOf ?Y2 . ?X headOf ?Y3 . ?X title ?Y4 . ?X officeNumber ?Y5 . ?X researchInterest ?Y6 . ?X name 'FullProfessor3' . ?X emailAddress ?Y7 . ?X age ?Y8 . ?X mastersDegreeFrom Department2 . ?X undergraduateDegreeFrom ?Y9 }
Q7 (4 MFSs)	SELECT * WHERE { ?X type Professor . ?X teacherOf Course2 . ?X name ?Y1 . ?X age ?Y2 . ?X emailAddress ?Y3 . ?X mastersDegreeFrom ?Y4 . ?X worksFor ?Y5 . ?Y5 subOrganizationOf ?Y6 . ?Y6 name ?Y7 . ?Y8 advisor ?X . ?Y8 mastersDegreeFrom ?Y4 . ?Y8 memberOf ?Y9 . ?Y8 emailAdress ?Y10 . ?Y8 takesCourse ?Y11 . ?Y8 name ?Y12 }

Experiment 1. We have first evaluated the scalability properties of MBS, O-MBS and F-MBS in comparison with our own implementation of the BFS algorithm proposed in [5]. This experiment has been run on Jena TDB with the LUBM100 dataset and k (the number of approximate answers) set to 50. Figures 4 and 5 show respectively the execution time and the number of executed queries for each workload query. For the MBS, O-MBS and F-MBS algorithms, these measures include both the computation of the MFSs and the execution of the relaxed queries.

In this experiment, BFS executes more queries than our algorithms. This difference increases with the size of the query. In particular, for queries Q6 and Q7 that have more than 10 triple patterns and 4 MFSs, this algorithm needs to explore a large part of the query relaxation graph to repair the MFSs. This result in more than 1000 executed queries. In the case of Q2, this difference in the number of executed queries does not imply a larger execution time as the relaxed queries have short execution times. But, for other queries, our fastest

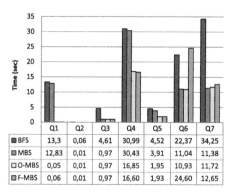

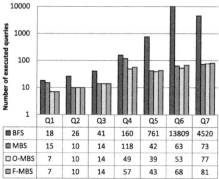

Fig. 4. Execution time (Jena)   **Fig. 5.** # Executed queries (log scale)

algorithm O-MBS outperforms BFS by more than a factor of 2 (average query times go from around 18 s to around 7 s).

Considering our proposed algorithms, O-MBS is better than the other strategies w.r.t minimizing the computation time as well as the number of executed queries in the majority of cases. Only the MBS algorithm executes less query than O-MBS for Q7 and it does not result in a larger execution time. For other queries O-MBS has better performance than MBS and in particular for Q1, Q4 and Q5. Let us consider Q1 to illustrate how O-MBS reduces the number of executed queries. Q1 has only 1 MFS: $(?X, title,' Dr')$ as there is not any *title* in the RDF database. MBS and O-MBS start by relaxing this triple pattern, for instance by $(?X, title, ?Y)$. Then the two algorithms differ in their strategies. O-MBS executes this triple pattern and find that it is still an MFS for the relaxed query. As a consequence it will relax again this triple pattern and find approximate answers. Conversely, as MBS only uses the MFSs of the initial query, this algorithm tries to relax the other triple pattern of the initial query, which will result in relaxed queries that fail. Thus, MBS will execute more queries than O-MBS and its execution time will be significantly longer.

Our last algorithm F-MBS always executes an equal or superior number of queries compared to O-MBS. This behavior is explained by the fact that the number of queries executed to find the MFSs of the relaxed queries is greater than the number of queries pruned in the relaxation graph (thanks to MFSs). We illustrate this fact with Q7. O-MBS executes 25 relaxed queries and 52 queries for computing the MFSs while F-MBS only executes 7 relaxed queries but 74 to compute all the MFSs. In some queries such as Q6, this difference impacts negatively the execution time of F-MBS in comparison with O-MBS.

Experiment 2. In the second experiment, we have evaluated the impact of the triplestore on the performance of our algorithms. Figure 6 presents the execution time of the different algorithms run on top of Virtuoso in the same conditions as the previous experiment. Again, we can observe that our algorithms perform better than BFS. For this latter algorithm, the query Q6 and Q7 took more

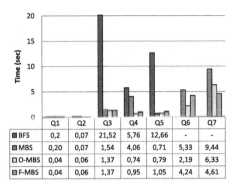

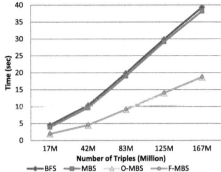

Fig. 6. Execution time (Virtuoso)   **Fig. 7.** Execution time vs data size (Jena)

than 1 hour to execute and thus their execution times are not shown. Even if the trends observed on Jena TDB are confirmed on Virtuoso, the execution times of the algorithms on Virtuoso differ significantly from the ones obtained with Jena TDB. For Q1 and Q4 the execution times are better on Virtuso and conversely for other queries. This is in agreement with the findings of several benchmarks (e.g., [14]) that no triplestore is superior for all queries.

Experiment 3. In the last experiment we have evaluated the scalability of our algorithms when the size of the dataset increases. This experiment was run on Jena TDB with k set to 50. Figure 7 presents the execution time of the different algorithms for the query Q5 when the dataset ranges from 17M to 167M triples. In this experiment, the algorithms scale almost linearly with the size of the repository. We obtained the same result for other queries. As the LUBM generates data that have proportionally similar statistics, our queries have the same MFSs on the different repositories and the same relaxed queries are executed in the same order. As the execution times of the queries scale linearly with the size of the datasets, this explain the result of this experiment.

6   Related Work

We provide here a review of the closest approaches related to our proposal both in the context of RDF and relational databases. In the first setting, Hurtado et al. [4] propose a relaxation approach based on the inferences rules of RDFS. This approach leverages the *subClassOf*, *subPropertyOf*, *domain* and *range* relationships of RDFS to relax a SPARQL query. The end-user can choose the triples that must be relaxed in a query by using the *RELAX* clause. Huang et al. [5] use the same relaxation techniques based on RDFS entailment. To ensure quality of alternative answers, they leverage a semantic similarity measure based on concept statistics. Several optimization techniques are also proposed to obtain the top-k approximate answers efficiently. Elbassuoni et al. [9] show a process for finding similar values of a precise value needed for a query relaxation.

In our previous work [6], we have proposed a set of primitive relaxation operators and have shown how these operators can be integrated in SPARQL in a simple or combined way. Cali et al. [7] have also extended a fragment of this language with query approximation and relaxation operators. As an alternative to query relaxation, *query auto-completion* techniques check the data during query formulation to avoid empty answers (e.g., [15]). It is worth noticing that none of the above approaches has considered the issue related to the causes of an RDF query failure and thus the issue of MFS computation. In our recent work [11], we have addressed this issue but only for providing users with some explanation about this failure.

As for relational databases, many approaches have been proposed for query relaxation (see Bosc et al. [16] for an overview). In particular, Godfrey [12] has defined the algorithmic complexity of the problem of identifying the MFSs of failing relational queries and developed the ISHMAEL algorithm for retrieving them. Jannach [17] studied the concept of MFS in the recommendation system setting. The MFSs computed are used to relax the query at hand by removing some parts of the query. Bosc et al. [16] and Pivert et al. [18] extended Godfrey's approach to the fuzzy queries context. In [16], to speed up the relaxation process, the authors attempt to leverage the MFSs when relaxing the failing query. Unfortunately, this approach does not work in all situations (e.g., when the set of MFSs of the relaxed query is not subsumed by the one of the original query).

As it can be seen, the MFS paradigm has never been used to guide the relaxation process in the context of RDF. To the best of our knowledge, this is the first attempt towards an MFS-driven relaxation of RDF queries.

7   Conclusion and Perspectives

In this paper, we have proposed three strategies to relax an RDF query. Their originalities is that they use different levels of information about the FCs of the initial query and its relaxed queries. In the first strategy, named MBS, only the FCs of the initial query are used to prune from the search space all the relaxed queries that include FCs. The second strategy named O-MBS extends the previous one by searching the FCs that remain in the relaxed query. In this strategy, all the FCs of a relaxed query are not necessarily discovered. Our last strategy named F-MBS fills this gap by using an optimized version of a previous work algorithm to find the FCs of the relaxed queries.

We have run several experiments on the LUBM benchmark with two triple-stores to compare these strategies with a state-of-the-art strategy, which consists in executing the relaxed queries in their ranking order. In these experiments, our best strategy O-MBS outperforms it by more than a factor of 2. O-MBS is a good compromise between MBS, which often does not use enough information about the FCs, and F-MBS which uses too much information about them.

This paper opens many perspectives. As our approach is defined for conjunctive RDF queries, we plan to extend it to support other SPARQL queries.

Studying the relevance of our strategies when they are applied on other query relaxation models that use different relaxation operators is another perspective. As our strategies gather a lot of information about the failure of many queries, we intend to design an interactive approach based on our strategies. Finally, we plan to investigate whether the FCs of a query could be used in conjunction with other cooperative techniques that aim at handling the empty-answer problem.

References

1. Lehmann, J., Isele, R., Jakob, M., Jentzsch, A., Kontokostas, D., Mendes, P.N., Hellmann, S., Morsey, M., van Kleef, P., Auer, S., Bizer, C.: DBpedia - a large-scale, multilingual knowledge base extracted from wikipedia. Semant. Web **6**(2), 167–195 (2015)
2. Dong, X., Gabrilovich, E., Heitz, G., Horn, W., Lao, N., Murphy, K., Strohmann, T., Sun, S., Zhang, W.: Knowledge vault: a web-scale approach to probabilistic knowledge fusion. In: ACM SIGKDD, pp. 601–610 (2014)
3. Saleem, M., Ali, M.I., Hogan, A., Mehmood, Q., Ngomo, A.-C.N.: LSQ: the linked SPARQL queries dataset. In: Arenas, M., et al. (eds.) ISWC 2015. LNCS, vol. 9367, pp. 261–269. Springer, Heidelberg (2015). doi:10.1007/978-3-319-25010-6_15
4. Hurtado, C.A., Poulovassilis, A., Wood, P.T.: Query relaxation in RDF. In: Spaccapietra, S. (ed.) Journal on Data Semantics X. LNCS, vol. 4900, pp. 31–61. Springer, Heidelberg (2008)
5. Huang, H., Liu, C., Zhou, X.: Approximating query answering on RDF databases. J. World Wide Web **15**(1), 89–114 (2012)
6. Fokou, G., Jean, S., Hadjali, A.: Endowing semantic query languages with advanced relaxation capabilities. In: Andreasen, T., Christiansen, H., Cubero, J.-C., Raś, Z.W. (eds.) ISMIS 2014. LNCS, vol. 8502, pp. 512–517. Springer, Heidelberg (2014)
7. Calì, A., Frosini, R., Poulovassilis, A., Wood, P.T.: Flexible querying for SPARQL. In: Meersman, R., Panetto, H., Dillon, T., Missikoff, M., Liu, L., Pastor, O., Cuzzocrea, A., Sellis, T. (eds.) OTM 2014. LNCS, vol. 8841, pp. 473–490. Springer, Heidelberg (2014)
8. Dolog, P., Stuckenschmidt, H., Wache, H., Diederich, J.: Relaxing RDF queries based on user and domain preferences. IJIIS **33**(3), 239–260 (2009)
9. Elbassuoni, S., Ramanath, M., Weikum, G.: Query relaxation for entity-relationship search. In: Antoniou, G., Grobelnik, M., Simperl, E., Parsia, B., Plexousakis, D., De Leenheer, P., Pan, J. (eds.) ESWC 2011, Part II. LNCS, vol. 6644, pp. 62–76. Springer, Heidelberg (2011)
10. Hogan, A., Mellotte, M., Powell, G., Stampouli, D.: Towards fuzzy query-relaxation for RDF. In: Simperl, E., Cimiano, P., Polleres, A., Corcho, O., Presutti, V. (eds.) ESWC 2012. LNCS, vol. 7295, pp. 687–702. Springer, Heidelberg (2012)
11. Fokou, G., Jean, S., Hadjali, A., Baron, M.: Cooperative techniques for SPARQL query relaxation in RDF databases. In: Gandon, F., Sabou, M., Sack, H., d'Amato, C., Cudré-Mauroux, P., Zimmermann, A. (eds.) ESWC 2015. LNCS, vol. 9088, pp. 237–252. Springer, Heidelberg (2015)
12. Godfrey, P.: Minimization in cooperative response to failing database queries. Int. J. Coop. Inf. Syst. **6**(2), 95–149 (1997)
13. Pérez, J., Arenas, M., Gutierrez, C.: Semantics and complexity of SPARQL. ACM Trans. Database Syst. **34**(3), 16:1–16:45 (2009)

14. Bizer, C., Schultz, A.: The Berlin SPARQL benchmark. Semant. Web Inf. Syst. **5**(2), 1–24 (2009)
15. Campinas, S.: Live SPARQL auto-completion. In: ISWC 2014 (Posters & Demos), pp. 477–480 (2014)
16. Bosc, P., Hadjali, A., Pivert, O.: Incremental controlled relaxation of failing flexible queries. JIIS **33**(3), 261–283 (2009)
17. Jannach, D.: Fast computation of query relaxations for knowledge-based recommenders. AI Commun. **22**(4), 235–248 (2009)
18. Pivert, O., Smits, G., Hadjali, A., Jaudoin, H.: Efficient detection of minimal failing subqueries in a fuzzy querying context. In: Eder, J., Bielikova, M., Tjoa, A.M. (eds.) ADBIS 2011. LNCS, vol. 6909, pp. 243–256. Springer, Heidelberg (2011)

CyCLaDEs: A Decentralized Cache for Triple Pattern Fragments

Pauline Folz[1,2(✉)], Hala Skaf-Molli[1], and Pascal Molli[1]

[1] LINA, Nantes University, Nantes, France
{pauline.folz,hala.skaf,pascal.molli}@univ-nantes.fr
[2] Nantes Métropole - Research,
Innovation and Graduate Education Department, Nantes, France

Abstract. The Linked Data Fragment (LDF) approach promotes a new trade-off between performance and data availability for querying Linked Data. If data providers' HTTP caches plays a crucial role in LDF performances, LDF clients are also caching data during SPARQL query processing. Unfortunately, as these clients do not collaborate, they cannot take advantage of this large decentralized cache hosted by clients. In this paper, we propose CyCLaDEs an overlay network based on LDF fragments similarity. For each LDF client, CyCLaDEs builds a neighborhood of LDF clients hosting related fragments in their cache. During query processing, neighborhood cache is checked before requesting LDF server. Experimental results show that CyCLaDEs is able to handle a significant amount of LDF query processing and provide a more specialized cache on client-side.

1 Introduction

Following Linked Data principles, data providers made billions of triples available on the web [4] and the number of triples is still growing [14]. A part of these data is available through public SPARQL endpoints maintained by data providers. However, public SPARQL endpoints have an intrinsic problem of availability as observed in [1]. The Linked Data Fragments (LDF) [16] tackles this issue by balancing the cost of query processing between data providers and data consumers. In Linked Data Fragments, data are hosted in Linked Data Fragments (LDF) servers providing low-cost publication of data, at the same time, SPARQL query processing is moved to the LDF clients side. This approach establishes a trade-off between data availability and performances leveraging the "pressure" on data providers. Consequently, a data provider can provide many datasets through one LDF server at a low-cost as demonstrated in WarDrobe [2] where more than 657,000 datasets are provided with few LDF servers[1].

Caching plays an important role in the performance of LDF servers [17]. Client-side SPARQL query processing using Triple-Pattern Fragments (TPF) generates many calls to LDF server. But as queries are decomposed into triple

---

[1] http://lodlaundromat.org/wardrobe/.

© Springer International Publishing Switzerland 2016
H. Sack et al. (Eds.): ESWC 2016, LNCS 9678, pp. 455–469, 2016.
DOI: 10.1007/978-3-319-34129-3_28

patterns, an important percentage of calls are intercepted by traditional HTTP caching techniques and leverage the pressure on LDF servers. However, HTTP caches are still on the charge of data providers and in the case of multiple datasets, the cache could be useless if a query does not belong to frequently accessed datasets.

During query processing, LDF clients are also caching data, a client replicates triple pattern fragments in its local cache. Unfortunately, as clients do not collaborate, they cannot take advantage of this large decentralized cache hosted by the clients. Building a decentralized cache on client-side has been already addressed by DHT-based approaches [9]. However, DHT-based approaches introduce high latency during the lookup of a content and can slow down the performance of the system. Behave [10] builds a behavioral cache for users browsing the web by exploiting similarities between browsing behaviors of users. Based on past navigation, the browser is directly connected to a fixed number of browsers with similar navigation profile. Consequently, a new requested URL could be checked in the neighborhood cache with a zero-latency connection. A behavior approach has not been applied in the context of the semantic web. Performing SPARQL queries and navigating on the web are different in terms of the number of HTTP calls per-second and clients profiling.

In this paper, we propose CyCLaDEs an approach that allows to build a behavioral decentralized cache hosted by LDF clients. More precisely, CyCLaDEs builds a behavioral decentralized cache based on Triple-Pattern Fragments (TPF). The main contributions of the paper are:

- We present CyCLaDEs an approach to build a behavioral decentralized cache on client-side. For each LDF client, CyCLaDEs builds a neighborhood of LDF clients hosting similar triple pattern fragments in their cache. A neighborhood cache is checked before requesting LDF server.
- We present an algorithm to compute clients profiles. The profile characterizes the content of the cache of LDF client at a given moment.
- We evaluate our approach by extending LDF client with CyCLaDEs. We experiment the extension in different setups, results show that CyCLaDEs reduces significantly the load on LDF server.

The paper is organized as follows: Sect. 2 summarizes related works. Section 3 describes the general approach of CyCLaDEs. Section 4 defines CyCLaDEs model. Section 5 reports our experimental results. Finally, conclusions and future works are outlined in Sect. 6.

2   Related Work

Improving SPARQL query processing with caching has been already addressed in the semantic web. Martin et al. [13] proposes to cache query results and manage cache replacement, Schmachtenberg [15] proposes a semantic query caching relying on queries similarity, and Hartig [11] proposes caching to improve efficiency and results completeness in link traversal query execution. All these approaches rely on a temporal locality where specific data are supposed to be reused again with a relatively small time duration, and caching resources are provided by data providers. CyCLaDEs relies on behavioral locality where clients with similar profiles are directly connected, and caching resources are provided by data consumers.

The Linked Data Fragments (LDF) [16,17] propose to shift complex query processing from servers to clients to improve availability and scalability of SPARQL endpoints. A SPARQL query is decomposed into triple patterns, an LDF server answers triple patterns and sends data back to the client. The client performs joins operations based on the nested loop operators, the triples patterns generated during the query processing are cached in the LDF client and in the traditional HTTP cache in front of the LDF Server. Although, a SPARQL query processing increases the number of HTTP requests to the server, a large number of requests are intercepted by the server cache reducing significantly the load on LDF server as demonstrated in [17]. LDF relies on a temporal locality, and the data providers have to provide resources for the data caching. Compared to other caching techniques in the semantic web, the LDF cache results of a triple pattern, increasing their usefulness for other queries, $i.e$, the probability of a cache hit is higher than the caching of a SPARQL query results. CyCLaDEs aims at discovering and connecting dynamically LDF clients according to their behaviors. CyCLaDEs makes the hypothesis that clients perform a limited query mix, consequently, a triple pattern of a query could be answered in a neighbor cache. To build a decentralized behavior cache, each LDF client must have a limited number of neighbors with a zero-latency access. During query processing, for each triple pattern subquery, CyCLaDEs checks if the triple pattern can be answered in the local cache, if not, in the cache of neighbors. A request is sent to LDF server only if the triple pattern cannot be answered neither in the local cache nor in the neighbors cache. CyCLaDEs improves LDF approach by hosting behavioral caching resources on the clients-side. Behavior cache reduces calls to an LDF server, especially, when the server hosts multiple datasets, the HTTP cache could handle frequent queries on a dataset but cannot absorb all calls. In other words, unpopular queries will not be cached in the HTTP cache and will be answered by the server. In CyCLaDEs, the neighborhood depends on fragments similarity which means that clients are gathered in communities depending on their past queries. By doing that unpopular or less frequent queries can be handled in the cache of the neighbors.

Decentralized cooperative caches were proposed in many research areas. Dahlin et al. [8] proposes a cooperative caching to improve file system read response time. By analyzing existing large-scale Distributed File Systems (DFS) workloads, Blaze [6] discovers that large proportion of "cache miss" is for files that are already copied in another client's cache. Blaze proposes dynamic hierarchical caching to reduce "cache miss" traffic for DFS and server's load. Research on peer-to-peer-oriented Content Delivery Networks (CDN) propose a decentralized web cache such as Squirrel [12], FlowerCDN [9] and Behave [10]. Squirrel and FlowerCDN use Distributed Hash Table (DHT) for indexing all content at all peers. If such approaches are relevant, querying the cache is expensive in term of latency. With n participants, a DHT requires $log(n)$ access to check the presence of a key in the DHT. As LDF query processing can generate thousands of sub-calls, DHT latency becomes a bottleneck and querying the DHT is considerably less performant than querying directly the LDF server.

Behave [10] is a decentralized cache for browsing the web. It is based on the Gossple approach [3]. The basic hypothesis is: if two users had visited the same web page in the past, they will likely to exhibit more common interests in the future. Based on this assumption, Behave relies on gossiping techniques to build dynamically a fixed-size neighboring for clients based on their profile, *i.e.*, their past HTTP access. When requesting a new URL, Behave is able to quickly checks if this URL is available in neighborhood. Compared to the DHT, the number of items available in Behave cache is smaller, but they are available with zero-latency, *i.e.*, with a direct socket or web socket connection. The available items are also personalized, they are based on the behavior of the client rather than a temporal locality.

In CyCLaDEs, we want to apply the general approach of Behave for LDF clients. However, compared to the human browsing, an LDF client could process a large number of queries per second and the local cache of the client could change quickly. We make the hypothesis that the clients processed same queries in the past will likely process similar queries in the future. We build a similarity metric by counting the number of predicates in the triple patterns on a sliding window. We demonstrate that this metric is efficient for building a decentralized cache for LDF clients.

3   CyCLaDEs Motivation and Approach

In CyCLaDEs, we make the assumption that clients who processed same queries in the past will likely process similar queries in the future. This is the case of a web applications proposing forms to the end-users and then executes parametrized SPARQL queries. The Berlin SPARQL Benchmark (BSBM) is built like that [5]. BSBM supposes a realistic web application where the users can browse products and reviews. BSBM generates a query mix based on 12 queries template and 40 predicates.

CyCLaDEs aims to build a behavioral decentralized cache for LDF query processing based on the similarities of LDF clients profiles. For each client, CyCLaDEs selects a fixed number of the best similar clients called *neighbors* and establishes a direct network connections with them. During a query processing on a given client, each triple pattern subquery is checked first on the local cache, next in the cache of the neighbors, before contacting the LDF server, if necessary. Because CyCLaDEs adds a new verification in the neighborhood, checking the cache of neighbors quickly is essential for the performance of CyCLaDEs. We expect that the orthogonality of behavioral cache hosted by the data consumers, and the temporal cache hosted by the data providers will reduce significantly the load on the LDF servers.

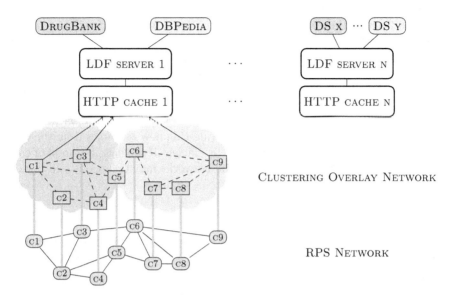

Fig. 1. c1–c9 represents LDF clients executing queries on LDF server 1. The RPS network connects clients in a random graph. CON network connects the same clients (red link) according to their queries. c1–c4 performs queries on DrugBank. c6–c9 perform queries on DBpedia. c5 performs queries on both. The total number of LDF servers is N (Color figure online).

In order to build a neighborhood and handle the dynamicity of the clients, we follow the general approach of Gossple [3]. CyCLaDEs builds two overlay networks on the client-side :

1. a *Random Peer Sampling (RPS)* overlay network that maintains the membership among connected clients. We rely on the Cyclon protocol [18] to maintain the network. Each client maintains a partial view on the entire network. The *view* contains a random subset of network nodes. Periodically, the client selects the oldest node from its view and they exchange parts of their views. This view is used to bootstrap and maintain the clustering network.
2. a *Clustering Overlay Network (CON)* builds on top of RPS, it clusters clients according to their profile. Each client maintains a second view, this view contains the *k-best* neighbors according to the similarity of their profile with the client profile. The maintenance of *k-best* neighbors is performed at RPS exchange time. To minimize the overhead of shuffling: (1) the profile informations have to be as small as possible, (2) the similarity metric has to be computed quickly in order to prevent slowing down the query engine.

Figure 1 shows LDF clients, clients $c1$–$c4$ performs queries on DrugBank, $c6$–$c9$ performs queries on DBpedia and $c5$ performs queries on both. The RPS network ensures that all clients are connected through a random graph, clients profiles make the clustered network converging towards two communities. $c1$–$c4$ will be highly connected because they access data over DrugBank, while $c6$–$c9$ will be grouped together due their interest in DBpedia. $c5$ could be connected to both communities because it performs query on DrugBank and DBpedia.

Thanks to the clustered overlay network, a client is now able to check the availability of triple patterns in its neighborhood before sending the request to the HTTP cache. Under the hypothesis of profiled clients, the behavioral cache should be able to handle a significant number of triple pattern queries and to scale with the number of clients without requesting new resources from the data providers. Of course, the behavioral cache is efficient if the neighborhood of each client is pertinent and the overhead of the networks maintenance is still low.

4   CyCLaDEs Model

In this section, we detail the model of the overlay networks built by CyCLaDEs on the client-side.

4.1   Random Peer Sampling

Random Peer Sampling (RPS) protocols [18] allow each client to maintain a view of a random subset of the network called *neighbors*. A view is a fixed-size table, associating a client ID to an IP address. The size of this view can be set to $log(N)$, where N is the total number of the node in the network. RPS protocols ensure that the network converges quickly to a random graph with no partitions, *i.e.*, a connected graph.

To maintain the connectivity of the overlay network, a client periodically selects the oldest node from its view and they exchange parts of theirs views. These periodic shuffling of the views of the clients ensures that each client view always contains a random subset of the network nodes and consequently maintains the clients connected through a random graph.

In CyCLaDEs, to ease the joining of the network, LDF server maintains a list of three last connected clients, *i.e*, called *bootstrap clients*. Each time a new client joins the network, *i.e.*, contacts the LDF server, the client receives automatically a list of the three last connected clients and add randomly one of them in its view. Periodic shuffling quickly re-establish the random graph property on the network including the new client.

4.2   LDF Client Profiles

The Clustering Overlay Network (CON) relies on the LDF client profile. A client profile has to characterize the content of the local cache of a client. At a given instant, the content of the cache is determined by the result of recent past processed queries.

The cache of a LDF client is a list of $(key, value)$ fixed-size LRU cache. The *key* is a triple pattern fragment where the predicate is a constant and the *value* is the set of triples that matches the fragment [16]. Each fragment matching a triple pattern fragment is divided into pages, each page contains 100 triples. The fragment is filled asynchronously and can be "incomplete", *e.g.*, a fragment $f1$ matches 1,000 triples, but currently only the first 100 triples has been retrieved from the server.

To illustrate, suppose that a LDF client is processing the following SPARQL query:

```
SELECT DISTINCT ?book ?author
WHERE {
    ?book  rdf:type  dbpedia−owl:Book;         tp1
           dbpedia−owl:author  ?author.         tp2
}
LIMIT 5
```

Listing 1.1. Q: Authors of books

The query is decomposed into triple pattern $tp1$ and $tp2$. The local cache will be asynchronously populated as described in Table 1. Because the number of matches of books (31,172) is smaller than those of authors (39,935), LDF client starts by retrieving books. Entry 0 contains $tp1$ with empty data, entry 1 contains $tp2$ with some data. LDF client starts by retrieving books and starts the nested loop to retrieve authors for a given book. Entries 2–9 contain all one triple as answer, but only the first five are needed to be retrieved to answer the query with *Limit* 5. Several strategies are possible to compute a profile of a LDF client:

– *Cache key*: we can consider a vector of keys of the cache as in Behave [10] and we reduce the dimension of the vector with a bloom filter. However, nested loop processing makes LDF quickly override the whole cache with the next query. If a new query searching for French authors is executed, then the nested loop will iterate on French authors instead of books and will completely rewrite the cache given in Table 1. Consequently, the state of the cache at a given time, do not reflect the near past.
– *Past queries*: we can analyze statically the past executed queries and extract processed predicates. Unfortunately, this does reflect the join ordering decided at run-time by LDF client and cannot take into account nested loops.
– *Count-min sketch*: we can use the count-min sketch [7] to analyze the frequency of processed predicates from the beginning of the session. However, count-min sketch does not forget and will not capture the recent past.

In CyCLaDEs, we want to define the profile in spirit of *count-min sketch* but with a short time memory, *i.e.*, a memory of the recent past. We denote the profile of a client c by $Pr(c) = \{(p, f_p, t)\}$, where Pr is a view of a fixed-size on the stream of the triple patterns processed by the client, p is a predicate in the triple pattern in the stream, f_q is the frequency and t is the timestamp of the last update of p. To avoid to mix predicates retrieved from different data sources, we concat the predicate and the provenance of predicate. For example, the general predicate *rdfs:label* retrieved from DBpedia should not be used with the same

Table 1. LDF client cache after execution of query in Listing 1.1

	key	triples
0	?book http:.../ontology/author ?author	⊔
1	'?book' http://www.w3.org/1999/02/22-rdf-syntax-ns#type http:.../ontology/Book	http:.../resource/%22...And_Ladies_of_the_Club%22 http://www.w3.org/1999/02/22-rdf-syntax-ns#type http:.../ontology/Book ... http:.../resource/%22K%22_Is_for_Killer http://www.w3.org/1999/02/22-rdf-syntax-ns#type http:.../ontology/Book
2	http:.../resource/%22...And_Ladies_of_the_Club%22 http:.../ontology/author ?author	http:.../resource/%22...And_Ladies_of_the_Club%22, http:.../ontology/author, http:.../resource/Helen_Hooven_Santmyer
3	http:.../resource/%22A%22_Is_for_Alibi http:.../ontology/author ?author	http:.../resource/%22A%22_Is_for_Alibi, http:.../ontology/author, http:.../resource/Sue_Grafton
4	http:.../resource/%22B%22_Is_for_Burglar, http:.../ontology/author, ?author	http:.../resource/%22B%22_Is_for_Burglar, http:.../ontology/author, http:.../resource/Sue_Grafton
5	http:.../resource/%22C%22_Is_for_Corpse, http:.../ontology/author, ?author	http:.../resource/%22C%22_Is_for_Corpse, http:.../ontology/author, http:.../resource/Sue_Grafton
6	http:.../resource/%22D%22_Is_for_Deadbeat, http:.../ontology/author, ?author	http:.../resource/%22D%22_Is_for_Deadbeat, http:.../ontology/author, http:.../resource/Sue_Grafton
7	http:.../resource/%22E%22_Is_for_Evidence, http:.../ontology/author, ?author	⊔
8	http:.../resource/%22F%22_Is_for_Fugitive, http:.../ontology/author, ?author	⊔
9	http:.../resource/%22G%22_Is_for_Gumshoe, http:.../ontology/author, ?author	⊔

Algorithm 1. ComputeProfile(s,w,t)

Require: : w: Window size, s: Stream of triples, t: timestamp
Ensure: : Pr : set of (predicate, frequency, timestamp) of size w
1: $Pr \rightarrow \varnothing$
2: **while** data stream continues **do**
3:   Receive the next streaming triple $tp = (s\ p\ o)$
4:   **if** $(tp.p, f_{p,\_})$ Pr **then**
5:     $Pr.update(tp.p,\ f_p + 1, t)$ {accumulate the frequency of the predicate p and update time}
6:   **else**
7:     Pr $(tp.p, 1, t)$ {add the new predicate p to the profile}
8:   **if** $|Pr| > w$ **then**
9:     $Pr \setminus (p_1, f_{p_1}, t_1) : (p_1, f_{p_1}, t_1)$ Pr ∄ (p_2, f_{p_2}, t_2) $Pr : t_2 < t_1$ {delete the oldest predicate from the profile}

---

predicate retrieved from DrugBank. In order to simplify notation, we just keep *predicate* that should be expanded to the couple *(provenance,predicate)*.

The Algorithm 1 presents CyCLaDEs profiling procedure. CyCLaDEs intercepts the stream of the processed triples and extracts the predicates. If the predicate belongs to the profile, the frequency of this predicate is incremented by one and the timestamp associated to this entry is updated. Otherwise, CyCLaDEs just insert a new entry in the profile. If the structure exceeds w entries, then CyCLaDEs removes the entry with the oldest timestamp. This profiling algorithm is designed to tolerate nested loops and forget predicates which are not used frequently. For the client whose cache is detailed in Table 1, after the entry 4 in the cache, the profile will be:

{(http://www.w3.org/1999/02/22-rdf-syntax-ns#type , 1) ,
 (http://dbpedia.org/ontology/author, 3)}

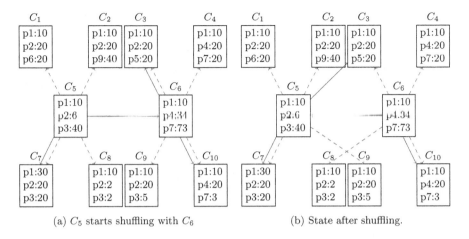

(a) C_5 starts shuffling with C_6 (b) State after shuffling.

Fig. 2. Partial CyCLaDEs network centred on C_5, C_6. Solid lines represent clients in RPS view (2). Dashed lines represent clients in CON view (4). Each client has a profile size of 3 defined as *predicate* : *frequency*.

4.3   Clustered Network and Similarity Metric

CyCLaDEs relies on a random peer sampling overlay network for managing memberships and on a clustered overlay network to manage the *k-best* neighbors. Concretely, the clustered network is just a second view on the network hosted by each client. This view is composed of the list of *k-best* neighbors with similar profiles. The view is updated during shuffling phase, when a client starts shuffling, it selects the oldest neighbor in its RPS view and they exchange profile informations, if the remote client has better neighbors in its view, then the local view is updated in order to keep the *k-best* neighbors.

To determine if a profile is better than another one, we use the generalized Jaccard similarity coefficient defined as:

$$J(x,y) = \frac{\sum_i min(x_i, y_i)}{\sum_i max(x_i, y_i)}$$

where x and y are two multi-sets and the natural numbers $x_i \geq 0$ and $y_i \geq 0$ are the multiplicity of item i in each multiset.

Figure 2 describes a CyCLaDEs network focused on C_5, C_6. The RPS view size is fixed to 2 and is represented as solid lines. The CON view size is fixed to 4 and is represented as dashed lines. Each client has a profile of size 3 that contains the last 3 mostly used predicates in the recent past. p_i represents a predicate and the associated integer indicates the frequency of this predicate in the recent past. Figure 2a illustrates the state of C_5, C_6 before C_5 triggered a shuffling with C_6. C_6 is chosen because it is the oldest client in RPS view of C_5. Figure 2b describes the state of C_5, C_6 after completion of shuffling. As we can see, only one RPS neighbor is changed for both C_5 and C_6. This is the result of exchanging half of RPS view between C_5 and C_6 as in Cyclon [18].

For CON views, C_5 integrated C_9 in its cluster while C_6 integrated C_8. During shuffling, C_5 retrieves the profiles of the CON view of C_6 including C_6. Next, it ranks all profiles according to the generalized Jaccard coefficient and keeps only the top-4. C_9 is more similar to C_5 than C_8 because $J(C_5, C_9) = 0,3$, and $J(C_5, C_8) = 0,25$ therefore, C_5 drops C_8 and integrates C_9. C_6 follows the same procedure by dropping C_9 and integrating C_8.

5   Experimental Study

The goal of the experimental study is to evaluate the effectiveness of CyCLaDEs. We measure mainly the *hit-ratio*; the fraction of queries answered by the decentralized cache.

5.1   Experimental Setup

We extended the LDF client[2] with the CyCLaDEs model presented in Sect. 4. CyCLaDEs source code is available at: https://github.com/pfolz/cyclades[3]. The setup environment is composed of an LDF server, a reverse proxy and different number of clients. Nginx is the reverse proxy with a cache set to 1 GB. We used Berlin SPARQL Benchmark (BSBM) [5] as in [16] with two datasets: 1M and 10M. We randomly generated 100 different query mix of the "explore" use-case of BSBM. Each query mix is composed of 25 queries and each client has its own query mix.

Table 2. Experimental parameters

Parameter	Values
Number of clients	10 - 50 - 100
RPS view	4 - 6 - 7
CON view	9 - 15 - 20
Local cache	100 - 1000 - 10000
Profile size	5 - 10 - 30
Shuffle time	10 s
Data sets	BSBM 1M - BSBM 10M
Queries	25 over BSBM

Table 2 presents the different parameters used in the experiment. We vary the value of parameters according to the objective of the experimentations as

---

[2] https://github.com/LinkedDataFragments/Client.js.
[3] The current implementation does not handle the introduction to the network and fragment transfer, *i.e*, data are retrieved from the LDF server eventually.

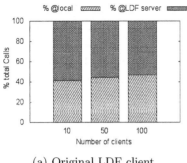

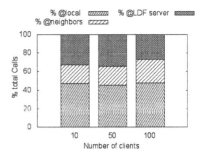

(a) Original LDF client   (b) LDF client with CyCLaDEs

Fig. 3. Impacts of clients number on hit-rate: (10 clients, $RPS_{view} = 4$, $CON_{view} = 9$), (50 clients, $RPS_{view} = 6$, $CON_{view} = 15$) and (100 clients, $RPS_{view} = 7$, $CON_{view} = 20$).

explained in the following sections. The shuffle time is fixed to 10 s for all experiments. For all experimentations, we first run a warm-up round and start the real round after a synchronization barrier. The warm-up round bootstraps the network, local cache and HTTP cache. In both rounds, each client executes its own query mix of 25 queries in the same order. Hit-ratio is measured during the real round.

Impact of the Number of the Clients on the Behavioral Cache. To study the impact of the number of clients on the behavior cache, we used BSBM dataset with 1M with a local cache of size 1,000 on each client. The RPS view size and CON view size are fixed to (4,9) for 10 clients, (6,15) for 50 clients, and (7,20) for 100 clients.

Figure 3a presents results of the LDF clients without CyCLaDEs. As we can see, ≈40 % of calls are handled by the local cache, regardless the number of clients. The flow of BSBM queries simulates a real user interacting with a web application. This behavior promotes the local cache.

Figure 3b describes the results obtained with CyCLaDEs activated. The performances of local cache is nearly the same ≈40 % of calls are handled by the local cache. However, ≈22 % of total calls are answered in the neighborhood, consequently, the number of calls to the LDF server are considerably reduced. Moreover, for 100 clients number of calls answered by the server and the neighborhood are nearly the same. Because the CON view size for 100 clients is larger than those of 10 or 50 clients.

Impact of the Size of the Data Sets on the Behavioral Cache. For this experimentation, we used two datasets, BSBM with 1M triples and BSBM with 10M triples, a local cache of 1,000, a profile view of size 10 and 10 LDF clients. $RPS_{view} = 4$ and $CON_{view} = 9$, as in previous experiment. Figure 4a shows the percentage of calls answered in the local cache, neighbors caches and in the

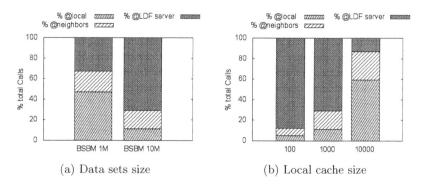

(a) Data sets size   (b) Local cache size

Fig. 4. Impacts of data sets size and local cache size on hit-rate. For 10 LDF clients with $RPS_{view} = 4$, $CON_{view} = 9$ and $Profile_{view} = 10$.

LDF server using the two datasets. As we can see, the calls to the local cache depends considerably on the size of the data, the percentage of hit-rate is 47 % in the case of BSBM with 1M, and it decreased to 11 % for BSBM with 10M. This is normal because the cache has a limited size and the temporal locality of the cache reduce its utility. However, the behavior cache calls stay stable with a hit-rate around 19 % for both data sets.

Impact of the Cache Size. We study the impact of the local cache size on the hit-rate of behavioral cache. We used the following parameters: BSBM 10M, 10 LDF clients, and $RPS_{view} = 4$ and $CON_{view} = 9$. Figure 4b shows that the number of calls answered by caches are proportional with the size of the cache. For local cache with 100 entries, the hit-rate of local cache and behavioral cache nearly equivalent 5 % for the local cache and 7 % for the behavioral cache. For local cache with 1,000 entries, the hit-rate behavioral cache is 18 %, greater than the hit-rate of the local cache of 11 %. Behavioral cache is more efficient than local cache. The situation changes for a local cache with 10,000 entries, in this case, the hit-rate of local cache is 59 % and 28 % for behavioral cache, only 13 % of calls are forwarded to the server.

Impact of the Profile Size on Cache-Hit. We run an experimentation with 2 different BSBM datasets of 1M, hosted on the same LDF server with 2 differents URLs. Each dataset has its own community of 50 clients running BSBM queries. As pointed out in Sect. 4.2, we use provenance to differentiate predicates in local cache of LDF clients. All clients run with $RPS_{view} = 6$, $CON_{view} = 15$ and a cache size of 1,000 entries.

We vary profile size to 5, 10 and 30 predicates. Figure 5 shows that performances of CyCLaDEs are quite similar. However, the performances with $profile_{size} = 5$ is less than $profile_{size} = 10$ or 30. The query mix of BSBM use often 16 predicates. Therefore, 5 entries in the profile is sometimes not enough to compute a good similarity.

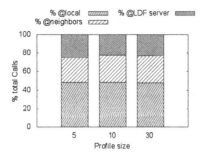

Fig. 5. Impacts of profile size on hit-rate for two datasets with 50 clients per dataset. $RPS_{view} = 6$ and $CON_{view} = 15$. $Profile_{size} = 5, 10$ and 30

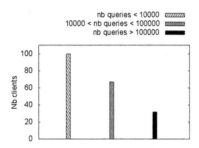

Fig. 6. Query distribution over clients

Query Load. As in the previous experimentation, we run a new experimentation with 2 different BSBM datasets of 1M hosted on the same LDF server with 2 different URLs.

Figure 6 shows the distribution of queries over clients. We want to verify if there is a hotspot, *i.e.*, one client receiving many cache queries from the others. As we can see, most of the clients handle 10,000 caches queries and a few handle more than 100,000 cache queries.

Impacts of the Profile Size on the Communities in the Clustering Overlay Network. As in the previous experimentations, we set up 2 BSBM datasets of 1M with 50 clients per dataset. We vary the profile size to 5, 10 and 30. In the Fig. 7, the directed graph represents the clustering overlay network where a node represents a client in the network and an edge represents the connection between clients. For example, an edge $1 \rightarrow 2$ means that the *client 1* has the *client 2* in its CON view. Figure 7 shows clearly that CyCLaDEs is able to build two clusters for both values of profile size. As we can see in Fig. 7b, a greater value of profile size promotes the clustering, *i.e.*, only clients with similar profiles will receive queries to retrieve fragments.

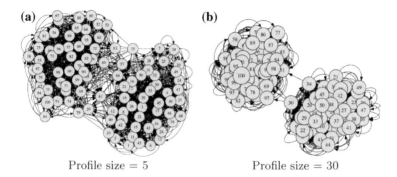

Profile size = 5   Profile size = 30

Fig. 7. Impacts of data profile size on the similarity in Clustering Overlay Network. Two distinct communities are discovered for two datasets.

6 Conclusion and Future Work

In this paper, we presented CyCLaDEs, a behavioral decentralized cache for LDF clients. This cache is hosted by clients and completes the traditional HTTP temporal cache hosted by data providers.

Experimental results demonstrate that a behavioral cache is able to capture a significant part of triple pattern fragment queries generated by LDF query processing, in the context of web applications as described in Sect. 3. Consequently, it leverages the pressure on data providers resources, by spreading the cost of query processing on clients. We proposed a cheap algorithm able to profile the subqueries processed on a client and gather the best neighbors for a client. This profiling has been proven effective in experiments. In this paper, we demonstrated how to bring data to queries with caching techniques, another approach could be to bring queries to data by choosing among neighbors, if a neighbor is able to process more than one triple pattern of a query. The promising results we obtained during experimentations encourage us to propose and experiment new profiling techniques that take into account the number of transferred triples and compare with the current profiling technique.

References

1. Buil-Aranda, C., Hogan, A., Umbrich, J., Vandenbussche, P.-Y.: SPARQL web-querying infrastructure: ready for action? In: Alani, H., et al. (eds.) ISWC 2013, Part II. LNCS, vol. 8219, pp. 277–293. Springer, Heidelberg (2013)
2. Beek, W., Rietveld, L., Bazoobandi, H.R., Wielemaker, J., Schlobach, S.: LOD laundromat: a uniform way of publishing other people's dirty data. In: Mika, P., et al. (eds.) ISWC 2014, Part I. LNCS, vol. 8796, pp. 213–228. Springer, Heidelberg (2014)
3. Bertier, M., Frey, D., Guerraoui, R., Kermarrec, A.-M., Leroy, V.: The gossple anonymous social network. In: Gupta, I., Mascolo, C. (eds.) Middleware 2010. LNCS, vol. 6452, pp. 191–211. Springer, Heidelberg (2010)

4. Bizer, C., Heath, T., Berners-Lee, T.: Linked data - the story so far. Int. J. Semant. Web Inf. Syst. **5**(3), 1–22 (2009)
5. Bizer, C., Schultz, A.: The Berlin SPARQL benchmark. Int. J. Semant. Web Inf. Syst. **5**(2), 1–24 (2009)
6. M. A. Blaze. Caching in large-scale distributed file systems. Ph.D. thesis, Princeton University, Princeton, NJ, USA (1993). UMI Order No. GAX93-11182
7. Cormode, G., Muthukrishnan, S.: An improved data stream summary: the count-min sketch and its applications. J. Algorithms **55**(1), 58–75 (2005)
8. Dahlin, M.D., Wang, R.Y., Anderson, T.E., Patterson, D.A.: Cooperative caching: using remote client memory to improve filesystem performance. In: 1st USENIX Conference on Operating Systems Design and Implementation (OSDI 1994), Berkeley, CA, USA (1994)
9. El Dick, M., Pacitti, E., Kemme, B.: Flower-CDN: a hybrid P2P overlay for efficient query processing in CDN. In: Proceedings of 12th International Conference on Extending Database Technology: Advances in Database Technology, EDBT 2009, pp. 427–438. ACM, New York (2009)
10. Frey, D., Goessens, M., Kermarrec, A.-M.: Behave: behavioral cache for web content. In: Magoutis, K., Pietzuch, P. (eds.) DAIS 2014. LNCS, vol. 8460, pp. 89–103. Springer, Heidelberg (2014)
11. Hartig, O.: How caching improves efficiency and result completeness for querying linked data. In: WWW 2011 Workshop on Linked Data on the Web, Hyderabad, India, 29 March 2011
12. Iyer, S., Rowstron, A., Druschel, P.: Squirrel: a decentralized peer-to-peer web cache. In: 21st Annual Symposium on Principles of Distributed Computing (PODC 2002), pp. 213–222. ACM, New York (2002)
13. Martin, M., Unbehauen, J., Auer, S.: Improving the performance of semantic web applications with SPARQL query caching. In: Aroyo, L., Antoniou, G., Hyvönen, E., ten Teije, A., Stuckenschmidt, H., Cabral, L., Tudorache, T. (eds.) ESWC 2010, Part II. LNCS, vol. 6089, pp. 304–318. Springer, Heidelberg (2010)
14. Schmachtenberg, M., Bizer, C., Paulheim, H.: Adoption of the linked data best practices in different topical domains. In: Mika, P., et al. (eds.) ISWC 2014, Part I. LNCS, vol. 8796, pp. 245–260. Springer, Heidelberg (2014)
15. Stuckenschmidt, H.: Similarity-based query caching. In: Christiansen, H., Hacid, M.-S., Andreasen, T., Larsen, H.L. (eds.) FQAS 2004. LNCS (LNAI), vol. 3055, pp. 295–306. Springer, Heidelberg (2004)
16. Verborgh, R., et al.: Querying datasets on the web with high availability. In: Mika, P., et al. (eds.) ISWC 2014, Part I. LNCS, vol. 8796, pp. 180–196. Springer, Heidelberg (2014)
17. Verborgh, R., Sande, M.V., Colpaert, P., Coppens, S., Mannens, E., de Walle, R.V.: Web-scale querying through linked data fragments. In: WWW Workshop on LDOW 2014 (2014)
18. Voulgaris, S., Gavidia, D., Van Steen, M.: Cyclon: inexpensive membership management for unstructured P2P overlays. J. Netw. Syst. Manag. **13**(2), 197–217 (2005)

LOTUS: Adaptive Text Search for Big Linked Data

Filip Ilievski$^{(\boxtimes)}$, Wouter Beek, Marieke van Erp,
Laurens Rietveld, and Stefan Schlobach

The Network Institute, VU University Amsterdam, Amsterdam, The Netherlands
{f.ilievski,w.g.j.beek,marieke.van.erp,l.j.rietveld,k.s.schlobach}@vu.nl

Abstract. Finding relevant resources on the Semantic Web today is a
dirty job: no centralized query service exists and the support for natural language access is limited. We present LOTUS: Linked Open Text
UnleaShed, a text-based entry point to a massive subset of today's Linked
Open Data Cloud. Recognizing the use case dependency of resource
retrieval, LOTUS provides an adaptive framework in which a set of
matching and ranking algorithms are made available. Researchers and
developers are able to tune their own LOTUS index by choosing and
combining the matching and ranking algorithms that suit their use case
best. In this paper, we explain the LOTUS approach, its implementation
and the functionality it provides. We demonstrate the ease with which
LOTUS enables text-based resource retrieval at an unprecedented scale
in concrete and domain-specific scenarios. Finally, we provide evidence
for the scalability of LOTUS with respect to the LOD Laundromat, the
largest collection of easily accessible Linked Open Data currently available.

Keywords: Findability · Text indexing · Semantic search · Scalable
data management

1 Introduction

A wealth of information is potentially available from Linked Open Data sources
such as those found in the LOD Cloud[1] or LOD Laundromat [3]. However,
finding relevant resources on the Semantic Web today is not an easy job: there
is no centralized query service and the support for natural language access is
limited. A resource is typically 'found' by memorizing its resource-denoting IRI.
The lack of a global entry point to resources through a flexible text index is a
serious obstacle for Linked Data consumption.

We introduce LOTUS: Linked Open Text UnleaShed,[2] a central text-based
entry point to a large subset of today's LOD Cloud. Centralized text search on

---

[1] http://lod-cloud.net/.

[2] An early version of LOTUS was presented at COLD 2015 [13]. This paper is a
significantly extended and updated version of that work.

© Springer International Publishing Switzerland 2016
H. Sack et al. (Eds.): ESWC 2016, LNCS 9678, pp. 470–485, 2016.
DOI: 10.1007/978-3-319-34129-3_29

the LOD Cloud is not new as Sindice[3] and LOD Cache[4] show. However, LOTUS differs from these previous approaches in three ways: (1) its scale (its index is about 100 times bigger than Sindice's was), (2) the adaptability of its algorithms and data collection, and (3) its integration with a novel Linked Data publishing and consumption ecosystem that does not depend on IRI dereferenceability.

LOTUS indexes every natural language literal from the LOD Laundromat data collection, a cached copy of a large subset of today's LOD Cloud spanning tens of billions of ground statements. The task of resource retrieval is a two-part process consisting of matching and ranking. Since there is no single combination of matching and ranking that is optimal for every use case, LOTUS enables users to customize the resource retrieval to their needs by choosing from a variety of matching and ranking algorithms. LOTUS is not a semantic search engine intended for end users, but a framework for researchers and developers in which semantic search engines can be developed and evaluated. The flexibility of the LOTUS approach towards resource retrieval shifts the adaptation task from the user to the retrieval system, making LOTUS attractive for a wide range of use cases including Information Retrieval, Entity Linking and Network Analysis. Existing Entity Linking systems such as NERD [17]), rely on a single or limited set of knowledge sources, typically DBpedia, and thus suffer from limited coverage. An adaptive linguistic entry point to the LOD Cloud, such as LOTUS, could inspire new research ideas for Entity Linking and facilitate Web of Data-wide search and linking of entities.

The remainder of this paper is structured as follows. In Sect. 2, we detail the problem of performing linguistic search on the LOD Cloud and we formalize it by defining an array of requirements. Section 3 presents relevant previous work on Semantic Web text search. Section 4 describes the LOTUS framework through its model, approach and initial collection of matching and ranking algorithms. The implementation of LOTUS is reported in Sect. 5. Scalability tests and typical usage scenarios of LOTUS are discussed in Sect. 6. We conclude by considering the key strengths, limitations and future plans for LOTUS in Sect. 7.

2   Problem Description

In this section, we detail the requirements for a global text-based entry point to linked open data and the strengths and weaknesses of current findability strategies with respect to these requirements.

2.1   Requirements

The Semantic Web currently relies on four main strategies to find relevant resources: datadumps, IRI dereferencing, Linked Data Fragments (LDF) and SPARQL, but these are not particularly suited to global text-based search.

Since it is difficult to memorize IRIs and structured querying requires prior knowledge of how the data is organized, text-based search for resources is an

---

[3] http://www.sindice.com/, discontinued in 2014.

[4] http://lod.openlinksw.com/.

important requirement for findability on the Semantic Web (Req1: Text-based). Furthermore, we also require text-based search to be resilient with respect to minor variations such as typos or spelling variations (Req2: Resilience). An important principle of the Semantic Web is that anybody can say anything about any topic (AAA). The findability correlate of this principle is not implemented by existing approaches: only what an authority says or links to explicitly can be easily found. We formulate the correlate requirement as "anybody should be able to find anything that has been said about any topic" (Req3: AFAA).

Decentralized data publishing makes text-based search over multiple data sources difficult. Not all sources have high availability (especially a problem for SPARQL) and results from many different sources have to be integrated on the client side (especially a challenge for IRI dereferencing and Linked Data Fragments). Hence, we set requirements on availability (Req4: Availability) and scalability (Req5: Scalability) for a text-based search service. While LDF provides better serviceability for singular endpoints than the other three approaches, it is still cumbersome to search for resources across many endpoints (Req6: Serviceability). Existing Semantic Web access approaches do not implement IR-level search facilities. In Sect. 3 some systems will be discussed that built some of this functionality on top of standard access methods. However, these systems focus on a single algorithm to work in each and every case. This may be suitable for an end-user search engine but not for a framework in which search engines are build and evaluated, bringing us to our last requirement of customizeability (Req7: customizeability). Below we iterate our requirements:

Req1. Text-based. Resource-denoting IRIs should be findable based on text-based queries that match (parts of) literals that are asserted about that IRI, possibly by multiple sources.

Req2. Resilience. Text-based search should be resilient against typo's and small variations in spelling (i.e., string similarity and fuzzy matching in addition to substring matching).

Req3. AFAA. Authoritative and non-authoritative statements should both be findable.

Req4. Availability Finding resources should not depend on the availability of all the original resource-publishing sources.

Req5. Scalability. Resources should be searchable on a Web scale, spanning tens of billions of ground statements over hundreds of thousands of datasets.

Req6. Serviceability. The search API must be freely available for humans (Web UI) and machines (REST) alike.

Req7. Customizability. Search results should be ranked according to a customizable collection of rankings to support a wide range of use cases.

2.2   Current State-of-the-Art

Datadumps implement a rather simple way of finding resource-denoting terms: one must know the exact Web address of a datadump in order to download and extract resource-denoting IRIs. This means that search is neither text-based

(Req1) nor resilient (Req2). Extraction has to be performed manually by the user resulting in low serviceability (Req6). Datadumps do not link explicitly to assertions about the same resource that are published by other sources (Req3).

IRIs dereference to a set of statements in which that IRI appears in the subject position or, optionally, object position. Which statements belong to the dereference result set is decided by the authority of that IRI, i.e., the person or organization that pays for the domain that appears in the IRI's authority component. Non-authoritative statements about the same IRI cannot be found. Non-authoritative statements can only be found accidentally by navigating the interconnected graph of dereferencing IRIs. As blank nodes do not dereference significant parts of the graph cannot be traversed. This is not only a theoretical problem as 7% of all RDF terms are blank nodes [11]. In practice, this means that non-authoritative assertions are generally not findable (Req3). Since only IRIs can be dereferenced, text-based access to the Semantic Web cannot be gained at all through dereferencing (Req1). Thus, it is not possible to find a resource-denoting IRI based on words that appear in RDF literals to which it is (directly) related, or based on keywords that bear close similarity to (some of the) literals to which the IRI is related (Req2).

Linked Data Fragments [21] (LDF) significantly increases the serviceability level for single-source Linked Data retrieval (Req6) by returning metadata descriptions about the returned data. E.g., by implementing pagination LDF allows all statements about a given resource to be extracted without enforcing arbitrary limits. This is not guaranteed by IRI dereferencing or SPARQL (both of which lack pagination). An extension to LDF [20] adds efficient substring matching, implementing a limited form of text-based search (Req1). Due to the reduced hardware requirements for running an LDF endpoint it is reasonable to assume that LDF endpoints will have higher availability than SPARQL endpoints (Req4). LDF does not implement resilient matching techniques such as fuzzy matching (Req2) and does not allow non-authoritative statements about a resource to be found (Req3).

SPARQL allows resources to be found based on text search that (partially) matches literal terms (Req1). More advanced matching and ranking approaches such as string similarity or fuzzy matching are generally not available (Req2). As for the findability of non-authoritative statements, SPARQL has largely the same problems as the other three approaches (Req3). There is also no guarantee that all statements are disseminated by some SPARQL endpoint. Endpoints that are not pointed to explicitly cannot be found by automated means. Empirical studies show that many SPARQL endpoints have low availability [4] (Req4).

Federation is implemented by both LDF and SPARQL, allowing queries to be evaluated over multiple endpoints. Federation is currently unable to implement Web-scale resource search (Req5) since every endpoint has to be included explicitly in the query. This requires the user to enter the Web addresses of endpoints, resulting either in low coverage (Req3) or low serviceability (Req6). Other problems are that the slowest endpoint determines the response time of the entire query and results have to be integrated at the client side. Since

Web-wide resource search has to span hundreds of thousands of data sources these problems are not merely theoretical.

3   Related Work

Several systems have implemented text-based search over Semantic Web data: Swoogle [8], SemSearch [14], Falcons [5], Semplore [22], SWSE [10], Hermes [18], Sindice/Sigma [19]. Swoogle allows keyword-based search of Semantic Web documents. Hermes performs keyword-based matching and ranking for schema resources such as classes and (object) properties. Falcons, Semplore, SWSE and Sindice search for schema and data alike.

Exactly how existing systems extract keywords from RDF data is largely undocumented. Most services use standard NLP approaches such as stemming and stopword removal to improve the keyword index. Furthermore, N-grams are used to take multi-word terms into account (Swoogle), WordNet is used by Hermes enrich keywords with synonyms. Many systems rely on off-the-shelf text indexing tools such as Lucene that perform similar keyword extraction tasks.

Many systems use metrics derived from the Information Retrieval field such as Term Frequency (TF), Inverse Document Frequency (IDF) and PageRank to rank results. Hermes, for example, implements a TF variant called Element Frequency (EF) that quantifies the popularity of classes/properties in terms of the number of instances they have. Sindice calculates a version of IDF by considering the distinctiveness of data sources from which instances originate, as well as the authority of a data sources: resources that appear in a data source that has the same host name are ranked higher than non-authoritative ones. In practice, TF and IDF are often combined in a single TF/IDF-based metric, in order to have a balanced measure of popularity and distinctiveness.

While the basic idea of adapting existing IR metrics such as TF/IDF and PageRank for text-based search of Semantic Web resources is a common ground for existing systems, they all implement this idea in different ways; using different metrics, the same metrics in different ways, or combining the various metrics in a different way. This makes it difficult to compare existing text-based Semantic Web search systems with one another. Also, the fact that the adapted algorithms differ between these systems provides evidence that there is no single retrieval algorithm that fits every use case. Acknowledging that combining existing approaches into a final ranking over end-results is highly application-dependent and is as much an art as a science, LOTUS takes a very different approach. LOTUS provides an environment in which multiple matching and ranking approaches can be developed and combined. This makes it much easier to evaluate the performance of individual rankers on the global end result.

Existing systems operate on data collections of varying size. Sindice, Falcons and Hermes are formally evaluated over hundreds of millions of statements, while Semplore is evaluated over tens of millions of statements. Falcons, Swoogle and Sindice have at some point in time been available as public Web Services for users to query. With Sindice being discontinued in 2014, no text-based Semantic Web search engine is widely available to the Semantic Web community today.

In addition to the work on semantic search engines, there have been multiple attempts to extend existing SPARQL endpoints with more advanced NLP tooling such as fuzzy string matching and ranking over results [9,12,15]. This improves text search for a restricted number of query endpoints but does not allow text-based queries to cover a large number of endpoints, as the problem of integrating ranked results from many endpoints at the client side has not been solved. Virtuoso's LOD Cache[5] provides public access to a text search-enriched SPARQL endpoint. It differs from LOTUS in that it does not allow the matching and ranking algorithms to be changed or combined in arbitrary ways by the user and as a commercial product its specific internals are not public.

4   LOTUS

LOTUS relates unstructured to structured data using RDF as a paradigm to express such structured data. LOTUS is integrated with a central architecture that exposes a large collection of resource-denoting terms and structured descriptions of those terms, all formulated in RDF. It indexes natural text literals that appear in the object position of RDF statements and allows the denoted resources to be findable based on approximate matching. LOTUS currently includes four different matching algorithms and eight ranking algorithms, which leverage both textual features and relational information from the RDF graph.

4.1   Model

Denoted Resources. RDF defines a graph-based data model in which resources can be described in terms of their relations to other resources. An RDF statement expresses that a certain relation holds between a pair of resources.

The textual labels denoting some of these resources provide an opening to relate unstructured to structured data. LOTUS does not allow every resource in the Semantic Web to be found through text search, as some resources are not denoted by a term that appears as a subject of a triple whose object term is a textual label. Fortunately, many Semantic Web resources are denoted by at least one textual label and as the Semantic Web adheres to the Open World Assumption, resources with no textual description today may receive one tomorrow.

RDF Literals. In the context of RDF, textual labels appear as part of *RDF literals*. We are specifically interested in literals that contain natural language text. However, not all RDF literals express – or are intended to express – natural language text. For instance, there are datatype IRIs that describe a value space of date-time points or polygons. Even though each dataset can define its own datatypes, we observe that the vast majority of RDF literals use RDF or XSD datatypes. This allows us to circumvent the theoretical limitation of not being able to enumerate all textual datatypes and focus on the datatypes `xsd:string` and `rdf:langString` [7]. Unfortunately, in practice we find that integers and

---

[5] http://lod.openlinksw.com/.

dates are also regularly stored under these datatypes. As a simple heuristic filter LOTUS only considers literals with datatype `xsd:string` and `xsd:langString` that contain at least two consecutive alphabetic Unicode characters.

4.2 Linguistic Entry Point to the LOD Cloud

LOD Laundromat. LOTUS is built on top of LOD Laundromat, a Semantic Web crawling and cleaning architecture and centralized data collection. LOD Laundromat spans hundreds of thousands of data documents and tens of billions of ground statements. Inherent to this design decision, LOTUS fulfills three of the requirements we set in Sect. 2: (1) the scale on which LOTUS operates is in range of billions and is 100 times bigger than that of previous semantic search systems that were made generally available for the community to use (Req5); (2) since the LOD Laundromat data is collected centrally, finding both authoritative and non-authoritative RDF statements is straightforward (Req3); (3) as a cached copy of linked data, LOD Laundromat allows its IRIs to be dereferenceable even when the original sources are unavailable (Req4).

Linguistic Entry Point. LOTUS allows RDF statements from the LOD Laundromat collection to be findable through approximate string matching on natural language literals. Approximate string matching [16] is an alternative to exact string matching, where one textual pattern is matched to another while still allowing a number of errors. In LOTUS, query text is approximately matched to existing RDF literals (and their associated documents and IRI resources) (Req1).

In order to support the approximate matching and linguistic access to LD through literals, LOTUS makes use of an inverted index. As indexing of big data in the range of billions of RDF statements is expensive, the inverted index of LOTUS is created offline. This also allows the approximation model to be efficiently enriched with various precomputed retrieval metrics.

4.3 Retrieval

Matching Algorithms. Approximate matching of a query to literals can be performed on various levels: as phrases, as sets of tokens, or as sets of characters. LOTUS implements four matching functions to cope with this diversity (Req2):

M1. Phrase matching: Match a phrase in an object string. Terms in each result should occur consecutively and in the same order as in the query.

M2. Disjunctive token matching: Match some of the query tokens in a literal. The query tokens are connected by a logical "OR" operator, expressing that each match should contain at least one of the queried tokens. The order of the tokens between the query and the matched literals need not coincide.

M3. Conjunctive token matching: Match all query tokens in a literal. The set of tokens are connected by a logical "AND" operator, which entails that all tokens from the query must be found in a matched literal. The order of the tokens between the query and the matched literals need not coincide.

M4. Conjunctive token matching with character edit distance: Conjunctive matching (with logical operator "AND") of a set of tokens, where a small Levenshtein-based edit distance on a character level is permitted. This matching algorithm is intended to account for typos and spelling mistakes.

To bring a user even closer to her optimal set of matches, LOTUS facilitates complementary filtering based on the language of the literals, as explicitly specified by the dataset author or automatically detected by a language detection library. While a language tag can contain secondary tags, e.g. to express country codes, LOTUS focuses on the primary language tags which denote the language of a literal and abstracts from the complementary tags.

Ranking Algorithms. Ranking algorithms on the Web of data operate on top of a similarity function, which can be content-based or relational [6].[6] The content-based similarity functions exclusively compare the textual content of a query to each potential result. Such comparison can be done on different granularity of text, leading to character-based (Levenshtein similarity, Jaro similarity, etc.) and token-based (Jaccard, Dice, Overlap, Cosine similarity, etc.) approaches. The content similarity function can also be information-theoretical, exploiting the probability distributions extracted from data statistics. Relational similarity functions complement the content similarity approaches by considering the underlying structure of the tree (tree-based) or the graph (graph-based).

We use this classification of similarity algorithms as a starting point for our implementation of three content-based (R1-R3) and five relational functions (R4-R8) in LOTUS, thus addressing Req7:

R1. Character length normalization: The score of a match is counter-proportional to the number of characters in the lexical form of its literal.

R2. Practical scoring function:[7] The score of a match is a product of three token-based information retrieval metrics: term frequency (TF), inverse-document frequency (IDF) and length normalization (inverse-proportional to the number of tokens).

R3. Phrase proximity: The score of a match is inverse-proportional to its edit distance with respect to the query.

R4. Terminological richness: The score of a match is proportional to the presence of controlled vocabularies, i.e. classes and properties, in the original document from which the RDF statement stems from.

R5. Semantic richness of the document: The score of a match is proportional to the mean graph connectedness degree of the original document.

R6. Recency ranking: The score of a match is proportional to the moment in time when the original document was last modified. Statements from recently updated documents have higher score.

---

[6] The reader is referred to this book for detailed explanation of similarity functions and references to original publications.

[7] This function is the default scoring function in ElasticSearch. Detailed description of its theoretical basis and implementation is available at https://www.elastic.co/guide/en/elasticsearch/guide/current/scoring-theory.html.

R7. Degree popularity: The score of a match is proportional to the total graph connectedness degree (indegree + outdegree) of its subject resource.

R8. Appearance popularity: The score of a match is proportional to the number of documents in which its subject appears.

5   Implementation

The LOTUS system architecture consists of two main components: the Index Builder (IB) and the Public Interface (PI). The role of the IB is to index strings from LOD Laundromat; the role of the PI is to expose the indexed data to users for querying. The two components are executed sequentially: data is indexed offline, after which it can be queried via the exposed public interface.

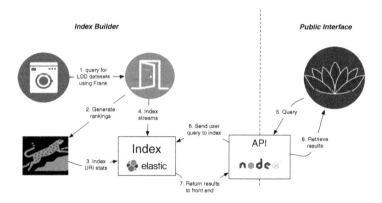

Fig. 1. LOTUS system architecture

5.1   System Architecture

Since our rankings rely on metadata about documents and resources which is reused across statements, we need clever ways to compute and access this metadata. For this purpose, we pre-store the document metadata needed for the ranking algorithms R4-R6, which includes the last modification date, the mean graph degree and the terminological richness coefficient of each LOD Laundromat document. We use the LOD Laundromat access tools (Frank [2] and the SPARQL endpoint[8]) to obtain these. The rankings R7 and R8 use metadata about resource IRIs. Computing, storage and access to this information is more challenging, as the number of resources in the LOD Laundromat is huge and their occurrences are scattered across documents. To resolve this, we store the graph degree of a resource and number of documents where it appears in RocksDB.[9]

---

[8] http://lodlaundromat.org/sparql/.
[9] http://rocksdb.org/.

Once the relational ranking data is cached, we start the indexing process over all data from LOD Laundromat through a batch loading procedure. This procedure uses LOD Laundromat's query interface, Frank (Step 1 in Fig. 1), to list all LOD Laundromat documents and stream them to a client script. Following the approach described in Sect. 4, we consider only the statements that contain a natural language literal as an object. The client script parses the received RDF statements and performs a bulk indexing request in ElasticSearch (ES),[10] where the textual index is built (Steps 2, 3 and 4 in Fig. 1).

As soon as the indexing process is finished, LOTUS contains the data it needs to perform text-based retrieval over the LOD Laundromat collection. Its index is only incrementally updated when new data is added in LOD Laundromat.[11]

For each RDF statement from the LOD Laundromat, we index: (1) *Information from the statement itself:* subject IRI, predicate IRI, lexical form of the literal ("string"), length of the "string" field (in number of characters), language tag of the literal and document ID; (2) *Metadata about the source document:* last modification date, terminological richness coefficient and semantic richness coefficient; (3) *Metadata about the subject resource:* graph degree and number of documents in which the resource appears.

We store the metadata for (2) and (3) in a numeric format to enable their straightforward usage as ranking scores by ElasticSearch.

5.2 Implementation of the Matching and Ranking Algorithms

While the matching algorithms we introduce are mainly adaptations of off-the-shelf ElasticSearch string matching functions, we allow them to be combined with relational information found in the RDF statement to improve the effectiveness of the matching process. The approximate matching algorithms M1-M4 operate on the content of the "string" field, storing the lexical form of a literal. This field is preprocessed ("analyzed") by ElasticSearch at index time, thus allowing existing ElasticSearch string matching functionalities to be put into practice for matching. We allow users to further restrict the matching process by specifying relational criteria: language tag of the literal ("langtag"), associated subject and predicate as well as LOD Laundromat document identifier.

Similarly to the matching algorithms, our ranking algorithms rely on both ElasticSearch functionality and relational information extracted from LOD Laundromat. Concretely, our rankings are based on: (1) *Scoring functions from Elastic-Search* (R1-R3); (2) *Document-level scores:* last modification date (R4), terminological richness (R5) and semantic richness (R6); (3) *Resource-level scores:* graph degree (R7) and number of documents that contain the resource (R8).

5.3 Distributed Architecture

In our implementation, we leverage the distributed features of ElasticSearch and scale LOTUS horizontally over 5 servers. Each server has 128 GB of RAM,

---

[10] https://www.elastic.co/products/elasticsearch.
[11] This procedure is triggered by an event handler in the LOD Laundromat itself.

6 core CPU with 2.40 GHz and 3 SSD hard disks with 440 GB of storage each. We enable data replication to ensure high runtime availability of the system.

5.4   API

Users can access the underlying data through an API. The usual query flow is described in steps 5–8 of Fig. 1. We expose a single query endpoint,[12] through which the user can supply a query, choose a combination of matching and ranking algorithms, and optionally provide additional requirements, such as language tag or number of results to retrieve. The basic query parameters are:[13]

- **string:** A natural language string to match in LOTUS
- **match:** Choice of a matching algorithm, one of *phrase, terms, conjunct, fuzzy-conjunct*
- **rank:** Choice of a ranking algorithm, one of *lengthnorm, psf, proximity, termrichness, semrichness, recency, degree, appearance*
- **size:** Number of best scoring results to be included in the response
- **langtag:** Two-letter language identifier.

LOTUS is also available as a web interface at http://lotus.lodlaundromat. org/ for human-friendly exploration of the data, thus fulfilling Req6 on usefulness for both humans and machines. Code of the API functions and data from our experiments can be found on github.[14] The code used to create the LOTUS index is also publicly available.[15]

6   Performance Statistics and User Scenarios

As LOTUS does not provide a one-size-fits-all solution, we present some performance statistics and scenarios in this section. We test LOTUS on a series of queries and show the impact of different matching and ranking algorithms.

6.1   Performance Statistics

Statistics over the indexed data are given in Table 1. LOD Laundromat contains over 12 billion literals, 8.81 billion of which are defined as a natural language string (`xsd:string` or `xsd:langString` datatype). According to our approach, 4.33 billion of these (∼35 %) express natural language strings. The initial LOTUS index was created in 67 h, consuming 509.81 GB of disk space. The current index consists of 4.33 billion entries, stemming from 493,181 distinct datasets.[16]

---

[12] http://lotus.lodlaundromat.org/retrieval.

[13] See http://lotus.lodlaundromat.org/docs for additional parameters and more detailed information.

[14] https://github.com/filipdbrsk/LOTUS_Search/.

[15] https://github.com/filipdbrsk/LOTUS_Indexer/.

[16] The number of distinct sources in LOTUS is lower than the number of documents in LOD Laundromat, as not every document contains natural language literals.

Table 1. Statistics on the indexed data

total # literals encountered	12,380,443,617
#xsd:string literals	6,205,754,116
#xsd:langString literals	2,608,809,608
# indexed entries in ES	4,334,672,073
# distinct sources in ES	493,181
# hours to create the ES index	67
disk space used for the ES index	509.81 GB
# in_degree entries in RocksDB	1,875,886,294
# out_degree entries in RocksDB	3,136,272,749
disk space used for the RocksDB index	46.09 MB

To indicate the performance of LOTUS from a client perspective we preformed 324,000 text queries. We extracted the 6,000 most frequent bigrams, trigrams and quadgrams (18,000 N-grams in total) from the source of *A Semantic Web Primer* [1]. Non-alphabetic characters were first removed and case normalization was applied. For each N-gram we performed a text query using one of three matchers in combination with one of six rankers. The results are shown in Fig. 2. We observe certain patterns in this Figure. Matching disjunctive terms (M2) is strictly more expensive than the other two matching algorithms. We also notice that bigrams are more costly to retrieve than trigrams and quadgrams. Finally, we observe that there is no difference between the response time of the relational rankings which is expected, because these rank results in the same manner, through sorting pre-stored integers in a decreasing order.

The performance and scalability of LOTUS is largely due to the use of ElasticSearch, which justifies our rationale in choosing ElasticSearch as one of our two main building blocks: it allows billions of entries to be queried within reasonable time constraints, even running on an academic hardware infrastructure.

6.2   Usage Scenarios

To demonstrate the flexibility and the potential of the LOTUS framework, we performed retrieval on the query "graph pattern". We matched this query as a phrase (M1) and iterated through the different ranking algorithms. The top results obtained with two of the different ranking modes are presented in Fig. 3.

The different ranking algorithms allow a user to customize her results. For example, if a user is interested in analyzing the latest changes in a dataset, the Recency ranking algorithm will retrieve statements from the most recently updated datasets first. A user who is more interested in linguistic features of a query can use the length normalization ranking to explore resources that match the query as precisely as possible. Users interested in multiple occurrences of informative phrases could benefit from the practical scoring function. When the popularity of resources is important, the degree-based rankings can be useful.

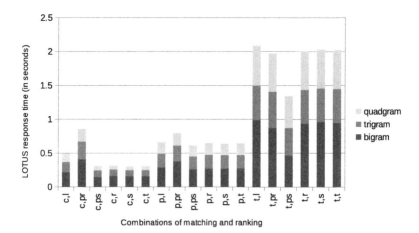

Fig. 2. LOTUS average response times in seconds for bi- tri- and quadgram requests. The horizontal axis shows 18 combinations of a matcher and a ranker. The matchers are *conjunct* (c), *phrase* (p) and *terms* (t). The rankers are *length normalization* (l), *proximity* (pr), *psf* (ps), *recency* (r), *semantic richness* (s) and *term richness* (t). The bar chart is cumulative per match+rank combination. For instance, the first bar indicates that the combination of conjunct matching and length normalization takes 0.20 s for bigrams, 0.15 s for trigrams, 0.15 s for quadgrams and 0.5 s for all three combined. The slowest query is for bigrams with terms matching and length normalization, which takes 1.0 s on average.

Fig. 3. Results of query "graph pattern" with terms-based matching and different rankings: (1) Semantic richness, (2) Recency.

Users can also vary the matching dimension. Suppose one is interested to explore resources with typos or spelling variation: fuzzy conjunctive matching would be the appropriate matching algorithm to apply.

7 Discussion and Conclusions

In this paper, we presented LOTUS, a full-text entry point to the centralized LOD Laundromat collection. We detailed the specific difficulties in accessing textual content in the LOD cloud today and the approach taken by LOTUS to address these. LOTUS allows its users to customize their own retrieval method by exposing analytically well-understood matching and ranking algorithms, taking into account both textual similarity and certain structural properties of the underlying data. LOTUS currently provides 32 retrieval options to be used in different use cases. Even though LOTUS is to some extent "connecting the dots" between the underlying infrastructures of LOD Laundromat and ElasticSearch, it is connecting a lot of these dots and due to the billions of statements that we are dealing with some of the connections are non-trivial.

In the current version of LOTUS we focus on context-free[17] ranking of results and demonstrate the versatility of LOTUS by measuring its performance and showing how the ranking algorithms affect the search results. A context-dependent ranking mechanism could make use of additional context coming from the query in order to re-score and improve the order of the results. To some extent, context-dependent functionality could be built into LOTUS. However, graph-wide integration with structured data would require a different approach, potentially based on a fulltext-enabled triplestore (e.g. Virtuoso).

Although further optimization is always possible, the current version of LOTUS performs indexing and querying in an efficient and scalable manner, largely thanks to the underlying distributed architecture. Since the accuracy of LOTUS is case-dependent, future work will evaluate the precision and recall of LOTUS on concrete applications, such as Entity Linking and Network Analysis.

Acknowledgments. The research for this paper was supported by the European Union's 7th Framework Programme via the NewsReader Project (ICT-316404) and the Netherlands Organisation for Scientific Research (NWO) via the Spinoza fund.

References

1. Antoniou, G., Groth, P., van Harmelen, F., Hoekstra, R.: A Semantic Web Primer, 3rd edn. The MIT Press, Cambridge (2012)
2. Beek, W., Rietveld, L.: Frank: algorithmic access to the LOD cloud. In: Proceedings of the ESWC Developers Workshop (2015)
3. Beek, W., Rietveld, L., Bazoobandi, H.R., Wielemaker, J., Schlobach, S.: Lod laundromat: a uniform way of publishing other peoples dirty data. ISWC **2014**, 213–228 (2014)
4. Buil-Aranda, C., Hogan, A., Umbrich, J., Vandenbussche, P.-Y.: SPARQL web-querying infrastructure: ready for action? In: Alani, H., Kagal, L., Fokoue, A., Groth, P., Biemann, C., Parreira, J.X., Aroyo, L., Noy, N., Welty, C., Janowicz, K. (eds.) ISWC 2013, Part II. LNCS, vol. 8219, pp. 277–293. Springer, Heidelberg (2013)
5. Cheng, G., Ge, W., Qu, Y.: Falcons: searching and browsing entities on the semantic web. In: Proceedings of the 17th International Conference on World Wide Web, WWW 2008, NY, USA, pp. 1101–1102 (2008). http://doi.acm.org/10.1145/1367497.1367676

---

[17] By "context-free", we mean that the retrieval process can not be directly influenced by additional restrictions or related information.

6. Christophides, V., Efthymiou, V., Stefanidis, K.: Entity Resolution in the Web of Data. Morgan and Claypool Publishers, San Rafael (2015)
7. Cyganiak, R., Wood, D., Lanthaler, M.: RDF 1.1 concepts and abstract syntax (2014)
8. Ding, L., Finin, T., Joshi, A., Pan, R., Cost, R.S., Peng, Y., Reddivari, P., Doshi, V., Sachs, J.: Swoogle: a search and metadata engine for the semantic web. In: Proceedings of the Thirteenth ACM International Conference on Information and Knowledge Management, CIKM 2004, NY, USA, pp. 652–659 (2004). http://doi.acm.org/10.1145/1031171.1031289
9. Feyznia, A., Kahani, M., Zarrinkalam, F.: Colina: a method for ranking sparql query results through content and link analysis. In: Proceedings of the 2014 International Conference on Posters & Demonstrations Track, ISWC-PD 2014, CEUR-WS.org, Aachen, Germany, vol. 1272, pp. 273–276 (2014). http://dl.acm.org/citation.cfm?id=2878453.2878522
10. Hogan, A., Harth, A., Umbrich, J., Kinsella, S., Polleres, A., Decker, S.: Searching and browsing linked data with swse: the semantic web search engine. Web Semant. Sci. Serv. Agents World Wide Web 9(4), 365–401 (2011). JWS special issue on Semantic Search. www.sciencedirect.com/science/article/pii/S1570826811000473
11. Hogan, A., Umbrich, J., Harth, A., Cyganiak, R., Polleres, A., Decker, S.: An empirical survey of linked data conformance. Web Semant. Sci. Serv. Agents World Wide Web 14, 14–44 (2012)
12. Ichinose, S., Kobayashi, I., Iwazume, M., Tanaka, K.: Ranking the results of DBpcdia retrieval with SPARQL query. In: Kim, W., Ding, Y., Kim, H.-G. (eds.) JIST 2013. LNCS, vol. 8388, pp. 306–319. Springer, Heidelberg (2014)
13. Ilievski, F., Beek, W., van Erp, M., Rietveld, L., Schlobach, S.: Lotus: linked open text unleashed. In: COLD workshop, ISWC (2015)
14. Lei, Y., Uren, V.S., Motta, E.: SemSearch: a search engine for the semantic web. In: Staab, S., Svátek, V. (eds.) EKAW 2006. LNCS (LNAI), vol. 4248, pp. 238–245. Springer, Heidelberg (2006)
15. Mulay, K., Kumar, P.S.: Spring: ranking the results of sparql queries on linked data. In: Proceedings of the 17th International Conference on Management of Data, COMAD 2011, Computer Society of India, Mumbai, India, pp. 12:1–12:10 (2011). http://dl.acm.org/citation.cfm?id=2591338.2591350
16. Navarro, G.: A guided tour to approximate string matching. ACM Comput. Surv. (CSUR) 33(1), 31–88 (2001)
17. Rizzo, G., Troncy, R.: NERD: a framework for unifying named entity recognition and disambiguation extraction tools. In: Proceedings of EACL 2012, pp. 73–76 (2012)
18. Tran, T., Wang, H., Haase, P.: Hermes: data web search on a pay-as-you-go integration infrastructure. Web Semant. Sci. Serv. Agents World Wide Web 7(3), 189–203 (2009). www.sciencedirect.com/science/article/pii/S1570826809000213. The Web of Data
19. Tummarello, G., Delbru, R., Oren, E.: Sindice.com: weaving the open linked data. In: Aberer, K., Choi, K.-S., Noy, N., Allemang, D., Lee, K.-I., Nixon, L.J.B., Golbeck, J., Mika, P., Maynard, D., Mizoguchi, R., Schreiber, G., Cudré-Mauroux, P. (eds.) ASWC 2007 and ISWC 2007. LNCS, vol. 4825, pp. 552–565. Springer, Heidelberg (2007)
20. Van Herwegen, J., De Vocht, L., Verborgh, R., Mannens, E., Van de Walle, R.: Substring filtering for low-cost linked data interfaces. In: Arenas, M. (ed.) ISWC 2015. LNCS, pp. 128–143. Springer, Switzerland (2015)

21. Verborgh, R., Hartig, O., De Meester, B., Haesendonck, G., De Vocht, L., Vander Sande, M., Cyganiak, R., Colpaert, P., Mannens, E., Van de Walle, R.: Querying datasets on the web with high availability. In: Mika, P., Tudorache, T., Bernstein, A., Welty, C., Knoblock, C., Vrandečić, D., Groth, P., Noy, N., Janowicz, K., Goble, C. (eds.) ISWC 2014, Part I. LNCS, vol. 8796, pp. 180–196. Springer, Heidelberg (2014)

22. Wang, H., Liu, Q., Penin, T., Fu, L., Zhang, L., Tran, T., Yu, Y., Pan, Y.: Semplore: a scalable IR approach to search the web of data. Web Semantics Science Services and Agents on the World Wide Web **7**(3), 177–188 (2009). www.sciencedirect.com/science/article/pii/S1570826809000262. The Web of Data

Query Rewriting in RDF Stream Processing

Jean-Paul Calbimonte[1]([⊠]), Jose Mora[2], and Oscar Corcho[2]

[1] Faculty of Computer Science and Communication Systems,
EPFL, Lausanne, Switzerland
`jean-paul.calbimonte@epfl.ch`
[2] Ontology Engineering Group, Universidad Politécnica de Madrid, Madrid, Spain
`j.mora@upm.es, ocorcho@fi.upm.es`

Abstract. Querying and reasoning over RDF streams are two increasingly relevant areas in the broader scope of processing structured data on the Web. While RDF Stream Processing (RSP) has focused so far on extending SPARQL for continuous query and event processing, stream reasoning has concentrated on ontology evolution and incremental materialization. In this paper we propose a different approach for querying RDF streams over ontologies, based on the combination of query rewriting and stream processing. We show that it is possible to rewrite continuous queries over streams of RDF data, while maintaining efficiency for a wide range of scenarios. We provide a detailed description of our approach, as well as an implementation, StreamQR, which is based on the kyrie rewriter, and can be coupled with a native RSP engine, namely CQELS. Finally, we show empirical evidence of the performance of StreamQR in a series of experiments based on the SRBench query set.

1 Introduction

Streams are currently one of the main sources of data on the Web, and are used in a number of applications, ranging from wearable devices for health monitoring to geospatial and environmental sensing. Some of the main challenges of managing this very dynamic type of data are linked to the *velocity* of the inputs, and the need for reactive processing. While these have been addressed to a large extent by the database community, through *Data Stream Management Systems* (DSMS) and *Complex Event Processors* (CEP), there is still a need for tackling the issues of heterogeneity, integration and interpretation of data streams on the Web. Semantic Web technologies and standards have provided fundamental concepts and tools to address these issues, such as ontologies, RDF triple stores and reasoners, although most of these are targeted towards stored data. The goal of RDF stream processing (RSP) is to apply and extend the Semantic Web models and languages for processing RDF data streams. Previous works have presented RSP engines [1,4,7,16,24], focusing on different aspects of query processing. However, most of them provide limited or no reasoning capabilities, nor use an ontology model (TBox) to provide inferences during query processing. Other works have also explored the field of *stream reasoning*, but they have centered their attention mainly on the materialization of streaming axioms in ontologies

© Springer International Publishing Switzerland 2016
H. Sack et al. (Eds.): ESWC 2016, LNCS 9678, pp. 486–502, 2016.
DOI: 10.1007/978-3-319-34129-3_30

[17, 23] and, to some extent, to query processing that takes into account material-ization rules [5]. The problem of answering queries over an ontology can be solved in different ways, and an important technique to do so is *query rewriting*. It is based on the idea of transforming the original query into an expanded query that captures the information of the ontology TBox [21]. Then this expanded query is evaluated over the ABox, providing answers that extract implicit knowledge of the data. This technique has been successfully used in scenarios such as OBDA (Ontology based data access) where data is stored in relational databases [22].

In this paper we address the problem of providing query answering over ontologies in RSP, through a novel approach that combines query rewriting techniques and RDF stream query processing. While most of the focus on query rewriting has been on OBDA for relational databases, we show that it is possi-ble to use it for rewriting continuous queries over streams of RDF data. RDF streams, understood as potentially infinite flows of timestamped triples, require reactive processing of queries, and therefore most RSP engines have focused on efficiency and high throughput. Our approach demonstrates that these engines can be coupled with a query rewriter, and still be efficient for a large range of sce-narios. Furthermore we implemented our solution, called StreamQR, extending the CQELS query processor with the kyrie [18] rewriting engine.

The paper is organized as follows. In Sect. 2 we introduce the notions related to RDF stream processing and query rewriting. In Sect. 3 we present our proposal for ontology query answering over data streams. Then, in Sect. 4 we describe the implementation of StreamQR. Section 5 provides details on the experimentation and applicability of this approach. In Sect. 6 we analyze and compare previ-ous work in the areas of stream reasoning, query rewriting and RDF stream processing. Finally we draw our conclusions in Sect. 7 and identify future areas of research.

2   Preliminaries

2.1   RDF Stream Processing

Data streams are infinite and time-varying sequences of data values [2], and can also be seen as more complex events whose patterns can be queried and processed [12]. In the case of RDF stream processing (RSP) the elements of the stream are RDF data, typically annotated with a timestamp. Managing stream-ing data differs significantly from classical stored data, given the potentially infinite nature of streams and the need for continuous evaluation of queries. Continuous processing changes the usual query execution model, since it is the data arrival that initiates query processing, and produces results as soon as streaming data matches the query criteria.

Most RSP systems define a data model based on timestamped triples to represent RDF streams. As in [4, 7], we define an RDF stream S as a sequence of pairs (T, t) where T is a triple $\langle s, p, o \rangle$ and t is a timestamp in the infinite set of non-decreasing timestamps $\mathbb{T}$:

$$S = \{(\langle s, p, o \rangle, t) \mid \langle s, p, o \rangle \in ((I \cup B) \times I \times (I \cup B \cup L)), t \in \mathbb{T}\}$$

where I, B and L are sets of IRIs, blank nodes and literals, respectively. The pairs (T, τ) are called a *timestamped triples*. An RDF stream can be identified by an IRI, which allows referencing a particular stream of tagged triples.

Most of the state-of-the-art RSP query languages and systems are to a large extent based on Data stream management systems (DSMS) and complex event processing (CEP) extensions to the standard SPARQL query language. These extensions include operators such as windows, the ability to input and output RDF streams, and the declaration of continuous queries. Examples of such RSP systems include C-SPARQL [4], SPARQL$_{\text{Stream}}$ [7] and CQELS [16], which incorporate these notions, although with some differences in syntax and semantics. As an example, the CQELS query in Listing 1 requests the maximum temperature and the sensor that reported it in the last hour, from the stream identified as http://example.org/stream (prefixes are omitted for brevity).

```
SELECT (MAX(?temp) AS ?maxtemp) ?sensor
WHERE {
  STREAM <http://example.org/stream> [RANGE 1 HOUR] {
  ?obs ssn:observationResult ?result;
       ssn:observedProperty cf-property:air_temperature;
       ssn:observedBy ?sensor.
  ?result ssn:hasValue ?obsValue.
  ?obsValue qu:numericalValue ?temp. }
} GROUP BY ?sensor
```

Listing 1. Query the maximum temperature in the last hour in CQELS.

Sliding windows such as the one in the previous example are available in almost all RSP engines. They limit the scope of triples to be considered by the query operators. In particular, a *window* can be defined as a function that takes a stream S and a time instant $t \in \mathbb{T}$ and produces an RDF graph of triples. A more particular case, the time window W is defined by parameters σ, δ where σ is the size and δ is the slide, such that: $W(S, t) = \{T_j \mid (T_j, t_j) \in S \text{ and } t - \sigma \leq t_j < t\}$ and $t_{\ell+1} - t_\ell = \delta$ for two consecutive windows $W(S, t_\ell)$ and $W(S, t_{\ell+1})$.

2.2   Query Rewriting

Query answering using ontologies provides the capability of extracting explicit and implicit knowledge from a data source. In general, an ontology $\mathcal{O}$ is composed of a *TBox* $\mathcal{T}$ containing intensional knowledge and an *ABox* $\mathcal{A}$ containing extensional knowledge [9]. Query rewriting can help answering queries over ontologies, by transforming –or rewriting– the original query into an expanded query that takes into account the TBox, and using this rewritten query to evaluate it against the ABox part of the ontology. This technique has been used in the past in different scenarios, most notably Ontology-based Data Access (OBDA). In OBDA, data from one (or more) data sources can be queried in terms of a high-level ontological model, hiding from the users the internal schema and storage details of the data sources [21].

In general, a *query rewriting* algorithm works in the following way [9]. First, given an input query q and an ontology $(\mathcal{T}, \mathcal{A})$, it transforms q using the TBox $\mathcal{T}$ into a query q', such that for every ABox $\mathcal{A}$, the set of answers that q' obtains from $\mathcal{A}$ is equal to the set of answers that are entailed by q over $\mathcal{T}$ and $\mathcal{A}$. The rewritten query q' can normally be unfolded and expressed as a union of conjunctive queries, as long as some restrictions are imposed on the expressivity of the ontology language. As an example, consider the query $q(x) \leftarrow$ Sensor(x), requesting instances of Sensor, and the TBox axiom: HumiditySensor $\sqsubseteq$ Sensor. Using this TBox assertion, query rewriting will produce the union of the following conjunctive queries, which will also request for all instances of HumiditySensor:

$$q'(x) \leftarrow \text{Sensor}(x) \qquad\qquad q''(x) \leftarrow \text{HumiditySensor}(x)$$

An important property for query rewriting is FOL-reducibility [9] or FO-rewritability [14], meaning that rewritten queries are first-order queries, what allows converting them to languages like SQL without using advanced features like recursion. It has been shown that the combined complexity of satisfiability on some of these logics is polynomial, while data complexity is AC^0 [3,14].

The ontology TBox can be described using different languages or logics. Depending on their expressiveness, the query rewriting process can be more or less expensive in terms of computation. A logic with special relevance in our case is $\mathcal{ELHIO}$, one of the most expressive logics currently used for query rewriting. $\mathcal{ELHIO}$ is not FOL-reducible (in the presence of certain cyclic axioms the rewriting produces a recursive datalog program) but remains tractable (PTIME-complete [21]) for the rewriting process. For the description of our TBoxes we will use acyclic $\mathcal{ELHIO}$.

3 RSP Query Rewriting

Our approach for querying RDF streams over ontologies stems from the combination of the query rewriting techniques described previously and RDF stream query processing. As described in Sect. 2.2, the first lack support for continuous query processing and the second, in general, do not take into account the TBox of an ontology and hence do not perform any inferences during querying.

3.1 RSP Query Evaluation Using the Ontology TBox

Most of the existing RSP engines do not make use an ontology TBox during query evaluation (excepting a materialization technique described in [5] which is not currently available in the C-SPARQL implementation). Consequently it might be the case that there are queries that return a reduced number of answers, or no answers at all, in contrast to what could be expected if TBox assertions were taken into account. For example, let's consider the following RDF graph G with triples of the form:

```
:obs1 rdf:type ssn:Observation .
:obs2 rdf:type ssn:Observation .
:obs3 rdf:type ssn:Observation .
```

If we pose the following query, requesting all instances that are observed by some other entity (e.g. a sensor):

$$q_G(x) \leftarrow \mathtt{ssn:observedBy}(x, y)$$

Without any other information, the evaluation of this query over the stream would produce no results, since we do not have any match to a corresponding triple in the stream, for the `ssn:observedBy` property. However if we take into account TBox assertions, the situation may change. For instance, the presence of the following TBox axiom:

$$\mathtt{ssn:Observation} \sqsubseteq \exists\, \mathtt{ssn:observedBy}$$

would mean that all subjects of the triples in the stream (i.e. obs1,obs2,obs3) are also implicitly observed by someone. By rewriting $q(x)$ with this TBox, we would obtain an expansion consisting of the two following queries:

$$q'_G(x) \leftarrow \mathtt{ssn:observedBy}(x, y)$$
$$q''_G(x) \leftarrow \mathtt{ssn:Observation}(x)$$

While q' is just the original queries and will not produce answers, q'' (containing `ssn:Observation(x)`) will match during the evaluation over the RDF stream.

The previous rewriting examples can be extended to handle continuous evaluation of RDF data streams. For instance, we can consider the following stream based on the dataset of the previous example[1]:

```
:obs1 rdf:type ssn:Observation .   [1]
:obs2 rdf:type ssn:Observation .   [3]
:obs3 rdf:type ssn:Observation .   [6]
...
```

As the stream is potentially infinite, the complete answer to a query theoretically takes infinite time to evaluate, but partial answers may be computed in a continuous fashion, e.g., applying a sliding window W of 3 time units $\sigma = 3, \delta = 3$. Then we can compute continuous queries of the form $q_{W(S,t)}$ for every evaluation time t, computed over the instantaneous graph produced by $W(S, t)$:

$$q_{W(S,t)}(x) \leftarrow \mathtt{ssn:observedBy}(x, y)$$

Then, for each instantaneous graph, all other SPARQL operators can be applied on top of the resulting operation. The time window operator is already incorporated into existing RDF stream languages such as CQELS, C-SPARQL, or SPARQL$_{\mathrm{Stream}}$, as described in Sect. 2.1.

---

[1] Turtle http://www.w3.org/TR/turtle/, extended with timestamps in square brackets.

3.2 Query Answering Semantics

We describe in this section the semantics of our approach of query answering for data streams, over $\mathcal{ELHIO}$ ontologies, one of the most expressive logics that can be currently handled in query rewriting. An ontology $\mathcal{O}$ is composed of a TBox $\mathcal{T}$ and an ABox $\mathcal{A}$, i.e. $\mathcal{O} = \langle \mathcal{T}, \mathcal{A} \rangle$. Given the inference capabilities provided by the TBox, the results to the queries that are posed to the ontology are called *certain answers*. These can be seen intuitively as the kind of answers that users would expect, i.e. the answers matching triples explicitly stated in the database and those entailed by the ontology. We constrain the queries in this work to *conjunctive queries* (CQ) of the form:

$$q_h(\boldsymbol{x}) \leftarrow p_1(\boldsymbol{x_1}) \wedge \ldots \wedge p_n(\boldsymbol{x_n})$$

where q_h is the head predicate of the query, $\boldsymbol{x}$ is a tuple of distinguished variables, $\boldsymbol{x_1} \ldots \boldsymbol{x_n}$ are tuples of variables or constants, and $p_i(\boldsymbol{x_i})$ are unary or binary atoms in the body of the query. Every variable $x_j \in \boldsymbol{x}$ in the head of the query also appears in the body of the query. The certain answers for this type of queries can be defined as the set:

$$cert(q, \mathcal{O}) = \{\boldsymbol{\alpha} \mid q \cup \mathcal{T} \cup \mathcal{A} \models q_h(\boldsymbol{\alpha})\}$$

where q_h is the head predicate of the query q over an ontology $\mathcal{O} = \langle \mathcal{T}, \mathcal{A} \rangle$.

The query rewriting process uses the TBox $\mathcal{T}$ to rewrite the query q into a new query q' such that:

$$cert(q, \mathcal{O}) = \{\boldsymbol{\alpha} \mid q' \cup \mathcal{A} \models q_h(\boldsymbol{\alpha})\}$$

In this work, q' is in fact a *union of conjunctive queries* (UCQ), which is a set of CQ with the same head. In the following, we describe: (i) the query rewriting semantics to obtain the UCQ q', and (ii) the adaptation of this semantics to a continuous evaluation.

Query Rewriting in $\mathcal{ELHIO}$. Query rewriting in our approach uses a non-recursive $\mathcal{ELHIO}$ ontology to rewrite the conjunctive query q into a union of conjunctive queries (UCQ) q' such that $cert(q, \mathcal{O}) = \{\boldsymbol{\alpha} \mid q' \cup \mathcal{A} \models q_h(\boldsymbol{\alpha})\}$. The details of the inference steps for the rewriting can be found in [18]. The transformation of the ontology to first order logic is performed according to the rules described in [21]. Afterwards, the UCQ is then converted into a union of basic graph patterns (BGPs). Since this conversion is merely syntactical, the semantics of the rules in $\mathcal{ELHIO}$ can be expressed over transformations on triple patterns, as detailed in Table 1.

Query Rewriting for Continuous Queries. In the context of this work, the ontology is not considered to be static at query evaluation time, but it is available as a stream. Assuming that that the ABox of the ontology is dynamic and that

Table 1. $\mathcal{ELHIO}$ axioms as rules over RDF triples. Converted from Table 2 in [20].

$\mathcal{ELHIO}$ axiom	antecedent $\rightarrow$ consequent
$A \sqsubseteq \{a\}$	`?x rdf:type ex:A` $\rightarrow$ `?x = ex:a`
$A1 \sqsubseteq A2$	`?x rdf:type ex:A1` $\rightarrow$ `?x rdf:type ex:A2`
$A1 \sqcap A2 \sqsubseteq A3$	`?x rdf:type ex:A1 ; rdf:type ex:A2` $\rightarrow$ `?x rdf:type ex:A3`
$A \sqsubseteq \exists P$	`?x rdf:type ex:A` $\rightarrow$ `?x ex:P []:fx`
$A1 \sqsubseteq \exists P.A2$	`?x rdf:type ex:A1` $\rightarrow$ `?x ex:P [rdf:type ex:A1]`
$A \sqsubseteq \exists P^-$	`?x rdf:type ex:A` $\rightarrow$ `[]:fx ex:P ?x`
$A1 \sqsubseteq \exists P^-.A2$	`?x rdf:type ex:A1` $\rightarrow$ `[ex:P ?x ; rdf:type ex:A2]`
$\exists P \sqsubseteq A$	`?x ex:P ?y` $\rightarrow$ `?x rdf:type ex:A`
$\exists P.A1 \sqsubseteq A2$	`?x ex:P [rdf:type ex:A1]` $\rightarrow$ `?x rdf:type ex:A2`
$\exists P^- \sqsubseteq A$	`?y ex:P ?x` $\rightarrow$ `?x rdf:type A`
$\exists P^-.A1 \sqsubseteq A2$	`[ex:P ?x ; rdf:type A1]` $\rightarrow$ `?x rdf:type A2`
$P \sqsubseteq S, P^- \sqsubseteq S^-$	`?x ex:P ?y` $\rightarrow$ `?x ex:S ?y`
$P \sqsubseteq S^-, P^- \sqsubseteq S$	`?x ex:P ?y` $\rightarrow$ `?y ex:S ?x`

the TBox does not change at query time, we can define an *instantaneous ABox* as $\mathcal{A}(t)$, for a given time t. An instantaneous ABox contains all assertions in the stream timestamped at time t. Then we can represent the stream as a sequence of ABoxes over time: $\mathcal{A}(0), \mathcal{A}(1), ..., \mathcal{A}(t_i), ...$ where t_i represents a time point. Furthermore, given a time window w with starting and ending times s and e, we can define $\mathcal{A}_w$ as the ABox consisting of the union of the instantaneous ABoxes $\mathcal{A}(t)$ such that $s \leq t < e$.

To provide continuous answers to queries on this stream of ABoxes, each query is executed over a sliding time window that limits the number of assertions of the stream. A slide parameter indicates how often this window is computed (as described in Sect. 2.1, but without loss of generality we can assume that the window size is equal to the slide). Then, if we assume t_0 as the initial evaluation time, given a query q_w over a window of size δ, it will be evaluated at times $t_0, t_0 + \delta, ..., t_0 + k\delta, ...$, with $k \in \mathbb{N}$. The certain answers for this query, for the k-th window can be defined as:

$$cert(q_{w_k}, \langle \mathcal{T}, \mathcal{A}_{w_k} \rangle) = \{\boldsymbol{\alpha} \mid q_w \cup \mathcal{T} \cup \mathcal{A}_{w_k} \models q_h(\boldsymbol{\alpha})\}$$

where the start and end times of the k-th window w_k define the contents of the ABox $\mathcal{A}_{w_k}$. This instantaneous query evaluation is compatible with the previous definition of certain answers given above, as we are referring to instantaneous snapshots of the ABox stream. Therefore, the query rewriting algorithms for static data sources can be used to compute the inferences on this instantaneous query. For the k-th window, the corresponding query q_{w_k} will be rewritten into a new query q'_{w_k} such that:

$$cert(q_{w_k}, \langle \mathcal{T}, \mathcal{A}_{w_k} \rangle) = \{\boldsymbol{\alpha} \mid q'_w \cup \mathcal{A}w_k \models q_h(\boldsymbol{\alpha})\}$$

As we will see in the next section, the time-window based query q' can be concretely implemented using an RDF stream processor that natively includes the window operator, as we have seen in Sect. 2.1. Given that queries are continuously evaluated, the rewriting process does not need to be performed on every window evaluation. Assuming that $\mathcal{T}$ does not change, the query rewriting can be executed only once.

4   RSP Rewriting Implementation

In this section we describe StreamQR, an implementation of the query rewriting approach presented in Sect. 3. This prototype is available as open-source code[2], and is based on two main components: (i) the query rewriter kyrie [18], which rewrites queries using $\mathcal{ELHIO}$ ontologies, and (ii) the RSP query engine CQELS [16]. The whole process is divided in a series of steps that are graphically summarized in Fig. 1.

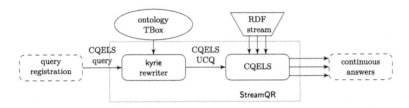

Fig. 1. High level architecture of StreamQR. The original query is rewritten by kyrie into a union of CQELS queries based on the ontology TBox. The rewritten queries are evaluated by the CQELS engine, on the incoming RDF stream.

RDF streams continuously feed StreamQR, specifically through the CQELS interface for consuming incoming streams, and allows registering queries specified in the CQELS language. In the following we describe the implementation of the rewriting and continuous execution.

As a running example, consider the following ontology TBox, including assertions about observations. For instance, temperature observations are observed by a temperature sensor, a thermistor is a type of temperature sensor, etc[3]:

$$met:TemperatureObservation \sqsubseteq \exists \; ssn:observedBy.aws:TemperatureSensor$$
$$met:AirTemperatureObservation \sqsubseteq met:TemperatureObservation$$
$$met:ThermistorObservation \sqsubseteq ssn:TemperatureObservation$$
$$aws:Thermistor \sqsubseteq aws:TemperatureSensor$$
$$aws:CapacitiveBead \sqsubseteq aws:TemperatureSensor$$

---

[2] Github StreamQR: https://github.com/jpcik/streapler/tree/streamQR.

[3] For brevity we use the prefixes aws: http://purl.oclc.org/NET/ssnx/meteo/aws, met: http://purl.org/env/meteo, ssn: http://purl.oclc.org/NET/ssnx/ssn.

Registration. In this stage StreamQR takes as input an ontology TBox and a registered CQELS query, to produce a union of conjunctive queries (UCQ). For instance the following query asks for all instances observed by a temperature sensor in the last 10 ms.

```
PREFIX ssn: <http://purl.oclc.org/NET/ssnx/ssn#>
PREFIX aws: <http://purl.oclc.org/NET/ssnx/meteo/aws#>
CONSTRUCT { ?o a :ObservedTemperature. }
WHERE {
  STREAM :stream1 [RANGE 10ms]  {
    ?o ssn:observedBy ?t .
    ?t a aws:TemperatureSensor.  }
}
```

Listing 2. CQELS query for the latest 10 ms of observations of temperature sensors.

Pre-processing. When this query is registered, the rewriting process is launched using the kyrie [18] module. kyrie uses $\mathcal{ELHIO}$ as the language for the ontologies, as it is one of the most expressive DLs that can be currently handled in query rewriting (see Sect. 6 for details). While the ontology is typically an OWL file, kyrie ignores assertions that go beyond the expressivity of $\mathcal{ELHIO}$. It also converts the ontology to Horn clauses and performs additional pre-processing. At this point, the system produces a conjunctive query from the basic graph patterns in the original CQELS query, while the time window definitions of the query (the query context) are preserved. For example, the following conjunctive query clauses can be obtained from the query in Listing 2:

```
Q(?0)   <-   aws:TemperatureSensor(?1), ssn:observedBy(?0,?1)
```

Saturation and Expansion. These clauses with the conjunctive query form a logic program that is saturated by using resolution with free selection (RFS) and including a series of optimizations, as described in [18]. This results in a number of clauses that is usually smaller than those produced by other similar techniques (e.g. REQUIEM [21]). After two saturation stages, we can remove functional terms to obtain a Datalog program or expand that program into a UCQ (as in our case without recursion). As an example, the previous conjunctive query is expanded to a union of the following conjunctive queries:

```
Q(?0)   <-   met:TemperatureObservation(?0)
Q(?0)   <-   met:AirTemperatureObservation(?0)
Q(?0)   <-   met:ThermistorObservation(?0)
Q(?0)   <-   aws:TemperatureSensor(?1), ssn:observedBy(?0,?1)
Q(?0)   <-   aws:Thermistor(?1), ssn:observedBy(?0,?1)
Q(?0)   <-   aws:CapacitiveBead(?1), ssn:observedBy(?0,?1)
```

Back to CQELS. The UCQ is then syntactically re-transformed back in CQELS using the context information from the original query. This includes the window definition, the query form and other modifiers. We refer to this query as the CQELS UCQ to avoid ambiguity with the original CQELS query. Following our example, after finishing the rewriting process we obtain the following union of queries (Listing 3), which already takes into account the axioms in the TBox. Given that CQELS does not allow unions on the stream clause, in practice this query is split into different CQELS queries with the same window and the union content as the pattern inside the stream clause.

```
PREFIX ssn: <http://purl.oclc.org/NET/ssnx/ssn#>
PREFIX aws: <http://purl.oclc.org/NET/ssnx/meteo/aws#>
PREFIX met: <http://purl.org/env/meteo#>
CONSTRUCT { ?o a :ObservedTemperature. }
WHERE { STREAM :stream1 [RANGE 10ms] {
    { ?o a met:TemperatureObservation } UNION
    { ?o a met:AirTemperatureObservation } UNION
    { ?o a met:ThermistorObservation } UNION
    { ?s a aws:TemperatureSensor . ?o ssn:observedBy ?s } UNION
    { ?s a aws:Thermistor . ?o ssn:observedBy ?s } UNION
    { ?s a aws:CapacitiveBead . ?o ssn:observedBy ?s }
}
```

Listing 3. The query of Listing 2 after rewriting.

The general description of the process is depicted in Fig. 2.

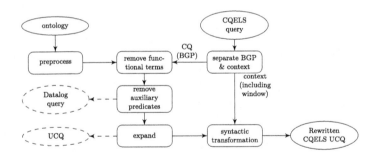

Fig. 2. Main steps of the query rewriting algorithm: preprocess, remove functional terms, auxiliary predicates and unfold (adapted from [18]). Added steps for syntactic conversion between CQELS and UCQ.

5   Evaluation

In this section we evaluate the performance of StreamQR in terms of the throughput of the query answering process, considering different input streaming rates and simultaneous continuous queries. More concretely, we compute the throughput of the system in terms of triples processed per unit of time, comparing

StreamQR with rewriting enabled, and StreamQR running CQELS without any rewriting. Then, we evaluate the throughput under different input data loads, considering that the query answering process depends not only on the input rate but also on the number of query matches produced in a given time span. Furthermore, we compare different queries whose rewriting produces different number of UCQs. Finally, we compare StreamQR with TrOWL [23], a state-of-the-art stream reasoner, which provides incremental reasoning over ontology streams, although not targeted towards query answering. We used a modified version of the SRBench [28] benchmark queries, as well as an ontology based on AWS (Ontology for Meteorological sensors[4]), which extends the W3C SSN ontology [11]. The ontology describes sensors, observations, features of interest, ans other weather-related concepts. The ontology is available online[5]. We have taken the SRBench queries and adapted them according to our extended ontology. The full set of queries is available in the Github repository, and additional information about the experiments can be found there as well[6]. All the experiments were performed on an Intel Core i7 3.1 GHz, 16 GB.

Comparing with CQELS without rewriting. A key indicator in stream processing is the system throughput, which can be measured in terms of the number of input elements processed per unit of time. Input rates, number of matching results and the number of concurrent queries are some of the main factors that impact throughput. In most evaluations concerning query rewriting, the rewriting time is also relevant. This is also the case in RSP, although to a lesser degree, because for continuous queries the rewriting is performed once, so its cost is not critical in the long run. In the first experiment we compared the throughput of StreamQR with rewriting enabled, to only evaluating the query with CQELS. We tested under different input rates, ranging from 10 to 100 K triples/s, as depicted in Fig. 3(a,b,c). We also tested under three different load conditions, i.e. in such a way that only 10 %, 50 % and 90 % of the input data matches the continuous query. As it can be seen in the results, the rewriting of StreamQR performs exactly as without rewriting for most of the input rates. Only under very high rates there is a considerable difference. This is the case for the three types of loads, although we can see considerable changes depending on the percentage of matching input triples. Notice that in these cases, up to 200 K triples/s., the behavior reaches the maximum expected throughput.

Variations in input matching. As we saw in the previous experiment the number of matching triples affects the overall throughput. The more matches, the more time the engine spends on evaluation. We performed a series of experiments under different input loads, with a varying distribution of the types of triples, in such a way that 10, 20, 50, 80 and 90 % of the triples match the query. As we can see in Fig. 4, up to 10 K triples/s., in almost all cases StreamQR is capable

---

[4] AWS: http://www.w3.org/2005/Incubator/ssn/ssnx/meteo/aws.

[5] http://jpcik.github.io/streapler/ontology/envsensors.owl.

[6] https://github.com/jpcik/streapler/wiki/StreamQR-experiments.

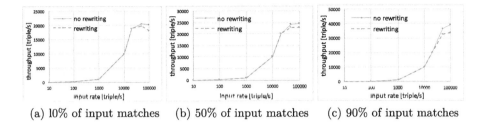

(a) 10% of input matches   (b) 50% of input matches   (c) 90% of input matches

Fig. 3. Throughput in StreamQR with and without rewriting.

of handling all the input. Beyond that, the throughput degrades until it reaches a limit. Notice that for each run, as StreamQR produces a UCQ, several queries are running simultaneously.

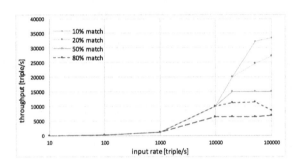

Fig. 4. Throughput in StreamQR for different distributions of input triples.

Query rewritings. Different queries may produce a UCQ with a different number of sub-queries. In this experiment we launched nine distinct queries that produce from 2 to over 180 sub-queries. As it was expected, in general the throughput decreases for queries that produce more rewritten queries (Fig. 5). Although this can also be affected by the complexity of the query, it is a limiting factor on the overall throughput. Existing techniques used in query rewriting and OBDA can be used to alleviate this, for instance by pruning queries that may not match any input. In stream processing this can be feasible in many cases as the data is often repetitive in terms of structure and can be deduced in the long run. Even then, for around 1 K triples/s, it still reaches maximum throughput.

Comparison with TrOWL. Finally, we compared the performance of StreamQR with TrOWL, which provides incremental reasoning for ABoxes. While the target of TrOWL is not query answering, but materialization, it is a state-of-the-art stream reasoner which can be used to populate an RDF store that can be periodically queried. We compared the throughput in three different settings. First,

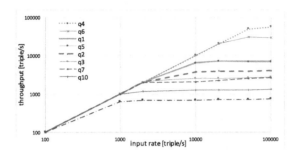

Fig. 5. Throughput for different queries with multiple rewritings, respectively 2, 2, 16, 18, 31, 36, 88, 51 and 185 sub-queries.

with TrOWL only consuming the data without performing any reasoning (no-reclassify), then activating the reasoning, but allowing only additions, and finally including removals as well. The removal operation is known to be expensive in incremental materialization. As we can see in Fig. 6, StreamQR sustains better throughput under fast input rates, even at the same level as TrOWL without any reasoning. With reasoning enable in TrOWL, this is even more noticeable. Under lower input rates both are able to reach maximum throughput. Given that the goal of TrOWL is not stream query answering, this comparison is only informative, showing that input throughput with materialization is lower than with query rewriting. A more systematic comparison of materialization vs. query rewriting is worth considering, although it is outside of the scope of this paper.

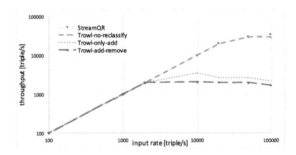

Fig. 6. Comparison with TrOWL, no reclassify, only additions, and with removals.

6   Related Work

Different DL languages have been explored and used for query rewriting and several systems have been implemented for these languages. The $DL-Lite$ family of languages [9], a first milestone in this area, derived into $DL-Lite_{\mathcal{R}}$ and $DL-Lite_{\mathcal{F}}$. $DL-Lite_{\mathcal{R}}$ includes ISA and disjointness assertions between roles and $DL-Lite_{\mathcal{F}}$ includes functionality restrictions on roles. These logics are first-order reducible with a tractable complexity [9], as done in Quonto and extended in Presto and Prexto [25,26]. The $OWL2\ QL$ profile was inspired by the $DL-Lite$ family and designed to keep the complexity of rewriting low, considering first-order rewritability. As a summary of a more extensive comparison [3], the main difference with $DL-Lite$ is related with the lack of the unique name assumption (UNA) Among the systems that address this expressiveness we can find Rapid [10].

The $\mathcal{ELHIO}^{\neg}$ logic [21] is more expressive. It extends the expressiveness of $DL-Lite_{\mathcal{R}}$ by including basic concepts of the form $\{a\}$, $\top$, and $B_1 \sqcap B_2$, as well as axioms of the form $\exists R.B \sqsubseteq C$. This logic does not preserve the first-order rewritability property, what means that depending on the query and the expressiveness in the ontology, the generated Datalog may contain recursive predicates. Thus some queries cannot be expressed as a union of conjunctive queries (UCQ) and must be rewritten to recursive Datalog. In spite of that, the computational complexity of the rewriting process remains tractable (PTIME-complete). Among the systems that can handle this logic we can find REQUIEM [21] and kyrie [18].

Some of the Datalog paradigms that ensure decidability are chase termination, guardedness or stickiness, extended to weak-stickiness by Calì et al. [8]. These paradigms limit the loops that can be present in some Datalog to ensure decidability of the unfolding and thus first-order rewritability. Among the systems that can handle this logic we can find Nyaya [14]. Finally, Horn-$\mathcal{SHIQ}$ includes role hierarchies and inverse roles as $\mathcal{ELHIO}$. It does also include universal restrictions and transitive roles ($\mathcal{S}$) axioms of the form $A \sqsubseteq \forall R.B$ and $trans(R)$. Among the systems that can handle this logic we can find Clipper [13].

With regards to RDF stream processing and reasoning, as described in Sect. 2.1, different approaches have surfaced in recent years, adding streaming support to SPARQL-based query processors. C-SPARQL [4] takes a hybrid approach that partially relies on a plug-in architecture that internally executes streaming queries with an existing DSMS. CQELS [16] implements a native RDF stream query engine with a focus on the adaptivity of streaming query operators and their ability to efficiently combine streaming and stored data. EP-SPARQL [1] adopts a perspective oriented to complex pattern processing, and includes sequencing and simultaneity operators. Other recent approaches focused on event processing based on the Rete algorithm for pattern matching include Sparkwave [15] and Instans [24]. There has also been a proposal for including rules from a knowledge base into C-SPARQL [5], although these are based on instantaneous materialization only for RDF-S, and are not available yet in the C-SPARQL software package. Concerning OBDA, the STARQL [19] framework introduced an ABox sequencing strategy which allows it to use unions of conjunctive queries combined with languages such as DL-Lite.

Previous efforts on stream reasoning have focused on ontology maintenance for streams, e.g. using *truth maintenance systems* [23] and approximate reasoning optimized for memory consumption, by eliminating unnecessary intermediate results. Other works have also proposed parallelization techniques for the materialization of inferences in streaming knowledge-bases, although limited only to a fragment of RDFS [27]. On a similar path, works on knowledge evolution [17] have used DL reasoning over *ontology streams* to detect and explain the nature of the changes on the ontology, as well as potential inconsistencies. Concerning theoretical results, the LARS framework [6] proposed a rule-based formalization that captures the semantics of stream reasoning engines.

7   Conclusions

In this paper we presented an approach for providing query answering over ontologies for RDF stream processors, through a novel approach that combines query rewriting techniques and an RSP engines. Furthermore, we implemented StreamQR, a system that incorporates the kyrie rewriter into an existing RSP engine, and that shows the feasibility of our approach. We also provided evidence that this implementation can still be efficient in terms of throughput, for a large range of scenarios, compared with an RSP engine with no rewriting or inferencing capabilities.

In the future, we plan to study other criteria such as correctness of the query answering process, which is known to be non trivial for data streams. Moreover, we are interested in exploring different expressiveness in order to find a good balance between efficiency and complexity. Finally, we believe that there is still large room for research in approaches that combine rewriting and incremental materialization for stream reasoners and query answering over ontologies.

Acknowledgments. Partially supported by the Nano-Tera.ch OpenSense2 and D1namo projects, evaluated by the SNSF. Supported by Ministerio de Economía y Competitividad (Spain) under the project 4V: Volumen, Velocidad, Variedad y Validez en la Gestión Innovadora de Datos (TIN2013-46238-C4-2-R).

References

1. Anicic, D., Fodor, P., Rudolph, S., Stojanovic, N.: EP-SPARQL: a unified language for event processing and stream reasoning. In: WWW, pp. 635–644 (2011)
2. Arasu, A., Babu, S., Widom, J.: The CQL continuous query language: semantic foundations and query execution. VLDB J. **15**(2), 121–142 (2006)
3. Artale, A., Calvanese, D., Kontchakov, R., Zakharyaschev, M.: The DL-Lite family and relations. J. Artif. Int. Res. **36**(1), 1–69 (2009)
4. Barbieri, D.F., Braga, D., Ceri, S., Della Valle, E., Grossniklaus, M.: C-SPARQL: SPARQL for continuous querying. In: WWW, pp. 1061–1062 (2009)
5. Barbieri, D.F., Braga, D., Ceri, S., Della Valle, E., Grossniklaus, M.: Incremental reasoning on streams and rich background knowledge. In: Aroyo, L., Antoniou, G., Hyvönen, E., ten Teije, A., Stuckenschmidt, H., Cabral, L., Tudorache, T. (eds.) ESWC 2010, Part I. LNCS, vol. 6088, pp. 1–15. Springer, Heidelberg (2010)
6. Beck, H., Dao-Tran, M., Eiter, T., Fink, M.: LARS: a logic-based framework for analyzing reasoning over streams. In: AAAI (2015)

7. Calbimonte, J.-P., Corcho, O., Gray, A.J.G.: Enabling ontology-based access to streaming data sources. In: Patel-Schneider, P.F., Pan, Y., Hitzler, P., Mika, P., Zhang, L., Pan, J.Z., Horrocks, I., Glimm, B. (eds.) ISWC 2010, Part I. LNCS, vol. 6496, pp. 96–111. Springer, Heidelberg (2010)

8. Calì, A., Gottlob, G., Pieris, A.: Query answering under non-guarded rules in datalog+/−. In: Hitzler, P., Lukasiewicz, T. (eds.) RR 2010. LNCS, vol. 6333, pp. 1–17. Springer, Heidelberg (2010)

9. Calvanese, D., De Giacomo, G., Lembo, D., Lenzerini, M., Rosati, R.: Tractable reasoning and efficient query answering in description logics: the DL-Lite family. J. Autom. Reason. **39**(3), 385–429 (2007)

10. Chortaras, A., Trivela, D., Stamou, G.: Optimized query rewriting for OWL 2 QL. In: Bjørner, N., Sofronie-Stokkermans, V. (eds.) CADE 2011. LNCS, vol. 6803, pp. 192–206. Springer, Heidelberg (2011)

11. Compton, M., Barnaghi, P., Bermudez, L., García-Castro, R., Corcho, O., Cox, S., Graybeal, J., Hauswirth, M., Henson, C., Herzog, A., Huang, V., Janowicz, K., Kelsey, W.D., Phuoc, D.L., Lefort, L., et al.: The SSN ontology of the W3C semantic sensor network incubator group. J. Web Semant. **17**, 25–32 (2012)

12. Cugola, G., Margara, A.: Processing flows of information: from data stream to complex event processing. ACM Comput. Surv. **44**(3), 15 (2011)

13. Eiter, T., Ortiz, M., Simkus, M., Tran, T.K., Xiao, G.: Query rewriting for horn-SHIQ plus rules. In: AAAI (2012)

14. Gottlob, G., Orsi, G., Pieris, A.: Ontological query answering via rewriting. In: Eder, J., Bielikova, M., Tjoa, A.M. (eds.) ADBIS 2011. LNCS, vol. 6909, pp. 1–18. Springer, Heidelberg (2011)

15. Komazec, S., Cerri, D., Fensel, D.: Sparkwave: continuous schema-enhanced pattern matching over RDF data streams. In: DEBS, pp. 58–68 (2012)

16. Le-Phuoc, D., Dao-Tran, M., Xavier Parreira, J., Hauswirth, M.: A native and adaptive approach for unified processing of linked streams and linked data. In: Aroyo, L., Welty, C., Alani, H., Taylor, J., Bernstein, A., Kagal, L., Noy, N., Blomqvist, E. (eds.) ISWC 2011, Part I. LNCS, vol. 7031, pp. 370–388. Springer, Heidelberg (2011)

17. Lécué, F.: Diagnosing changes in an ontology stream: a DL reasoning approach. In: AAAI (2012)

18. Mora, J., Corcho, O.: Engineering optimisations in query rewriting for OBDA. In: I-SEMANTICS, pp. 41–48 (2013)

19. Özçep, Ö.L., Möller, R., Neuenstadt, C.: A stream-temporal query language for ontology based data access. In: KI, pp. 183–194 (2014)

20. Pérez-Urbina, H.: Tractable query answering for description logics via query rewriting. Ph.D. thesis (2009)

21. Pérez-Urbina, H., Horrocks, I., Motik, B.: Efficient query answering for OWL 2. In: Bernstein, A., Karger, D.R., Heath, T., Feigenbaum, L., Maynard, D., Motta, E., Thirunarayan, K. (eds.) ISWC 2009. LNCS, vol. 5823, pp. 489–504. Springer, Heidelberg (2009)

22. Poggi, A., Lembo, D., Calvanese, D., De Giacomo, G., Lenzerini, M., Rosati, R.: Linking data to ontologies. In: Spaccapietra, S. (ed.) Journal on Data Semantics X. LNCS, vol. 4900, pp. 133–173. Springer, Heidelberg (2008)

23. Ren, Y., Pan, J.Z.: Optimising ontology stream reasoning with truth maintenance system. In: CIKM, pp. 831–836 (2011)

24. Rinne, M., Törmä, S., Nuutila, E.: SPARQL-based applications for RDF-encoded sensor data. In: SSN, vol. 904, pp. 81–96 (2012)

25. Rosati, R.: Prexto: query rewriting under extensional constraints in DL-Lite. In: Simperl, E., Cimiano, P., Polleres, A., Corcho, O., Presutti, V. (eds.) ESWC 2012. LNCS, vol. 7295, pp. 360–374. Springer, Heidelberg (2012)
26. Rosati, R., Almatelli, A.: Improving query answering over DL-Lite ontologies. In: KR (2010)
27. Urbani, J., Margara, A., Jacobs, C., van Harmelen, F., Bal, H.: Dynamite: parallel materialization of dynamic RDF data. In: Alani, H., et al. (eds.) ISWC 2013, Part I. LNCS, vol. 8218, pp. 657–672. Springer, Heidelberg (2013)
28. Zhang, Y., Duc, P.M., Corcho, O., Calbimonte, J.-P.: SRBench: a streaming RDF/SPARQL benchmark. In: Cudré-Mauroux, P., et al. (eds.) ISWC 2012, Part I. LNCS, vol. 7649, pp. 641–657. Springer, Heidelberg (2012)

Services, APIs, Processes and Cloud Computing Track

Linking Data, Services and Human Know-How

Paolo Pareti[1(⊠)], Ewan Klein[1], and Adam Barker[2]

[1] University of Edinburgh, Edinburgh, UK
p.pareti@sms.ed.ac.uk, ewan@inf ed.ac.uk
[2] University of St Andrews, St Andrews, UK
adam.barker@st-andrews.ac.uk

Abstract. An increasing number of everyday tasks involve a mixture of human actions and machine computation. This paper presents the first framework that allows non-programmer users to create and execute workflows where each task can be completed by a human or a machine. In this framework, humans and machines interact through a shared knowledge base which is both human and machine understandable. This knowledge base is based on the PROHOW Linked Data vocabulary that can represent human instructions and link them to machine functionalities. Our hypothesis is that non-programmer users can describe how to achieve certain tasks at a level of abstraction which is both human and machine understandable. This paper presents the PROHOW vocabulary and describes its usage within the proposed framework. We substantiate our claim with a concrete implementation of our framework and by experimental evidence.

1 Introduction

This paper addresses the largely unexplored problem of enabling non programmers to specify the type of behaviour they require from machines, as opposed to being constrained by their pre-defined functionalities. Applications such as IFTTT[1] (If-This-Than-That) have demonstrated that, given the right tools, users are capable of composing different components, such as triggers and actions, to program certain desired behaviours on a system. IFTTT for example, allows users to define conditions, such as "if there is a chance of rain tomorrow", in order to automatically trigger actions, such as "notify me".

At the opposite end of the spectrum from computer programs, human instructions can be seen as ways to "program" human behaviours. Humans are capable of specifying complex workflows based on human actions. Online instructions, such as those available on the wikiHow[2] website, often include different methods, steps, loops, and conditions. The main disadvantage of human instructions is that they are not machine understandable. Consequently, any useful functionality that a human could benefit from while following a set of instructions has to be manually accessed and triggered by the user.

---

[1] http://ifttt.com/.
[2] http://www.wikihow.com/.

© Springer International Publishing Switzerland 2016
H. Sack et al. (Eds.): ESWC 2016, LNCS 9678, pp. 505–520, 2016.
DOI: 10.1007/978-3-319-34129-3_31

The main contributions of this paper are two. The first is a framework that allows non-programmers to define and execute human-machine workflows. Our approach overcomes the limitations of traditional human instructions by allowing machines to automatically detect when certain actions are needed, and consequently to actively execute them. Section 4 presents the components of our framework and describes how they achieve its objective. An implementation of this framework is described in Sect. 6 and its usability by non-programmer users is evaluated in Sect. 7.

The second contribution is the PROHOW Linked Data vocabulary that we use to represent this knowledge base. This vocabulary not only describes human-machine workflows but can also trigger and describe their execution. Thus Linked Data, in our framework, acts both as a data representation and as a programming language. The details of this vocabulary, along with its intended logical interpretation, are given in Sect. 5.

2 Motivating Example

As a running example, we consider a scenario where a non-programmer human is writing instructions that include some automatable steps. In this scenario, Jane is an employee of a company who is in charge of curating content on the company website. Today Jane has gained access to a text file containing a long list of cities that the company has worked with. Each line of the file contains the name and Wikipedia page of the city in the following format:

```
<London: https://en.wikipedia.org/wiki/London>
```

Jane's job is to display the name and official website of each city as a list on the company's website. To achieve this, she wants to transform every entry in the dataset into an HTML list element in this format:

```
<li>London: https://london.gov.uk/</li>
```

Jane decides to delegate this task (that we will denote t) to John, one of her collaborators, and she gives him these instructions:

Step 1: Remove < and > from the string of text.
Step 2: Substitute the wikipedia URL with the official website of the city.
Step 3: Enclose the string of text with <li> </li>.

Jane notices that some steps in her instructions could be easily automated by a machine. However, Jane's problem is that nobody in her company is a programmer. Consequently, although some of the required functionalities are available in their computers, the whole task might have to be completed manually.

Our proposed solution for Jane's problem is to let her use a system that automatically translates her natural language instructions into machine understandable data and that semi-automatically links them to machine functionalities. This system might analyse the natural language description of steps 1 and 3 and detect that they fall within its capabilities. Given this configuration, when task t is started with a new string of text as input, the system could immediately execute the first step of the instructions. Then, as soon as the second step is complete, the third step will also be automatically executed.

The system will store all the information about Jane's task as Linked Data. This will allow the system to publish this information to other systems, and maybe discover that there is an external service that can also automate the second step of Jane's instructions. For example, an external service might be able to automate Jane's second step by querying the DBpedia[3] dataset, which contains information about the official websites of a large number of cities.

An important observation that can be derived from this example is that it is inefficient to create ad hoc systems to achieve these types of tasks, especially when they are small in scale, and when they can occur frequently but with variations. For example, other problems might require the same functionalities required by t, but combined together into a different workflow. Instead of creating rigid ad hoc solutions for human-machine collaboration, our approach to automation makes use of simple and generic machine capabilities that can be integrated with several different human-made instructions.

Many such capabilities can be imagined. For example, a trigger capability could instruct a machine to start the task "organise lunch in the park" if the weather is sunny and if no other commitment is scheduled for lunchtime in the user's calendar. If the user decides to do this activity, the calendar might also be updated automatically. Alternatively, a machine could assist the organiser of an event who decides to "send a message" to a large number of invitees. A machine might know two methods to automatically "send a message" to a person: by email and by mobile text. The machine could automatically send the message to all invitees whose email or mobile number is known. The remaining invitees could then be manually contacted by the event organiser using other channels. This scenario highlights the flexibility of human-machine collaboration, since a purely manual approach would be inefficient, and a purely automatic approach would be infeasible. More examples of machine capabilities that are being used by non-programmer users can be found on the IFTTT website.

3   Problem Description

The concept of *tasks* refers to things that can be accomplished. As such, they can be used to define goals, namely things that a person wants to accomplish. The main use of the concept of tasks is to provide a layer of abstraction over the actual *actions* that are performed to accomplish them. The types of tasks that humans can describe is very broad, and when interacting with machines, the level of abstraction at which they are described plays an important role. At the opposite ends of the abstraction spectrum we can find very abstract tasks, such as "Behave well", and very specific tasks, such as "Increment variable X by 1 unit". Typically, machines struggle to understand abstract tasks, while they can often accomplish specific tasks more efficiently than humans. Humans, on the other hand, can easily

---

[3] http://wiki.dbpedia.org/.

describe tasks in abstract terms, but struggle to define them in a very specific and rigorous way. Tasks that are too abstract for machines to automate, or too specific for humans to describe, are outside the focus of this paper. Our hypothesis is that there is a non-empty intersection T between the tasks that can be easily defined by (non-programmer) humans H, and those that are understandable by machines M. Lacking a previous baseline to compare to, we define a task to be *easily* definable if the majority of (non-programmer) humans can define it at a sufficient level of abstraction for it be machine understandable.

Our objective is to allow (non-programmer) humans to make better use of machine functionalities on the Web by allowing machines to understand which of their services is needed and when. This type of human-machine collaboration can be achieved by having humans and machines interact through a shared knowledge base. To this end, the problem that we address is how to allow a (non-programmer) web user to share (1) human-made instructions and (2) information on the progress made at completing the tasks described in those instructions, in a format that is both human and machine understandable.

4   Methodology

At the core of our approach for allowing human-machine collaboration is a knowledge base that is shared between humans and machines. Humans and machines collaborate with each other by interacting with this knowledge base, as depicted in Fig. 1. The contents of the knowledge base rely on the PROHOW vocabulary. This is a Linked Data vocabulary that we developed for key concepts that occur in human-made instructions (such as steps and requirements) or that describe aspects of their execution (such as information as to which steps have been completed).

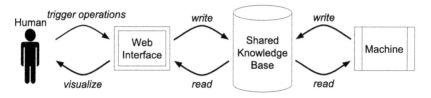

Fig. 1. Schema of the main components of our framework. Humans and machines interact with each other through a shared knowledge base.

In our approach, communication is performed indirectly, by modifying the shared knowledge base. This can be seen as a type of *stigmergy* [6], namely indirect communication through modification of the environment. For example, if a machine wants to communicate to humans (or to other machines) that a particular step has been automated, this will be done by adding this information to the knowledge base. This type of indirect communication avoids the problem of how to implement direct communication between human and machines. Instead, it casts human-machine collaboration as a knowledge sharing problem, and as such it is amenable to being handled by Semantic Web technologies.

Human and machines interact with the knowledge base in two different ways. Machines can directly access it to query or modify its contents, since the knowledge base is represented as an RDF[4] graph. Moreover, machines can understand this knowledge base by following its logical interpretation, as defined by the PRO-HOW vocabulary. This logical interpretation allows machines to make inferences over this knowledge base. For example, a machine could infer that a certain task has been implicitly accomplished because all of its steps have been completed, or it could infer that it is not yet time to complete a certain task because some of its requirements are still incomplete. The logical interpretation of the PROHOW vocabulary will follow in Sect. 5.

Humans, on the other hand, interact with this knowledge base through an intuitive Web interface. This interface allows humans to "read" the knowledge base by providing it in a human-readable format. For example, while a machine could query the knowledge base to retrieve all the steps of a given task, a human could visualize the list of steps in an HTML page. This interface also allows humans to "write" to the knowledge base. This can be done, for example, by parsing a user's natural language instructions into RDF, or by interpreting a user action of ticking off a certain step as an indication that the step has been completed. An implementation of this interface will be presented in Sect. 6.

Once humans and machines interact through a shared knowledge base, they can collaborate on the execution of tasks. This collaboration can be divided into three phases: (1) know-how acquisition, (2) know-how linking and (3) execution. We will now describe the typical workflow of our framework across these phases.

4.1 Know-How Acquisition

In the first phase of our framework, know-how is converted into Linked Data. Our hypothesis is that humans can write instructions in semi-structured format. Websites like wikiHow, for example, require users to explicitly divide their instructions into steps, methods and requirements. Further support for our hypothesis is provided by the existence of large instructional websites that contain instructions with this level of structure. Evidence of this has been provided by a large scale conversion of over 200,000 instructions from the wikiHow and Snapguide[5] websites into an RDF format using the PROHOW vocabulary [7].

This existing structure can be extracted and represented in RDF. For example, consider the following natural language instructions for the string transformation task :t described in Sect. 2:

```
Step 1: Remove < and > from the string of text.
Step 2: Substitute the wikipedia page of the city in the
    string of text with official homepage of the city.
Step 3: Enclose the string of text with <li> </li>.
```

---

[4] http://www.w3.org/TR/rdf11-concepts/.
[5] https://snapguide.com/.

Table 1. The RDF namespaces used in this document.

Prefix	Namespace
prohow:	http://w3id.org/prohow
rdfs:	http://www.w3.org/2000/01/rdf-schema
:	http://example.org/

By exploiting the implicit structure of this text, it is possible to automatically generate the following RDF graph. RDF graphs listed in this paper are serialised in Turtle[6] format, and use the namespaces defined in Table 1.

```
:t prohow:has_step :1, :2, :3 .
:1 rdfs:label "Remove < and > from the string of text.".
:2 rdfs:label "Substitute the wikipedia page of the city in the
    string of text with official homepage of the city.".
:3 rdfs:label "Enclose the string of text with <li> </li>.".
:2 prohow:requires :1 .
:3 prohow:requires :2 .
```

This RDF graph explicitly represents the subdivision of task :t into three steps (:1, :2 and :3) using the prohow : has_step relation. The correct ordering of the steps is specified by the prohow : requires relations.

4.2   Know-How Linking

In the second phase of our methodology, an existing set of instructions is linked with automatable functions. For example, we can imagine a machine :x capable of removing specific characters from strings of text. This machine can describe its capability in terms of the task it can accomplish. For example, the following RDF graph describes the task :t1 of "Remove a character from a string of text". This task specifies the requirements :r1 "The string of text to modify" and :r2 "The character to remove" which should be known before the task can be automated.

```
:t1 rdfs:label "Remove a character from a string of text".
:r1 rdfs:label "The string of text to modify" .
:r2 rdfs:label "The character to remove" .
:t1 prohow:requires :r1, :r2 .
```

PROHOW allows functionalities to be defined at the input/output level. A functionality :f with a set of inputs I and a set of outputs O is described with a set of prohow : requires and prohow : has_method relations. A prohow : requires link from :f to one of its inputs :i represents the dependency between :f and :i, thus making sure that the functionality will not be executed before its input is available. A prohow : has_method link from an output :o to :f represents the fact that one way to obtain :o is to perform :f.

---

[6] http://www.w3.org/TR/turtle/.

It can be observed that the **prohow : requires** relation can be used to represent both the dependency between (1) a task and an input and (2) a task and a step that needs to be done beforehand. This is a result of the fact that in the domain of human know-how the distinction between actions and objects can be blurred. A cooking recipe, for example, could mention the ingredient "eggs" as an input, or it could mention "get eggs" as one of its steps. In such a situation, the choice of whether to represent something as an object or as an action is arbitrary, and it should not lead to semantically different formalisations. Therefore, the PROHOW vocabulary diverges from typical process formalisations in that it does not enforce a distinction between actions and objects.

Going back to our example, we can imagine a step :t2 of a procedure that requires the character "<" to be removed from string :s. This step can be linked to the functionality offered by machine :x with the following triples:

```
:t2 prohow:has_method :t1 .
:t2 prohow:has_constant :c1 .
:c1 rdfs:label "<" .
:r1 prohow:binds_to :s .
:r2 prohow:binds_to :c1 .
```

Once this link has been created, machine :x will detect that its capability of accomplishing task :t1 can also be used to accomplish task :t2. When task :t2 needs to be completed, machine :x will try to accomplish it by executing :t1. Bindings between tasks can be used to specify which particular parametrisation of a task can be used to accomplish another task. In this scenario, for example, the character to remove :r2 is bound to the constant "<". These types of links can connect a single machine functionality, or one of its parametrisations, to any number of more abstract tasks that this functionality can accomplish. For example, task :t1 could be linked to any string modification task that specifies a string and a character to remove from that string.

The discovery of these kind of links can be seen as a form of subsumption matching, since the set of possible ways of accomplishing the more abstract task subsume the set of possible ways of accomplishing the more specific one. This discovery process can be performed in different ways. In a previous experiment, we automated the creation of links between different sets of instructions from wikiHow (in the PROHOW format) using Natural Language Processing and Machine Learning [7]. This showed that creating links between sets of PROHOW instructions can be achieved with high accuracy. Indeed, the number and precision of the discovered links was shown to be superior to the equivalent human-generated HTML links already present in wikiHow. In this paper we consider instead a semi-automated approach. Whenever a set of instructions is created, an artificial system can select from a large number of available resources the ones that seem to be most related to each part of the instructions. Humans will then be asked to verify whether a link should be created or not.

4.3 Execution

When a human or a machine intends to execute a task, an RDF graph is created that declares a new execution of that task. For example, the following triple is sufficient to declare a new attempt :en to accomplish task :t.

```
:en prohow:has_goal :t .
```

Declaring this intention could be as simple as pressing a "Do it!" button available on the same web page that describes task :t.

After the creation of this triple, it is possible to retrieve information about :en in order to view (and if necessary modify) the current state of the execution. For a human user, a visualization of :en could display, for example, which steps of the procedure have already been completed, and which still need to be completed. We can represent the fact that the execution :ex1 of the first step :1 of the instructions :t has been completed with the following graph:

```
:ex1 prohow:has_task :1 .
:ex1 prohow:has_result prohow:complete .
:ex1 prohow:has_environment :en .
```

After this information is stored in the shared knowledge base, all the humans and machines collaborating on this task will be able to discover that the first step :t1 has been completed. They would then infer that it no longer needs to be carried out and could decide to execute the following step instead.

5 Logical Interpretation of the PROHOW Vocabulary

This section describes the logical interpretation of the PROHOW vocabulary that enables machines to understand the shared knowledge base of our framework. The terms of this vocabulary are listed in Table 2, and follow the namespaces described in Table 1.

The most important concept in the PROHOW vocabulary is the concept of *task* (prohow : task). The same task can be accomplished multiple times. For example, in the example scenario introduced in Sect. 2, the string transformation task would need to be completed once for each string of text that needs to be transformed. The concept of *environment* (prohow : environment) is used to group together all the information about a specific intention to achieve a task. We use the logical expression has_goal(e, y) to indicate that the main task (or goal) of environment e is task y.

With the term *execution* (prohow : execution) we refer to a particular attempt to complete a task. A task y is complete (prohow : complete) in an environment e if there is an execution i in that environment that has succeeded (Formula 1); it is said to be failed (prohow : failed) if that execution has failed instead (Formula 2). Intuitively, the completion of a task means that a satisfactory result has been reached, and there is no need for completing the same task in the same environment again. If an execution has failed instead, it is possible to attempt another execution of the same task in the same environment.

$$\text{complete}(y, e) \Longleftarrow \exists i.\text{has\_env}(i, e) \wedge \text{has\_task}(i, y) \wedge \text{success}(i) \tag{1}$$

$$\text{failed}(y, e) \Longleftarrow \exists i.\text{has\_env}(i, e) \wedge \text{has\_task}(i, y) \wedge \text{failure}(i) \tag{2}$$

An environment is said to be finished when its goal is complete:

$$\text{finished}(e) \Longleftarrow \exists g.\text{has\_goal}(e, g) \wedge \text{complete}(g, e) \tag{3}$$

Tasks are not only used to denote actions, such as "Remove all the < and > characters from a string of text" but they are also used to refer to objects and data, such as "the string of text" that needs to be modified. In this last case, completing a task means obtaining the object or discovering its value as a variable. The particular object/value associated with a completed execution might be specified using the has_value relation. With the following formula we state that an object-task y has a value of z in environment e if there is a complete execution in that environment that refers to task y and has value z:

$$\begin{aligned}\text{value}(z, y, e) &\Longleftarrow \exists i.\text{has\_env}(i, e) \\ &\wedge \text{has\_task}(i, y) \wedge \text{success}(i) \wedge \text{has\_value}(i, z)\end{aligned} \tag{4}$$

A common property of tasks is the set of requirements that need to be completed before the execution of the task is ready to start. A task y is ready in an environment e if all of its requirements (if any) are complete in the same environment (Formula 5). If a task is not ready, than no attempt at performing it, or any of its sub-tasks, should occur. Intuitively, this means that a task can be started only when all of its requirements are complete. This relation can also be used to order (totally or partially) the steps of a task.

$$\text{ready}(y, e) \Longleftarrow \forall x.\text{requires}(y, x) \rightarrow \text{complete}(x, e) \tag{5}$$

If a task y has at least one step, and all of its steps are complete in an environment e, then task y is also complete in that environment (Formula 6). Intuitively, this means that a task can be accomplished by accomplishing all of its steps.

$$\text{complete}(y, e) \Longleftarrow \exists x.\text{has\_step}(y, x) \bigwedge \forall x.\text{has\_step}(y, x) \rightarrow \text{complete}(x, e) \tag{6}$$

If a method x of a task y is complete in a sub-environment of e, then task y is complete in environment e (Formula 7). Intuitively, this means that a task can be completed by completing any of its methods.

$$\text{complete}(y, e) \Longleftarrow \exists x, a.\text{has\_method}(y, x) \wedge \text{complete}(x, a) \wedge \text{sub\_env}(a, e) \tag{7}$$

Environments that are connected with the sub-environment relation are said to be related environments:

$$\begin{aligned}\text{related}(a, e) &\Longleftarrow \text{sub\_env}(a, e) \vee \text{sub\_env}(e, a) \\ \text{related}(a, e) &\Longleftarrow \exists x.\text{related}(a, x) \wedge \text{related}(x, e)\end{aligned} \tag{8}$$

Unless bindings have been specified, a task that is complete in one environment is not necessarily complete in its related environments. A task x is complete in an environment e if it has a binding with another task y which is complete is an environment a related to e (Formula 9). Values can be shared in a similar way, as defined in Formula 10.

$$\text{complete}(x, e) \Longleftarrow \exists y, a.\text{binds}(x, y) \wedge \text{complete}(y, a) \wedge \text{related}(a, e) \qquad (9)$$

$$\text{value}(z, x, e) \Longleftarrow \exists y, a.\text{binds}(x, y) \wedge \text{value}(z, y, a) \wedge \text{related}(a, e) \qquad (10)$$

Bindings can be used to choose which tasks and values can be shared between environments. For example, the task "Remove a character from a string of text" could be completed in two different (but related) environments to remove two different characters (e.g. $<$ and $>$) from the same string of text. In this scenario, the string of text will be shared between the two environments, while the characters to remove will not.

Values that can be shared between unrelated environments are called constants. If a task y has a constant x, then x will be automatically considered accomplished in any environment where y is being executed:

$$\begin{aligned}\text{complete}(x, e) \wedge \text{value}(x, x, e) \Longleftarrow \exists i.\text{has\_env}(i, e)\\ \wedge\, \text{has\_task}(i, y) \wedge \text{has\_constant}(y, x)\end{aligned} \qquad (11)$$

It should be noted that the expressiveness of the PROHOW vocabulary goes beyond simple step sequences. In fact, this vocabulary can express all the basic control flow patterns defined by Van der Aalst et al. [11]. For example, tasks are by default non-ordered and any partial ordering of the tasks can be expressed using `prohow : requires` relations. This implements the *sequence, parallel split* and *synchronisation* patterns. The `prohow : has_method` relation can describe a choice point where only one out of multiple paths needs to be followed. Other tasks can be made to wait until one such path has been completed. This implements the *exclusive choice* and *simple merge* patterns.

6   Human Interfaces to the PROHOW Vocabulary

Unlike machines, humans cannot directly interact with PROHOW data. Their interactions need to be mediated through human-understandable interfaces that allow users to both visualise and modify this data. For example, such an interface might present entities organised into familiar structures, such as an ordered list of steps or to-do checklists. In order to allow users to modify the data, several functionalities need to be implemented. In particular, users should be able to create new sets of instructions, create links between them, start new executions of a task and update their progress.

As a proof of concept, we have implemented one such interface as an online service[7] which allows users to follow the three main phases defined in Sect. 4.

---

[7] http://w3id.org/prohow/editor.

Table 2. The concepts and relations of the PROHOW vocabulary.

Concept	Description of the concept
prohow : task	a task that can be accomplished
prohow : execution	an attempt to perform a task
prohow : environment	a collection of executions to achieve a goal
prohow : complete	the positive result of accomplishing a task
prohow : failed	the negative result of accomplishing a task
Relation	Logical definition (with subject y and object x)
prohow : requires	requires(y, x)
prohow : has_step	has_step(y, x)
prohow : has_method	has_method(y, x)
prohow : has_task	has_task(y, x)
prohow : has_goal	has_goal(y, x)
prohow : binds_to	binds$(y, x) \wedge$ binds(x, y)
prohow : has_constant	has_constant(y, x)
prohow : has_value	has_value(y, x)
prohow : has_result	success(y) if x is prohow : complete
	failure(y) if x is prohow : failed
prohow : has_environment	has_env(y, x)
prohow : sub_environment_of	sub_env(y, x)

This interface includes an online editor that translates free text into RDF using the PROHOW vocabulary. This editor parses a user's input and looks for keywords such as "Step" and "Requires" to identify steps and requirements. After parsing the text into an RDF graph, a visualization is provided to the user. If the user is satisfied with the machine interpretation of the instructions, a save button stores the generated know-how as an RDF graph in the knowledge base of the system, minting new URIs for each component of the instructions. Those URIs are dereferenceable, and users can use them to request a human-readable visualization. The same URI can be used by machines to obtain a machine-readable version. This is done by content negotiation, for example by setting the *Accept* header of an HTTP request to *application/rdf+xml*. The graphical interface also supports users during the know-how linking and execution phases. It allows them to create or link steps, methods and requirements and to initiate and update task executions. The user can choose whether to complete all tasks manually, or to let the system automate them whenever possible.

7 Experiments

The objective of this experiment is to support our hypothesis that the majority of (non-programmer) Web users can define certain tasks at a level of abstraction

which is understandable by machines. We evaluated this hypothesis with respect
to the three phases of our framework described in Sect. 4: know-how acquisition,
know-how linking and execution. This experiment is also meant to demonstrate
the application of our approach in a concrete scenario. Human participation in this
experiment has been obtained through the crowdsourcing platform Crowdflower.[8]
Participants were asked information about their computer skills to exclude answers
from computer experts.

7.1 Evaluation of the Know-How Acquisition Phase

During the know-how acquisition phase, our hypothesis is that humans can write
semi-structured procedures which can be automatically parsed into an RDF representation. We evaluated this by asking 10 workers to solve the example task
described in Sect. 2 through an online survey.[9] Workers submitted their instructions in a text-box in natural language. To improve the quality of their submissions, workers were asked to follow certain rules, such as to clearly divide their
instructions into steps, and were offered a bonus compensation for creating high
quality instructions. All the original submissions are available online.[10] Of the
10 submissions we received, we rejected 3 of them as non-genuine attempts. Of
the remaining submissions, we considered the 5 best solutions for the next part
of the experiment.

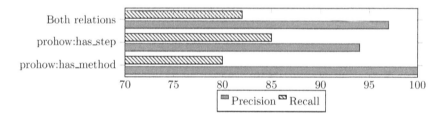

Fig. 2. Precision and recall of the links generated by the workers.

7.2 Evaluation of the Know-How Linking Phase

To judge whether non-programmer humans can correctly link instructions to
automatable functions, we have paired each of the 16 steps of the five best
sets of instructions with 10 different machine functionalities. All automatable
steps have been paired with one or more relevant functionalities, as well as
unrelated ones. For each step-functionality pair, we have then asked workers to
judge whether a particular functionality (in the survey called *action*), such as

---

[8] http://www.crowdflower.com/.
[9] http://w3id.org/prohow/r1/survey.
[10] http://w3id.org/prohow/r1/survey_results.

"Remove every occurrence of a particular character from the string of text" is relevant for the execution of a particular step (in the survey called *goal*), such as "Remove < and > from the string of text". Each worker was asked to choose one of the following three answers: (1) "YES, this action can completely achieve the goal", (2) "YES, BUT the action can only achieve part of the goal" or (3) "NO, the action is unrelated with the goal". For some functionalities, workers were also asked to provide information on how the function should be completed, for example by answering the question: "Which is the character to remove?".

For each step-function pair we have asked the judgement of 10 different workers, and then we have chosen the judgement given by the majority of the workers. For all questions, the most common answer was always chosen by more than 50 % of the workers. To judge whether an answer is correct or not, we interpret the first answer as the creation of a `prohow : has_method` link between the step and the functionality; the second answer as the creation of a `prohow : has_step` relation, and the third answer as no relation. We have manually evaluated the links generated by the majority of the workers and the precision and recall of those links is shown in Fig. 2. The result of this evaluation shows that the majority of the workers correctly chose to create a link between a step and a functionality 97 % of the time, discovering 82 % of all possible correct links.

7.3 Demonstration of the Execution Phase

To enable collaborative human-machine execution we have developed the machine functionalities which workers previously created links to. Our system listens to changes to its knowledge base to detect when and how its functionalities are needed. When this happens, the system will execute the functionality and modify the knowledge base accordingly, so as to allow the human user to notice that a task has been accomplished. As a result of this experiment, all the five sets of instructions are now available online,[11] and each of them contains at least one automatable function.

8   Related Work

We frame our work at the intersection of several related areas, which highlight its interdisciplinary nature. The idea of combining human and machine efforts to solve tasks that neither humans or machines alone could solve efficiently is central to the field of Human-Computation [8]. In Human Computation systems, humans typically play a subordinate role, as they have no direct control over the computation, and are not in charge of initiating it. For example, users of the Galaxy Zoo[12] project are asked to detect patterns in sky images to help machines to classify galaxies. However, they have no control over the machine computation that utilizes their contributions. More control is given to users

---

[11] http://w3id.org/prohow/r1/instructions.
[12] http://www.galaxyzoo.org/.

of Human-Provided Services (HPS) [10]. For example, HPS users can actively define and advertise the services they want to offer, and manage their interactions with other users. In general, Human Computation is better suited to accomplish large and complex tasks, while HPS can better address dynamic tasks with rich user interactions. However, both of them require expert intervention to define task workflows and therefore neither of them can effectively address the goals of individual users. The main objective of our work, instead, is to put humans in control of the computation. In our framework, humans participate to solve tasks that they have defined and that they are directly interested in accomplishing.

In our approach, non-programmer users define workflows by providing natural language instructions. While we limit the analysis of these instructions at the structural level (e.g. steps and requirements), the possibility of extending this analysis by means of Natural Language Processing has been investigated in the literature. For example, an approach has been developed to translate if-this-than-that constructs from natural language into executable programs [9]. Other approaches are more domain specific, such as focusing on cooking recipes [4]. Similarly to Controlled Languages, most of those approaches rely on certain structural or lexical properties of the instructions to extract their meaning.

Several languages have been created to describe processes in different fields, most notably OWL-S [5] in the Semantic Web community. While the majority of these languages focuses on describing automated functionalities, such as Web Services, some languages also include human participation in the computation. For example, the CompFlow [3] ontology allows the definition of workflows that can interleave both human and machine computation. None of these languages, however, is meant to be understandable by generic Web users, and experts are required to define the specific workflows. Moreover, most of these formalisations are too domain specific or logic heavy to conveniently represent human know-how. For example, human tasks are incompatible with OWL-S, since this ontology defines a process as a "specification of the ways a client may interact with a service" [5]. It should be noted, however, that integrations between PRO-HOW and other languages are possible. For example, the URI of a PROHOW task could point to an OWL-S service that might accomplish the task once invoked.

In our framework, humans and machines interact through the modification of a shared RDF knowledge base. This type of indirect communication resembles Blackboard Systems (BS) [1]. In BS, multiple agents collaborate to compute the solution to a problem by modifying a shared resource (the blackboard). From BS originates another related communication mechanism: Triple Space Computing (TSC) [2]. In the TSC approach, coordination between Semantic Web Services is achieved indirectly by publishing and reading RDF resources on the Web, which are organised into *Triple Spaces*. TSC includes several functionalities, such as the possibility to create, advertise or subscribe to particular Triple Spaces. Although useful in a more generic setting, such functionalities are not currently included in our framework, which can be imagined as having a single Triple Space.

9   Conclusion

In this paper we have addressed the problem of enabling non-programmer humans to specify the type of behaviour they require from machines, as opposed to being constrained by pre-defined functionalities. To solve this problem we have presented the first framework that allows non-programmers to define and execute human-machine workflows that is, workflows that combine human and machine actions to achieve a common goal.

Human-machine collaboration is achieved by indirect communication through a shared Linked Data knowledge base defined with our PROHOW vocabulary. The logical interpretation associated with this vocabulary makes the knowledge base machine understandable, allowing machines to infer when and how their functionalities are required. At the same time, this knowledge base is made human understandable by a direct visualization of its contents through an intuitive web interface. Using this interface, user-generated instructions in natural language are automatically translated into Linked Data.

Unlike a modelling language, the main objective of our vocabulary is not to describe how humans and machines collaborate, but rather to enable this collaboration in practice. To demonstrate this, we presented an implementation of our framework which we evaluated in a concrete test scenario. The results of this experiment support our hypothesis that non-programmers can specify certain types of instructions at the level of detail required for machine understanding.

References

1. Corkill, D.D.: Blackboard systems. AI Expert **6**(9), 40–47 (1991)
2. Fensel, D., Facca, F.M., Simperl, E., Toma, I.: Triple space computing for semantic web services. In: Semantic Web Services, pp. 219–249 (2011)
3. Luz, N., Pereira, C., Silva, N., Novais, P., Teixeira, A., Oliveira e Silva, M.: An ontology for human-machine computation workflow specification. In: Polycarpou, M., de Carvalho, A.C.P.L.F., Pan, J.-S., Woźniak, M., Quintian, H., Corchado, E. (eds.) HAIS 2014. LNCS, vol. 8480, pp. 49–60. Springer, Heidelberg (2014)
4. Malmaud, J., Wagner, E.J., Chang, N., Murphy, K.: Cooking with semantics. In: Proceedings of the ACL 2014 Workshop on Semantic Parsing, pp. 33–38 (2014)
5. Martin, D., Burstein, M., Hobbs, J., et al.: OWL-S: Semantic markup for web services. W3C Member Submission (2004)
6. Omicini, A., Ricci, A., Viroli, M., Castelfranchi, C., Tummolini, L.: Coordination artifacts: environment-based coordination for intelligent agents. In: Proceedings of the Third International Joint Conference on Autonomous Agents and Multiagent Systems, vol. 1, pp. 286–293 (2004)
7. Pareti, P., Testu, B., Ichise, R., Klein, E., Barker, A.: Integrating know-how into the linked data cloud. In: Janowicz, K., Schlobach, S., Lambrix, P., Hyvönen, E. (eds.) EKAW 2014. LNCS, vol. 8876, pp. 385–396. Springer, Heidelberg (2014)
8. Quinn, A.J., Bederson, B.B.: Human computation: a survey and taxonomy of a growing field. In: Proceedings of the SIGCHI Conference on Human Factors in Computing Systems, pp. 1403–1412 (2011)

9. Quirk, C., Mooney, R., Galley, M.: Language to code: learning semantic parsers for if-this-then-that recipes. In: Proceedings of the 53rd Annual Meeting of the Association for Computational Linguistics (ACL–2015), pp. 878–888 (2015)
10. Schall, D.: Service-Oriented Crowdsourcing: Architecture Protocols and Algorithms, chapter Human-Provided Services, pp. 31–58 (2012)
11. Van Der Aalst, W.M.P., Ter Hofstede, A.H.M., Kiepuszewski, B., Barros, A.P.: Workflow patterns. Distrib. Parallel Databases **14**(1), 5–51 (2003)

Smart Cities, Urban and Geospatial Data Track

VOLT: A Provenance-Producing, Transparent SPARQL Proxy for the On-Demand Computation of Linked Data and its Application to Spatiotemporally Dependent Data

Blake Regalia$^{(\boxtimes)}$, Krzysztof Janowicz, and Song Gao

STKO Lab, University of California, Santa Barbara, USA
{blake,sgao}@geog.ucsb.edu, janowicz@ucsb.edu

Abstract. Powered by Semantic Web technologies, the Linked Data paradigm aims at weaving a globally interconnected graph of raw data that transforms the ways we publish, retrieve, share, reuse, and integrate data from a variety of distributed and heterogeneous sources. In practice, however, this vision faces substantial challenges with respect to data quality, coverage, and longevity, the amount of background knowledge required to query distant data, the reproducibility of query results and their derived (scientific) findings, and the lack of computational capabilities required for many tasks. One key issue underlying these challenges is the trade-off between storing data and computing them. Intuitively, data that is derived from already stored data, changes frequently in space and time, or is the result of some workflow or procedure, should be computed. However, this functionality is not readily available on the Linked Data cloud with its current technology stack. In this work, we introduce a proxy that can transparently run on top of arbitrary SPARQL endpoints to enable the on-demand computation of Linked Data together with the provenance information required to understand how they were derived. While our work can be generalized to multiple domains, we focus on two geographic use cases to showcase the proxy's capabilities.

Keywords: Linked data · Semantic web · SPARQL · Geo-data · Cyber-infrastructure · Geospatial semantics · VOLT

1 Introduction and Motivation

Linked Data described the paradigm for a Web of densely interconnected yet distributed data. It provided methods and tools that dramatically ease the publication, retrieval, sharing, reuse, and integration of semantically rich data across heterogeneous sources. Over the last few years, we have witnessed a rapid increase in available data sources on the Linked Data cloud and a fast uptake of the involved technologies in academia, governments, and industry. Nonetheless, several key issues remain to be addressed in order to enable the full potential of Linked Data. One of these issues is the trade-off between storing data and

© Springer International Publishing Switzerland 2016
H. Sack et al. (Eds.): ESWC 2016, LNCS 9678, pp. 523–538, 2016.
DOI: 10.1007/978-3-319-34129-3_32

computing them. To give a concrete example, if the population and area of a county are available, should the population density be stored as well or should it be computed on-demand as it depends on already stored properties? Storing such data is often problematic or even impossible for multiple reasons. Keeping the population density in sync with a changing population is just one example. Consequently, such statements should be computed. However, this functionality is not readily available on the Linked Data cloud and is not fully supported by existing query languages, endpoints, or APIs.

Recently, a variety of approaches [1,4,5,9,10] have been proposed to address this and related issues. Here, we argue why these approaches alone are not sufficient and propose a framework inspired by a combination of their findings. Essentially, we propose a proxy[1] that can transparently run on top of any SPARQL 1.1 compliant endpoint while providing a framework for the on-demand computation and caching of Linked Data. Going beyond existing work, our approach also provides the provenance information required to make sense of the (cached) results, thereby improving reproducibility. Essentially, all (derived) data together with the procedures used to compute them are stored as RDF in separate graphs.

In the following, and as space permits, we highlight key aspects of the VOLT[2] proxy and framework by example. Instead of focusing on technical (implementation) aspects alone, we showcase VOLT's capabilities by discussing two use cases in detail. These use cases also serve as the evaluation of our work, e.g., they demonstrate how to improve the data quality of DBpedia and reduce storage size at the same time. While our work can be generalized to multiple domains, both use cases focus on geo-data. We believe that the challenges introduced by spatiotemporal data are ideal for discussing the need for provenance information on the procedural (workflow) level, the difficulties resulting from keeping *dependent* data in sync, and the problem that allegedly *raw* data was created by using some latent assumptions that now hinder reproducibility and thus interoperability.

2   The VOLT Framework and Proxy

Work that aims at bringing API-like features to the Semantic Web typically does so by either suggesting ways to extend SPARQL or by providing additional functionality outside of the typical Semantic Web layer cake; see Sect. 4. Implementing such solutions often requires a custom SPARQL engine or the adoption of future W3C recommendations. Furthermore, running non-standard SPARQL engines threatens Linked Data interoperability and reusability of federated and non-federated queries alike. For these reasons, we often fail to see widespread use of experimental technologies. Finally, most of these technologies are not transparent, i.e., they require additional knowledge or at least awareness by the end user. To overcome these issues, we strive to develop a transparent framework that embraces the existing technology stack without any changes to SPARQL. Our approach functions as a *transparent proxy* [3] to any existing SPARQL 1.1

---

[1] A working VOLT proxy prototype is available at: http://demo.volt-name.space/.
[2] VOLT: **V**OLT **O**ntology and **L**inked data **T**echnology.

engine and thereby acts as a legitimate endpoint. When a query is issued to the proxy, it triggers a series of interactions with the underlying, encapsulated SPARQL endpoint before forwarding the results back to the client. In other words, the client does not notice any difference to a regular endpoint.

In this section we introduce the general VOLT architecture, highlight important aspects such as transparency, and give an overview of the implementation.

2.1 SPARQL as an API

The idea of using triples within the basic graph pattern of a query to invoke computation is referred to by iSPARQL as *virtual triples* [5]. A virtual triple uses the predicate to identify a procedure and effectively treats each subject and object of the triple as an input or output to the procedure. Like virtual triples and the *magic properties*[3] of Apache's ARQ, we use the triple's predicate as a way to identify a user-defined procedure. However, we make a distinction between the various ways in which these special patterns are used in our framework:

Firstly, *computable properties* simply represent an existential relation between two named entities. For example, consider a computable property named `udf:intersects` that tests for the spatial intersection between two individuals. A client may trigger computation on the individuals `:A` and `:B` by issuing a SPARQL ASK query with the basic graph pattern `:A udf:intersects :B`. Alternatively, a client may find all things that intersect with `:A` via a SELECT query `:A udf:intersects ?other`, where the object of the previous triple has been replaced by the variable `?other`. Yet another style allows the client to test multiple computable properties on the same triple by using a variable in place of the predicate along with a triple that constrains the variable to a specific `rdf:type`. For instance, a client may discover all topological relations between two particular regions by issuing the SELECT query `:A ?relation :B. ?relation a udf:RegionConnectionRelation`. In this variation, the triple that constrains the variable `?relation` functions as a *computable property trigger*. It indicates the client's intention to invoke testing on an entire class of computable properties.

Secondly, *functional triples* act as interfaces for calling user-defined procedures with named inputs and outputs. To invoke a procedure, functional triples expect a primary *root triple* where the subject is anonymous and the object is a blank node. The blank node object acts as a hashmap for both the input arguments and output variables to the procedure. Whereas EVT[4] functions accept an ordered list of input arguments and return a single RDF term, functional triples accept an unordered set of named input arguments, are capable of returning multiple output bindings, and allow both inputs and outputs to be either RDF terms or RDF graphs. Once the functional triple call is executed, the entire graph constructed within the blank node is saved (either temporarily or persistently) to a graph in the triplestore. Doing so enables auxiliary pattern groups within the same query to work as if the entire functional triple's blank node was

---

[3] https://jena.apache.org/documentation/query/extension.html#property-functions.
[4] Extensible Value Testing.

matched to an existing set of triples. The subject of a root triple must be an unbounded variable or a top-level blank node (that is, anonymous) in the query as the entire functional triple will be materialized and the subject will become a URI suffixed by a UUID[5]. A functional triple example will be shown in Listing 8.

Thirdly, *pattern rewriters* perform special expansions to the SPARQL query at runtime for patterns that may be otherwise impossible to write in a single query, such as subqueries that construct RDF graphs. A pattern rewriter is invoked by a functional triple which identifies the rewriter's procedure along with its input arguments. A group of query patterns gets associated to the rewriter by exploiting the GRAPH keyword in SPARQL. Consider an example where we want to select only the first valid object matched by a list of acceptable predicates that are semantically equivalent (in a certain context). Say we want to count the sum of populations given by the DBpedia dbp:population predicate for some distinct places. Since we do not want to count the same subject twice, we only want to match a single value for each subject. If a subject does not have a valid numeric literal belonging to the primary predicate dbp:population, then we opt for a secondary predicate, dbp:populationTotal. One can perform this in a regular SPARQL query as depicted in Listing 1.

```
select (sum(?population) as ?totalPopulation) where {
    {   ?s dbp:population ?population .
        filter(isNumeric(?population))
    } union {
        ?s dbp:populationTotal ?population .
        filter(isNumeric(?population))
        filter not exists {
            ?s dbp:population ?primary_population .
            filter(isNumeric(?primary_population))}}}
```

Listing 1. Select the sum of population counts using a preferred order of predicates in a query to a regular SPARQL endpoint.

As the number of predicates to test for increases, so does the number of FILTER NOT EXISTS blocks in each new UNION group. Furthermore, if we wanted to use a list of predicate IRIs from an RDF collection found in a triplestore, then this selection would be impossible to perform in a single query. Employing a pattern rewriter, we can automate building such queries in addition to having their bindings projected onto the surrounding query level; see Listing 2.

```
select (sum(?population) as ?totalPopulation) where {
    ?matcher volt:firstMatch [
        input:onVariable "?p"^^volt:Variable ;
        input:useValuesFrom (dbp:population dbp:populationTotal) ;
        input:sampleFromVariables ("?population"^^volt:Variable) ] .
    graph ?matcher {
        ?s ?p ?population .
        filter(isNumeric(?population))   }}
```

Listing 2. The ?matcher variable can be thought of as binding to the URI of a named, transient graph. In reality, the pattern rewriter's procedure will transform the patterns within the GRAPH group into a new subquery. This code snippet along with the expanded query can be seen in its entirety at https://git.io/v2Nxb .

---

[5] Universally Unique Identifier.

2.2   Transparency and Reproducibility

A key limitation of previous approaches has been with the client's inability to inspect the source code behind an API function. Functions are not always trivial and their algorithms may overlook cornercases or depend on undocumented assumptions – leading to a breakdown in *semantic interoperability*. Our approach is to make the source code for all procedures readily accessible to the client by storing everything in the triplestore as RDF. Each procedure is serialized according to the VOLT ontology[6] and stored in the *model graph*. In order to execute a procedure, the proxy downloads a segment of the model graph and evaluates each step from the procedure's sequence of instructions. A simple example of an instruction is the assignment of a variable to an expression, e.g., `?x = ?y + ?z`, which applies the addition operator to the values stored in the variables `?y` and `?z`, then puts the result in the locally-scoped variable `?x`. In the model graph, this expression is serialized as an abstract syntax tree; shown in Listing 3.

```
... [ a volt:Assignment ;
    volt:name "?x"^^volt:Variable ;
    volt:gets [ a volt:BinaryOperation ;
        volt:operator "+"^^volt:Operator ;
        volt:lhs "?y"^^volt:Variable ;
        volt:rhs "?z"^^volt:Variable ]] ...
```

Listing 3. Abstract syntax tree of an assignment instruction for a VOLT procedure.

In taking this approach, we are able to statically evaluate the validity of a procedure's RDF serialization by using an ontology. Another example of a procedural instruction might be a SPARQL query, which has the benefit of referential integrity in its serialized form. This implies that a client can discover a procedure that depends on a particular IRI by querying the model graph for that IRI in the object position of a triple. E.g., we can discover any procedures that depend on the `geo:geometry` predicate by using the query shown in Listing 4.

```
describe ?procedure from named volt:graphs where {
    graph volt:graphs { ?modelGraph a volt:ModelGraph }
    graph ?modelGraph {
        ?procedure rdf:type/rdfs:subClassOf volt:Procedure .
        ?procedure (!</>)+ geo:geometry . }}
```

Listing 4. Discover any VOLT procedures that depend on the `geo:geometry` predicate by using the nexus property path `(!</>)+`.

Thus, VOLT is transparent in two ways: (1) the proxy sits on top of a regular endpoint without a client noticing any difference, i.e., computed Linked Data behave as if they were stored in the underlying triplestore [3], and (2) procedures (defined by users or providers) are open for inspection.

2.3   Provenance

During execution of a procedure, all SPARQL queries and function calls are analyzed and recorded. Any information used during the evaluation of these transactions gets serialized as RDF triples and stored into a *provenance graph*.

---

[6] https://github.com/blake-regalia/volt.

Those details are used to associate a cached triple to the inputs and expected outputs of SPARQL queries and function calls which led to that result. This offers two advantages: (1) the provenance of a cached triple is stored and remains available for inspection by which a client has the means to review source information that led a procedure to its conclusion and (2) it enables the invalidation of *stale cache*.

2.4   Caching and Cache Invalidation

To improve the performance of matching query patterns against computable properties and functional triples, we make use of caching. When caching is enabled by the proxy's host, each cacheable result is diverted to a persistent *output graph* instead of a temporary *results graph*. The input query is ultimately executed on the union of the source graph(s), results graph, and output graph, known collectively as the *content graph*. Determining whether or not a result should be cached depends on the ontological definition of the procedure that was used. Caching will only take place on a result when the procedure allows it. However, a client can bypass caching any results for the entire duration of a query's execution by including `optional {[] volt:ignoreCache true}` in the input query. Using the OPTIONAL keyword ensures that the query is reusable against arbitrary SPARQL engines, e.g., ones that do not run the proxy.

Each time a new triple is cached for a computable property, that triple runs the risk of being obsolete for future queries if the contents of its original source graph were to change. To detect this issue and protect against stale cache, we embed a cache invalidation feature within the framework. For procedures that use simple SPARQL queries, this may just involve confirming the existence of triples. In these cases, a cached result may be validated by a single query directed at the actual SPARQL engine. However, procedures that use more complex queries can employ patterns such as property paths or aggregate functions which can only be validated by executing those queries in full. We realize the need for an ontology that enables serializing, with various levels of complexity, methods of result validation for outputs of function calls and SPARQL queries given their inputs.

2.5   VOLT Procedures

The VOLT framework supports several types of user-defined procedures; each type serves a different purpose. In the model graph, a user may define procedures for EVT functions, computable properties, functional triples, and pattern rewriters. For each of these mediums, there is an ontological class that defines how an associated procedure must be encoded as RDF in the user-defined model graph. For example, a VOLT EVT function must have at least one member of type `volt:ReturnStage` in the RDF collection object pointed to by the `volt:stages` predicate within the procedure's set of defining triples.

The user-defined model exists as an RDF graph which encodes each procedure definition as a sequence of instructions. These instructions are limited to the basics, such as: operational expressions, control flow, SPARQL queries, and so on. To provide developers with the full flexibility of a programming language, the user-defined model can be extended by scripting plugins. Plugins are ideal for handling tasks such as complex calculations, networking, and I/O. They are treated as namespaced modules. A single plugin may host an array of functions. For instance, we created a plugin that uses PostGIS[7] to handle geographic calculations; see Listing 5 for a call to the EVT function `postgis:azimuth`.

Under the hood, each plugin registers a specific namespace with the proxy by inserting RDF statements about itself into the model graph. This information includes metadata such as the path of the binary to execute, the path or URL of the source code if available, the namespace IRI, and process-related configuration. Anytime an EVT function call has a registered namespace, it will trigger the corresponding plugin. If the plugin is not already running, the proxy will load it into memory by spawning a child process. Once a plugin is running, the proxy pipes the function name and input arguments, serialized as JSON over stdin, to that child process. A single process may be used to run multiple tasks in series and multiple process of a plugin may be spawned in order to run tasks in parallel. Idle and busy processes may be terminated at the discretion of the proxy.

```
prefix postgis: <http://stko.geog.ucsb.edu/volt-plugins/postgis/#>
select ?angle where {
        dbr:Santa_Barbara geo:geometry ?wktFrom .
        dbr:Ventura geo:geometry ?wktTo .
        bind( postgis:azimuth(?wktFrom, ?wktTo) as ?angle ) }
```

Listing 5. Calls the user-defined EVT function 'azimuth' in the PostGIS plugin.

2.6   Query Flow Overview

To give a brief overview of how the proxy works, we examine VOLT's computable property feature. In Case Study I, we will demonstrate the use of such a property to determine the cardinal direction between Santa Barbara and Ventura. Figure 1 depicts the process of executing the procedure for `stko:east` as a flowchart.

3   Case Studies

This section discusses two geographic use cases to showcase VOLT in action. Each use case highlights a different capability of the framework.

3.1   Case Study I: Cardinal Directions

The four cardinal directions North (N), South (S), East (E), and West (W) are among the most common means to express directional relations. The equal

---

[7] http://postgis.net/.

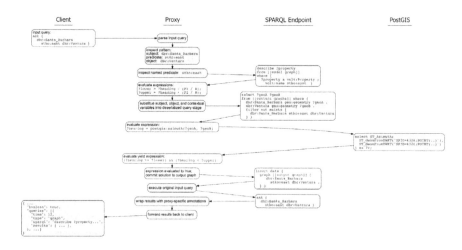

Fig. 1. The execution of computable property `stko:east` represented by a flowchart.

directional divisions of a compass rose are known as the four intercardinal directions, i.e., Northeast (NE), Southeast (SE), Southwest (SW) and Northwest (NW). In this section, *cardinal directions* will refer to all eight directions. Figure 2 shows how the bearing span ω for a cardinal direction is represented. The directionality is determined by testing if the azimuth between the point geometries of two places falls within ω from the primary angle of the direction. For the 8 cardinal directions, ω is set to $\pi/8$. For example, SE (*stko:southeast* here) covers the range $5\pi/8$ to $7\pi/8$ which is measured from the positive y-axis.

According to a SPARQL query for all resources of type `dbo:Place` or having a `geo:geometry`, there are over 1 million places in DBpedia.[8] Nearly 35,000 of them are associated to at least one triple with a cardinal direction predicate, leading to a total of 108,818 distinct triples involved. While this number is large, it is only a small portion ($\approx 1.2\%$) of the potential amount of cardinal direction relations among all places if merely storing a single triple per direction, e.g., only storing the nearest place to the North, South, and so forth. Trying to store all cardinal directions between all places would lead to a combinatorial explosion.

The entities contained in these triples vary widely and include macro-scale types such as *Mountain Range* or *Country*, meso-scale types, such as *City* or *River*, and micro-scale place types such as *Hospital*. Interestingly, types such as *Person* also show up, likely confusing persons with the places they were buried; e.g., `dbr:Saint Mechell` is `dbp:north` of `dbr:Tref Alaw`. Intuitively, and leaving cases such as headlands and meandering rivers aside, there should be only one cardinal direction relation between two places. Surprisingly, there are 3,411 places (involving $\approx$ 17,000 triples) with more than one cardinal direction to the same entity. For instance, Chicago, IL is both *dbp:east* and *dbp:west* of

---

[8] All queries & experiments were performed on the stable DBpedia 2015-04 version.

Lincolnwood, Rosemont, and Schiller Park, which is controversial. Consequently, we are compelled to test the accuracy of cardinal directions in DBpedia.

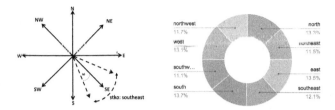

Fig. 2. The eight primary/inter-cardinal directions and their range (left) and the proportional distribution of mismatched directions normalized by categorical count (right).

In order to compute the cardinal direction accuracy between entities of type `dbo:Place` that have one or more `geo:geometry` property in DBpedia, we selected all combinations of geometries between two places[9]. Our selection yielded 136,964 results of which 91,890 matched correctly, leaving 45,084 rows (33 %) marked as incorrect. To validate that our computational representation of the cardinal directions does not introduce bias, we show in Fig. 2 that each of the eight cardinal directions have roughly equal portions of incorrect relations. If we consider all 133,941 cardinal direction triples in DBpedia, we find that 55,928 (42 %) of them have a subject or object lacking `geo:geometry`, or are not of type `dbo:Place`. In fact, 17,957 triples have cardinal direction relations to RDF literals, 537 of which are of datatype `xsd:integer`. Most importantly, our argument is that given the few correct existing cardinal direction triples, a Linked Data user has to wonder why these specific relations are present in DBpedia and not a comprehensive set of cardinal directions between all places. This, however, would far exceed the total number of triples in DBpedia today. The imbalanced cardinal direction distribution becomes immediately clear by inspecting Table 1.

Another challenging issue is the computation of cardinal directions between polygonal representations of places. It is straightforward to compute point-to-point cardinal direction results on-the-fly if centroids are taken as the representations of regions. Depending on the polygons and the representativeness of centroids, there may be a varying degree of uncertainty associated with a cardinal direction relation between two regions. For example, according to DBpedia, the city of *Ventura* is linked to the city of *Santa Barbara* via the `dbp:northwest` relation, i.e., Ventura should be located to the *southeast* of Santa Barbara. This may be true for a certain point-feature representation of the cities but is not correct for all points inside the city boundaries. In fact, by taking the OpenStreetMap polygons for Santa Barbara and Ventura and defining a regular point grid of 1×1 km, we can compute the probability of grid points contained in Ventura to locate in the *southeast* of Santa Barbara (grid points). We filter out those points

---

[9] Please note that some places have more than one geometry.

Table 1. Cardinal direction accuracies of the top 20 places with the most relations.

Place	Matches	Total	Accuracy	Place	Matches	Total	Accuracy
Wrexham	47	71	0.66	Karimnagar	27	38	0.71
Dolgellau	49	58	0.84	Ranchi	27	38	0.80
Ruthin	27	57	0.47	Shrewsbury	34	35	0.97
Bradford	29	53	0.55	Brothertoft	30	34	0.88
Bala, Gwynedd	22	47	0.47	Burton-upon-Trent	34	34	1.0
Orlando, Florida	27	47	0.57	Boston, Lincolnshire	25	33	0.76
Lichfield	43	46	0.93	Kirkby	29	33	0.88
Corwen	26	43	0.60	Ford, Shropshire	17	31	0.55
Aberystwyth	31	43	0.72	Mansfield	26	31	0.84
Derby	22	42	0.52	Glensanda	24	30	0.80

which are either outside of the city boundary or in the ocean. In total, we get 79 representative points for Santa Barbara and 88 points for Ventura. As depicted in Fig. 3, we compute cardinal direction relations between 6,952 pairs of points in total. Our result shows that *southeast* is only the correct relation in 7.6 % of the cases while it is *east* in 92.4 % of the cases. That is to say that the DBpedia statement of Ventura being southeast of Santa Barbara is merely true for 527 point pairs, while east is the correct relation for 6,425 other pairs. The situation would be even more complex if we consider fuzzy-set representation typically used for cognitive regions, e.g., *downtown*.

The last issue that remains to be discussed is performance. Clearly, computing cardinal directions takes longer than retrieving stored triples. A SPARQL query for all cardinal directions of the top 20 places takes about 3.3 s on DBpedia's public endpoint. A *cold*, i.e., non-cached, VOLT prototype computes the same relations and returns its results (but does not yield erroneous data as does DBpedia) in about 18 s on a modern laptop. This number should be taken into perspective by comparing it to the cache-enabled VOLT which takes only 6.9 s after an initial run. Finally, it is important to remember that queries typically ask for the cardinal direction between a place and other geographic features and not for hundreds of directions among 20 random places. In such real-world cases, however, the overhead introduced by computation is relatively small.

Summing up, DBpedia currently only stores a very small, and from an end user's perspective, arbitrary fraction of cardinal direction relations. Approximately 33 % of these relations are defective and many other need an understanding of the involved uncertainties to make use of them in a reproducible setting. For instance, there is no way for a user to understand what is returned by a SPARQL query for cardinal directions: are the results about the closest entity in a given direction, multiple entities, entities of the same type (e.g., the city north of LA), and so forth. Using the VOLT proxy, cardinal directions between

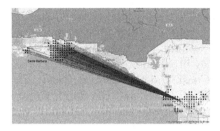

Fig. 3. Uncertainty in cardinal directions for Santa Barbara and Ventura.

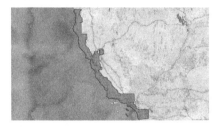

Fig. 4. Union of coastal counties computed as adjacent to Pacific Ocean.

all places can be computed on-demand along with provenance records that document how the computation was done and based on which formal definitions.

3.2   Case Study II: Counting Regional Population

For the second case study, let us assume that a client wants to count the total population of California's coastal counties. She discovers the DBpedia resources for: North Coast of California, Central Coast of California and South Coast of California; each of which embodies counties along the coast. Intuitively, the user expects these three regions to be spatially disjoint and after inspecting the page for the Central Coast, naively devises the SPARQL query shown in Listing 6.

```
select (sum(?regionalPopulation) as ?coastalPopulation) where {
    ?region dbp:population ?regionalPopulation .
    values ?region {
        <http://dbpedia.org/resource/North_Coast_(California)>
        <http://dbpedia.org/resource/Central_Coast_(California)>
        <http://dbpedia.org/resource/South_Coast_(California)> }}
```
Listing 6. Select the sum of population counts for all three CA coastal regions.

At the time of this writing, the query from Listing 6 returns a ?costalPopulation of 2,249,558 - the same number as the population property given by the DBpedia resource for the Central Coast. In fact, the South Coast was not included since its population value is the literal "~ 20 million" and the North Coast does not have a population property to begin with. Therefore, since the query does not check if each region was matched to a triple, and since the sum aggregate function in SPARQL *silently* ignores non-numeric values, the result of this query is misleading. Even more, the three coastal regions are neither continuous nor disjoint. For example, there are two coastal counties, San Francisco County and San Mateo County, which do not belong to any of the three coastal regions in California; they break the continuity of these regions by making a gap in between the North Coast and the Central Coast. The regions are also not disjoint because the Central Coast and the South Coast both include Ventura County; this could lead to counting the population of Ventura County twice.

Clearly, the client needs a better way to select the coastal counties of California and should be able to validate the accuracy of their operation by inspecting the provenance of constituent population values. By modifying our data to be GeoSPARQL-conformant, we can build a better query as shown in Listing 7.

```
# count the population of coastal counties in California
select (sum(?countyPopulation) as ?coastalPopulation) where {
    # get geometry of Pacific Coast as WKT
    data:PacificCoast geo:hasGeometry/geo:asWkt ?pacificCoastWkt .
    # use a subquery to group by place; avoid counting same place twice
    { select ?county (sample(?population) as ?countyPopulation) {
        # select all California counties and geometries as WKT
        ?county a yago:CaliforniaCounties .
        ?county geo:hasGeometry/geo:asWKT ?countyWkt .
        # make sure the county geometry is a polygon
        filter(regex(?countyWkt, '^(<[^>]*>)?(MULTI)?POLYGON', 'i'))
        # filter for coastal counties only
        filter(geof:sfTouches(?countyWkt, ?pacificCoastWkt))
        # get population of each county using best valid property name
        {   # best property to use is 'dbo:populationTotal'
            ?county dbo:populationTotal ?population .
            filter(isNumeric(?population))
        } union {
            # next best property is 'dbp:populationTotal'
            ?county dbp:populationTotal ?population .
            filter(isNumeric(?population))
            # block counties that have the preferred property
            filter not exists {
                ?county dbo:populationTotal ?best_population .
                filter(isNumeric(?best_population))
        }}} group by ?county }}
```

Listing 7. Use GeoSPARQL to count the population of California's coastal counties.

While the GeoSPARQL query is more likely to yield an accurate result, the user cannot perform aggregate spatial operations. In order to check if the entire coast was accounted for, she would have to issue a separate query in which ?countyWkt is selected without any aggregate functions and then plot each geometry on a map. With the VOLT framework however, we provide namespaced aggregate functions that construct temporary RDF graphs in the SPARQL query from a list of results for a single variable. By keeping only the county selection patterns in the subquery and aggregating those counties into an RDF Set, we can then call the user-defined stko:sumOfPlaces method to sum the values of the population properties as shown in Listing 8. Additionally, the user-defined method can construct a single geometry feature that is the union of all coastal counties in California. We then plot this geometry feature on a map to inspect the areas included in our population count, as shown in Fig. 4.

```
prefix volt:   <http://volt-name.space/ontology/>
prefix input:  <http://volt-name.space/vocab/input#>
prefix output: <http://volt-name.space/vocab/output#>
prefix stko:   <http://stko.geog.ucsb.edu/vocab/>
# count the population of coastal counties in California
select ?population ?area where {
    # in a subquery, aggregate all California's coastal counties into a set
    { select (volt:cluster(?county) as ?setOfCounties) {
        # select only California counties
        ?county a yago:CaliforniaCounties .
        # ...that are 'along' the Pacific Coast (refers to a computable property)
        ?county stko:along data:PacificCoast . } }
    # let 'sumOfPlaces' method compute the total population of coastal counties
    [] stko:sumOfPlaces [
        input:places ?setOfCounties ;
        input:propertyList (dbo:populationTotal dbp:populationTotal) ;
        output:sum ?population ;
        output:coveredArea ?area ; ] }
```

Listing 8. Computes the sum of values for the first valid numeric property from
`dbo:populationTotal` or `dbp:populationTotal` for all coastal counties in California.

The `stko:sumOfPlaces` method is stored in the model graph as RDF triples. To
simplify the process of programming user-defined procedures in RDF, we developed the VOLT syntax and its compiler[10]. The language allows inline embedding
of SPARQL query fragments, dynamically-scoped variables, operational expressions, and basic flow control. The VOLT source code for the `stko:sumOfPlaces`
method is shown in Listing 9. At runtime, the population example will cause this
method to generate the SPARQL query shown in Listing 10. Note that the VOLT
language does not invalidate our claim of the proxy being transparent and only
depending on well established W3C technologies. The language is only used to
simplify the production of ontologically-compatible RDF statements which define
custom functions and optionally their connections to external systems such as
PostGIS. This language is not used for querying or any other functionality exposed
to the client. As explained before, each procedure is serialized to RDF and stored
in the model graph where it is available for public inspection.

```
method stko:sumOfPlaces {
    input ?places decluster into ?place
    input ?propertyList(list)
    select ?sum=sum(?value) ?placeGeomsWkt=volt:collect(?placeWkt) {
        ?matcher volt:firstMatch [
            input:forVariable "?property"^^volt:Variable ;
            input:useValuesFrom ?propertyList ;
            input:sampleFromVariables ("?value"^^volt:Variable) ] .
        graph ?matcher {
            ?place ?property ?value .
            filter(isNumeric(?value)) }
        ?place geo:hasGeometry/geo:asWKT ?placeWkt }
    output ?sum  # shorthand for 'output [output:sum ?sum]'
    if object has output:placeGeometries {
        output [output:placeGeometries ?placeGeomsWkt] }
    if object has output:coveredArea {
        ?coveredArea = postgis:union(?placeGeomsWkt)
        output ?coveredArea }
    if object has output:overlap {
        ?overlap = postgis:union(postgis:intersectionAmong(?placeGeomsWkt))
        output ?overlap } }
```

Listing 9. User-defined **sumOfPlaces** method in VOLT syntax. It accepts two inputs:
(1) a set of places whose properties should be summed and (2) a list of property IRIs
ordered by the most preferred property value to match each distinct **?place**.

---

[10] https://github.com/blake-regalia/volt.

```
select (sum(?value) as ?sum)
    (group_concat(?_n3_placeWkt; separator='\n') as ?placeGeomsWkt)
where {
    # volt:firstMatch for variable ?property, use values from: (dbo:populationTotal
    ↪    dbp:populationTotal). sample from variable ?value
    { select ?place ?property (sample ?_sample_value as ?value)
        where {
            {    ?place ?property ?_sample_value .
                 filter(isNumeric(?_sample_value))
                 values ?property { dbo:populationTotal }
            } union {
                 ?place ?property ?_sample_value .
                 filter(isNumeric(?_sample_value))
                 values ?property { dbp:populationTotal }
                 filter not exists {
                     ?place dbo:populationTotal ?_0_value .
                     filter(isNumeric(?_0_value))
            }}} group by ?place ?property }
    ?place geo:hasGeometry/geo:asWKT ?placeWkt .
    # decluster ?places into ?place
    values ?place { dbr:Alameda_County    dbr:Contra_Costa_County  dbr:Del_Norte_County ... }
    # volt:collect(?placeWkt)
    bind( if(isBlank(?placeWkt), concat('_:', struuid()),
        if(isIri(?placeWkt), concat('<', str(?placeWkt), '>'),
        if(isLiteral(?placeWkt),
            concat('"',
                replace(
                    replace(str(?placeWkt), '"', '\\\"'),
                    '\n', '\\\n' ), '"',
                if(lang(?placeWkt) = '',
                    concat('^^<', str(datatype(?placeWkt)), '>'),
                    concat('@', lang(?placeWkt)) )),
        concat('?', struuid()) )))
        as ?_n3_placeWkt   ) }
```

Listing 10. A SPARQL query issued by the proxy on behalf of the client's input query. The client invokes the `stko:sumOfPlaces` method that substitutes values and subquery selection results into its own SELECT stage, ultimately yielding this SPARQL query.

Summing up, this second use case highlights the difficulties in naively querying Linked Data and the misleading results that commonly result from such queries. We use it to showcase VOLT's capabilities with respect to user (or provider) defined methods and the provenance information that allows others to inspect how the returned query results came to be.

4   Related Work

In this section we introduce work that is either related in terms of common goals, similar technological approaches, similar target domain, i.e., geo-data, or inspired and informed our thinking while developing the VOLT framework.

Linked Data Services (LIDS) [9] describes a formalization for connecting SPARQL queries to RESTful Web APIs by enabling a service layer behind query execution. Service calls have named inputs and outputs in the query. VOLT provides functional triples which also use named inputs and outputs in the query to make API calls to registered plugins. Plugins execute asynchronously and may perform networking tasks such as requests to RESTful Web APIs.

Linked Open Services (LOS) [7] sets forth the principles on how to establish interoperability between RESTful resources and Linked Open Data by semantically lifting flat content to RDF.

The **Linked Data API** (LDA) [8] is used to create RESTful APIs over RDF triple stores to streamline the process of web applications consuming Linked Data. Similar to LDA, VOLT also runs as a SPARQL proxy and dynamically generates SPARQL queries on behalf of the client.

Linked Data Fragments (LDF) [11] highlight the role of clients for scaling query engines by offloading partial execution to the web browser. Since our prototype is implemented in JavaScript, the proxy also runs as a standalone instance in the browser. The framework only needs a connection to a SPARQL endpoint over HTTP, or a locally emulated one such as the LDF client. In this regard, we aim to achieve Web-Scale querying as described by Verborgh [11].

SCRY [10] is a SPARQL endpoint that allows a client to invoke user-defined services by using named predicates in SPARQL queries. It simply identifies which service to execute and forwards the appropriate arguments given by the associated triple. SCRY's current implementation requires services to be implemented as Python modules or as command-line executables. Compared to VOLT, it does not provide the means for a client to inspect the source of user-defined services.

iSPARQL is a *virtual triple approach* [5] to invoking custom functions similar to the concept of *magic properties*. It extends the SPARQL grammar with a *SimilarityBlockPattern* production to distinguish between basic graph pattern triples and triple-like function calls having the form *?v apf:funct ArgList* [5].

The **SPIN** [1] framework generates entailments by issuing SPARQL queries to perform inferencing. The framework consists of a set of vocabularies that enable the serialization of user-defined rules, input as SPARQL queries, directly into an RDF graph; a technique that preserves IRI referential integrity. VOLT also serializes SPARQL fragments and graph patterns into RDF to use as inference rules. However, SPIN requires use of a proprietary extension of the SPARQL language to explicitly invoke computation while VOLT is designed to automatically recognize the need for computation on regular SPARQL queries that are issued as if the patterns are simply being matched to existing triples.

Logical Linked Data Compression [4] proposes a lossless compression technique which benefits large datasets when storage and sharing may be an issue. Similar to their compression, VOLT reduces the number of triples by using procedures to generate statements that can be deduced from source triples. However, our approach increases the total size of a dataset when caching is enabled. With caching disabled, one can instead opt for computing such statements on-demand thus saving storage space at the cost of query execution time.

5   Conclusions

In this work we introduced the transparent VOLT proxy for SPARQL endpoints. We outlined its core features, highlighted selected implementation details, and presented use cases that demonstrate the proxy's capabilities in addressing key shortcomings that we believe prevent the wide usage of Linked Data in science. Instead of storing triples that depend on already stored data, we propose to compute results on-demand and then cache them. Our work goes beyond merely reducing the amount of stored triples but also addresses quality issues as the *dependent* triples have to be kept in sync with their source data, e.g., when storing population densities in addition to population and areal data. We also address issues of provenance and the reproducibility of results by making the VOLT functions available and inspectable and by storing all data and procedures that were used to arrive at certain results in a separate graph. Finally, we discuss two use cases to demonstrate the difficulty in querying Linked Data, quality issues in Linked Data, and the need for the implemented VOLT capabilities.

Future work will focus on improving our current prototype and making it easier to extend and customize by others. We will also work on improving the proxy's performance and an alignment of our provenance and model graphs with ontologies such as PROV-O [6] and (semantic) workflow models in general [2].

Acknowledgements. This work was partially funded by NSF under award 1440202 and the USGS *Linked Data for the National Map* award. The authors would also like to thank Johannes Gross from NASA/JPL for his comments.

References

1. SPIN - SPARQL Inferencing Notation (2011). http://spinrdf.org/
2. Gil, Y., Deelman, E., Ellisman, M., Fahringer, T., Fox, G., et al.: Examining the challenges of scientific workflows. IEEE Comput. **40**(12), 26–34 (2007)
3. Janowicz, K., Schade, S., Bröring, A., Keßler, C., Maué, P., Stasch, C.: Semantic enablement for spatial data infrastructures. Trans. GIS **14**(2), 111–129 (2010)
4. Joshi, A.K., Hitzler, P., Dong, G.: Logical linked data compression. In: Cimiano, P., Corcho, O., Presutti, V., Hollink, L., Rudolph, S. (eds.) ESWC 2013. LNCS, vol. 7882, pp. 170–184. Springer, Heidelberg (2013)
5. Kiefer, C., Bernstein, A., Stocker, M.: The fundamentals of iSPARQL: a virtual triple approach for similarity-based semantic web tasks. In: Aberer, K., et al. (eds.) ASWC/ISWC 2007. LNCS, vol. 4825, pp. 295–309. Springer, Heidelberg (2007)
6. McGuinness, D., Lebo, T., Sahoo, S.: PROV-O: the PROV Ontology. Technical report, W3C Recommendation. 30 April 2013
7. Norton, B., Krummenacher, R.: Consuming dynamic linked data. In: COLD (2010)
8. Reynolds, D., Tennison, J., Dodds, L.: Linked Data API (2012). https://github.com/UKGovLD/linked-data-api
9. Speiser, S., Harth, A.: Integrating linked data and services with linked data services. In: Antoniou, G., Grobelnik, M., Simperl, E., Parsia, B., Plexousakis, D., De Leenheer, P., Pan, J. (eds.) ESWC 2011, Part I. LNCS, vol. 6643, pp. 170–184. Springer, Heidelberg (2011)
10. Stringer, B., Meroño-Peñuela, A., Loizou, A., Abeln, S., Heringa, J.: To SCRY linked data: extending SPARQL the easy way. In: Diversity++, ISWC 2015 (2015)
11. Verborgh, R., Vander Sande, M., Colpaert, P., Coppens, S., Mannens, E., Van de Walle, R.: Web-scale querying through linked data fragments. In: Proceedings of the 7th Workshop on Linked Data on the Web (2014)

Learning to Classify Spatiotextual Entities in Maps

Giorgos Giannopoulos[(✉)], Nikos Karagiannakis, Dimitrios Skoutas,
and Spiros Athanasiou

IMIS Institute, "Athena" Research Center, Athens, Greece
giann@imis.athena-innovation.gr

Abstract. In this paper, we present an approach for automatically recommending categories for spatiotextual entities, based on already existing annotated entities. Our goal is to facilitate the annotation process in crowdsourcing map initiatives such as OpenStreetMap, so that more accurate annotations are produced for the newly created spatial entities, while at the same time increasing the reuse of already existing tags. We define and construct a set of training features to represent the attributes of the spatiotextual entities and to capture their relation with the categories they are annotated with. These features include spatial, textual and semantic properties of the entities. We evaluate four different approaches, namely SVM, kNN, clustering+SVM and clustering+kNN, on several combinations of the defined training features and we examine which configurations of the algorithms achieve the best results. The presented work is deployed in OSMRec, a plugin for the JOSM tool that is commonly used for editing content in OpenStreetMap.

1 Introduction

The Semantic Web and Linked Data practices have been gaining increasing interest the last years and are being adopted by crowdsourcing and mapping initiatives. In conjunction with the widespread use of smartphones and GPS enabled devices, this has resulted in a large number of RDF datasets containing geospatial information, which is of high importance in several application scenarios, such as navigation, tourism, and location-based social media.

In particular, OpenStreetMap (OSM) is an initiative for crowdsourcing map information from users. It is based on a large and active community contributing both data and tools that facilitate the constant enrichment and enhancement of OSM maps. An important feature of OSM is a large hierarchy of categories[1] for annotating spatial entities on the map. Its Linked Data counterpart, Linked-GeoData, serves the whole OSM dataset as RDF data adhering to a respective ontology[2] which maps OSM categories into equivalent OWL classes.

---

[1] http://wiki.openstreetmap.org/wiki/Map_features.
[2] http://linkedgeodata.org/ontology.

© Springer International Publishing Switzerland 2016
H. Sack et al. (Eds.): ESWC 2016, LNCS 9678, pp. 539–555, 2016.
DOI: 10.1007/978-3-319-34129-3_33

One of the most prominent tools for editing OSM data is JOSM[3], a graphical tool that allows users to create and edit spatial entities in OSM. Through its graphical user interface, the user can draw the geometry of a spatial entity. Then, she can annotate the entity with categories (alt. classes, tags), which are represented in the form of key-value pairs and assign semantics to the entity. Each entity may belong to multiple categories; for example, a "building" can be further characterized as "school" or "house".

The categories used to semantically annotate spatial entities on the map can either be selected from the already existing hierarchy of categories mentioned above or be defined for the first time by the user. Although this provides a lot of flexibility, which is an important requirement when relying on crowdsourcing, it also increases the complexity of maintaining the taxonomy of classes. Ideally, already existing categories should be reused as much as possible when creating new entities, while new categories should only be introduced when a new entity appears that cannot be appropriately classified and characterized by the existing ones. However, manually browsing through the class hierarchy to determine which category(-ies) is the appropriate one for an entity is a time consuming task, especially for a non-expert user who is not already familiar with the OSM taxonomy. Thus, it may often result in choosing a category poorly (e.g., selecting a more general one, although a more specific one exists) or introducing a new category while in fact an appropriate one (e.g., a synonym) already exists.

To deal with these shortcomings, in this work we propose a process that trains recommendation models on existing, annotated spatiotextual datasets in order to subsequently recommend categories automatically for newly created entities. The main contribution of our work lies on defining and implementing specific training features in order to capture the relations between the spatial and textual properties of each spatial entity with the categories/classes that characterize it. Essentially, this way, the proposed framework takes into account the similarity of the new spatial entities to existing ones that are already annotated with categories. This similarity is specified at several levels: *spatial similarity*, e.g. the number of nodes of the feature's geometry; *textual similarity*, e.g. common important keywords in the names of the features; and *semantic similarity*, i.e. common or related categories that characterize already annotated entities.

To evaluate the proposed methodology, we perform an extensive experimental evaluation that assesses the effectiveness of several feature subsets deployed in the frame of two classification algorithms: (i) Support Vector Machines (SVM) and (ii) k-Nearest Neighbors (kNN). In addition, we assess two hybrid solutions: (iii) clustering+SVM and (iv) clustering+kNN. The experimental results show that a proper combination of classification algorithm and training features can achieve recommendation precision of more than 90 %, rendering the proposed approach suitable for deployment on real-world use cases. Indeed, to further validate it in real-world scenarios, the proposed method is implemented and made

---

[3] https://josm.openstreetmap.de/.

available as a JOSM plugin[4] [9], allowing the real-time and effective annotation of newly created spatial entities into OSM.

The rest of the paper is organized as follows. In the next section, we briefly review related work. Then, Sect. 3 describes our proposed method, including the defined training features and the assessed algorithms. Section 4 presents the evaluation of training features and algorithms in terms of recommendation performance, while Sect. 5 concludes the paper.

2 Related Work

During the past years, the amount of Volunteered Geographic Information is constantly increasing, while its value and importance in numerous applications and services is constantly becoming more recognized and prominent. Hence, the quality of this content is becoming a crucial factor. Nevertheless, so far few works address the problem of semantically enriching crowdsourced spatiotextual data.

A recent work for addressing the semantic heterogeneity of OSM data proposing a tag recommendation plugin for JOSM is presented in [1]. The main idea of this approach is to recommend additional similar categories for a spatial entity in OSM, when the user has already inserted a category. The recommendation process is held by constructing a semantic network for OSM. This network holds the scores for semantic similarities between pairs of OSM tags and the recommendation process is based upon these scores. The authors also focus on evaluating the effectiveness of the tool and user satisfaction by performing a thorough user study. Thus, they solve a very similar problem to ours, but from another perspective. Given that, the two approaches could complement each other in order to further increase the recommendation effectiveness of the two individual systems.

The work presented in [2] proposes a machine learning based solution for assessing the semantic quality of OSM data. This work focuses on classifying road segments in OSM, thus it specializes only on geometrical and topological features of the specific entities and reduces the space of recommendation categories from more than 1000 to only 21.

In a broader context, relating to recommendations of geospatial data, a Semantic Web Knowledge System is described in [3] that is able to recommend Points of Interest (POIs) to drivers using a content-based recommendation approach. Specifically, they utilize a kNN classifier that takes into account historical POIs the driver has visited to recommend new POIs. In [4], a method is presented for recommending tags for photos, exploiting geospatial proximity, image similarity and several other estimators between photos. The authors feed these estimators into several machine learning algorithms, training this way classifiers that recommend tags for new photos. In [5], the authors utilize Gaussian mixture models to represent the geospatial profile of a user, extracted by microblog posts on music listening events. Considering either user geographic positions or geographic neighbourhoods of the user, they exploit these models into a collaborative filtering approach for identifying similar users and recommending music.

---

[4] http://wiki.openstreetmap.org/wiki/JOSM/Plugins/OSMRec.

The works in [6,7] perform analyses on OSM annotation statistics and on semantic relations between point geometries in OSM respectively. For the latter, the results of their analysis can be exploited from systems that perform category recommendations or quality assessment of the annotation of the spatial entities. Finally, [8] presents a thorough overview of recommendation concepts and methods on location based social networks.

3   Recommendation Models

3.1   Problem Definition and Method Outline

We cast the problem as a multilabel classification task. Given a set $\mathcal{E}$ of spatial entities $e_i \in \mathcal{E}$ and a set $\mathcal{T}$ of categories $t_n \in \mathcal{T}$: (i) properly represent each entity as a feature vector $v_{e_i} = <g_i, w_i, c_i>$, where g_i, w_i and c_i are sets of geospatial, textual and semantic attributes of the entity; (ii) learn a function $\mathcal{F}(v_{e_i}, t_n) => \{true, false\}$ that maps entities to categories.

We further break down this task into three distinct sub-tasks: (a) analyse spatial entities into meaningful attributes (training features) that properly describe the entities and capture their latent relation to the categories that annotate them; (b) train a machine learning algorithm to "learn" relations between attributes of the spatiotextual entities and the respective categories and (c) input the new (test) entities into the trained algorithm to produce category recommendations for them.

In Sect. 3.2 we describe the defined features that correspond to geometric, textual and semantic properties of the entities. The next step is to feed training entities, expressed through their features, into a classification algorithm that utilizes them to classify new entities to categories. We applied both model based (Support Vector Machines) and memory based (k Nearest Neighbour) state-of-the-art classification methods[5]. Further, we tested two hybrid solutions that first create clusters of similar training spatial entities and then apply either SVM or kNN respectively. Each of the algorithms is described in Sect. 3.3. During the third step, depending on the algorithm, either the trained model is applied on the new entity, or the new entity is matched to the training ones (using the cosine similarity of the respective feature vectors of the entities), in order to produce category recommendations. In the case where the similarity between a cluster and an (external) entity needs to be computed, this is done by considering the average feature vector of all entities contained in the cluster and, similarly, applying the cosine similarity with the vector of the (external) entity. The general rule used for producing the average vector of each cluster is by applying an OR operation on the boolean training features (since we wanted the cluster to be characterized with the specific training feature value if at least one of the clusters

---

[5] More sophisticated or recent algorithms could have been tested, however, the focus of this work lies mainly on defining meaningful and effective training features. Thus, these two algorithms were selected as intuitive representatives of two categories of classification algorithms that have shown to be effective in several scenarios.

entities was characterized by it) and computing the average value of the training features represented by double values.

3.2   Feature Selection

In our scenario, the entities to be classified are the spatiotextual entities that exist on the map (e.g. buildings, roads, parks) and the variable to be predicted is the category(ies) of the entities. Note that an entity may belong to multiple categories. To learn a classifier for assigning entities to categories, we represent each spatiotextual entity by a feature vector. We consider features that capture spatial, textual and semantic properties of the entities, as listed next.

- *Spatial properties*
  - *Type*. Six distinct geometry types are considered: Point, LineString, Polygon, LinearRing, Circle and Rectangle. Each type is treated as a boolean feature, so six positions in the feature vector are used.
  - **Geometry**
    * *Points*. This denotes the number of points that the geometry of the entity comprises when represented as a point set. It provides an indication of how complex the geometry is.
    * *Area*. This represents the size of the entity's geometry.
    * *Mean*. Mean edge length is a feature used to capture properties of the shape of the entity.
    * *Variance*. Variance of edge lengths is also related to the entity's shape.
- *Textual properties*
  - *Text*. For each entity of the training set, we extract the textual description of its name. We consider each word separately and count their frequency within the dataset. Then, we sort the list of words by their frequency, filtering out words with frequency below a certain threshold (set to 20 in our experiments). Finally, we apply a stopword list[6], removing this way words without any particular meaning. What remains are special meaning identifiers, such as "Road", "Avenue", "Park", "Court", etc. Each of these keywords is used as a separate boolean feature, realizing, thus, a bag-of-words model for representing the text of the entities.
- *Semantic properties*
  - *Categories*. This is a set of $1,421$ boolean features, corresponding to each of the OSM categories. This feature is used in the scenario where an entity has been previously annotated with one or more categories, and we wish to recommend additional relevant categories for this entity[7].

---

[6] https://code.google.com/archive/p/stop-words/.

[7] These features are only mentioned here for completeness, since they are utilized in the implemented OSMRec prototype - JOSM plugin. However, they are not included in the experimental evaluation that is described next.

We should note here that we select different representations for the numeric features (double precision number or set of boolean features) to comply with the functionality of the different models and similarity functions we apply. Namely, we consider boolean features in the case of SVM, since it is well known that such models perform better when the input feature vectors are vertically normalized within (or at least close) to the interval [0,1]. Thus, for example, for the feature "Area of Geometry", defining several area ranges corresponding to separate feature positions, we allow the model to relate these different areas (feature positions) with different training classes. On the other hand, in the case of clustering or applying kNN, where similarity measures such as the cosine similarity are applied, using the exact value of a feature (e.g. area of geometry) is preferable in order to better quantify the difference between two entities.

3.3   Algorithms

SVM. The first algorithm applies multilabel SVM classification, using the $SVM^{multiclass}$ implementation[8], considering as training items the spatiotextual entities themselves and as labels the categories that characterize them. The method maps the training entities into a multidimensional feature space and aims at finding the optimal hyperplanes that discriminate the entities belonging to different categories. The optimality of the hyperplane depends on the selected parameter C, which adjusts the trade-off between misclassified training entities and optimal discrimination of correctly classified entities. The output of the training process is a model (essentially a weight vector) that is able to map the feature vector of a new, unannotated entity to a set of categories, providing, also, a matching score for each category. This way, one can obtain the top-n most fitting categories for a new spatial entity.

kNN. The second algorithm searches for k-Nearest Neighbours in the set of training entities. The algorithm compares the new entity with each one of the training entities and recommends the categories that characterize the most similar training entities. As similarity measures we consider *cosine similarity* and *Euclidean distance*. The two similarity functions were initially applied to the feature vectors of the respective entities. However, we empirically observed that, in the specific setting, boosting the vector-calculated similarity score with the area-based and point-based similarity scores between two entities improves the precision of the algorithms. Specifically, we use the following formula in our experiments to calculate the similarity score S between two entities u, v:

$$S_{cos} = cosSim + 2 * (1 - areaDist) + 2 * (1 - pointDist)$$

$$\text{where } areaDist = \frac{|area_u - area_v|}{\max area_u, area_v} \tag{1}$$

$$\text{and } pointDist = \frac{|points_u - points_v|}{\max points_u, points_v}$$

---

[8] https://www.cs.cornell.edu/people/tj/svm_light/svm_multiclass.html.

where $cosSim$ is the cosine similarity on the whole feature vector of the two entities and is interchanged in our experiments with the similarity that is based on the euclidean distance

$$euSim = 1 - euDist / \max euDist \qquad (2)$$

The similarity of a trained entity with the new entity is propagated to the categories that annotate the training entity. So, for each candidate category, we get an aggregate score from all matching training entities that are annotated by it. This way, we can get the top-n most fitting categories for a new spatial entity.

clustering+SVM. The third algorithm, SVM on clustered entities, first clusters the training entities, to identify groups of similar entities, based on the training features that represent them. Then, the entities assigned to each cluster go through the SVM training process in separate groups. This procedure produces a number of SVM models equal to the number of clusters. The rationale of this approach is that the clustering step is expected to produce subsets of the training entities containing more homogeneous entities, w.r.t. to the defined training features. Thus, the SVM training step is performed on more coherent groups of entities, with the expectation to produce more specialized SVM models. Then, a test entity is matched only with some of these models (corresponding to the most similar training entities) and the eventually used category recommendation models are trained only by entities similar to the test entity.

We applied an Expectation Maximization clustering algorithm (using the implementation provided by Weka[9]). The algorithm starts by assigning initial probabilities to items belonging to clusters and then it iteratively re-defines clusters and re-assigns items to clusters, until it converges. The specific implementation provides the option either to define a fixed number of clusters to be formed, or to allow the system to result to the most fitting number of clusters. Thus, we performed our experiments using both options.

In order to match a new entity to appropriate cluster(s), we calculate the cosine similarity of this entity's feature vector against *all* the entities contained in the training set. Based on the K[10] most similar training entities e_j, $j \in [1, K]$ that emerged from this process and, using their cosine similarity scores $cos(e_j, e_{new})$ with the new entity's feature vector e_{new}, we assign similarity weights between each cluster and the new entity. Specifically, the value of a cluster weight w_{c_i} is calculated by the frequency of the cluster's appearance in the set of K most similar training entities and additionally, the similarity of each training entity (belonging to the cluster) with the new entity:

$$w_{c_i} = \sum_{j=1}^{K} cos(e_j, e_{new}) * \mathbb{1}_{(e_j \in c_i)} \qquad (3)$$

---

[9] http://www.cs.waikato.ac.nz/ml/weka/.
[10] Experimentally tuned to $K = 50$ by testing the recommendation results while changing its value in the interval $[10, 100]$ with step 10.

with $\mathbb{1}$ being the indicator function.

Then, we apply each of the M matched SVM models (corresponding to the matched clusters above) to produce a ranking of categories for the new entity. Let $r_{c_i,j}$ denote the ranking of the j-th category by the ranking function of cluster c_i; the final ranking score between a category j and the new entity is given by linearly weighting the cluster rankings with the cluster importance w_{c_i}:

$$S(e_{new}, j) = \sum_{i=1}^{M} w_{c_i} * r_{c_i,j}. \tag{4}$$

Finally, we consider the top-n categories ranked according to the process above.

clustering+kNN. The fourth algorithm, kNN on clustered entities, first clusters the training entities to identify groups of similar entities (using the same aforementioned Expectation Maximization clustering algorithm), based on the training features that represent them. Then, it performs a kNN algorithm that, for each new spatiotextual entity, finds the most similar cluster. To do so, each cluster is represented by its average feature vector, as described in Sect. 3.1. The same rationale described in the clustering+SVM case applies here. Through clustering, we expect to specialize the groups of training entities on which the kNN algorithm is applied, only to those training entity groups that are found more similar to the test entity.

Upon clustering, the process is identical to the one followed in kNN algorithm with the difference that the training items are now clusters of entities and their labels are the total of the categories that characterize the entities of the cluster. The similarity function compares a new spatial entity with the average vector of each cluster, and assigns it a ranked list of categories, based on the similarity score of the entity with each cluster. Defining the set of all categories N that belong to a ranked list of matching clusters c_i, $i \in [1, \mathcal{C}]$, that is all the categories that characterize entities belonging to the respective list of clusters and $w_{c_i,n}$ the matching score of the cluster c_i's centroid (where category n belongs to cluster c, with normalized frequency $f_{c_i,n}$) to the new entity, the aggregate matching score of each category with the new entity is calculated as follows:

$$S_n = \sum_{i=1}^{K} w_{c_i,n} * f_{c_i,n} \tag{5}$$

In our experiment, we tried out $K = 1$ and $K = 5$, i.e. we considered the 1 and the 5 Nearest Neighbours-clusters of the new entity for detecting matching categories[11].

---

[11] In the case of clustering+kNN, our experiments consistently showed that the best results were achieved with a low number of clusters, specifically, 14 and 28 for Athens and London respectively. Thus, given this upper limit for k, we indicatively evaluated the 1-NN and 5-NN cases.

4   Experimental Evaluation

Next, we present the evaluation of the proposed methods w.r.t. the recommendation precision they achieve. First, we describe the dataset used and the evaluation methodology. Then, we compare the four algorithms and discuss the results.

4.1   Dataset and Evaluation Methodology

We performed our evaluation on two distinct subsets of OSM data covering parts of: (i) Athens, Greece and (ii) London, UK, which we exported through the Overpass API[12] from the OSM website. Each of the two datasets contains a total of about 20,000 spatiotextual entities which were properly divided into training, validation and test sets, as will be described next. Table 1 presents some statistics on both datasets.

Each dataset is partitioned into five subsets of similar sizes. Then, combining each time different subsets to make (a) the training (3 subsets), (b) the validation (1 subset) and (c) the test set (1 subset), we create five different arrangements for five-fold cross-validation. In every fold, the validation set was used to tune the parameters of the classification model and the test set was the one where the actual evaluation of the method was performed. In this evaluation, we considered the setting where test entities were clear of annotations, that is, they were to be annotated for the first time. Thus, we do not consider the Semantic properties mentioned in Sect. 3.2.

We experimented varying the following parameters. Parameter C of the SVM algorithm was set with the following values: {0.001, 0.01, 0.1, 1.0, 10.0, 100.0, 1000.0}. For Cl, the number of created clusters for the two hybrid approaches, we set the values: {10, 50, 100, 200} and we also considered the automatic cluster creation option of the algorithm that decided the optimal number of clusters to be created. Finally, for the number k of Nearest Neighbours, we set 10 for the kNN algorithm and experimented with 1 and 5 for the clustering-kNN approach.

Table 1. Dataset statistics.

Statistics	Athens	London
Distinct classes	186	306
Classes per entity	1.0	1.1
Majority class #1 (Building/Building) frequency	7133	8036
Majority class #2 (ResidentialHighway/Footway) frequency	3850	1942
Majority class #3 (Footway/UnclassifiedHighway) frequency	1122	1083
Categories with frequency $= 1$	20 %	25 %
Categories with frequency ≤ 2	33 %	38 %

---

[12] http://overpass-api.de/.

As evaluation measure, we consider the precision of category recommendations, i.e. the ratio of correct recommendations to the total recommendations:

$$P = \frac{\#correct\_category\_recommendations}{\#total\_category\_recommendations} \tag{6}$$

We consider three variations of the measure, depending on how strictly we define the recommendation correctness:

P^1: In this case, a recommendation is considered correct if the recommended category with the highest rank from the recommendation model is indeed a category that characterizes the test entity.

P^5: In this case, a recommendation is considered correct if one of the five recommended categories with the highest rank from the recommendation model is indeed a category that characterizes the test entity.

P^{10}: Similarly, a recommendation is considered correct if one of the ten recommended categories with the highest rank from the recommendation model is indeed a category that characterizes the test entity.

The rationale behind the definition of these three measures is that, in practice, it is common to provide the top-N recommendations. So, using these measures, we wanted to evaluate some representative cases, setting $N = 1, 5, 10$. Note that the above measures are slight variations of the commonly used measure of *Precision@N* which measures the number of correct results in the top-N rank positions. Our measures are defined in a boolean manner, indicating whether *at least one* correct category is found within the recommended ones to annotate the entity. Given that, whether the recommendation set consists of 1, 5 or 10 categories, the recommendation is considered successful if at least one category from it indeed characterizes the respective entity. Another reason for this adaptation is the very low average number of categories that annotate each entity in our datasets (and in OSM in general), that lies a little above 1 category per entity. This means that measuring *Precision@N* with $N > 1$ would be meaningless.

Finally, we note that when we used a validation step, the best performing configuration was chosen according to the highest P^1 **validation** value (since this is the strictest of the applied measures), while when no validation was applicable, then we considered the P^1 **test** value. Thus, the overall precision Table 8 presents the best sets of P^1, P^5 and P^{10} values considering the best P^1 as described above.

4.2   Algorithm Comparison and Discussion

We first present individual analysis results for each algorithm separately.

SVM. Table 2 presents the test values on precision of the SVM algorithm, for several training feature configurations. A general observation is that Spatial properties (either geometry types or geometry values) highly contribute to the precision of the model. On the other hand, Text features result to very low quality recommendations, at least when used on their own. The best precision in terms of P^1 comes by combining Spatial properties and Text. Type features used on their own provide the best values for P^{10} in both datasets.

Table 2. SVM feature combinations for Athens and London.

Features	Athens test set				London test set			
	C	P^1	P^5	P^{10}	C	P^1	P^5	P^{10}
Type	0.1	59.25	**82.85**	**91.06**	0.001	52.15	73.54	**83.25**
Points and area	100	47.76	72.65	79.10	100	47.30	59.20	62.50
Mean and variance	0.01	53.73	77.66	87.39	0.1	44.71	69.32	78.19
Points, area & type	0.01	59.15	76.55	89.95	100	47.26	58.55	67.44
Text	1	6.78	9.85	9.94	1	14.05	23.27	24.93
Points, area, type & text	10	59.49	70.55	75.67	1000	**59.59**	73.20	80.02
Spatial properties & text	1000	**59.99**	81.61	89.56	1000	55.44	**75.92**	81.43
Spatial properties	1000	59.50	79.69	88.42	1000	50.34	66.09	75.90

Table 3. kNN feature combinations for Athens and London.

Features	Athens test set			London test set		
	P^1	P^5	P^{10}	P^1	P^5	P^{10}
Type	42.57	53.62	54.96	44.67	54.50	56.40
Points and area	40.06	51.00	52.25	43.2	52.74	54.55
Mean and variance	57.42	69.32	71.56	51.88	63.06	65.40
Points, area & type	42.66	53.66	54.93	44.60	54.42	56.35
Points, area, type & text	46.91	58.06	59.46	54.13	64.04	66.29
Spatial properties & text	**59.98**	**71.57**	**73.45**	**58.81**	**70.15**	**72.39**
Spatial properties	56.58	68.13	69.92	52.01	62.77	64.94
Points, area, type - double values	42.57	53.62	54.96	44.67	54.51	56.40

kNN. Table 3 presents the recommendation precision values for the kNN algorithm. Spatial properties and Text combined give the highest precision in both datasets. However, these values are lower than the respective ones produced by SVM. Further, the difference in precision increases while moving from P^1 (negligible) to P^{10} (high). The overall better performance of SVM compared to kNN can be probably attributed to the generalization properties of the SVM model: kNN compares straightforwardly training and test entities and utilizes only categories from matching training entities. On the other hand, SVM, produces a model that assigns a score to *every available category* for the respective test entity, based on the importance of the entity's features for the trained model. Essentially, by finding optimal hyperplanes that discriminate different classes-categories in the training feature space, SVM can effectively handle both the potential sparseness of the training data and outlier data (i.e. incorrect entity annotation cases) that would otherwise negatively affect the model's training.

Table 4. Clustering+SVM feature combinations for Athens and London for the automatic cluster selection mode.

Features	Athens test set					London test set				
	C	Clust.	P^1	P^5	P^{10}	C	Clust.	P^1	P^5	P^{10}
Type	0.01	14	**31.21**	**68.25**	**75.71**	0.01	21	33.57	**57.10**	**66.16**
Points and area	0.01	14	28.02	46.66	59.54	0.001	21	26.37	36.60	50.93
Mean and variance	0.1	14	28.98	47.23	59.41	0.001	28	26.74	37.57	52.75
Points, area & type	0.001	14	27.86	46.69	60.01	0.01	28	26.61	37.50	51.97
Points, area, type & text	100	14	28.41	48.80	57.94	1000	28	35.18	45.87	51.12
Spatial properties & text	1000	14	29.37	45.28	57.94	1000	28	**38.02**	45.14	54.62
Spatial properties	100	14	28.65	41.76	55.65	1000	28	30.02	37.35	48.33

Table 5. Clustering+SVM feature combinations for Athens and London with manual selection of clusters.

Features	Athens test set					London test set				
	C	Clust.	P^1	P^5	P^{10}	C	Clust.	P^1	P^5	P^{10}
Points, area & type	1000	10	44.07	56.99	**68.91**	0.001	10	25.26	44.06	55.20
Spatial properties & text	1000	10	40.98	53.97	59.53	0.001	10	23.38	37.36	49.16
Spatial properties	1000	10	44.46	57.22	64.78	0.01	10	22.76	36.29	48.21
Points, area, type & text	1000	10	43 45	**63.48**	64.75	1000	10	25.30	40.37	48.21
Type	0.1	10	44.70	60.33	65.20	0.1	10	29.60	44.37	57.20
Points, area & type	0.01	50	43.37	45.18	59.03	0.01	50	**43.37**	50.18	**60.71**
Points, area, type & text	0.01	50	43.37	45.18	59.03	0.001	50	23.56	33.60	48.08
Spatial properties	1000	50	42.39	50.33	61.31	0.01	50	23.28	32.80	46.81
Spatial properties & text	1000	50	43.06	51.43	62.02	0.01	50	23.98	32.82	47.34
Type	0.01	50	38.02	39.37	41.67	0.01	50	25.01	34.40	38.56
Points, area & type	0.01	100	40.15	45.10	59.70	0.001	100	19.76	35.44	56.01
Points, area, type & text	0.001	100	**45.26**	58.34	68.32	1000	100	25.51	42.79	48.68
Spatial properties	0.01	100	40.18	51.36	61.44	0.01	100	26.14	37.77	57.51
Spatial properties & text	0.001	100	40.18	51.35	61.46	1000	100	32.71	**53.68**	59.58
Type	-	100	-	-	-	-	100	-	-	-
Points, area & type	0.001	200	30.39	39.04	50.60	0.001	200	-	-	-
Points, area, type & text	0.001	200	**45.26**	58.34	68.32	0.001	200	-	-	-
Spatial properties	0.1	200	26.91	39.97	48.01	0.001	200	23.68	32.08	43.12
Spatial properties & text	100	200	38.75	44.76	53.40	0.01	200	23.97	23.60	43.80
Type	-	200	-	-	-	-	100	-	-	-

clustering+SVM. Tables 4 and 5 present recommendation precision values for the hybrid clustering+SVM approach we propose. The first table presents results using the automatic number of cluster selection of the EM algorithm for Athens and London, while the second one the respective results when we manually provide several values for the number of clusters parameter. The best values are produced by manual selection of the number of clusters and, in most cases, for low values of this parameter (10 or 50 clusters). However, the results are rather poor, compared to the SVM and kNN algorithms. A possible explanation is that the clustering process is not performed with the proper objective in order to facilitate and enhance the upcoming SVM training process. Namely, the clusters are created using an improper clustering criterion, w.r.t. the specific task. Another issue might lie on the selection of the matching process, where we match test entities with training entities straightforwardly and not with a representation of the clusters they belong to, as we do in the clustering+kNN case. Nevertheless, since this general approach of creating ensembles of classifiers and properly combining them to obtain better results has worked in several other classification settings, we intend to further investigate it.

clustering+kNN. Tables 6 and 7 present recommendation precision values for the hybrid clustering+kNN approach we propose for applying 1-NN and 5-NN on clusters respectively. In this case, we report only results produced by the automatic number of clustering selection option of EM algorithm. The manual setting of the parameter produced consistently worse results and these results are, thus, omitted, due to lack of space. This hybrid solution produces very good recommendation precision, that even surpasses the simple SVM model for P^5 and P^{10}. Here, in contrast with the clustering+SVM case, the clustering step facilitates the upcoming kNN classification process. This can be intuitively explained as follows: by considering clusters of similar entities, instead of individual entities, we basically enhance the "Nearest Neighbours" of test entities; the test entities are still matched with similar training entities, but now, through the matched clusters, a more rich/diverse set of categories can be mapped to the test entities. So, when the precision regards the top-5 or top-10 recommended entities, finding correctly recommended categories becomes more probable.

Another observation is that the 1-NN configuration provides much better results than the 5-NN. This can be explained by the fact that each cluster already contains several categories to be recommended. So, if we consider many clusters, the recommendation set of categories gets too diverse, hurting, thus, the recommendation precision. Also, we can observe that relatively small numbers of clusters achieve the best precision values (14 for Athens - 28 for London). Finally, again, the geometry related features are the ones providing the highest precision; specifically, the Type for Athens and the Spatial properties for London.

All Methods. Table 8 presents the best achieving configuration for each tested algorithm, w.r.t. P^1. SVM achieves the highest accuracy for P^1 and clustering+kNN

Table 6. Clustering+kNN (1-NN) feature combinations for Athens and London.

Features	Athens test set				London test set			
	Clusters	P^1	P^5	P^{10}	Clusters	P^1	P^5	P^{10}
Spatial properties & text	14	51.22	71.14	79.14	21	47.54	61.43	69.71
Points, area, type & text	14	58.69	80.51	89.13	21	44.35	55.12	62.03
Mean and variance	14	44.43	55.52	60.00	28	45.92	55.81	61.83
Points, area & type	14	31.61	53.41	61.37	28	50.38	71.23	81.46
Points, area & type - double val.	14	39.62	51.04	56.51	28	45.24	56.64	65.65
Type	14	**59.15**	**82.53**	**91.08**	28	**51.67**	71.44	82.85
Spatial properties	14	55.95	79.36	87.89	28	51.24	**72.93**	**84.03**
Points and area	14	55.94	79.18	87.68	28	31.37	47.90	56.04

Table 7. Clustering+kNN (5-NN) feature combinations for Athens and London.

Features	Athens test set				London test set			
	Clusters	P^1	P^5	P^{10}	Clusters	P^1	P^5	P^{10}
Spatial properties & text	14	39.13	58.74	62.56	21	41.62	52.37	58.79
Points, area, type & text	14	38.69	58.16	61.64	21	41.62	48.77	56.05
Mean and variance	14	38.52	49.37	54.39	28	41.62	49.06	54.55
Points, area & type	14	50.53	**65.82**	72.05	28	**43.96**	56.27	63.90
Points, area & type - double val.	14	38.35	44.02	46.90	28	41.62	46.70	49.92
Type	14	**50.73**	63.86	**76.12**	28	41.62	52.89	61.29
Spatial properties	14	50.24	61.54	70.20	28	40.72	**60.17**	**67.00**
Points and area	14	38.46	60.21	65.15	28	41.62	45.12	50.89

for P^{10} in both datasets, while P^5 is a "tie" for the two methods considering both datasets. kNN comes third, with its performance degrading from P^1 to P^{10}. clustering+SVM provides the worst results of all. These results indicate that SVM is preferable when we want to provide very few recommendations, but as accurate as possible. On the other hand, when the setting allows us to provide more recommendations, with the hope that a few of them will be correct, then clustering+kNN works better.

Further, as a naive baseline, we consider the case where the majority class, i.e. the category with the highest frequency in each dataset is constantly (i.e. for each test entity) recommended, w.r.t. the P^1 measure. Similarly, we consider the cases where the 5 and 10 most frequent classes are constantly recommended for P^5 and P^{10}. The precision values are provided in the last line of Table 8 and demonstrate that the precision of the best solutions are consistently much higher than the naive baseline, with a difference ranging from 10% to 24%.

Table 8. Overall recommendation precision for Athens and London.

Algorithms	Athens test set			London test set		
	P^1	P^5	P^{10}	P^1	P^5	P^{10}
SVM	**59.99**	81.61	89.56	**59.59**	**73.20**	80.02
kNN	59.98	71.57	73.45	58.81	70.15	72.39
Clustering+SVM	43.45	63.48	64.75	43.37	50.18	60.71
Clustering+kNN	59.15	**82.53**	**91.08**	51.67	71.44	**82.85**
Majority class recommendation	35.65	68.33	75.76	40.17	63.07	72.97

Further, in general, it seems that applications of Geometry features or Spatial properties features give the best results, enhanced in occasions by being combined with Text features. However, in some cases (clustering+kNN), Type features seem to work very well by their own. Text features, seem to have the less effect on precision, probably due to their sparseness.

Another observation is the difference in recommendation precision between Athens and London, with Athens having slightly better P^1 values and considerably better P^5 and P^{10} values. This can be explained by examining some simple statistics from the two datasets. First, Athens is based on a more compact annotation set of 186 distinct classes, while the same number of spatial entities (20000 entities) in London are annotated by a set of 300 classes, resulting to higher heterogeneity in annotations. Second, the London dataset contains relatively more classes that do essentially hurt the evaluation results, due to their very low frequency. For example, categories with only one appearance in the dataset comprise 20 % of Athens but 25 % of London dataset, while the respective numbers for categories with at most two appearances are 33 % and 38 % respectively. Conclusively, the compactness of the annotation set and the lack of outlier categories (which is the case of the specific Athens dataset, relatively compared to the London one) seem to favour the recommendation precision.

With respect to the importance of the achieved values for the three evaluation measure variations, reaching a precision of around 60 % in a multilabel classification task, and for recommending *just one* category, is a rather important achievement. However, the other two measures are also important, considering a real world deployment of the recommendation algorithms: recommending 5 or 10 categories

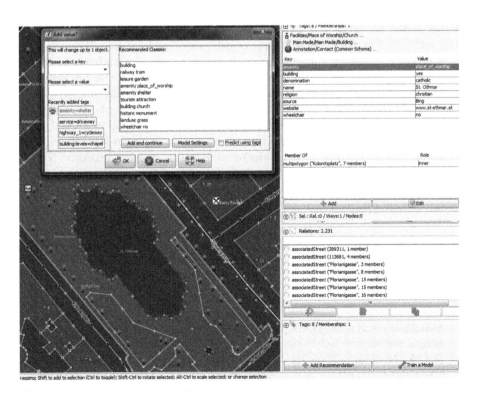

Fig. 1. Recommendations example

to the user to choose from is a realistic option, and the proposed algorithms can achieve precision up to 91 % for the 10 categories recommendation setting. This means that most times, the system will be able to recommend at least one useful category to the user.

To intuitively demonstrate the above conclusion, we present an exemplary recommendation of the JOSM plugin we have implemented using our approach, OSMRec [9]. In Fig. 1, we can see an example of recommendations for a church in Vienna. The already existing, manual annotations set (right top of the screen), contains the specific annotation pairs *"amenity => place_of_worship"*, *"denomination => catholic"* and *"religion => christian"*. Out of these three annotations, the most *geospatially* meaningful is the first one, *"amenity => place_of_worship"*, which is identified by our recommendation model (left top panel). Further, our model recommends an even more accurate annotation, *"building => church"*, that does not exist in the initial, manual annotation set. Apart from that, our model recommends a more diverse annotation, *"historic => monument"*, which is very probable to also characterize the specific entity.

5   Conclusions

In this paper, we presented a framework for producing category recommendations on spatial entities, based on previously annotated entities. We defined a set of problem specific training features that where applied with four classifiers, two state of the art and two hybrid solutions we proposed, and we reported on their recommendation precision for two real OpenStreetMap datasets from Athens and London. Specifically, we showed that the recommendation precision is high enough to be used in real world applications, such as JOSM OSM editing tool, where our method is already implemented as a plugin (OSMRec). Our future work includes further generalizing and testing the implemented framework with more geospatial semantic datasets and evaluating the effectiveness of Category features by using existing categories of an entity to predict new, correct annotation categories for the entity.

Acknowledgments. This work was partially supported by EU projects GeoKnow (GA no. 318159) and City.Risks (H2020-FCT-2014-653747).

References

1. Arnaud, V., Rodolphe, D.: Improving volunteered geographic information quality using a tag recommender system: the case of OpenStreetMap. In: Arsanjani, J., Zipf, A., Mooney, P., Helbich, M. (eds.) OpenStreetMap in GIScience, pp. 59–80. Springer, Cham (2015)
2. Jilani, M., Corcoran, P., Bertolotto, M.: Automated highway tag assessment of OpenStreetMap road networks. In: Proceedings of the SIGSPATIAL 2014 (2014)

3. Parundekar, R., Oguchi, K.: Learning driver preferences of POIs using a semantic web knowledge system. In: Simperl, E., Cimiano, P., Polleres, A., Corcho, O., Presutti, V. (eds.) ESWC 2012. LNCS, vol. 7295, pp. 703–717. Springer, Heidelberg (2012)

4. Silva, A., Martins, B.: Tag recommendation for georeferenced photos. In: LBSN (2011)

5. Schedl, M., Vall, A., Farrahi, K.: User geospatial context for music recommendation in microblogs. In: SIGIR (2014)

6. Mooney, P., Corcoran, P.: Annotating spatial features in OpenStreetMap. In: GIS-RUK (2011)

7. Mülligann, C., Janowicz, K., Ye, M., Lee, W.-C.: Analyzing the spatial-semantic interaction of points of interest in volunteered geographic information. In: Egenhofer, M., Giudice, N., Moratz, R., Worboys, M. (eds.) COSIT 2011. LNCS, vol. 6899, pp. 350–370. Springer, Heidelberg (2011)

8. Bao, J., Zheng, Y., Wilkie, D., Mokbel, M.: Recommendations in location-based social networks: a survey. Geoinformatica **19**, 525–565 (2015)

9. Karagiannakis, N., Giannopoulos, G., Skoutas, D., Athanasiou, S.: OSMRec tool for automatic recommendation of categories on spatial entities in OpenStreetMap. In: RecSys (2015)

Supporting Geo-Ontology Engineering Through Spatial Data Analytics

Gloria Re Calegari[✉], Emanuela Carlino, Irene Celino, and Diego Peroni

CEFRIEL – Politecnico of Milano, Via Fucini 2, 20133 Milano, Italy
{gloria.re,emanuela.carlino,irene.celino,diego.peroni}@cefriel.it

Abstract. Geo-ontologies are becoming first-class artifacts in spatial data management because of their ability to represent places and points of interest. Several general-purpose geo-ontologies are available and widely employed to describe spatial entities across the world. The cultural, contextual and geographic differences between locations, however, call for more specialized and spatially-customized geo-ontologies. In order to help ontology engineers in (re)engineering geo-ontologies, spatial data analytics can provide interesting insights on territorial characteristics, thus revealing peculiarities and diversities between places.

In this paper we propose a set of spatial analytics methods and tools to evaluate existing instances of a general-purpose geo-ontology within two distinct urban environments, in order to support ontology engineers in two tasks: (1) the identification of possible location-specific ontology restructuring activities, like specializations or extensions, and (2) the specification of new potential concepts to formalize neighborhood semantic models. We apply the proposed approach to datasets related to the cities of Milano and London extracted from LinkedGeoData, we present the experimental results and we discuss their value to assist geo-ontology engineering.

1 Introduction and Motivation

Imagine you live in the UK but you are currently on vacation in Italy for the first time. In the morning, when you are at home, you are used to buy newspapers and cigarettes in the small store around the corner; where can you get the same items now, since none of the shops in the vicinity of your B&B looks like a convenience store? Then, after having walked all day long throughout the old town, you finally reach the recommended restaurant in your tourist guide to taste Italian cuisine; unluckily it is closed and you seem to be in a residential area. You'd better find a district with a wide choice of eating places and – why not? – also at walking distance to nightlife venues; is there in town such a neighborhood, which is both rich in restaurants and clubs?

When we experience the environment surrounding us, we tend to elaborate what we see through the spatial categories we are used to; however, when we are in a less familiar place, we recognize that our conceptualizations do not perfectly apply and, even when they do, to better get oriented we make use of guides classifying for us the territory.

© Springer International Publishing Switzerland 2016
H. Sack et al. (Eds.): ESWC 2016, LNCS 9678, pp. 556–571, 2016.
DOI: 10.1007/978-3-319-34129-3_34

This semantic diversity has consequences on the conceptual models that knowledge engineers can build to describe spatial objects. When modeling a geo-ontology of urban points of interest, creating a general-purpose yet correct conceptualization is a challenge, because the ontology engineer has to decide the most suitable level of abstraction [1]. If the geo-ontology is generic, the risk is that it does not provide enough details to describe and distinguish between spatial objects with similar characteristics or functions in different places. On the other hand, if the engineering process produces a very rich and exhaustive conceptualization, some spatial objects' types could be too peculiar because of cultural diversity and apply only to specific geographic areas.

Many popular geospatial ontologies, adopted throughout different geographic regions, prefer to remain at a quite high level of abstraction, focusing on the basic concepts, the primitive notions [2]. A successful example is the LinkedGeoData ontology [3], which results from the ontologization of OpenStreetMap descriptive tags. How would it be possible to easily add location-specific extensions to a general-purpose geospatial ontology? Is there any analytics technique that could help in finding spatial concepts with a similar or diverging meaning at different locations? Would it be feasible to also characterize a territory by identifying its emerging semantic neighborhoods from the proximity analysis of different spatial object types?

Those are the questions that we address in our work, by proposing a set of spatial analytics techniques that, by processing the instances of a general-purpose geo-ontology, can provide hints to ontology engineers to support their modeling activities to create location-specific ontology extensions. The remainder of the paper is organized as follows: Sect. 2 describes the background of our approach in the context of related work; the main objectives and the experimental data set-up is described in Sect. 3; we detail the proposed approach, its empirical application and the obtained results in Sect. 4 at spatial feature level and in Sect. 5 at neighborhood level; finally, in Sect. 6 we offer our concluding remarks.

2   Background and Related Work

Geographic information usually describes locations as simple coordinates which are point-like, ubiquitous and precise. Beyond this straightforward interpretation of geo-information as geographic coordinates, there is the concept of place, which is the human way to understand and refer to space. Places are not point-like and have fuzzy boundaries determined by physical, cultural, and cognitive processes, such as the concept of "downtown" [4]. The identity and meaning of places cannot be captured considering only the spatial component of data, for which the semantic aspect is essential.

In this sense the analysis of geographic information in terms of its semantics is crucial and, nowadays, the modelling of geo-ontologies is becoming a great challenge [2]. Since geographic information spans across different domains, geo-ontology engineering needs to move from the top-down development of a small number of global ontologies, to the creation of a higher number of local ontologies that reflect location-specific perspectives and are developed in a bottom-up fashion [5].

An important aspects in ontology building is the choice of the level of abstraction: ontologies can contain both top-level concepts that apply across many or all domains and bottom-level concepts that apply only within a specific domain. The same holds for geospatial datasets, which can be semantically described at various levels to convey their meaning [1]. Since geo-ontologies are strictly related to the context, domain-specific concepts can mean situated concepts, i.e. conceptualizations dependent on specific processes (natural, social, scientific, or possibly machine) and whose instances are entities within a specific spatio-temporal context [6].

In our experiments we extract data from LinkedGeoData [3], which is described with a general-purpose ontology, and we aim at finding possible location-specific specializations or extensions to its ontology. LinkedGeoData uses the information collected by the OpenStreetMap project with the aim of providing a rich integrated and interlinked geographic dataset for the Semantic Web.

Since OpenStreetMap is a prominent example of volunteered geographic information (VGI) [7], LinkedGeoData knowledge reflects the way in which the environment is experienced [8]. This peculiarity implies that this type of information can be effectively employed to characterize a specific area not only in terms of geolocation [9]. LinkedGeoData information can highlight whether differences between regions exist [10] and can be useful to discover semantic dissimilarities of point features [11].

In this paper, starting from the LinkedGeoData ontology, we use a set of spatial analytics methods on OpenStreetMap data to check how to assist some steps of geo-ontology engineering [12], like evolution, repair and specialization.

The combination of the ontology, which represents the semantics of the data, and the sheer spatial analysis, which considers only the geographic coordinates, has been used for various purposes: to guide the choice of suitable data and clustering method for the task of locating shopping malls [13]; to characterize citizens behavior through location-based social network [14]; to create geographic summaries using social media [15]; to extract urban land use and support smart cities planning activities [16].

In the following, we use this combination of geo-ontologies and spatial analytics to provide a semantic characterization of the territory, in order to give citizens and visitors a thorough knowledge of a region in terms of different semantics areas.

3   Objectives, Data Preparation and Assumptions

In this paper, we present our approach to analyze the instances of a general-purpose geospatial ontology in two different urban environments, Milano and London. We apply a set of spatial analytics methods to derive helpful insights on their "urban semantics" to support the ontology engineers' re-engineering efforts on that ontology, in the specialization and extension phases (as defined in [12]). The two main objectives of our work are:

O1. Identifying concepts that play a different role in the two cities, thus possibly indicating different cultural or pragmatic meanings; we address this goal, by analyzing the pattern distribution of each spatial feature

O2. Highlighting new potential concepts to characterize the urban neighborhoods of the two cities; we address this goal by analyzing the co-occurring aggregations of different spatial features.

Our approach is inspired by the observation-driven framework proposed in [5], with the following differences: we do not aim to build a new ontology but to identify improvement points in a pre-existing conceptualization; we base our analysis on volunteered geographic information instead of sensor data; we focus on spatial primitives, touching geo-ontology design patterns only to a limited extent with the neighborhoods characterization.

We apply the approach described in the following Sects. 4 and 5 to a dataset derived from LinkedGeoData [3]. More specifically, our spatial objects are individuals in LinkedGeoData; we focus our investigation on the instances of the `lgdo:Amenity` concept and its sub-classes (shops, restaurants, hotels, offices, etc.). Each spatial object is described in terms of a semantic feature – its LinkedGeoData ontology class, which represent the place's category – and a spatial characterization; regarding the latter, in case of a point (named "node" in OpenStreetMap terminology) we take its latitude-longitude pair, in case of a polygon ("way") we compute its centroid and consider the coordinate pair of that centroid.

The experimental dataset consists of the spatial objects included in two reference zones: the Milano municipality border (around 200 km^2) and the Central London sub-region, the innermost boroughs of the UK capital city (around 130 km^2). The experiments illustrated in this paper are based on a dataset extracted at the end of October 2015; this means that the Milano objects include also the pavilions of the World Exposition in the EXPO 2015 area. In total, we collected around 13,000 spatial objects in Milano and 30,000 in London; those objects are instances of around 180 LinkedGeoData ontology classes (our spatial features).

The assumptions we make on the considered dataset are as follows. LinkedGeoData is derived from OpenStreetMap and OpenStreetMap is an open, collaborative bottom-up effort for collecting this large-scale spatial knowledge base. Therefore, the data cannot be expected to be complete and it is hard to give an estimate of its actual coverage [17]; nonetheless, since the data is the result of manual annotation, we can consider that OpenStreetMap volunteer editors add the most relevant and characterizing spatial objects, which are indeed the features that we wish to analyze to derive some insights on the urban space. Still, the mapping can be inhomogeneous (some zones can be more detailed annotated than others). Since we decided to focus on Milano and London, however, we can discard this potential issue: our direct knowledge of the city of Milano let us affirm that the spatial objects mapping is quite good and homogeneous throughout the city; OpenStreetMap coverage in the London area was evaluated in [18] and shown to be quite accurate in comparison to official sources. Furthermore, according to global OpenStreetMap statistics[1], Italy and UK are ranked 7th and 10th for number of created spatial objects, and 4th and 5th for density of created spatial objects per square kilometer. More details and further experimental results are available at http://swa.cefriel.it/geo/eswc2016.html.

4 Objective 1: Re-Engineering Spatial Features

If a particular spatial feature is condensed in a specific region of a larger area, that feature can be considered as a relevant element to characterize that region. Starting from this point of view, the first step of our experiments consists in conducting a spatial analytics exploration to find which spatial features can be considered prominent to tipify our reference areas and to verify if the same concept plays a different role in the different places. This is the first kind of support we can provide to geo-ontology engineers.

To reach this goal, we resort to two different statistical analysis: the analysis of the spatial patterns and aggregations and the points density analysis of each spatial feature. Naturally, the two analyses move into the same direction and are helpful to discover similarities and dissimilarities between different areas.

4.1 Analyzing Spatial Objects' Distribution

Regarding the first statistical analysis, we adopt some spatial analytics metrics to evaluate the tendency of each spatial feature to show spatial patterns and to create aggregations. The indicators we choose are the Morishita index and the Moran index.

The Morishita index [19] is a statistical measure of dispersion; we use it to detect spatial point patterns based on quadrat counts: the geographic area is divided into a regular grid with cells of equal size and shape and the numbers of points falling in each cell are counted and compared. If the point pattern is completely random, the index should be approximately equal to 1; values greater than 1 suggest that a spatial aggregation exists. The trend of the Morishita diagram and the value of the index indicate if a spatial aggregation can occur. Some examples of Morishita diagrams are reported on the companion website.

The other indicator is the Moran index [20], which is a measure of spatial autocorrelation: this index values range from –1 (perfect dispersion) to +1 (perfect correlation); a zero value indicates a random spatial pattern. The Moran and the Morishita indexes can be jointly used as indicators of spatial autocorrelation and aggregation.

We compute the Morishita and the Moran indexes for all spatial features, i.e. for all selected LinkedGeoData classes. The instances of some spatial features, like lgdo:FuelStation, lgdo:Supermarket, lgdo:Police – which are usually

---

[1] Cf. http://osmstats.neis-one.org/?item=countries.

homogeneously spread throughout a city, without specific aggregations – present low values for both the Morishita and the Moran Index, while those features that are normally grouped in specific areas, like lgdo:Clothes or lgdo:Cafe in a big touristic city, show higher values for both indicators. These considerations, although simple and immediate, confirm the reliability of the two adopted indexes, that we use to select the features on which we apply density based clustering, as explained in the following section.

4.2   Clustering Spatial Objects

In the second step, we start from the consideration that through clustering we can highlight how spatial objects, described only by their latitude-longitude coordinates, form agglomerations in a specific region. Density-based clustering methods [21] are used to this end: they find clusters based on the density of points in regions and they are able to identify clusters of arbitrary shapes. The key idea of density-based clustering is that, for each object of a cluster, its neighborhood within a given radius ϵ has to contain at least a minimum number $minpts$ of other objects, i.e. the cardinality of the object's neighborhood has to exceed a threshold.

We decided to adopt the OPTICS algorithm [22], which is the extended hierarchical version of the more famous DBSCAN method [21] to detect clusters with different densities. Both methods group points that are closely packed together and mark as outliers those points that lay in low-density regions. We prefer to adopt OPTICS because, since it is a hierarchical clustering, it is possible to "cut" its reachability plot at a given threshold ϵ_{cut}, thus revealing clusters with the same density.

Figure 1a shows the reachability plot of the instances of lgdo:Restaurant in Milano (shown in Fig. 1b); cutting the plot at ϵ_{cut} (represented by the red horizontal line) discriminates between the restaurant clusters and the more isolated points. Figure 1c shows the restaurant clusters as colored points on the map, while the white spots represents the outliers that are not clustered (i.e. less than $minpts$ restaurants in a radius of ϵ_{cut}).

To apply the OPTICS algorithm, it is necessary to set the three parameters, ϵ, $minpts$ and ϵ_{cut}. The first two parameters are used to compute the distances between each point and its nearest one, thus building the reachability plot; than the ϵ_{cut} parameter defines the desired level of clusters density.

We decide to perform OPTICS clustering on each spatial feature separately (i.e., lgdo:Restaurant, lgdo:Pub, lgdo:Clothes, etc.), in order to identify areas characterized by a high density of a specific class. To ensure effective results comparability, we decide to adopt the same parameters' values for all spatial features and for all locations.

Since we are working with points characterized only by their spatial component (the latitude-longitude pair), we choose parameters with a precise physical meaning: we consider that points instances of a specific spatial feature constitute a cluster if there are at least 5 points of the same class in a radius of 150 m (i.e. $minpts = 5$ and $\epsilon_{cut} = 150$). Imposing this constraint implies that not all the

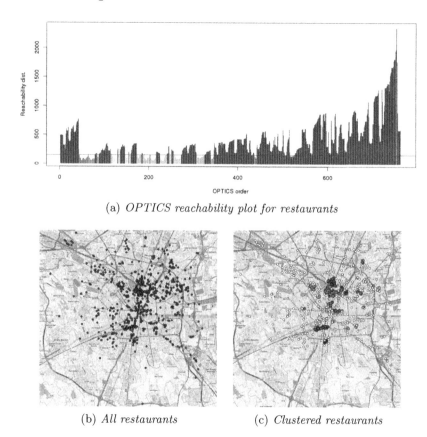

(a) *OPTICS reachability plot for restaurants*

(b) *All restaurants*   (c) *Clustered restaurants*

Fig. 1. Results of OPTICS clustering applied on instances of `lgdo:Restaurant` in Milano, with $minpts = 5$ and $\epsilon_{cut} = 150\,\text{m}$. (Color figure online)

considered spatial features result in clusters. This circumstance is useful to better understand the semantic characterization of each single place and to compare the two environments finding any dissimilarities.

Analyzing the results deriving from this clustering step, London and Milano outcomes are similar for some spatial features, but appear considerably different for other ones. In terms of spatial features which produce at least one cluster, the difference between the two city is remarkable: in London 56 different amenity types are sufficiently dense in at least one region, while in Milano only 25. Results on `lgdo:Hotel` amenities clustering highlight the difference of the considered areas in the two cities: in London we take only the innermost boroughs, characterized by a strongly touristic component, and OPTICS clustering reflects this consideration: 40 % of points in this category were clustered; conversely, in Milano we consider both central and peripheral regions, so the share of clustered hotels is much lower (14 %).

To highlight the possible existence of different semantic interpretations of the same concept in the two cities, let's consider two classes: lgdo:Telephone and lgdo:Pub. The red telephone box in London is a hallmark, while in Milano it is simply an installation useful for citizens located in the main places of displacement. This distinction is remarked by our clustering results: in London telephone clusters are discovered and mostly in the touristic areas; in Milano the OPTICS algorithm does not identify any cluster, because points are spread all over the city.

Pub is another concept with a different meaning for London and Milano communities: in UK pubs are widespread around a city, because people go there to eat and drink both at lunch- and dinner-time; in Italy people usually go to pubs after dinner and only to have a drink, so they are concentrated in popular nightlife areas. In London only 7 % of points are clustered, while in Milano this percentage reaches 20 %.

4.3 Support to Ontology Engineers

The instance analysis of an existing conceptualization can provide useful suggestions to the ontology engineer, who would like to evaluate the validity and applicability of a general geo-ontology in different locations. The approach outlined above with spatial distribution and spatial density analysis already ends up with some hints.

If a spatial feature corresponding to an ontological concept displays the same behavior in different places, this means that the concept is valid throughout the different territories and does not require any ontology re-engineering intervention (e.g. lgdo:Restaurant or lgdo:Clothes in our analysis of Milano and London).

On the other hand, whenever a spatial feature shows different values of the Morishita and Moran indicators and/or different tendency to form clusters, that sign can indicate the need to intervene: maybe location-specific extensions to the geo-ontology are required to provide a more precise conceptualization. For example, the lgdo:Pub concept could be split to take into account the two different meanings that spatial feature has in UK and Italy; similarly, the lgdo:Telephone concept could become a sub-class of lgdo:TourismThing in a London-specific ontology extension.

It is worth noting that also a pure numerical analysis of the spatial objects can provide interesting insights: it could reveal concepts whose instances in a place are "outliers" with regards to the instances of the same concept in other areas. This is for example the case of lgdo:Convenience in Milano: only a few instances of this class are present (instead of the 750 in central London); indeed, in Italy this type of shop is usually simply considered a supermarket, while cigarettes are only sold by tobacconists. A pure numerical outlier feature could further be investigated through the above spatial analysis: indeed lgdo:Convenience instances have a zero Moran index (which means random spatial pattern) and do not cluster in Milano, while they have a Moran index of 0.39 and form 18 clusters in London.

5   Objective 2: Specifying Spatial Neighborhoods

After identifying the features that best characterize the two cities, and those that exhibit different meanings in different places, our second goal is to answer the following questions: which are the spatial features that occur together? Is it possible to semantically characterize a region according to its semantic features co-occurence? The second step of our experiments is organized into three sub-steps, characterized by different, but strongly interrelated, statistical analysis.

Starting with the spatial feature clusters computed for each category (result-ing from the approach illustrated in Sect. 4), first we analyze their spatial co-occurence using again clustering techniques to discover emerging neighborhoods. Then, we further investigate the feature composition of those neighborhoods in terms of a new spatial indicator, inspired by the popular *tf-idf* score, that we introduce to identify the amenities that better define the urban space. Lastly, we characterize neighborhoods with new district-specific concepts, representing shopping, cultural or residential areas, by building spatio-semantic "queries" that incorporate spatial features' co-occurrence.

5.1   Identifying Neighborhoods

For each cluster of spatial objects obtained with the OPTICS algorithm, we define the convex hull polygon containing all the points belonging to that cluster and we compute its centroid. In this way we obtain a set of points representing all the categories' hotspot locations, which are urban areas with a high number of amenity of a given type.

Then we apply a hierarchical agglomerative clustering [23] technique (with Ward minimum variance method) on these centroids with the aim of dividing the urban space in neighborhoods, according to the simultaneous presence of spatial features. In this case, we do not use a density-based clustering algorithm because the centroids are artificially-created points that actually represent sets of points, hence the centroid density is not a meaningful indicator; consequently, a pure hierarchical clustering is a better fit for our goal. Using this technique, we obtain 12 clusters for both Milano and London, as shown in Fig. 2.

By comparing the obtained clusters with the official boundaries of the two cities' boroughs/districts, we discovered some interesting matches. In Milano (cf. Fig. 2a) we can identify, in the city centre, Duomo area, Centrale and Garibaldi railway stations and the area around Navigli and, in the outer parts, the uni-versity area – Città studi –, the most important industrial sites – Bicocca and Bovisa – and the area that hosted EXPO 2015. In London (cf. Fig. 2b) some famous boroughs are highlighted too: the town centre consisting in the City of London, West End, St. James and Covent Garden; the characteristic districts of Camden Town, Hyde Park, Kensington and Chelsea.

5.2   Characterizing Neighborhoods

Starting from the neighborhoods resulting from the latter clustering on both cities, we further investigate our clustering results to discover which spatial features best

MILANO

○ Garibaldi	○ Navigli	○ Città studi	● Mecenate
● Centrale	● Milano East	○ Bicocca	○ Expo
● Duomo	○ Pta Romana	● Bovisa	● Milano South

LONDON

● West End	○ Islington South	○ Southwark	● Chelsea
● St.James-Covent G.	○ Warwick	○ Camden Town	● Hampstead
● City of London	○ Brixton	○ Hyde Park-Kensinghton	● Islington North

(a) *Milano*   (b) *London*

Fig. 2. Centroid clustering to identify emerging neighborhoods. (Color figure online)

characterize each city region. To reach this goal, we define a spatial-specific version of the popular *tf-idf* score widely used in information retrieval and text mining. The classical version of this indicator is a numerical statistics used to highlight how important a word is to a document in a collection or corpus, taking into account the number of times a word appears in the document, weighted by its frequency in the corpus (*tf* part) and penalized if it appears very frequently in all documents (*idf* part).

Referring to our experiments, we want to highlight how important a spatial feature is to a specific neighborhood. Therefore, similarly to [24], for each neighborhood and for each feature we define the two components of our index as follows. The spatial object frequency *sof* of a spatial feature in a neighborhood is defined as:

$$sof = \frac{|n \cap f|}{|n|}$$

where n is the set of clustered points in the neighborhood and f is the set of all spatial points with that feature. Similarly, we define the inverse neighborhood frequency *inf* of a spatial feature as:

$$inf = 1 + log\frac{|N|}{|\{n : n \cap f \neq \emptyset\}|}$$

where $|N|$ is the total number of neighborhoods (24 in our experiments, because we have 12 districts in each city) and the denominator is the number of neighborhoods in which the spatial feature is represented by at least a point.

Table 1. Top three *sof-inf* scores in each neighborhood

Garibaldi	Centrale	Duomo	Navigli	MiEast	Pt.Romana	CittàStudi	Bicocca	Bovisa	Mecenate	Expo	MiSouth
office	hotel	shoes	pub	school	bank	university	indust.	indust.	indust.	fast_food	indust.
(1.28)	(3.30)	(4.18)	(4.18)	(4.18)	(0.37)	(2.90)	(0.63)	(1.48)	(0.34)	(1.39)	(0.05)
bar	bar	clothes	bar	parking	parking	bicycle_p	office	office	office	indust.	parking
(1.16)	(1.04)	(2.78)	(0.49)	(0.46)	(0.15)	(0.60)	(0.50)	(1.06)	(0.20)	(0.25)	(0.03)
restaur.	bank	bank	bank	office	office	bar	univ.	bicycle_p	bicycle_p	office	/
(0.71)	(0.70)	(1.90)	(0.41)	(0.25)	(0.15)	(0.49)	(0.24)	(0.31)	(0.06)	(0.08)	

(a) *Milano*

WestEnd	St.James	City	Isl.South	Warwick	Brixton	Southwark	Camden	HydePark	Chelsea	Hampstead	Isl.North
shoes	theatre	pub	school	hotel	greengr.	atm	charity	antiques	embassy	school	comm_c
(2.18)	(4.18)	(1.31)	(2.32)	(0.59)	(3.49)	(0.96)	(4.18)	(2.80)	(2.63)	(1.85)	(3.08)
art	musical	office	indust.	indust.	housew.	university	tattoo	hotel	shoes	parking	conven.
(2.15)	(2.39)	(1.01)	(2.26)	(0.58)	(2.57)	(0.70)	(2.24)	(2.40)	(0.86)	(0.25)	(0.65)
clothes	pub	bank	conven.	atm	butcher	bicycle_p	bar	conven.	clothes	conven.	/
(2.00)	(2.40)	(0.99)	(1.12)	(0.44)	(1.13)	(0.28)	(0.57)	(0.90)	(0.53)	(0.14)	

(b) *London*

Finally, we obtain our spatial object frequency–inverse neighborhood frequency index *sof-inf* by multiplying the two components.

To analyze neighborhoods composition in terms of their most relevant spatial features, we sort the *sof-inf* scores of all features in a neighborhood, thus ranking the distinctive amenity categories. Table 1a and b show Milano and London results respectively.

Considering Milano, we can offer our considerations based also on our direct knowledge of the environment. As highlighted previously, Bicocca and Bovisa are the most important industrial sites and this is confirmed by the highest *sof-inf* scores of the lgdo:Industrial spatial feature. The Garibaldi, Centrale and Duomo districts present higher values for those features strongly related to tourists: lgdo:Bar, lgdo:Restaurant, lgdo:Clothes, lgdo:Hotel. Navigli, the area known for its dynamic nightlife, is the only neighborhood in which the lgdo:Pub spatial feature achieves a considerable *sof-inf* score. The last consideration applies to the university area Città-Studi: the *sof-inf* analysis confirms the youthful look of the area, which is dense of typical structures related to student lifestyle (lgdo:University, lgdo:BicycleParking and lgdo:Bar). It is evident that there is a clear difference between the central areas, characterized by features related to shopping and to the daily/night life, and peripheral region where features as industrial and office are the most frequent ones.

Similar considerations can be made on London, even if only the very centre of the city is explored. West End and St. James are the most touristic areas (higher values for lgdo:Shoes, lgdo:Clothes, lgdo:Art and lgdo:Theatre). The City of London business area is marked by clusters of lgdo:Office and lgdo:Bank instances. The *sof-inf* analysis also highlights two peculiar areas, that we cannot find in Milano: Camden Town and Chelsea. The former area is the only one characterized by an "alternative" category like lgdo:Tattoo: it actually is the best known area in London for its "quirks". The latter area is one of the most prestigious, which is distinguished by presence of lgdo:Embassy clusters, i.e. structures normally situated in the most rich regions of a city.

5.3 Semantically Querying Neighborhoods

The third analysis is aimed to create neighborhood concepts that express the co-occurrence of different spatial features in the same area. To this end, we define combinations of LinkedGeoData amenities that we expect to characterize a specific class of districts, as follows (we omit the `lgdo:` prefix for simplicity):

Shopping: `Clothes`, `Shoes`, `Chemist` and `BeautyShop`
Nightlife: `Restaurant`, `Pub`, `Bar` and `Theatre`
Residential: `School`, `Atm`, `Butcher`, `Greengrocer`, `Convenience` and `Parking`
Culture: `Theatre`, `ArtShop`, `BookShop`, `MusicalInstruments` and `College`
Alternative: `Erotic`, `Tattoo`, `Charity`, `CommunityCentre` and `ArtShop`

In a sense, we aim to define neighborhood types with an ontology design pattern [25] that implies the simultaneous presence of a set of spatial features.

Those combinations of spatial features can be considered our *spatio-semantic queries*, similarly to a list of terms to be retrieved in a document corpus. We compute the *sof-inf* score defined in the previous section also for each query; then we compute the cosine similarity between the query *sof-inf* vector and each neighborhood *sof-inf* vector, thus evaluating the match between the classes of districts and the urban neighborhoods. The intrinsic difference of the two cities is reflected in our semantic queries' results, as shown in Fig. 3.

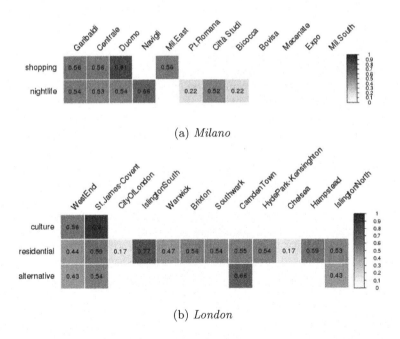

(a) *Milano*

(b) *London*

Fig. 3. Similarity scores between *sof-inf* query and neighborhood vectors.

In Milano it is reasonable that the *shopping* query presents the highest match (0.81) within the Duomo neighborhood, the principal touristic area. Milano *nightlife* is mainly concentrated in the area around Navigli (0.66), which is known as a lively area full of restaurants and pubs. The similarity scores of the same queries get uniform values in London across all districts, indicating that shopping and nightlife are more equally spread; it is worth reminding that in our experiment we analyze only the central boroughs of London, while in Milano area the suburbs are also included.

The cosmopolitan character of London is reflected by the existence of two neighborhoods strongly connoted as *cultural* and *alternative*: respectively, St. James–Covent Garden area (0.90), full of theatres, museums and art centres, and Camden Town (0.66), famous for its peculiarity and eccentric culture. As regards *residential* areas, it is interesting to focus on the districts with the lowest similarity scores: the City of London, the business core, and Chelsea, the most exclusive and prestigious area.

Our method would work on any other city and can prove effective also to compare similar cities; an example to compare Milano and London is reported at http://swa.cefriel.it/geo/eswc2016.html.

5.4   Support to Ontology Engineers

This second set of spatial analytics techniques is oriented to provide ontology engineers with some evidence of emerging neighborhood conceptualizations; a possible result then could be the specification of a new set of concepts to synthesize the "semantics" of urban districts.

While we use the clustering of spatial feature clusters only as a means to split the area of interest in regions with a specific characterization (as opposed to official and administrative boundaries), the definition of the *sof-inf* score, inspired by information retrieval practice, is an important step to allow a knowledge engineer to select the most prominent spatial features that characterize a neighborhood, similarly to [9]. This technique can help in effectively describing spatial regions by summarizing their distinctive categories.

Finally, the semantic query approach presented in Sect. 5.3 is the method we propose to support the ontological specification of neighborhood concepts as a design pattern combining spatial features: adopting this technique, not only it is possible to verify the actual "instantiation" of such concepts in different geographic areas, but the ontology engineer can also test different hypotheses, computing the similarity scores for different feature compositions. A possible extension to the proposed procedure can introduce spatial features' weights in the *sof-inf* computation of the spatio-semantic query.

This approach can prove useful also when the target neighborhood ontology must fit a specific level of abstraction: if the conceptualization is expected to be general-purpose, the semantic query should get homogeneous similarity scores in different geographical areas; conversely, if the new concepts are location-specific, the combination of spatial features can be selected based on the maximization of its similarity score with the desired regions.

6   Conclusions

Even when applying proper methodologies, ontology engineering largely remains an art that requires a deep domain knowledge. In the case of geo-ontologies, the understanding of location-specific knowledge is key; whatever the kind of spatial objects to be described, their geographic distribution can be investigated through spatial analytics to gain hints and suggestions to support the ontology engineering of their thematic characterization.

In this paper we employed a set of state-of-the-art data analytics techniques to study and compare geospatial objects from LinkedGeoData and provide additional insights to guide and support the (re-)engineering of the respective ontology. We showed how the plain analysis of objects coordinates can reveal cultural and location-specific differences between different cities. Our contribution also included the definition of a spatial variant of the tf-idf index, to illustrate how the combined analysis of semantic and spatial information can support the specification of neighborhood characterization.

The adopted methods are generic enough to be applied to different spatial datasets, since our experiments demonstrated that the diverse number of spatial objects and the possible dataset incompleteness do not negatively impact their analysis. Nonetheless, we will apply the proposed approach to a larger set of heterogeneous cities to further test, refine and select the best spatial metrics and indicators to reveal location-specific semantics. We would like to further investigate the generality and applicability of our techniques, by analyzing different types of spatial entities, possibly also within non-urban contexts.

The main limitation of our approach is that we do not provide any automated means to geo-ontology re-engineering, because our approach's findings bring only supporting insights. In addition, our analyses can confirm already-known characteristics, or they can discover unknown specificities that are hard to interpret and that require further investigation by the engineer. The natural next step is therefore to more tightly integrate spatial analytics within ontology engineering processes and tools.

Acknowledgments. This work was supported by the PROACTIVE project (id 40723101) co-funded by Regione Lombardia (POR-FESR 2007–2013).

References

1. Frank, A.U.: Chapter 2: ontology for spatio-temporal databases. In: Sellis, T.K., et al. (eds.) Spatio-Temporal Databases. LNCS, vol. 2520, pp. 9–77. Springer, Heidelberg (2003)
2. Janowicz, K., Scheider, S., Pehle, T., Hart, G.: Geospatial semantics and linked spatiotemporal data-past, present, and future. Semant. Web J. **3**(4), 321–332 (2012)

3. Stadler, C., Lehmann, J., Höffner, K., Auer, S.: Linkedgeodata: a core for a web of spatial open data. Semant. Web **3**(4), 333–354 (2012)
4. Montello, D.R., Goodchild, M.F., Gottsegen, J., Fohl, P.: Where's downtown?: behavioral methods for determining referents of vague spatial queries. Spat. Cogn. Comput. **3**(2–3), 185–204 (2003)
5. Janowicz, K.: Observation-driven geo-ontology engineering. Trans. GIS **16**(3), 351–374 (2012)
6. Brodaric, B., Gahegan, M.: Experiments to examine the situated nature of geoscientific concepts. Spat. Cogn. Comput. **7**(1), 61–95 (2007)
7. Goodchild, M.: Citizens as sensors: the world of volunteered geography. GeoJournal **69**, 211–221 (2007)
8. Brodaric, B.: Geo-pragmatics for the geospatial semantic web. Trans. GIS **11**(3), 453–477 (2007)
9. Tomko, M., Purves, R.S.: Venice, city of canals: characterizing regions through content classification. Trans. GIS **13**(3), 295–314 (2009)
10. Mooney, P., Corcoran, P.: The annotation process in OpenStreetMap. Trans. GIS **16**(4), 561–579 (2012)
11. Mülligann, C., Janowicz, K., Ye, M., Lee, W.-C.: Analyzing the spatial-semantic interaction of points of interest in volunteered geographic information. In: Egenhofer, M., Giudice, N., Moratz, R., Worboys, M. (eds.) COSIT 2011. LNCS, vol. 6899, pp. 350–370. Springer, Heidelberg (2011)
12. Suárez-Figueroa, M.C., Gómez-Pérez, A., Motta, E., Gangemi, A.: Ontology Engineering in a Networked World. Springer, Heidelberg (2012)
13. Wang, X., Hamilton, H.J.: Towards an ontology-based spatial clustering framework. In: Kégl, B., Lee, H.-H. (eds.) Canadian AI 2005. LNCS (LNAI), vol. 3501, pp. 205–216. Springer, Heidelberg (2005)
14. Noulas, A., Scellato, S., Mascolo, C., Pontil, M.: Exploiting semantic annotations for clustering geographic areas and users in location-based social networks. In: The Social Mobile Web (2011)
15. Rizzo, G., Falcone, G., Meo, R., Pensa, R.G., Troncy, R., Milicic, V.: Geographic summaries from crowdsourced data. In: Presutti, V., Blomqvist, E., Troncy, R., Sack, H., Papadakis, I., Tordai, A. (eds.) ESWC Satellite Events 2014. LNCS, vol. 8798, pp. 477–482. Springer, Heidelberg (2014)
16. Calegari, R.G., Carlino, E., Peroni, D., Celino, I.: Extracting urban land use from linked open geospatial data. ISPRS Int. J. Geo-Inf. **4**(4), 2109–2130 (2015)
17. Mooney, P., Corcoran, P., Winstanley, A.C.: Towards quality metrics for OpenStreetMap. In: Proceedings of the 18th SIGSPATIAL International Conference on Advances in Geographic Information Systems, pp. 514–517. ACM (2010)
18. Haklay, M., et al.: How good is volunteered geographical information? a comparative study of OpenStreetMap and ordnance survey datasets. Environ. Plann. B Plan. Des. **37**(4), 682 (2010)
19. Morisita, M.: Measuring of the dispersion of individuals and analysis of the distributional patterns. Mem. Fac. Sci. Kyushu Univ. Ser. E **2**(21), 5–235 (1959)
20. Gittleman, J.L., Kot, M.: Adaptation: statistics and a null model for estimating phylogenetic effects. Syst. Biol. **39**(3), 227–241 (1990)
21. Ester, M., Kriegel, H.P., Sander, J., Xu, X.: A density-based algorithm for discovering clusters in large spatial databases with noise. In: KDD, pp. 226–231 (1996)
22. Ankerst, M., Breunig, M.M., Kriegel, H.P., Sander, J.: Optics: ordering points to identify the clustering structure. In: Proceedings of the 1999 ACM SIGMOD International Conference on Management of Data, pp. 49–60. ACM (1999)

23. Rokach, L., Maimon, O.: Clustering methods. In: Maimon, L., Rokach, L. (eds.) Data Mining and Knowledge Discovery Handbook, pp. 321–352. Springer, Heidelberg (2005)
24. Walker, A.R., Moody, M.P., Pham, B.L.: A spatial similarity ranking framework for spatial metadata retrieval (2006)
25. Gangemi, A., Presutti, V.: Ontology design patterns. In: Staab, S., Studer, R. (eds.) Handbook on Ontologies, pp. 221–243. Springer, Heidelberg (2009)

Trust and Privacy Track

Provenance Management for Evolving RDF Datasets

Argyro Avgoustaki[1,2(✉)], Giorgos Flouris[2], Irini Fundulaki[2],
and Dimitris Plexousakis[1,2]

[1] Department of Computer Science, University of Crete, Heraklion, Greece
[2] Institute of Computer Science, FORTH, Heraklion, Greece
{argiro,fgeo,fundul,dp}@ics.forth.gr

Abstract. Tracking the provenance of information published on the Web is of crucial importance for effectively supporting trustworthiness, accountability and repeatability in the Web of Data. Although extensive work has been done on computing the provenance for SPARQL queries, little research has been conducted for the case of SPARQL updates. This paper proposes a new provenance model that borrows properties from both *how* and *where* provenance models, and is suitable for capturing the triple and attribute level provenance of data introduced via SPARQL INSERT updates. To the best of our knowledge, this is the first model that deals with the provenance of SPARQL updates using algebraic expressions, in the spirit of the well-established model of *provenance semirings*. We present an algorithm that records the provenance of SPARQL update results, and a reconstruction algorithm that uses this provenance to identify a SPARQL update that is *compatible* to the original one, given only the recorded provenance. Our approach is implemented and evaluated on top of Virtuoso Database Engine.

1 Introduction

During the last few years, we have witnessed an explosion in the volume of semantic data available on the Web. These data are usually published using the RDF data model[1], where information is represented using *triples*, organized in *named graphs* [6], thereby forming *quadruples*. Querying and updating RDF data is performed using the W3C standards SPARQL[2] and SPARQL Update[3] respectively.

Nowadays semantic data is the most prominent example of large scale data where one could create new *datasets* (sets of quadruples) by integrating existing ones. In this setting, recording the *provenance* of such data, i.e., their *origin*, which describes from *where* [5] and *how* [12] the data was obtained, is of crucial importance for supporting trustworthiness, accountability and repeatability. This is necessary due to the open and unconstrained nature of the Web of

---

[1] http://www.w3.org/TR/rdf11-primer/.
[2] http://www.w3.org/TR/sparql11-overview/.
[3] http://www.w3.org/TR/sparql11-update/.

© Springer International Publishing Switzerland 2016
H. Sack et al. (Eds.): ESWC 2016, LNCS 9678, pp. 575–592, 2016.
DOI: 10.1007/978-3-319-34129-3_35

Data and the growing tendency to populate scientific data warehouses through SPARQL updates offered by SPARQL endpoints.

In this work we deal with the problem of *capturing and managing the provenance of quadruples constructed through SPARQL updates*. More specifically, we focus on SPARQL INSERT operations (we refer to them as INSERT updates) used to add newly created triples in a target named graph (i.e., quadruples). The purpose of computing the provenance for such operations is to record from *where* and *how* each quadruple was constructed, thereby allowing us to determine the quadruples and the SPARQL operators that were used to produce it.

The problem of managing provenance information has received considerable attention [5,9–13,15,16,20,21], but most works deal with query provenance. W3C published a recommendation [18] concerning the interchange of provenance information, which, however, focuses on providing a syntactic means to represent provenance rather than providing a method for identifying or computing it. Algebraic expressions have been used to capture (query) provenance in varying levels of detail [11,12,15]. In the RDF context, provenance is often represented using named graphs [6,7,10,15].

However, the unique requirements associated with SPARQL update provenance do not allow a direct reuse of such approaches. One problem is that the named graph component of a quadruple is defined by the user in the INSERT update, so triples with different origin may be added to the same named graph. Thus, the standard approach of capturing provenance through the named graph of a quadruple is not sufficient, and provenance should be defined for quadruples, rather than triples (as in most works).

In addition, quadruples created via INSERT updates could be the result of combining values found in different quadruples through different SPARQL operators. This creates a unique challenge, because each attribute of a quadruple may have a different provenance. Thus, fine-grained, *attribute level* provenance models are called for, and more expressive models that go beyond the named graphs approach are needed.

Another challenge stems from the persistence of a SPARQL update result, which implies that when a quadruple is accessed, the SPARQL update that generated the quadruple is no longer available. This makes standard *how* provenance models unsuitable for recording provenance at a fine-grained level in this setting. As an example, standard how-provenance approaches will record that a join was used to generate a quadruple, but will not record the components of the quadruples that were joined to produce the result; even though this information is easily available during queries (via the SPARQL query), this is not the case for SPARQL updates (where the SPARQL update is not available). Recording the INSERT update is not an efficient remedy for the situation, because (a) the syntactic form of the actual INSERT update is irrelevant and (b) the INSERT update is no longer relevant, as the dataset has evolved.

Therefore, more fine-grained forms of how-provenance are called for. We define this more demanding form of how-provenance in an indirect manner, by introducing the notion of *reconstructability*, which refers to the ability of using the provenance information for *reconstructing* an INSERT update that is *compatible* (see Definition 4) with the INSERT update that generated this quadruple.

We show that, to satisfy the requirement of reconstructability, the provenance of a quadruple should be expressive enough to identify: (a) the quadruples that contributed to its creation (*where provenance* [8]), and (b) how these quadruples were used (via joins and unions) to generate the new one (*how provenance* [12]), under the more demanding form of how-provenance explained above.

The main contributions of this paper[4] are:

- The introduction of a *fine-grained and expressive provenance model* that borrows from both *where* and *how* provenance models, is suitable for encoding both *triple* and *attribute* level provenance of quadruples obtained via INSERT updates, and allows the *reconstructability* of such updates from their provenance.
- The provision of algorithmic support for our model via the *provenance construction* and *update reconstruction* algorithms. The former is used for computing and recording the provenance of the result of an INSERT update based on the proposed model, whereas the latter exploits the expressiveness of our model to report on the generation process of a quadruple.
- The implementation, theoretical analysis, and experimental evaluation of these algorithms on top of Virtuoso Database Engine.

2   Preliminaries

We consider provenance in the context of an *RDF dataset* (denoted by D); for simplicity, we assume that an RDF dataset is composed of a set of *quadruples* of the form *(s,p,o,n)*, where *(s,p,o)* is a triple belonging in a *named graph n*.

SPARQL 1.1 is the official W3C recommendation for querying and updating RDF datasets, and is based on the concept of matching patterns against such graphs. Patterns are defined via *quad patterns* which are like quadruples but allow *variables* (prefixed with ?) in the subject, property or object position. Quad patterns can be combined using *SPARQL operators* to form *graph patterns*. In this work, we focus on *union* (UNION) and *join* (".") operators only, ignoring *optional* and *filters* (we plan to deal with these operators in future work). Thus, the considered INSERT updates are of the form: $U := $ INSERT $\{qp_{ins}\}$ WHERE $\{gp\}$, where qp_{ins} is a quad pattern and gp is a graph pattern formed as a union of individual graph patterns, gp^1 UNION ... UNION gp^k. Each gp^i is of the form qp_1^i . qp_2^i . .... . qp_m^i. Note that all INSERT updates containing only union and join operators can be equivalently written in the above form [19]. Note also that INSERT DATA operations can be defined in terms of INSERT [2].

In addition, we require that for each qp_j^i there is a sequence $\langle qp_{j_1}^i, \ldots \rangle$ of quad patterns from gp^i, such that each element in the sequence has a common

---

[4] Detailed presentation of our approach including the source code of our implementation can be found in http://www.ics.forth.gr/isl/provenance.

variable with the previous element in the sequence, whereas the first element has a common variable with qp_{ins}. This restriction is necessary to "strip" the graph pattern in the WHERE clause from quad patterns that play no essential role in its evaluation [19].

SPARQL Update specifications do not fully clarify the principles governing transactions with multiple updates [13]; here, we focus on transactions consisting of single atomic updates. Further details on SPARQL are omitted (see [1,2,19]).

3   Motivating Example

We provide an example from the medical domain to motivate our approach. Note that this example is used for illustration purposes only and any consequences pertaining to data privacy are out of the scope of our paper. Table 1 shows a dataset D_1 containing four quadruples (with identifiers $c_1, \ldots, c_4$), each with a certain provenance ($p_1, \ldots, p_4$). These quadruples describe treatments for hypertension that have been provided by different doctors.

Table 1. Dataset D_1

	S	P	O	N	PROV
c_1	<hypertension>	<treatedWith>	<diuretics>	<Diabetologist>	p_1
c_2	<hypertension>	<treatedWith>	<diuretics>	<Pathologist1>	p_2
c_3	<hypertension>	<treatedWith>	<diuretics>	<Pathologist2>	p_3
c_4	<hypertension>	<treatedWith>	<b_blockers>	<Pathologist2>	p_4

Now suppose that a patient visits the hospital and a young doctor diagnoses hypertension. To decide on the proper treatment, he checks the system for previous treatments of hypertension, paying special attention to those proposed by the diabetologist, because the patient's history includes diabetes and some medications may raise the blood sugar levels, a dangerous condition for a diabetic. The result of his query needs also to be recorded in the database, as it will be his suggested treatment, so he executes U:

$$\text{INSERT } \{qp_{ins}\} \text{ WHERE } \{qp_1^1 \text{ UNION } qp_1^2 \; . \; qp_2^2\}$$

where: qp_{ins}: (<hypertension>, <treatedWith>, $?o$, <YoungDoctor>)

$\quad\quad qp_1^1$: (<hypertension>, <treatedWith>, $?o$, <Diabetologist>)

$\quad\quad qp_1^2$: (<hypertension>, <treatedWith>, $?o$, <Pathologist1>)

$\quad\quad qp_2^2$: (<hypertension>, <treatedWith>, $?o$, <Pathologist2>)

The application of U upon D_1 leads to the insertion of c_5, forming dataset D_2, shown in Table 2. The expression p_5 below is used to describe the provenance of c_5:

$$p_5 \; : \quad \{(\bot, \; \bot, \; {}_{qp_1^1.o}(c_1)) \quad \oplus \quad (\bot, \; \bot, \; {}_{qp_1^2.o}(c_{2\{qp_1^2.o\}} \odot_{\{qp_2^2.o\}} c_3))\}$$

Some explanations on p_5 are in order. First, each operand of $\oplus$ indicates a different way through which c_5 occurred (due to the existence of UNION). The first operand $(\perp, \perp,\ _{qp_1^1.o}(c_1))$ resulted from the prescription of the diabetologist; in particular, its subject and property values were dictated by the corresponding constants in qp_{ins} (indicated by the value $\perp$), whereas its object resulted by "copying" the object value (o) of c_1 due to the quad pattern qp_1^1 (denoted by $_{qp_1^1.o}(c_1)$).

Similarly, the second operand's subject and property were dictated by the constants of qp_{ins}, whereas the object resulted from the agreeing prescriptions of the two pathologists. In particular, the object value was the result of a join (indicated by $\odot$) between two quadruples, namely c_2, c_3; the join happened between the object position of qp_1^2 (i.e., $qp_1^2.o$) and the object position of qp_2^2 (i.e., $qp_2^2.o$), hence the left and right subscript of $\odot$. Finally, the result of this join was projected over the object position as indicated by the outer subscript $qp_1^2.o$.

Table 2. Dataset D_2

	S	P	O	N	PROV
c_1	\<hypertension\>	\<treatedWith\>	\<diuretics\>	\<Diabetologist\>	p_1
c_2	\<hypertension\>	\<treatedWith\>	\<diuretics\>	\<Pathologist1\>	p_2
c_3	\<hypertension\>	\<treatedWith\>	\<diuretics\>	\<Pathologist2\>	p_3
c_4	\<hypertension\>	\<treatedWith\>	\<b_blockers\>	\<Pathologist2\>	p_4
c_5	**\<hypertension\>**	**\<treatedWith\>**	**\<diuretics\>**	**\<YoungDoctor\>**	p_5

The created expression (p_5) is inspired by standard how-provenance expressions [12,15] used in abstract provenance models, but contains additional information not present in such expressions. In particular, we include, for each attribute of a result quadruple:

– a subscript denoting the quad pattern position in the WHERE clause that the element's value is taken from (arbitrarily we set this to be the first matching position).
– two subscripts in the provenance join operator $(_{\{\}}\odot_{\{\}})$ to describe the positions of the quad patterns where the joins take place; the first subscript is written to the left of the join operator and refers to the first operand of the join, whereas the second is written to the right and refers to the second operand. This information is important for understanding how c_5 found its way in the dataset (reconstructability).

4   Abstract Provenance Model

Standard *abstract provenance models* are comprised of *abstract identifiers* and *abstract operators* [12,15]. Abstract identifiers (*quadruple identifiers* in our case, denoted by c_i) are uniquely assigned to RDF quadruples, whereas abstract operators describe the computations performed on quadruples to derive a result quadruple. We additionally introduce the notion of *quad pattern positions*, which are used to describe the position of the occurrence of a constant or a variable in a quad pattern (we provide more details below). Using this infrastructure, RDF quadruples are annotated with complex expressions that involve the identifiers, the operators and the quad pattern positions:

Definition 1. *The provenance p of a quadruple q is defined as $p := \{cpe_1, \ldots, cpe_k\}$. A cpe is a complex provenance expression defined as $cpe := pe^1 \oplus pe^2 \oplus \ldots \oplus pe^m$, where $m \geq 1$, pe^j is a simple provenance expression and $\oplus$ is the provenance operator for union. An expression pe is of the form $(prov_s, prov_p, prov_o)$, where $prov_{pos}$ is the provenance of the attribute pos (described in detail in Definition 2).*

In the above definition, p is the full provenance of the quadruple. Since a quadruple can be the result of more than one INSERT updates applied over the course of time, we use cpe_i to record each such update. As explained in Sect. 3, each pe^i corresponds to one operand of a UNION operator that leads to the generation of the quadruple, whereas each $prov_{pos}$ describes how the current attribute resulted. Note that $prov_{pos}$ allows the identification of the origin of each element-attribute individually (attribute-level provenance [4]). We are not interested in the provenance of the graph component (the fourth element of a quadruple), as this is explicitly defined by the INSERT update.

Example 1. In our running example (Sect. 3), $p_5 = \{cpe_1\}$ and $cpe_1 = pe^1 \oplus pe^2$, where $pe^1 = (\bot, \bot, \ _{qp_1^1.o}(c_1))$ and $pe^2 = (\bot, \bot, \ _{qp_1^2.o}(c_{2\{qp_1^2.o\}} \odot_{\{qp_2^2.o\}} c_3))$; each pe^i results from one operand of the UNION. In pe^1, $prov_s = prov_p = \bot$, whereas $prov_o = \ _{qp_1^1.o}(c_1)$. □

Now let's see how the simple provenance expression pe is constructed. For reasons that will be made apparent later, it is necessary to refer to each individual variable or constant of an update. For this purpose, we arbitrarily number:

a. graph patterns, gp^i $(i \geqslant 1)$ indicates the ith graph pattern of the WHERE clause.
b. quad patterns, qp_j^i $(j \geqslant 1)$ indicates the jth quad pattern in the graph pattern gp^i.

Moreover, we refer to the quad pattern in the INSERT clause as qp_{ins}.

Using this identification mechanism, each variable or constant in a quad pattern can be uniquely identified by a *quad pattern position*, i.e., $qp_j^i.x$ (or $qp_{ins}.x$), where qp_j^i (qp_{ins}) is the corresponding quad pattern and x is one of s, p, o, to indicate one of the these positions in a quad pattern (e.g., $qp_2^1.o$ denotes the object of the 2th quad pattern of the 1st graph pattern).

Definition 2. *The provenance of attribute pos (pos $\in \{s, p, o\}$), namely $prov_{pos}$, is defined as $prov_{pos} := \bot \mid \ _{varSub}(spe)$, where $\bot$ is a special label, varSub is the var subscript (a quad pattern position) and spe is a standard* *provenance expression. spe is defined as* $spe := (c_i\ _{joinSub^1} \odot\ _{joinSub^2}\ c_j \cdots$ $_{joinSub^{r-1}} \odot\ _{joinSub^r}\ c_k)$, *where* c_x *is a quadruple identifier,* $joinSub^x$ *is a join subscript (quad pattern position IDs) and* $\odot$ *is the provenance operator of join.*

As proposed in [4,20], the special label $\perp$ is used to record the case where the INSERT update constructs an element of the new quadruple using a constant, e.g., $prov_s$, $prov_p$ in pe^1 and pe^2 expressions of p_5 in our motivating example. This is the case where the corresponding position in qp_{ins} contains a constant.

If a quad pattern position in qp_{ins} (say $qp_{ins}.pos_1$) contains a variable, then the corresponding value is copied by a quadruple in the dataset, or generated via SPARQL joins. This is recorded using the form $_{varSub}(spe)$, where $varSub$ determines the quad pattern position $qp_j^i.pos_2$ that the value should originate from (this position contains the same variable as $qp_{ins}.pos_1$), and spe describes the operation (join or simple "copy") that created it. When there is a copy (in the sense of [4]), spe records the quad pattern position ID from where the value is taken. When there is a join, spe records the joined quadruples, and the positions in said quadruples that were joined (via the left and right subscripts of $\odot$). This is similar to [15], except that [15] does not record the joined positions, which is critical for reconstructability.

Example 2. In our example, $qp_{ins}.s$ and $qp_{ins}.p$ are constants, so the s,p positions of pe^1, pe^2 are set to $\perp$. For the o position, the expression pe^1 contains the var subscript $qp_1^1.o$ because this is the position where the variable $?o$ appears in the first operand of the UNION. In this case, the value is taken directly from the corresponding quadruple (c_1), so in pe^1, $prov_o = \ _{qp_1^1.o}(c_1)$. Similarly, for pe^2, the corresponding var subscript is $qp_2^2.o$; note that $qp_2^2.o$ contains the same variable, but we take, by convention, the first valid appearance of said variable. The actual value of the quadruple is generated through a join between the o positions of qp_1^2, qp_2^2, hence the subscripts of the $\odot$ operand; the joined quadruples are c_2 and c_3. Thus, for pe^2, $prov_o = \ _{qp_1^2.o}(c_{2\{qp_1^2.o\}} \odot_{\{qp_2^2.o\}} c_3))$.   □

5   Provenance Algorithms

5.1   Provenance Construction Algorithm

The *provenance construction* algorithm (Algorithm 1) is used to record the provenance of quadruples resulting from an INSERT update. This algorithm takes as input an INSERT update U and a dataset D, and returns a provenance expression p_k to associate with each newly created quadruple q_k. Due to space limitations, we will present a simplified version of the algorithm, where the INSERT update generates only one result quadruple; the interested readers can see the full algorithm in [2].

Computing p_k amounts to computing the new cpe_r (resulting from U) to be added; the actual addition happens in line 22 (line 21 determines the corresponding quad q_k). The computation of cpe_r proceeds as follows: the outer FOR (lines 1–20) computes all pe^i (one for each operand of UNION), which are added to

cpe_r (line 19), whereas the inner FOR (lines 2–17) computes $prov_s, prov_p, prov_o$ which are composed to form pe^i (line 18). The value of each $prov_{pos}$ is determined by the corresponding $qp_{ins}.pos$: if it is a constant, then $prov_{pos} = \bot$ (line 15); otherwise (if it is a variable), the computation is more complex and is performed in lines 4–13.

Algorithm 1. Provenance Construction Algorithm

Input: An INSERT update U, a dataset D
Output: The provenance p_k of a result quadruple q_k, P
1: **for all** $(gp^i \in$ WHERE clause) **do**
2:      **for all** $qp_{ins}.pos$ **do**
3:         **if** $qp_{ins}.pos \in \mathbb{V}$ **then**
4:            Create the set $MatchingPatterns$ $\{mp_1, \ldots mp_z\}$
5:            $spe =$ FINDIDS(mp_1) ▷ "Copy" case
6:            $j = 2$
7:            **while** $mp_j \neq null$ **do**   ▷ Join case
8:               Create $joinSub^x$ and $joinSub^{x+1}$
9:               $spe = spe \;_{joinSub^1} \odot \;_{joinSub^2}$ FINDIDS(mp_j)
10:               j++
11:            **end while**
12:            Create the $varSub$
13:            $prov_{pos} = {}_{varSub}(spe)$
14:         **else**
15:            $prov_{pos} = \bot$
16:         **end if**
17:      **end for**
18:      $pe^i = (prov_s, prov_p, prov_o)$
19:      $cpe_r = cpe_r \oplus pe^i$
20: **end for**
21: $q_k =$ GETQUAD(cpe_r, qp_{ins})
22: $p_k = p_k \cup cpe_r$
23: **return** (q_k, p_k)

In the latter case, we first compute the ordered set $MatchingPatterns$ (line 4), which contains all quad pattern identifiers that belong in gp^i and are *related* to the evaluation of the variable in $qp_{ins}.pos$. A quad pattern is related if it contains the specific variable, or if it joins (possibly via another variable) with another related quad pattern.

Then, spe is initially set to be equal to the quad pattern identifier that matches the first item in MatchingPatterns (line 5). If MatchingPatterns has a single item, then we have no joins, i.e., we have a "copy"; lines 7–11 will be skipped, $varSub$ will be computed in line 12, and $prov_{pos}$ in line 13.

If, however, MatchingPatterns has more than one items, then there is one or more joins, which have to be taken into account in the computation of spe. Each join is identified in line 8 (by iterating over the quad patterns and recording the positions where the joins take place by looking at their common variables), line 9 enhances spe with the new join (and the respective quadruple identifier) and the process (lines 7–11) continues until no more MatchingPatterns exist.

It should be noted that there may be more than one quadruple identifiers matching a given quad pattern. In this case, all the different valid combinations are considered by FINDIDS, and each combination results to a different spe and $prov_{pos}$.

Example 3. In our example, $qp_{ins}.s$, $qp_{ins}.p$ are constants, $prov_s = prov_p = \bot$; on the other hand, $qp_{ins}.o = ?o$. For the graph pattern gp^1, we have *Matching-Patterns* $= \{qp_1^1\}$ (which contains the variable $?o$); line 5 will set $spe = c_1$ and line 12 will set $varSub = qp_1^1.o$; the final result (line 18) will be: $pe^1 = (\bot, \bot, qp_1^1.o(c_1))$.

For gp^2, the set $MatchingPatterns = \{qp_1^2, qp_2^2\}$, indicating that there was a join between qp_1^2, qp_2^2 that created this quadruple. Line 5 sets $spe = c_2$, line 8 identifies the common variable(s) between these two quad patterns ($qp_1^2.o, qp_2^2.o$), and line 9 computes the final $spe = c_2{}_{\{qp_1^2.o\}} \odot_{\{qp_2^2.o\}} c_3$. Note that the evaluation of qp_2^2 also matches c_4, but we ignore it since it does not join with c_2. As before, line 12 will set $varSub = qp_1^2.o$, and the result will be: $pe^2 = (\bot, \bot, qp_1^2.o(c_2 {}_{\{qp_1^2.o\}} \odot_{\{qp_2^2.o\}} c_3)$. $\qquad\square$

5.2 Update Reconstruction Algorithm

Algorithm 2 exploits the rich semantics of the provenance expression of a quadruple in order to determine how the quadruple found its way in the dataset. It takes as input a complex provenance expression cpe that is part of the provenance of the input quadruple q and a dataset D, and returns another INSERT update U'; as we will show below, U' is *compatible* with the original INSERT update that led to the creation of q, i.e., the same in most relevant aspects. The reason why Algorithm 2 takes as input cpe, rather than the full provenance, is that each cpe is the result of one INSERT update operation. Before presenting the algorithm, we provide some formal definitions:

Definition 3. *Let gp and gp' be graph patterns. We say that gp' is filter-compatible to gp (denoted $gp \sim gp'$) iff gp' differs from gp only in the filters that it may employ.*

Note that Definition 3 refers also to implicit filters created by a constant value in the WHERE clause, e.g., <hypertension> in qp_1^1, qp_1^2, qp_2^2 of our motivating example.

Definition 4. *Let U and U' be INSERT updates. We say that U' is compatible to U (denoted $U \leadsto U'$) if there is a renaming of variables in U', such as $qp_{ins} = qp'_{ins}$ and for each gp' in U' there is a filter-compatible gp in U.*

Intuitively, Definition 4 says that U' is compatible to U iff U contains a subset of the graph patterns in U', modulo filters and variable renaming. As a consequence of Definition 4, the following theorem can be deduced:

Theorem 1. *Let U and U' be UNION-free INSERT updates. If U' is compatible to U ($U \leadsto U'$), then U is also compatible to U' ($U' \leadsto U$).*[5]

The Algorithm 2 can be split in three parts, each of which computes a different component of the output $U' = $ INSERT $\{qp'_{ins}\}$ WHERE $\{gp'\}$. In particular, lines 1–8 compute qp'_{ins}; lines 9–34 compute gp' and line 35 combines the above to form U'.

---

[5] Proofs for all theorems can be found in [2].

For the first part, the graph position (n) of qp'_{ins} is determined by the graph attribute of the input q (line 1). For the s, p, o positions, we exploit the fact that, if $prov_{pos}$ of pe^1 is equal to $\bot$, then the corresponding quadruple attribute was created by a constant, so we set $qp'_{ins}.pos = q.pos$ (note also that in this case, the $prov_{pos}$ of all pe^i will be equal to $\bot$); otherwise, $qp'_{ins}.pos$ is associated with a new variable.

The main part of the algorithm (lines 9–34) contains one FOR loop which computes the graph patterns (gp'^i), each corresponding to one pe^i in cpe; each loop computes all the quad patterns qp'^i_j of gp^i, composes them using join in line 32 (to form gp'^i), and uses the result to progressively built the final graph pattern gp' (line 33).

To construct gp'^i, we progressively fill the positions of each quad pattern in gp'^i with variables, taking special care to use the same variables in positions that are joined, and also to reuse the variables already in qp'_{ins} when appropriate.

Initially, we compute the size of gp^i, i.e., the number of quad patterns in gp^i (line 10), by scanning all quad pattern identifiers found in the var or join subscripts of pe^i.

Line 11 deals with the fourth attribute of quad patterns, which does not accept variables, so its value is taken directly by the fourth attribute of the corresponding quadruple. Finding the corresponding quadruple is easy: for a quad pattern appearing in a var subscript, its corresponding quadruple is the first that appears in the respective spe, whereas for join subscripts we take the quadruple in the respective "side" of the join.

The most important task is done in lines 12–31, where the s, p, o positions of quad patterns are filled. Lines 12–15 are the starting point: we "read" the $varSub$ of each $prov_{pos} \neq \bot$, in order to identify where each position in qp'_{ins} took its value from. Line 13 finds j (i.e., the proper quad pattern qp'^i_j in gp^i) and the position in said quad pattern (pos'), whereas line 14 fills this position with the variable found in $qp'_{ins}.pos$.

Lines 16–28 essentially "follow the chain of joins" that is recorded in the join subscripts, so as to assign common variable names where appropriate, reusing the variables in qp'_{ins}, or introducing new ones. Recall that each join contains two join subscripts; the number of quad pattern positions in each subscript of the pair depends on the number of positions in which the join is applied. In our algorithm, $AllJoinSubs$ is an ordered list of all such join subscripts (easily found by scanning $prov_{pos}$); by construction, $joinSub^1$ and $joinSub^2$ appear in the same join (same for $joinSub^3$ and $joinSub^4$ and so on). Each $joinSub^r$ is a sequence of quad patterns $\langle jp^r_1, \ldots, jp^r_k \rangle$.

If $AllJoinSubs$ is empty, then we have a "copy"; lines 19–27 will be skipped, and the only variable assignment necessary is the one already performed in line 14. In the more complex case where $AllJoinSubs$ contains some elements, these are processed in pairs, as indicated by the WHILE in line 19 and the increment in line 26. The idea is to put the same variable in positions that are joined, i.e., the same variable in jp^r_k, jp^{r+1}_k for all pairs $r, r+1$ (for r an odd number). If jp^r_k has already an assigned variable, this was created either by line 14, or by a previous execution of line 24, so this value is copied in jp_k^{r+1}; if not, a "fresh" variable is assigned to jp_k^r (line 22) and the process continues normally. The assumption that the quad pattern position appearing in the var subscript is the first one that matches is critical for this process, because it guarantees that we will not assign a fresh variable when the variables should be taken from qp'_{ins}.

Any unbound quad pattern positions (i.e., positions with no assigned variables) remaining after the execution of lines 16–28, are filled with "fresh" variables (line 30).

Algorithm 2. Update Reconstruction Algorithm

Input: A *cpe* expression of the form $pe^1 \oplus \ldots \oplus pe^k$, a quadruple q (s, p, o, n), a dataset D

Output: An INSERT update U', such that $U' \rightsquigarrow U$

1: $qp'_{ins} = (qp'_{ins}.s, qp'_{ins}.p, qp'_{ins}.o, n)$
2: **for all** $pos \in \{s, p, o\}$ **do**
3:     **if** $prov_{pos}$ of pe^1 is equal to $\perp$ **then**
4:         $qp'_{ins}.pos = q.pos$
5:     **else**
6:         $qp'_{ins}.pos = \text{NEWVAR}(\ )$
7:     **end if**
8: **end for**
9: **for all** pe^i in *cpe* **do**
10:     $l = \text{COMPUTEGPSIZE}(pe^i)$
11:     $\text{ASSIGNGRAPHS}(pe^i)$
12:     **for all** $prov_{pos}$ in pe^i, such that $prov_{pos} \neq \perp$ **do**
13:         $(j, pos') = \text{GETPOSFROMVARSUB}(prov_{pos})$
14:         $qp_j^i.pos' = qp'_{ins}.pos$
15:     **end for**
16:     **for all** $prov_{pos}$ in pe^i, such that $prov_{pos} \neq \perp$ **do**
17:         Set $AllJoinSubs = \langle joinSub^1, \ldots, joinSub^x \rangle$, where $joinSub^r = \langle jp_1^r, \ldots, jp_k^r \rangle$
18:         $r = 1$
19:         **while** $joinSub^r \neq \emptyset$ **do**
20:             **for all** $jp_k^r \in joinSub^r$ **do**
21:                 **if** $jp_k^r = $ null **then**
22:                     $jp_k^r = \text{NEWVAR}(\ )$
23:                 **end if**
24:                 $jp_k^{r+1} = jp_k^r$
25:             **end for**
26:             $r = r + 2$
27:         **end while**
28:     **end for**
29:     **for all** unbound $qp_j^i.pos$ **do**
30:         $qp_j^i.pos = \text{NEWVAR}(\ )$
31:     **end for**
32:     $gp^i = qp_1^i \cdot qp_2^i \cdot \ldots \cdot qp_l^i$
33:     $gp' = gp' \text{ UNION } gp^i$
34: **end for**
35: **return** $U' = \text{INSERT } \{qp'_{ins}\} \text{ WHERE } \{gp'\}$

Example 4. Now, we will explain how Algorithm 2 works for our motivating example. We first determine the graph attribute of qp'_{ins} (<YoungDoctor>), taken from c_5 (line 1). Then (lines 2–8), we note that the s, p values of c_5 resulted from a constant (see pe^1, pe^2), whereas the o value resulted from a "copy" or join; thus, $qp'_{ins} = $ (<hypertension>, <treatedWith>, $?v0$, <YoungDoctor>).

Subsequently, the FOR loop in line 9 is called for each pe^i. For pe^1, only qp_1^1 appears, whose named graph is the one of c_1, i.e., <Diabetologist>. The o position of $qp_1'^1$ is taken from $qp'_{ins}.o$, as indicated by the var subscript (so $qp_1'^1.o = ?v0$. There are no joins, so the block in lines 16–28 has no effect, and fresh variables are assigned in the other positions in line 30. Thus, $qp_1'^1 = (?v1, ?v2,$ $?v0$, <Diabetologist>).

Similarly, for pe^2, we have two quad pattern identifiers, qp_1^2, qp_2^2, whose named graph attributes are taken from c_2, c_3 respectively (lines 10–11). The value of $qp_1'^1.o$ is set to $?v0$ (equal to $qp_{ins}'.o$, as indicated by the var subscript of pe^2). In this case, we have a join, so in line 24 we will copy the value of $qp_1'^1.o$ (i.e., $?v0$) to $qp_2'^1.o$; this is due to the form of the join $(_{\{qp_1^1.o\}} \odot _{\{qp_2^1.o\}}$, which will set $AllJoinSubs = \langle\langle qp_1^1.o\rangle, \langle qp_2^1.o\rangle\rangle)$. There are no further joins to process, so we put fresh variables in the unbound positions of $qp_1'^1, qp_2'^1$, resulting into: $qp_2'^1 = (?v3, ?v4, ?v0, <\text{Pathologist1}>)$, $qp_2'^1 = (?v5, ?v6, ?v0, <\text{Pathologist2}>)$.
After the composition of the above quad patterns in lines 32, 33, 35 we get U':

INSERT $\{qp_{ins}'\}$ WHERE $\{qp_1'^1$ UNION $qp_1'^2. qp_2'^2\}$

where: qp_{ins}': $(<\text{hypertension}>, <\text{treatedWith}>, ?v0, <\text{YoungDoctor}>)$

   $qp_1'^1$: $(?v1, ?v2, ?v0, <\text{Diabetologist}>)$

   $qp_1'^2$: $(?v3, ?v4, ?v0, <\text{Pathologist1}>)$

   $qp_2'^2$: $(?v5, ?v6, ?v0, <\text{Pathologist2}>)$

Note that U' differs from U only in the (implicit) filters that U employs $(<\text{hypertension}>, <\text{treatedWith}>)$ in its quad patterns, as well as in the variable names.   □

The following theorem proves the correctness of our algorithms:

Theorem 2. *Let U be an INSERT update evaluated on a dataset D, q a result quadruple and cpe a complex provenance expression in the provenance of q as computed by Algorithm 1. Assume that we run Algorithm 2 with input (cpe, q, D) and we get as output the INSERT update U'. Then, U' returns q among other quadruples and $U \rightsquigarrow U'$.*

Theorem 2 proves that the output of Algorithm 2 is compatible with the original INSERT update that created the input quadruple; thus, the intended semantics of a provenance expression, as given in Sect. 4, are correctly recorded by Algorithm 1, and interpreted by Algorithm 2 to reconstruct the original INSERT update.

5.3   Complexities

The time complexity of Algorithm 1 is *linear* with respect to the update size (number of quad patterns in the WHERE clause). To see this, note that lines 2–17 will be executed three times, each run costing $O(m_i)$, where m_i is the number of joined quad patterns in gp^i; thus, the total cost is $O(3 \cdot \sum_i m_i) = O(m)$, where m is the update size. Algorithm 1 is also of *logarithmic* complexity with respect to the dataset size (number of quadruples), say R. Specifically, the dataset is accessed in two occasions: to find the quadruple identifiers (lines 5, 9), and to get the attribute values of q_k (line 21). Each access costs $O(\log R)$ time (assuming appropriate indexes), and happens a constant number of times (assuming a constant update size), so the total cost is $O(\log R)$. The above complexities are related to the cost of annotating the result of the INSERT update with its provenance, and do not include the cost of computing the result itself;

an obvious conclusion is that the overhead imposed by the provenance algorithm is negligible.

The time complexity of Algorithm 2 is *linear* with respect to the *size of the cpe*. Lines 9–34 run once for each pe^i, each run costing $O(m_i)$ time (where m_i is the number of quad patterns in pe^i), because each part of provenance is accessed a constant number of times. Hence, the complexity is $O(\sum_i m_i) = O(m)$, where m is the total number of quad patterns in the WHERE clause (i.e., update size). As with Algorithm 1, Algorithm 2 only accesses the dataset in specific points (lines 4, 11), each being run a constant number of times (for a fixed update size) and costing $O(\log R)$ (assuming adequate indexes) over a dataset of size R. Thus, the complexity of Algorithm 2 is *logarithmic* with respect to the number of quadruples in D.

Regarding space complexity, we note that the size of provenance is analogous to the size of the input U (and vice-versa), and that all temporarily stored information is no larger in size than the size of the update/provenance (respectively) in either algorithm. Thus, the space complexity of both algorithms is *linear* with respect to U/cpe.

6   Implementation and Evaluation

6.1   Implementation and Storage (Relational Schema)

Existing SPARQL engines do not support the kind of complex provenance information proposed by our model. Thus, we used Virtuoso Database Engine as our triple store, where quadruples and provenance expressions are stored in relational tables. On top of Virtuoso we built a main memory Java implementation of our algorithms. The quadruples and the related provenance expressions are stored in a relational schema, which uses two tables: *Quads(qid,s,p,o,n)* and *Prov(qid, cpeNo, peNo, prov_s, prov_p, prov_o)*. Table *Quads* stores the quadruple's (ID, subject, property, object, named graph). Table *Prov* stores the provenance information of a quadruple: *qid* is the quadruple ID, *cpeNo* and *peNo* are the IDs of *cpe* and *pe* expressions, while $prov_s$, $prov_p$ and $prov_o$ contain the provenance of the corresponding attribute related to the specific *cpe* and *pe*.

6.2   Experiments

In our experiments we used real data that were taken from the Billion Triple Challenge (BTC) dataset (small crawl)[6]. The BTC dataset contains 10 million quadruples, but we used smaller excerpts containing 100, 250 and 500 thousand unique quadruples. Due to the absence of a standard benchmark for provenance, we used our own custom synthetic set of INSERT updates[7]. All experiments were conducted on a Dell OptiPlex 755 desktop with CPU Intel® Core™ 2 Duo CPU Q6600 at 2.40 GHz, 6 GB of memory, running Windows 7 Professional x86_64.

---

[6] https://km.aifb.kit.edu/projects/btc-2009/.
[7] http://www.ics.forth.gr/isl/provenance/updates.pdf.

We conducted three experiments. EXPERIMENT 1 measures the time required to compute the results of an INSERT update along with their provenance information, whereas EXPERIMENT 2 considers the time required to compute only the result quadruples. The difference in time of EXPERIMENT 1 and EXPERIMENT 2 indicates the overhead for computing the provenance. EXPERIMENT 3 computes the time needed for reconstructing a compatible INSERT update based on a quadruple's provenance.

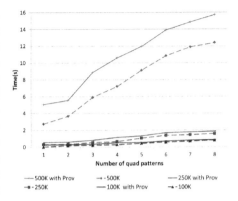

Fig. 1. EXPERIMENT 1, 2

Our experiments confirm the theoretical complexity results above. In particular, the evaluation time depends on the number of quad patterns, and is also affected by the number of quadruples in the dataset, the number of quad patterns in the WHERE clause (i.e., update size), and, of course, the applied SPARQL operators (join, union).

Figure 1 shows the computation time for executing the INSERT update with and without the provenance computation. The graph shows that the provenance computation time increases linearly with the number of quad patterns, and that it is, in all cases, only a fraction of the time required for evaluating the INSERT update. Moreover, note that the dataset's size has a great impact on the evaluation time of both experiments.

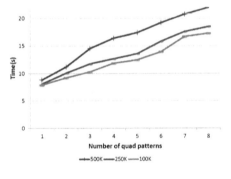

Fig. 2. EXPERIMENT 3

Figure 2 shows how the performance of Algorithm 2 scales as the complexity of the INSERT update increases, for the considered datasets. We note that the evaluation time increases linearly with respect to the number of quad patterns in the WHERE clause, and that performance is not seriously affected by the dataset size.

7   Related Work

Data provenance has been widely studied in several different contexts such as databases, distributed systems, Semantic Web etc. In [16], Moreau explores the different aspects of provenance in the Web. Likewise, Cheney et al. [8] provide an extended survey that considers the provenance of query results in relational databases regarding the most popular provenance models.

Research on data provenance can be categorized depending on whether it deals with *updates* [3,4,10,13,20] or *queries* [4,8–12,15,20,21]; compared to querying, the problem of provenance management for updates is less well-understood.

Another important classification is based on the underlying data model, SQL [5,12,20] or RDF [9–11,13,15,21], which determines whether the model deals with the relational or SPARQL algebra operators respectively. Despite its importance, only a few works deal with the problem of update provenance, and even fewer consider the problem in the context of SPARQL updates [13].

A third categorization stems from the expressive power of the employed provenance model, e.g., *how*, *where* and *why* among others. Since our proposed model is based on how and where provenance models, we discuss them thoroughly here. *Where provenance* is a popular provenance model [3,5,10,20] that describes where a piece of data is copied from, i.e., which quadruples contributed to produce a result quadruple in our context. *How provenance* describes not only the quadruples used for producing an output, but also how these source quadruples were combined (through operators) to derive it. In [12], *provenance semirings* are used to record *how provenance* for the relational setting through polynomials; whereas [11,15] showed how to apply provenance semirings for the RDF/SPARQL setting.

An important work on update provenance for the relational setting is [4], which focuses on the *copy* and *modify* operations. The proposed formalization is based on "tagging" tuples using "colors" propagated along with their data item during the computation of the output. The provenance of the output is the provenance propagated from the input item(s). Our model follows this approach to capture the provenance of a quadruple attribute, but uses identifiers instead of colors, as well as a more expressive provenance model.

In the context of SPARQL update provenance, there are no works that consider abstract provenance models. Instead, RDF named graphs are used to represent both past versions and changes to a graph [13]. This is achieved by modelling the provenance of an RDF graph as a set of history records, including a special provenance graph and additional auxiliary versioning named graphs.

Moreover, our work builds on [15]. This work presents how popular relational data provenance models such as (*how*, *why*) can be adapted to capture the provenance of the results of positive SPARQL queries (i.e., without SPARQL OPTIONAL clauses). More specifically, the authors investigate how provenance models for the positive fragment of the relational algebra (like [12]) can be adapted for unions of conjunctive SPARQL queries. The present paper extends this model in order to address the extra challenges associated with provenance management of SPARQL updates (as opposed to queries).

Another major line of work deals with the different ways in which provenance can be serialized and modelled in an ontology in the form of Linked Data [14, 17,18]. In [14], Hartig proposes a provenance model that captures information about Web-based data access as well as information about the creation of data. Moreau et al. created the Open Provenance Model [17] that supports the digital representation of provenance for any "thing", no matter how it was produced. In this context, PROV was released as a W3C reccomendation [18]. The goal of

PROV is to enable the wide publication and interchange of provenance on the Web and other information systems. PROV can exhibit provenance information using widely available formats such as RDF and XML.

8   Conclusions

As the volume of data made available in the Web is continuously increasing, the need for capturing and managing the provenance of such data becomes all the more important. Our work addresses this problem for RDF data, by proposing a novel, fine-grained and expressive provenance model to record the triple and attribute-level provenance of RDF quadruples generated through SPARQL INSERT updates.

Our work follows the approach of [10,15], where the use of abstract identifiers and operators is proposed; we build upon the novel notion of *quad pattern positions* in order to provide a richer set of operators, that allow the identification of the attributes of quad patterns that were involved in a join or "copy" operation. Our model is richer than standard query provenance models since it captures fine-grained provenance both at triple and attribute level.

Our model supports the feature of update *reconstructability*. Reconstructability prescribes that the information stored in the provenance of a quadruple allows the identification of an INSERT update that is almost identical (in the sense of *compatibility*) to the original one that was used to create said quadruple. This is a stronger form of *how provenance*. On the algorithmic side, we introduce two algorithms that allow recording the provenance information, as well as interpreting it to identify how the quadruple found its way in the dataset, through the identification of a compatible INSERT update as described above. The overhead imposed by these algorithms in the execution of an INSERT update is negligible. We implemented the *provenance construction* and the *update reconstruction* algorithms on top of Virtuoso Database Engine and conducted a preliminary set of experiments that verified the complexity of the proposed algorithms.

In the future, we plan to consider FILTER and non-monotonic SPARQL operators (OPTIONAL) as well as SPARQL functions. In addition, we will study the SPARQL DELETE, CREATE and DROP operations since all SPARQL update operations can be written as a combination of INSERT, DELETE, CREATE and DROP statements. Furthermore, we plan to take under consideration benchmarks supporting update operations and will try to extend them in order to compute the provenance information using our model.

We also intend to explore the use of PROV approach for representing our model in the form of Linked Data. As a long term plan we aim at working towards a provenance aware triple store in the spirit of TripleProv [21].

References

1. Arenas, M., Gutierrez, C., Perez, J.: On the Semantics of SPARQL. In: De Virgilio, R., Giunchiglia, F., Tanca, L. (eds.) Semantic Web Information Management: A Model-Based Perspective, pp. 281–307. Springer, Heidelberg (2009)
2. Avgoustaki, A.: Provenance management for SPARQL updates. Master's thesis, University of Crete (2014). http://www.icu.forth.gr/isl/provenance/provenance.pdf
3. Buneman, P., Chapman, A., Cheney, J.: Provenance management in curated databases. In: Chaudhuri, S., Hristidis, V., Polyzotis, N. (eds.) ACM SIGMOD International Conference on Management of Data, pp. 539–550 (2006)
4. Buneman, P., Cheney, J., Vansummeren, S.: On the expressiveness of implicit provenance in query and update languages. In: Schwentick, T., Suciu, D. (eds.) ICDT 2007. LNCS, vol. 4353, pp. 209–223. Springer, Heidelberg (2006)
5. Buneman, P., Khanna, S., Tan, W.-C.: Why and where: a characterization of data provenance. In: Bussche, J., Vianu, V. (eds.) ICDT 2001. LNCS, vol. 1973, pp. 316–330. Springer, Heidelberg (2000)
6. Carroll, J.J., Bizer, C., Hayes, P., Stickler, P.: Named graphs. J. Web Semant. 3(4), 247–267 (2005)
7. Carroll, J.J., Bizer, C., Hayes, P.J., Stickler, P.: Named graphs, provenance and trust. In: International Conference on World Wide Web, pp. 613–622 (2005)
8. Cheney, J., Chiticariu, L., Tan, W.-C.: Provenance in databases: why, how, and where. Found. Trends Databases 1(4), 379–474 (2009)
9. Damásio, C.V., Analyti, A., Antoniou, G.: Provenance for SPARQL queries. In: Cudré-Mauroux, P., Heflin, J., Sirin, E., Tudorache, T., Euzenat, J., Hauswirth, M., Parreira, J.X., Hendler, J., Schreiber, G., Bernstein, A., Blomqvist, E. (eds.) ISWC 2012, Part I. LNCS, vol. 7649, pp. 625–640. Springer, Heidelberg (2012)
10. Flouris, G., Fundulaki, I., Pediaditis, P., Theoharis, Y., Christophides, V.: Coloring RDF triples to capture provenance. In: Bernstein, A., Karger, D.R., Heath, T., Feigenbaum, L., Maynard, D., Motta, E., Thirunarayan, K. (eds.) ISWC 2009. LNCS, vol. 5823, pp. 196–212. Springer, Heidelberg (2009)
11. Geerts, F., Karvounarakis, G., Christophides, V., Fundulaki, I.: Algebraic structures for capturing the provenance of SPARQL queries. In: International Conference on Database Theory, pp. 153–164 (2013)
12. Green, T.J., Karvounarakis, G., Tannen, V.: Provenance semirings. In: Principles Of Database Systems, pp. 31–40 (2007)
13. Halpin, H., Cheney, J.: Dynamic provenance for SPARQL updates. In: Mika, P., Tudorache, T., Bernstein, A., Welty, C., Knoblock, C., Vrandečić, D., Groth, P., Noy, N., Janowicz, K., Goble, C. (eds.) ISWC 2014, Part I. LNCS, vol. 8796, pp. 425–440. Springer, Heidelberg (2014)
14. Hartig, O.: Provenance information in the web of data. In: Proceedings of the 2nd Linked Data on the Web Workshop at the World Wide Web Conference (2009)
15. Karvounarakis, G., Fundulaki, I., Christophides, V.: Provenance for linked data. In: Tannen, V., Wong, L., Libkin, L., Fan, W., Tan, W.-C., Fourman, M. (eds.) Buneman Festschrift 2013. LNCS, vol. 8000, pp. 366–381. Springer, Heidelberg (2013)
16. Moreau, L.: The foundations for provenance on the web. Found. Trends Web Sci. 2(2–3), 99–241 (2010)

17. Moreau, L., Clifford, B., Freire, J., Futrelle, J., Gil, Y., Groth, P.T., Kwasnikowska, N., Miles, S., Missier, P., Myers, J., Plale, B., Simmhan, Y., Stephan, E.G., den Bussche, J.V.: The open provenance model core specification (v1.1). Future Gener. Comput. Syst. **27**(6), 743–756 (2011)
18. Moreau, L., Missier, P.: PROV-DM: the PROV data model. W3C Recommendation (2013)
19. Pérez, J., Arenas, M., Gutierrez, C.: Semantics and complexity of SPARQL. In: Cruz, I., Decker, S., Allemang, D., Preist, C., Schwabe, D., Mika, P., Uschold, M., Aroyo, L.M. (eds.) ISWC 2006. LNCS, vol. 4273, pp. 30–43. Springer, Heidelberg (2006)
20. Vansummeren, S., Cheney, J.: Recording provenance for SQL queries and updates. IEEE Data Eng. Bull. **30**(4), 29–37 (2007)
21. Wylot, M., Cudré-Mauroux, P., Groth, P.T.: Tripleprov: efficient processing of lineage queries in a native RDFstore. In: International Conference on World Wide Web, pp. 455–466 (2014)

Private Record Linkage: Comparison of Selected Techniques for Name Matching

Pawel Grzebala[✉] and Michelle Cheatham

DaSe Lab, Wright State University, Dayton, OH 45435, USA
{grzebala.2,michelle.cheatham}@wright.edu

Abstract. The rise of Big Data Analytics has shown the utility of analyzing all aspects of a problem by bringing together disparate data sets. Efficient and accurate private record linkage algorithms are necessary to achieve this. However, records are often linked based on personally identifiable information, and protecting the privacy of individuals is critical. This paper contributes to this field by studying an important component of the private record linkage problem: linking based on names while keeping those names encrypted, both on disk and in memory. We explore the applicability, accuracy and speed of three different primary approaches to this problem (along with several variations) and compare the results to common name-matching metrics on unprotected data. While these approaches are not new, this paper provides a thorough analysis on a range of datasets containing systematically introduced flaws common to name-based data entry, such as typographical errors, optical character recognition errors, and phonetic errors.

1 Introduction and Motivation

Data silos, in which organizations keep their data tightly isolated from other systems, are a major barrier to the effective use of data analytics in many fields. Unfortunately, when the data in question involves information about people, integrating it often necessitates querying or joining based on personally identifiable information (PII) that could be used to explicitly identify an individual. As recent security breaches at organizations ranging from Target to the United States Postal Service have made clear, it is important to protect PII, both while it is at rest on a system and when it is read into memory. The goal of this effort is to explore the applicability, accuracy, and speed of existing algorithms for querying and joining databases while keeping the PII within those databases protected.

This work focuses particularly on the situation in which a data provider maintains a database which authorized subscribers are able to query. For instance, consider a company that maintains a database containing its customer data. The company wishes to allow third party entities who have contracted with it to access the information in this database.[1] At the same time, the company wants

---

[1] A standard access control system to allow authorized consumers to query the database while preventing unauthorized users from doing so is assumed to be in place.

© Springer International Publishing Switzerland 2016
H. Sack et al. (Eds.): ESWC 2016, LNCS 9678, pp. 593–606, 2016.
DOI: 10.1007/978-3-319-34129-3_36

to limit its vulnerability to data breaches by keeping the data encrypted as much as possible, including while it is stored in the database and when it is loaded into memory to do query processing. For instance, if an attacker gets access to the system on which the database resides, he should not be able to see the raw data values, either on disk or in memory.

Even though this situation occurs frequently, research on private record linkage tends to focus more on a different use case, in which the data provider and the data consumer do not fully trust one another. This typically leads to solutions involving trusted third parties or asymmetric cryptography that go beyond the requirements of this ubiquitous application scenario, and these additional, unneeded capabilities negatively impact performance. For instance, because access control mechanisms are already in place, this work is not concerned about the data consumer (who has paid to access the information in the database) gaining knowledge of any, or even all, of the records in the database. Furthermore, this project is not concerned about the data consumer preventing the data provider from gaining knowledge about what queries are being made. Rather, the present use case allows a system in which the data consumer submits a query containing the raw PII values, these values are encrypted using symmetric key cryptography[2], and the encrypted values are then used to query the database.

This work focuses on supporting privacy-preserving querying and merging on string attributes and does not consider numeric data. While private record linkage based on numeric fields of is course an important capability to establish, the techniques involved for this are distinctly different than for string-based linking. Furthermore, string attributes, in particular person names, are a particularly common linkage point between datasets. We therefore leave the challenge of numeric attributes for future work and focus on name-based linking here. The requirements of our target application scenario require PRL methods that support encryption and do not need to act directly on the raw field values, so approaches that utilize the original string values at any stage in the process are not suitable in this case. Because names are frequently misspelled, mispronounced, or mistyped, it is important for the approach to support fuzzy (approximate) matching as well as exact matching. This fuzzy matching should be particularly tailored to support the types of lexical variations specific to names. No data should be decrypted, even in memory, until a match is ensured. In this paper we analyze the accuracy and efficiency of several metrics that meet these requirements and compare those results to that of standard name-matching methods employed on unencrypted data. The paper focuses entirely on technical considerations of the targeted use case. Laws and regulations also have a bearing on this application, but that aspect is not addressed here due to wide variance between legal jurisdictions and the authors' lack of legal expertise.

Note that nothing in this application scenario places any restrictions upon the infrastructure in which the data records are stored. In particular, the results presented here can be applied directly, with no modification, to data stored as

---

[2] Note that this exposes the raw PII values in memory, though only those in the query, not those in every database record.

RDF triples in accordance with the linked data principles. This work therefore joins a growing body of literature regarding how linked data can be secured while retaining its utility for those authorized to access it [5, 7].

The main contributions of this paper are:

- The usage of an advanced name matching benchmark generation tool to analyze the performance of several different name-based similarity metrics in a nuanced way. In our analysis we consider numerous realistic sources of errors and study the effect of the threshold value applied to each of the metrics.
- The accuracy of the privacy-preserving similarity metrics is compared to that of standard string metrics on unprotected data in order to establish the accuracy lost in support of data privacy.
- The computational efficiency of the privacy-preserving similarity metrics is also compared to that of standard string metrics.

The rest of this paper is organized as follows: In Sect. 2 we provide an overview of some related work and briefly discuss the challenges that make record linkage on names difficult. Section 3 introduces the metrics and algorithms used to perform record linkage in this study. This includes the string similarity metrics for unencrypted data which are used as a baseline for comparison purposes and the metrics relevant to private record linkage. Section 4 analyzes and evaluates the performance of the algorithms mentioned in Sect. 3 in terms of accuracy and computational efficiency. Finally, Sect. 5 concludes the paper by summarizing the results and provides an outlook to future work.

2 Background

There have been numerous approaches to solving the problem of record linkage based on person names. A comprehensive overview of several name matching techniques was provided by Snae in [9]. Snae describes four different types of name matching algorithms and compares them in terms of accuracy and execution time: spelling analysis based algorithms (Guth and Levenshtein), phonetic based algorithms (Soundex, Metaphonez, and Phonex), composite algorithms (combination of sound and spelling based methods, e.g. ISG), and hybrid algorithms (combination of phonetic and spelling based approaches, e.g. LIG). The hybrid algorithms were recommended for many name based record linkage applications because of their flexibility that allows them to be easily tuned for specific use cases. However, the results indicated that there is no single best method for name matching. In the conclusion, the author suggests that the choice of the name matching algorithm should depend on the specific application needs. Moreover, this work doesn't take into consideration the important aspect of our study, which is linking records while keeping them encrypted.

As mentioned previously, many existing techniques for private record linkage assume that the two parties involved do not want to reveal their data to the other party. One way this is commonly achieved is by developing algorithms that avoid directly comparing the records to be linked. For example, in the two-party protocol presented by Vatsalan and his colleagues in [11], two database owners compute similarity values between the records in their dataset and public reference values. Then, the similarity values are binned into intervals and the bins are exchanged between the two database owners. Based on the exchanged bins the protocol uses the reverse triangular inequality of a distance metric to compute the similarity values of two records without revealing the records themselves. Another, somewhat similar, two-party protocol was proposed by Yakout et al. in [12]. In this approach, each database owner converts all of their records into vector representations that are later mapped to points in a complex plane. The planes are then exchanged between the owners in order to identify pairs of points that are in proximity of each other. To calculate similarity values between the candidate vectors, the Euclidean distance of two records is computed using a secure distance computation. These two approaches are typical of many existing PRL techniques and, like the majority of those techniques, they implicitly assume that the records to be linked are not encrypted. We now turn our attention to examples of the few approaches that do not make this assumption.

One important thing to note is that not all string similarity metrics can be applied to the problem of name-based private record linkage. In order for a metric to be usable in this scenario, the metric must not require access to individual characters within this string. This is because any such metric would have to "encrypt" a string character-by-character, which is essentially a classical substitution cipher that is not at all secure. This eliminates common metrics such as Levenshtein and Monge Elkan from consideration.

Among the techniques that support approximate matching for linking records are the Soundex and q-gram string similarity metrics. The Soundex metric was originally designed as a phonetic encoding algorithm for indexing names by sound. [1] Soundex encodes a string representing a name into a code that consists of the first letter of the name followed by three digits, by applying a set of transformation rules to the original name. When two Soundex encodings are compared, the comparison is an exact match rather than approximate comparison but common name mispronunciations will not cause the algorithm to miss a match[3]. To use Soundex for private record linkage, both the name and the phonetic encoding are stored in the database in encrypted form for each record, but the encrypted phonetic encoding is the one used to respond to queries. The comparison is still an exact rather than fuzzy comparison, but because it is now being done on a phonetic encoding, common misspellings or other slight

---

[3] In 1990 Lawrence Philips created a phoenetic algorithm called Metaphone that improves upon Soundex by considering numerous situations in which the pronunciation of English words differs from what would be anticipated based on their spelling [8]. Metaphone was not considered for this effort because the extensions that it makes beyond Soundex are primarily intended to improve the performance on regular words rather than on names; however, the metric does fit the requirements for use in this application, and will be considered during our future work on this topic.

differences will not cause the algorithm to miss matching records. This was the approach suggested in [3].

Another of the string similarity metrics that *can* be used is *q*-grams. A *q*-gram is created by splitting a string into a set of substrings of length *q*. An example of a *q*-gram, given $q = 2$ and the input string *Alice*, is { "*Al*", "*li*", "*ic*", "*ce*"}. As with the Soundex approach, in order to use *q*-grams for name based private record linkage additional information must be stored with each record. In the case of *q-grams*, the person's name is divided into *q*-grams, each of the substrings in the set of *q*-grams is encrypted, and those encrypted substrings are also stored as part of the record. The amount of similarity between two records is then computed as the degree of overlap between these set of encrypted *q*-grams for each record. Each individual substring is compared based on exact match. The degree of overlap is computed using a traditional set similarity metric such as Jaccard or Dice, which are calculated as follows:

$$\text{Jaccard} = \frac{\text{grams}_{\text{common}}}{\text{grams}_1 + \text{grams}_2 - \text{grams}_{\text{common}}}$$

$$\text{Dice} = \frac{2 \times \text{grams}_{\text{common}}}{\text{grams}_1 + \text{grams}_2},$$

where $\text{grams}_{\text{common}}$ corresponds to the number of *q*-grams that are common to both strings, grams_1 to the number of *q*-grams in the first string, and grams_2 to the number of *q*-grams in the second string. The intuition behind using *q*-grams to compare two names is that a typo, misspelling, or other variation will only impact a limited number of substrings and therefore similar strings will still have a high degree of overlap and thus a high similarity value. The downside is that the order of the substrings is not considered, so it is possible for two very different strings, such as "stop" and "post" to have very high similarity according to this metric.

A more in-depth review of techniques proposed to achieve private record linkage can be found in [2].

In this work, we have evaluated the performance of the Soundex and *q*-gram algorithms for name-based private record linkage in the scenario described in the introduction. Because it is unrealistic to expect a privacy-preserving record linkage algorithm to perform better than a linkage method that does not provide any protection for the data, we have compared the performance of Soundex and *q*-gram to the performance of some traditional string similarity metrics on unencrypted data. Specifically, we have used Levenshtein and Jaro-Winkler, two of the most commonly used string similarity metrics, as a baseline. Levenshtein is an edit distance metric. It simply counts the number of edits (insertions, deletions, or substitutions) that must be applied to one string in order to transform it into another one. For example, the Levenshtein distance between "Michelle" and "Micheal" is 2. Jaro-Winkler is based on the Jaro metric, which counts the number of "common" characters of two strings. Characters are considered common when the difference between their indexes is no greater than half of the length of the longer string. Jaro also takes into consideration the number of character transpositions. The Jaro-Winkler version of the algorithm increases the similarity value returned by Jaro if the two strings begin with the same sequence of characters and differences appear only in the middle or at the end of string [2].

Name matching has been researched for many years and numerous studies have proven that it is not an easy task. This is because a name's spelling can be malformed in a wide variety of ways, including punctuation, abbreviation, pronunciation, spelling, the order of writing, use of prefixes, typos, or optical recognition errors to name a few. In addition, privacy concerns have made it very difficult to find publicly available data that can be used for benchmark purposes, particularly a collection of names that accurately reflects worldwide name distribution rather than being US-centric. This lack of suitable benchmarks was a considerable challenge during this study, leading to the use of a newly-available name matching benchmark generation system, described in Sect. 4.

3   Approach

We analyzed the performance of several string similarity metrics for linking encrypted records. The metrics considered were Soundex and several variations of the q-gram technique. This performance was compared against those of Jaro and a normalized version of Levenshtein on the unencrypted data. The data and Java source code are available from https://github.com/prl-dase-wsu/prl-technique-comparison.

We used two metrics based on q-grams. The first is q-grams with $q = 2$ (also called bigrams). Because studies have shown [6] that padding the input string with a special character (one that never appears as part of any string in the dataset) at the beginning and the end of string can increase the accuracy when comparing two different q-grams we also tried padded q-grams with $q = 2$. Both q-grams and padded q-grams were compared using two different similarity coefficient methods, Jaccard and Dice.

The string metrics that were used on unencrypted data, Jaro and Levestein, were introduced in Sect. 2. To formally define both algorithms, the similarity value of two strings returned by the Jaro algorithm is calculated as follows:

$$Jaro = \frac{\frac{c}{s_1} + \frac{c}{s_2} + \frac{c - t}{c}}{3},$$

where c is the number common characters in both strings, s_1 is the length of the first string, s_2 is the length of the second string, and t is the number of transpositions (the number of common characters that are not in sequence order, divided by 2). Since all of the metrics used in this study return a value between 0.0 (when strings are completely different) and 1.0 (when string are the same) we modified the original Levenshtein algorithm so that it returns a similarity value that falls in the same range. The Normalized Levenshtein formula is defined as follows:

$$NormalizedLevenshtein = 1 - \frac{Levenshtein}{max(s_1, s_2)},$$

where *Levenshtein* is the number of replacements needed to transform the first string into the second string, s_1 is the length of the first string, and s_2 is the length of the second string.

The benchmark datasets used in this study were created by using the advanced personal data generation tool called "GeCo" and developed by K-N. Tran et al. [10] The tool was created to address the issue of lack of publicly available data that contains PII information. GeCo has two main functionalities: data generation and data corruption. The data generation module provides the user with an interface capable of producing records with five different attribute generation mechanisms. The first two can be used to generate individual attributes such as credit card number, social security number, name, age, etc. The attribute values are created by either user-defined functions or based on frequency look-up files that specify the set of all possible values of an attribute and their relative frequencies. The other three types of attribute generation mechanisms allow the user to produce compound attributes where the attributes' values depend on each other. For example, a compound attribute with fields such as: city, gender, and blood pressure can be created, where the value of the blood pressure depends on the previously generated city and gender values. The second module of GeCo provides users with a sophisticated interface allowing them to corrupt the generated data using six different corruption techniques that simulate real-world errors that can occur during data processing. Those techniques include introducing: (1) missing values (one of the record's fields gets lost), (2) character edits (a random character of a string attribute is inserted, deleted, substituted, or transposed), (3) keyboard edits (simulates a human mistake during typing), (4) optical character recognition (OCR) errors (simulates OCR software mistakes), (5) phonetic edits (replaces substrings with their corresponding phonetic variations), and (6) categorical value swapping (replaces an attribute value with one of its possible variations). The user can also specify numerous other parameters such as: the number of records to corrupt, the number of corruptions applied to a record or single attribute, or the probability of corruption of a particular attribute.

For benchmark purposes we generated a dataset of 10,000 records where each of the records had the following attributes: first name, last name and credit card number. Then, we used the GeCo tool to introduce various types of realistic corruption to the generated dataset. The corrupted datasets produced by the GeCo tool were categorized using three parameters: type of applied corruption technique (Character Edit, Keyboard Edit, OCR Edit, Phonetic Edit, or mix of all), the percentage of original record corruption (high - 10 %, medium - 5 %. or low - 2 %), and the number of corruptions applied to either the first name, last name, or both (1 or 2). This resulted in 30 variations of the dataset. Once the datasets were corrupted we added additional attributes to each of the records from all datasets to be able to perform record linkage using encrypted q-grams and Soundex encodings. Each q-gram array and Soundex encoding were encrypted using 256-bit AES password-based encryption.

To evaluate the performance of the string metrics, the uncorrupted dataset was cross joined with each of the corrupted datasets using each of the string metrics discussed in the previous section. During the join operation only the pairs of records with the highest similarity score that exceeded a threshold value were joined. If the an individual pair of joined records corresponded to the same individual we counted it as a "Correct" join, otherwise the join was counted as "False Positive". If none of the scores returned by a string metric exceeded a threshold value we incremented the "False Negative" count by 1 to indicate that a corresponding record was not found in the other dataset. In a special case, when more than one pair of records had the same highest score, the pair of records that corresponded to the same individual was marked as "Correct" and the rest of the pairs were counted as "False Positive".

4   Evaluation and Analysis

Performing record linkage using each of the seven string metrics (q-grams compared using Jaccard coefficient, q-grams compared using Dice coefficient, padded q-grams compared using Jaccard coefficient, padded q-grams compared using Dice coefficient, Soundex, Levenshtein, and Jaro) between the uncorrupted dataset and the 30 corrupted datasets resulted in a massive amount of statistical data. Instead of presenting the outcome of every single cross join operation, this section summarizes our key findings, with an emphasis on practical advice related to selecting a metric, setting the threshold, and conveying the type of performance that can be expected by someone attempting to do name-based private record linkage.

4.1   Observation 1: Soundex is Not Viable, but (Padded) q-grams are

Figure 1 Shows the results of all of the string metrics on a version of the data in which 10 % of the names have had one (the solid lines) or two (the dotten lines) characters edited. In the single edit case, all versions of the q-gram metric are able to achieve the same, nearly perfect, accuracy on the encrypted data that Levenshtein and Jaro achieve on the encrypted data. The performance of all metrics is lower for the two character edit case, with a top accuracy of 90 % rather than the completely accurate results possible in the single edit situation. However, we again see that the performance of at least the padded versions of the q-gram approach on the ciphertext can match that of Levenshtein and Jaro on the plaintext.

The results for the Soundex metric are not included in Fig. 1 because the results showed that comparing names based on encrypted Soundex encodings is not viable in most of the cases as a record linkage technique. The only exception was noted when the datasets containing records with the phonetic type of record corruption were joined. Still, in the best case scenario only 60.71 % of corrupted data was successfully matched using this technique. Table 1 presents the accuracy of record linkage on all types of corrupted datasets using the Soundex technique.

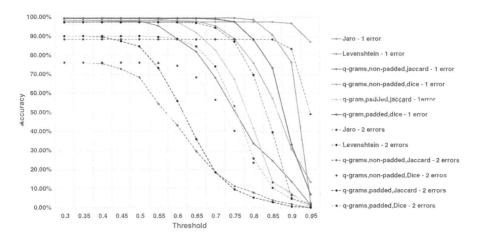

Fig. 1. Illustration of the decrease in accuracy of record linkage of selected string metrics. Solid lines correspond to accuracy when linkage was performed on datasets corrupted with 1 Character Edit, dotted lines with 2 Character Edits.

Table 1. Performance of record linkage based on encrypted Soundex encodings. The percentage values reflect the number of corrupted records that were successfully matched with their uncorrupted versions.

Corruption type	Number of corruptions per record	
	1	2
Character Edit	47.24 %	24.06 %
Keyboard Edit	48.06 %	21.94 %
OCR Edit	38.29 %	13.71 %
Phonetic Edit	60.71 %	43.88 %
Mix	50.82 %	25.47 %

4.2 Observation 2: Dice is Preferable to Jaccard for Calculating q-gram Similarity

Out of the four similarity metrics based on q-grams, the ones using the Dice coefficient to measure the similarity between the sets of encrypted q-grams were more accurate. This was the case with q-grams as well as padded q-grams. This can be explained by the fact that Dice favors the occurrences of common q-grams more than Jaccard. As a result, a pair of similar records is likely to have a higher similarity score when calculated using Dice coefficient. To illustrate this, in Table 2 we provide a sample results from record linkage performed against a dataset with the phonetic type of corruption, where 10 % of original records had two phonetic errors introduced. Similar results were recorded for datasets with other types of corruptions.

Table 2. Sample results of record linkage performed against phonetically corrupted dataset showing the performance of q-grams based string similarity metrics

Threshold	Unpadded q-grams						Padded q-grams					
	Jaccard			Dice			Jaccard			Dice		
	Correct	FP	FN	Correct	FP	FN	Correct	FP	FN	Correct	FP	FN
0.30	9837	60	163	9837	60	163	9949	45	51	9949	45	51
0.35	9837	60	163	9837	60	163	9949	45	51	9949	45	51
0.40	9834	57	166	9837	60	163	9948	45	52	9949	45	51
0.45	9814	56	186	9837	60	163	9937	43	63	9949	45	51
0.50	9778	48	222	9837	60	163	9910	38	90	9949	45	51
0.55	9614	29	386	9834	57	166	9790	27	210	9949	45	51
0.60	9506	25	494	9826	57	174	9589	18	411	9946	45	54
0.65	9329	21	671	9778	48	222	9337	18	661	9910	38	90
0.70	9198	18	802	9637	32	363	9170	18	830	9792	27	208
0.75	9101	18	899	9467	22	532	9081	17	919	9538	18	462
0.80	9046	18	954	9264	19	736	9032	17	968	9233	18	767
0.85	9023	18	977	9125	18	875	9011	17	989	9097	17	903
0.90	9009	18	991	9036	18	964	9002	17	998	9021	17	979
0.95	9002	18	998	9009	18	991	9000	17	1000	9002	17	998

4.3   Observation 3: Lower Thresholds are Better for q-grams

Figure 1 illustrates that the threshold value for the Levenshtein and Jaro metrics can be set relatively high without sacrificing accuracy when linking the unencrypted data, which was not the case when the q-gram techniques were used to link the encrypted data. For instance, to achieve an accuracy of 99.5 % when performing linkage against datasets where records contain one corruption of any type, the threshold value applied to the Jaro or Levenshtein metric was set to 0.8 whereas the threshold value applied to q-grams based metrics needs to be set to a value between 0.55 and 0.75 to achieve the same result, depending on the type of corruption applied to the datasets.

Table 2 makes the point that the padded versions of the q-gram metric in particular have better performance when the threshold value is kept low, which as explained in the previous paragraph is the optimal approach. For threshold values up to 0.7 for the Jaccard coefficient and 0.8 for the Dice coefficient, padding the q-grams produces better results. For higher threshold values, the unpadded version is slightly better. The reason behind this is that similarity scores calculated using padded q-grams are higher when the differences between the strings used to generate the q-grams appear in the middle of the strings. [2] When the differences appear at the beginning or at the end of strings the similarity scores are lower because the number of common q-grams is smaller. Statistically, the differences between strings appear more often in the middle, which explains why the padded q-grams can produce higher similarity scores for the majority of corrupted data. This pattern occurred in all of the results produced during this study.

4.4   Observation 4: Some Types of Errors are Worse than Others

Out of all corrupted datasets the worst performance in terms of accuracy and the number of false positives found was "achieved" when the datasets with OCR Edits were linked. This is most likely due to the fact that some of the mistakes that OCR Edits introduce are replacements of two characters in place of one character, or vice versa. For instance, character "m" can be replaced with "rn" and the string "cl" can be replaced by the character "d". Those kind of replacements can have a significant negative impact on the string similarity scores produced by all of the metrics. The best performance results were recorded when the datasets corrupted with Character Edits were linked, those are presented in Fig. 1. Figure 2 illustrates the accuracies of linking datasets corrupted with OCR Edits errors. The accuracies of datasets corrupted with Keyboard Edits, Phonetic Edits, and a mix of all types of edits fall in between the accuracies presented in Figs. 1 and 2.

Another pattern common for the results obtained from linking all types of corrupted datasets was a significant drop in accuracy when the corrupted records contained more than one error of any type. For instance, for $Threshold = 0.85$ the accuracies of the Jaro, Levenshtein, unpadded q-grams compared using the Dice coefficient, and padded q-grams compared using the Dice coefficient were 97.5 %, 90.79 %, 57.46 %, and 73.37 % respectively when there was only one error of the Character Edit type per record. When the number of errors per corrupted record increased to two, the accuracies decreased to 88.39 %, 39.54 %, 13.31 %, and 10.41 %. Figure 1 presents a full overview of the accuracy degradation for datasets corrupted with Character Edits where 10 % of all original records were corrupted.

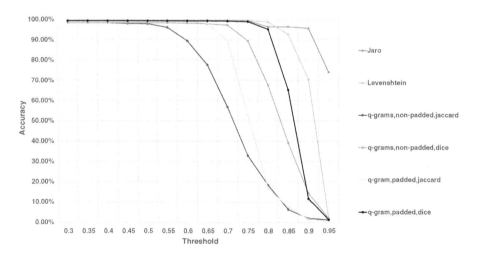

Fig. 2. Accuracy of string metrics used to perform record linkage on dataset with 10 % of the records corrupted using OCR Edits with 1 corruption per record.

4.5   Observation 5: The Efficiency Penalty for These Privacy-Preserving String Similarity Metrics is Small

The Jaro, Levenshtein, and Jaccard and Dice variants of the q-grams metric all have a $O(nm)$ time complexity, where n and m are the lengths of the strings to be compared. Because the Soundex metric is only checking for equality of the Soundex representation of the strings, its time complexity just $O(n)$. When determining whether a particular name is in the dataset, the query name is compared against all of the names in the dataset. It should be noted that the Soundex algorithm, because it is an exact rather than fuzzy match, could be made more efficient by indexing the database on the Soundex representation of the name. Also, there has been some work on eliminating the need to consider all names in the dataset when querying using the Jaro metric through the user of character and length-based filters to quickly determine if it is possible for a particular name to match the query within a specified threshold [4]. Neither of these optimizations were considered in this work.

While most of the string metrics considered have the same computational complexity, constant factors differ between the approaches. For example, because Jaro only looks for matching characters within a window that is half the length of the longer string, it is generally faster than Levenstein. To evaluate the computational efficiency of each of the string metrics in a practical setting, the time taken to perform the join operation between the original dataset and the corrupted dataset was measured. We explored the impact on the performance when the number of characters in names increases. In this case, the datasets always consisted of 10,000 records but the number of characters in each name was equal to 10, 15, or 20. These tests were done on datasets with a record corruption of 10 %, where the records were corrupted using the Character Edit technique and contained one corruption per record. The results are shown in Fig. 3. The timing

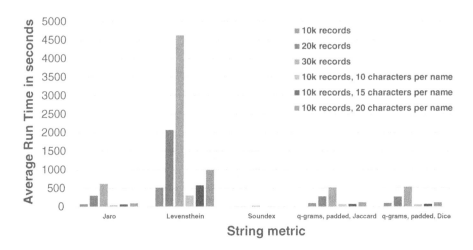

Fig. 3. The average time of record linkage using selected string metric techniques.

results of linkage performed on the other corrupted datasets were very similar to the ones presented in this figure.

The results show that the q-grams approaches are very slightly faster than Jaro in these tests, and significantly faster than Levenshtein. The average time taken to perform the join operation on the datasets using Levenshtein was more than five times the magnitude of the time taken by the other string metrics. Of course, the best speed was observed when the datasets were linked using the Soundex metric. In those cases the linkage was performed almost instantly, averaging only about one second.

Additionally, we have investigated the impact on the performance when the number of records increases. Three datasets of different volumes (10, 20, and 30 thousand records) were linked to conduct the tests. The results shown in Fig. 3 indicate that as the number of records to be linked increases, the time required to link all the records is again very similar for Jaro and the q-grams techniques, significantly greater for Levenstein, and very low for Soundex.

5 Conclusions and Future Work

In this work we evaluated the accuracy and speed of selected string metrics that support approximate matching for querying and joining databases based on encrypted names. An advanced benchmark generation tool, "GeCo", was used to produce sample datasets with records containing common mistakes in name spelling such as typographical errors, optical recognition errors, and phonetic errors. The performance of several string metrics that support approximate matching on encrypted data (four variations of q-grams based techniques and one technique based on encodings produced by the Soundex algorithm), was compared against commonly used string metrics, such as Jaro and Levenshtein, employed on unencrypted data.

Joining databases based on Soundex encodings did not prove to be a feasible option since it failed to find a correct match for more than 50 % of records when the name in a corrupted record contained one error, and for almost 75 % when corrupted records could contain two errors in a single name. Q-grams based techniques seem to be viable option for joining databases on encrypted names. While their performance in terms of precision is slightly worse than the performance of metrics such as Jaro or Levenshtein on unencrypted data, this can be easily dealt with by adjusting the threshold value that determines when two q-grams are likely to correspond to the same name.

In future work we plan to extend the range of attribute types that can be used to perform record linkage. In this study we focused on linking records based only on an individual's first and last name. However other types of attributes, such as numeric or categorical ones, can also carry PII. We want to be able to integrate those kind of attributes into private record linkage queries. Finally, we want to address the potentially significant security vulnerability of the encrypted q-grams approach, on which a frequency attack based on common q-grams can be launched, by investigating possible ways to make the encrypted q-grams resilient to those kind of attacks.

Acknowledgments. This work was partially supported by the LexisNexis corporation.

References

1. Christen, P.: A comparison of personal name matching: techniques and practical issues. In: Sixth IEEE International Conference on Data Mining Workshops, ICDM Workshops 2006, pp. 290–294. IEEE (2006)
2. Christen, P.: Data Matching: Concepts and Techniques for Record Linkage, Entity Resolution, and Duplicate Detection. Springer, Heidelberg (2012)
3. Churches, T., Christen, P.: Some methods for blindfolded record linkage. BMC Med. Inform. Decis. Mak. **4**(1), 9 (2004)
4. Dreßler, K., Ngomo, A.C.N.: Time-efficient execution of bounded jaro-winkler distances. In: Proceedings of the 9th International Conference on Ontology Matching, vol. 1317, pp. 37–48. CEUR-WS. org (2014)
5. Giereth, M.: On partial encryption of RDF-graphs. In: Gil, Y., Motta, E., Benjamins, V.R., Musen, M.A. (eds.) ISWC 2005. LNCS, vol. 3729, pp. 308–322. Springer, Heidelberg (2005)
6. Keskustalo, H., Pirkola, A., Visala, K., Leppänen, E., Järvelin, K.: Non-adjacent digrams improve matching of cross-lingual spelling variants. In: Nascimento, M.A., de Moura, E.S., Oliveira, A.L. (eds.) SPIRE 2003. LNCS, vol. 2857, pp. 252–265. Springer, Heidelberg (2003)
7. Muñoz, J.C., Tamura, G., Villegas, N.M., Müller, H.A.: Surprise: user-controlled granular privacy and security for personal data in smartercontext. In: Proceedings of the 2012 Conference of the Center for Advanced Studies on Collaborative Research, pp. 131–145. IBM Corp. (2012)
8. Philips, L.: Hanging on the metaphone. Comput. Lang. **7**(12) (1990)
9. Snae, C.: A comparison and analysis of name matching algorithms. Int. J. Appl. Sci. Eng. Technol. **4**(1), 252–257 (2007)
10. Tran, K.N., Vatsalan, D., Christen, P.: Geco: an online personal data generator and corruptor. In: Proceedings of the 22nd ACM International Conference on Conference on Information & Knowledge Management, pp. 2473–2476. ACM (2013)
11. Vatsalan, D., Christen, P., Verykios, V.S.: An efficient two-party protocol for approximate matching in private record linkage. In: Proceedings of the Ninth Australasian Data Mining Conference, vol. 121, pp. 125–136. Australian Computer Society, Inc. (2011)
12. Yakout, M., Atallah, M.J., Elmagarmid, A.: Efficient private record linkage. In: IEEE 25th International Conference on Data Engineering, ICDE 2009, pp. 1283–1286. IEEE (2009)

Vocabularies, Schemas, Ontologies Track

An Ontology-Driven Approach for Semantic Annotation of Documents with Specific Concepts

Céline Alec[✉], Chantal Reynaud-Delaître, and Brigitte Safar

LRI, Univ. Paris-Sud, CNRS, Université Paris-Saclay, 91405 Orsay, France
{celine.alec,chantal.reynaud,brigitte.safar}@lri.fr

Abstract. This paper deals with an ontology-driven approach for semantic annotation of documents from a corpus where each document describes an entity of a same domain. The goal is to annotate each document with concepts being too specific to be explicitly mentioned in texts. The only thing we know about the concepts is their labels, i.e., we have no semantic information about these concepts. Moreover, their characteristics in the texts are incomplete. We propose an ontology-based approach, named SAUPODOC, aiming to perform this particular annotation process by combining several approaches. Indeed, SAUPODOC relies on a domain ontology relative to the field under study, which has a pivotal role, on its population with property assertions coming from documents and external resources, and its enrichment with formal specific concept definitions. Experiments have been carried out in two application domains, showing the benefit of the approach compared to well-known classifiers.

Keywords: Ontology-driven approach · Ontology population · Ontology enrichment with specific concepts

1 Introduction

Nowadays, many Semantic Web applications use ontologies as rich conceptual schemas to give meanings to terms used as annotations of web contents. This paper deals with such an ontology-based approach addressing semantic annotation of documents when annotations are specific concepts. The only thing we know about these concepts is their labels. They have no definitions. We face three difficulties: (1) specific concepts are not explicitly mentioned in the documents under consideration, (2) specific concepts are not defined, even if the designer of the system knows their meaning and what kind of information is relevant to define them, (3) textual documents are incomplete, i.e., some characteristics describing these specific concepts are missing. For example, the concept "Destination where one can do Water sports during Winter" (DWW) is a very specific concept. Advertising descriptions of holiday destinations do not mention explicitly if a destination matches this concept. However, the designer is able to say whether a given textual description of a destination has to be annotated with this specific concept or not. Moreover, he knows DWW refers to a place

© Springer International Publishing Switzerland 2016
H. Sack et al. (Eds.): ESWC 2016, LNCS 9678, pp. 609–624, 2016.
DOI: 10.1007/978-3-319-34129-3_37

with the possibility of doing water sports and warm enough in winter, so that water sports are practicable. He also knows that terms referring to water sports can be found in documents but with no mention of the values of the winter temperatures. The missing information has to be searched in an external resource. In this paper, we show that an ontology can be enriched and then be used to annotate documents in such a constrained context.

The automation of this annotation process needs to formally define each specific concept, when it may be hard to do it. For instance, what is a place *warm enough* in winter? A solution is to automatically learn definitions provided that the designer is able to manually annotate documents as positive or negative examples for a given concept. Once a definition is learned, new descriptions of the same domain can be automatically annotated. Definitions are crucial in our work, which is currently used for a Business to Consumer application whose goal is to provide users with entities matching their needs in a particular domain. In that context, formal definitions can be used by reasoning to deliver partial satisfactory proposals when totally satisfactory ones do not exist. For example, if a user wants to go on vacation during winter and do water sports, entities annotated with DWW have to be proposed. If the DWW definition has lots of constraints, it is possible that no available destinations match this concept, implying no proposals to the user. This must be avoided. Close destinations have to be proposed, e.g., with a slightly lower temperature in winter.

This paper focuses on how an ontology can be populated and enriched in order to annotate documents. We investigate how several approaches can be combined in order to jointly contribute to address this annotation problem. Our contribution is then the SAUPODOC (Semantic Annotation Using Population of Ontology and Definitions of Classes) approach, relying on a domain ontology relative to the field under study, which has a pivotal role, on its population with property assertions, and on automatic generation of formal concept definitions from the populated ontology.

The remainder of the paper is organized as follows. Section 2 presents some related work. Section 3 describes the general aspect of our approach while Sect. 4 presents the various tasks involved. Section 5 presents experiments to evaluate the approach. Section 6 concludes and outlines future work.

2   Related Work

In this section, we review some existing literature about semantic annotation and highlight the need, in our context, for an ontology-based approach. Semantic annotation is a large area of the Semantic Web [20]. Annotating implies to attach data to some other pieces of data. Methods of semantic annotation of documents can be classified into two categories [23]: (1) pattern-based methods based on an initial set of entities and/or a set of patterns and (2) machine learning-based methods based on probability or induction. The idea is to look, in documents, for textual fragments that mention an entity. Named entity recognition [19] is part of this kind of annotation. However, our goal is a bit different.

We want to annotate a whole document, i.e., the entity described in it, not the elements mentioned in it. A few similar works deal with this idea. Their goal is to evaluate the proximity between a description of an entity and more specific elements (other documents, instances of concepts, concepts). [14] wants to match a job offer with candidates (CV, cover letters). Both are textual documents. Documents from both side are represented as vectors and various similarity measures are used to match them. [5] is to match hotel services. The hotel manager gives a description of the services of his hotel to be matched with a pre-existing list of services. The proximity is based on an n-gram calculation. Finally, [3] wants to annotate product catalogs with very fine-grained concepts. As similarity measures are not possible in this context, a first manual annotation is performed by an expert helped with an annotation tool. Then, machine learning techniques apply. Like these works, we want to match a description with specific concepts, i.e., to annotate descriptions with specific concepts. However, we face one more stake. We want comprehensive annotations. This means the process of annotation cannot be a black box. Indeed, when a concept asked by a user is not associated with any descriptions, we would like to make some refinements, i.e., to generalize its definition to be able to have some answers. This is what makes our work original.

Since the concepts used for annotation are not explicitly mentioned in the descriptions to be analyzed, a (at least) two-time process needs to be done. The first step is a classic extraction process while the second step is a reasoning one over the results of the first step. These two phases can be observed in concept-based approaches [8], such as concept-level sentiment analysis [7], which focuses also on text analysis relying on features considered as implicit because not represented in the texts but reasoning is here very specific. To the best of our knowledge, two works with a research purpose close to ours follow this idea. Both use ontologies. In the BOEMIE system [21], concepts from an ontology are divided into primitive and composite concepts. The first ones are populated via classical information extraction techniques. For the composite concepts, it is not that easy since they cannot be found in texts. However, their properties can be found. In this way, they can be defined in terms of primitive concepts. Composite concepts are populated via a reasoning performed on primitive instances. [27] aims to extract facts from texts using an ontology and natural language processing tools. New facts, which are not explicitly mentioned in texts, are then learned from the extracted facts and ontology knowledge. This learning is done via inference rules, written manually, relying on background knowledge. Compared to [21, 27], we do not have a manual definition of our concepts. These definitions need to be learned. This is one of our contributions.

3 The Saupodoc Approach: An Ontology-Based Approach

In this section, we present the general idea of the approach. The main point is the use of a domain ontology progressively populated with information extracted from documents under consideration and from external resources. Then, concept definitions are learned based on this enhanced ontology and on some documents manually annotated. The last step is a reasoning one, where definitions apply in order to generate annotations of new documents. The approach, with the tasks as defined in the paper, is automatic and domain-independent. For example, it has been used to annotate destination descriptions as well as film descriptions.

3.1   Inputs of the Approach

Each domain needs its own inputs: (1) a domain ontology; (2) the list of concepts used to annotate, called target concepts; (3) a corpus of documents from which some of them have to be manually annotated as positive or negative examples for each target concept; (4) a specification of correspondences between the ontology properties and properties of external resources.

The Ontology. It defines the domain. It is a guide to analyze documents, to search for missing information in external resources and to reason with definitions. It contains all elements defining entities in the application domain. It is approach-independent. Indeed, the only constraints imposed by the approach are described in this section. The ontology can therefore be largely reused, or (semi-) automatically built, however, we do not focus on this point in this paper. More formally, the ontology $\mathcal{O}$ is an OWL ontology defined as a tuple $(\mathcal{C}, \mathcal{P}, \mathcal{I}, \mathcal{A})$ where $\mathcal{C}$ is a set of classes, $\mathcal{P}$ a set of (datatype, object and annotation) properties characterizing the classes, $\mathcal{I}$ a set of individuals and property assertions, and $\mathcal{A}$ a set of axioms including constraints on classes and properties: subsumption, equivalence, type, domain/range, characteristics (functional, transitive, etc.), disjunction.

Figure 1 shows an excerpt of an ontology in the domain of holiday destinations. The classes Activity, Environment and FamilyType are respectively the roots of a hierarchy, e.g., Environment expresses the natural environment (Aquatic, Desert, etc.) or its quality (Beauty, View). Some object properties represented on the figure have subproperties, not represented here. Datatype properties are represented under their domain class. Individuals are not represented.

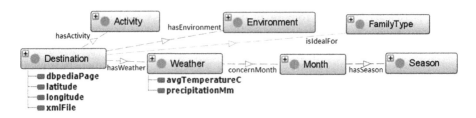

Fig. 1. The structure of the destination ontology

$\mathcal{C}$ groups two types of classes. ***The main class*** corresponds to the general type of entities described in the corpus, e.g., Destination. ***Descriptive classes*** are all other classes, useful to define the main class, e.g., Activity.

$\mathcal{P}$ is the set of properties characterizing the classes, datatype or object. A property assertion is a triple $<s,\ p,\ o>$ which links an individual s to an other individual or a literal o via a property p of $\mathcal{O}$. For example, if "Destination has-Activity Activity" is an axiom in $\mathcal{A}$, if d and a are respectively instances of Destination and Activity, then $<d$, hasActivity, $a>$ may be a property assertion. No object/datatype property assertions are initially expressed. Our aim is to collect them.

$\mathcal{I}$ initially contains instances of descriptive classes, e.g., _rainForest is an instance of Forest (descendant of Environment) and dense forest is one of its labels.

The Target Concepts. They are simple names of concept like "Destinations where one can do Water sports during Winter" (DWW), listed by the designer. Target concepts will be introduced in the ontology as specialized classes of the main class. One target concept will be denoted by tc in the following.

The Corpus of Documents. These are XML documents describing a domain entity, with very little structure. The structure of documents highlights the name of the entity and its textual description (containing labels of instances of descriptive classes). In our context, documents may be extracted from advertising catalogs, praising the assets of the entity described. In any cases, they describe the main features of entities and few negative expressions are present. However, names of target concepts are not explicitly mentioned. In our approach, some documents have to be manually annotated by the designer, for all the target concepts, as positive or negative examples of target concepts, i.e., either by tc or by *not tc*. It is not very time-consuming since the designer associates target concepts to a whole document. As an expert of the domain, he can provide the annotations based on his own background knowledge. He is not obliged to analyze precisely the document content or to seek information from external resources.

The Correspondences Between the Ontology and External Resources. We distinguish two types of properties, *document properties* and *external properties*. Documents from one corpus are supposed to be complete w.r.t. *document properties*. For example, the documents describing destinations mention the activities that one can do in this destination. When an activity is not mentioned, we suppose it cannot be practiced in this destination. Nevertheless, documents are incomplete w.r.t. *external properties*. This means these properties are not mentioned at all in the documents, e.g., the weather in the destination corpus.

Document properties can be asserted from each document. However, *external properties* need to be asserted from external resources. In this paper, we focus our completion with data from LOD (Linked Open Data). The designer indicates the *external properties* and selects the most relevant LOD datasets to populate them. This task can be performed manually because the number

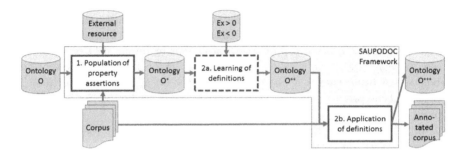

Fig. 2. The SAUPODOC workflow

of *external properties* is not too large: these properties deal only with precise information not included in the documents (like numerical data) or with misinterpretation of documents. However, the ontology and the LOD datasets differ in regards to vocabulary terms and structure. Complex correspondences have to be established. We propose a model in Sect. 4.2 to support their specification.

3.2 Functional Description

The SAUPODOC approach is based on four tasks guided by the ontology. The first two tasks aim to populate the ontology (step 1) with property assertions. The next two tasks (step 2) are two alternative reasoning tasks. Step 2a discovers formal definitions of target concepts while Step 2b populates the ontology classes corresponding to those definitions. This population is equivalent to an annotation of the documents. Indeed, if a document d is (respectively is not) an instance of a class corresponding to a target concept tc, then d is annotated with tc (respectively *not tc*). We call tc a positive annotation and *not tc* a negative annotation w.r.t. the target concept tc. The annotation process is executed off-line.

Figure 2 describes the workflow of the approach. The initial ontology O is progressively populated and enriched. First, each document entity is introduced as an instance of the main class. Each textual description is used to populate the ontology with property assertions expressing the features of each entity. These assertions are then completed thanks to information found in external resources (O^+). Target concepts are inserted into the ontology as classes, called target classes, which are specializations of the main class. Their definition is learned based on manually annotated examples (O^{++}). Finally, definitions apply in order to populate target classes (O^{+++}) and annotate the corpus with target concepts.

4 Tasks of the Approach

This section presents the various tasks exploiting data at different abstraction levels (classes and individuals) and having to cooperate to reach the final goal.

The pivotal role of the ontology is central in the approach. A preliminary task creates, for each document, an instance of the main class representing the entity described in the document. For example, in the destination domain, an individual Dominican_Republic is created from the document describing it. This individual is created such as <Dominican_Republic rdf:type Destination>. For each entity, the two tasks of step 1 populate the ontology with information that will be used by the two reasoning tasks of step 2. In what follows, we present how the ontology is a guide for each of these tasks.

4.1   Data Extraction from Texts

The first task of step 1 extracts data from documents. Its goal is to enhance the ontology with assertions of *document properties*. This extraction is guided by the ontology, more particularly by its terms related to instances of descriptive classes being ranges of the *document properties*. If there is a match between a term and a document, then a property assertion is added in the ontology. For example, the constraint <Destination, hasActivity, Activity> requires that the range value of the property hasActivity belongs to the extension of the class Activity. From this constraint, if the text describing an entity e contains a term of an instance a of Activity, then the assertion <e, hasActivity, a> is built. Figure 3 represents an excerpt of a document describing the Dominican Republic. The expressions scuba divers and diving are terms of the individual _diving, which is an instance of a class specializing WaterSport (subclass of Activity). So, the assertion <Dominican_Republic, hasActivity, _diving> is added. Note that in the two study cases of our experiments, we specialized property ranges to avoid having two properties with the same range. The approach should be extended in a future work to take into account cases where such a specialization is impossible.

In our work, we use GATE [6,10], an open source software performing a lot of text processing tasks. The GATE resource OntoRoot Gazetteer, in combination with other generic GATE resources, can produce label-

especially loved by scuba divers. Over 20 exiting diving sites and 3 old shipwrecks are waiting to be discovered.

Fig. 3. Excerpt of the document on the Dominican Republic annotated by GATE

ing over textual documents, called lookups, w.r.t. an ontology given as input. GATE was chosen for its ability of using an external ontology as input, unlike other tools such as Open Calais [1] which annotates with named entities, facts or events but cannot be used with an external ontology. GATE can be used with a JAPE transducer, which applies JAPE (Java Annotation Patterns Engine) rules. In our context, JAPE rules transform lookups into property assertions. They are automatically created from one pattern, used whatever the ontology. The pattern is instantiated for all *document properties*, creating as many JAPE rules as *document properties*.

Note that we are in a context where documents describe the main features of entities and do not include negative expressions that could disrupt the process. Thus, a simple information extraction task like this one is appropriate.

4.2   Data Completion with External Resources

Textual descriptions are often short and do not contain all the necessary information. For instance, defining a DWW requires to know temperature and precipitation during winter for every destination. This data is not mentioned in descriptions. Data collection has to be enriched exploiting available on line resources.
This is the second task of step 1. Again, this task is guided by the ontology.
It involves to find a RDF resource dealing with entities of the corpus and to
identify in this resource what properties correspond to those required by the
ontology (*external properties*).

We chose to work with DBpedia [4] and we use DBpedia Spotlight [18], a
tool able to automatically annotate a text with references to DBpedia entities.
Applied on the entity name of each document, it gives a direct access to the
DBpedia resource corresponding to the entity, unlike other tools like Wikifier
[9,22] or AIDA [28], which return Wikipedia pages.

In this section, we present a model of data acquisition used by the designer
to express (i) correspondences and (ii) access paths. The model of acquisition is
currently used to extract information from DBpedia (but can be used for other
LOD datasets) and insert property assertions into the ontology.

Correspondences. Since the vocabulary from the ontology and from the
resource can differ, mechanisms have to be conceived to establish correspondences between the required elements and those from the resource. A model,
briefly presented here, allows the designer to express these correspondences and
supports automatic generation of SPARQL queries. A correspondence consists
in associating an *external property* from the ontology with a property expression
from the target data source, called PE_t. This PE_t is not necessarily explicitly represented in the target data source, instead, it may result from complex
treatments.

A property expression in the target source (PE_t) is a property p or its inverse
p^{-1}, or an expression (f) using one or several property expressions in the target
source. A PE_t may include constraints (*Constr*).

$$PE_t = p \mid p^{-1} \mid f(PE_t) \mid f(PE_t, PE_t) \mid PE_t.Constr$$

By recursion, a PE_t can be a function of n PE_t. The function f is specified by
the designer. He has to indicate whether it is a function of aggregation (minimum,
average, etc.), of transformation (mathematical calculation, concatenation, etc.),
or a set-theoretic operation (union, difference), and to clarify its nature. *Constr*
represents any domain or range constraints.

For example, the *external property* precipitation_in_January is expressed by six
different properties in DBpedia (janPrecipitationMm, janRainMm, janPrecipitation-
Inch, janRainInch, janPrecipitationIn and janRainIn). So, the correspondence of pre-
cipitation_in_January can be expressed as the union of the values of these six properties, or even the average value of their union.

Access Paths. Since external resources are incomplete (like DBpedia), some properties from a PE_t may be missing. In our settings, having complete information is essential in order to achieve the best results as output of the entire process. Consequently, we propose an alternative way to get missing data, by browsing close resources. The idea is to have an approximation of the data, which is better than nothing.

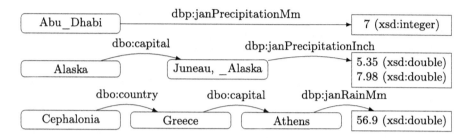

Fig. 4. Access paths in DBpedia

This mechanism is based on the composition of properties and allows the designer to establish access paths, to reach resources containing the required information. For example, Fig. 4 shows two examples (Alaska and Cephalonia) where the values for January precipitation are not available for the destination but where an approximation can be found for a close entity (the capital here).

These complex correspondences with their access paths are specified by the designer. CONSTRUCT SPARQL queries are automatically built based on these specifications, allowing to collect data via SPARQL endpoints in a transparent way and to insert ontology assertions in the source ontology. This process is not the focus of the paper. It will not be discussed here.

4.3   Learning the Definitions of Target Concepts

The first task of step 2 is a reasoning task, which is executed only once. It aims to learn the definitions of target concepts, based on the manual annotations of documents provided by the designer and the data collected in step 1.

Most machine learning tools do not take into account explicit specifications of relations (subsumption, object/datatype properties) between features as it is expressed in an ontology. Three tools, YinYang [11], DL-FOIL [12] and DL-Learner [15], use inductive logic programming (ILP) on Description Logics (DL) to perform concept learning. We chose to use DL-Learner, an open-source software capable of learning definitions of classes expressed in DL, from expert-provided examples, using an ontology as input. It allows us to get an explicit definition for all target concepts, an important point in concrete applications. The DL-Learner definitions are conjunctions and disjunctions of elements. An element can be a class (Destination) or an expression using object properties (hasActivity some Nightlife), numerical datatype properties (avgTemperatureC some double[>= 23.0]), or cardinality constraints (hasCulture min 3 Culture). Ranges are conjunctions and disjunctions of elements. For example, the definition of DWW that can be learned by DL-Learner, could be something like this:

(Destination and (hasActivity some Watersport)
    and (hasWeather min 2 ((concernMonth some (hasSeason some MidWinter))
        and (avgTemperatureC some double[>= 23.0])
        and (precipitationMm some double[<= 70.0]))))).

DL-Learner parameters were chosen based on the user manual and discussions with the software developers. We use the CELOE algorithm [16] announced as the best class learning algorithm currently available within DL-Learner, and the default reasoner, called fast instance checker, making the closed-world assumption (CWA). However, learned definitions have to be incorporated in an OWL ontology where reasoning is based on the open world assumption (OWA). To be able to learn and exploit minimum cardinality constraints, e.g., (hasActivity min 3 Activity), instances are automatically expressed as disjoint (Unique Name Assumption) to avoid being linked by owl:sameAs. Moreover, we disable the negation operator (NOT), the operator of universal restriction (ONLY), and operator of the maximum cardinality restriction (MAX), so that the obtained definitions are applicable under OWA.

Some DL-Learner parameters have been set making a compromise between the expressiveness of the definition and the execution time. Hence, we allow statements with a cardinality value set maximum at 10 instead of 5 (default value) such as definitions like hasObjectProperty min 10 class_name can be learned and the maximum execution time is set at 200 s. These parameters will be used whatever the application. An other important parameter is the noise percentage, i.e., the percentage of positive examples not covered by a definition. We have proceeded by trial and error to set it. Hence, we have developed a methodology from the conducted experiments, where 5 different values for the noise percentage (5–15–25–35–45 %) are tested, and tuned using test experiments. Moreover, in case of really complex and far from easy to learn definitions, we set up parameters to apply a search heuristic to get longer definitions. When these latter parameters are activated, we call it the complex configuration, when they are not, the basic one. This means 10 configurations are tested: the basic and complex configurations both with 5 different values for the noise percentage. For each test, the highest ranked solution which is the best one in terms of accuracy and length is automatically kept. For each target concept, the best definition from the 10 tests is chosen.

4.4 Reasoning to Populate Target Classes and Annotate Documents

The second task of step 2 consists in applying the learned definitions to populate the target classes in the ontology. This task is done any time new descriptions have to be annotated. We chose to use FaCT++ [26], an OWL-DL reasoner, which scales well with a large number of instances, unlike HermiT [24] and Pellet [25] according to our experiments. Indeed, these two reasoners have never terminated the process when used on 10,000 documents. FaCT++ relies on the definitions of target concepts to identify the document entities that should be annotated with a target concept. For each target concept tc, if the entity described in a document d fits the definition of tc, then this entity becomes an instance of the class representing tc. In that way, the document d is annotated by tc. On the opposite, if the entity does not fit the definition of tc, then this entity does not become an instance of the tc class and d is annotated by *not tc*. Doing this sort of annotation relies on a closed world assumption (CWA), whereas the reasoning with OWL is generally based on an open world assumption (OWA). Nevertheless, our context is particular and allows us to simulate the CWA in each step. In step 1, for each entity, if a property assertion is not created, then we consider it does not exist. Indeed, the extraction of property assertions from documents operates under CWA since documents are supposed to be complete for all *document properties*. Moreover, the extraction of property assertions from external resources nearly operates under CWA too thanks to the model of acquisition presented in Sect. 4.2. Indeed, access paths providing approximate values are good ways to overcome incompleteness. For step 2a, as stated in Sect. 4.3, definitions respect the CWA.

5   Experimental Evaluation

5.1   Procedure

To evaluate the annotation process, we compare our approach with classification approaches. Classification approaches are able to annotate documents in the same way as SAUPODOC, i.e., with a positive annotation (tc) or negative annotation (*not tc*) for each target concept tc. In this section, we assess the quality of the annotation. To do that, a set of annotated documents is used. This set is split. The training set (2/3) provides positive and negative examples given in inputs to learn definitions. The testing set (1/3) is used to compare the annotations obtained by each process with the correct annotations. The comparison is based on several metrics: precision, recall, accuracy and F-measure.

$$Precision = \frac{TP}{TP + FP} \qquad Recall = \frac{TP}{TP + FN}$$

$$Acc. = \frac{TP + TN}{TP + FP + TN + FN} \qquad F\text{-}measure = \frac{2 \times Precision \times Recall}{Precision + Recall}$$

To make a fair assessment between SAUPODOC and classification approaches, we consider the same domain terminology. Indeed, SAUPODOC is based on an ontology but classifiers are not. The input of classifiers consists of documents represented with a list of features and a label. The label is binary w.r.t. a target concept (the annotation of the document with this target concept is true or false). To make the list of features, we consider the domain terminology of the ontology, i.e., the set of labels and key-expressions associated to each individual, as a domain dictionary. We call a *word* of the dictionary, a keyword or keyphrases from the domain dictionary. Two expressions referring to a same idea (same individual) are put into a same word of the dictionary, e.g., scuba divers and diving. All the words of the dictionary are lemmatized. Each document corresponds to a vector of features (Vector Space Model). A bag-of-words method is used. This means that each element of the vector (each feature) corresponds to a word of the dictionary. Documents are lemmatized. If a word of the dictionary is found, the value in its vector for its element is the TF-IDF value, otherwise 0. In summary, for a corpus, a list of vectors is made, one vector for each document representing its content w.r.t. the list of features, and a label w.r.t. a target concept. For each target concept, we launch the classifier with the same features but we change the binary label, depending on whether the annotation is true or false.

Two classifiers are tested: an SVM classifier and a decision tree classifier. We use several parameters. Indeed, several values are tested for kernel (PolyKernel with several exponents, RBF with several gammas) and complexity for SVM; and several confidence factors for decision trees. We only keep the best results, i.e., the results with the best accuracy on the test set.

5.2   Versions Used in the Evaluation

The experimental evaluation has been performed using the following versions of the SAUPODOC components: GATE 8.0, DBpedia 2014, DL-Learner 1.0, FaCT++ 1.6.2.

For classifiers, we used Stanford NLP 3.4.1 [17] to lemmatize and Weka 3.6.6 [13] to execute the classifiers.

5.3   The Two Tested Corpora

Experiments were conducted in two application domains, each one having different characteristics. The objective is to see whether the size of the corpus and the ontology richness affect the quality of results.

The Destination Corpus. The corpus of destinations contains 80 documents, which have been automatically extracted from the Thomas Cook catalog [2]. Each document describes a specific place (country, region, island or city). The documents are promotional, i.e., they express the qualities of destinations and have hardly any negative expressions. Geolocation and meteorological data are missing. This information will be extracted from DBpedia thanks to the model of acquisition (Sect. 4.2).

The main class of the ontology is Destination. There are 161 descriptive classes, which express the nature of the environment (46 classes), the activities that can be done (102 classes), the kind of family that should go there, e.g., people with kids, couples, etc. (6 classes) and the weather information (7 classes), e.g., the seasons. The individuals, which are instances of these descriptive classes, have terminological forms via annotation properties (label, isDefinedBy). For example, the terms archaeology, archaeological, acropolis, roman villa, excavation site, mosaic are terminological forms of the individual archaeology.

In all, 39 target concepts are under study. Every destination of the corpus is annotated by the designer as a positive or negative example for each target concept.

The Film Corpus. The film corpus contains 10,000 documents. These documents have been created via an automatic extraction from DBpedia. Each document corresponds to a film described in a DBpedia resource. The document contains the DBpedia URI from which it was extracted and the abstract of the film. The abstract of film fits our context, i.e., there are hardly any negative expressions. Since the DBpedia URI is provided, we do not need to use DBpedia Spotlight to get it. Duration of the film as well as languages and countries are extracted from DBpedia. Indeed, duration is missing in the description. Languages and countries might be mentioned but a misinterpretation is possible, this is the reason why we prefer to trust DBpedia than documents for these properties. For instance, the presence of the term "French" may have different meanings. The film may be in French (language), or it may be a French film (country), or it can tell the story of a French person.

The film ontology is basic. It contains the main class Film and five descriptive classes expressing characteristics about films. The ontology only contains the classes needed for defining the target concepts in our experiments. It should be completed in case of new target concepts.

In total, 12 target concepts are taken into consideration. They correspond to some DBpedia categories. Annotated examples are automatically generated for films: a film f is a positive example for a target concept corresponding to a category c if $<f$ dcterms:subject $c>$ according to DBpedia, otherwise it is a negative example.

Let us note that, for our approach, the domain terminology may be partial, for which external resources are used. For example, there are no terms about either languages or countries in our film ontology but the DBpedia step can add instances of them. To be fair for classifiers, terms from these external resources are also added into the domain dictionary of classifiers.

5.4   Validation of the Approach

Table 1 shows the results for the testing set, i.e., the part not used in the process of construction of annotations. We can see that all the three approaches give good results in terms of accuracy, even if SAUPODOC has slightly better ones. However, accuracy is not the only metric to take into account. Most of the target concepts have many negative examples and few positive examples. Hence, if a classifier predicts negative for each document given in input, then the accuracy is high, e.g., 91.76 % on average for target concepts of the film corpus. Moreover, no instances of each target classes is found, which means all documents are annotated with negative annotations. As already said, our approach is currently used for a Business to Consumer application whose goal is to provide users with entities matching their needs in a particular domain. This means it is important to obtain positive annotations. This is why alternative metrics like precision, recall and f-measure are needed. They allow us to evaluate the positive annotations, which are central in this settings. From Table 1, we can observe that SAUPODOC significantly outperforms classifiers for these three metrics.

Table 1. Average results for target concepts

Metric (%)	Accuracy			F-measure		
Approach / Corpus	SAU-PODOC	SVM	Decision Tree	SAU-PODOC	SVM	Decision Tree
Destination (39 target concepts)	95.89	84.52	86.23	72.23	54.14	63.22
Film (12 target concepts)	95.46	94.41	94.32	75.65	61.74	61.40

Metric (%)	Precision			Recall		
Approach / Corpus	SAU-PODOC	SVM	Decision Tree	SAU-PODOC	SVM	Decision Tree
Destination (39 target concepts)	73.95	58.10	64.23	71.58	55.32	65.89
Film (12 target concepts)	76.27	69.90	67.72	77.76	57.59	58.99

Moreover, for all the target concepts $(39 + 12 = 51)$ from the two corpora, SAUPODOC generates 8 target concept definitions that assign every input of the test set as a negative example. To avoid this, as definitions are intelligible, the designer can easily refine them in order to have some positive examples. This refinement is mandatory in contexts such as Business to Consumer applications, for which positive examples are needed for all target concepts.

Classifiers encounter the same problem. However, a refinement is impossible, since there is no explicit definitions. Indeed, SVM classifiers create a model, which is not comprehensive for human. Decision tree classifiers are more comprehensive since trees can be seen as sets of rules. But here, these rules deal with the TF-IDF number associated with a dictionary word, which is hard to be interpreted by humans. For example, the decision tree for the target concept coastal destinations is given in Fig. 5 (accuracy of 96.30 %). It means positive annotations are made when the TF-IDF value of the dictionary word urban is less than 0.18893 and beach is less than 0 and sea is more than 0.005502, or if urban is less than 0.18893 and beach is more than 0. Such a tree is difficult to be adjusted by the designer in case a refinement is needed.

```
_urban <= 0.018893
|   _beach <= 0
|   |   _sea <= 0.005502: 0
|   |   _sea > 0.005502: 1
|   _beach > 0: 1
_urban > 0.018893: 0
```

Fig. 5. The decision tree for coastal destinations

6   Conclusion and Future Work

We proposed an ontology-driven approach dealing with semantic annotations of documents describing entities of a same domain. Annotation is performed with a list of concepts, which are not explicitly mentioned into the documents. The approach, called SAUPODOC, is based on an ontology, and combines population and enrichment steps. It makes tasks cooperate both at individual level and at concept level. Innovative mechanisms have been implemented to exploit the LOD. Property complex correspondences between the ontology and a data set can be defined and alternatives to missing data are provided. Results clearly show the benefit over well-known classifiers and the relevance of an ontology-based approach relying on a particular combination of various techniques to semantically annotate documents.

This work has been validated by the Wepingo company. It is a part of a wider approach to answer user's queries whose keywords are specific concepts by delivering documents related to instances of these concepts. Future work will be devoted to the integration of this semantic annotation process, extended with automatic generation of SPARQL queries to easily access LOD data, into a framework supporting the overall approach.

Acknowledgements. This work has been funded by the PORASO project, in the setting of a collaboration with the Wepingo company.

References

1. http://www.opencalais.com/
2. http://www.thomascook.com/
3. Alec, C., Reynaud-Delaître, C., Safar, B., Sellami, Z., Berdugo, U.: Automatic ontology population from product catalogs. In: Janowicz, K., Schlobach, S., Lambrix, P., Hyvönen, E. (eds.) EKAW 2014. LNCS, vol. 8876, pp. 1–12. Springer, Heidelberg (2014)
4. Auer, S., Bizer, C., Kobilarov, G., Lehmann, J., Cyganiak, R., Ives, Z.G.: DBpedia: a nucleus for a web of open data. In: Aberer, K., Choi, K.-S., Noy, N., Allemang, D., Lee, K.-I., Nixon, L.J.B., Golbeck, J., Mika, P., Maynard, D., Mizoguchi, R., Schreiber, G., Cudré-Mauroux, P. (eds.) ASWC 2007 and ISWC 2007. LNCS, vol. 4825, pp. 722–735. Springer, Heidelberg (2007)
5. Béchet, N., Aufaure, M.A., Lechevallier, Y.: Construction et peuplement de structures hiérarchiques de concepts dans le domaine du e-tourisme. In: IC, pp. 475–490 (2011)
6. Bontcheva, K., Tablan, V., Maynard, D., Cunningham, H.: Evolving GATE to meet new challenges in language engineering. Nat. Lang. Eng. **10**(3/4), 349–373 (2004)
7. Cambria, E., Fu, J., Bisio, F., Poria, S.: AffectiveSpace 2: enabling affective intuition for concept-level sentiment analysis. In: AAAI, pp. 508–514 (2015)
8. Cambria, E., White, B.: Jumping NLP curves: a review of natural language processing research [review article]. IEEE Comp. Int. Mag. **9**(2), 48–57 (2014)
9. Cheng, X., Roth, D.: Relational inference for wikification. In: EMNLP (2013)
10. Cunningham, H., et al.: Text Processing with GATE. University of Sheffield Department of Computer Science, Sheffield (2011)

11. Esposito, F., Fanizzi, N., Iannone, L., Palmisano, I., Semeraro, G.: Knowledge-intensive induction of terminologies from metadata. In: McIlraith, S.A., Plexousakis, D., van Harmelen, F. (eds.) ISWC 2004. LNCS, vol. 3298, pp. 441–455. Springer, Heidelberg (2004)

12. Fanizzi, N., d'Amato, C., Esposito, F.: DL-FOIL concept learning in description logics. In: Železný, F., Lavrač, N. (eds.) ILP 2008. LNCS (LNAI), vol. 5194, pp. 107–121. Springer, Heidelberg (2008)

13. Hall, M., Frank, E., Holmes, G., Pfahringer, B., Reutemann, P., Witten, I.H.: The WEKA data mining software: an update. SIGKDD Explor. 11(1), 10–18 (2009)

14. Kessler, R., Béchet, N., Roche, M., Moreno, J.M.T., El-Bèze, M.: A hybrid approach to managing job offers and candidates. Inf. Process. Manage. 48(6), 1124–1135 (2012)

15. Lehmann, J.: DL-Learner: learning concepts in description logics. J. Mach. Learn. Res. 10, 2639–2642 (2009)

16. Lehmann, J., Auer, S., Bühmann, L., Tramp, S.: Class expression learning for ontology engineering. J. Web Seman. 9, 71–81 (2011)

17. Manning, C.D., Surdeanu, M., Bauer, J., Finkel, J., Bethard, S.J., McClosky, D.: The stanford CoreNLP natural language processing toolkit. In: 52nd ACL: System Demonstrations, pp. 55–60 (2014)

18. Mendes, P.N., Jakob, M., García-Silva, A., Bizer, C.: DBpedia spotlight: shedding light on the web of documents. In: I-Semantics, pp. 1–8. ACM, New York (2011)

19. Nadeau, D., Sekine, S.: A survey of named entity recognition and classification. Linguisticae Investigationes 30, 3–26 (2007)

20. Oren, E., Möller, K., Scerri, S., Handschuh, S., Sintek, M.: What are Semantic Annotations? Technical report, DERI Galway (2006)

21. Petasis, G., Möller, R., Karkaletsis, V.: BOEMIE: reasoning-based information extraction. In: LPNMR, pp. 60–75. A Corunna, Spain (2013). CEUR-WS.org

22. Ratinov, L., Roth, D., Downey, D., Anderson, M.: Local and global algorithms for disambiguation to wikipedia. In: ACL (2011)

23. Reeve, L.: Survey of semantic annotation platforms. In: ACM Symposium on Applied Computing, pp. 1634–1638. ACM Press (2005)

24. Shearer, R., Motik, B., Horrocks, I.: HermiT: a highly-efficient OWL reasoner. In: OWLED, vol. 432 (2008). CEUR-WS.org

25. Sirin, E., Parsia, B., Grau, B.C., Kalyanpur, A., Katz, Y.: Pellet: a practical OWL-DL reasoner. J. Web Seman. 5(2), 51–53 (2007)

26. Tsarkov, D., Horrocks, I.: FaCT++ description logic reasoner: system description. In: Furbach, U., Shankar, N. (eds.) IJCAR 2006. LNCS (LNAI), vol. 4130, pp. 292–297. Springer, Heidelberg (2006)

27. Yelagina, N., Panteleyev, M.: Deriving of thematic facts from unstructured texts and background knowledge. In: Klinov, P., Mouromtsev, D. (eds.) KESW 2014. CCIS, vol. 468, pp. 208–218. Springer, Heidelberg (2014)

28. Yosef, M.A., Hoffart, J., Bordino, I., Spaniol, M., Weikum, G.: AIDA: an online tool for accurate disambiguation of named entities in text and tables. PVLDB 4(12), 1450–1453 (2011)

Qanary – A Methodology for Vocabulary-Driven Open Question Answering Systems

Andreas Both[1]([✉]), Dennis Diefenbach[2], Kuldeep Singh[3], Saedeeh Shekarpour[4], Didier Cherix[5], and Christoph Lange[3,4]

[1] Mercateo AG, Munich, Germany
andreas.both@mercateo.com
[2] Laboratoire Hubert Curien, Saint-Etienne, France
dennis.diefenbach@univ-st-etienne.fr
[3] Fraunhofer IAIS, Sankt Augustin, Germany
kuldeep.singh@iais.fraunhofer.de, langec@cs.uni-bonn.de
[4] University of Bonn, Bonn, Germany
shekarpour@uni-bonn.de
[5] FLAVIA IT-Management GmbH, Kassel, Germany
didier.cherix@gmail.com

Abstract. It is very challenging to access the knowledge expressed within (big) data sets. Question answering (QA) aims at making sense out of data via a simple-to-use interface. However, QA systems are very complex and earlier approaches are mostly singular and monolithic implementations for QA in specific domains. Therefore, it is cumbersome and inefficient to design and implement new or improved approaches, in particular as many components are not reusable.

Hence, there is a strong need for enabling best-of-breed QA systems, where the best performing components are combined, aiming at the best quality achievable in the given domain. Taking into account the high variety of functionality that might be of use within a QA system and therefore reused in new QA systems, we provide an approach driven by a core QA vocabulary that is aligned to existing, powerful ontologies provided by domain-specific communities. We achieve this by a methodology for binding existing vocabularies to our core QA vocabulary without recreating the information provided by external components.

We thus provide a practical approach for rapidly establishing new (domain-specific) QA systems, while the core QA vocabulary is re-usable across multiple domains. To the best of our knowledge, this is the first approach to open QA systems that is agnostic to implementation details and that inherently follows the linked data principles.

Keywords: Semantic web · Software reusability · Question answering · Semantic search · Ontologies · Annotation model

1 Introduction

Data volume and variety is growing enormously on the Web. To make sense out of this large amount of data available, researchers have developed a number of

© Springer International Publishing Switzerland 2016
H. Sack et al. (Eds.): ESWC 2016, LNCS 9678, pp. 625–641, 2016.
DOI: 10.1007/978-3-319-34129-3_38

domain-specific monolithic question answering systems (e.g., [5,6,11,23]). These QA systems perform well in their specific domain, but find limitation in their reusability for further research due to specific focus on implementation details. Hence, creating new question answering systems is cumbersome and inefficient; functionality needs to be re-implemented and the few available integrable services each follow different integration strategies or use different vocabularies. Hence, an ecosystem of components used in QA systems could not be established up to now. However, component-oriented approaches have provided high values in other research fields (like service-oriented architectures or cloud computing) while increasing efficiency. This is achieved by establishing exchangeability and isolation in conjunction with interoperability and reusability. Increased efficiency of both creating new question answering systems as well as establishing new reusable services would be the major driver for a vital and accelerated development of the QA community in academics and industry.

However, currently the integration of components is not easily possible because the semantics of their required parameters as well as of the returned data are either different or undefined. Components of question answering systems are typically implemented in different programming languages and expose interfaces using different exchange languages (e.g., XML, JSON-LD, RDF, CSV). A framework for developing question answering systems should not be bound to a specific programming language as it is done in [14]. Although this reduces the initial effort for implementing the framework, it reduces the reusability and exchangeability of components. Additionally, it is not realistic to expect that a single standard protocol will be established that subsumes all achievements made by domain-specific communities. Hence, establishing just one (static) vocabulary will not fulfill the demands for an open architecture. However, a standard interaction level is needed to ensure that components can be considered as isolated actors within a question answering system while aiming at interoperability. Additionally this will enable the benchmarking of components as well as aggregations of components ultimately leading to best-of-bread domain-specific but generalized question answering systems which increases the overall efficiency [8]. Furthermore it will be possible to apply quality increasing approaches such as ensemble learning [7] with manageable effort.

Therefore, we aim at a methodology for open question answering systems with the following attributes (requirements): *interoperability*, i.e., an abstraction layer for communication needs to be established, *exchangeability and reusability*, i.e., a component within a question answering system might be exchanged by another one with the same purpose, *flexible granularity*, i.e., the approach needs to be agnostic the processing steps implemented by a question answering system, *isolation*, i.e., each component within a QA system is decoupled from any other component in the QA system.

In this paper we describe a methodology for developing question answering systems driven by the knowledge available for describing the question and related concepts. The knowledge is represented in RDF, which ensures a self-describing message format that can be extended, as well as validated and reasoned upon using off-the-shelf software. Additionally, using RDF provides the advantage of retrieving or updating knowledge about the question directly via SPARQL. In previous work, we have already established a QA system vocabulary qa [19]. qa is a core vocabulary that represents a standardized view on concepts that existing QA systems have in common, on top of an annotation framework. The main focus of this paper is to establish a methodology for integrating external components into a QA system. We will eliminate the need to (re)write adapters for sending pieces of information to the component (service call) or custom interpreters for the retrieved information (result). To this end, our methodology binds information provided by (external) services to the QA systems, driven by the qa vocabulary. Because of the central role of the qa vocabulary, we call our methodology Qanary: **Q**uestion **a**nswering vocabul**ary**. The approach is enabled for question representations beyond text (e.g., audio input or unstructured data mixed with linked data) and open for novel ideas on how to express the knowledge about questions in question answering systems. Using this approach, the integration of existing components is possible; additionally one can take advantage of the powerful vocabularies already implemented for representing knowledge (e.g., DBpedia Ontology[1], YAGO[2]) or representing the analytics results of data (e.g., NLP Interchange Format [9], Ontology for Media Resources[3]). Hence, for the first time an RDF-based methodology for establishing question answering systems is available that is agnostic to the used ontologies, available services, addressed domains and programming languages.

The next section motivates our work. Section 3 reviews related work. In Sect. 4 the problem is will be broken down to actual requirements. Thereafter, we present our approach (Sect. 5) followed by a methodology to align existing vocabularies to our qa vocabulary in Sect. 6. In Sect. 7, we present a case study where a QA system is created containing actual reusable and exchangeable components. Section 8 concludes, also describing future research tasks.

2   Motivation

QA systems can be classified by the domain of knowledge in which they answer questions, by supported types of demanded answer (factoid, boolean, list, set, etc.), types of input (keywords, natural language text, speech, videos, images, plus possibly temporal and spatial information), data sources (structured or unstructured), and based on traditional intrinsic software engineering challenges (scalability, openness, etc.) [13].

Closed domain QA systems target specific domains to answer a question, for example, medicine [1] or biology [2]. Limiting the scope to a specific domain or ontology makes ambiguity less likely and leads to a high accuracy of answers, but closed domain systems are difficult or costly to apply in a different domain.

---

[1] http://dbpedia.org/services-resources/ontology.

[2] YAGO: A High Quality Knowledge Base; http://www.mpi-inf.mpg.de/departments/databases-and-information-systems/research/yago-naga/yago/.

[3] W3C Recommendation 09 February 2012, v1.0, http://www.w3.org/TR/mediaont-10/.

Open domain QA systems either rely on cross-domain structured knowledge bases or on unstructured corpora (e.g., news articles). DBpedia [3], and Google's non-public knowledge graph [20] are examples of semantically structured general-purpose *Knowledge Bases* used by open domain QA systems. Recent examples of such QA systems include PowerAqua [11], FREyA [6], QAKiS [5], and TBSL [23]. QuASE [21] is a corpus-based open domain QA system that mines answers directly from Web documents.

Each of these QA systems addresses a different subset of the space of all possible question types, input types and data sources. For example, PowerAqua finds limitation in linguistic coverage of the question, whereas TBSL overcomes this shortcoming and provides better results in linguistic analysis [23]. It would thus be desirable to combine these functionalities of [23] and [12] into a new, more powerful system.

For example, the open source web service DBpedia Spotlight [15] analyzes texts leading to named entity identification (NEI) and disambiguation (NED), using the DBpedia ontology (cf., Subsect. 3.2). AIDA [10] is a similar project, which uses the YAGO ontology (cf., Subsect. 3.2). AGDISTIS [24] is an independent NED service, which, in contrast to DBpedia Spotlight and AIDA, can use any ontology, but does not provide an interface for NEI. The PATTY system [17] provides a list of textual patterns that can be used to express properties of the YAGO and DBpedia ontologies. As these components have different levels of granularity and as there is no standard message format, combining them is not easy and demands the introduction of a higher level concept and manual work.

3 Related Work

We have already reviewed the state of the art of QA systems in Sect. 2. Work that is related to ours in a closer sense includes other frameworks that aim at providing an abstraction of QA systems, as well as other ontologies used by QA systems.

3.1 Abstract QA Frameworks

The QALL-ME framework [8] is an attempt to provide a reusable architecture for multilingual, context aware QA. QALL-ME uses an ontology to model structured data of a specific domain at a time. However, it focuses on closed domain QA, and finds limitation to get extended for heterogeneous data sources and open domain QA systems.

openQA [14] on other hand is an extensible framework for answering questions using open domain knowledge. openQA has a pipelined architecture to incorporate multiple external QA systems such as SINA [18] and TBSL to answer questions. openQA requires all components of the pipeline to be implemented in Java. The OKBQA Hackathon[4], on the other hand, is a collaborative effort to

---

[4] OKBQA Hackathon: http://2015.okbqa.org/development/documentation (last accessed: 2016-03-04).

develop knowledge bases and question answering systems that are generic and independent of programming languages.

3.2 Ontologies for Question Answering

Ontologies play an important role in question answering. First they can be used as a knowledge source to answer the questions. Prominent examples are the DBpedia Ontology and YAGO. DBpedia is a cross domain dataset of structured data extracted from Wikipedia articles (infoboxes, categories, etc.). The DBpedia Ontology is "a shallow, cross-domain ontology, which has been manually created based on the most commonly used infoboxes within Wikipedia".[5]

The YAGO ontology unifies semantic knowledge extracted from Wikipedia with the taxonomy of WordNet. The YAGO Knowledge Base contains more than 10 million entities and more than 120 million facts about these entities. YAGO links temporal and spatial dimensions to many of its facts and entities.

Ontologies can also be used to model the search process in a question answering system. For example, the research presented in [22] describes a search ontology that abstracts a user's question. One can model complex queries using this ontology without knowing a specific search engine's syntax for such queries. This approach provides a way to specify and reuse the search queries. However, the approach focuses on textual queries, i.e., it is not completely agnostic to possible question types (e.g., audio, image, ...). Search Ontology also does not cover other possibly useful properties, such as the dataset that should be used for identifying the answer.

However, so far no ontology has been developed that would provide a common abstraction to model the whole QA process.

4 Problem Statement, Requirements and Idea

Our work is motivated by existing QA systems not being sufficiently interoperable and their components not being sufficiently reusable, as pointed out in Sect. 2. Related work on abstract frameworks and QA ontologies has not yet solved the interoperability problem fully, as explained in Sect. 3. In this section, we provide a precise statement of the problem, from which we derive requirements for an abstraction, and finally present our idea for an ontology that can drive extensible QA infrastructures.

4.1 Problem Statement

Question answering systems are complex w.r.t. the components needed for an adequate quality. Sophisticated QA systems need components for NEI, NED, semantic analysis of the question, query building, query execution, result analysis, etc. Integrating multiple such components into a QA system is inconvenient

---

[5] http://wiki.dbpedia.org/services-resources/ontology (last accessed: 2016-03-08).

and inefficient, particularly considering the variety of input and output parameters with the same or similar semantics (e.g., different terms for referring to "the question", or "a range of text", or "an annotation with a linked data resource", or just plain string literals where actual resources are used). As no common vocabulary for communicating between components exists, the following situation is observable for components that need to be integrated: (1) a (new) vocabulary for input values is established, (2) a (new) vocabulary for the output values is established, (3) input or output values are represented without providing semantics (e.g., as plain text, or in JSON or XML with an ad hoc schema). Confronted with these scenarios, developers of QA systems have the responsibility to figure out the semantics of the components, which is time-consuming and error-prone. Hence, efficiently developing QA systems is desirable for the information retrieval community in industry and academics. We observed in Sects. 2 and 3 that the number of reusable (components of) QA systems is negligible so far.

4.2   Requirements

From the previous problem statement and our observations, we derived the following requirements for a vital ecosystem of QA system's components:

Req. 1 (Interoperability). *Components of question answering systems are typically implemented in different programming languages and expose interfaces using different exchange languages (e.g., XML, JSON-LD, RDF, CSV). It is not realistic to expect that a single fixed standard protocol will be established that subsumes all achievements made by domain-specific communities. However, a consistent standard interaction level is needed. Therefore, we demand a (self-describing) abstraction of the implementation.*

Req. 2 (Exchangeability and Reusability). *Different domains or scopes of application will require different components to be combined. Increasing the efficiency for developers in academia and industry requires a mechanism for making components reusable and enable a best-of-breed approach.*

Req. 3 (Flexible Granularity). *It should be possible to integrate components for each small or big step of a QA pipeline. For example, components might provide string analytics leading to Named Entity Identification (NEI) (e.g., [15]), other components might target the Named Entity Disambiguation (NED) only (e.g., [24]) and additionally there might exist components providing just an integrated interface for NEI and NED in a processing step.*

Req. 4 (Isolation). *Every component needs to be able to execute their specific step of the QA pipeline in isolation from other components. Hence, business, legal and other aspects of distributed ownership of data sources and systems can be addressed locally per component. This requirement targets the majority of the QAS platform, to enable benchmarking of components and the comparability of benchmarking results. If isolation of components is achieved, ensemble learning or similar approaches are enabled with manageable effort.*

No existing question answering system or framework for such systems fulfills these requirements. However, we assume here that fulfilling these requirements will provide the basis for a vital ecosystem of question answering system components and therefore unexpectedly increased efficiency while building question answering systems.

4.3 Idea

In this paper we are following a two step process towards integrating different components and services within a QA system.

1. On top of a standard annotation framework, the Web Annotation Data Model (WADM[6]), the qa vocabulary is defined. This generalized vocabulary covers a common abstraction of the data models we consider to be of general interest for the QA community. It is extensible and already contains properties for provenance and confidence.
2. Vocabularies used by components for question answering systems for their input and output (e.g., NIF for textual data annotations, but also any custom vocabulary) are aligned with the qa vocabulary to achieve interoperability of components. Hence, a generalized representation of the messages exchanged by the components of a QA system is established, independently of how they have been implemented and how they natively represent questions and answers.

Thereafter, the vocabulary qa provides the information needed by the components in the implemented question answering system, i.e., a self-describing, consistent knowledge base is available – fulfilling Req. 1. Hence, any component can use the vocabulary for retrieving previously annotated information and to annotate additional information (computed by itself), i.e., each component is using this knowledge base as input and output. This fact and the alignment of the component vocabularies fulfills Req. 2, as each component can be exchanged by any other component serving the same purpose with little effort, and any component can be reused in a new question answering system. Following this process might result in a message-driven architecture (cf., Sect. 7) as it was introduced earlier for search-driven processes on hybrid federated data sources (cf., [4]). However, the methodology might be implemented by different architectures.

5 Approach

5.1 Web Annotation Framework

The Web Annotation Data Model (WADM), currently a W3C Working Draft, is a framework for expressing annotations. A WADM annotation has at least a target and a body. The target indicates the resource that is described, while the body indicates the description. The basic structure of an annotation, in Turtle syntax, looks as follows:

---

[6] W3C Working Draft 15 October 2015, http://www.w3.org/TR/annotation-model.

```
<anno>  a                oa:Annotation  ;
        oa:hasTarget  <target>      ;
        oa:hasBody    <body>        .
```

Additionally the oa vocabulary provides the concept of selectors, which provide access to specific parts of the annotated resource (here: the question). Typically this is done by introducing a new oa:SpecificResource, which is annotated by the selector:

```
<mySpTarget>  a                   oa:SpecificResource  ;
              oa:hasSource    <URIQuestion>  ;
              oa:hasSelector  <mySelector>  .
<mySelector>  a                   oa:TextPositionSelector  ;
              oa:start     "n"^^xsd:nonNegativeInteger  ;
              oa:end       "m"^^xsd:nonNegativeInteger  .
```

Moreover one can indicate for each annotation the creator using the oa:annotatedBy property and the time it was generated using the oa:annotatedAt property.

5.2   Vocabulary for Question Answering Systems

In [19] we introduced the vocabulary for the Qanary approach. Following the data model requirements of question answering systems, this vocabulary – abbreviated as qa – is used for exchanging messages between components in QA systems [19].

Qanary extends the WADM such that one can express typical intermediate results that appear in a QA process. It is assumed that the question can be retrieved from a specific URI that we denote with URIQuestion. This is particularly important if the question is not a text, but an image, an audio file, a video or data structure containing several data types. URIQuestion is an instance of an annotation class called qa:Question. The question is annotated with two resources URIAnswer and URIDataset of types qa:Answer and qa:Dataset respectively. All of these new concepts are subclasses of oa:Annotation. Hence, the minimal structure of all concepts is uniform (provenance, service URL, and confidence are expressible via qa:Creator, oa:annotatedBy, and qa:score) and the concepts can be extended to more precise annotation classes.

These resources are further annotated with information about the answer (like the expected answer type, the expected answer format and the answer itself) and information about the dataset (like the URI of an endpoint expressing where the target data set is available). This model is extensible since each additional information that needs to be shared between components can be added as a further annotation to existing classes. For example, establishing an annotation of the question is possible by defining a new annotation class qa:AnnotationOfQuestion (using OWL Manchester Syntax):

```
Class: qa:AnnotationOfQuestion
EquivalentTo: oa:Annotation that oa:hasTarget some qa:Question
```

For additional information about the vocabulary we refer to [19].

5.3 Integration of (External) Component Interfaces

Following the Qanary approach, existing vocabularies should not be overturned. Instead, any information that is useful w.r.t. the task of question answering will have to be aligned to Qanary to be integrated on a logical level, while the domain-specific information remains available. Hence, we provide a standardized interface for interaction while preserving the richness of existing vocabularies driven by corresponding communities or experts. Existing vocabularies will be aligned to Qanary via axioms or rules. These alignment axioms or rules will typically have the expressiveness of first-order logic and might be implemented using OWL subclass/subproperty or class/property equivalence axioms as far as possible, using SPARQL *CONSTRUCT* or *INSERT* queries, or in the Distributed Ontology Language DOL, a language that enables heterogeneous combination of ontologies written in different languages and logics [16]. The application of these alignment axioms or rules by a reasoner or a rule engine will translate information from the Qanary knowledge base to the input representation understood by a QA component (if it is RDF-based), and it will translate the RDF output of a component to the Qanary vocabulary, such that it can be added to the knowledge base. Hence, after each processing step a consolidated representation of the available knowledge about the question is available.

Each new annotation class (with a specific semantics) can be derived from the existing annotation classes. Additionally, the semantics might be strengthened by applying restrictions to `oa:hasBody` and `oa:hasTarget`.

6 Alignment of Component Vocabularies

Our goal in this section is to provide a methodology for binding the `qa` vocabulary to existing ones used by QA systems. Of course, it is not possible to provide a standard solution for bindings of all existing vocabularies due to the variety of expressing information. However, here we provide three typical solution patterns matching standard use cases and presenting the intended behavior.

As running example we consider an implemented exemplary question answering system with a pipeline of three components (NEI + NED, relation detection, and query generation and processing; cf., Sect. 7). In the following the components are described briefly and also a possible alignment implementation of the custom vocabulary to `qa`.

6.1 NE Identification and Disambiguation via DBpedia Spotlight

DBpedia Spotlight [15] provides the annotated information via a JSON interface. An adapter was implemented translating the untyped properties DBpedia Spotlight is returning into RDF using NIF. On top of this service we developed a reusable service that aligns the NIF concepts with the annotations of `qa`. First we need to align the implicit NIF selectors defining the identified named entities with the `oa:TextPositionSelector` while aligning the `oa:TextPositionSelector` with `nif:String` on a logical level iff `nif:beginIndex` and `nif:endIndex` exist. This is expressed by the following first-order rule:

$$\text{rdf:type}(?s, \text{nif:String}) \wedge \text{nif:beginIndex}(?s, ?b) \wedge \text{nif:endIndex}(?s, ?e)$$
$$\implies (\exists ?x \bullet \text{rdf:type}(?x, \text{oa:TextPositionSelector}) \wedge \text{oa:start}(?x, ?b) \wedge \text{oa:end}(?x, ?e)) \tag{1}$$

Additionally the identified resource of the named entity (`taIdentRef` of the vocabulary `itsrdf`) needs to be constructed as annotation. We encode this demanded behavior with the following rule:

$$\text{itsrdf:taIdentRef}(?s, ?NE) \wedge \text{nif:confidence}(?s, ?conf)$$
$$\implies \text{rdfs:subClassOf}(\text{qa:AnnotationOfEnitites}, \text{oa:AnnotationOfQuestion}) \wedge$$
$$(\exists ?sp \bullet \text{rdfs:type}(?sp, \text{oa:SpecificResource}) \wedge \text{oa:hasSource}(?sp, < \text{URIQuestion} >) \wedge \tag{2}$$
$$\text{oa:hasSelector}(?sp, ?s)) \wedge (\exists ?x \bullet \text{rdfs:type}(?x, \text{oa:AnnotationOfNE}) \wedge$$
$$\text{oa:hasBody}(?x, ?NE) \wedge \text{oa:hasTarget}(?x, ?sp) \wedge \text{qa:score}(?x, ?conf))$$

Figure 1 shows our SPARQL implementations of this rule. After applying this rule, named entities and their identified resources are available within the qa vocabulary.

```
PREFIX itsrdf: <http://www.w3.org/2005/11/its/rdf#>
PREFIX nif:    <http://persistence.uni-leipzig.org/nlp2rdf/ontologies/nif-core#>
PREFIX qa:     <http://www.wdaqua.eu/qa#>
PREFIX oa:     <http://www.w3.org/ns/openannotation/core/>

INSERT {
    ?s a oa:TextPositionSelector .
    ?s oa:start ?begin .
    ?s oa:end ?end .
    ?x a qa:AnnotationOfNE .
    ?x oa:hasBody ?NE .
    ?x oa:hasTarget [  a     oa:SpecificResource;
                       oa:hasSource     <URIQuestion>;
                       oa:hasSelector   ?s  ] .
    ?x qa:score ?conf .
    ?x oa:annotatedBy 'DBpedia_Spotlight_wrapper' .
    ?x oa:annotatedAt ?time
} WHERE { SELECT ?x ?s ?NE ?begin ?end ?conf
          WHERE { graph <http://www.wdaqua.eu/qa#tmp> {
                       ?s itsrdf:taIdentRef ?NE .
                  ?s nif:beginIndex ?begin .
                  ?s nif:endIndex ?end .
                  ?s nif:confidence ?conf .
                  BIND (IRI(CONCAT(str(?s),'#',str(RAND())))) AS ?x) .
                  BIND(now() as ?time) .
} } };
```

Fig. 1. Aligning identified NE to a new **qa** annotation using SPARQL

6.2   Relation Detection Using PATTY Lexicalization

PATTY [17] can be used to provide lexical representation of DBpedia properties. Here we created a service that uses the lexical representation of the properties to detect the relations in a question. The service adds annotations of type `qa:AnnotationOfEntity`. Consequently, the question is annotated by a selector and a URI pointing to a DBpedia resource comparable to the processing in Fig. 1.

For example, the question "Where did Barack Obama graduate?" will now contain the annotation:

```
PREFIX dbo: <http://dbpedia.org/ontology/>
 <urn:uuid:a...> a oa:TextPositionSelector ;
          oa:start "24"^^xsd:nonNegativeInteger ;
          oa:end  "33"^^xsd:nonNegativeInteger ;
 <urn:uuid:b...> a qa:AnnotationOfEntity ;
          oa:hasBody dbo:almaMater ;
          oa:hasTarget [  a                oa:SpecificResource  ;
                          oa:hasSource     <URIQuestion> ;
                          oa:hasSelector   <urn:uuid:a...> ] ;
          qa:score "23"^^xsd:decimal ;
          oa:annotatedBy <http://wdaqua.example/Patty> ;
          oa:annotatedAt "2015-12-19T00:00:00Z"^^xsd:dateTime .
```

In our use case the PATTY service just extends the given vocabulary. Hence, components within a QA system called after the PATTY service will not be forced to work with a second vocabulary. Additionally, the service might be replaced by any other component implementing the same purpose (Reqs. 2 and 4 are fulfilled).

6.3   Query Construction and Query Execution via SINA

SINA [18] is an approach for semantic interpretation of user queries for question answering on interlinked data. It uses a Hidden Markov Model for disambiguating entities and resources. Hence, it might use the triples identifying entities while using the annotation of type `qa:AnnotationOfEntity`, e.g., for "Where did Barack Obama graduate?" the entities http://dbpedia.org/resource/Barack_Obama and http://dbpedia.org/ontology/almaMater are present and can be used. The SPARQL query generated by SINA as output is a formal representation of a natural language query. We wrap SINA's output into RDF as follows:

```
PREFIX sparqlSpec: <http://www.w3.org/TR/sparql11-query/#>
<urn:uuid:...> sparqlSpec:select "SELECT * WHERE {
  <http://dbpedia.org/resource/Barack_Obama>
  <http://dbpedia.org/ontology/almaMater> ?v0 . }".
```

As this query, at the same time, implicitly defines a *result set*, which needs to be aligned with the `qa:Answer` concept and its annotations. We introduce a new annotation `oa:SparqlQueryOfAnswer`, which holds the SPARQL query as its body.

sparqlSpec:select($?x, ?t$) $\land$ rdf:type($?t$, xsd:string)

$\implies$ rdfs:subClassOf(oa:SparqlQueryOfAnswer, oa:AnnotationOfAnswer) $\land$

$(\exists?x \bullet$ rdfs:type($?x$, oa:SparqlQueryOfAnswer) $\land$ oa:target($?x$, <URIAnswer>)$\land$

$$\tag{3}$$

oa:body($?x$, "SELECT ... "))

The implementation of this rule as a SPARQL INSERT query is straight-forward and omitted due to space constraints. Thereafter, the knowledge base of the question contains an annotation holding the information which SPARQL query needs to be executed by a query executor component to obtain the (raw) answer.

6.4 Discussion

In this section we have shown how to align component-specific QA vocabularies. Following our Qanary approach each component's knowledge about the current question answering task will be aligned with the qa vocabulary. Hence, while using the information of the question answering system for each component there is no need of knowing other vocabularies than qa. However, the original information is still available and usable. In this way Req. 4 is fulfilled, and we achieving Req. 2 by being able to exchange every component.

Note that the choice of how to implement the alignments depends on the power of the triple store used. Hence, more elegant vocabulary alignments are possible but are not necessarily usable within the given system environment (e.g., an alternative alignment for Sect. 6.1, implemented as an OWL axiom, is given in the online appendix[7]).

Here our considerations finish after the creation of a *SELECT* query from an input question string. A later component should execute the query and retrieve the actual resources as result set. This result set will also be used to annotate URIAnswer to make the content available for later processing (e.g., HCI compo-nents).

7   Case Study

In this section we present a QA system that follows the idea presented in Sect. 4.3. Note that in this paper our aim was not to present a pipeline that performs better by quantitative criteria (e.g., F-measure) but to show that the alignment of isolated, exchangeable components is possible in an architecture derived from the Qanary methodology. In this paper, we have extended the vocabulary proposed in [19] to align individual component vocabularies together to integrate them into a working QA architecture. Without such an alignment, these components cannot be integrated easily together because of their heterogeneity.

---

[7] Alternative alignment: https://goo.gl/hdsaq4.

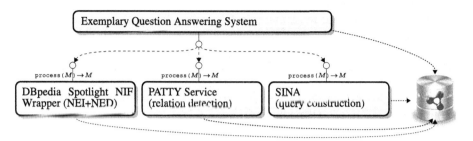

Fig. 2. Architecture of the exemplary question answering system.

Our exemplary QA system consists of three components: DBpedia Spotlight for named entity identification and disambiguation, a service using the relational lexicalizations of PATTY for relation detection, and the query builder of SINA. All information about a question is stored in a named graph of a triple store using the QA vocabulary. As a triple store, we used Stardog[8].

The whole architecture is depicted in Fig. 2. Initially the question is exposed by a web server under some URI, which we denote by URIQuestion. Then a named graph reserved for the specific question is created. The WADM and the qa vocabularies are loaded into the named graph together with the predefined annotations over URIQuestion described in Sect. 5.2. Step by step each component receives a message M (cf., Fig. 2) containing the URI where the triple store can be accessed and the URI of the named graph reserved for the question and its annotations. Hence, each component has full access to all the messages generated by the previous components through SPARQL *SELECT* queries and can update that information using SPARQL *UPDATE* queries. This in particular allows each component to see what information is already available. Once a component terminates, a message is returned to the question answering system, containing the endpoint URI and the named graph URI (i.e., the service interface is defined as process(M) $\rightarrow$ M). Thereafter, the retrieved URI of the triple store and the name of the named graph can be passed by the pipeline to the next component.

Now let us look into detail about the working of each component.

The first component wraps DBpedia Spotlight and is responsible for linking the entities of the question to DBpedia resources. First it retrieves the URI of the input question from the triple store and then downloads the question from that URI. It passes the question to the external service DBpedia Spotlight by using its REST interface. The DBpedia Spotlight service returns the linked entities. The raw output of DBpedia Spotlight is transformed using the alignment from Subsect. 6.1 to update the information in the triple store with the detected entities.

The second component retrieves the question from the URI and analyses of all parts of the question for which the knowledge base does not yet contain

---

[8] http://stardog.com/, community edition, version 4.0.2.

annotations. It finds the most suitable DBpedia relation corresponding to the question using the PATTY lexicalizations. These are then updated in the triple store (cf., Sect. 6.2).

The third component ignores the question and merely retrieves the resources with which the question was annotated directly from the triple store. The query generator of SINA is then used to construct a SPARQL query which is then ready for sending to the DBpedia endpoint.

We implemented the pipeline in Java but could have used any other language as well. The implementation of each component requires just a few lines of code (around 2–3 KB of source code); in addition, we had to implement wrappers for DBpedia Spotlight and PATTY (4–5 KB each) to adapt their input and output (e.g., to provide DBpedia Spotlight's output as NIF). Note that this has to be done just once for each component. The components can be reused for any new QAS following the Qanary approach.

Overall, it is important to note that the output of each component is not merely passed to the next component just like other typical pipeline architecture, but every time when an output is generated, the triple store is enriched with the knowledge of the output. Hence, it is a message-driven architecture built on-top of a self-describing blackboard-style knowledge base containing valid information of the question. Each component fetches the information that it needs from the triple store by itself.

In conclusion, the case study clearly shows the power of the approach. The knowledge representation is valid and consistent using linked data technology. Moreover, each component is now isolated (cf., Req. 4), exchangeable and reusable (cf., Req. 2), as the exchanged messages follow the qa vocabulary (cf., Req. 1), which contains the available pieces of information about the question, and their provenance and confidence. The components are independent and lightweight, as the central triple store holds all knowledge and takes care of querying and reasoning. As Qanary does not prescribe an execution order or any other processing steps, Req. 3 is also fulfilled.

The case study is available as online appendix[9].

8   Conclusion and Future Work

We have presented an extensible, generalized architecture for question answering systems. The idea is driven by the observation that, while many question answering systems have been created in the last years, the number of reusable components among them are still negligible. Hence, the creation of new question answering systems is cumbersome and inefficient at the moment. Most of the created QA systems are monolithic in their implementation; neither the systems nor their components can be reused. To overcome this problem, our approach follows the linked data paradigm to establish a self-describing vocabulary for messages exchanged between the components of a QA system. Qanary – the question

---

[9] https://github.com/WDAqua/Pipeline.

answering vocabulary – covers the requirements of open question answering systems and their integrated components. However, our goal is not to establish an independent solution. Instead, by using the methodology of annotations, Qanary is designed to enable the alignment with existing/external vocabularies, and it provides provenance and confidence properties as well.

On the one hand, developers of the components for question answering (e.g., question analyses, query builder, ...) can now easily use our standard vocabulary and also have descriptive access to the knowledge available for the question via SPARQL. Additionally, aligning the knowledge of such components with our vocabulary and enabling them for broader usage within question answering systems is now possible. Fulfilling the requirements (cf., Reqs. 1–4) this ultimately sets the foundation for rapidly establishing new QA systems. A main advantage of our approach are the reusable ontology alignments, increasing the efficiency and the exchangeability in an open QA system.

Our contribution to the community is a vocabulary and a methodology, which take into account the major problems while designing (complex) question answering systems. Via alignments, our vocabulary is extensible with well-known vocabularies while preserving standard information such as provenance. This enables best-of-breed QA approaches where each component can be exchanged according to considerations about quality, domains or fields of application. Additionally, meta approaches such as ensemble learning can be applied easily. Hence, the approach presented in this paper provides a clear advantage in comparison to earlier closed monolithic approaches. Eventually, for the first time the foundations for a vital ecosystem for components of question answering systems is on the horizon. The paper already provides some components and the alignment of their vocabulary to the Qanary vocabulary. In the future, we will integrate further available components by implementing wrappers for them and specifying ontology alignments. Checking the logical consistency of alignments is also a future issue. Additionally an extension for benchmarking and a corresponding framework is planned to be established.

Acknowledgments. Parts of this work received funding from the European Union's Horizon 2020 research and innovation programme under the Marie Skłodowska-Curie grant agreement No. 642795, project: Answering Questions using Web Data (WDAqua). We would like to thank the anonymous peer reviewers for their constructive feedback.

References

1. Abacha, A.B., Zweigenbaum, P.: Medical question answering: translating medical questions into SPARQL queries. In: ACM IHI (2012)
2. Athenikos, S.J., Han, H.: Biomedical question answering: a survey. Comput. Methods Programs Biomed. **99**(1), 1–24 (2010)

3. Auer, S., Bizer, C., Kobilarov, G., Lehmann, J., Cyganiak, R., Ives, Z.G.: DBpedia: a nucleus for a web of open data. In: Aberer, K., Choi, K.-S., Noy, N., Allemang, D., Lee, K.-I., Nixon, L.J.B., Golbeck, J., Mika, P., Maynard, D., Mizoguchi, R., Schreiber, G., Cudré-Mauroux, P. (eds.) ASWC 2007 and ISWC 2007. LNCS, vol. 4825, pp. 722–735. Springer, Heidelberg (2007)
4. Both, A., Ngomo, A.-C.N., Usbeck, R., Lukovnikov, D., Lemke, C., Speicher, M.: A service-oriented search framework for full text, geospatial and semantic search. In: SEMANTiCS (2014)
5. Cabrio, E., Cojan, J., Aprosio, A.P., Magnini, B., Lavelli, A., Gandon, F.: QAKiS: an open domain QA system based on relational patterns. In: Proceedings of the ISWC 2012 Posters & Demonstrations Track (2012)
6. Damljanovic, D., Agatonovic, M., Cunningham, H.: FREyA: an interactive way of querying linked data using natural language. In: García-Castro, R., Fensel, D., Antoniou, G. (eds.) ESWC 2011. LNCS, vol. 7117, pp. 125–138. Springer, Heidelberg (2012)
7. Dietterich, T.G.: Ensemble learning. In: Arbib, M.A. (ed.) The Handbook of Brain Theory and Neural Networks. The MIT Press, Cambridge (2002)
8. Ferrández, Ó., Spurk, C., Kouylekov, M., Dornescu, I., Ferrández, S., Negri, M., Izquierdo, R., Tomás, D., Orasan, C., Neumann, G., Magnini, B., González, J.L.V.: The QALL-ME framework: a specifiable-domain multilingual Question Answering architecture. J. Web Sem. 9(2), 137–145 (2011)
9. Hellmann, S., Lehmann, J., Auer, S., Brümmer, M.: Integrating NLP using linked data. In: Alani, H., Kagal, L., Fokoue, A., Groth, P., Biemann, C., Parreira, J.X., Aroyo, L., Noy, N., Welty, C., Janowicz, K. (eds.) ISWC 2013, Part II. LNCS, vol. 8219, pp. 98–113. Springer, Heidelberg (2013)
10. Ibrahim, Y., Yosef, M.A., Weikum, G.: Aida-social: entity linking on the social stream. In: Exploiting Semantic Annotations in Information Retrieval (2014)
11. Lopez, V., Fernández, M., Motta, E., Stieler, N.: PowerAqua: supporting users in querying and exploring the semantic web. Semant. Web 3(3), 249–265 (2011)
12. Lopez, V., Motta, E., Sabou, M., Fernandez, M.: PowerAqua: a multi-ontology based question answering system-v1. OpenKnowledge Deliverable D8.4 (2007)
13. Lopez, V., Uren, V., Sabou, M., Motta, E.: Is question answering fit for the semantic web? a survey. Semant. Web 2(2), 125–155 (2011)
14. Marx, E., Usbeck, R., Ngomo, A.-C.N., Höffner, K., Lehmann, J., Auer, S.: Towards an open question answering architecture. In: SEMANTiCS (2014)
15. Mendes, P.N., Jakob, M., García-Silva, A., Bizer, C.: DBpedia spotlight: shedding light on the web of documents. In: I-SEMANTICS (2011)
16. Mossakowski, T., Kutz, O., Lange, C.: Three semantics for the core of the Distributed Ontology Language. In: Formal Ontology in Information Systems (2012)
17. Nakashole, N., Weikum, G., Suchanek, F.M.: PATTY: a taxonomy of relational patterns with semantic types. In: EMNLP-CoNLL (2012)
18. Shekarpour, S., Marx, E., Ngomo, A.-C.N., Auer, S.: SINA: semantic interpretation of user queries for question answering on interlinked data. Web Semant. Sci. Serv. Agents WWW 30, 39–51 (2015)
19. Singh, K., Both, A., Diefenbach, D., Shekarpour, S.: Towards a message-driven vocabulary for promoting the interoperability of question answering systems. In: Proceedings of the 10th IEEE International Conference on Semantic Computing (ICSC) (2016)
20. Singhal, A.: Introducing the knowledge graph: things, not strings. Official Google Blog, May 2012

21. Sun, H., Ma, H., Yih, W.-T., Tsai, C.-T., Liu, J., Chang, M.-W.: Open domain question answering via semantic enrichment. In: WWW (2015)
22. Uciteli, A., Goller, C., Burek, P., Siemoleit, S., Faria, B., Galanzina, H., Weiland, T., Drechsler-Hake, D., Bartussek, W., Herre, H.: Search ontology, a new approach towards semantic search. In: Workshop on Future Search Engines (2014)
23. Unger, C., Bühmann, L., Lehmann, J., Ngomo, A.-C.N., Gerber, D., Cimiano, P.: Template-based question answering over RDF data. In: WWW (2012)
24. Usbeck, R., Ngonga Ngomo, A.-C., Röder, M., Gerber, D., Coelho, S.A., Auer, S., Both, A.: AGDISTIS - graph-based disambiguation of named entities using linked data. In: Mika, P., Tudorache, T., Bernstein, A., Welty, C., Knoblock, C., Vrandečić, D., Groth, P., Noy, N., Janowicz, K., Goble, C. (eds.) ISWC 2014, Part I. LNCS, vol. 8796, pp. 457–471. Springer, Heidelberg (2014)

Test-Driven Development of Ontologies

C. Maria Keet[1(✉)] and Agnieszka Ławrynowicz[2]

[1] Department of Computer Science, University of Cape Town,
Cape Town, South Africa
mkeet@cs.uct.ac.za
[2] Institute of Computing Science, Poznan University of Technology,
Poznań, Poland
agnieszka.lawrynowicz@cs.put.poznan.pl

Abstract. Emerging ontology authoring methods to add knowledge to an ontology focus on ameliorating the validation bottleneck. The verification of the newly added axiom is still one of trying and seeing what the reasoner says, because a systematic testbed for ontology authoring is missing. We sought to address this by introducing the approach of test-driven development for ontology authoring. We specify 36 generic tests, as TBox queries and TBox axioms tested through individuals, and structure their inner workings in an 'open box'-way, which cover the OWL 2 DL language features. This is implemented as a Protégé plugin so that one can perform a TDD test as a black box test. We evaluated the two test approaches on their performance. The TBox queries were faster, and that effect is more pronounced the larger the ontology is.

1 Introduction

The process of ontology development has progressed much over the past 20 years, especially by the specification of high-level, information systems-like methodologies [8,25], and both stand-alone and collaborative tools [9,10]. But support for effective low-level *ontology authoring*—adding the right axioms and adding the axioms right—has received some attention only more recently. Processes at this 'micro' level of the development may use the reasoner to propose axioms with FORZA [13], use Ontology Design Patterns (ODPs) [6], and repurpose ideas from software engineering practices, notably exploring the notion of unit tests [27], eXtreme Design with ODPs [3], and Competency Question (CQ)-based authoring using SPARQL [23].

However, testing whether a CQ can be answered does not say how to add the knowledge represented in the ontology, FORZA considers simple object properties only, and eXtreme Design limits one to ODPs that do not come out of the blue but have been previously prepared. Put differently, there is no systematic testbed for ontology engineering, other than manual efforts by a knowledge engineer to add or change something and running the reasoner to check its effects. This still puts a high dependency on expert knowledge engineering, which ideally should not be in the realm of an art, but be rather at least a systematic process for good practices.

© Springer International Publishing Switzerland 2016
H. Sack et al. (Eds.): ESWC 2016, LNCS 9678, pp. 642–657, 2016.
DOI: 10.1007/978-3-319-34129-3_39

We aim to address this problem by borrowing another idea from software engineering: *test-driven development* (TDD) [2]. TDD ensures that what is added to the program core (here: ontology) does indeed have the intended effect specified upfront. Moreover, TDD in principle is cognitively a step up from the 'add stuff and lets see what happens'-attitude, therewith deepening the understanding of the ontology authoring process and the logical consequences of an axiom.

There are several scenarios of TDD usage in ontology authoring:

I. *CQ-driven TDD* Developers (domain experts, knowledge engineers etc.) specify CQs. A CQ is translated automatically into one or more axioms. The axiom(s) are the input of the relevant TDD test(s) to be carried out. The developers who specify the CQs could be oblivious to the inner workings of the two-step process of translating the CQ and testing the axiom(s).

II-a. *Ontology authoring-driven TDD - the knowledge engineer* The knowledge engineer knows which axiom s/he wants to add, types it, which is then fed directly into the TDD system.

II-b. *Ontology authoring-driven TDD - the domain expert* As there is practically a limited amount of 'types' of axioms to add, one could create templates, alike the notion of the "logical macro" ODP [22], which then map onto *generic*, domain-independent tests (as will be specified in Sect. 3). For instance, a domain expert could choose the all-some template from a list, i.e., an axiom of the form $C \sqsubseteq \exists R.D$. The domain expert instantiates it with relevant domain entities (e.g., Professor $\sqsubseteq \exists$teaches.Course), and the TDD test for the $C \sqsubseteq \exists R.D$ type of axiom is then run automatically. The domain expert need not know the logic, but behind the usability interface, what gets sent to the TDD system is that axiom.

While in each scenario the actual testing can be hidden from the user's view, it is necessary to specify what actually happens during such testing and how it is tested. Here, we assume that either the first step of the CQ process is completed, or the knowledge engineer adds the axiom, or that the template is populated, respectively; i.e., that we are at the stage where the axioms are fed into the TDD test system. To realise the testing, a number of questions have to be answered:

1. Given the TDD procedure in software engineering—check that the desired feature is absent, code it, test again (*test-first* approach)—then what does that mean for ontology testing when transferred to ontology development?
2. TDD requires so-called *mock objects* for 'incomplete' parts of the code; is there a parallel to it in ontology development, or can that be ignored?
3. In what way and where (if at all) can this be integrated as a methodological step in existing ontology engineering methodologies that are typically based on waterfall, iterative, or lifecycle principles?

To work this out for ontologies, we take some inspiration from TDD for conceptual modelling. Tort et al. [26] essentially specify 'unit tests' for each feature/possible addition to a conceptual model, and test such an addition against sample individuals. Translating this to OWL ontologies, such testing is possible by means of ABox individuals, and then instead of using an ad hoc algorithm, one can avail of the automated reasoner. In addition, for ontologies, one can avail of a query language for the TBox, namely, SPARQL-OWL [15], and most of the tests can be specified in that language as well. We define TBox and ABox-driven TDD tests for the basic axioms one can add to an OWL 2 DL ontology. To examine practical feasibility for the ontology engineer and determine which TDD strategy is the best option, we implemented the TDD tests as a Protégé plugin and evaluated it on performance by comparing TBox and ABox TDD tests for 67 ontologies. The TBox TDD tests outperform the ABox ones except for disjointness and this effect is more pronounced with larger ontologies. Overall, we thus add a new mechanism and tool to the ontology engineer's 'toolbox' to enable systematic development of ontologies in an agile way.

The remainder of the paper is structured as follows. Section 2 describes related works on TDD in software and ontology development. Section 3 summarises the TDD tests and Sect. 4 evaluates them on performance with the Protégé plugin. We discuss in Sect. 5 and conclude in Sect. 6. Data, results, and more detail on the TDD test specifications is available at https://semantic.cs. put.poznan.pl/wiki/aristoteles/doku.php.

2   Related Work

To 'transfer' TDD to ontology engineering, we first summarise preliminaries about TDD from software engineering and subsequently discuss related works on tests in ontology engineering.

TDD in Software Development. TDD was introduced as a software development methodology where one writes new code only if an automated test has failed [2]. TDD permeates the whole development process, which can be summarised as: (1) Write a test for a piece of functionality (that was based on a requirement), (2) Run all tests to check that the new test fails, (3) Write relevant code that passes the test, (4) Run the specific test to verify it passes, (5) Refactor the code, and (6) Run all tests to verify that the changes to the code did not change the external behaviour of the software (regression testing) [24]. The important difference with unit tests, is that TDD is a *test-first* approach rather than *test-last* (design, code, test). TDD results in being more focussed, improves communication, improves understanding of required software behaviour, and reduces design complexity [17]. Quantitatively, TDD produced code passes more externally defined tests—i.e, better software quality—and involves less time spent on debugging, and experiments showed that it is significantly more productive than test-last [11].

TDD has been applied to conceptual data modelling, where each language feature has its own test specification in OCL that involves creating the objects that should, or ought not to, instantiate the UML classes and associations [26]. Tort and Olivé's tool was evaluated with modellers, which made clear, among others, that more time was spent on developing and revising the conceptual model to fix errors than on writing the test cases [26].

Tests in ontology engineering. In ontology engineering, an early explorative work on borrowing the notion of testing from software engineering is described in [27], which explores several adaptation options: testing with the axiom and its negation, formalising CQs, checks by means on integrity constraints, autoepistemic operators, and domain and range assertions. Working with CQs has shown to be the most popular approach, notably [23], who analyse CQs and their patterns for use with SPARQL queries that then would be tested against the ontology. Their focus is on individuals and the formalisation stops at *what* has to be tested, not *how* that can, or should, be done. Earlier work on CQs and queries include the OntologyTest tool for ABox instances, which specifies different types of tests, such as "instantiation tests" (instance checking) and "recovering tests" (query for a class' individuals) and using mock individuals where applicable [7]; other instance-oriented test approaches is RDF/Linked Data [16]. There is also an eXtreme Design NeON plugin with similar functionality and ODP rapid design [3,21], likewise with RapidOWL [1], which lacks the types of tests, and a more basic variant exists in the EFO Validator[1]. Neither are based on the principle of TDD. The only one that aims to zoom in on unit tests for TBox testing requires the tests to be specified in Clojure and the ontology in Tawny-Owl notation, describes subsumption tests only [28], and the tests are tailored to the actual ontology rather than reusable 'templates' for the tests covering all OWL language features.

Related notions have been proposed in methods for particular types of axioms, such as disjointness [5] and domain and range constraints [13]. Concerning methodologies, none of the 9 methodologies reviewed by [8] are TDD-based, nor is NeON [25]. The Agile-inspired OntoMaven [20] has OntoMvnTest with 'test cases' only for the usual syntax checking, consistency, and entailment [20].

Thus, full TDD ontology engineering has not been proposed yet. While the idea of unit tests—which potentially could become part of TDD tests—has been proposed, there is a dearth of actual specifications as to what exactly is, or should be, going on in such as test. Even when one were to specify basic tests for each language feature, it is unclear whether they can be put together in a modular fashion for the more complex axioms that can be declared with OWL 2. Further, there is no regression testing to check that perhaps an earlier modelled CQ—and thus a passed test—conflicts with a later one.

3   TDD Specification for Ontologies

First the general procedure and preliminaries are introduced, and then the TBox and RBox TDD tests are summarised.

3.1   Preliminaries on Design and Notation of the TDD Tests

The generalised TDD test approach is summarised as follows for the default case:

---

[1] http://www.ebi.ac.uk/fgpt/sw/efovalidator/index.html.

1. input: CQ into axiom, axiom, or template into axiom.
2. given: axiom x of type X to be added to the ontology.
3. check the vocabulary elements of x are in ontology O (itself a TDD test)
4. run TDD test twice:
   (a) the first execution should fail (check $O \nvDash x$ or not present),
   (b) update the ontology (add x), and
   (c) run the test again which then should pass (check that $O \models x$) and such that there is no new inconsistency or undesirable deduction
5. Run all previous successful tests, which should pass (i.e., regression testing).

There are principally two options for the TDD tests: a test at the TBox-level or always using individuals explicitly asserted in the ABox. We specify tests for both approaches, where possible. For the test specifications, we use the OWL 2 notation for the ontology's vocabulary: $C, D, E, ... \in V_C$, $R, S, ... \in V_{OP}$, and $a, b, ... \in V_I$, and SPARQL-OWL notation [15] where applicable, as it conveniently reuses OWL functional syntax-style notation merged with SPARQL's queried objects (i.e., ?x) for the formulation of the query. For instance, $\alpha \leftarrow$ SubClassOf (?x D) will return all subclasses of class D. Details of SPARQL-OWL and its implementation are described in [15].

Some TBox and all ABox tests require additional classes or individuals for testing purposes only, which resembles the notion of *mock objects* in software engineering [14,18]. We shall import this notion into the ontology setting, as *mock class* for a temporary OWL class created for the TDD test, *mock individual* for a temporary ABox individual, and *mock axiom* for a temporary axiom. These mock entities are to be removed from the ontology after completion of the test.

Steps 3 and 4a in the sequence listed above may give an impression of epistemic queries. It has to be emphasised that there is a fine distinction between (1) checking when an element is in the vocabulary of the TBox of the ontology (in V_C or V_{OP}) versus autoepistemic queries, and (2) whether something is *logically* true or false versus a *test* evaluating to true or false. In the TDD context, the epistemic-sounding 'not asserted in or inferred from the ontology' is to be understood in the context of a *TDD test*, like whether an ontology has some class C in its vocabulary, not whether it is 'known to exist' in one's open or closed world. Thus, an epistemic query language is not needed for the TBox tests.

3.2   Generic Test Patterns for TBox Axioms

The tests are introduced in pairs, where the primed test names concern the tests with individuals; they are written in SPARQL-OWL notation. They are presented in condensed form due to space limitations. The TDD tests in algorithm-style notation are available in an extended technical report of this paper [12].

Class subsumption, $Test_{cs}$ or $Test'_{cs}$. When the axiom to add is of type $C \sqsubseteq D$, with C and D named classes, then $O \models \neg(C \sqsubseteq D)$ should be true if it were not present. Logically, then in the tableau, $O \cup \neg(\neg(C \sqsubseteq D))$ should be inconsistent, i.e., $O \cup (\neg C \sqcup D)$. Given the current Semantic Web technologies, it is easier to query the ontology for the subclasses of D and to ascertain that

C is not in query answer α rather than create and execute tailor-made tableau algorithms:

Test$_{cs}$ $=\alpha \leftarrow$ SubClassOf(?xD). *IfC $\notin \alpha$, then $C \sqsubseteq D$ is neither asserted nor entailed in the ontology; the test fails.* ◂

After adding $C \sqsubseteq D$ to the ontology, the same test is run, which should evaluate to $C \in \alpha$ and therewith $Test_{cs}$ returns 'pass'. The TTD test with individuals checks whether an instance of C is also an instance of D:

Test$'_{cs}$ $=$ *Create a mock object a and assert $C(a)$. $\alpha \leftarrow$ Type(?x D). If $a \notin \alpha$, then $C \sqsubseteq D$ is neither asserted nor entailed in the ontology.* ◂

Class disjointness, $Test_{cd}$ or $Test'_{cd}$. One can assert the complement, $C \sqsubseteq \neg D$, or disjointness, $C \sqcap D \sqsubseteq \bot$. Let us consider the former first (test $Test_{cd_c}$), such that then $\neg(C \sqsubseteq \neg D)$ should be true, or $T(C \sqsubseteq \neg D)$ false (in the sense of 'not be in the ontology'). Testing for the latter only does not suffice, as there are more cases where $O \not\vDash C \sqsubseteq D$ holds, but disjointness is not really applicable— being classes in distinct sub-trees in the TBox—or holds when disjointness is asserted already, which is when C and D are sibling classes. For the complement, we simply can query for it in the ontology:

Test$_{cd_c}$ $= \alpha \leftarrow$ ObjectComplementOf(C ?x). *If $D \notin \alpha$, then $O \not\vDash C \sqsubseteq \neg D$; hence, the test fails.* ◂

For $C \sqcap D \sqsubseteq \bot$, the test is:

Test$_{cd_d}$ $= \alpha \leftarrow$ DisjointClasses(?x D). *If $C \notin \alpha$, then $O \not\vDash C \sqcap D \sqsubseteq \bot$.* ◂

The ABox option uses a query or classification; availing of the reasoner only:

Test$'_{cd}$ $=$ *Create individual a, assert $C(a)$ and $D(a)$. ostate $\leftarrow$ Run the reasoner. If ostate is consistent, then either $O \not\vDash C \sqsubseteq \neg D$ or $O \not\vDash C \sqcap D \sqsubseteq \bot$ directly or through one or both of their superclasses (test fails). Else, the ontology is inconsistent (test passed); thus either $C \sqsubseteq \neg D$ or $C \sqcap D \sqsubseteq \bot$ is already asserted among both their superclasses or among C or D and a superclass of D or C, respectively.* ◂

Further, from a modelling viewpoint, it would make sense to also require C and D to be siblings. The sibling requirement can be added as an extra check in the interface to alert the modeller to it, but not be enforced from a logic viewpoint.

Class equivalence, $Test_{ce}$ and $Test'_{ce}$. When the axiom to add is of the form $C \equiv D$, then $O \vDash \neg(C \equiv D)$ should be true before the edit, or $O \not\vDash C \equiv D$ false. The latter is easier to test—run $Test_{cs}$ twice, once for $C \sqsubseteq D$ and once for $D \sqsubseteq C$—or use one SPARQL-OWL query:

Test$_{ce}$ $= \alpha \leftarrow$ EquivalentClasses(?x D). *If $C \notin \alpha$, then $O \not\vDash C \equiv D$; the test fails.* ◂

Note that D can be complex here, but C cannot. For class equivalence with individuals, we can extend $Test'_{cs}$:

Test$'_{ce}$ $=$ *Create a mock object a, assert $C(a)$. Query $\alpha \leftarrow$ Type(?x D). If $a \notin \alpha$, then $O \not\vDash C \equiv D$ and the test fails; delete $C(a)$ and a. Else, delete $C(a)$, assert $D(a)$. Query $\alpha \leftarrow$ Type(?x C). If $a \notin \alpha$, then $O \not\vDash C \equiv D$, and the test fails. Delete $D(a)$ and a.* ◂

Simple existential quantification, $Test_{eq}$ or $Test'_{eq}$. The axiom pattern is $C \sqsubseteq \exists R.D$, so $O \not\vDash \neg(C \sqsubseteq \exists R.D)$ should be true, or $O \vDash C \sqsubseteq \exists R.D$ false (or: not asserted) before the ontology edit. One could do a first check that D is not a descendant of R but if it is, then it may be the case that $C' \sqsubseteq \exists R.D$, with C a different class from C'. This still requires one to confirm that C is not a subclass of $\exists R.D$. This can be combined into one query/TDD test:

Test$_{eq}$ $= \alpha \leftarrow$ SubClassOf(?x ObjectSomeValuesFrom(R D)). If $C \notin \alpha$, then $O \nvDash C \sqsubseteq \exists R.D$, hence the test fails. ◄

If $C \notin \alpha$, then the axiom is to be added to the ontology, the query run again, and if $C \in \alpha$, then the test cycle is completed.

From a modelling viewpoint, desiring to add a CQ that amounts to $C \sqsubseteq \exists R.\neg D$ may look different, but $\neg D \equiv D'$, so it amounts to testing $C \sqsubseteq \exists R.D'$, i.e., essentially the same pattern. This also can be formulated directly into a SPARQL-OWL query, encapsulated in a TDD test:

Test$_{eq_{nd}}$ $= \alpha \leftarrow$ SubClassOf(?x ObjectSomeValuesFrom(R ObjectComplementOf(D))). If $C \notin \alpha$, then $O \nvDash C \sqsubseteq \exists R.\neg D$; hence, the test fails. ◄

It is slightly different for $C \sqsubseteq \neg \exists R.D$. The query with TDD test is as follows:

Test$_{eq_{nr}}$ $= \alpha \leftarrow$ SubClassOf(?x ObjectComplementOf(ObjectSomeValuesFrom(R D))). If $C \notin \alpha$, then $O \nvDash C \sqsubseteq \neg \exists R.D$, and the test fails. ◄

The TDD test $Test'_{eq}$ with individuals only is as follows:

Test$'_{eq}$ $=$ Create mock objects a, assert $(C \sqcap \neg \exists R.D)(a)$. $ostate \leftarrow$ Run the reasoner. If $ostate$ is consistent, then $O \nvDash C \sqsubseteq \exists R.D$; test fails. Delete $(C \sqcap \neg \exists R.D)(a)$, and a. ◄

This holds similarly for $C \sqsubseteq \exists R.\neg D$ ($Test'_{eq_{nd}}$). Finally, for $C \sqsubseteq \neg \exists R.D$:

Test$'_{eq_{nr}}$ $=$ Create two mock objects, a and b; assert $C(a)$, $D(b)$, and $R(a, b)$. $ostate \leftarrow$ Run the reasoner. If $ostate$ is consistent, then $O \nvDash C \sqsubseteq \neg \exists R.D$, hence, the test fails. Delete $C(a)$, $D(b)$, $R(a, b)$, a, and b. ◄

Simple universal quantification, $Test_{uq}$ *or* $Test'_{uq}$. The axiom to add is of the pattern $C \sqsubseteq \forall R.D$, so then $O \nvDash \neg(C \sqsubseteq \forall R.D)$ should hold, or $O \models C \sqsubseteq \forall R.D$ false (not be present in the ontology), before the ontology edit. This has a similar pattern for the TDD test as the one for existential quantification,

Test$_{uq}$ $= \alpha \leftarrow$ SubClassOf(?x ObjectAllValuesFrom(R D)). If $C \notin \alpha$, then $O \nvDash C \sqsubseteq \forall R.D$, hence, the test fails. ◄

which then can be added and the test ran again. The TDD test for $Test'_{uq}$ is alike $Test'_{eq}$, but then the query is $\alpha \leftarrow$ Type(?x, ObjectAllValuesFrom(R D)).

3.3 Generic Test Patterns for Object Properties

TDD tests for object properties (the RBox) do not lend themselves well for TBox querying, though the automated reasoner can be used for the TDD tests.

Domain axiom, $Test_{da}$ *or* $Test'_{da}$. The TDD needs to check that $\exists R \sqsubseteq C$ that is not yet in O, so $O \models \neg(\exists R \sqsubseteq C)$ should be true, or $O \models \exists R \sqsubseteq C$ false. There are two options with SPARQL-OWL. First, one can query for the domain:

Test$_{da}$ $= \alpha \leftarrow$ ObjectPropertyDomain(R ?x) If $C \notin \alpha$, then $O \nvDash \exists R \sqsubseteq C$; test fails. ◄

Alternatively, one can query for the superclasses of $\exists R$ (it is shorthand for $\exists R.\top$), where the TDD query is: $\alpha \leftarrow$ SubClassOf(SomeValuesFrom(R Thing) ?x). Note that $C \in \alpha$ only will be returned if C is the only domain class of R or when $C \sqcap C'$ (but not if it is $C \sqcup C'$, which is a superclass of C). The ABox test is:

Test$'_{\mathbf{da}}$ $=$ *Check $R \in V_{OP}$ and $C \in V_C$. Add individuals a and topObj, add $R(a, topObj)$. Run the reasoner. If $a \notin C$, then $O \nvDash \exists R \sqsubseteq C$ (also in the strict sense as is or with a conjunction); hence the test fails. Delete a and topObj.* ◄
If the answer is empty, then R does not have any domain specified yet, and if $C \notin \alpha$, then $O \nvDash \exists R \sqsubseteq C$, hence, it can be added and the test run again.

Range axiom, Test$_{ra}$ or Test$'_{ra}$. Thus, $\exists R^- \sqsubseteq D$ should not be in the ontology before the TDD test. This is similar to the domain axiom test:

Test$_{\mathbf{ra}}$ $= \alpha \leftarrow$ ObjectPropertyRange(R ?x). *If $D \notin \alpha$, then $O \nvDash \exists R^- \sqsubseteq D$; test fails.* ◄
Or one can query $\alpha \leftarrow$ SubClassOf(SomeValuesFrom(ObjectInverseOf(R) Thing) ?x). Then $D \in \alpha$ if $O \models \exists R^- \sqsubseteq D$ or $O \models \exists R^- \sqsubseteq D \sqcap D'$, and only owl:Thing $\in \alpha$ if no range was declared for R. The test with individuals:

Test$'_{\mathbf{ra}}$ $=$ *Check $R \in V_{OP}$ and $D \in V_C$. Add individuals a and topObj, add $R(topObj, a)$. If $a \notin D$, then $O \nvDash \exists R^- \sqsubseteq D$. Delete $R(topObj, a)$, a, topObj.* ◄

Object property subsumption and equivalence, Test$_{ps}$ and Test$_{pe}$, and Test$'_{ps}$ and Test$'_{pe}$. For property subsumption, $R \sqsubseteq S$, we have to test that $O \models \neg(R \sqsubseteq S)$, or that $R \sqsubseteq S$ fails. This is simply:

Test$_{\mathbf{ps}}$ $= \alpha \leftarrow$ SubObjectPropertyOf(?x S) *If $R \notin \alpha$, then $O \nvDash R \sqsubseteq S$; test fails.* ◄
Regarding the ABox variant, for $R \sqsubseteq S$ to hold given the OWL semantics, it means that, given some individuals a and b, that if $R(a, b)$ then $S(a, b)$:

Test$'_{\mathbf{ps}}$ $=$ *Check $R, S \in V_{OP}$. Add individuals a, b, add $R(a, b)$. Run the reasoner. If $S(a, b) \notin \alpha$, then $O \nvDash R \sqsubseteq S$; test fails. Delete $R(a, b)$, a, and b.* ◄
Upon the ontology update, it should infer $S(a, b)$. There is no guarantee that $R \sqsubseteq S$ was added, but $R \equiv S$ instead. This can be observed easily with the following test:

Test$'_{\mathbf{pe}}$ $=$ *Check $R, S \in V_{OP}$. Add mock individuals a, b, c, d, add $R(a, b)$ and $S(c, d)$. Run the reasoner. If $S(a, b) \in \alpha$ and $R(c, d) \notin \alpha$, then $O \models R \sqsubseteq S$ (hence the ontology edit was correct); test fails. Else, i.e. $\{S(a, b), R(c, d)\} \in \alpha$, so $O \models R \equiv S$; test passes. Delete $R(a, b)$ and $S(c, d)$, and a, b, c, d.* ◄
For object property equivalence at the Tbox level, i.e., $R \equiv S$, one could use $Test_{ps}$ twice, or simply use the EquivalentObjectProperties:

Test$_{\mathbf{pe}}$ $= \alpha \leftarrow$ EquivalentObjectProperties(?x S) *If $R \notin \alpha$, then $O \nvDash R \equiv S$; test fails.* ◄

Object property inverses, Test$_{pi}$ and Test$'_{pi}$. There are two options since OWL 2: explicit inverses (e.g., teaches with its inverse declared as taught by) or 'implicit' inverse (e.g., teaches and teaches$^-$). For the failure-test of TDD, only the former case can be tested. Also here there is a TBox and an ABox approach; their respective tests are:

Test$_{\mathbf{pi}}$ $= \alpha \leftarrow$ InverseObjectProperties(?x S) *If $R \notin \alpha$, then $O \nvDash R \sqsubseteq S^-$; test fails.* ◄

Test$'_{\mathbf{pi}}$ $=$ *Check $R, S \in V_{OP}$. Assume S is intended to be the inverse of R (with R and S having different names). Add mock individuals a, b, and add*

$R(a, b)$. Run the reasoner. If $O \not\models S(b, a)$, then $O \not\models R \sqsubseteq S^-$; hence, the test fails. Delete a, b. ◄

Object property chain, $Test_{pc}$ or $Test'_{pc}$. The axiom to be added is one of the permissible chains (except for transitivity; see below), such as $R \circ S \sqsubseteq S$, $S \circ R \sqsubseteq S$, $R \circ S_1 \circ ... \circ S_n \sqsubseteq S$ (with $n > 1$). This is increasingly more cumbersome to test, because many more entities are involved, hence, more opportunity to have incomplete knowledge represented in the ontology and thus more hassle to check all possibilities that lead to not having the desired effect. Aside from searching the owl file for `owl:propertyChainAxiom`, with the relevant properties included in order, the SPARQL-OWL-based TDD test is:

Test$_{\mathbf{pc}}$, for $R \circ S \sqsubseteq S = \alpha \leftarrow$ SubObjectPropertyOf(ObjectPropertyChain(R S) ?x). If $S \notin \alpha$, then $O \not\models R \circ S \sqsubseteq S$, and the test fails. ◄

and similarly with the other permutations of property chains. However, either option misses three aspects of chains: (1) a property chain is pointless if the properties involved are never used in the intended way, (2) this cannot ascertain that it does only what was intended, and (3) whether the chain does not go outside OWL 2 due to some of them being not 'simple'. For $O \models R \circ S \sqsubseteq S$ to be interesting for the ontology, also at least one $O \models C \sqsubseteq \exists R.D$ and one $O \models D \sqsubseteq \exists S.E$ should be present. If they all were, then a SPARQL-OWL query $\alpha \leftarrow$ SubClassOf(?x ObjectSomeValuesFrom(S E)) will have $C \in \alpha$. If either of the three axioms are not present, then $C \notin \alpha$. The ABox TDD test is more cumbersome:

Test$'_{\mathbf{pc}}$, for $R \circ S \sqsubseteq S =$ Check $R, S \in V_{OP}$ and $C, D, E \in V_C$. If $C, D, E \notin V_C$, then add the missing class(es) (C, D, and/or E) as mock classes. Run the test $Test_{eq}$ or $Test'_{eq}$, for both $C \sqsubseteq \exists R.D$ and for $D \sqsubseteq \exists S.E$. If $Test_{eq}$ is false, then add $C \sqsubseteq \exists R.D$, $D \sqsubseteq \exists S.E$, or both, as mock axiom. If $O \models C \sqsubseteq \exists S.D$, then the test is meaningless, for it would not test the property chain. Then add mock class C', mock axiom $C' \sqsubseteq \exists R.D$. Verify with $Test_{eq}$ or $Test'_{eq}$. $\alpha \leftarrow$ SubClassOf(?x ObjectSomeValuesFrom(S E)). If $C' \notin \alpha$, then $O \not\models R \circ S \sqsubseteq S$; test fails. Else, i.e., $O \not\models C \sqsubseteq \exists S.D$: $\alpha \leftarrow$ SubClassOf(?x ObjectSomeValuesFrom(S E)). If $C \notin \alpha$, then $O \not\models R \circ S \sqsubseteq S$; test fails. Delete all mock entities. ◄

Assuming that the test fails, i.e., $C \notin \alpha$ (resp. $C' \notin \alpha$) and thus $O \not\models R \circ S \sqsubseteq S$, then add the chain and run the test again, which then should pass (i.e., $C \in \alpha$). The procedure holds similarly for the other permissible combinations of object properties in a property chain/complex role inclusion.

Object property characteristics, $Test_{p_x}$. TDD tests can be specified for the ABox approach, but only transitivity and local reflexivity have a TBox test.

R is functional, $Test'_{p_f}$, i.e., an object has at most one R-successor:

Test$'_{\mathbf{p_f}}$ $=$ Check $R \in V_{OP}$ and $a, b, c \in V_I$; if not present, add. Assert mock axioms $R(a, b)$, $R(a, c)$, and $b \neq c$, if not present already. Run reasoner. If O is consistent, then $O \not\models$ Func(R), so the test fails. (If O is inconsistent, then the test passes.) Remove mock axioms and individuals, as applicable. ◄

R is inverse functional, $Test'_{p_{if}}$. This is as above, but then in the other direction, i.e., $R(b, a)$, $R(c, a)$ with b, c declared distinct. Thus:

Test$'_{\mathbf{p_{if}}}$ $=$ Check $R \in V_{OP}$ and $a, b, c \in V_I$; if not present, add. Assert mock axioms $R(b,a)$, $R(c,a)$, and $b \neq c$, if not present already. Run reasoner. If O is consistent, then $O \nvDash \mathsf{InvFun}(R)$, so the test fails. (If O is inconsistent, then $\mathsf{InvFun}(R)$ is true.) Remove mock axioms and individuals, as applicable. ◀

R is transitive, $Test_{p_t}$ or $Test'_{p_t}$. As with object property chains ($Test_{pc}$), transitivity is only 'interesting' if there are at least two related axioms so that one obtains a non-empty deduction; if the relevant axioms are not asserted, they have to be added. The TBox and ABox tests are as follows:

$\mathbf{Test_{p_t}}$ = Check $R \in V_{OP}$ and $C, D, E, \in V_C$. If $C, D, E, \notin V_C$, then add the missing class(es) (C, D, and/or E as mock classes). If $C \sqsubseteq \exists R.D$ and $D \sqsubseteq \exists R.E$ are not asserted, then add them to O. Query $\alpha \leftarrow \mathsf{SubClassOf}(?x$ $\mathsf{ObjectSomeValuesFrom}(R\ E))$. If $C \notin \alpha$, then $O \nvDash \mathsf{Trans}(R)$, so the test fails. Remove mock classes and axioms, as applicable. ◀

$\mathbf{Test'_{p_t}}$ = Check $R \in V_{OP}$, $a, b, c \in V_I$. If not, introduce mock a, b, c, $R(a,b)$, and $R(b,c)$, if not present already. Run reasoner. If $R(a,c) \notin \alpha$, then $O \nvDash \mathsf{Trans}(R)$, so the test fails. Remove mock entities. ◀

R is symmetric, $Test'_{p_s}$, $\mathsf{Sym}(R)$, so that with $R(a,b)$, it will infer $R(b,a)$. The test-to-fail—assuming $R \in V_{OP}$—is as follows:

$\mathbf{Test'_{p_s}}$ = Check $R \in V_{OP}$. Introduce a, b as mock objects ($a, b \in V_I$). Assert mock axiom $R(a,b)$. $\alpha \leftarrow \mathsf{ObjectPropertyAssertion}(R\ x?\ a)$. If $b \notin \alpha$, then $O \nvDash \mathsf{Sym}(R)$, so the test fails. Remove mock assertions and individuals. ◀

Alternatively, one can check in the ODE whether $R(b,a)$ is inferred.

R is asymmetric, $Test'_{p_a}$. This is easier to test with its negation, i.e., assert objects symmetric and distinct, then if O is not inconsistent, then $O \nvDash \mathsf{Asym}(R)$:

$\mathbf{Test'_{p_a}}$ = Check $R \in V_{OP}$. Introduce a, b as mock objects and assert mock axioms $R(a,b)$ and $R(b,a)$. Run reasoner. If O is not inconsistent, then $O \nvDash \mathsf{Asym}(R)$, so the test fails. Remove mock axioms and individuals. ◀

R is reflexive, $Test'_{p_{rg}}$ or $Test'_{p_{rg}}$. The object property can be either globally reflexive ($\mathsf{Ref}(R)$), or locally ($C \sqsubseteq \exists R.Self$). Global reflexivity is uncommon, but if the modeller does want it, then the following test should be executed:

$\mathbf{Test'_{p_{rg}}}$ = Check $R \in V_{OP}$. Add mock object a. Run the reasoner. If $R(a,a) \notin O$, then $O \nvDash \mathsf{Ref}(R)$, so the test fails. Remove mock object a. ◀

Adding $\mathsf{Ref}(R)$ will have the test evaluate to true. Local reflexivity amounts to checking whether $O \vDash C \sqsubseteq \exists R.Self$. This is essentially the same as $Test_{eq}$ but then with Self cf. a named D, so there is a TBox and an ABox TDD test:

$\mathbf{Test_{p_{rl}}}$ = $\alpha \leftarrow \mathsf{SubClassOf}(?x\ \mathsf{ObjectSomeValuesFrom}(R\ \mathsf{Self}))$. If $C \notin \alpha$, then $O \nvDash C \sqsubseteq \exists R.Self$, so the test fails. ◀

$\mathbf{Test'_{p_{rl}}}$ = Check $R \in V_{OP}$. Introduce a as mock objects ($a \in V_I$). Assert mock axiom $C(a)$. $\alpha \leftarrow \mathsf{Type}(?x\ C)$, $\mathsf{PropertyValue}(a\ R\ ?x)$. If $a \notin \alpha$, then $O \nvDash C \sqsubseteq \exists R.Self$, so the test fails. Remove $C(a)$ and mock object a. ◀

R is irreflexive, $Test'_{p_{ir}}$. As with asymmetry, the TDD test exploits the converse:

$\mathbf{Test'_{p_i}}$ = Check $R \in V_{OP}$, and add $a \in V_I$. Add mock axiom $R(a,a)$. Run reasoner. If O is consistent, then $O \nvDash \mathsf{Irr}(R)$; test fails. (Else, O is inconsistent, and $\mathsf{Irr}(R)$) is true. Remove mock axiom and individual, as applicable. ◀

This concludes the basic tests. While the logic permits that a class on the left-hand side of the inclusion axiom is an unnamed class, we do not consider this here, as due to the tool design of the most widely used ODE, Protégé, the class on the left-hand side of the inclusion is typically a named class.

4   Evaluation with the Protégé Plugin for TDD

In order to support ontology engineers in performing TDD, we have implemented a Protégé plugin, TDDOnto, which provides a view where the user may specify the set of tests to be run. After their execution, the status of the tests is displayed. One also can add a selected axiom to the ontology (and re-run the test).

The aim of the evaluation is to answer *Which TDD approach—queries or mock objects—is better?*, as performance is likely to affect user opinion of TDD. To answer this question, we downloaded the TONES ontologies from OntoHub [https://ontohub.org/repositories], of which 67 could be used (those omitted were either in OBO format or had datatypes incompatible with the reasoner). The ontologies were divided into 4 groups, based on the number of axioms: up to 100 (n−20), 100–1000 axioms (n=35), 1000–10,000 axioms (n=10), and over 10,000 (n=2) to measure effect of ontology size. The tests were generated randomly, using the ontology's vocabulary, and each test kind was repeated 3 times to obtain more reliable results as follows. For each axiom kind of the basic form (with C and D as primitive concepts) there is a fixed number of "slots" that can be replaced with URIs. For each test, these slots were randomly filled from the set of URIs existing in the ontology taking into account whether an URI represents a class or a property. The tested axioms with the result of each test are published in the online material. The test machine was a Mac Book Air with 1.3 GHz Intel Core i5 CPU and 4 GB RAM. The OWL reasoner was HermiT 1.3.8, which is the same that is built-in into OWL-BGP to ensure fair comparison.

The first observation during our experiments was that not all the features of OWL 2 are covered by OWL-BGP, in particular the RBox axioms (e.g., subPropertyOf and property characteristics). Therefore, we only present the comparative results of the tests that could be run in both settings: ABox tests and TBox tests with use of the SPARQL-OWL query answering technology implemented in the OWL-BGP tool.

The performance results per group of ontologies are presented in Fig. 1. Each box plot has the median m (horizontal line); the first and third quartile (bottom and top line of the box); the lowest value above $m - 1.5 \cdot IQR$ (horizontal line below the box), and the highest value below $m + 1.5 \cdot IQR$ (horizontal line above the box), where IQR (interquartile range) is represented with the height of the box; outliers are points above and below of the short lines. It is evident that TBox (SPARQL-OWL) tests are generally faster than the ABox ones, and these differences are larger in the sets of larger ontologies. A comparison was done also between two alternative technologies for executing a TBox test—based on SPARQL-OWL and based on OWL API with the reasoner—showing even better performance of the TBox based TDD tests versus ABox based ones (results

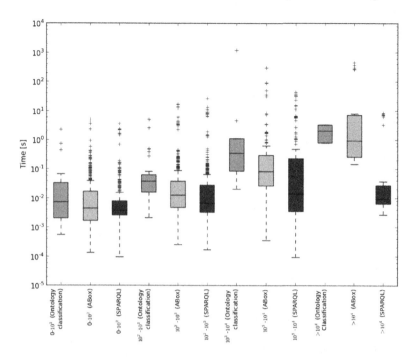

Fig. 1. Performance times by ontology size (four groups, with lower and upper number of the axioms of the ontologies in that group), and classification and test type for each.

available in the online material). Before running any test on an ontology, we also measured ontology classification time, which is also included in Fig. 1: it is higher on average in comparison to the times of running the test. Performance by TDD test type and the kind of axiom is shown in Fig. 2, showing the better general performance of the TBox approach in more detail, except for disjointness.

5   Discussion

The current alternative to TDD tests is browsing the ontology for the axiom. This is problematic, for then one does not know the implications it is responsible for, it results in cognitive overload that hampers ontology development, and one easily overlooks something. Instead, TDD can manage this in one fell swoop. In addition, the TDD tests also facilitate regression testing.

On Specifying and Implementing a TDD Tool. TBox tests can be implemented in different ways; e.g., in some instances, one could use the DL query tab in Protégé; e.g., T_{cs}'s as: D and *select* Sub classes, without the hassle of unnamed classes (complex class expressions) on the right-hand-side of the inclusion axiom (not supported by BGP [15]). However, it lacks functionality for object property tests (as did all others, it appeared during evaluation); one still can test the sequence 'manually' and check the classification results, though.

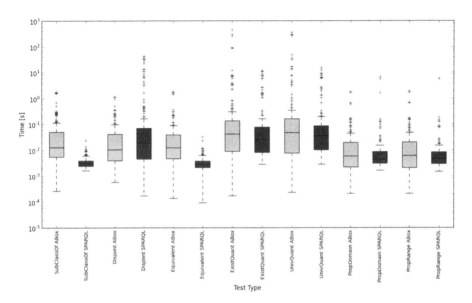

Fig. 2. Test computation times per test type and per the kind of the tested axiom.

The core technological consideration, however, is the technique to obtain the answer of a TDD test: SPARQL SELECT-queries, SPARQL-OWL's BGP (with SPARQL engine and HermiT), or SPARQL-DL with ASK queries and the OWL API. Neither could do all TDD tests in their current version. Regarding performance, the difference between the ABox and TBox tests are explainable—the former always modifies the ontology, so requires an extra classification step—though less so for disjointness or the difference being larger (subsumption, equivalence) or smaller (queries with quantifiers). Overall performance is likely to vary also by reasoner [19], and, as observed, by ontology size. This is a topic of further investigation.

A related issue is the maturity of the tools. Several ontologies had datatype errors, and there were the aforementioned RBox tests limitations. Therefore, we tested only what could be done with current technologies (the scope is TDD evaluation, not extending other tools), and infer tendencies from that so as to have an experimentally motivated basis for deciding which technique likely will have the best chance of success, hence, is the best candidate for extending the corresponding tool. This means using TBox TDD tests, where possible.

A Step Toward a TDD Ontology Engineering Methodology. A methodology is a structured collection of methods and techniques, processes, people having roles possibly in teams, and quality measures and standards across the process (see, e.g., [4]). A foundational step in the direction of a TDD ontology development methodology that indicates where and how it differs from the typical waterfall, iterative, or lifecycle-based methodologies is summarised in Fig. 3, adapting the software development TDD procedure. One can refine these steps,

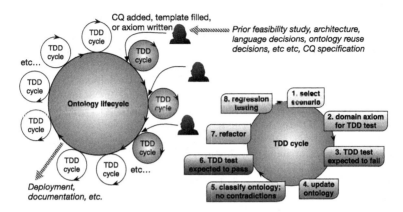

Fig. 3. Sketch of a possible ontology lifecycle that focuses on TDD, and the typical, default, sequence of steps of the TDD procedure summarised in key terms.

such as managing the deductions following from the ontology update and how to handle an inconsistency or undesirable deduction due to contradictory CQs. Refactoring could include, e.g., removing an explicitly declared axiom from a subclass once it is asserted for its superclass. These details are left for future work. Once implemented, a comparison of methodologies is also to be carried out.

6   Conclusions

This paper introduced 36 tests for *Test-Driven Development* of ontologies, specifying what has to be tested, and how. Tests were specified both at the TBox-level with queries and for ABox individuals, using mock entities. The implementation of the main tests demonstrated that the TBox test approach performs better, which is more pronounced with larger ontologies. A high-level 8-step process for TDD ontology engineering was proposed.

Future work pertains to extending tools to also implement the remaining tests, elaborate on the methodology, and conduct use-case evaluations.

Acknowledgments. This research has been supported by the National Science Centre, Poland, within the grant number 2014/13/D/ST6/02076.

References

1. Auer, S.: The RapidOWL methodology-towards agile knowledge engineering. In: Proceedings of WETICE 2006. pp. 352–357. IEEE Computer Society, June 2006
2. Beck, K.: Test-Driven Development: By Example. Addison-Wesley, Boston (2004)
3. Blomqvist, E., Seil Sepour, A., Presutti, V.: Ontology testing - methodology and tool. In: ten Teije, A., Völker, J., Handschuh, S., Stuckenschmidt, H., d'Acquin, M., Nikolov, A., Aussenac-Gilles, N., Hernandez, N. (eds.) EKAW 2012. LNCS, vol. 7603, pp. 216–226. Springer, Heidelberg (2012)

4. Cockburn, A.: Selecting a project's methodology. IEEE Softw. **17**(4), 64–71 (2000)
5. Ferré, S., Rudolph, S.: Advocatus diaboli – exploratory enrichment of ontologies with negative constraints. In: ten Teije, A., Völker, J., Handschuh, S., Stuckenschmidt, H., d'Acquin, M., Nikolov, A., Aussenac-Gilles, N., Hernandez, N. (eds.) EKAW 2012. LNCS, vol. 7603, pp. 42–56. Springer, Heidelberg (2012)
6. Gangemi, A., Presutti, V.: Ontology design patterns. In: Staab, S., Studer, R. (eds.) Handbook on Ontologies, pp. 221–243. Springer, Berlin (2009)
7. García-Ramos, S., Otero, A., Fernández-López, M.: Ontologytest: a tool to evaluate ontologies through tests defined by the user. In: Omatu, S., Rocha, M.P., Bravo, J., Fernández, F., Corchado, E., Bustillo, A., Corchado, J.M. (eds.) IWANN 2009, Part II. LNCS, vol. 5518, pp. 91–98. Springer, Heidelberg (2009)
8. Garcia, A., O'Neill, K., Garcia, L.J., Lord, P., Stevens, R., Corcho, O., Gibson, F.: Developing ontologies within decentralized settings. In: Chen, H., et al. (eds.) Semantic e-Science. Annals of Information Systems, vol. 11, pp. 99–139. Springer, New York (2010)
9. Gennari, J.H., et al.: The evolution of Protégé: an environment for knowledge-based systems development. Int. J. Hum Comput Stud. **58**(1), 89–123 (2003)
10. Ghidini, C., Kump, B., Lindstaedt, S., Mahbub, N., Pammer, V., Rospocher, M., Serafini, L.: Moki: the enterprise modelling wiki. In: Aroyo, L., et al. (eds.) ESWC 2009. LNCS, vol. 5554, pp. 831–835. Springer, Heidelberg (2009)
11. Janzen, D.S.: Software architecture improvement through test-driven development. In: Companion to ACM SIGPLAN 2005, pp. 240–241. ACM Proceedings (2005)
12. Keet, C.M., Ławrynowicz, A.: Test-driven development of ontologies (extended version). Technical report 1512.06211, arxiv.org, December 2015. http://arxiv.org/abs/1512.06211
13. Keet, C.M., Khan, M.T., Ghidini, C.: Ontology authoring with FORZA. In: Proceedings of CIKM 2013, pp. 569–578. ACM Proceedings (2013)
14. Kim, T., Park, C., Wu, C.: Mock object models for test driven development. In: Proceedings of SERA2006. IEEE Computer Society (2006)
15. Kollia, I., Glimm, B., Horrocks, I.: SPARQL query answering over owl ontologies. In: Antoniou, G., Grobelnik, M., Simperl, E., Parsia, B., Plexousakis, D., De Leenheer, P., Pan, J. (eds.) ESWC 2011, Part I. LNCS, vol. 6643, pp. 382–396. Springer, Heidelberg (2011)
16. Kontokostas, D., Westphal, P., Auer, S., Hellmann, S., Lehmann, J., Cornelissen, R., Zaveri, A.: Test-driven evaluation of linked data quality. In: Proc. of WWW'2014. pp. 747–758. ACM Proceedings (2014)
17. Kumar, S., Bansal, S.: Comparative study of test driven development with traditional techniques. Int. J. Softw. Comput. Eng. **3**(1), 352–360 (2013)
18. Mackinnon, T., Freeman, S., Craig, P.: Endo-testing: unit testing with mock objects. In: Extreme Programming Examined, pp. 287–301. Addison-Wesley, Boston (2001)
19. Parsia, B., Matentzoglu, N., Goncalves, R., Glimm, B., Steigmiller, A.: The OWL Reasoner Evaluation (ORE) 2015 competition report. In: Proceedings of SSWS 2015. CEUR-WS, Bethlehem, USA, vol. 1457, 11 October 2015
20. Paschke, A., Schaefermeier, R.: Aspect OntoMaven - aspect-oriented ontology development and configuration with OntoMaven. Technical report 1507.00212v1, Free University of Berlin, July 2015. http://arxiv.org/abs/1507.00212
21. Presutti, V., Daga, E., et al.: Extreme design with content ontology design patterns. In: Proceedings of WS on OP 2009, CEUR-WS, vol. 516, pp. 83–97 (2009)

22. Presutti, V., et al.: A library of ontology design patterns: reusable solutions for collaborative design of networked ontologies. NeOn deliverable D2.5.1, NeOn Project, ISTC-CNR (2008)
23. Ren, Y., Parvizi, A., Mellish, C., Pan, J.Z., van Deemter, K., Stevens, R.: Towards competency question-driven ontology authoring. In: Presutti, V., d'Amato, C., Gandon, F., d'Aquin, M., Staab, S., Tordai, A. (eds.) ESWC 2014. LNCS, vol 8465, pp. 752–767. Springer, Heidelberg (2014)
24. Shrivastava, D.P., Jain, R.: Metrics for test case design in test driven development. Int. J. Comput. Theory Eng. **2**(6), 952–956 (2010)
25. Suárez-Figueroa, M.C., et al.: NeOn methodology for building contextualized ontology networks. NeOn Deliverable D5.4.1, NeOn Project (2008)
26. Tort, A., Olivé, A., Sancho, M.R.: An approach to test-driven development of conceptual schemas. Data Knowl. Eng. **70**, 1088–1111 (2011)
27. Vrandečić, D., Gangemi, A.: Unit tests for ontologies. In: Meersman, R., Tari, Z., Herrero, P. (eds.) OTM 2006 Workshops. LNCS, vol. 4278, pp. 1012–1020. Springer, Heidelberg (2006)
28. Warrender, J.D., Lord, P.: How, What and Why to test an ontology. Technical report 1505.04112, Newcastle University (2015). http://arxiv.org/abs/1505.04112

In-Use & Industrial Track

Semantically Enhanced Quality Assurance in the JURION Business Use Case

Dimitris Kontokostas[1]([✉]), Christian Mader[2], Christian Dirschl[3], Katja Eck[3], Michael Leuthold[3], Jens Lehmann[1], and Sebastian Hellmann[1]

[1] Institut für Informatik, AKSW, Universität Leipzig, Leipzig, Germany
{kontokostas,lehmann,hellmann}@informatik.uni-leipzig.de
[2] Semantic Web Company, Vienna, Austria
c.mader@semantic-web.at
[3] Wolters Kluwer Germany, Munich, Germany
{cdirschl,keck,mleuthold}@wolterskluwer.de
http://aksw.org

Abstract. The publishing industry is undergoing major changes. These changes are mainly based on technical developments and related habits of information consumption. Wolters Kluwer already engaged in new solutions to meet these challenges and to improve all processes of generating good quality content in the backend on the one hand and to deliver information and software in the frontend that facilitates the customer's life on the other hand. JURION is an innovative legal information platform developed by Wolters Kluwer Germany (WKD) that merges and interlinks over one million documents of content and data from diverse sources such as national and European legislation and court judgments, extensive internally authored content and local customer data, as well as social media and web data (e.g. DBpedia). In collecting and managing this data, all stages of the Data Lifecycle are present – extraction, storage, authoring, interlinking, enrichment, quality analysis, repair and publication. Ensuring data quality is a key step in the JURION data lifecycle. In this industry paper we present two use cases for verifying quality: (1) integrating quality tools in the existing software infrastructure and (2) improving the data enrichment step by checking the external sources before importing them in JURION. We open-source part of our extensions and provide a screencast with our prototype in action.

Keywords: RDF quality · Linked Data · Enrichment

1 Introduction

The publishing industry is - like many other industries - undergoing major changes. These changes are mainly based on technical developments and related habits of information consumption[1]. The world of customers has changed dramatically and as an information service provider, Wolters Kluwer wants to meet

---

[1] For example: http://hmi.ucsd.edu/pdf/HMI_2009_ConsumerReport_Dec9_2009.pdf.

© Springer International Publishing Switzerland 2016
H. Sack et al. (Eds.): ESWC 2016, LNCS 9678, pp. 661–676, 2016.
DOI: 10.1007/978-3-319-34129-3_40

these changes with adequate solutions for customers and their work environment. For a couple of years, Wolters Kluwer has already engaged in new solutions to meet these challenges and to improve processes for generating good quality content in the backend on the one hand and to deliver information and software in the frontend that facilitates the customers life on the other hand.

One of these frontend applications is a platform called JURION.[2] JURION is a legal information platform developed by Wolters Kluwer Germany (WKD) that merges and interlinks over one million documents of content and data from diverse sources, such as national and European legislation and court judgments, extensive internally authored content and local customer data; as well as social media and web data (e.g. from DBpedia). In collecting and managing this data, all stages of the Data Lifecycle are present – extraction, storage, authoring, interlinking, enrichment, quality analysis, repair and publication. On top of this information processing pipeline, the JURION development teams add value through applications for personalization, alerts, analysis and semantic search.

Based on the FP7 LOD2 project[3], parts of the *Linked Data stack*[4] have been deployed in JURION to handle data complexity issues. LOD2 aimed at developing novel, innovative Semantic Web technologies and also at the expansion and integration of openly accessible and interlinked data on the web. More detailed information can be found in [1]. WKD acted as a use case partner for these technologies, supported the development process of semantic technologies and integrated them to support the expansion of linked data in business environments. The software development process and data life cycle at WKD are highly independent from each other and require extensive manual management to coordinate their parallel development, leading to higher costs, quality issues and a slower time-to-market. This is why the JURION use case presented here is situated within both the Software Engineering as well as in the Data Processing area.

Through the ALIGNED project[5], JURION focuses on closing the gap between Software & Data Engineering. This paper describes the JURION results of the first phase of the project. In this phase, we concentrated mainly on the enhancement of data quality and repair processes. We created novel approaches for integrating RDF tools in the existing software engineering tool stack and created bindings to widely used Java libraries. We additionally created a link validation service for cleaning up external metadata residing in our databases. As a proof of concept, we open sourced some of our extensions and provide a screencast of the prototype implementation.

This industry paper is structured as follows: Sect. 2 provides a detailed description of JURION and its architecture. The existing tools that are used and enhanced are detailed in Sect. 3. Section 4 describes the challenges that drove this development. Section 5 provides the detailed approach we took for tackling each challenge. We provide an in-depth evaluation in Sect. 6 and conclude in Sect. 7.

---

[2] See JURION website https://www.jurion.de/de/home/guest.

[3] http://lod2.eu/Welcome.html.

[4] http://stack.linkeddata.org/.

[5] http://aligned-project.eu.

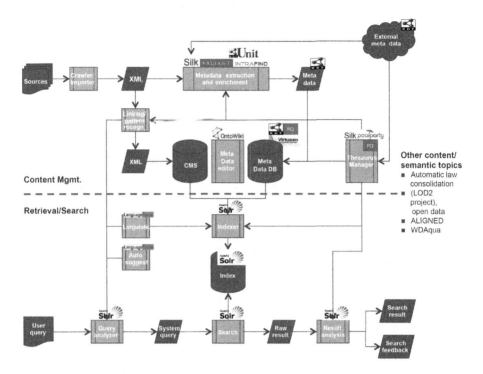

Fig. 1. JURION content pipeline and semantic search

2 The JURION Business Use Case

WKD is a leading knowledge and information service provider in the domains of law, companies and tax and offers high quality business information for professionals. This information is more and more integrated in digital solutions and applications. When using these solutions, customers can make critical decisions more efficiently and they can enhance their productivity in a sustainable way. Wolters Kluwer n.v. is based in more than 40 countries and serves customers in more than 175 countries worldwide.

JURION is the legal knowledge platform developed by WKD. It is not only a legal search platform, but considers search for legal information as an integrated part of the lawyer's daily processes. JURION combines competencies in the areas of legal publishing, software, portal technology and services, which cover all core processes of the lawyer within one single environment by connecting and integrating many different internal and external data sources.

The main goal of JURION is not to be yet another search engine. On the one hand, because Google as a reference application has made major progress in recent times, even in search environments dedicated to professionals. On the other hand, legal research is just one part of the lawyers main and daily tasks. So the higher the coverage of core processes of a digital offering, the more added-value on the customers side is generated and the higher the willingness to pay for that service will be. In addition, the more touchpoints between vendor and customer exist, the lower is the possibility for the service provider to be replaced by others.

Figure 1 describes the overall JURION content processing and retrieval infrastructure. Within the content pipeline, metadata is extracted from the proprietary WKD XML schema and transformed in RDF. In the thesaurus manager, controlled vocabularies and domain models based on SKOS standard are created, maintained and delivered for further usage. The indexing process of a search engine includes more and more additional information on top of the pure text. Queries are analyzed for legal references and keywords, which are matched against existing data in the metadata management systems. Once there are matches, the semantic relations are shown in the results overview by specific references to texts and related knowledge visualizations.

Since most of these new service offerings like workflow support at the lawyers desk are not document and content driven, the current paradigm of using pure XML as the only major data format had to be given up. Data and metadata are driving internal processes and therefore most of the features and functionalities in the JURION application. So, this data must not be locked in DTDs or XML schemas anymore. Conversion of this data in traditional RDBMS would have been possible, but the main benefits of these systems like high performance and robustness were not the major requirements in this setting. Instead, the data needed to be stored and maintained in a much more flexible and inter-connectable format, so that new data format requirements like adding new metadata or a new relationship type could be processed in a more or less real-time fashion. Semantic web technologies were chosen to meet that need, since e.g. their triple store technology supports this flexibility and since high performance is as already mentioned not a major requirement in a CMS environment. In addition, due to government initiatives on a national and European level, quite a lot of valuable data in the legal field was available in SKOS and RDF format, so that the integration effort for this data was rather limited, as soon as the basic technical infrastructure was laid. Once the internal and integrated external data was available, new features like personalization based on domain preferences (e.g. boosting labor law documents in the result list, based on current search behavior), context-sensitive disambiguation dialogues to resolve query issues (e.g. contract as labor contract or sales contract) or the sustainable linking to external data sources like EU directives and court decisions with their own legal contexts, e.g. across languages and countries could be established.

2.1   Related Work

In industrial settings, architectural system details are most times kept hidden due to conflicts of interest. However, major companies in the media & publishing sector are using semantic web technologies. In 2013, BBC published their internal knowledge graph[6]. BBC keeps an open position on Linked Data and

---

[6] http://www.bbc.co.uk/things/.

publishes many of their semantic-web-enabled features[7]. Thomson Reuters provides B2B semantic solutions with OpenCalais[8] and PermID[9]. Guardian uses Linked Data to augment the data they provide behind a paid API[10]. At the end of 2015, Springer announced a semantic wrapper of their data[11]. Nature Publishing Group has been an early adopter on linked data[12]. Finally, Pearson is also observed to use semantic web technologies[13].

3   Use Case Tools

In this paper, we focus on the use and enhancement of two tools: RDFUnit, an RDF Unit Testing suite and PoolParty Thesaurus Server (PPT) which is part of the PoolParty Semantic Suite.[14]

RDFUnit [3] is an RDF validation framework inspired by test-driven software development. In software, every function should be accompanied by a set of unit tests to ensure the correct behaviour of that function through time. Similarly, in RDFUnit, every vocabulary, ontology, dataset or application can be associated by a set of data quality test cases. Assigning test cases (TCs) in ontologies results in tests that can be reused by datasets sharing the same ontology.

The test case definition language of RDFUnit is SPARQL, which is convenient to directly query for identifying violations. For rapid test case instantiation, a pattern-based SPARQL-Template engine is supported where the user can easily bind variables into patterns. RDFUnit has a *Test Auto Generator* (TAG) component. TAG searches for schema information and automatically instantiates new test cases. Schema information can be in the form of RDFS[15] or OWL axioms that RDFUnit translates into SPARQL under Closed World Assumption (CWA) and Unique Name Assumption (UNA). These TCs cover validation against: domain, range, class and property disjointness, (qualified) cardinality, (inverse) functionality, (a)symmetricity, irreflexivity and deprecation. Other schema languages such as *SHACL, IBM Resource Shapes* or *Description Set Profiles* are also supported. RDFUnit can check an RDF dataset against multiple schemas but when this occurs, RDFUnit does not perform any reasoning/action to detect inconsistencies between the different schemas.

The PoolParty Semantic Suite[16] is a commercial product developed by Semantic Web Company.[17] For the JURION business case, PPT is of main

---

[7] http://www.bbc.co.uk/blogs/internet/tags/linked-data.

[8] http://www.opencalais.com/.

[9] https://permid.org/.

[10] http://www.theguardian.com/open-platform/blog/linked-data-open-platform.

[11] http://lod.springer.com/wiki/bin/view/Linked+Open+Data/About.

[12] http://www.nature.com/ontologies/.

[13] https://www.semantic-web.at/pearson.

[14] https://www.poolparty.biz/.

[15] http://www.w3.org/TR/rdf-schema/.

[16] https://www.poolparty.biz/.

[17] https://www.semantic-web.at/.

relevance. With PPT, taxonomists can develop thesauri in a collaborative way, either from scratch or supported by extraction of terms from a document corpus. The created thesauri fully comply to the 5-star Open Data principles[18] by using RDF and SKOS as the underlying technologies for representing the data. Using Linked Data technologies, it is possible to automatically retrieve potential additional concepts for inclusion into the thesauri by querying SPARQL endpoints (e.g. DBpedia). Furthermore, PPT provides interfaces to conveniently identify and link to related resources that are either defined in other PPT projects that reside on the same server or located on the Web. Depending on configuration, taxonomies developed with PPT can be made available as data dumps in various RDF serialization formats or directly queried by using the taxonomy's SPARQL endpoint. When additional semantics are required that exceed that of SKOS, PPT supports the creation of custom schemas. Taxonomies can use them do define their own classes and relation types by using elements of RDFS such as `rdfs:subClassOf` or `rdfs:domain` and `rdfs:range`.

Users of PPT are at any time supported by automated quality assurance mechanisms. Currently, three methodologies are in place: first, conformance with a custom schema or the SKOS model is ensured by the user interface. One for instance, cannot create two preferred labels in the same language. Second, the enforcement level of some quality metrics can be configured by the user so that it is, e.g., possible to get an alert if circular hierarchical relations [4] are introduced. Third, a taxonomy can be checked "as a whole" against a set of potential quality violations, displaying a report about the findings.

4   Challenges

JURION merges and interlinks over one million documents of content and data from diverse sources. Currently, the software development process and data life cycle are highly independent from each other and require extensive manual management to coordinate their parallel development, leading to higher costs, quality issues and a slower time-to-market. The higher flexibility of the data model as well as the shortened time-to-market for data features can only be materialized when most of the testing and QA effort – after data & chema changes are introduced – are tackled in a semi-automatic fashion. Thus, benefits like flexibility and scalability only materialize when it is not necessary to do all the quality checks manually, or involving expensive domain experts.

As depicted in Fig. 1, within the content pipeline, metadata is extracted from the proprietary WKD XML schema and transformed in RDF. Due to regular changes in the XML format, the correct transformation process based on existing XSLT scripts must be secured, so that no inconsistent data is fueled into the metadata database. In the thesaurus manager, controlled vocabularies and domain models based on SKOS standard are created, maintained and delivered for further usage. The integrity of the knowledge management system as a whole needs to be ensured. Therefore, regular local and global quality checks need to

---

[18] http://5stardata.info/en/.

be executed, so that e.g. inconsistencies across different controlled vocabularies can be detected and resumed.

Through the ALIGNED project, we target to enable JURION to address more complex business requirements that rely on tighter coupling of software and data. In this paper, we focus on improving the metadata extraction process as well as inconsistencies across controlled vocabularies and in particular external links coming from the JURION enrichment phase.

4.1   Metadata RDF Conversion Verification

At the top of Fig. 1 of the content pipeline, metadata is extracted from the proprietary WKD XML schema and transformed to RDF. Due to regular changes in the XML format, the correct transformation process based on existing XSLT scripts must be secured, so that no inconsistent data is fuelled into the metadata database. The main challenge of this task is to reduce the error in data transformation and accelerate the delivery of metadata information to JURION.

Approach & Goals. Based on the schema, test cases should automatically be created, which are run on a regular basis against the data that needs to be transformed. The errors detected will lead to refinements and changes of the XSLT scripts and sometimes also to schema changes, which impose again new automatically created test cases. This approach provides: 1. better control over RDF metadata 2. streamlined transformation process from XML to RDF 3. early detection of errors in RDF metadata, since the resulting RDF metadata are a core ingredient for many subsequent process steps in production and application usage, and 4. more flexibility in RDF metadata creation.

Impact. Continuous high quality triplification of semi-structured data is a common problem in the information industry, since schema changes and enhancements are routine tasks, but ensuring data quality is still very often purely manual effort. So any automation will support a lot of real-life use cases in different domains.

Existing Infrastructure. As part of the core CMS tasks within JURION each WKD XML document is checked-in through internal workflow functionality and gets converted to RDF which is based on the "Platform Content Interface" (PCI) ontology. The PCI ontology is a proprietary schema that describes legal documents and metadata in OWL. Due to change requests and new use cases for the RDF metadata in the ontology, the conversion logic or both the conversion logic and ontology need amendments. In these cases we need to make sure that the RDF data that is generated from the WKD XML documents still complies with the PCI ontology for quality assurance.

Quality Assurance. As a gatekeeper to avoid loading flawed data into the triple store, each result of the conversion from WKD XML into PCI RDF was sent to a dedicated, proprietary validation service that inspects the input and verifies compliance with the ontology. This approach assured that the conversion results are verified but came with major issues which makes it unsuitable for ad-hoc testing and quick feedback. The three most important ones are:

- the current service only processes entire PCI-packages, i.e. several datasets; this makes error detection on single data units quite difficult and service errors block the whole processing pipeline
- the service is a SOAP based web service that operates asynchronously with many independent process steps, which imposes high complexity on its usage
- it depends on other services and requires permanent network access and therefore is potentially unstable.

To improve these issues, we want to implement unit-test scenarios that can be run directly coupled to the development environment of the conversion projects and is therefore seamlessly integrated into the workflow. The tests should be run both automatically on every change in the project, but also be able to be manually triggered. Tests should be easily extendable and expressive enough to effectively spot issues in the conversion process. The feedback loop should be coupled as tightly as possible to the submitted change.

4.2 Quality Control in Thesaurus Management

At the middle right of Fig. 1 of the JURION content pipeline, controlled vocabularies and domain models are created, maintained and delivered for further usage, based on the SKOS standard. The integrity of the knowledge management system as a whole needs to be ensured. Therefore, regular local and global quality checks need to be executed, so that e.g. inconsistencies across different controlled vocabularies can be detected and resumed. The domain models and controlled vocabularies within and beyond the system are partly dependant on each other; sometimes there is even an overlap, e.g. concerning domains and concepts. This dependency must be transparent and consistency must be maintained. In addition, versioning issues must be addressed, since subsequent processes need to be aware of model changes, especially deletions.

Until now we had no effective overview over the validity of linked sources. This for example, causes problems in frontend applications where links do not resolve. The only way to evaluate the quality was to analyze the frontend representations of the linked sources or to follow a link to detect a missing source. There was in general no process in place to control the validity of external sources.

Approach & Goals. WKD is already using the thesaurus management system PoolParty for several vocabularies. With the increasing operational use, amount of content and the extended functionality to define custom schemas, we encounter various pre-existing and new challenges: 1. Transparency of vocabulary dependencies, 2. Consistency of vocabulary dependencies, 3. Versioning issues due to model changes, deletions etc., 4. Tracking of subsequent changes performed by different users, 5. Process definition for the maintenance of vocabularies, 6. Usability related to the understanding of the data models, 7. Ambiguities and doublets. Our goal is to deploy prototypical approaches in the operational system of WKD and to investigate approaches to ensure data quality, enhance the transparency and consistency of dependencies, resolve versioning issues, deploy tracking functionalities, deploy a maintenance process, identify and encounter ambiguities, and deploy a solution for dealing with doublets.

Impact. The creation and maintenance of knowledge models is gaining importance in the Web of Data. These tasks are increasingly being executed by SME's in the domain, not in knowledge modelling and IT as such. Therefore, better automatic support of these processes will directly help achieving quality and efficiency gains.

Existing Infrastructure. As the number of controlled vocabularies and custom schemas we are integrating in the metadata management tool PoolParty is increasing, we need solutions that give an overview of existing relations between projects and external data and schemas. Besides, the number of user roles is growing so that we also need a solution that provides an overview for a number of different users with different purposes. By different queries and enhancements we want to get an impression about the relations between projects and the usage of specific custom schemas. Connections between projects and schemas are not easily traceable. Owners of vocabularies need to provide documentation so that others can also understand the projects and their scope and context. Without this documentation, it is hard to analyse the different projects. Within the tool the user can only analyse the individual concepts for relations to investigate any relations with schemas. For linking to other projects it is possible to get a list of links. This list does not provide the number of links and specific numbers for different kinds of linking. These figures need to be searched manually.

5   Semantically Enhanced Quality Assurance

The JURION challenges described in Sect. 4 are targeted by the following extensions to the JURION workflow and its components: (a) verification of correct metadata conversion, (b) integration of repeatable tests into the development environment and (c) automation of external link clean-up. In the following sections we present our approach for tackling each challenge.

A six-minute screencast video has been developed which showcases the demonstrator described in this paper. The screencast shows some background context on the JURION use case and the prototype implementation in action. It is available on YouTube[19] and is linked through the ALIGNED project website[20]. RDFUnit [3], including the extensions developed for this demonstrator is available as open source code. PoolParty is a commercial closed-source

---

[19] https://youtu.be/6aLXK7N7wFE.
[20] http://aligned-project.eu/jurion-demonstration-system/.

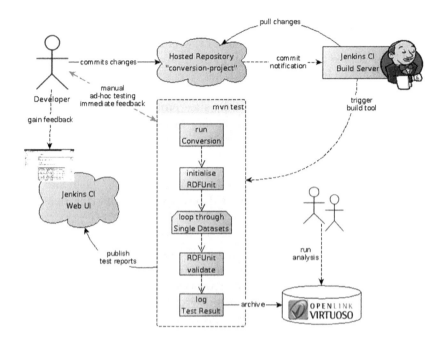

Fig. 2. RDFUnit integration in JURION

product[21] and the extensions developed for this demonstrator will be folded into future releases of the product when this is commercially viable.

XML to RDF Conversion Verification. To allow comparable and reproducible tests with short execution times, a number of WKD XML reference documents have been selected, against which the actual conversion into PCI RDF is executed. Each resulting RDF dataset is then verified individually. The prototyped solution (cf. Figure 2) integrates RDFUnit as the core driver of the tests. It is set upon the JUnit-API[22] so that it integrates seamlessly into the development tool chain and the software build process in general. For instance, a developer can trigger the test chain manually on his local workstation to retrieve direct feedback at any time and any change in the conversion project automatically leads to full test run which is performed by the build system.

All executed tests are based on RDFUnit's auto-generators, which derive test cases at runtime from the latest version of the PCI-ontology. As a proof of concept RDFUnit's test results (the validation model based on the Test Driven Data Validation Ontology [2]) linked to this test is stored into a Virtuoso triplestore to enable future analysis/reviews of historical data.

---

[21] https://www.poolparty.biz/.
[22] http://junit.org/javadoc/latest/.

Early and quick feedback on changes to the project is very valuable to assure that the project is in good health and existing functionality meets the defined expectations. Good coverage with automated tests prevents bugs from slipping in released functionality which may have side effects on other parts of the system. RDFUnit enables possibilities but still needs a tighter integration as a library with our existing toolchain to improve reporting capabilities and make its feedback even more useful. RDFUnit proves as very useful and will be a fixed component of the operational tech stack within WKD JURION from now on. We will provide further requirements to improve RDFUnit's integration into our development pipeline. At a later point in time, we will utilize RDFUnit to enable monitoring the existing data store to implement quality assurance on operational side.

We additionally integrated RDFUnit with JUnit. JUnit is a unit testing framework for Java, supported by most Java developer IDEs and build tools (e.g. maven, Eclipse). JUnit allows to execute repeatable tests as part of the development workflow in line with the test-driven development paradigm. As part of the described use cases we contributed the integration of the RDFUnit-JUnit-module[23]. We added specific Java annotations on JUnit classes that can define the input dataset and the schema that RDFUnit can test. Each test generated by RDFUnit TAGs is translated in a separate JUnit test and reported by JUnit. This approach facilitates simpler setups and can verify specific input files. The benefit is the immediate integration of RDF dataset testing on existing software development tool stacks.

Verifying Available Instance Data for Linked External Metadata. We currently implemented two different kinds of statistical metrics and integrated them into the PoolParty UI, (i) checking for external links validity and (ii) links to other PoolParty projects on the server. These metrics differ in the methodology they are evaluated. Checking the validity of external links cannot be done using SPARQL and requires external tool support (e.g., Java code, see Sect. 6 on external link validity). Reporting links to PoolParty projects can be achieved in a similar way than checking for data consistency violations [3]. Each statistical property can be formulated as a SPARQL query, which is executed on the relevant project data, i.e., the current project data and metadata as well as all linked project data and metadata.

6   Evaluation

Our methodology for baseline data collection is divided into three categories: productivity, quality and agility. The analysis is based on measured metrics and the qualitative feedback of experts and users. Participants of the evaluation study were selected from WKD staff in the fields of software development and

---

[23] https://github.com/AKSW/RDFUnit/tree/master/rdfunit-junit.

data development. There were seven participants in total: four involved in the expert evaluation and three content experts involved in the usability/interview evaluation.

6.1 Productivity

Collection Methods & Metrics. We collected content expert evaluations for the metadata extraction verification, a test suite was set up to measure metrics of productivity for all implemented features and finally, interviews were conducted to obtain feedback from prospective users of PoolParty functionalities. For the RDFUnit integration, we measured the total time for quality checks and error detection, as well as the need for manual interaction. For the external link validation we measured the number of checked links, the number of violations and the total time.

RDFUnit enabled us to develop automated tests that provide tight feedback and good integration into the existing toolchain. It enabled error messages, which point exactly to the offending resource, making bug fixing much easier. Depending on the size of the document and size of the ontology, total time to execute a test-suite varies, but can be indicated with 1 ms to 50 ms per single test. With this approach, the feedback is as close to real-time as possible; currently a couple of minutes. Since it is possible to trigger the quality checks manually at any time through the existing developers IDE menus, speedy performance is desirable to avoid developer idle time. Though quality checks can be triggered by manual execution, they are always verified automatically by the build system, which sends a notification if an error occurs. Due to the development process it is guaranteed that with every change the whole set of quality checks is executed and reported automatically. The current setup generates and runs about 44000 tests with a total duration of 11 min which may scale-up easily when parallelized or clustered. The details-section reports each violated RDFUnit test individually with it's corresponding error message and a list of failing resources.

The external link validation is a new feature that evaluates the links to external sources and informs the user in case the sources are not available anymore. Previously, it was only possible to check the links manually in random samples. We evaluated a sample of four WKD projects that were checked for external relations. The results are provided in Table 1. This validation process can considerably improve time spent on error reduction, so that external links can be maintained efficiently and corrected at short notice. The table shows that the validation of the links of the Arbeitsrechtsthesaurus took a relatively long time. The presentation of the results was well understood. In general, the tool was received well by the experts, which was reflected by their feedback in the interviews.

6.2 Quality

Collection Methods & Metrics. We set up a test suite to measure quality metrics for all features, we additionally used content expert dataset evaluations. For the

Table 1. External link validation of several Wolters Kluwer thesauri

External linking	Concepts	Checked links	Violations	Time
Thesaurus	6510	0	0	0,5 min
Gesetzesthesaurus	1307	0	0	0,001 s
Arbeitsrechtsthesaurus	1728	1868	60	10 min
Gerichtsthesaurus	1503	1434	81	3 min 4 s

metadata extraction verification we collected the number of detected error categories, the test coverage and expert evaluation. For the external link validation we measured the correctness of results and usability aspects.

In the metadata extraction verification we had the following questions: "What kind of errors can be detected" and "is categorization possible"? We used the RDFUnit supported axioms to categorize the errors wherever possible. As stated in Sect. 3, RDFUnit supports many RDFS & OWL axioms. Regarding test coverage, RDFUnit provides test coverage metrics. However, we did not yet integrate the test coverage metrics in our operational tool stack. This is a next (crucial) step, as we need to evaluate the relevance of individual tests to the tested dataset. Ideally, we would need to get the percentage of the input dataset that is covered by tests and how many of these tests actually measured features of the input dataset. From our expert evaluation we concluded that it is helpful to spot errors introduced by changes, since issues spotted in this way can be assumed to point to really existing errors; the causes of which can be identified and addressed. In contrast, successful tests are less significant as we are not yet able to evaluate whether and how the measurements taken correspond to target measures and these tests do not point to concrete errors. To resolve this, we will proceed to integrate measures that help evaluating the test cases on the one hand, and the input datasets on the other hand.

To analyse the quality of the external links validation, we evaluated results of two thesauri. Figure 3 depicts the number of concepts that are detected. In the Labor Law Thesaurus there are 40 correctly detected broken image links. The courts thesaurus provides results of better quality. All of the detected violations are indeed broken links and are corrected. The presentation of external link validation results works well in general. The number of checked and incorrect links are shown in the overview. The interface, in its prototype state, still lacks some usability. The interview strengthens this impression of the expert evaluation, and gives suggestions to improve the interface. A next step will be to add the related preferred label of the concept that links to the broken external source to enhance the usability and the quality evaluation, as well as an improvement of the general structure to ease also the reading of the results.

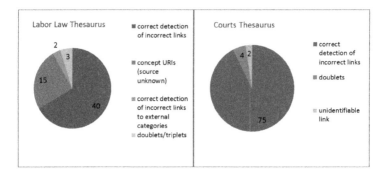

Fig. 3. Incorrect external links of the Labor Law Thesaurus and the Courts Thesaurus

6.3   Agility

Collection Methods & Metrics. We collected evaluations of content experts for the RDFUnit integration and evaluations of technical experts for the PoolParty tool. We collected metrics for time to include new requirements in RDFUnit integration and for the external links validation scope of external link checks, possibility of integration, configuration time and extension.

With respect to the XML-to-PCI conversion verification, including new constraints or adapting existing constraints is a convenient process. The procedure works by adding new reference documents to the input dataset to make the test environment as representative as possible. As the process of generating tests and testing is fully automated, it adapts very easily to changed parameters. However, adding more documents to the input dataset increases the total runtime of the test-suite, which affects the time to feedback. Therefore, one must be careful with the selection of proper reference documents.

Regarding the scope of external link checks, the current solution resolves all links that point to hosts which are not identified as the local host. While this is certainly useful for getting an overview, in many cases it is desired to limit the link lookups and adapt the way links to external datasets are detected. Possibilities are, for instance, to use the current projects base URI or regular expression-based techniques. Internal link checks are not yet implemented. Determining which URIs should be resolved can be done either directly with SPARQL or within the Java resolution algorithm. In each case, the effort for change is low, allowing for agile reaction on changed requirements. However, changes to the current configuration require recompilation and redeployment of PoolParty, which reduces agilty. In order to address performance issues, our future plans are to delegate link checking to an external application. This task would be suitable to be performed by UnifiedViews[24], which we also envision to take a central role in the PoolParty Semantic Suite. This also requires to have a method in place that enables UnifiedViews to report the created statistics back to PoolParty, which is on the roadmap for future work.

---

[24] https://github.com/UnifiedViews.

6.4   Analysis

The evaluation of the prototype shows clearly that during the first phase of prototype development, we have achieved our aim to improve the productivity and quality of data processes within the data lifecycle. With the presented features, these improvements could be shown. Performance and quality/error rates of the test results were reasonable. In addition to the tests, there will be new scope for data repair processes to correct the detected errors of the dataset, as we gained new insights into data violations (e.g. more categories of violated external links than we expected). Nonetheless, there need to be further improvements, especially with regard to usability, performance, integration of functionalities and required details that are not yet fully working.

In summary, the productivity of data processes is clearly improved by the initial prototype. The statistics and external link validation functionalities can help to save much time by replacing time consuming manual work by efficient data overviews.

Concerning the quality of the prototype functionalities, the results are very satisfying. For notifications and external link validation there are only few issues. For the data transformation with RDFUnit there needs to be further investigations to enable comprehensive and extensive data testing results. Usability issues need to be tackled in all of the features for a better operational implementation. As this is only an initial prototype, usability was less of a focus.

The testers feedback for agility of features is quite positive. The agility of RDFUnit is seen as satisfying, as the automated service allows the implementation of new requirements easily. External link validation has a reasonable agility and is planned to be done by an external application to address performance issues.

7   Conclusions & Future Work

In this paper, we described an industrial use case of RDF technologies with JURION. We managed to weaken the gap between software and data development by integrating quality checks in our existing software tool stack. We provided a screencast of our prototype and contributed bindings between RDFUnit and JUnit as open source.

In future work we plan to improve the RDFUnit integrations in JURION. Further research is needed for test coverage reports as well as the generation test analytics. For example, time to fix a bug, identification of regressions, etc. In regard to PoolParty we will improve the user experience of the extenal metadata checking tool. As far as functionality is concerned, we plan to support internal link checking and provide advanced configuration for the end users.

Acknowledgments. This work was supported by grants from the EU's H2020 Programme ALIGNED (GA 644055). WKD and JURION is a use case partner in the ALIGNED project.

References

1. Auer, S., Bryl, V., Tramp, S. (eds.): Linked Open Data - Creating Knowledge Out of Interlinked Data, vol. 8661, 1st edn. Springer, Heidelberg (2014)
2. Kontokostas, D., Brümmer, M., Hellmann, S., Lehmann, J., Ioannidis, L.: NLP data cleansing based on linguistic ontology constraints. In: Presutti, V., d'Amato, C., Gandon, F., d'Aquin, M., Staab, S., Tordai, A. (eds.) ESWC 2014. LNCS, vol. 8465, pp. 224–239. Springer, Heidelberg (2014)
3. Kontokostas, D., Westphal, P., Auer, S., Hellmann, S., Lehmann, J., Databugger, R.C.: A test-driven framework for debugging the web of data. In: WWW Companion 2014, pp. 115–118 (2014)
4. Mader, C., Haslhofer, B., Isaac, A.: Finding quality issues in SKOS vocabularies. In: Zaphiris, P., Buchanan, G., Rasmussen, E., Loizides, F. (eds.) TPDL 2012. LNCS, vol. 7489, pp. 222–233. Springer, Heidelberg (2012)

Adaptive Linked Data-Driven Web Components: Building Flexible and Reusable Semantic Web Interfaces

Ali Khalili[✉], Antonis Loizou, and Frank van Harmelen

Department of Computer Science, Vrije Universiteit Amsterdam,
Amsterdam, The Netherlands
{a.khalili,a.loizou,frank.van.harmelen}@vu.nl

Abstract. Due to the increasing amount of Linked Data openly published on the Web, user-facing Linked Data Applications (LDAs) are gaining momentum. One of the major entrance barriers for Web developers to contribute to this wave of LDAs is the required knowledge of Semantic Web (SW) technologies such as the RDF data model and SPARQL query language. This paper presents an adaptive component-based approach together with its open source implementation for creating flexible and reusable SW interfaces driven by Linked Data. Linked Data-driven (LD-R) Web components abstract the complexity of the underlying SW technologies in order to allow reuse of existing Web components in LDAs, enabling Web developers who are not experts in SW to develop interfaces that view, edit and browse Linked Data. In addition to the modularity provided by the LD-R components, the proposed RDF-based configuration method allows application assemblers to reshape their user interface for different use cases, by either reusing existing shared configurations or by creating their proprietary configurations.

1 Introduction

With the growing number of structured data published, the Web is moving towards becoming a rich ecosystem of machine-understandable Linked Data[1]. Semantically structured data facilitates a number of important aspects of information management such as information retrieval, search, visualization, customization, personalization and integration [11]. Despite all these benefits, Linked Data Applications (LDAs) are not yet adopted by the large community of Web developers outside the Semantic Web domain and, causally, by the end-users on the Web. The usage of semantic data is still quite limited and most of the currently published Linked Data is generated by a relatively small number of publishers [6] which points to entrance barriers for the wide-spread utilization of Linked Data [2].

---

[1] lodlaundromat.org recently (17.12.2015) reported approx. 38.6 billion triples published on the Web.

© Springer International Publishing Switzerland 2016
H. Sack et al. (Eds.): ESWC 2016, LNCS 9678, pp. 677–692, 2016.
DOI: 10.1007/978-3-319-34129-3_41

The current communication gap between Semantic Web developers and User Experience (UX) designers, caused by the need to bear Semantic Web knowledge, prevents the streamlined flow of best practices from the UX community into Linked Data user interface (UI) development. The resulting lack of adoption and standardization often makes current LDAs inconsistent with user expectations and impels more development time and costs on LDA developers. In this situation, more time is spent in re-designing existing UIs rather than focusing on innovation and creation of sophisticated LDAs.

This paper presents the concept of *Adaptive Linked Data-driven Web components* together with its open source implementation available at http://ld-r.org to build flexible and reusable Semantic Web UIs. *Web Components* are a set of W3C standards [5] that enable the creation of custom, reusable user interface widgets or components in Web documents and Web applications. The *Resource Description Framework* (RDF), on the other hand, provides a common data model that allows data-driven components to be created, shared and integrated in a structured way across different applications. Linked Data-driven (LD-R) Web components as defined in this paper are a species of Web components that employ the RDF data model for representing their content and specification (i.e. metadata about the component). LD-R components are supported by a set of predefined core Web components, each representing a compartment of the RDF data model on the Web UI. Thus, the Semantic Web nature of an LDA can be encapsulated in LD-R components thereby allowing UX designers and Web developers outside the Semantic Web community to contribute to LDAs. The components also provide current Semantic Web developers with a mechanism to reuse existing Web components in their LDAs. Furthermore, LD-R components exploit the power and flexibility of the RDF data model in describing and sharing resources to provide a mechanism to adapt the Web interfaces based on the meaning of data and user-defined rules.

The LD-R approach offers many benefits that we will describe in the remainder of the paper. Among them are:

- *Bootstrapping LDA UIs.* LD-R components exploit best practices from modern Web application development to bring an exhaustive architecture to perform separation of concerns and thereby bootstrapping an LDA by only selecting a minimal relevant configuration. For example, a developer only needs to set the URL of his in-use SPARQL endpoint and start developing the viewer components without dealing with the underlying connection adapters and data flow mediators in the system.
- *Standardization and Reusability of LDA UIs.* Instead of creating an LDA UI from scratch, in the component-based development of LDA UIs, application assemblers choose from a set of standard UIs which will reduce the time and costs associated with the creation of LDAs. For example, to render DBpedia resources of type 'Location', a standard map can be reused.
- *Customization and Personalization of LDA UIs.* The RDF-based nature of LD-R components allow application assemblers to reshape their user interface based on the meaning of data or user context. For example, for all the

resources of type foaf:Person, the content can be rendered with a 'Contact-Card' component.

– *Adoption of LDA UIs by non-Semantic Web developers and end-users.* Most of the current Linked Data interfaces fall into the *Pathetic Fallacy of RDF* [10] where they display RDF data to the users as a graph because the underlying data model is a graph. Abstracting the complexity of RDF and graph-based data representation provides more *Affordances* [16] for non-Semantic Web users to contribute to Linked Data UIs. Engaging more UX designers and Web developers into LDA UIs will also result in more affordances on the end-user's side to better understand the possible actions and advantages of the LDAs.

2   The Current Status of Linked Data UI Development

In order to understand the current pitfalls of LDA UI design, we conducted a survey targeting active Semantic Web developers[2]. The participants where selected from the community of Semantic Web (SW) developers on Github who have had at least one active SW-related repository. Github is currently the most popular repository for open source code and its transparent environment implies a suitable basis for evaluating reuse and collaboration among developers [21]. We used Github APIs to search[3] for SW repositories and to collect contact information for the corresponding contributors when available. The search, after removing organizations and invalid email addresses, resulted in 650 potential SW developers. We then contacted the candidates to ask them about the current pitfalls in developing LDA UIs. In our enquiry, we clearly mentioned to skip the questionnaire if they have not developed any SW application so far. We used a minimal set of 7 questions to attract more responses and also used inline forms for GMail users to allow filling out the questionnaire in the same time as reading the enquiry email. We collected 79 responses to our questionnaire, which is a considerable number of participants (almost 12 % of the potential candidates). Figure 1 shows the main results of our survey.

Participants. Based on their LDA development experience, we divided the participants into three groups: basic (less than 2 applications), intermediate (3–5 applications) and advanced (more than 5 applications) developers. The result showed that the majority (62 %) of participants were intermediate and advanced developers. In addition to their development experience, developers were asked about their knowledge of Semantic Web to compare their practical and conceptual experience. As results revealed, the majority of participants (63 %) had proficient (4–5 years) and expert (more than 5 years) knowledge of Semantic Web and Linked Data which makes a good sample for our evaluation.

Questions addressed the following topics:

---

[2] Results are available at https://goo.gl/cltqhv.
[3] Keywords: "Semantic Web" OR "Linked Data" OR "RDF" OR "SPARQL".

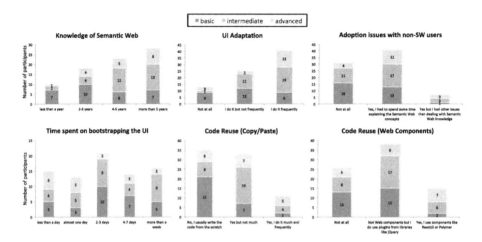

Fig. 1. Results of our user study on the current status of LDA UI development.

– *Amount of time spent on bootstrapping LDA UIs.* Before designing the UIs in an LDA, developers need to spend some time on creating the skeleton of their application where querying data and the business logic of the application is handled. The results confirm that developers spend a lot of time (on average more than 2 days) on bootstrapping their LDAs before they can start working on the UI.
– *Reuse of code by Semantic Web developers.* Developers usually reuse sections of code, templates, functions, and objects to save time and resources when developing LDAs. We asked participants about two types of reuse: reuse by copy/pasting code from existing LDAs and reuse by employing current Web components. Reuse by copy/pasting code can be seen as an indicator of the state of standardization, modularity and reusability of current LDAs. The results indicate that a considerable amount of users (46 %), prefer to write the code from scratch instead of reusing code from existing Semantic Web projects. This situation is more pronounced for basic developers who still prefer to write the code from scratch although they have less experience in programming LDAs. Furthermore, the results on reuse of Web components give an insight on the adoption of current Web Components by Semantic Web developers. The results indicate that despite the prevalence of Web Components solutions, only 19 % of the participants (mainly advanced users) were employing them in their applications. Interestingly, the majority of participants (49 %) were already reusing other component-like libraries which shows an attitude and capacity towards adopting the Web components.
– *Adaptation of LDA UIs.* Most of the modern Web applications provide a mechanism to customize and personalize their user interfaces based on the type of data and the information needs of their end-users. Proactive user interface adaptation allows the application to act more like a human and consequently, more intelligently [9]. As our study shows, within the current LDA developers, 52 %

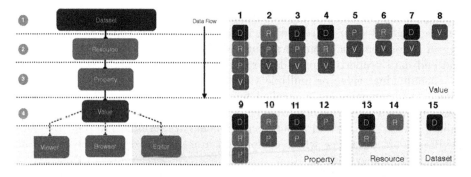

Fig. 2. (Left) core LD-R Web components. (Right) LD-R scopes based on the permutation of dataset, resource, property and value identifiers.

had experience adapting the user interface of their applications frequently. There were also 32 % that were doing the UI adaptation but not frequently.

- *Adoption issues with non-Semantic Web developers.* In order to examine if there is a communication gap between UI designers and Semantic Web developers, we asked the participants about their experience when collaborating with a non-SW developer. Among the participants, 51 % had communication issues with non-Semantic Web developers to familiarize them with Semantic Web concepts before they can start contributing to the application. The distribution of this issue among more experienced developers (57 % of the intermediate and advanced users) further emphasizes the importance of this communication gap.

3   Adaptive Linked Data-Driven Web Components

In order to streamline the process of UI development in LDAs, we propose an architecture of adaptive LD-R Web components – Web components enriched by the RDF data model. The proposed architecture addresses LDA UI reusability and flexibility by incorporating RDF-based Web components and scopes. In the following sections, the main elements of the architecture are described:

3.1   LD-R Web Components

As depicted in Fig. 2, there are four core component levels in an LD-R Web application. Each core component abstracts the actions required for retrieving and updating the graph-based data and provides a basis for user-defined components to interact with Linked Data in three modes: view, edit and browse.

The data-flow in the system starts from the *Dataset* component which handles all the events related to a set of resources under a named graph identified by a URI. The next level is the *Resource* component which is identified by a URI and indicates what is described in the application. A resource is described by a set of properties which are handled by the *Property* component. Properties can be either individual or aggregate when combining multiple features of a resource (e.g. a component that combines longitude and latitude properties; start date and end date properties for a date range, etc.). Each property is instantiated by an individual value or multiple values in case of an aggregate object. The value(s) of properties are controlled by the *Value* component. In turn, Value components invoke different components to view, edit and browse the property values. *Viewer*, *Editor* and *Browser* components are terminals in the LD-R single directional data flow where customized user-generated components can be plugged into the system.

User interactions with the LD-R components are controlled by a set of configurations defined on one or more selected component levels known as scopes.

3.2   Scopes and Configurations

LD-R Web components provide a versatile approach for context adaptation. A context can be a specific domain of interest, a specific user requirement or both. In order to enable customization and personalization, the LD-R approach exploits the concepts of *Scope* and *Configuration*. A scope is defined as a hierarchical permutation of Dataset, Resource, Property and Value components (cf. Fig. 2). Each scope conveys a certain level of specificity on a given context ranging from 1 (most specific) to 15 (least specific). Scopes are defined by using either the URIs of named graphs, resources and properties, or by identifying the resource types and data types. A configuration is defined as a setting which affects the way the LDA and Web components are interpreted and rendered (e.g. render a specific component for a specific RDF property or enforce a component to display Wikipedia page URIs for DBpedia resources). UI adaptation is handled by traversing the configurations for scopes, populating the configurations and overwriting them when a more specific applicable scope is found.

Scopes can also be defined on a per user basis, facilitating the versioning and reuse of user-specific configurations. User-Specific configurations provide different views on components and thereby data, based on the different personas dealing with them. In addition to the fine-grained component customization, LD-R Web applications provide a fine-grained access control over the data through the component scopes. For example, an application developer can restrict access to a specific property of a specific resource in a certain dataset and on a specific interaction mode.

3.3   Semantic Markup for Web Components

The innate support of RDF in LD-R Web components enable the automatic creation of semantic markup on the UI level. Lower semantic techniques such as *RDFa*, *Mircodata* and *JSON-LD* can be incorporated in the core LD-R components to expose structured data to current search engines which are capable of parsing semantic markup. For example, an LD-R component created based on the *Good Relations* or schema.org ontologies, can automatically expose the product data as Google Rich Snippets for products which will provide better visibility of the data on Web search results (i.e. SEO). In addition to automatic annotation of data provided by the LD-R Web components, the approach offers semi-automatic markup of Web components by creating component metadata. Component metadata consists of two categories of markup:

- Automatic markup generated by parsing component package specification – metadata about the component and its dependencies. It includes general metadata such as name, description, version, homepage, author as well as technical metadata on component source repository and dependencies.
- Manual markup created by component authors which exposes metadata such as component level (dataset, resource, property, value), granularity (individual, aggregate), mode (view, edit, browse) and configuration parameters specification.

Similar to content markup, Component markup can utilize commonly-known ontologies such as schema.org in order to improve the visibility of LD-R components and enable application assemblers to better understand the intended usage and capabilities of a given component.

3.4   Stackeholders and Life Cycle

As shown in Fig. 3, the LD-R components lifecycle encompasses four primary types of stakeholders:

- *Linked Data Provider.* Since the LD-R approach focuses mainly on Linked Data applications, the provision of RDF-compliant data is an essential phase in developing the LD-R components. There are different stages [1] in Linked Data provision, including data extraction, storage, interlinking, enrichment, quality analysis and repair which should be taken into account by data scientists and Linked Data experts. Once the data and schemata are provided

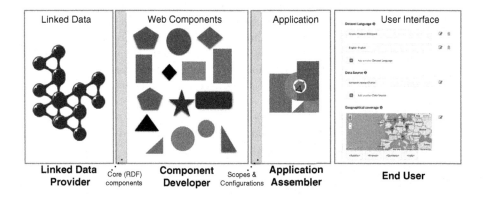

Fig. 3. LD-R components life cycle.

to the LD-R component system, the system can bring a reciprocal value to
Linked Data providers to better understand and curate the data when needed.
For example, in the case of geo-coordinates, a map component can enable data
providers to easily curate the outlier data (e.g. ambiguous entities) within a
certain geo boundary in a visual manner.

– *Component Developer.* Component developers are UX designers and Web programmers who are involved in component fabrication. There are two types of
Web components developed in this step: (a) *Core components* (cf. Fig. 2) which
abstract the underlying RDF data model. These components are built-in to
the system, however can still be overwritten by developers who have profi-
ciency in Semantic Web and Linked Data. (b) *Community-driven components*
which exploit the core components. These components are either created from
scratch or by remixing and repurposing existing Web components found on
the Web.

– *Application Assembler.* The main task of application assemblers is to identify
the right components and configurations for the application; and to combine
them in a way which fits the application requirements. Within the LD-R
component system, the metadata provided by each Web component facilitates
the discovery of relevant components. Having shared vocabularies on Linked
Open Data allows assemblers to not only reuse components but also reuse
the existing configurations and scopes published on the Web. For example, if
there is already a suitable configuration for RP scope which uses `foaf:Person`
as resource type and `dcterms:description` as property URI, the assembler
can reuse that configuration within his application.

– *End-User.* End-users experience working with the components to pursue goals
in a certain application domain. As such, they may request the development
of new components or configurations in order to fulfil their requirements and
are expected to provide feedback on existing components.

4 Implementation

In order to realize the idea of adaptive Linked Data-driven Web components,
we implemented an open-source software framework called *Linked Data Reactor
(LD-Reactor)* which is available online at http://ld-r.org. LD-Reactor utilizes
Facebook's ReactJS[4] components, the Flux[5] architecture, Yahoo!'s Fluxible[6]
framework for isomorphic Web applications (i.e. running the components code
both on the server and the client) and the Semantic-UI[7] framework for flexible UI themes. The main reasons we chose *React* components over other Web
Components solutions (e.g. *Polymer, AngularJS, EmberJS*, etc.) were the maturity and maintainability of the technology, the native multi-platform support,
the number of developer tools/components/applications, and the efficiency of
its underlying virtual DOM approach.

---

[4] https://facebook.github.io/react/.
[5] https://facebook.github.io/flux.
[6] http://fluxible.io/.
[7] http://semantic-ui.com/.

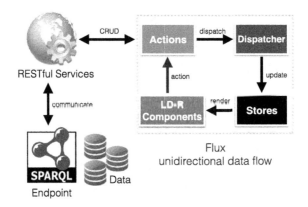

Fig. 4. Data flow in the LD-Reactor framework.

As shown in Fig. 4, LD-Reactor follows the Flux architecture which eschews MVC (Model-View-Controller) in favour of a unidirectional data flow. When a user interacts with a React component, the component propagates an action through a central dispatcher, to the various stores that hold the application's data and business logic, and updates all affected components. The component interaction with SPARQL endpoints to retrieve and update Linked Data occurs through the invocation of RESTful services in actions.

In order to allow the bootstrapping of LDA UIs, LD-Reactor provides a comprehensive framework that combines the following main elements:

- A set of RESTful Web services that allow basic CRUD operations on Linked Data using SPARQL queries[8].
- A set of core components called *Reactors* which implement core Linked Data components (see Fig. 2) together with their corresponding actions and stores.
- A set of default components which allow basic viewing, editing and browsing of Linked Data.
- A set of minimal viable configurations based on the type of data and properties from commonly-used vocabularies (e.g. foaf, dcterms and SKOS).
- A basic access control plugin which allows restricting read/write access to data.

LD-Reactor implementation is compliant with *Microservices Architecture* [13] where the existing ReactJS components can be extended by complementary LD services. In contrast to the centralized monolithic architecture, the microservices architecture allows placing the main functionalities of the LDA into separate decoupled services and scale by distributing these services across

---

[8] The framework is compliant with the SPARQL 1.1 standard. However, we have identified certain inconsistencies between OpenRDF Sesame and OpenLink Virtuoso RDF stores, which did not allow the execution of syntactically identical queries across both systems. Thereby, specific adaptors have been implemented for each of these two RDF stores.

servers, replicating as needed. This architectural style also helps to minimize the redeploying of the entire application when changes in components were requested.

Semantic markup of data (as discussed in Sect. 3.3) is supported natively within the framework by embedding Microdata annotations within the LD-R Web components. Additionally, in order to facilitate the creation of component metadata, we developed a tool[9] which automatically generates the general metadata about the components in JSON-LD, using schema.org's SoftwareApplication schema.

5   Use Cases

The LD-Reactor framework is already in use within the RISIS[10] and Open PHACTS[11] projects.

5.1   RISIS

The RISIS project aims to provide an infrastructure for research and innovation, targeting researchers from various science and technology domains. The LD-Reactor framework was utilized in RISIS to help data providers with no Linked Data experience to provide RDF metadata about their datasets[12]. This metadata is then used to allow researchers to search, identify, and request access to the data they are interested in[13].

In the following, we present the main requirements for configurations and components, together with their representation in the LD-Reactor framework[14].

Configurations:

– The UI should be able to render metadata properties in different categories.
– The labels for properties should be changeable in the UI especially for technical properties (e.g. RDF dump) that are unknown to researchers outside the Semantic Web domain.
– There should be a hint for properties to help metadata editors to understand the meaning of the property.
– Instead of showing the full URIs, the output UI should render either a shortened URI or a meaningful string linked to the original URI.
– Whenever a DBpedia URI is provided, display the corresponding Wikipedia URI enabling users to retrieve human readable information.
– When a dropdown menu is provided, there should be the ability to accommodate user-defined values which are not listed in the menu.

---

[9] https://github.com/ali1k/ld-r-metadata-generator.
[10] http://risis.eu.
[11] http://www.openphacts.org.
[12] http://sms.risis.eu.
[13] http://datasets.risis.eu.
[14] See the complete configuration file at http://github.com/risis-eu/sms-platform.

Components:

- A component for `dcterms:spatial` values to allow searching and inserting resources from DBpedia based on the entity type (e.g. Place, Organization, etc.).
- A component for `dcterms:subject` values to allow inserting and viewing DBpedia URIs as subject.
  A component for `dcterms:language` values to allow inserting and viewing languages formatted in ISO 639-1 using standard URIs (e.g. http://id.loc.gov/vocabulary/iso639-1/en).
- A component for `dcat:byteSize` values to allow inserting and viewing file size specified by a unit.
- A component for `dcterms:format` values to allow inserting and viewing mime types.

In accordance to the LD-Reactor microservices architecture (cf. Sect. 4), we built a *DBpediaGMap* viewer component where we reused the current react-google-maps together with DBpedia lookup and query services to retrieve the coordinates for the recognized DBpedia resource values.

5.2 Open PHACTS

The Open PHACTS Discovery Platform has been developed to reduce barriers to drug discovery, by collecting and integrating a large number of prominent RDF datasets in the pharmacology domain. The platform provides a uniform RESTful API for application developers to access the integrated data. In collaboration with the data providers, the Open PHACTS consortium has created a comprehensive dataset description specification[15] based on the Vocabulary of Interlinked Datasets (VoID)[16]. The metadata provided in this context enables (among others) exposing the detailed provenance for each result produced by the platform, the location of source files for each dataset, and example resources.

The provision of VoID dataset descriptors that adhere to the specification proved to be a non-trivial challenge, even for data providers that are well versed in providing RDF distributions of their core data. A series of UIs were therefore created to facilitate the creation of the VoID descriptors. However, as the specification evolved over the first 2 years of the project, any changes or additions made had to be reflected in the UI source code as well; a cumbersome process. Due to the inevitable delay between specification changes and UI development, users often found themselves having to edit large RDF files using text editors, which resulted in frequent syntax errors being made.

A new version of the VoID editor implemented using the LD-Reactor framework is now available online[17]. Though the import/export capabilities of the editor are still not implemented at the time of writing, we have received very positive feedback from the community for a number of reasons:

---

[15] http://www.openphacts.org/specs/2013/WD-datadesc-20130912/.
[16] http://www.w3.org/TR/void/.
[17] http://void.ops.labs.vu.nl/ Source: http://github.com/openphacts/ld-r.

Fig. 5. (Left) screenshot of RISIS VoID editor. (Right) screenshot of the BasicCalendarInput component for editing datetime values in Open PHACTS.

– UI updates: As the UI is generated based on the underlying data, the process of staying up to date with the current specification becomes trivial. The RDF example provided by the specification can be simply loaded into the RDF store and the changes are immediately visible in the UI through the default core components. Users are then able to adapt the example VoID description to their dataset. The resulting VoID file can be downloaded by exporting all triples in the named graph corresponding to the dataset.

– Dataset releases: A large number of property values remain the same across releases. Similarly with using the example from the Dataset Description Specification, users are able to upload their old VoID description and only edit the outdated values.

– Access control: Using the built–in user authentication mechanism of LD-Reactor we were able to ensure that only the owner(s) of a particular dataset are able to edit its metadata.

– Non-standard properties: Some data providers elect to include additional properties that are not prescribed by the specification. The visualisation of such properties is supported easily using the core LD-R components.

– Intuitive navigation: Typically, each dataset consists of a number of subsets, which are also of type `void:Dataset`, and may have further subsets themselves. Displaying all the information together can easily become confusing for the user; instead the LD-Reactor framework was used to provide navigation through the subset links, displaying only a single dataset (or subset) at a time.

– Datetime component: Manually typing datetimes in the required format (e.g. '2015-02-12T16:48:00Z') can be an error prone process. Instead, we have been able to reuse the `react-bootstrap-datetimepicker` component, to create a new LD-R value editor for datetimes (BasicCalendarInput[18]) with a graphical interface as shown in Fig. 5.

In addition to the significant improvements over previous versions of the VoID editor outlined above, we were able to develop the LD-Reactor version in a fraction of the time that was required for earlier versions.

6   Related Work

Component-based software engineering (CBSE) has been an active research area since 1987 with numerous results published in the research literature [22]. Within the Semantic Web community, the main focus has been on enriching current service-oriented architectures (SOAs) with semantic formalisms and thereby providing Semantic Web services as reusable and scalable software components [23]. There have also been a few attempts to create Semantic Web Components by integrating existing Web-based components with Semantic Web technology [3,7].

When it comes to component-based development of LDAs, the works typically fall into software application frameworks that address building scalable LDAs in a modular way. The survey conducted by [8] identified the main issues in current Semantic Web applications and suggested the provision of component-based software frameworks as a potential solution to the issues identified. The Semantic Web Framework [4] was one of the first attempts in that direction to decompose the LDA development requirements into an architecture of reusable software components. In most of the current full-stack LDA frameworks such as Callimachus[19] and LDIF[20] the focus is mainly on the backend side of LDAs and less attention is paid on how Linked Data is consumed by the end-user. There are also more flexible application frameworks such as OntoWiki [6] which provide UI widgets and extensions to expose Linked Data to non-SW end-users.

Besides these generic LDA frameworks, there are also approaches that focus on the development of user interfaces for LDAs. WYSIWYM (What You See Is What You Mean) [12] is a generic semantics-based UI model to allow integrated visualization, exploration and authoring of structured and unstructured data. Our proposed approach utilizes the WYSIWYM model for binding RDF-based data to viewer, editor and browser UIs. Uduvudu [14] is another approach to making an adaptive RDF-based UI engine to render LD. Instead of adopting Web components, Uduvudu employs a set of flexible UI templates that can be combined to create complex UIs. Even though the static templates do not provide enough interactions for editing and browsing data (in contrast to Web components), we believe that algorithms for automatic selection of templates employed in Uduvudu can be reused in the LD-Reactor framework for automatic generation of configurations. Another similar approach is SemwidgJS [20]

---

[18] https://github.com/openphacts/ld-r/blob/master/components/object/editor/individual/BasicCalendarInput.js.

[19] http://callimachusproject.org/.

[20] http://ldif.wbsg.de/.

which brings a semantic Widget library for the rapid development of LDA UIs. SemwidgJS offers a simplified query language to allow the navigation of graph-based data by ordinary Web developers. The main difference between LD-R and SemwidgJS is that LD-Reactor suggests a more interactive model which is not only for displaying LD but also for providing user adaptations based on the meaning of data. LD-Viewer [15] is another related Linked Data presentation framework particularly tailored for the presentation of DBpedia resources. In contrast to LD-Reactor, LD-Viewer builds on top of the traditional MVC architecture and its extensions rely heavily on the knowledge of RDF which is a burden for developers unfamiliar with SW technologies.

In addition to the LDA UI frameworks, there are several ad-hoc tools for Linked Data visualization and exploration such as Balloon Synopsis [18] and Sgvizler [19] which can be utilized as Web components within the LD-Reactor framework. [17] provides an extensive list of these tools aiming to make Linked Data accessible for common end-users who are not familiar with Semantic Web.

Overall, what distinguishes LD-Reactor from the existing frameworks and tools is its modern isomorphic component-based architecture that addresses reactive and reusable UIs as its first class citizen.

7   Conclusion and Future Work

This paper presented adaptive Linked Data-driven Web components as a solution to increase the usability of current Linked Data applications. The proposed component-based solution emphasizes the reusability and separation of concerns in respect to developing Linked Data applications. The RDF-based UI adaptation mechanism aims to provide better customization and personalization based on the meaning of data. Furthermore, employing standard Web components aspires to bring a better communication between UX designers and Semantic Web developers in order to reuse best UI practices within Linked Data applications.

We argue that bridging the gap between Semantic Web Technologies and Web Components worlds brings mutual benefits for both sides. On one hand, SW technologies provide support for richer component discovery, interoperability, integration, and adaptation on the Web. On the other, Web Components bring the advantages of UI standardization, reusability, replaceability and encapsulation to current Semantic Web applications.

As our future plan, we envisage creating a cloud infrastructure for sharing and reusing LD-R scopes and configurations as well as LD-R Web components without the need to install the framework. We also plan to make a user interface to facilitate creation of the LD-R scopes and configurations. Another direction for future research is developing mechanisms for the automatic configuration and composition of Web components based on the semantic markup provided.

Acknowledgement. We would like to thank prof. Peter van den Besselaar from the faculty of Social Sciences and our colleagues from the Knowledge Representation & Reasoning research group at Vrije Universiteit Amsterdam for their helpful comments during the development of the LD-Reactor framework. This work was supported by a grant from the European Union's 7th Framework Programme provided for the project RISIS (GA no. 313082).

References

1. Auer, S., Lehmann, J., Ngonga Ngomo, A.-C., Zaveri, A.: Introduction to linked data and its lifecycle on the web. In: Rudolph, S., Gottlob, G., Horrocks, I., Harmelen, F. (eds.) Reasoning Web 2013. LNCS, vol. 8067, pp. 1–90. Springer, Heidelberg (2013)
2. Benson, E., Karger, D.R.: End-users publishing structured information on the web: an observational study of what, why, and how. In: 32nd Conference on Human Factors in Computing Systems, CHI 2014, pp. 1265–1274. ACM (2014)
3. Casey, M., Pahl, C.: Web components and the semantic web. Electr. Notes Theor. Comput. Sci. **82**(5), 156–163 (2003)
4. Castro, R.G., Pérez, A.G., Óscar, M.-G.: The semantic web framework: a component-based framework for the development of semantic web applications. In: 19th International Conference on Database and Expert Systems Application, DEXA 2008, pp. 185–189. IEEE Computer Society, Washington, D.C (2008)
5. Cooney, D.: Introduction to web components (2014). http://www.w3.org/TR/components-intro/
6. Frischmuth, P., Martin, M., Tramp, S., Riechert, T., Auer, S.: OntoWiki–an authoring, publication and visualization interface for the data web. Semant. Web **6**(3), 215–240 (2015). doi:10.3233/SW-140145
7. Hartig, O., Kost, M., Freytag, J.C.: Designing component-based semantic web applications with DESWAP. In: Bizer, C., Joshi, A. (eds.) Poster and Demonstration Session at the ISWC 2008, CEUR Workshop Proceedings, vol. 401 (2008). CEUR-WS.org
8. Heitmann, B., Kinsella, S., Hayes, C., Decker, S.: Implementing semantic web applications: reference architecture and challenges. In: 5th International Workshop on Semantic Web-Enabled Software Engineering (2009)
9. Hervás, R., Bravo, J.: Towards the ubiquitous visualization: adaptive user-interfaces based on the semantic web. Interact. Comput. **23**(1), 40–56 (2011)
10. Karger, D., Schraefel, M.: The pathetic fallacy of RDF. Position Paper for SWUI06 (2006)
11. Khalili, A., Auer, S.: User interfaces for semantic authoring of textual content: a systematic literature review. Web Seman. Sci. Serv. Agents World Wide Web **22**, 1–18 (2013)
12. Khalili, A., Auer, S.: WYSIWYM - integrated visualization, exploration and authoring of semantically enriched un-structured content. Semant. Web **6**(3), 259–275 (2015). doi:10.3233/SW-140157
13. Lewis, J., Fowler, M.: Microservices (2014). http://martinfowler.com/articles/microservices.html
14. Luggen, M., Gschwend, A., Bernhard, A., Cudre-Mauroux, P.: Uduvudu: a graph-aware and adaptive UI engine for linked data. In: Bizer, C., Auer, S., Berners-Lee, T., Heath, T. (eds.) Workshop on Linked Data on the Web (LDOW), CEUR Workshop Proceedings, Aachen, vol. 1409 (2015)

15. Lukovnikov, D., Stadler, C., Lehmann, J.: LD viewer - linked data presentation framework. In: Proceedings of the 10th International Conference on Semantic Systems, SEM 2014, pp. 124–131. ACM, New York (2014)
16. Norman, D.A.: The Design of Everyday Things: Revised and Expanded Edition. Basic Books Inc., New York (2013)
17. Ojha, S.R., Jovanovic, M., Giunchiglia, F.: Entity-centric visualization of open data. In: Abascal, J., Barbosa, S., Fetter, M., Gross, T., Palanque, P., Winckler, M. (eds.) INTERACT 2015. LNCS, vol. 9298, pp. 149–166. Springer, Heidelberg (2015)
18. Schlegel, K., Weißgerber, T., Stegmaier, F., Granitzer, M., Kosch, H.: Balloon synopsis: a jquery plugin to easily integrate the semantic web in a website. In: Verborgh, R., Mannens, E. (eds.) ISWC Developers Workshop, CEUR Workshop Proceedings, vol. 1268, pp. 19–24 (2014). CEUR-WS.org
19. Skjæveland, M.G.: Sgvizler: a JavaScript wrapper for easy visualization of SPARQL result sets. In: 9th Extended Semantic Web Conference (ESWC 2012), May 2012
20. Stegemann, T., Ziegler, J.: SemwidgJS: a semantic widget library for the rapid development of user interfaces for linked open data. In: 44. Jahrestagung der Gesellschaft für Informatik, Informatik 2014, Big Data - Komplexität meistern, 22–26 September 2014, pp. 479–490, Stuttgart (2014)
21. Tsay, J., Dabbish, L., Herbsleb, J.: Let's talk about it: evaluating contributions through discussion in github. In: 22nd ACM SIGSOFT International Symposium on Foundations of Software Engineering, FSE 2014, pp. 144–154. ACM (2014)
22. Vale, T., Crnkovic, I., de Almeida, E.S., Neto, P.A.D.M.S., Cavalcanti, Y.C., de Lemos Meira, S.R.: Twenty-eight years of component-based software engineering. J. Syst. Softw. **111**, 128–148 (2016). doi:10.1016/j.jss.2015.09.019
23. Wang, H.H., Gibbins, N., Payne, T., Patelli, A., Wang, Y.: A survey of semantic web services formalisms. Concurrency Comput. Pract. Exp. **27**(15), 4053–4072 (2015)

Building the Seshat Ontology for a Global History Databank

Rob Brennan[1]([⊠]), Kevin Feeney[1], Gavin Mendel-Gleason[1],
Bojan Bozic[1], Peter Turchin[2], Harvey Whitehouse[3],
Pieter Francois[3,4], Thomas E. Currie[5], and Stephanie Grohmann[3]

[1] KDEG and ADAPT, School of Computer Science and Statistics,
Trinity College Dublin, Dublin, Ireland
{rob.brennan,kevin.feeney,
mendelgg,bozicb}@scss.tcd.ie
[2] Department of Ecology and Evolutionary Biology,
University of Connecticut, Mansfield, USA
peter.turchin@uconn.edu
[3] Institute of Cognitive and Evolutionary Anthropology,
University of Oxford, Oxford, UK
{harvey.whitehouse,pieter.francois,
stephanie.grohmann}@anthro.ox.ac.uk
[4] History Group, School of Humanities,
University of Hertfordshire, Hatfield, UK
[5] Centre for Ecology and Conservation, Biosciences,
University of Exeter, Penryn, UK
T.Currie@exeter.ac.uk

Abstract. This paper describes OWL ontology re-engineering from the wiki-based social science codebook (thesaurus) developed by the Seshat: Global History Databank. The ontology describes human history as a set of over 1500 time series variables and supports variable uncertainty, temporal scoping, annotations and bibliographic references. The ontology was developed to transition from traditional social science data collection and storage techniques to an RDF-based approach. RDF supports automated generation of high usability data entry and validation tools, data quality management, incorporation of facts from the web of data and management of the data curation lifecycle.

This ontology re-engineering exercise identified several pitfalls in modelling social science codebooks with semantic web technologies; provided insights into the practical application of OWL to complex, real-world modelling challenges; and has enabled the construction of new, RDF-based tools to support the large-scale Seshat data curation effort. The Seshat ontology is an exemplar of a set of ontology design patterns for modelling uncertainty or temporal bounds in standard RDF. Thus the paper provides guidance for deploying RDF in the social sciences. Within Seshat, OWL-based data quality management will assure the data is suitable for statistical analysis. Publication of Seshat as high-quality, linked open data will enable other researchers to build on it.

Keywords: Ontology engineering · Ontology design patterns · Cliodynamics

© Springer International Publishing Switzerland 2016
H. Sack et al. (Eds.): ESWC 2016, LNCS 9678, pp. 693–708, 2016.
DOI: 10.1007/978-3-319-34129-3_42

1   Introduction

The success of linked data has seen semantic web technology widely deployed. However in many domains such as social sciences, despite a strong tradition of quantitative research, linked data has made little headway. This stems partially from a lack of social sciences research ICT infrastructure but also from the challenges of describing human systems with all their uncertainties and disagreements in formal models.

Here we describe re-engineering an OWL ontology from the structured natural language codebook (thesaurus) developed by the international Seshat: Global History Databank initiative[1] [1]. This evolving codebook consists of approximately 1500 variables used to study human cultural evolution at a global scale from the earliest societies to the modern day. Each variable forms a time series and represents a single fact about a human society such as identifying the capital city, the capital's population or the presence of infrastructure such as grain storage sites. The variables are grouped – measures of social complexity, warfare, ritual, agriculture, economy and so on. However the historical and archaeological record is incomplete, uncertain and disagreed upon by experts. All these aspects, along with annotations need to be recorded. An example variable definition in the codebook is: "**Polity territory** in squared kilometers". An instance of this variable, showing uncertainty and temporal scoping of values is "**Polity territory** 5,300,000: 120bce-75bce; 6,100,000:75bce-30ce".

Current data collection in Seshat uses a wiki based on the natural language codebook. This is unsustainable as data quality assurance is impossible and better tools are required to manage the collection, curation and analysis of the dataset. In addition it is desired to publish the dataset as linked data to enable other scholars to build upon the Seshat work. The new tools will be RDF-based using the Dacura data curation platform developed at Trinity College Dublin[2] as part of the ALIGNED H2020 project[3].

This paper investigates the research question: what is a suitable structure in RDF to represent the Seshat codebook that will support data quality assurance? Our technical approach is to develop an OWL ontology describing the codebook based on a set of design patterns for Seshat variables that capture the requirements for variable uncertainty, temporal scoping, annotations and provenance while producing a compact, strongly typed data model that is suitable for quality assurance in a very large dataset.

The contributions of this paper are: an identification of challenges for converting social science codebooks to RDF, a description of the Seshat ontology, new ontology design patterns for uncertainty and temporal scoping, a case study of the Seshat ontology deployed in a data curation system and finally the lessons learned.

The paper structure is: Sect. 2 background on Seshat, Sect. 3 ontology re-engineering challenges, Sect. 4 the Seshat ontology and design patterns Sect. 5 deployment of the ontology in the RDF-based data collection infrastructure, Sect. 6

---

[1] http://seshatdatabank.info/.

[2] http://dacura.scss.tcd.ie.

[3] http://www.aligned-project.eu.

lessons learned for social sciences ontology development, Sect. 7 surveys related work and Sect. 8 is conclusions & future work.

2 Background – Seshat: The Global History Databank

The study of past human societies is currently impeded by the fact that existing historical and archaeological data is distributed over a vast and disparate array of databases, archives, publications, and the notes and minds of individual scholars. The scope and diversity of accumulated knowledge makes it impossible for individual scholars, or even small teams, to engage with the entirety of this data. The aim of 'Seshat: The Global History Databank' is therefore to systematically organize this knowledge and make it accessible for empirical analysis, by compiling a vast repository of structured data on theoretically relevant variables from the past 10.000 years of human history [1]. In this way, it becomes possible to test rival hypotheses and predictions concerning the 'Big Questions' of the human past, for example the evolution of social complexity[4], the deep roots of technologically advanced areas[5], or the role axial age religions play in explaining social inequality[6].

Seshat data is currently manually entered either by domain experts (historians, archaeologists and anthropologists), or by research assistants whose work is subsequently reviewed and validated by domain experts. The aim is to move to quality assured data collection facilitated by customized software that can automatically import data from existing web resources such as DBpedia. A central requirement for the Seshat information architecture is a flexible and agile system that allows for the continuous development of the Codebook (which structures the data), the adaptation of variables to different research interests and theoretical approaches, and the participation of a large number of additional researchers and teams.

The databank's information structure comprises of a range of units of analysis, including polities, NGAs (i.e. 'Natural Geographic Areas'), cities and interest groups [2]. These are associated with temporally-scoped variables to allow for a combination of temporal and spatial analyses. Each variable currently consists of a value, typically marking a specific feature "absent/present/unknown/uncoded", and indicating levels of inference, uncertainty or scholarly disagreement about this feature. In addition to the values, which are used for statistical analysis, variables contain explanatory text as well as references to secondary literature. Where it is not possible to code variables due to missing or incomplete source data, variables are sometimes coded by inference (for example, if it cannot be ascertained if a given feature was present for a certain time period, but it is known to be present in the time periods immediately before and after, the feature

---

[4] 'Ritual, Community, and Conflict' research project funded by the ESRC/UK (http://www.esrc.ac.uk/research/our-research/ritual-community-and-conflict/).

[5] 'The Deep Roots of the Modern World: Investigating the Cultural Evolution of Economic Growth and Political Stability', funded by the Tricoastal Foundation/US (http://seshatdatabank.info/seshat-projects/deep-roots-economic-growth/).

[6] 'Axial-Age Religions and the Z-curve of Human Egalitarianism', funded by the John Templeton Foundation; (http://seshatdatabank.info/seshat-projects/axial-age-egalitarianism/).

would be coded 'inferred present'). By linking descriptions of past societies to both sources and coded data amenable to statistical analysis, the databank thus combines the strengths of traditional humanistic and scientific approaches.

In the initial stages of the project, the database was implemented in a Wiki, however, as the number of coded variables has been rapidly growing, it was decided to move the Seshat data to an RDF-based triplestore. Based on the Dacura data curation platform, this will facilitate all steps of the Seshat research process, from data gathering, validation, storage, querying and exporting down to analysis and visualization.

3 Seshat Codebook to Ontology Re-Engineering Challenges

The purpose of creating the Seshat ontology was not simply to translate or uplift an existing dataset to RDF for publication as linked data. Instead we wished to use the ontology at the heart of a set of RDF-based tools that would produce a step change in the data collection and curation capabilities of the Seshat consortium by improving data quality, productivity and agility (Fig. 1). The primary goal of the formal OWL model is to enable data quality management as even uncertain facts can be omitted, mistyped, duplicated, inconsistent and so on. This creates a huge data cleaning overhead before statistical processing in the pre-OWL system. Later we hope to extend the utility of DL reasoning to support inference, fact reuse and other advanced features.

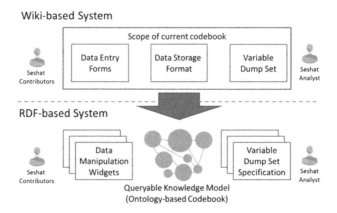

Fig. 1. The Seshat codebook re-engineering vision

The characteristics of the Seshat codebook that made this re-engineering process challenging were as follows:

1. **The codebook was specified in semi-formal structured natural language** designed for human consumption. While a common approach in social sciences it is not often studied in ontology engineering, e.g. the methodology for ontology re-engineering from non-ontological resources [3] doesn't consider it.

2. **The ontology must not depend on custom reasoning or triple-stores.** Rather than moving beyond RDF triples to specify qualified relations or temporal scoping it must be possible to use standard, state of the art, scalable triple-stores.
3. **The ontology must be expressive enough to support data quality validation.** The flexibility of wiki-based collection means that the data collected needed extensive cleanup before analysis. The ontology must eliminate this workload.
4. **Every historical fact (Seshat variable value) recorded was potentially subject to uncertainty.** The historical and archeological record often does not permit definite statements of the sort normally recorded by RDF triples.
5. **Each Seshat variable assertion is temporally scoped.** This is because historical facts are typically only true for a certain period of time.
6. **Each temporal scoping was potentially subject to uncertainty.** Many historical dates are unknown or only have known ranges of values.
7. **Time-series variables must support human-readable annotations in addition to data-values.** Seshat is primarily data-oriented but the data collection and expert verification process depends upon the availability of a flexible annotation scheme.
8. **Efficiency of representation for storage and query.** The Seshat dataset is going to be very large. Hence it is desirable to create a tight data model.
9. **Seshat variables do not represent a full model of the domain.** Each Seshat variable is a time series that is conceptually linked to other variables in the codebook based on social science concerns. However there are many missing relations between variables or unifying concepts that only reside in the minds of the domain experts that constructed the codebook and perform analysis on the collected data.
10. **Dataset will be sparse, sampling rates not fixed.** History does not provide sufficient data to populate a classical data cube, there are too many gaps and it is necessary to record data when available rather than imposing a rigid sampling scheme.
11. **Hierarchical structures present in the codebook are often arbitrary.** The hierarchical patterns used to organize variables within the Seshat codebook serve purposes such as navigation, templating or grouping of items for data entry.
12. **Data provenance important but cannot overload infrastructure.** In addition in the RDF-based data curation platform will use provenance to record activities, agents and entities within the platform.
13. **Representing time from circa 10,000BC to the present day.** Typical IT applications and date-time formats do not deal with >4 character BC dates well.

The next section describes our solutions in the Seshat Ontology for each challenge.

4   The Seshat Ontology

In this section we introduce the Seshat ontology[7], describe the development process and describe the key design patterns deployed in the ontology.

---

[7] http://www.aligned-project.eu/ontologies/seshat.

4.1   Overview

The Seshat codebook is primarily aimed at collecting geo-temporally scoped time series variable values describing two main units of analysis – the Polity, representing an independent historical human culture or society and the natural geographical region (NGA) which is a unit of data collection or analysis defined spatially by a polygon drawn on a map. In the RDF-based approach we use three named graphs to represent the dataset: V, the data value graph which is described by the Seshat ontology; A, the annotation graph (based on Open Annotation) where textual annotations of data values are held and P, the provenance graph (challenge 12, Sect. 3) where W3C PROV statements are recorded that describe the annotation and variable value lifecycles as they travel through the data curation system (Fig. 1). The Seshat ontology extends the set of units of analysis by creating a hierarchical structure of entity classes as seen in Fig. 2. Each of these entities has a set of Seshat variables associated with it. Each variable value for an entity is associated with geographical and temporal scoping information.

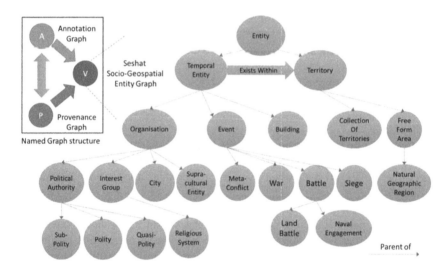

Fig. 2. Seshat named graph structure and Seshat ontology geo-temporally scoped entities

In order to model the additional context required by the qualified nature of a Seshat variable, each is modelled as an OWL class and a property pointing from the appropriate Seshat entity to that class (challenge 2, Sect. 3). In order to keep the data model compact a large number of data pattern upper classes are defined for each variable. By exploiting multiple inheritance and OWL DL's complete class definitions it is possible to overload the class definition to provide a toolbox of assertions which can be automatically classified and constrained by an appropriate OWL reasoner (challenge 8, Sect. 3). Each value type associated with a variable is either an XSD datatype or a custom OWL class definition, often with a declared set of allowed values. At the variable definition level in the Seshat ontology it is possible to associate a unit of measure with data values (Fig. 3).

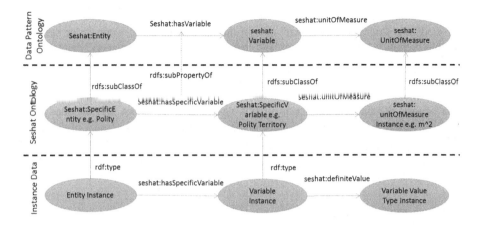

Fig. 3. Seshat ontology variable structure - modelled as a qualified relation

4.2   Development Methodology

The ontology has been developed at Trinity over the last 18 months. No formal ontology engineering process has been followed exactly. We used an iterative development model where the domain was explored in group sessions and requirements established. Then individual knowledge engineers worked on surveying the literature and generating solutions for specific aspects of the model. Then new versions of the combined model were developed. Then hand-coding of instance data was done to evaluate the consequences of designs. The ontology was primarily written in turtle in a syntax-highlighting text editor. Using Protégé for editing has several drawbacks – turtle comments on development are silently dropped, the import of a file often reduces properties to annotations if Protégé cannot understand them, additional meta-data and comments were generated. RDF validation has been periodically performed with the rdf2rdf[8] command line tool. More recently the ontology has been validated by the Dacura Quality Service [4], a custom OWL/RDFS reasoner that can check an ontology for a wider range of logical, typographical and syntactic errors. In addition the ontology has been used for testing the Dacura data curation tools being developed for Seshat. The ontology was split into an upper part containing basic patterns and a lower part containing the ontology of the Seshat codebook based on those patterns.

Close collaboration with the domain experts that developed the codebook was necessary. Several workshops have been held to understand their modelling concerns and describe our approach. Developing a common understanding and hence appropriate model of data unreliability and uncertainty was the most conceptually challenging topic. Three separate sources of uncertainty were identified: (1) within the codebook there was a syntax defined for variable bags of values or ranges (2) some apparently boolean variables were assigned enumerated values of "uncoded, present, inferred present, absent, inferred absent, unknown", and (3) the codebook syntax

---

[8] http://www.l3s.de/~minack/rdf2rdf/.

allowed multiple experts to disagree on a value. It was discovered that the use of "inferred" and "uncoded/unknown" tags in the dataset instances went wider than the variable definitions of the codebook and hence these represented generic patterns that needed to be available for all variables, not just those specified as an enum. Modelling of values, bags and ranges was straightforward (Sect. 4.3). The concept of an "inferred" value was added as an attribute for any value to indicate a human researcher had gone beyond the direct evidence to infer a value. Both unknown and uncoded were collapsed into one concept that of epistemic incompleteness – a statement of the limits of human knowledge about the past, given the expertise of the person asserting it (in the Seshat wiki a research assistant would put uncoded and an expert unknown but our PROV logs could distinguish these cases).

4.3   Design Patterns

In this section we use description logic and commentary to describe how each ontology re-engineering challenge is overcome by using the basic patterns of the Seshat ontology. In the following description logic we define $\uplus$ as the disjoint union operator where $A \uplus B \equiv A \sqcup B$ where $A \sqcap B \sqsubseteq \bot$.

Representing Uncertain Time. Two main references were used as a basis for representing time - the W3C draft Time Ontology in OWL (henceforth owltime) and the W3C PROV-O ontology. Owltime is attractive since it makes explicit the granularity of representation, for example in cases where the historical record only records a year but no month or day, whereas PROV-O uses a simpler structure for time whereby activities are directly linked to an xsd:datetime value using the prov:hasBeginning and prov:hasEnd properties. In contrast owltime uses 4 intermediate nodes for each time value in an interval. Neither specification has any support for uncertainty in time assertions or non-Gregorian calendars (although Cox [5] has recently extended owltime to handle this).

Our approach, based on triple efficiency concerns, has been to re-use the expressive owltime:DateTimeDescription directly linked to a qualified variable object via the atDatetime, hasEnd and hasBeginning properties in the PROV-O pattern. i.e.

$$Instant \equiv (= 1 \ \forall atDateTime^-.DateTimeDescription)$$
$$Interval \equiv (= 1 \forall hasEnd^-.DateTimeDescription) \sqcup ($$
$$= 1 \forall hasBeginning^-.DateTimeDescription$$

We have then extended the definition of an InstantValue to be either an Instant or UncertainInstant, which is defined as a thing having two or more assertions of the atDateTime property:

$$InstantValue \equiv Instant \sqcup UncertainInstant \text{ where } Instant$$
$$\sqcap UncertainInstant \sqsubseteq \bot$$
$$UncertainInstant \equiv (\geq 2 \ \forall atDateTime^-.DateTimeDescription)$$

Then we generalized an IntervalValue to be either an Interval or an Uncertainterval which is defined as the disjoint union of the three types of temporal uncertainty:

$$IntervalValue \equiv Interval \uplus UncertainInterval$$

$UncertainInterval$

$$\equiv UncertainEndInterval \uplus UncertainBeginInterval \uplus UncertainBothInterval$$

$$UncertainEndInterval \equiv (\geq 2\, \forall hasEnd^{-}.DateTimeDescription) \sqcup ($$
$$= 1\, \forall hasBeginning^{-}.DateTimeDescription)$$
$$UncertainBeginInterval \equiv (= 1\, \forall hasEnd^{-}.DateTimeDescription) \sqcup ($$
$$\geq 2\, \forall hasBeginning^{-}.DateTimeDescription)$$
$$UncertainBothInterval \equiv (\geq 2\, \forall hasEnd^{-}.DateTimeDescription) \sqcup ($$
$$\geq 2\, \forall hasBeginning^{-}.DateTimeDescription)$$

This gives a flexible and compact notation (challenge 8, Sect. 3) for defining certain or uncertain temporal scopes (challenge 6, Sect. 3). We currently use Gregorian dates, which we project back in time using the common interpretation of ISO 8601 that allows for greater than 4 digit dates if preceded by a minus sign (challenge 13, Sect. 3).

Representing Uncertain Data Values. A key feature of Seshat is that many uncertain facts must be recorded (challenge 4, Sect. 3). We deal with this through the intermediate qualification node in a Seshat variable value. From this we define four properties: definiteValue, valuesFrom, maxValue and minValue. This enables a given variable to have a single value, a bag or a range:

$$DefiniteValue \equiv (= 1\, \forall definiteValue^{-}.\top)$$
$$BagOfValues \equiv (\geq 1\, \forall valuesFrom^{-}.\top)$$
$$RangeMaxValueRestriction \equiv (= 1\, \forall maxValue^{-}.\top)$$
$$RangeMinValueRestriction \equiv (= 1\, \forall minValue^{-}.\top)$$
$$Range \equiv RangeMaxValueRestruction \sqcap RangeMinValueRestriction$$

One special type of value in the Seshat codebook is one that is inferred from the historical record by the person entering the data, rather than by reference to a historical source. This is modelled as a new type but it is always a form of definite value:

$$InferredValue \equiv Inferred \sqcap DefiniteValue$$

When a Value is present it is always a member of the disjoint union of definite values, bags or ranges:

$$Value \equiv DefiniteValue \uplus BagOfValues \uplus Range$$

However in addition to these types of uncertainty it is important for Seshat data collectors to be able to express the presence of epistemic incompleteness, i.e. that a search has been performed and that, to the extent of the current author's knowledge, the data value is not present in the historical record. In this case we set the variable to UnknownValue which carries these semantics and record the author in the PROV graph. This leads to the full definition of an UncertainVariable in Seshat:

$$UncertainVariable \equiv Value \uplus UnknownValue$$

In fact due to OWL's inability to create properties that have a range of both datatypes and objects it is necessary for us to create 4 additional properties named definiteDataValue, dataValuesFrom, maxDataValue and minDataValue and parallel class definitions (DefiniteDataValue etc.) to the above to allow variables to have data or object properties. The base range for data values is rdfs:Literal rather than owl:Thing.

Temporal Constraints. The final pattern needed is the ability to express temporal constraints as part of the qualification of a Seshat variable (challenge 5, Sect. 3). To do this we build upon our uncertain representation of time above to add scoping properties to the variable qualification class. Hence we first define the TemporalScoping as the disjoint union of the temporal types:

$$TemporalScoping$$
$$\equiv Instant \uplus Interval \uplus UncertainInstant \uplus UncertainInterval$$

Then we construct a TemporalScopedVariable as the intersection of uncertain-variables and things with a defined temporal scoping.

$$TemporalScopedVariable \equiv UncertainVariable \sqcap TemporalScoping$$

Finally we have our Seshat variable qualifier base class the UncertainTemporalVariable which can pick and mix both certain and uncertain temporal scoping and values:

$$UncertainTemporalVariable$$
$$\equiv UncertainVariable \sqcup TemporalScopedVariable$$

Again it is necessary to have a parallel definition of an UncertainTemporalDataVariable for variables that refer directly to xsd:datatypes instead of OWL classes. These parallel definitions are all available in the online version of the Seshat ontology.

Example Seshat Datatype Variable Definition. To illustrate the use of the previous sections we define here an example Seshat datatype variable based on xsd:dateTime. In order to enable quality analysis and constraint checking we need to make this as strongly typed as possible. This means that all our data accessor properties must be restricted to using a single datatype (xsd:dateTime in this example) and the base type of UncertainTemporalVariable. We do this by declaring the 4 restriction classes (one for each data accessor property) and the intersection of these with our base type:

$$DateDataValueRestriction \equiv (= 1 \ \forall definiteDataValue^-.XsdDateTime)$$
$$DateBagOfDataValuesRestriction$$
$$\equiv (\geq 1 \ \forall dataValuesFrom^-.XsdDateTime)$$
$$DateRangeMinDataValueRestriction$$
$$\equiv (= 1 \ \forall minDataValue^-.XsdDateTime)$$
$$DateRangeMaxDataValueRestriction$$
$$\equiv (= 1 \ \forall maxDataValue^-.XsdDateTime)$$
$$UncertainDateTimeVariable \equiv UncertainTemporalDataVariable \sqcap$$
$$DateDataValueRestriction \sqcap DateBagOfDataValuesRestriction \sqcap$$
$$DateRangeMinDataValueRestriction \sqcap$$
$$DateRangeMaxDataValueRestriction$$

This is a full, usable Seshat variable and we would follow the same pattern if we had defined a custom OWL Class to hold our variable value. In practice we have defined all the common xsd:datatypes in this way as part of our base ontology and when a specific Seshat variable is based on a specific datatype we declare a sub-property in the Seshat ontology to declare specific annotation properties (rdfs:comment, rdfs:name) and meta-properties such as the units of measure for that variable.

5 Application and Use Case

The Seshat ontology is deployed in the pilot Seshat data curation system[9] based on the Dacura platform developed within the H2020 ALIGNED project. This platform allows Seshat users to enter data, manage the structure and quality of the entered data and output it in standard formats. In the pilot system, four of the components from Fig. 4 are used: (1) The wiki data entry/validation tools (top left in figure); (2) The schema management tools; (3) The data quality controls (lower middle of figure) which perform schema and data integrity checks; and (4) the data export tool which can transform Seshat data into the TSV dumps required by statistical analysts. The Seshat ontology in this system is used by all our tools and enables more structured information to be captured than the original Seshat wiki, data validation at the point of entry and triple-store data integrity enforcement by the Dacura Quality Service.

6 Lessons Learned

The exercise of re-engineering the Seshat codebook into an OWL DL ontology has provided us with valuable experiences in the areas of social science codebook translation, data uplift to RDF, OWL modelling and Linked Data publishing. Each of these is summarized in Table 1 and further discussed below.

The overwhelming experience of developing the Seshat ontology from the wiki-based codebook is that taking a semantic web approach will add a lot of value.

---

[9] For a video demonstration see https://www.youtube.com/watch?v=OqNtpSClczU.

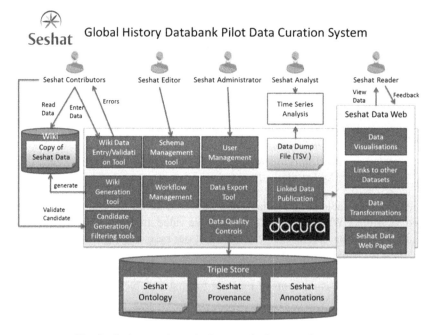

Fig. 4. Seshat ontology deployment in data curation system

Table 1. Lessons learned

Area/Issue	Resolution/Impact	OWL Adv[a]
1. Codebook Translation		
1.1 Implicit data patterns in codebook	Required manual design of new data patterns	Y
1.2 Implicit semantics of blank values	Explicit modelling of epistemic incompleteness	Y
1.3 Lack of data-typing	Defined variables as xsd:floats, ints or unsigned ints	P
1.4 Domain model incomplete	Attached OWL classes to variable definitions	P
1.5 Atomic concepts evolve	Require patterns for composite and inferred variables	Y
1.6 Support mandatory annotations	Model at the variable definition level	P
1.7 Measurement unit definitions	Model in variable definition, link to units ontology	Y
2. OWL Modelling		
2.1 RDFS insufficient for data quality	Moved to OWL to express constraints	Y

(Continued)

Table 1. (*Continued*)

Area/Issue	Resolution/Impact	OWL Adv[a]
2.2 Minimizing number of properties creates complex OWL restrictions	Knowledge model complexity increases faster than an interface specification as properties are reused	P
2.3 OWL data/object property split	Parallel definitions for owl:Thing and rdf:Literal	N
2.4 Compact data representation	OWL disjoint unions to access a palette of properties	Y
2.5 OWL Restriction classes verbose	Automated generation of OWL from design patterns	N
2.6 Intermediate logical classes needed	Additional classes defined, hide from users	N
2.7 Constraints for xsd:datatypes	OWL restrictions provide excellent property reuse	Y
3. Linked Data		
3.1 Open Annotation Inconsistent	OA imports 64 vocabularies, hard to work with as OWL (see also [6])	–
3.2 Time vocabulary	Compromised between owltime and W3C PROV-O	–
3.3 GeoSPARQL	Badly named specification, not clear is an ontology	–
4. Uplift/Import of Wiki		
4.1 Seshat coding sheet variations	Need flexible uplift mappings	N
4.2 OWL model drift from codebook	The more complex the knowledge model, the harder the uplift and dump as TSV	P
4.3 Modelling inter-entity relations	Important to provide support for text-based links as well as true relations	Y

[a]Was OWL an advantage for resolving this issue, especially wrt the wiki: Y = yes, P = partial, N = no.

However given the emphasis on fixing the data quality issue in the wiki it has proved necessary to move to OWL for the ontology rather than using a linked data/RDFS approach. This is because the demands of data validation and the imprecision of what Gómez-Pérez terms "Frankenstein" linked data ontologies were ultimately incompatible. In general the process has helped the domain experts too as they have had to clarify and make explicit the semantics embedded in the codebook. The biggest hurdles in terms of OWL modelling have been the lack of support for a property top that spans both object and datatypes. This has created a doubling-up of the data patterns required. In terms of the future, by moving to a natively RDF-based system it is hoped to be able to automate the exploitation of the vast quantity of structured data produced by the semantic web community and of course this would not be possible in a manual approach based on the wiki without a lot of brittle, custom development.

7 Related Work

The major influences on this work have been Dodds and Davis' catalogue of design patterns [6], especially the modelling patterns section, the W3C PROV ontology [7] and Open Annotation [8]. In terms of ontology engineering process, the many works of Gómez-Pérez, e.g. [2], have been influential. Our treatment of uncertainty is inspired by the work of the W3C Uncertainty Reasoning for the World Wide Web group [9]. The works of Horrocks, Patel-Schneider and their collaborators, e.g. [10], have been vital in shaping our understanding of OWL DL. Finally the survey of Zaveri et al. [11] has been instrumental in guiding the development of a Seshat ontology that is suitable for data quality assurance.

There have been many initiatives that tackle the challenge of representing historical data using semantic web technology. One important standard is CIDOC CRM [12] published by ISO. It has the broad remit of defining an ontology for cultural heritage information. In contrast to Seshat, its primary role is to serve as a basis for mediation between local representations of cultural heritage resources such as museum collections. Hence the term definitions and subsumption hierarchy are incomplete, there is no full grounding of datatypes, for example as xsd:datatypes but instead the lowest level is abstract types such as string. The RDFS-based ontology definition the standard includes is not the primary reference but a derived one. Nonetheless the FP7 ARIADNE infrastructure project[10] has made progress with using it as a basis for linked data publication and interworking between collections. There is great potential for future collaboration with the Seshat consortium in terms of data sharing.

DBpedia [13] of course contains many historical facts that are of interest to Seshat and it is hoped that by leveraging the work already done there it will be possible to quickly import candidate data for Seshat, to be then curated by the Seshat research assistants and domain experts. Nontheless the current DBpedia data is not in a format suitable for processing as time series and does not comply with the conceptual models underlying the Seshat codebook so mapping techniques will have to be employed. Through the ALIGNED project we are collaborating with the AKSW group at the University of Leipzig and it is planned to establish a virtuous circle whereby DBpedia extracts crowd-sourced facts from Wikipedia, Seshat uses those facts as input to their historical time-series, the Seshat team curates and refines the facts and publishes them as high quality linked data which in turn is available to DBpedia+, the new multi-source, improved quality version of DBpedia in development by the DBpedia community. This integration will be trialled in year 3 of ALIGNED (2017).

There are also a large number of other curated RDF datasets describing historical locations and facts such as Pleiades[11] that focuses on ancient names, places and locations. Nonetheless these datasets are typically based on controlled vocabularies rather than formal semantic data models and RDF is provided as a dump that transforms the internal representation. This gap presents an opportunity for Seshat as a

---

[10] http://www.ariadne-infrastructure.eu/.

[11] http://pleiades.stoa.org/home.

provider of high quality native linked data with strong consistency assurances. Once again it is hoped that Seshat will work with these other dataset publishers in the future.

Finally there are a wide range of historical time series data collection efforts in the social sciences that are not RDF-based or publishing linked data. Most of these have much more limited scope than Seshat. For example Sabloff's datasets describing the limited geographic region of Mongolia throughout time [14] or the Database of Religious History [15] that has similar geo-temporal scope to Seshat but deals only with religion rather than all aspects of human cultural evolution.

8   Conclusions and Future Work

Our ambition for the Seshat ontology goes beyond constraining, structuring and classifying the uncertain and sparse (although voluminous) historical time series data that forms the basis of the Seshat: Global History Databank. In future work we will enrich the knowledge model by adding semantic relationships between Seshat time-series variables to support domain knowledge-based quality assurance. This will enable, for example, the identification of inconsistent statements about a historical society's military metal technology and the metals used for agricultural tools.

The current ontology reflects the modelling foci in the original Seshat codebook and several areas would benefit from generalization or extension. Two high priority areas are (1) the creation of richer models of the politico-geographical relationships between historical societies as this will add greater flexibility to the model and (2) adding support for inferred variable values in addition to collected values as this will reduce data collection effort and improve consistency. Similarly the ontology will be extended for publication as linked data. For example, creating interlinks between Seshat and the web of data or mapping Seshat to common linked data vocabularies like GeoSPARQL to make it more easily consumed.

In addition to data validation and quality assurance, a key use of the ontology within Seshat is the generation of customised, dataset-specific, high usability user interfaces for data entry, import, interlinking, validation and domain expert-based curation. This requires the development of form generation tools for presenting ontology elements and widgets that streamline data entry and constrain the entered data to be syntactically and semantically correct. As this form generation technology develops it may produce new design patterns for the structure of the Seshat ontology.

Acknowledgements. This work was supported by a John Templeton Foundation grant, "Axial-Age Religions and the Z-Curve of Human Egalitarianism," a Tricoastal Foundation grant, "The Deep Roots of the Modern World: The Cultural Evolution of Economic Growth and Political Stability," an ESRC Large Grant, "Ritual, Community, and Conflict" (REF RES-060-25-0085), European Union Horizon 2020 research and innovation programme (grant agreement No 644055 [ALIGNED, www.aligned-project.eu]) and the ADAPT Centre for Digital Content Technology, SFI Research Centres Programme (Grant 13/RC/2106) co-funded by the European Regional Development Fund. We gratefully acknowledge the contributions of our team of research assistants, post-doctoral researchers, consultants, and experts. Additionally, we have received invaluable assistance from our collaborators. Please see the Seshat website for a full list of private donors, partners, experts, and consultants and their respective areas of expertise.

References

1. Turchin, P., Brennan, R., Currie, T., Feeney, K., Francois, P., Hoyer, D., et al.: Seshat: the global history databank. Cliodynamics J. Quant. Hist. Cult. Evol. **6**, 77–107 (2015)
2. Francois, P., Manning, J., Whitehouse, H., Brennan, R., Currie, T., Feeney, K., Turchin, P.: A macroscope for global history. Seshat Global History Databank: a methodological overview (Submitted)
3. Villazón-Terrazas, B., Gómez-Pérez, A.: Reusing and re-engineering non-ontological resources for building ontologies. In: Suárez-Figueroa, M.C., Gómez-Pérez, A., Motta, E., Gangemi, A. (eds.) Ontology Engineering in a Networked World, pp. 107–145. Springer, Heidelberg (2012). doi:10.1007/978-3-642-24794-1_6
4. Mendel-Gleason, G., Feeney, K., Brennan, R.: Ontology consistency and instance checking for real world linked data. In: Proceedings of the 2nd Workshop on Linked Data Quality co-located with 12th Extended Semantic Web Conference (ESWC 2015), Portorož, Slovenia, 1 June 2015
5. Cox, S.J.D.: Time ontology extended for non-gregorian calendar applications. Seman. Web J. (2015, to appear). http://www.semantic-web-journal.net/content/time-ontology-extended-non-gregorian-calendar-applications-0
6. Feeney, K., Mendel-Gleason, G., Brennan, R.: Linked data schemata: fixing unsound foundations. Submission to Seman. Web J. (2015). http://www.semantic-web-journal.net/content/linked-data-schemata-fixing-unsound-foundations
7. Lebo, T., Sahoo, S., McGuinness, D. (eds.): PROV-O: The PROV Ontology, W3C Recommendation, 30 April 2013
8. Sanderson, R., Ciccarese, P., Van de Sompel, H. (eds.): Open Annotation Data Model, Community Draft, 08 February 2013. http://www.openannotation.org/spec/core/
9. W3C Uncertainty Reasoning for the World Wide Web XG, UncertaintyOntology (2005). http://www.w3.org/2005/Incubator/urw3/wiki/UncertaintyOntology.html
10. Grau, B.C., Horrocks, I., Motik, B., Parsia, B., Patel-Schneider, P., Sattler, U.: OWL 2: The next step for OWL. J. Web Seman. **6**(4), 309–322 (2008). doi:10.1016/j.websem.2008.05.001
11. Zaveri, A., Rula, A., Maurino, A., Pietrobon, R., Lehmann, J., Auer, S.: Quality assessment for linked data: a survey. Seman. Web **7**, 1–31 (2015). doi:10.3233/SW-150175
12. ISO 21127: 2014 Information and documentation – a reference ontology for the interchange of cultural heritage information, 2nd edn., ISO (2014)
13. Bizer, C., Lehmann, J., Kobilarov, G., Auer, S., Becker, C., Cyganiak, R., Hellmann, S.: DBpedia-a crystallization point for the web of data. J. Web Semant. Sci. Serv. Agents World Wide Web **7**, 154–165 (2009)
14. Sabloff, P.L.W.: Mapping Mongolia: Situating Mongolia in the World from Geologic Time to the Present. University of Pennsylvania Press, Pennsylvania (2011). ISBN 978-1-934536-18-6
15. Slingerland, E., Sullivan, B.: Durkheim with data: the database of religious history (DRH). J. Am. Acad. Relig. (2015, in press)

RMLEditor: A Graph-Based Mapping Editor for Linked Data Mappings

Pieter Heyvaert[1]([⊠]), Anastasia Dimou[1], Aron-Levi Herregodts[2],
Ruben Verborgh[1], Dimitri Schuurman[2],
Erik Mannens[1], and Rik Van de Walle[1]

[1] Data Science Laboratory, Ghent University - iMinds, Ghent, Belgium
`pheyvaer.heyvaert@ugent.be`
[2] Ghent University – iMinds – MICT, Ghent, Belgium

Abstract. Although several tools have been implemented to generate Linked Data from raw data, users still need to be aware of the underlying technologies and Linked Data principles to use them. Mapping languages enable to detach the mapping definitions from the implementation that executes them. However, no thorough research has been conducted on how to facilitate the editing of mappings. We propose the RMLEditor, a visual graph-based user interface, which allows users to easily define the mappings that deliver the RDF representation of the corresponding raw data. Neither knowledge of the underlying mapping language nor the used technologies is required. The RMLEditor aims to facilitate the editing of mappings, and thereby lowers the barriers to create Linked Data. The RMLEditor is developed for use by data specialists who are partners of (i) a companies-driven pilot and (ii) a community group. The current version of the RMLEditor was validated: participants indicate that it is adequate for its purpose and the graph-based approach enables users to conceive the linked nature of the data.

1 Introduction

Semantic Web technologies rely on data which is interlinked and whose semantically enriched representation is available, the so-called Linked Data [1]. Most of the current Linked Data stems originally from (semi-)structured formats. Mappings specify in a declarative way how Linked Data is generated from such raw data. Nevertheless, defining and executing them still remains complicated, despite the significant number of tools implemented for this scope. At first, most approaches that map raw data to its RDF representation [2] incorporated the mappings in the implementation that executes them. Thus, not only knowledge of Semantic Web and Linked Data is required. However, also dedicated software development cycles for creating, updating and extending the implementations,

The described research activities were funded by Ghent University, iMinds, the Institute for the Promotion of Innovation by Science and Technology in Flanders (IWT), the Fund for Scientific Research Flanders (FWO Flanders), and the European Union.

© Springer International Publishing Switzerland 2016
H. Sack et al. (Eds.): ESWC 2016, LNCS 9678, pp. 709–723, 2016.
DOI: 10.1007/978-3-319-34129-3_43

whenever new or updated semantic annotations are desired, are needed. Mapping languages, such as R2RML [3] and RML [4], enable *to detach the mapping definitions from the implementation that executes them.* Besides knowledge of the underlying mapping language that is required to define the mappings, manually editing and curating them requires a substantial amount of human effort [5]. Moreover, data specialists are not Semantic Web experts or developers. Thus, the task of editing mappings should be addressed independently, and disassociated from the corresponding mapping language and/or underlying technology used. Facilitating the editing of mappings further lowers the barriers of obtaining Linked Data and, thus stimulates the adoption of Semantic Web technologies. Nevertheless, dedicated environments that support users to intuitively edit mappings were not thoroughly investigated yet, as it occurred with applications that actually execute them and deliver its RDF representation. *Step-by-step* wizards prevailed, e.g., fluidOps editor [6], as an easy-to-reach solution. However, such applications restrict data publishers' editing options, hamper altering parameters in previous steps, and detach mapping definitions from the overall knowledge modeling, since related information is separated in different steps. We propose the RMLEditor[1], an editing environment for specifying mappings of raw data to their RDF representation based on graph visualizations, without requiring knowledge of the underlying mapping language. The RMLEditor is developed to support partners of (i) a companies-driven pilot for sharing and integrating the RDF representations of their data and co-develop third-party applications, and of (ii) a community-group-driven bootstrap for showcasing the advantages of Linked Data and Semantic Web technologies. The tool is available to interested parties under custom licensing conditions. We performed an exploratory user validation of our proposed solution, which showed that the RMLEditor achieves the goal it was implemented for. 82 % of the participants found the use of graphs beneficial for editing mappings using the RMLEditor and 70 % could better conceive that a relationship exists between multiple data sources. The remainder of the paper is structured as follows: Sect. 2 outlines existing mapping languages and mapping editors. Section 3 describes the mapping process without and with the use of a mapping editor. Section 4 presents our proposed solution, the RMLEditor. Section 5 outlines the use cases and explains the exploratory user validation and presents the results. Last, Sect. 6 discusses the results and presents the conclusions of our solution.

2   Related Work

In this section, we discuss existing mapping languages. Moreover, we elaborate on existing mapping editors, with a distinction between editors either supporting homogeneous or heterogeneous data sources.

---

[1] http://rml.io/RMLeditor.

2.1 Mapping Languages

Mapping languages specify in a declarative way how Linked Data is generated from raw data. At first, existing formalizations were considered as mapping languages, such as XPath [7] or XQuery language [8]. Nevertheless, there were also languages defined for this particular task. R2RML [3] is the W3C recommended language to define mappings to generate RDF from data derived from relational databases. Besides R2RML, other *format-specific* languages were defined, such as X3ML[2] for XML data. There are also *query-oriented* languages such as XSPARQL [9], which combines XQuery and SPARQL to map XML data, and Tarql[3], for data in CSV. However, these languages only support homogeneous data sources. A number of tools were developed supporting mappings from heterogeneous data sources to RDF, such as Datalift[4], RDFizers[5] and Virtuoso Sponger[6]. However, those tools actually employ separate *source-centric* approaches for each format they support, which does not allow the interlinking between sources in different formats. The RDF Mapping Language (RML) [4] circumvents this, by enabling the generation of data in RDF representation based on multiple heterogeneous data sources, e.g., XML and JSON.

2.2 Mapping Editors

Despite the significant number of mapping languages, the number of corresponding editors that support users to define the mappings is not comparable. Similar to the mapping languages, a distinction can be made between tools supporting homogeneous data sources and tools supporting heterogeneous data sources.

Homogeneous Data Sources. The fluidOps editor [6] is a browser application that provides an intuitive user interface for editing mappings. The underlying mapping language is R2RML. The fluidOps editor relies on a single *step-by-step* workflow. There are six successive steps, similar to actually creating a mapping document. Although its Graphical User Interface (GUI) aims to hide the R2RMLvocabulary, it still strongly focuses on concepts and terminology introduced by R2RML(e.g., subject maps and object maps). Therefore, knowledge of the language is required to use the editor. This decreases its adoption by non-Semantic Web experts and lowers the GUI's reusability potential. Additionally, only once the users reach the final step, they are able to preview the mappings in R2RMLsyntax and identify possible inconsistencies. Consequently, they need to restart the workflow to update the definitions of the previous steps. Pinkel et al. [5] adapted the original fluidOps editor to overcome flexibility limitations imposed by the database-driven step-by-step workflow. Their extension supports

---

[2] https://github.com/delving/x3ml/blob/master/docs/x3ml-language.md.
[3] https://tarql.github.io/.
[4] http://datalift.org/.
[5] http://simile.mit.edu/wiki/RDFizers.
[6] http://virtuoso.openlinksw.com/dataspace/doc/dav/wiki/Main/VirtSponger.

the *ontology-driven* approach. With the former approach creating the mappings starts with the data in the databases, semantic annotations are added afterwards. With the latter approach the mappings are created based on an existing ontology. Next, the mappings are complemented with data fractions from the databases. sheet2RDF [10] is a platform that uses a PEARL [11] document to map data in spreadsheets to RDF. Its GUI allow users to view the source data, define the mappings by editing the PEARLdocument directly, and view the resulting RDF through a tabular-structure. However, the adoption of the tool decreases because users need knowledge about PEARLto edit the mappings. An alternative approach is proposed by Rodrıguez-Muro et al. [12] who introduced -ontopPro-[7], a plugin for Protégé [13]. It allows users to generate RDF based on data from database(s) and an ontology. Tabs are provided to manage the databases and the mappings. Users need to write SPARQL-like templates to define how to map the original data. However, data sources are limited to databases and users need to understand the template's custom syntax.

Heterogeneous Data Sources. Karma[8] differentiates from aforementioned tools because it supports heterogeneous sources, such as databases, delimited text files (e.g., CSV files), JSON, XML, Microsoft Excel and web APIs. It uses Global-Local-As-View [14] rules to perform the mappings. These rules can be exported using R2RMLor D2RQ [15]. When displaying the mappings to the users, Karma takes a data-centric approach: users can only view the input data. Users are not able to follow the ontology-driven approach, as it occurs with the fluidOps editor. DataOps [16] uses the latter to support heterogeneous data formats. However, users are still confronted with the syntax of the used languages. RDF123 [17] is similar to sheet2RDF, however, it also supports CSV files, and it uses custom *map graphs* to represent the mappings, which are converted to a custom map function that produces RDF based on the input data. Consequently, users are not required to understand or know about the map function to define mappings, as it occurs with PEARLfor sheet2RDF. Additionally, they offer a web service that uses a link to a Google Spreadsheet or a CSV file, and generates RDF based on the mappings defined with the application. Therefore, the maps can be used outside the desktop application. However, their use is limited to that web service, as a custom map function is used and not a mapping language. TopBraid Composer[9] supports heterogeneous sources. It allows data integration from databases, XML, UML, RSS, spreadsheets and RDF data backends. The data can be reconciled with DBpedia [18]. However, as in the case of Karma, interlinking data from different sources is not possible. To set up the mapping process, the GUI offers a *data-driven step-by-step* workflow. However, an ontology-driven approach is not possible. Similar to the original fluidOps editor, a more simplified wizard has been built on top of RML, instead of R2RML. This form-based

---

[7] http://ontop.inf.unibz.it/components/sample-page/.

[8] http://www.isi.edu/integration/karma/.

[9] http://www.topquadrant.com/tools/modeling-topbraid-composer-standard-edition/.

browser application[10] supports heterogeneous data sources, because it has RMLas its underlying mapping language, compared to only homogeneous sources (i.e., databases) which are supported by the fluidOps editor. Last, OpenRefine[11] is a browser-based tool for cleansing raw data, changing the data format and incorporating external data using Web services. The RDF Refine extension[12] [19] allows users to export the data in RDF. A RDF graph is used to visualize the mappings. However, the RDF graph is forced in a hierarchy-layout, which weakens the advantages of using a graph representation. Additionally, this extension supports reconciliation services that offer HTTP interfaces. User intervention is needed to assess the quality of each reconciliation. A preview of candidate entities is available to help the user.

3   Mapping Process

The mapping process is a series of steps performed in order to generate Linked Data from raw data. There are two variations of the process, depending on whether a mapping editor is used or not.

Mappings Without Editor. Generating RDF from (semi-)structured data, using a mapping language without a mapping editor, consists of two consecutive steps: First, the mapping is created (see Fig. 1a). In this step, the user needs to be aware of the input data and has to have knowledge about the domain and the *mapping language's specification.* The latter is, in principle, possessed by *Semantic Web experts.* In most cases, a text editor is used to create the mappings. Statements in the mapping language follows this step. In the second

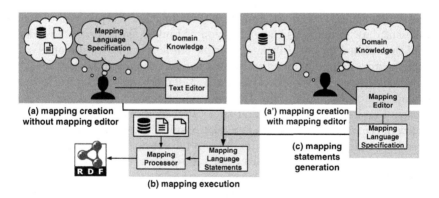

Fig. 1. The difference in the mapping process if during the mapping creation a mapping editor is used or not.

---

[10] http://pebbie.org/mashup/rml.
[11] http://openrefine.org/.
[12] http://refine.deri.ie/.

step the statements together with the input data sources are used by the mapping processor to generate the RDF triples (see Fig. 1b). The mapping processor is a tool that knows how to interpret the mapping language's specification to generate the RDF representation of the corresponding data, taking into consideration the statements derived from the previous step.

Mappings with Editor. When the mapping process is done with a mapping editor, the users do not longer need to be aware of the *mapping language's specification* (see Fig. 1a'). This allows *non-Semantic Web experts* to define the mappings. Additionally, the text editor is replaced by the mapping editor. The mapping editor knows how to interpret the mappings created by the user using the editor based on the language's specification. Subsequently, it generates the mapping language statements (see Fig. 1c). Important to note is that in this step no user knowledge about the specification is needed. Subsequently, the mapping is executed, and RDF triples are generated.

4   RMLEditor

The RMLEditor is a browser-based GUI with the goal to support users in production environments to define, in a uniform way, mappings that specify how to generate Linked Data represented using the prevalent RDF framework. In previous work [20], we listed 7 desired features of a GUI for uniform mapping editors. First, it should be independent of the underlying mapping language, so that users are able to create mappings without knowledge of the language's syntax. Second, it should allow users to execute the mappings outside of the editor, because it is only meant to create the mappings. Third, it should enable users to map multiple data sources at the same time, as it might occur that data is spread across multiple sources. Fourth, the editor should support data sources in different data formats, as the generation of Linked Data should be independent of the original format. Fifth, as multiple ontologies and vocabularies can be used to create a mapping, an editor should support the use of both existing and customs ontologies and vocabularies. Sixth, it should allow multiple alternative modeling approaches, as certain use cases might benefit from using a specific approach. Finally, by supporting non-linear workflows, users are able to keep an overview of the mapping model and its relationships. The GUI of the RMLEditor is designed to implement these features. The GUI uses graphs to visualize the mappings. Manipulation of these graphs results in creating, updating and extending the mappings, which can be done without any knowledge of the underlying mapping language or other used technologies. The graphs express how the raw data will be represented as RDF. However, this expressiveness is independent of the language's expressiveness. The RMLEditor triggers the mapping processor which executes the mappings exported by the RMLEditor and generates RDF statements. For the RMLEditor, we chose RMLwhich can support mappings derived from a GUI that covers all of the aforementioned features. However, any other mapping language could be used instead, if it allows to implement the features.

In Sect. 4.1 we discuss the RMLEditor's architecture. In Sect. 4.2, we elaborate on how the features are implemented in the GUI. In Sect. 4.3, we explain how the RMLProcessor, the mapping processor for RML, is used with the RMLEditor. In Sect. 4.4, we present two real-life use cases of the RMLEditor.

4.1 Architecture

The RMLEditor's high-level architecture is based on the *multilayered architecture pattern* [21]. This allows to separate the presentation and the logic of the mappings, using the *presentation layer* and *application layer*, respectively. The loading of mappings and data sources is done using the *data access layer*. The latter only communicates with the application layer. Communication between the presentation layer and the data access layer is not possible, as the architecture prohibits communication between layers that are not directly under or above each other. For the presentation layer, the Webix JavaScript library[13] is used to build the GUI, in cooperation with the d3.js library [22] for the presentation of the graphs. The communication with application layer is facilitated by the *Model-View-Controller* pattern [23]. The Graph Markup Language (GraphML) [24] is used to represent the graph visualization of the mappings independently of the underlying mapping language. This allows users to export the graphs in an application-independent format. Additionally, the GraphML-version of the graphs are used to generate the corresponding RMLstatements. Users are able to load graphs by instructing the RMLEditor to load the corresponding GraphML document. When RMLstatements need to be loaded, the statements are first converted to a GraphML document, then interpreted as graph elements, and shown in the GUI.

4.2 Graphical User Interface

The graphical user interface of the RMLEditor allows users to define mappings on existing data. To implement the aforementioned features for the GUI, the RMLEditor offers three panels to the users: *Input Panel*, *Modeling Panel* and *Results Panel* (see Fig. 2). They are aligned next to each other, however, when users want to focus on a specific panel, they are able to hide the other panels. The *Input Panel* shows the data sources to users (Feature 3). Each data source is assigned with a unique color. Depending on the data format, an adequate visualization is chosen (Feature 4). The *Modeling Panel* shows the mappings using a graph representation. The color of each node and edge depends on the data source that is used in that specific mapping, if any. It offers the means to manipulate the nodes and edges of the graphs in order to update the mappings. Semantic annotations can be added using multiple vocabularies and ontologies (Feature 5). The Linked Open Vocabularies[14] (LOV) can be consulted via the GUI to get suggestions on which classes, properties and datatypes

---

[13] http://webix.com/.
[14] http://lov.okfn.org/dataset/lov/.

to use. As the graphs offer a generic representation of the mappings, because they do not depend on the underlying mapping language, this panel addresses Feature 1. Additionally, the graph representation and the RMLstatements can be exported (Feature 2), allowing the execution of the mappings outside the RMLEditor. The *Results Panel* shows the resulting RDF dataset when the mappings defined in the *Modeling Panel* are executed on the data in the *Input Panel*. For each RDF triple of the dataset it shows the subject, predicate and object. The functionality and the interaction between the panels supports the different mapping generation approaches, as we described in previous work [20, 25] (Feature 6). The *data-driven* approach uses the input data sources as the basis to construct the mappings. The classes, properties and datatypes of the schemas are then assigned to the mappings. When users start with the vocabularies and ontologies to generate the mappings, the *schema-driven* approach is followed. Next, data fractions from the data sources can be associated to the mappings. Additionally, by not restricting users in when to interact with which panels – as would be the case for linear workflows – the RMLEditor supports non-linear workflows (Feature 7).

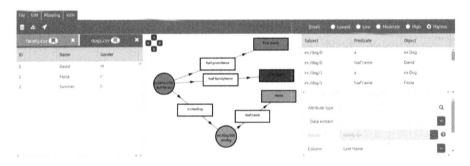

Fig. 2. The RMLEditor with the *Input Panel* on the left, the *Modeling Panel* in the center and the *Results Panel* on the right

4.3   RMLProcessor Server

As the RMLEditor uses RML, it needs the functionality of the RMLProcessor[15], a Java application which generates RDF based on provided RMLmapping documents. However, the processor is not needed to define the mappings. For the RMLEditor's needs, the RMLProcessor's functionality is offered through a Web API. The Web API is developed using Node.js[16] and offers three functions: (i) executing a mapping document on a set of data sources; (ii) converting a GraphML document to RMLto execute the mappings using the RMLProcessor, and (iii) converting RMLstatements to a GraphML document to visualize the mappings that are loaded in the RMLEditor.

---

[15] https://github.com/RMLio/RML-Mapper.
[16] https://nodejs.org/.

4.4   Real-Life Use Cases

The RMLEditor is developed for a pilot (COMBUST[17]), initiated by companies in Flanders who aim to build their collaboration network on top of their inter-linked data. The RMLEditor covers their need to generate the RDF representation of their raw data to be further used in other third-party applications. Moreover, the RMLEditor supports the partners of the Open Tourism working group of the Belgian Chapter of Open Knowledge Foundation (OKFN)[18] to semantically anno-tate their data with the Open Standard for Tourism Ecosystems Data[19] in the frame of the *Sustainable Mobile Guides for Tourism (in Flanders)* bootstrap. In both cases, the partners of the project have the raw data and the required knowledge about the domain. However, they have no understanding of map-ping languages, and without the use of the RMLEditor they need assistance of a Semantic Web expert. Additionally, for every update and execution of the map-pings the expert is consulted, which introduces significant overhead. We collected feedback during the deployment of the RMLEditor and concluded that they were able to create mappings for their data using their own domain knowledge. This is confirmed during our exploratory user validation, which involved members of the aforementioned pilot (see Sect. 5). Furthermore, the RMLEditor is the topic of a number of workshops and tutorials for interested groups and users[20].

5   Exploratory User Validation

We performed an exploratory user validation to assess the RMLEditor's adequacy to support users in defining mappings that generate RDF datasets. In Sect. 5.1 we discuss the two use cases that the participants completed. In Sect. 5.2 we explain the two groups of participants, the apparatus and the procedure followed during the validation. In Sects. 5.3 and 5.4 we elaborate on both the subjective and objective aspect of the validation. Finally, in Sect. 5.5 we discuss the results.

5.1   Use Cases

In this section, two use cases are outlined and it is explained in more detail how the RMLEditor is used to generate the mappings. Each use case covers a differ-ent mapping generation approach: *data-driven* or *schema-driven* [25]. The first use case involves data about employees and projects they work on, originally in two different data sources. The goal is to semantically annotate them, and link the employees to the projects they work on. The *data-driven* approach is recommended for editing the mappings to RDF. Users start by loading the two data sources into the RMLEditor. Next, mappings are generated based on data

---

[17] http://www.iminds.be/en/projects/2015/03/11/combust.
[18] http://www.openknowledge.be/.
[19] http://tourism.openknowledge.be:8080/spec/.
[20] http://rml.io/RMLevents.

fractions from the sources. Subsequently, users semantically annotate the mappings. Identifying suitable classes, properties and datatypes is supported by the LOV. The second use cases involves data about movies and their directors. The goal is to semantically annotate them, and to interlink each movie to the person that directed it. The *schema-driven* approach, supported by the RMLEditor, is recommended to the users for editing the mappings. Users start by generating mappings reusing movie concepts from the DBpedia Ontology[21] and person concepts from FOAF[22]. When the modeling is completed, the data sources are loaded. Subsequently, the mappings are updated with the data fractions to be used for generating the final RDF dataset.

5.2   Method

Participants. 15 participants from both Ghent University and the COMBUST network took part. They were divided in two major groups: (i) *Semantic Web experts* with 10 participants and (ii) *non-Semantic Web experts* with 5 participants. A *Semantic Web expert* is a user who has knowledge about Semantic Web technologies and standards, including Linked Data and RDF. A *non-Semantic Web expert* is not aware of the Semantic Web technologies, including the editor's underlying language RML. The *Semantic Web experts* are further distinguished in two sub-categories: (i) 5 experts with experience in Linked Data publishing, and (ii) 5 experts with no previous experience in Linked Data publishing.

Apparatus. The evaluation was carried out using the current version of the RMLEditor. Participants used their own computer with Chrome[23]. Both the RMLEditor and the RMLProcessor were hosted on the same physical server. Each participant participated in both use cases described in Sect. 5.1.

Procedure. The following steps were followed during the evaluation:

Step 1. We gave a presentation[24] to the participants where basic concepts, such as Linked Data, RDF and schemas were explained. In order for them to understand the purpose of the RMLEditor. Additionally, the user interface's panels were described. However, no introductory tutorial regarding how to use the RMLEditor was given.

Step 2. We conducted a pre-assessment by asking the participants to fill in a questionnaire.

Step 3. Half of each group's participants started with Use Case A, and the other half with Use Case B. This was done to eliminate the influences of the use case-specific data and editing approach on the final results of the

---

[21] http://dbpedia.org/ontology/.
[22] http://xmlns.com/foaf/0.1/.
[23] Version 45.0.2454.101 or higher; https://www.google.com/chrome/.
[24] http://www.slideshare.net/secret/vI6AO0ywqzyk1m.

evaluation. Subsequently, they completed the other use case. During the execution of the use cases, the experts were able to ask questions when they had problems with the RMLEditor. It was recorded when intervention was needed. For the *non-Semantic Web experts*, closer observation was conducted and the think-aloud method [26] was taken into consideration. This allowed for the identification of specific usability issues related to the RMLEditor. After each use case, we conducted an assessment to inquire about the difficulty of the use case.

Step 4. (Experts with prior RMLknowledge only) Half of the experts, with knowledge of the underlying mapping language RML, also had to create a mapping using a plain text editor, to observe differences and preferences between use and non-use of the RMLEditor. However, this could not be done by all participants, as they do not possess the required knowledge.

Step 5. At the end of each use case, we collected the created mappings. Additionally, a second and third questionnaire were filled in by the participants after their first and second use case, respectively.

Step 6. We conducted a post-assessment by providing the participants with a fourth questionnaire to fill in.

5.3 Subjective Validation

We conducted a subjective validation of the RMLEditor by presenting the participants with four questionnaires during the validation in step 2, 5 and 6. With the pre-assessment we assess the participants expectations before interacting with the RMLEditor. For the pre-assessment's questionnaire, we used the System Usability Scale (SUS) scale [27]. The SUS statements were translated to the use of the RMLEditor. The intermediate questionnaires of step 5 allowed for a subjective self-assessment of the previously completed use case, and more specific the approach of the specific use case. With the post-assessment we assess the participants' experience with the RMLEditor and determine what the positive and negative aspects are of the RMLEditor. For the intermediate questionnaires and the post-assessment's questionnaire, we used the 7-point Likert scale [28] from 'not difficult at all' to 'very difficult' to measure difficulty, from 'not useful at all' to 'very useful' to measure usefulness, and from 'not agree at all' to 'very much agree' to measure the degree of agreement to a given statement.

5.4 Objective Validation

After each completed use case, we collected each participant's mapping. We conducted a objective validation of the RMLEditor by comparing all mappings to the baseline mapping that we created ourselves. Additionally, if a participant also created a mapping using RMLdirectly, it was compared with the participant's mapping created using the RMLEditor.

5.5   Results

During the pre-assessment we measured that both groups of participants had high expectations regarding the ease-of-use and the improvements the RMLEditor would bring to the mapping process. Interesting to note is that *Semantic Web experts* expected the tool to be hard to use if no introduction tutorial is provided, in contradiction to the *non-Semantic Web experts*.

Learning Curve. After having completed the first use case in row, we measured via the subjective validation that 60 % of both the Semantic Web experts and non-experts found it difficult to use the RMLEditor to define the mappings that generate the RDF representation. After having completed the second use, all non-experts found it easier to use the RMLEditor. However, 30 % of the experts still found it difficult to use the RMLEditor over all. This is partially due to the fact that they all had high expectations of the RMLEditor. All participants needed at least once help during the first in row use case. However, during the participants their second use case, 40 % of the *Semantic Web experts* did not need any help at all anymore. However, all *non-Semantic Web experts* still needed help during their second use case. Nevertheless, over all, a significant lower level of intervention by us was needed. We deduced that the completeness and accuracy of the mappings increased when users performed their second use case, compared to their first one. This shows that there is a learning curve to use the RMLEditor, however, once users learn how to use it, the RMLEditor is adequate for its scope. The latter is verified via subjective validation by the fact that 47 % of the participants found that the overall usage of the RMLEditor was not difficult. However, still 20 % found it too difficult to use. Again, this is partially due to the fact that the participants had high expectations of the RMLEditor. Even though the RMLEditor aims to eliminate language and RDF-specific terminology, through the observation during the validation we found that 40 % of the participants had trouble with the used terminology in the GUI, such as 'child' and 'parent' when data sources need to be interlinked, and 'templates' when the URIs of the entities are constructed based on a data fraction from the data source. However, once explained to the users, 90 % of the users could use the terminology as intended.

Editing Approaches. 67 % of the participants were able to start editing the mappings using the given approach with no assistance at all. Through oral feedback during the evaluation, participants stated their preferred approach. Because of that preference, when they were asked to follow the other approach, they stated it was counterintuitive. If users prefer the *data-driven* approach, the *schema-driven* approach is counterintuitive, because no data is loaded initially. If users prefer the *schema-driven* approach, the *data-driven* is counterintuitive, because no predefined schema to be used is given.

Graph Visualizations. During the post-assessment, participants were asked to what extend they agree with the statement that the use of graph is beneficial for editing mappings, and the statement that the graphs make the linked nature of the final RDF dataset clear. Through the subjective validation, we found that 82 % of all participants found the use of graphs beneficial for editing mappings, and that graph-based visualizations are adequate for conceiving how the final RDF dataset will be. The objective validation showed that in the case of *Semantic Web experts*, in only 15 % of the total twenty use cases there were incomplete or inaccurate mappings, regarding the modeling of the domain. In 67 % of the use cases they were able to create mappings as complete and accurate as expected.

Linking Data Sources. In 33 % of the use cases, participants missed the interlinking between multiple data sources when using RML, when we compared the mapping created with the RMLEditor and the mapping created directly using RML, via the objective validation. Through the post-assessment, we concluded that the links were missing because of the difficulties with the terminology of the language, or because the participants just forgot the interlinking. With the RMLEditor this only happened in 10 % of the use cases. However, it still occurred because not all terminology was well covered in the GUI, as the participants made clear in the post-assessment.

6   Discussion and Conclusions

During the assessment both groups of participants were able to generate Linked Data through the RMLEditor's GUI, with the graph-based visualization as the most important contributor. Therefore, the RMLEditor fulfills its initial goal to supports users to define, in a uniform way, mappings that specify how to generate Linked Data. We present four main findings. First, non-Semantic Web experts are able to generate Linked Data when using the RMLEditor. Additionally, for experts, the quality of the Linked Data is better when using the RMLEditor instead of RMLdirectly, by looking at the mapping created with the RMLEditor and the mapping created by using RMLdirectly. Therefore, second, when Semantic Web experts generate Linked Data using the RMLEditor, the resulting RDF dataset is of at least the same quality as the dataset generated when using directly the RMLlanguage, if quality metrics, defined by the Semantic Web experts, are kept constant. Furthermore, for example, when a relationship between two data sources exists, however, not made explicit by the user, this is reflected in the graph: a link between the resource nodes belonging to the sources is not present. Using RMLdirectly, finding the missing relationship is more difficult. Therefore, third, graph-based visualizations of mappings improve the interlinking between multiple data sources during the mapping creation, compared to the use of no visualizations. Nevertheless, improvements to the RMLEditor, and more specifically to the GUI can be made during the next development sprints. As the main aspect of the RMLEditor is the use of graphs, it is important that manipulations on them are made as easy as possible. Although the required functionality for the manipulations is available, improving even more its accessibility will benefit the mapping process. Additionally, allowing users to determine the size of the graphs allows them to select the desired detail-level of their mappings. As the participants where presented with two use cases, they were able to perceive a learning curve. Future evaluations need to be conducted to determine how steep this learning curve is. However, as the fourth finding, by providing a set of tutorials, where the features of the RMLEditor are discussed will already improve the initial use of the RMLEditor by new users. Topic of future research is the validation of these observations during further large-scale user testing. Moreover, the RMLEditor supports both the *data-driven* and *schema-driven* approach. Whether a user starts with the use case using the *data-driven* or the *schema-driven* approach, the second use case in row is less difficult. Therefore, there is no approach enforced on the user by the RMLEditor, and the preferred approach depends on the person and the circumstances [5]. Both Semantic Web experts and non-Semantic Web experts have different expectations from the RMLEditor. Non-Semantic Web experts expect that they can generate Linked Data without any knowledge about a mapping language or Semantic Web technology. Semantic Web experts can already generate Linked Data, as they can use RMLdirectly. Therefore, they expect that they can at least do the same as with RML. Additionally, they have a set of requirements for their Linked Data, for which they look in the RMLEditor. These requirements improve, according to them, the quality of their generated Linked Data. In the long run, we want to see these expectations united. Non-Semantic Web experts do not only generate Linked Data. They are also concerned with the quality of their Linked Data, in order to further improve its adoption.

References

1. Bizer, C., Heath, T., Berners-Lee, T.: Linked data-the story so far. In: Emerging Concepts, Semantic Services, Interoperability and Web Applications (2009)
2. Brickley, D., Guha, R.: RDF Schema 1.1. Working group recommendation, W3C, February 2014. http://www.w3.org/TR/rdf-schema/
3. Das, S., Sundara, S., Cyganiak, R.: R2RML: RDB to RDF Mapping Language. Working group recommendation W3C, September 2012. http://www.w3.org/TR/r2rml/
4. Dimou, A., Vander Sande, M., Colpaert, P., Verborgh, R., Mannens, E., Van de Walle, R.: RML: a generic language for integrated RDF mappings of heterogeneous data. In: Workshop on Linked Data on the Web (2014)
5. Pinkel, C., Binnig, C., Haase, P., Martin, C., Sengupta, K., Trame, J.: How to best find a partner? An evaluation of editing approaches to construct R2RML mappings. In: Presutti, V., d'Amato, C., Gandon, F., d'Aquin, M., Staab, S., Tordai, A. (eds.) ESWC 2014. LNCS, vol. 8465, pp. 675–690. Springer, Heidelberg (2014)
6. Sengupta, K., Haase, P., Schmidt, M., Hitzler, P.: Editing R2RML mappings made easy. In: Proceedings of the 12th International Semantic Web Conference (2013)
7. Berglund, A., Boag, S., Chamberlin, D., Fernández, M.F., Kay, M., Robie, J., Siméon, J.: XML path language (XPath). World Wide Web Consortium (W3C) (2003)
8. Boag, S., Chamberlin, D., Fernández, M.F., Florescu, D., Robie, J., Siméon, J., Stefanescu, M.: XQuery 1.0: An XML query language (2002)
9. Bischof, S., Decker, S., Krennwallner, T., Lopes, N., Polleres, A.: Mapping between rdf, xml with xsparql. J. Data Seman. **1**(3), 147–185 (2012). ISSN 1861–2032

10. Fiorelli, M., Lorenzetti, T., Pazienza, M.T., Stellato, A., Turbati, A.: Sheet2RDF: a flexible and dynamic spreadsheet import&lifting framework for RDF. In: Ali, M., Kwon, Y.S., Lee, C.-H., Kim, J., Kim, Y. (eds.) IEA/AIE 2015. LNCS, vol. 9101, pp. 131–140. Springer, Heidelberg (2015)
11. Pazienza, M.T., Stellato, A., Turbati, A.: Pearl: Projection of annotations rule language, a language for projecting (uima) annotations over rdf knowledge bases. In: LREC (2012)
12. Rodriguez-Muro, M., Hardi, J., Calvanese, D.: Quest: efficient SPARQL-to-SQL for RDF and OWL. In: 11th International Semantic Web Conference ISWC, p. 53. Citeseer (2012)
13. Noy, N.F., Sintek, M., Decker, S., Crubézy, M., Fergerson, R.W., Musen, M.A.: Creating semantic web contents with protege-2000. IEEE Intell. Syst. **2**, 60–71 (2001)
14. Friedman, M., Levy, A.Y., Millstein, T.D., et al.: Navigational plans for data integration. In: AAAI/IAAI, pp. 67–73 (1999)
15. Bizer, C., Seaborne, A.: D2RQ - treating non-RDF databases as virtual RDF graphs. In: Proceedings of the 3rd International Semantic Web Conference (ISWC 2004), vol. 2004. Citeseer, Hiroshima (2004)
16. Pinkel, C., Schwarte, A., Trame, J., Nikolov, A., Bastinos, A.S., Zeuch, T.: DataOps: seamless end-to-end anything-to-RDF data integration. In: Gandon, F., Guéret, C., Villata, S., Breslin, J., Faron-Zucker, C., Zimmermann, A. (eds.) ESWC 2015. LNCS, vol. 9341, pp. 123–127. Springer, Heidelberg (2015)
17. Han, L., Finin, T., Parr, C., Sachs, J., Joshi, A.: RDF123: from spreadsheets to RDF. In: Sheth, A., Staab, S., Dean, M., Paolucci, M., Maynard, D., Finin, T., Thirunarayan, K. (eds.) ISWC 2008. LNCS, vol. 5318, pp. 451–466. Springer, Heidelberg (2008)
18. Bizer, C., Lehmann, J., Kobilarov, G., Auer, S., Becker, C., Cyganiak, R., Hellmann, S.: DBpedia - a crystallization point for the web of data. Web Seman. Sci. Serv. Agents World Wide Web **7**(3), 154–165 (2009)
19. Maali, F., Cyganiak, R., Peristeras, V.: Re-using cool URIs: entity reconciliation against LOD hubs. In: LDOW, vol. 813 (2011)
20. Heyvaert, P., Dimou, A., Verborgh, R., Mannens, E., Van de Walle, R.: Towards a uniform user interface for editing mapping definitions. In: Proceedings of the 4th Workshop on Intelligent Exploration of Semantic Data, October 2015
21. Richards, M.: Software Architecture Patterns. O'Reilly, Sebastopol (2015)
22. Bostock, M., Ogievetsky, V., Heer, J.: D^3 data-driven documents. IEEE Trans. Vis. Comput. Graph. **17**(12), 2301–2309 (2011)
23. Osmani, A.: Learning JavaScript Design Patterns. O'Reilly Media, Sebastopol (2012)
24. Brandes, U., Eiglsperger, M., Herman, I., Himsolt, M., Marshall, M.S.: GraphML progress report layer proposal. In: Mutzel, P., Jünger, M., Leipert, S. (eds.) GD 2001. LNCS, vol. 2265, pp. 501–512. Springer, Heidelberg (2002)
25. Heyvaert, P., Dimou, A., Verborgh, R., Mannens, E., Van de Walle, R.: Approaches for generating mappings to RDF. In: Proceedings of the 14th International Semantic Web Conference: Posters and Demos, October 2015
26. Charters, E.: The use of think-aloud methods in qualitative research an introduction to think-aloud methods. Brock Educ. J. **12**(2), 68–82 (2003)
27. Brooke, J.: SUS - a quick and dirty usability scale. Usability Eval. Ind. **189**(194), 4–7 (1996)
28. Likert, R.: A technique for the measurement of attitudes. Arch. Psychol. **22**, 1–55 (1932)

Enriching a Small Artwork Collection
Through Semantic Linking

Mauro Dragoni[1(✉)], Elena Cabrio[2], Sara Tonelli[1], and Serena Villata[3]

[1] FBK, Trento, Italy
{dragoni,satonelli}@fbk.eu
[2] University of Nice Sophia Antipolis, Nice, France
elena.cabrio@unice.fr
[3] CNRS, I3S Laboratory, Sophia Antipolis, France
villata@i3s.unice.fr

Abstract. Cultural heritage institutions have recently started to explore the added value of sharing their data, opening to initiatives that are using the Linked Open Data cloud to integrate and enrich metadata of their cultural heritage collections. However, each museum and each collection shows peculiarities, which make it difficult to generalize this process and offer one-size-fits-all solutions. In this paper, we report on the integration, enrichment and interlinking activities of metadata from a small collection of verbo-visual artworks in the context of the Verbo-Visual-Virtual project. We investigate how to exploit Semantic Web technologies and languages combined with natural language processing methods to transform and boost the access to documents providing cultural information, i.e., artist descriptions, collection notices, information about technique. We also discuss the open challenges raised by working with a small collection including little-known artists and information gaps, for which additional data can be hardly retrieved from the Web.

1 Introduction

In the last years, cultural heritage institutions have been involved in several initiatives in order to exploit digital means to increase their visibility. Galleries, libraries, archives and museums (GLAM) typically own rich and structured datasets developed over many years and organized by domain, which in principle could be easily connected with the databases of other institutions and then made available online to a larger audience. However, several issues need to be faced, for instance the need for a standard format for data sharing, and the lack of technical skills especially in small museums, so that data manipulation and conversion can hardly be achieved. Relevant standardization efforts such as that of Europeana[1], a framework at European level to publish and link cultural heritage metadata through a unified data model, go in this direction and contribute

---

This work has been partially carried out in the framework of the VVV Project supported by Fondazione Cassa di Risparmio di Trento e Rovereto.
[1] http://www.europeana.eu.

H. Sack et al. (Eds.): ESWC 2016, LNCS 9678, pp. 724–740, 2016.
DOI: 10.1007/978-3-319-34129-3_44

to raise museums' awareness on the importance of knowledge sharing among cultural heritage institutions. The advantages include driving users to new content, stimulating collaboration in the cultural heritage domain, enabling new scholarship through the availability of new digital content, and more generally increasing the relevance of cultural heritage institutions [1].

In this work, we present the process performed to map the metadata from the Verbo-Visual-Virtual Project [2] to the Linked Open Data[2] (LOD) cloud and the related data enrichment. Although the work was largely inspired by past efforts by other cultural heritage institutions [3–5], we face new challenges, partly related to the small size of the collection, with little-known artists and few information available from other online sources, and partly to the integration of Natural Language Processing (NLP) techniques to enrich the metadata. On the one hand, we show that linking metadata to DBpedia[3] contributes to improving the quality and richness of the data owned by the museum. On the other hand, that small collections with little-known artists present specific issues, e.g. the limited coverage of external resources, that need to be addressed in a semi-automatic way. We make available both the developed ontology and the RDF data set containing the enriched metadata of our collection.

The remainder of the paper is as follows. Section 2 describes the Verbo-Visual-Virtual cultural heritage collection, Sect. 3 presents the VVV ontology we have defined to represent our cultural heritage data, and details about the data interlinking and enrichment steps we have addressed to enrich the available information. The approach is evaluated in Sect. 4 to prove its feasibility, and the encountered difficulties are discussed. Conclusions end the paper.

2   The Verbo-Visual-Virtual Project

The Verbo-Visual-Virtual Project (VVV) started in 2013 as a joint effort between two museums in Trentino-Alto Adige and a technological partner, with the goal to create a unified virtual collection of "Archivio di Nuova Scrittura" (ANS) [6]. The collection, albeit international, is mainly centered around the artworks of Italian artists active between 1950 and 1990, and finds its origin in the collecting activity of Paolo Della Grazia, an entrepreneur with a passion for interdisciplinary forms between art and poetry. Towards the end of the nineties, Della Grazia decided to donate to a public institution the archive, which had been steadily growing and needed an appropriate site. He decided first to contact MUSEION[4], the Museum of Contemporary Art in Bolzano. However, since the museum did not have enough space available, Della Grazia decided to split the collection into two parts, one assigned to MUSEION through a long-term loan, and the other to MART[5], the Museum of Modern and Contemporary Art in Rovereto. For this reason, a collection which was originally conceived as a single archive is now

---

[2] http://lod-cloud.net/.
[3] http://www.dbpedia.org/.
[4] http://www.museion.it.
[5] http://www.mart.trento.it.

divided in two and hosted by two different institutions. ANS fulcrum is represented by works linked to concrete poetry, visual poetry, Fluxus, and conceptual art. The main features of the collection are its small size (around 5,000 artworks in total) and its homogeneity, i.e. all works were created in a limited time span by a relatively small number of artists.

Given the possibility offered by current digital technologies to access art collections online, VVV was launched to create a unified virtual collection of ANS works, where all information about the collection would be semantically enriched and made available through a web-based platform. The Digital Humanities group at Fondazione Bruno Kessler has therefore worked in the last two years to make all the work records consistent, possibly add information automatically retrieved from the Web, and implement a navigation platform to display and search works in a virtual exhibition. The project will end in Spring 2016, when the VVV platform will be made accessible.

2.1 Project Challenges

The VVV project brought about several challenges to address:

1. The *quality* of information available in databases describing small collections exposed in local museums is a critical aspect, especially when archives have been created incrementally and at different time points. Information could be inconsistent even for important elements such as the artwork title or author. Here, experts have to be supported in the management of such items by providing facilities to interlink information about collection items with as many external data sources as possible, to improve the qualitative description of each item. In VVV, this is particularly relevant to guarantee that the two portions of the archive are curated with the same quality.
2. Besides quality, also the *quantity* of available information is an important issue. Local authors are a good example scenario. Generally, it is not easy to find information about them on the Web, and, when it is available, it should be manually extracted for enriching the related content in the knowledge base. Addressing this challenge implies the design and implementation of information extraction approaches supporting experts in the enrichment of artworks information.
3. The exposure of information in semantic formats requires artworks to be classified through a classification schema. For this reason, the use of ontologies in the cultural heritage domain has gained a lot of attention in the last years. However, well-known classification schemas (for example the Europeana Data Model[6]) may be too generic to capture the peculiarities of minor collections, e.g., the local museum in a village in which a specific battle of World War I took place. In this case, a *conceptual modeling* activity is needed to tailor existing models to the specific needs of a collection.

---

[6] http://www.europeana.eu/.

4. Finally, disseminating the cultural heritage of local environments is effective only when information is published in languages currently spoken in the territory. In Trentino-Alto Adige, where the VVV use cases are located, Italian and German are the official languages, while English is widely used for tourism purposes. For this reason, it is crucial that artworks preserved in local museums are described at least in these three languages. When necessary, a *translation* task has to be performed for breaking the language barriers and yielding an effective publication of information.

2.2   Data Description

As a first step in this direction, in this paper we investigate the possibility to perform automatic enrichment of data through semantic interlinking and information extraction from the Web, based on the data split belonging to MART. We focus first on this part of the collection because the information stored at MART is more stable, while the records at MUSEION are still being updated. Nevertheless, the framework introduced in this paper (Sect. 3) is designed to support multiple sources and to fuse them through a domain ontology. MART has adopted a well-known record management system called *MuseumPlus*[7], that is used by the museum personnel to fill information about the artworks and curate them, as a knowledge base of the museum objects. Since the system is used by several people, and required information is mostly filled as free text, some inconsistencies are present in the data, especially spelling errors.

   In order to perform semantic enrichment, we first export the VVV database stored at MART as raw data in CSV format. Since *MuseumPlus* offers the possibility to input a variety of information about each artwork and each artist, which however are not present for all entries of the VVV collection, we export only a subset of fields, which domain experts consider mandatory to identify a work of art, that is:

`title,inventory_code, dimension, date, technique,`
`author_name, author_surname, author_born_place,`
`author_born_year`

   Overall the raw dataset contains 592 artworks created by 287 artists. However, prints are often collected in portfolios, which are also recorded as artworks. If we merge the works that belong to the same portfolio, we obtain 495 entries. Another issue is related to vague information in the fields: 187 works have no title ("Senza titolo"), not because of missing information but due to the artist's choice. Besides, 27 works have been created by an unknown artist.

3   The Framework

In this section, we describe the steps carried out to transform VVV raw data into the semantically enriched VVV data set. Figure 1 shows the proposed framework.

---

[7] http://www.zetcom.com/en/products/museumplus/.

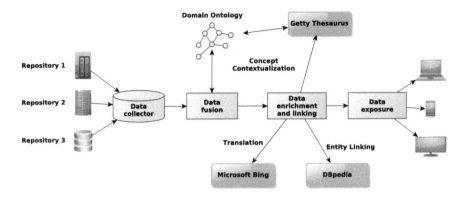

Fig. 1. The platform developed in the context of the VVV project.

Data can be collected from diverse repositories connected to the platform. As mentioned earlier among the challenges of the project, raw data are affected by problems like incomplete records and missing information about authors. Interoperability is promoted by the fusion and annotation of input data with semantic information modeled through an ontology, described in Sect. 3.1. After the merging and annotation process of all collected information, each record is enriched in two different ways:

1. Recognized entities are linked with information available in the Linked Open Data (LOD) cloud. More precisely, in the current version of the system only DBpedia[8] is exploited for interlinking (Sect. 3.2).
2. Natural Language Processing methods are used for extracting information from web pages containing relevant details for the record that needs to be enriched. For example, if for an author only the name is available, extraction patterns are applied for collecting information such as the birth place, the birth date, and so on (Sect. 3.3).

Finally, the last component of the pipeline consists in exposing the created knowledge base, by using LOD formats, to make it available to third-party services. Data are exposed through a RESTful interface providing requested information in Turtle format[9].

Such a pipeline, specifically designed and implemented for the VVV project, is generally applicable to any process whose goal is to convert raw cultural heritage data into an enriched version. Each task of the pipeline is run when a new raw dataset is imported. Given that the final purpose of the process is to expose an enriched dataset, and the fact that collected data are automatically translated and linked by the different components of the pipeline, we provide a tool that implements facilities enabling the manual verification and refinement of all information. More precisely, the described pipeline has been implemented

---

[8] http://wiki.dbpedia.org/.
[9] http://www.w3.org/TR/turtle/.

in a collaborative knowledge management tool called MoKi [7,8]. MoKi[10] is a collaborative MediaWiki-based[11] tool for modeling ontological and procedural knowledge in an integrated manner[12]. MoKi is grounded on three main pillars, which we briefly illustrate: *(i)* each basic entity of the ontology, i.e., concepts, object and datatype properties, and individuals, is associated with a wiki page; *(ii)* each wiki page describes an entity by means of both unstructured, e.g., free text, images, and structured, e.g., OWL axioms, content; and *(iii)* a multi-mode access to the page contents is provided to support easy usage by users with different skills and competencies. In the VVV project, the tool has been customized for supporting the manual refinement activity performed by experts after the automatic execution of the entire pipeline[13].

3.1 The VVV Ontology

In the cultural heritage domain, a number of ontologies has been proposed to represent the semantics of cultural heritage data. The most known ontologies proposed in this context are CIDOC-CRM[14] and the Europeana Data Model (EDM)[15].

CIDOC Conceptual Reference Model (CIDOC-CRM) claims to be a "formal ontology intended to facilitate the integration, mediation and interchange of heterogeneous cultural heritage information" [9]. It is developed by International Counsel of Museums (ICOM), and is accepted as an ISO Standard. CIDOC-CRM is one of the most recommended models for cultural heritage. It combines knowledge about artworks together with all the events concurring to its creation. CIDOC-CRM is defined with the aim to facilitate the identification and sharing of knowledge about cultural heritage data, and the interoperability among the different sources of cultural heritage data and their own data representation models. This ontology mainly defines general concepts, allowing for ontology extensions with the concepts and properties needed by each source.

EDM is a data model that has been defined by a set of European museums in the Europeana project. The aim of Europeana is to build a computational library including the cultural heritages from various European museums. It allows to access to the collections of galleries, museums and libraries of all types (including images and audiovisual resources). The British Library in London, the Rijksmuseum in Amsterdam and the Louvre Museum in Paris are among the 1500 institutions that have participated in the construction of this cultural library. The EDM data model is constructed based on the CIDOC-CRM model, i.e., it inherits some concepts and properties of CIDOC-CRM. This reuse of properties and

---

[10] http://moki.fbk.eu.

[11] http://www.mediawiki.org.

[12] Though MoKi allows to model both ontological and procedural knowledge, here we will limit our description only to the features for building ontologies.

[13] A read-only version, but with all functionalities available, of the MoKi instance described in this paper is available at https://dkmtools.fbk.eu/moki/3_5/vvv/.

[14] http://www.cidoc-crm.org.

[15] http://pro.europeana.eu/edm-documentation.

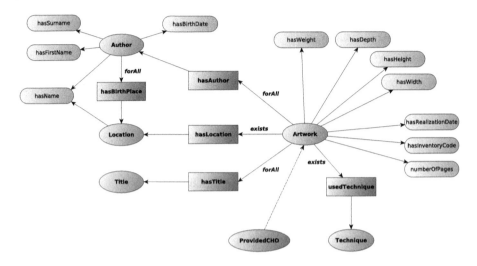

Fig. 2. The ontology of the VVV project.

concepts from CIDOC-CRM supports interoperability with other data sources represented through the CIDOC-CRM ontology.

In analyzing these two models with the purpose of selecting an existing ontology (if any) that suits the requirements of the VVV project, we choose to take EDM as the basis of our ontology[16], so as to reuse EDM general concepts and properties and further extend it with the additional information we need to represent the VVV data semantics in a compliant way. More precisely, our data contain a set of mandatory fields reported in Sect. 2. Moreover, we need to address the challenge of managing several missing data about the items in local museums. For these reasons, we design a new ontology, called VVV, reflecting a specific model that fits well our project needs. As illustrated in Fig. 2, our ontology extends the EDM ontology with new concepts and properties.

The VVV ontology reuses two concepts from the EDM ontology. Such concepts have been redefined in terms of label for accommodating the schema readability by the domain experts. The Author concept is used for representing artists, and it is aligned with the concept "Agent" contained in the EDM, while the concept Location is used for representing either the places where an artwork is exposed or the birth place of an author. This concept is aligned with the concept "Place" contained in the EDM.

In addition, the VVV ontology introduces the following concepts: (i) Artwork, defined as a child of the concept ProvidedCHO modeled in the EDM ontology, is used for instantiating artworks in the knowledge base; (ii) Technique represents the style adopted for realizing a certain art work; (iii) Title defines the caption of an artwork.

---

[16] CIDOC-CRM model is not appropriate to represent the granularity of the information we are interested in.

The definition of the concept `Title`, instead of adopting a simpler annotation property, is required since such information may be missing in the knowledge base. Indeed, our intention is to have a conceptual modeling of the missing title of an artwork, and not only mark the absence of an annotation. This fact is modeled through the use of the `hasTitle` object property. The same situation occurs, always for the concept `Artwork`, when data about its author are missing. Thus, the concept `Author` is linked to the concept `Artwork` through the `hasAuthor` object property. Both object properties have been modeled using the universal quantifier.

Moreover, among the object properties, also the `usedTechnique`, `hasLocation`, and `hasBirthPlace` object properties are defined. Such properties allow to model the artistic technique adopted for realizing a certain artwork, the location is which the artwork is exhibit, and the birth place of an author, respectively.

Finally, the ontological model is completed with the definition of the following annotation properties:

- `hasWeight`, artwork weight;
- `hasDepth`, artwork depth;
- `hasWidth`, artwork width;
- `hasHeight`, artwork height;
- `hasRealizationDate`, the date when an artwork has been realized;
- `hasInventoryCode`, the inventory code of an artwork in the exposition where it is preserved;
- `numberOfPages`, in case of literary artworks, the number of pages;
- `hasName`, the full name of an author or of a location;
- `hasSurname`, the surname of an author;
- `hasFirstName`, the first name of an author;
- `hasBirthDate`, the birth date of an author.

The described ontology is automatically populated while importing raw data from the external databases. Information is initially available only in the Italian language. One of the challenges of the project is to provide the ontology in two additional languages, i.e. English and German. Therefore, the tool has been equipped with a component connecting MoKi with the Microsoft Bing Translation service[17]. Through the interface, the experts are able to correct the proposed translations with most appropriate ones if needed.

3.2 Data Interlinking

In this section, we describe how we address the interlinking of our dataset with external knowledge bases in order to retrieve further information to contextualize our data. The entire interlinking activity is performed in two steps:

---

[17] http://www.bing.com/translator.

1. when a new dataset is imported into the pipeline, an automatic procedure is performed to retrieve the candidate links from external resources;
2. considering that all data provided through the platform should contain high quality information, a further manual refinement is carried out by experts to validate the linked information.

   In the VVV project, two interlinking activities have to be performed: *(i)* to contextualize the technique used for creating an artwork with respect to a domain-specific thesaurus, and *(ii)* to interlink the entities, like authors and locations, with conceptual nodes defined on external knowledge bases such as DBpedia.

Linking with the Getty Thesaurus. As already mentioned in Sect. 2.1, it is very important to translate the concept labels as a preliminary activity to the interlinking step. Since in our case all information in the database is available in Italian only, all records are first automatically translated into English through Microsoft Bing Translation service. After that, the *Getty* thesaurus is used to map modeled concepts with information on the techniques contained in our dataset. Such thesaurus is currently one of the most widely used linguistic resources in cultural heritage projects, with high-quality, manually curated domain information.

The mapping operation consists in querying the Getty web service[18] with the English label describing the used technique. Each query may return more than one result. Thus, we decided to show to the experts a selection of the top five candidate concepts, based on the confidence score provided by the web service, that can be used for defining a new mapping. Such a strategy has been already applied in [10] demonstrating its suitability for this kind of scenario. Once results are shown to experts, they are able to select which concept, from the Getty thesaurus, to map with the ones defined in our platform. In this way, an imported dataset can be semantically connected with external knowledge bases in a semi-automatic fashion.

Linking with DBpedia. A similar approach has been adopted for linking information about authors and locations with DBpedia. This activity has been performed by exploiting the lookup service connected with DBpedia[19]. Such a service works with a REST interface receiving a query containing, for instance, a named entity label, and by returning a ordered rank of candidate DBpedia nodes that can be linked with the source label. Similarly to the linking with the Getty thesaurus, also in this case we provided five candidate suggestions to the domain experts, and let them choose the final alignment.

---

[18] http://vocab.getty.edu/.
[19] https://github.com/dbpedia/lookup.

3.3   Data Enrichment

Given that some of the artists of the VVV collection are little-known, they may be missing in DBpedia (and therefore, it is not possible to connect them to such a resource). For this reason, in order to provide additional information also for these entities to enrich our knowledge base, we resort to a semi-automatic process combining manual selection of texts and Natural Language Processing techniques. The methodology comprises the following steps:

1. *Manual search of online information about VVV artists:* The entire enrichment process cannot be performed completely in an automatic way due to the high risk of retrieving wrong information from the Web. Thus, we first search the Web for extracting textual snippets or paragraphs describing artists (in particular their biography or career), from sources that may contain relevant and correct information about artists (for instance, art collections websites). Texts are collected independently by the language used to write them. Indeed, most of the documents about the artists contained in the VVV collection, are expressed in the language of the artist birth country. For example, the biography of the artist "Massimo Pompeo", can be found at http://www.zam.it/biografia_Massimo_Pompeo only in Italian.
2. *Automatic content translation:* For the reasons expressed above, each text written in a language other than English is automatically translated into English by using the Microsoft Bing Translation service.
3. *Part-of-speech tagging:* Information contained in the pages retrieved online is typically unstructured and presented in natural language. For enabling the detection of potential significant relations, text is preprocessed and tokens are tagged with the TreeTagger [11] library on the English translation of the text.
4. *Relation extraction:* We apply a standard method for the automatic acquisition of relations based on hand-crafted rules [12]. In particular, we define a set of basic patterns (Table 1) expressed as regular expressions, and retrieve in the text the matching strings. These are assigned to the extracted relations with the corresponding properties.

MoKi supports the enrichment process with facilities (i) for inserting new rules to be exploited for extracting significant relations from text, and (ii) for selecting, among the suggested relations, the ones that can be used for enriching the ontology. Figure 3 shows the interface used by the domain expert for analyzing the list of the extracted relations and for choosing which ones can be used for enriching the ontology. In the upper part of the picture, we may see how the expert is able to invoke the relation extraction component. In the middle part, we may see how suggestions are displayed to the experts; while, in the bottom part, how the selected relations are shown in the entity mask.

Table 1. Hand written rules for relation extraction.

Example patterns	Relation	Property	Examples of matched sentence
born in DATE; (LOC, DATE; born on DATE)	birthdate	`rdaa:dateOfBirth`	He was born in Brno in 1946
died in LOC	place of death	`dbpedia:deathPlace`	He died in Paris
author of .*	author of	`vvv:authorOf`	He is author of numerous children's and other radio plays
influenced by .*	influenced by	`dbpedia:influencedBy`	The first works were influenced by the metaphysical painting

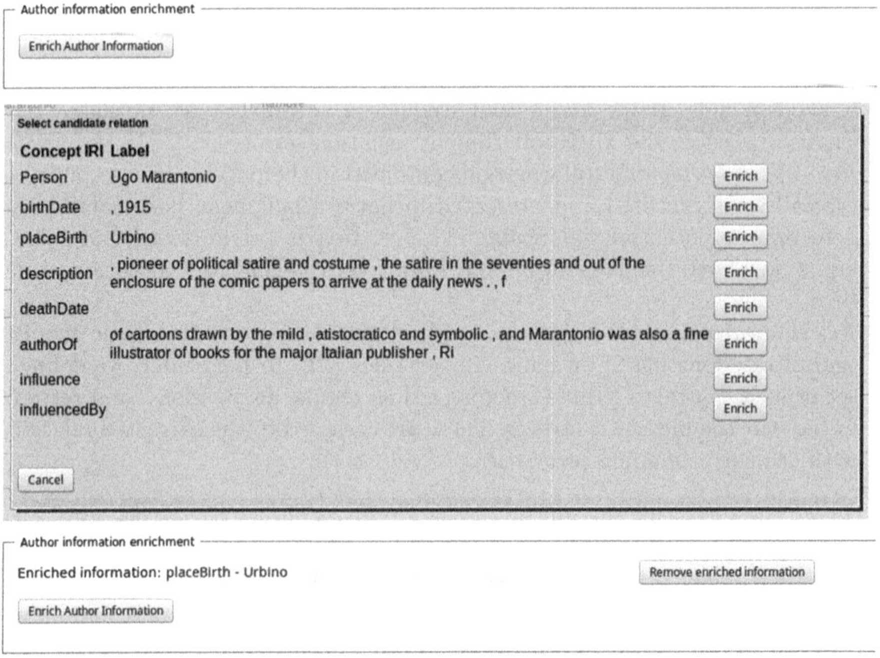

Fig. 3. The enrichment mask implemented in MoKi.

4   Evaluation

In this section, we present the evaluation of the platform proposed in the VVV-project. Such an evaluation has been conducted from two perspectives:

- Quantitative evaluation (Sect. 4.1): we report on the numeric results about the effectiveness of the data interlinking and the data enrichment algorithms. These values allow to measure how much support is provided to experts by the automatic modules in the pipeline.
- Lessons learned (Sect. 4.2): we discuss the lessons learned from the experience in applying Semantic Web technologies to the use case of data enrichment and integration of small cultural heritage collections. We include some suggestions provided by two art curators, expert of the VVV collection, during a demo session.

4.1   Quantitative Evaluation

The quantitative evaluation step includes the measure of the effectiveness of the DBpedia lookup service for covering and linking the modeled knowledge base, the accuracy of the enrichment information provided by the NLP component in charge of analyzing external textual resources, and the precision of the candidate alignments with the Getty thesaurus suggested by the system.

The raw data from the VVV collection contain works from 287 artists and for only 139 of them a DBpedia page can be linked. The interlinking algorithm we implemented has been executed automatically and we manually verified the correctness of the created links. For none of the 148 artists with no DBpedia page, links have been created. For the 139 artists that have a DBpedia page, 26 artists have been linked to the wrong page (i.e. the links pointed to entities that are homonyms of the considered artist). Thus, 113 VVV entities were correctly linked to DBpedia by obtaining an accuracy of 0.812.

Concerning data enrichment, we applied the proposed approach to the 148 artists with no DBpedia page and for 93 of them we were able to find online a description of the author biography and artistic career. Hand-written rules have been applied for extracting the following relations: "place of birth", "date of birth", "date of death", "author of", "influenced by", "description". As mentioned in Sect. 3.3, the set of rules can be changed dynamically by adding further rules or by refining the existing ones. To evaluate the accuracy of the relations suggested by the extraction algorithm, the texts have been manually annotated to build a gold standard. Results obtained for relation extraction of the six above mentioned properties are: *Precision: 0.93, Recall: 0.89, and F-measure: 0.90.* Finally, concerning the alignment between our ontology and the Getty thesaurus, the candidate mappings service implemented in our platform obtained an accuracy of 0.976.

From these results, we can infer a couple of practical lessons useful for possible future inclusion of further collections. First, when dealing with little-known artists, a methodology based on metadata enrichment by linking artists' names to DBpedia is not satisfactory. Artists' contextual information are needed for improving the disambiguation capability of the linking approach. Secondly, the adoption of a semi-automatic enrichment approach based on (i) a manual retrieval of textual snippets from the Web and (ii) the analysis of such snippets for suggesting further information about artists, is recommended to ensure a good balance between curation time and quality.

4.2   Lessons Learned and Future Work

Here, we sum up our experience in using Semantic Web technologies for supporting the management and the enrichment of a little-known collection in the context of the VVV project. As already mentioned in this paper, the main issue when working with little-known collections is the quality of the raw data. Different from large, established collections, entries are often recorded once and rarely cross-checked, causing a lack of neatness and everlasting mistakes. In these cases, linking is a useful strategy both for performing a quality check and for spotting possible inconsistencies in the data. For example, in our collection, we found that the date of birth of four artists available in the VVV repository did not match with the ones published on DBpedia. We asked a domain expert for a final adjudication, and she further checked the dates in a third database, the Virtual International Authority (VIAF) File[20]. We found that: (i) in two cases DBpedia contained the correct information, (ii) in one case the VVV repository was right, and (iii) in another case VIAF did not match with any of the two other sources. This example shows that the quality of information stored by museums is not always more accurate than the one available, for example, in the Linked Open Data cloud. Thus, the challenge of providing accurate information to users is still an open challenge due to the necessity of double-checking all exposed data.

The problem of inconsistency in raw data affects also other parts of the collection. For instance, in some cases, the same field contains incompatible information. An example is the "author" field, which sometimes contains the author of the artwork, while in other cases it contains the editor of the artwork. These problems are caused by the inaccurate design of the data management system used for storing the raw data. However, in our data fusion activity, such issues are easily spotted and corrected automatically.

Other useful hints for current improvement and future developments were provided by the art curators during a demo session, in which the MoKi interface with the VVV records was presented. The curators were impressed by the power of such a knowledge management system and provided a generally positive feedback, praising in particular the time they could spare to complete records through automatic linking. The possibility to manually correct translations or to choose the best piece of information among a set of options was deemed to be very useful, since art experts would never trust a completely automatic enrichment process. A possible improvement, which they consider groundbreaking, would be the possibility to enrich information starting from an image, which should be linked automatically to related images or to additional information through image processing techniques. This would be, however, computationally very demanding and we will not be able to deliver it in the framework of the

---

[20] http://viaf.org.

VVV project. Another improvement suggested by the domain experts is to connect MoKi with the platform to navigate the collection, so that all the changes and corrections performed by them would be automatically shown to the visitors of the platform in real time, without the need to update periodically the database underlying the navigation platform. This suggestion is technically less demanding on the short run and we consider implementing it before the end of the project.

The only negative feedback provided by the curators was related to the usability of the discussion facility embedded in MoKi. Such an interface has been perceived as less intuitive with respect to the others included facilities. Curators suggested possible improvements that will be addressed in the next version of the tool.

Concerning future work, the effort for improving the results obtained in the VVV project will be focused on two directions. On the tool side, the workflow for managing data modeled in the knowledge base will be extended with an approval mechanism allowing (i) the management of a more fine-grained set of roles assigned to the users for enabling them to modify only parts of knowledge bases, and (ii) the possibility by the curators manager to decide which changes can be carried out on the knowledge base and which cannot. On the evaluation side, it is already planned to extend the evaluation to further collections coming from other local museums, in order to validate the interoperability level and the effort needed for migrating the entire platform to a different environment.

5   Related Work

Given the growing interest in enriching cultural heritage data using Semantic Web languages and techniques, a number of works have been proposed to address this issue. Hyvonen [13] presents an overview on why, when, and how Linked (Open) Data and Semantic Web technologies can be employed in practice in publishing cultural heritage collections and other content on the Web.

An example is provided by Szekely et al. [3], where the authors define a specific ontology which extends the EDM ontology. This extension is motivated by the lack of properties needed to represent specific data of the Smitsonian museum. The data is then linked to knowledge bases such as DBpedia, the Getty vocabulary "Ulan" (Union List of Artist Names), and the list of artists of the museum Rijksmuseum.

de Boer and colleagues [4,5] present the Amsterdam Museum Linked Open Data project, where the data of the museum has been transformed into RDF data in an automated way. The authors used the EDM ontology for defining the semantics of the data coupled with vocabularies such as the Dublin Core vocabulary.

On the one side, we share with these works the idea of starting from raw cultural heritage data available in different formats, e.g., tables, texts, CSV, and then transforming this data into semantic data. We all rely on the EDM ontology to start and we extend it depending on the purpose of our data translation task.

We also adopt other vocabularies like Dublin Core and FOAF to represent additional features. Moreover, a data interlinking step with DBpedia is addressed in these works as well. On the other side, several differences arise. More precisely, we do not only translate the raw data into RDF data and interlink it with DBpedia instances, but we address also an enrichment step where we extracted from the textual resources available on the Web structured information to be integrated with our starting data, e.g., information about the artists. Such an enrichment step has been achieved using Natural Language Processing techniques.

Other approaches deal with the definition of more specific ontologies and annotation tools. Benjamins and colleagues [14] present an ontology of Humanities then exploited into a semi-automatic tool for the annotation of cultural heritage data to ease the knowledge acquisition task.

Other related work has been proposed concerning multilingual access to cultural heritage information. Dannells and colleagues [15] address the problem of multilingual access to cultural heritage by adopting Semantic Web languages. They process museum data extracted from two distinct sources and making this data accessible in natural language, thanks to a grammar-based system designed to generate coherent texts from Semantic Web ontologies in 15 languages. Here, natural language processing techniques have been applied, coupled with Semantic Web technologies, similarly to our approach, even if the final goal is different from our one.

Besides the cultural heritage domain, annotations has been adopted also in the digital libraries field. In [16] the authors discuss a framework about the use of semantic annotations for enriching the content of digital archives.

Finally, the LOD cloud contains some examples of semantic data from museums. Among others, the British Museum[21] has published its data collection using the CIDOC-CRM ontology to allow data manipulation and reuse, and it provides a SPARQL access point to query the available data. The main difference with respect to this initiative is in the features of the two collections, i.e., a small and not well documented collection in the case of VVV, and a well documented and huge collection in the case of the British Museum. For these reasons, the interlinking of our data with other information sources in the LOD cloud is less efficace than in the case of the British Museum, as discussed in Sect. 4.2.

6   Conclusions

In this work, we described the process to link the metadata of a collection of verbo-visual art to DBpedia, and to enrich such a collection with additional information retrieved from the Web. We showed that the workflow proposed in the past for other important collections was applicable to our case only to a limited extent. In particular, the fact that the collection and the involved artists are in many cases little known affects the coverage of the interlinking and the amount of additional information retrieved from the Web. Besides, other issues

---

[21] http://collection.britishmuseum.org.

related to the management and the curation of the raw data, as well as those related to specific characteristics of VVV sample data, have to be addressed.

In the future, we plan to extend the enrichment process to the whole collection, and to make the data available. Besides, we will enrich the data also with information about the artwork content, since around 50 % of verbo-visual works contain some texts that can be transcribed and analyzed automatically with NLP techniques.

References

1. Oomen, J., Baltussen, L.B., van Erp, M.: Sharing cultural heritage the linked open data way: Why you should sign up. In: Proceedings of Museums and the Web (2012)
2. Marchetti, A., Tonelli, S., Sprugnoli, R.: The verbo-visual virtual platform for digitizing and navigating cultural heritage collections. In: Proceedings of the 2nd Annual Conference on Collaborative Research Practices and Shared Infrastructures for Humanities Computing (AIUCD-2013) (2013)
3. Szekely, P., Knoblock, C.A., Yang, F., Zhu, X., Fink, E.E., Allen, R., Goodlander, G.: Connecting the smithsonian american art museum to the linked data cloud. In: Cimiano, P., Corcho, O., Presutti, V., Hollink, L., Rudolph, S. (eds.) ESWC 2013. LNCS, vol. 7882, pp. 593–607. Springer, Heidelberg (2013)
4. de Boer, V., Wielemaker, J., van Gent, J., Hildebrand, M., Isaac, A., van Ossenbruggen, J., Schreiber, G.: Supporting linked data production for cultural heritage institutes: The Amsterdam museum case study. In: Simperl, E., Cimiano, P., Polleres, A., Corcho, O., Presutti, V. (eds.) ESWC 2012. LNCS, vol. 7295, pp. 733–747. Springer, Heidelberg (2012)
5. de Boer, V., Wielemaker, J., van Gent, J., Oosterbroek, M., Hildebrand, M., Isaac, A., van Ossenbruggen, J., Schreiber, G.: Amsterdam museum linked open data. Semant. Web 4(3), 237–243 (2013)
6. Ferrari, D.: Archivio di Nuova Scrittura Paolo della Grazia. Storia di una Collezione/Geschichte einer Sammlung. Silvana Editoriale, Milan, Italy (2012)
7. Dragoni, M., Bosca, A., Casu, M., Rexha, A.: Modeling, managing, exposing, and linking ontologies with a wiki-based tool. In: Proceedings of the Ninth International Conference on Language Resources and Evaluation (LREC-2014), ELRA, pp. 1668–1675 (2014)
8. Ghidini, C., Rospocher, M., Serafini, L.: Modeling in a wiki with moki: Reference architecture, implementation, and usages. Int. J. Adv. Life Sci. 4, 111–124 (2012)
9. Crofts, N., Doerr, M., Gill, T., Stead, S., Stiff, M. (eds.): Definition of the CIDOC Conceptual Reference Model. ICOM/CIDOC CRM Special Interest Group (2009)
10. Dragoni, M.: Multilingual ontology mapping in practice: a support system for domain experts. In: Arenas, M., et al. (eds.) ISWC 2015. LNCS, vol. 9367, pp. 169–185. Springer, Heidelberg (2015). doi:10.1007/978-3-319-25010-6_10
11. Schmid, H.: Improvements in part-of-speech tagging with an application to German. In: Proceedings of the ACL SIGDAT-Workshop (1995)
12. Hearst, M.A.: Automatic acquisition of hyponyms from large text corpora. In: Proceedings of the 14th Conference on Computational Linguistics, COLING-1992, pp. 539–545. ACL (1992)

13. Hyvönen, E.: Publishing and Using Cultural Heritage Linked Data on the Semantic Web. Synthesis Lectures on the Semantic Web. Morgan & Claypool Publishers, Palo Alto (2012)

14. Benjamins, V.R., Contreras, J., Blázquez, M., Dodero, J.M., García, A., Navas, E., Hernandez, F., Wert, C.: Cultural heritage and the semantic web. In: Bussler, C.J., Davies, J., Fensel, D., Studer, R. (eds.) ESWS 2004. LNCS, vol. 3053, pp. 433–444. Springer, Heidelberg (2004)

15. Dannells, D., Ranta, A., Enache, R., Damova, M., Mateva, M.: Multilingual access to cultural heritage content on the semantic web. In: 7th Workshop on Language Technology for Cultural Heritage, Social Sciences, and Humanities, pp. 107–115 (2013)

16. Agosti, M., Ferro, N.: Annotations: enriching a digital library. In: Koch, T., Sølvberg, I.T. (eds.) ECDL 2003. LNCS, vol. 2769, pp. 88–100. Springer, Heidelberg (2003)

Ontology-Based Data Access for Maritime Security

Stefan Brüggemann[1], Konstantina Bereta[2], Guohui Xiao[3(✉)],
and Manolis Koubarakis[2]

[1] Airbus Defence and Space, Bremen, Germany
Stefan.s.brueggemann@airbus.com
[2] National and Kapodistrian University of Athens, Athens, Greece
konstantina.bereta@di.uoa.gr
[3] Free University of Bozen-Bolzano, Bolzano, Italy
koubarak@di.uoa.gr

Abstract. The maritime security domain is challenged by a number of
data analysis needs focusing on increasing the maritime situation awareness, i.e., detection and analysis of abnormal vessel behaviors and suspicious vessel movements. The need for efficient processing of dynamic
and/or static vessel data that come from different heterogeneous sources
is emerged. In this paper we describe how we address the challenge of
combining and processing real-time and static data from different sources
using ontology-based data access techniques, and we explain how the
application of semantic web technologies increases the value of data and
improves the processing workflow in the maritime domain.

Keywords: Maritime security · EMSec · RMSAS · Geospatial ·
OBDA · *Ontop* · SPARQL · R2RML · OWL

1 Introduction

The maritime security domain is challenged by a number of data analysis needs
with a focus on increasing the maritime situation awareness. Vessel movements
are of major importance for maritime data analysts and decision makers. Abnormal vessel behaviors and suspicious vessel movements need to be detected and
understood to properly increase the maritime domain awareness. The project
EMSec[1] (real-time services for the maritime security) has the aim to support
the maritime security by improving the availability and accessibility of relevant data and information ashore and offshore. The central data management
component of EMSec is the "Real-time Maritime Situation Awareness System"
(RMSAS), which is in charge of integrating various types of data from different
sources.

This paper focuses on the integration and analysis of data for vessel movements in RMSAS. For the proper analysis a storage capability is needed to

---

[1] http://www.emaritime.de/projects/emsec/.

© Springer International Publishing Switzerland 2016
H. Sack et al. (Eds.): ESWC 2016, LNCS 9678, pp. 741–757, 2016.
DOI: 10.1007/978-3-319-34129-3_45

properly analyze vessel data and to identify vessel trajectories and their stops and movements for the last days. However, the vessel data integration and management task in RMSAS is challenged by the following requirements: *(a)* The vessel data is heterogeneous and in particular consists of dynamic position data or static metadata. *(b)* There is a need for integrating third party data, i.e., open data like GeoNames and OpenStreetMap. *(c)* The size of the data is large, deriving from the acquisition and processing of large radar and satellite images. *(d)* The data about vessels are produced in real-time, i.e., approximately 1000 vessel positions are acquired per second.

The motivation of this work was to address the above needs for combining data from heterogeneous sources, such as in-situ data, AIS data, and open data developing an automated solution avoiding manual work as much as possible. We considered that the conceptualized model offered by ontologies would meet the requirements of that purpose, so we used state-of-the-art technologies and tools into this direction. The rationale behind our design choices was to avoid the creation of replicas of the same data in other formats and also to avoid storing natively data that are already available as open data. For the first, we opted for a well-known ontology-based data access (OBDA) system, *Ontop*, instead of using a triple store, to avoid the cost (in disk space and response time) of materializing our frequently updated data to RDF and storing them natively. For the second, we used federation to query data coming from different endpoints (e.g., in-house data and linked data like geonames). In the same direction, we used the web-based tool Sextant to visualize geospatial data coming from different SPARQL endpoints (*Ontop* endpoints, triple stores, etc.) creating composite maps, instead of storing everything natively in a geospatial relational database and visualizing them using GIS tools.

The RMSAS system has been implemented on top of these state-of-the-art Semantic Web techniques and tools. The evaluation has shown that RMSAS eases the data analysis by using virtual triples and standardized vocabularies. Next, the integration of several heterogeneous data sources is a benefit for maritime decision makers and the maritime security. Finally, the approach contributes significantly to the detection of routine traffic and abnormal vessel behavior.

The rest of the paper is structured as follows: Sect. 2 describes the data sources that are available in the German maritime research project EMSec and focus on the data integration requirements. Section 3 defines a maritime domain ontology. Section 4 shows how to properly analyze these data using semantic technologies, focusing on how the concept of ontology based data access can be used to add a level of semantics on these data and how to access and spatially analyze relational data. Section 5 shows how OBDA approach can add a value to the analysis of vessel movements. Section 6 concludes the paper with a summary and discussion.

2 The Maritime Context

The maritime security is currently challenged by several influences: More than two-thirds of the overall volume of cargo worldwide is transported seaborne. This massively increases the number of ships traveling on the seas. Next, the continuously increasing number of offshore wind parks has an impact on the security of the citizens. The energy supply must be ensured even without fossil fuels. This turns offshore wind park into assets with a strong demand for protection. Moreover, industrial nations worldwide use the potential of the seas, but are threatened by pirates and terrorists. Finally, besides the danger of criminal actions, disasters and storm floods also challenge the maritime security [4].

In Germany, it is the task of several institutions from the federal government and the federal states to ensure the maritime security and to mitigate risks. For that reason, several detailed information are needed to effectively gain maritime domain awareness [11] and to analyze maritime emergency situations. Data from different data sources are needed to support these tasks and to create a common overview of the maritime situation.

The German federal ministry of education and research funded project EMSec (real-time services for the maritime security) is aiming at providing a consistent and user-oriented access to data and information from different data sources. These data can be satellite images, aerial images, weather information like wind, rain, drift, or else. All these data shall be displayed in a flexible manner to ensure the situation awareness of the end users, combining data from several sources. A faster and more detailed provision of these data shall enable responsible organizations and decision makers to early recognize and avoid critical situations. In an emergency situation or within a criminal activity, action forces can benefit from accessing detailed information in real time to handle the situation efficiently.

2.1 RMSAS: Real-Time Maritime Situation Awareness System

As previously mentioned, a system is needed inside EMSec that is capable of integrating data coming from several data sources. Since the concrete information needs of the user are not known and are highly situation-dependent, a flexible system and an agile iterative design approach is needed. A distributed federated system is needed to face the diversity of the maritime players and governmental constraints like laws, IT-security, or else. A system of systems approach has been identified as being able to cope with these challenges in the maritime domain [12]. Consequently, with RMSAS, a real-time maritime situation awareness system is implemented in EMSec as a system of systems to:

- integrate vessel data coming from various sensors,
- enrich these data with data from other sources (e.g. open data),
- harmonize these data using established maritime standards,
- retrieve new information from these integrated data,
- infer knowledge from this information,

- retrieve and deliver this knowledge in near real-time,
- create a maritime domain awareness for the end user, and
- enable maritime decision makers to handle maritime situations more efficiently.

A service oriented architecture using semantic web technologies has been evaluated to support these tasks. The approach presented in [12] will be used and extended to challenge IT-security constraints.

Situation-aware data shall be presented to the user in near real-time. To achieve this, data have to be integrated and consolidated from several different sources. This allows for properly displaying the combination of this data to the user via an application. Providing data faster and in further detail shall allow the involved parties to identify critical situations better and earlier, to avoid these situations, and to manage them efficiently.

Being the central data management component in EMSec, the RMSAS aims at integrating and consolidating data. RMSAS uses the "System of Systems" approach and implements a federated information system based on separate services (SOA). Data are integrated in RMSAS in near real-time, next they are consolidated based on semantic data models and techniques and provided to the end user as information products. Ontologies are used in the consolidation of these heterogeneous data.

2.2   Data Sources

In the following we describe some typical data sources that are used in EMSec. The Automatic Identification System (AIS)[2] is a common data source for the maritime navigation used worldwide. These data serve as a basis for several maritime applications and also serve as reference data in EMSec. AIS is a cooperative system and calls for the active participation of every vessel. However, it has only a limited trustworthiness due to the fact that every vessel owner can manipulate the system or can completely switch it off. Transferring and receiving the data can also be manipulated or hindered. Terrestrial AIS uses coast based receiver with good availability but with limited range and coverage. Satellite based AIS is also available, but with limited good reliability these days. EMSec analyzes the benefits of additional data sources that are based on Earth Observation from satellites and airborne systems. Here it is important to extend the spatial and temporal resolution of maritime data to add a value to the maritime security.

The EMSec partners provide the following data for the integration in RMSAS:

AIS. AIS messages are not only used as reference data, they are additionally used for quality inspections and – wherever possible – analyzed to identify certain movement pattern, for instance for ferries. Several AIS types are available:

---

[2] https://de.wikipedia.org/wiki/Automatic_Identification_System.

Terrestric AIS, satellite AIS[3], and the AIS signal that comes from the Columbus-module of the ISS[4]. In EMSec we receive AIS data about 800–1000 vessels in the German bay every 1–3 s.

Satellite SAR. TerraSAR-X provides satellite-based synthetic aperture radar (SAR) and creates radar images with a high resolution. Algorithms can be used to detect objects (e.g., vessels) and to link these detected objects with previously collected AIS messages. The radar images can also be analyzed to extract wind and wave information and connect them with conventional secondary weather information.

Airborne Systems. The EMSec consortium utilizes an airplane that comes with an AIS receiver and a radar system. The AIS messages are used as described before. The radar system provides objects and their movements as plots and tracks. Next, another airborne system provides optical images that are used to detect vessels in these images. RMSAS is capable of providing these object detections together with weather information and geospatial information to the end user applications.

2.3 Request Management in EMSec

This section describes the concept of managing and answering requests in EMSec. Figure 1 describes that these requests may come from a user, a SOA-architecture or else. The main concept is that requests are formulated using the Top level ontology (TLO) and are posed using SPARQL. This enables the end user to use the described high level semantics of the TLO. The semantic data processing component utilizes *Ontop* to translate the queries to SQL queries in order to be evaluated in the underlying RDBMS, e.g., a PostgreSQL database. This paper focuses on relational data sources, so other input data formats such as CSV will not be discussed here.

2.4 Scenarios in EMSec

The validation of the created methods, architectures, algorithms and concepts will be done in a campaign, where two maritime security scenarios are executed. First, a concrete satellite mission is utilized. Second, both airborne missions are requested and executed. The generated data are transferred to RMSAS in near real time and integrated, analyzed, consolidated and finally transferred to the user. Several maritime regions are deserving protection. Restricted areas can be off-shore platforms, wind parks, or preserved areas. These call for limited vessel traffic with certain restrictions. *Geographic fences* can be created to analyze the vessel traffic focusing in specific areas of interest. Possible scenarios are to check

---

[3] http://www.esa.int/Our_Activities/Space_Engineering_Technology/
ESA_satellite_receiver_brings_worldwide_sea_traffic_tracking_within_reach.

[4] http://www.esa.int/Our_Activities/Space_Engineering_Technology/
Space_Station_keeps_watch_on_world_s_sea_traffic.

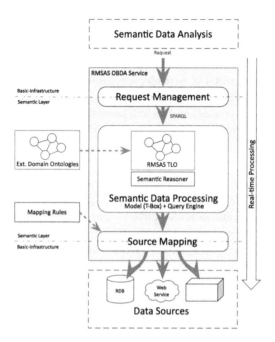

Fig. 1. Request management within EMSec

that the speed over ground is within a limited range in these regions, that certain vessel types like oil-tanker may not pass these regions, or that under certain sea conditions no vessel traffic is allowed.

2.5   Categorization of Data

RMSAS processes several types of data. These data can be categorized as follows:

Data Streams. Data streams are data which are created continuously in real-time. These data streams can be AIS-data that continuously report new arriving vessels in the German bay.

Static Data. Data are static when they are stored in databases, on FTP-servers, or in external systems. These can be metadata about vessels as for example the vessel type, cargo, port of departure, and historical data about previous routes. After transmission to the earth, satellite data are made available as packages.

Open Data. Open data are data coming from the linked open data cloud or from other external data sources. Using these data can improve certain kinds of analysis. For example, these data sources contain information about real existing harbors (as in GeoNames[5]), or information about certain points of interest

---

[5] http://geonames.org.

(as in DBpedia[6] or OpenStreetMap[7]), or that contain weather data (as in Open-WeatherMap[8]).

GeoNames is a gazetteer that collects both spatial and thematic information for various place names around the world. GeoNames data is available through various Web services but it is also published as linked data. The features in GeoNames are interlinked with each other defining regions that are inside the underlined feature (children), neighboring countries (neighbors) or features that have certain distance with the underlined feature (nearby features).

OpenStreetMap (OSM) maintains a global editable map that depends on users to provide the information needed for its improvement and evolution. OpenStreetMap datasets are available in RDF format from the LinkedGeoData project[9]. However, it was more convenient for us to download the most up-to-date original OpenStreetMap data about Bremen, available as Shapefiles[10]. We imported the Shapefiles into a PostGIS database and created virtual geospatial RDF views on top of them using *Ontop*-spatial, as described at https://github.com/ConstantB/ontop-spatial/wiki/Shapefiles.

3 Ontologies for the Maritime Domain

In this section we present the central knowledge representation and data management in maritime component of RMSAS. It includes a model of the maritime domain, the logical data model and the ontology. The last part of the section is devoted to interlinking the RMSAS data with the linked open data (LOD) cloud.

3.1 Modeling the Maritime Domain

Maritime domain models are results of several research projects, both national and international. The CoopP-project has created the CISE-ontology [1], which is reused in RMSAS and adopted to meet the project's specific requirements.

Object. Objects can be any involved parts of the maritime domain. They can be physical elements that are airborne, onshore and offshore, such as vessels, containers, planes, icebergs, or satellites. Vessels are central elements of interest and modeled in greatest detail with a special focus on the information that are available in AIS.

Geometry. Geometry is dedicated to deal with information about space and geographical localizations of the maritime objects. The geometries contained in our data are encoded in WKT format, which is an OGC standard for the serialization of geometries. The geometries encountered in our dataset are mainly polygons and

---

[6] http://dbpedia.org.
[7] http://openstreetmap.org/.
[8] http://openweathermap.org/.
[9] http://linkedgeodata.org.
[10] http://download.geofabrik.de/europe/germany/bremen.html.

points. The geometries of areas, for example, are represented as polygons. These areas can be marked regions ashore, for instance. Dimension describes the specifics of an object like length, width, or height. A location describes places with a geographical name like cities or harbours. They can be identified using a URI which makes it possible to interlink them with external sources like GeoNames or DBPedia. Movements describe the track of an object including its course and speed over ground and optionally its rate of turn. Points describe a dedicated geographical point described using its geographical coordinates and its height. A position then is a point combined with a timestamp.

Time. Time is used to describe timestamps that can be used to model positions of objects, to label data during data integration and to support temporal data analysis.

3.2   The RMSAS Movement Ontology

In order to model our data, we have constructed an ontology that is shown in Fig. 2. In this paper we focus on the aspects of vessel movements and trajectories.

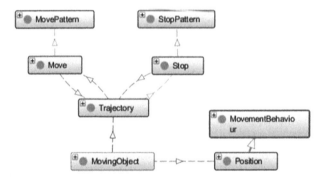

Fig. 2. RMSAS movement ontology

The movement ontology defines the necessary structures for modeling object movements like vessel, satellites or aircrafts. The ontology allows for enriching native position data with semantics. This allows to model vessel positions as being moves or stops. Any moving object has position data and consists of trajectories that reflect the historic positions of an object. The use of semantics to these positions facilitates the monitoring of the status of the moving object, i.e. whether it has stopped or was moving.

4   Semantic Data Analysis

In this section we describe how RMSAS uses the Semantic Web technologies mentioned in the introduction in order to achieve the following goals:

- Transparent integration of different, geospatial and thematic data sources using ontologies.
- Processing of in-house dynamic and static data, enriching them with information already available on the web (linked open data).
- Avoid replicating the same data as much as possible (e.g., materializing data to RDF, storing data from scratch when a SPARQL endpoint for them is already available) using OBDA techniques and federation.
- Visualization of the data and creation of persistent, web accessible maps, with no need to load the datasets or issue the queries again every time we want to populate the existing databases/endpoints with fresh data.

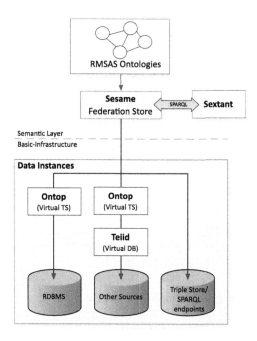

Fig. 3. Abstract architecture of RMSAS

We illustrate the abstract architecture of RMSAS in Fig. 3. RMSAS uses the OBDA system *Ontop* and *Ontop-spatial* to expose the data we need from the relational databases as SPARQL endpoints. For accessing non-relational data sources, RMSAS first wraps these sources into relational ones by Teiid, and then uses *Ontop* [5,8] to access them. For federating third party SPARQL endpoints like GeoNames, Sesame is used for the SPARQL 1.1 federated query answering. Finally, Sextant is used for visualizing the results on temporally-enabled maps combining geospatial and temporal results from different (Geo)-SPARQL endpoints.

4.1   Linking RMSAS Data to the RMSAS Ontology

The relational data in RMSAS can be faithfully mapped to the ontology using the ontology-based data access (OBDA) approach. We use *Ontop* and the its extension *Ontop-spatial* for this purpose. As illustrated in Fig. 3, *Ontop* allows for querying relational data sources through a conceptual representation of the domain of interest, provided in terms of an ontology, to which the data sources are mapped. *Ontop* answers the SPARQL queries by translating them into SQL queries over the database and avoids materializing triples. *Ontop*-spatial is an extension of *Ontop* with geospatial features.

```
mappingId  Vessel
target     :Vessel-{v.id} a :Vessel ; :hasName {v.name} .
source     SELECT v.id, v.name FROM Vessel v

mappingId  VesselPosition
target     :Vessel-{v.id} :hasLocation :Position-{vp.position_id}.
source     SELECT v.id, vp.position_id FROM Vessel v,
                         Vessel_position vp WHERE v.id=vp.vessel_id

mappingId  Position
target     :Position-{id} a :Position ; :hasLatitude {latitude} ;
             :hasLongitude {longitude} ; :hasDateTime {ts} ;
             :hasGeometry geos:Geometry-{id} .
source     SELECT id, latitude, longitude, ts FROM position

mappingId  Geometry
target     geos:Geometry-{id} a geos:Geometry ;
                         geos:asWKT {geom}^^geos:wktLiteral .
source     SELECT id, geom FROM position
```

Fig. 4. Example mappings in RMSAS

Ontop uses declarative mappings to encode how relational data are mapped to the respective RDF terms. *Ontop* supports W3C R2RML mapping language [6] and its native *Ontop* mapping languages. In this paper, we use the native syntax because it is more compact. An *Ontop* mapping consists of three fields: *mappingId*, *source* and *target*. The *mappingId* is an identifier for mapping; the *source* is an arbitrary SQL query over the database; and the *target* is a triple template written in Turtle syntax that contains placeholders referencing column names mentioned in the source query.

For example, all information about the positions of vessels are stored in a spatially enabled PostGIS database. In Fig. 4, we present mappings related to vessels in *Ontop* native syntax. The coordinates of the mappings that are stored in the respective columns named `longitude` and `latitude` in the database in textual form are mapped into RDF literals, as objects of the respective virtual triples as indicated in the mapping assertion with mappingId "Position". The respective geometries that represent the vessels positions are also stored in the well-known binary format (WKB) in a separate column, named **geom**. The mapping assertion **Geometry** indicates how this information is mapped to RDF: The binary geometry of the database is exported as a well-known text literal (WKT), following the OGC GeoSPARQL standard [2].

4.2 SPARQL Queries

In the following we present two example SPARQL queries that we used in order to process our data using OBDA technologies and combine them with other sources.

```
PREFIX : <http://www.rmsas.de/DMARitime#>
PREFIX geos: <http://www.opengis.net/ont/geosparql#>
SELECT DISTINCT ?x ?z ?g ?timestamp
WHERE {
  ?x rdf:type :Vessel.  ?x :hasName "Vesselname"^^xsd:string.
  ?x :hasLocation ?z. ?z :hasDateTime ?timestamp .
  ?z geos:asWKT ?g. }
ORDER BY DESC(?timestamp)
```

Fig. 5. SPARQL query retrieving positions of a vessel through time

The query described in Fig. 5 retrieves geometries of the locations of vessels (ordered by the timestamps) that are stored in binary (WKB) format in the relational database. Objects of this datatype are internally handled by *Ontop-spatial* and are eventually transformed into RDF literals of WKT datatype, as specified the OGC standard GeoSPARQL and indicated by the mappings that we presented in the previous section. This is the template of the queries we posed to retrieve the locations of ferries to three German islands (Langeoog, Spiekeroog, and Wangerooge). Figure 8(a) presents the visualization of results using Sextant.

```
PREFIX geof: <http://www.opengis.net/def/function/geosparql/>
PREFIX osm: <http://linkedgeodata.org/ontology#>
PREFIX geo: <http://www.opengis.net/ont/geosparql#>
   SELECT DISTINCT ?lu ?geo WHERE {
     ?x osm:landUse lgd:port . ?x geo:asWKT ?geo .
     ?x1 geo:asWKT ?geo1 . ?x1 osm:landUse ?lu .
     FILTER (geof:sfIntersects(?geo,?geo1)) }
```

Fig. 6. SPARQL query retrieving locations of ports and land use of intersecting areas

The query described in Fig. 6 retrieves the geometries that represent the locations of ports and the land use of areas that they intersect with (e.g., farmyards, commercial/religious areas).

4.3 SPARQL Federation

For federating third party SPARQL endpoints like GeoNames, RMSAS relies on the SPARQL 1.1 federated query [10] implemented in Sesame [3]. In the query described in Fig. 7, we use "SERVICE" function in order to combine information coming from different endpoints exposed by *Ontop*. The first endpoint (Position-Store) contains dynamic data about the locations of vessels stored in a PostGIS database. The second endpoint (ObjectStore) contains static metadata about vessels, such as dimensions, name, etc. The query retrieves all available information about a specific vessels combining both *Ontop* endpoints in a federated store.

```
PREFIX : <http://www.rmsas.de/DMARitime#>
PREFIX geos: <http://www.opengis.net/ont/geosparql#>
SELECT ?vessel ?location ?geometry ?wkt ?mmsi ?length ?height
WHERE {
  SERVICE <http://www.rmsas.de/openrdf-sesame/PositionStore> {
    ?vessel rdf:type :Vessel .
    ?vessel :hasLocation ?location. ?vessel :hasName "388328333".
    ?location :hasGeometry ?geometry. ?geometry geos:asWKT ?wkt.
    OPTIONAL {
      SERVICE <http://www.rmsas.de/openrdf-sesame/ObjectStore> {
        ?vessel :hasMMSI ?mmsi ; :hasName "388328333".
        ?vessel :hasLength ?length ; :hasHeight ?height .}}}}
```

Fig. 7. SPARQL federation: finding locations of a vessel and their static metadata

4.4 Visualization of Results

For the visualization of the geospatial results presented above, we used the tool Sextant [9][11], which is a web based and mobile ready platform for visualizing, exploring and interacting with linked geospatial data. Sextant is mainly used to create thematic maps by combining geospatial and temporal information that exists in a number of heterogeneous data sources ranging from standard SPARQL endpoints, to SPARQL endpoints following the OGC standard GeoSPARQL, or well-adopted geospatial file formats, like KML, GML and GeoTIFF.

More specifically, we use the capabilities of Sextant to issue queries to remote GeoSPARQL endpoints and project geometries that are included in the result set on a map. Every layer on that map corresponds to results (i.e., geometries) retrieved from a SPARQL or GeoSPARQL query. By this way, we can combine and visualize different geospatial sources. Another Sextant capability that is useful in this use-case is the *timeline* capability. As the location of vessels is associated with a timestamp, the results of a GeoSPARQL query that retrieves

---

[11] http://sextant.di.uoa.gr/.

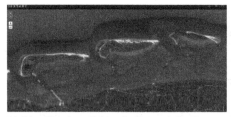 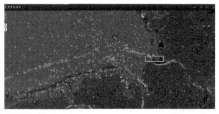

(a) Identification of three ferries to three German islands, Langeoog, Spiekeroog, Wangerooge.

(b) The red polygon represents a geofence that observes all incoming and leaving vessels via the river Elbe

Fig. 8. Utilization of Sextant to display WKT-data that is made available by Ontop-spatial

both the geo-location of the vessel and the timestamp can be visualized in both the map and the timeline respectively; as the user scrolls the timeline band of Sextant, they can see where the vessel was at that time. The temporal features of Sextant are described at [9].

(a) Displaying OSM information about ports

(b) The position of all vessels in the German bay at a certain time point

Fig. 9. Ports and vessels visualized in Sextant

As *Ontop* can be used as a standard SPARQL endpoint, thus, *Ontop-spatial* can be used as a GeoSPARQL endpoint, so we used *Ontop-spatial* endpoints as some of the source endpoints of Sextant. Then, we posed geospatial queries like the ones described in the previous section and the geometries that were included in the result set of each query were displayed on the map, creating one layer for each one of the geospatial queries posed. Screen shots of the query results visualized using Sextant are provided in Figs. 8, 9 and 10.

5   Evaluation

In the context of the project EMSec we have developed an approach for integrating data that comes from various data sources in the RMSAS system. The

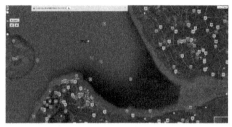

(a) Visualization of further information from GeoNames and DBpedia.

(b) Identifying vessels nearby a certain point of interest

Fig. 10. GeoNames, DBpedia, and GeoSPARQL are used with Sextant

main focus in this work was on analyzing data about vessels. The benefits of the approach that we presented in this paper are explained below.

Improved Data Analysis Using Virtual Triples. The data given in this project mainly exists in databases and data streams and is modeled with respect to different data models. The use of OBDA techniques facilitates the process of data analysis as these data are mapped to the ontology that has been created for RMSAS. This allows decision makers to formulate queries against a standardized ontology instead of articulating different queries in different languages against different data sources like the ones described in Sect. 2.2.

Benefits of Data Integration for Maritime Decision Makers. Compared to the old workflow with respect to information exchange and integration that was identified in the beginning of the EMSec project, where maritime staff had to exchange data often in very traditional ways like email, USB-sticks, mail, paper, or else, the current workflow is significantly improved. With the presented technologies in place, maritime decision makers have all the desired information at hand in near real-time, integrated from different data sources. This increasing having an overview on the maritime security and having a better maritime situational awareness.

Detection of Routine Traffic and Abnormal Vessel Behavior. In the process on data analysis, SPARQL queries and SWRL rules [7] have been used as a good means (w.r.t. expressivity and efficiency) to detect routine traffic and abnormal vessel behavior. Since we cannot display these rules here for confidentiality reasons, we can state that vessel movements can be easily classified using the movement ontologies that were introduces in Sect. 3.2 and that vessel behavior can be classified using the introduced movement pattern. Having combined this with the OBDA approach and with the utilization of (Geo-)SPARQL functionalities, this has strong benefits regarding the detection of routine traffic and abnormal vessel behavior.

6   Summary, Lessons Learned and Future Work

6.1   Summary

In this paper we described challenges for the maritime security and how a German project named EMSec addresses these challenges by introducing a system called RMSAS. This system has been developed to incorporate data from different data sources and in different formats. An ontology was introduced to model the maritime domain and to provide a common view on the different data sources. The concept of ontology based data access was used to map the original data to this ontology. This allows maritime decision makers to efficiently pose queries against the high level ontology. We have then shown how these queries are articulated by the *Ontop* framework and how they are translated to SPARQL.

We have further introduced the movement ontology as a concrete use case on identifying and analyzing vessel movements and detecting abnormal vessel behavior. Open geospatial data like OpenStreetMap and Geonames have been used to put vessel data and information regarding ports in relation to the publicly available data. This allowed for a comparison of the data acquired about vessels with the data that have been publicly created and reviewed.

6.2   Lessons Learned

Looking back the design and implementation of RMSAS which heavily replies on OBDA techniques for *virtual* data integration, it is clearly that these new techniques introduced some learning curve for developers. However, we emphasize that this curve would be bigger if we had to use ETL tools for converting and storing RDF data or if we did this without Semantic Web technologies since relational databases do not offer the conceptual model for data integration that RDF/OWL could offer.

Regarding the new languages and tools in RMSAS, at this moment, mapping designers are comfortable with the *Ontop* mapping language since it is easy to understand and write. Meanwhile a more friendly GUI tool for assisting mapping construction would still be helpful for improving productivity. *Ontop* Sesame workbench comes with a basic and simple interface for configuring SPARQL endpoints; however it currently lacks functionalities like version control or more fine-grained configurations.

Another limitation of the Semantic Web technologies that were used in this use case was the lack of geospatial federation support. Although *Ontop-spatial* supports spatial filters in queries, federated queries that perform spatial joins spanning different geospatial endpoints are not supported in any federated system to the best of our knowledge. Sextant is a user-friendly tool which provides another approach of data integrating by visualization of open linked geospatial data and its comparison with in-situ data. We would also like to be able to pose federated geospatial queries and project their results in Sextant but due to the reasons described above, this is not supported yet.

6.3   Future Work

The RMSAS system is already scheduled to be deployed this autumn outside of the lab according to the project management and there is no technical barrier for the deployment. Future work will focus on clustering abnormal vessel behavior and the creation of early pre-warning systems. This would allow to concentrate the situational awareness to potentially conspicuous vessels. Another field of future work is to manage uncertainties that are due to missing or inconsistent data and where rules like SWRL rules may fail. We will also carry out an extensive performance evaluation.

Acknowledgement. Brüggemann is supported by the German BMBF project EMSec; Bereta and Xiao are supported by the EU under the large-scale integrating project (IP) Optique (Scalable End-user Access to Big Data), grant agreement n. FP7-318338.

References

1. Towards the integration of maritime surveillance: a common information sharing environment for the EU maritime domain. Technical report (2009). http://eur-lex.europa.eu/legal-content/EN/TXT/HTML/?uri=URISERV:pe0011&from=EN
2. Open Geospatial Consortium: OGC GeoSPARQL - A geographic query language for RDF data. OGC Candidate Implementation Standard, February 2012. http://www.opengeospatial.org/standards/geosparql
3. Broekstra, J., Kampman, A., van Harmelen, F.: Sesame: a generic architecture for storing and querying RDF and RDF schema. In: Horrocks, I., Hendler, J. (eds.) ISWC 2002. LNCS, vol. 2342, pp. 54–68. Springer, Heidelberg (2002)
4. Brueggemann, S., Foerster, S.: User requirements for real-time services for the maritime security. In: Proceedings of ISIS 2014: International Symposium on Information on Ships (2014)
5. Calvanese, D., Cogrel, B., Komla-Ebri, S., Kontchakov, R., Lanti, D., Rezk, M., Rodriguez-Muro, M., Xiao, G.: Ontop: answering SPARQL queries over relational databases. Semant. Web J. (2016, to appear)
6. Das, S., Sundara, S., Cyganiak, R.: R2RML: RDB to RDF mapping language. W3C Recommendation, World Wide Web Consortium (2012). http://www.w3.org/TR/r2rml/
7. Horrocks, I., Patel-Schneider, P., Boley, H., Tabet, S., Grosof, B., Dean, M.: SWRL: a semantic web rule language combining OWL and RuleML. W3C Member Submission, World Wide Web Consortium (2004)
8. Kontchakov, R., Rezk, M., Rodríguez-Muro, M., Xiao, G., Zakharyaschev, M.: Answering SPARQL queries over databases under OWL 2 QL entailment regime. In: Mika, P., et al. (eds.) ISWC 2014, Part I. LNCS, vol. 8796, pp. 552–567. Springer, Heidelberg (2014)
9. Nikolaou, C., Dogani, K., Bereta, K., Garbis, G., Karpathiotakis, M., Kyzirakos, K., Koubarakis, M.: Sextant: visualizing time-evolving linked geospatial data. Web Semant. Sci. Serv. Agents World Wide Web **35**, Part 1:35–52 (2015). Geospatial Semantics

10. Prud'hommeaux, E., Buil-Aranda, C.: SPARQL 1.1 Federated query. W3C Recommendation, World Wide Web Consortium, March 2013. www.w3.org/TR/sparql11-federated-query/
11. Vance, G., Vicente, P.: Maritime domain awareness. Proc. Mar. Saf. Secur. Counc. **63**(3), 6–8 (2006)
12. Willems, N., Robert, W., Hage, V., Vries, G.D.: An integrated approach for visual analysis of a multi-source moving objects knowledge base. Int. J. Geogr. Inf. Sci. **24**(9), 1–16 (2010)

WarSampo Data Service and Semantic Portal for Publishing Linked Open Data About the Second World War History

Eero Hyvönen$^{(\boxtimes)}$, Erkki Heino, Petri Leskinen, Esko Ikkala, Mikko Koho, Minna Tamper, Jouni Tuominen, and Eetu Mäkelä

Semantic Computing Research Group (SeCo), Aalto University, Espoo, Finland
{eero.hyvonen,erkki.heino,petri.leskinen,esko.ikkala,mikko.koho,
minna.tamper,jouni.tuominen,eetu.makela}@aalto.fi
http://seco.cs.aalto.fi/

Abstract. This paper presents the WarSampo system for publishing collections of heterogeneous, distributed data about the Second World War on the Semantic Web. WarSampo is based on harmonizing massive datasets using event-based modeling, which makes it possible to enrich datasets semantically with each others' contents. WarSampo has two components: First, a Linked Open Data (LOD) service WarSampo Data for Digital Humanities (DH) research and for creating applications related to war history. Second, a semantic WarSampo Portal has been created to test and demonstrate the usability of the data service. The WarSampo Portal allows both historians and laymen to study war history and destinies of their family members in the war from different interlinked perspectives. Published in November 2015, the WarSampo Portal had some 20,000 distinct visitors during the first three days, showing that the public has a great interest in this kind of applications.

1 Motivation: Second World War on the Semantic Web

Many websites publish information about the Second World War (WW2), the largest global tragedy in human history[1]. Such information is of great interest not only to historians but to potentially hundreds of millions of citizens globally whose relatives participated in the war actions, creating a shared trauma all over the world. However, WW2 information on the web is typically meant for human consumption only, and there are hardly any web sites that serve *machine-readable data* about the WW2 for digital humanists [3,5] and end-user applications to use. It is our belief that by making war data more accessible our understanding of the reality of the war improves, which not only advances understanding of the past but also promotes peace in the future.

The goal of this paper therefore is to (1) initiate and foster large scale LOD publication of WW2 data from distributed, heterogeneous data silos and (2) demonstrate and suggest its use in applications and research. We introduce the

---

[1] http://ww2db.com, http://www.world-war-2.info, Wikipedia, etc.

© Springer International Publishing Switzerland 2016
H. Sack et al. (Eds.): ESWC 2016, LNCS 9678, pp. 758–773, 2016.
DOI: 10.1007/978-3-319-34129-3_46

LOD service WarSampo Data[2] and the semantic WarSampo Portal[3] on top of it. WarSampo is to our best knowledge the first large scale system for serving and publishing WW2 LOD on the Semantic Web.

World war history makes a promising use case for Linked Data (LD) because war data is by nature heterogeneous, distributed in different countries and organizations, and written in different languages. WarSampo is based on the idea of creating a shared, open semantic data repository with a sustainable "business model" where everybody wins [8]: When an organization contributes to the WW2 LOD cloud with a piece of information, say a photograph, its description is automatically connected to related data, such as persons or places depicted. At the same time, the related pieces of information, provided by others, are enriched with links to the new data.

In the following, we first present the WarSampo Data service, and then the WarSampo Portal with six different application perspectives enriching each other via data linking and shared addressing practices. In conclusion, contributions of the system are summarized and related work discussed.

2   WarSampo Datasets, Conceptual Model, and Data Service

Datasets. The WarSampo Data Service contains datasets related to the Finnish Winter War 1939–1940 against the Soviet attack, the Continuation War 1941–1944, where the occupied areas of the Winter War were temporarily regained by the Finns, and the Lapland War 1944–1945, where the Finns pushed the Germans out of Lapland. The datasets in use are presented in Table 1. The casualties data (1) includes data about the deaths in action during the wars. War diaries (2) are digitized authentic documentations of the troop actions in the frontiers. Photos and films (3) were taken during the war by the troops of the Defense Forces. The Kansa Taisteli magazine (4) was published in 1957–1986; its articles contain mostly memoirs of the men that fought on the fronts. Karelian places (5) and maps (6) cover the war zone area in pre-war Finland that was ultimately annexed by the Soviet Union. Senate atlas (7) contains historical maps of Southern Finland, and the municipalities data (8) contains the Finnish municipalities that existed during the wartime. Organization cards (9), written after the war, document events of military units during the war. National Biography (10) contains over 6,300 biographies of Finnish national figures. In WarSampo the data related to 500 persons active during the war is utilized. Data about wartime events (11), persons (12), and army units (13) were collected from various war history text books. The RDF data in WarSampo contains at the moment 7,176,900 triples.

Table 1. Central datasets of WarSampo.

#	Name	Providing organization	Size
1	Casualties of WW2	National Archives	94,700 death records
2	War diaries	National Archives	13,000 war diaries of troops
3	Photos & films	Defence Forces	160,000 photos & films
4	Kansa Taisteli magazine articles	The Assoc. for Military History in Finland & Bonnier	3,400 articles of veteran soldiers
5	Karelian places	Jyrki Tiittanen / National Land Survey	32,400 places of the annexed Karelia
6	Karelian maps	National Land Survey	47 wartime maps of Karelia
7	Senate atlas	National Archives	404 historical maps of Finland
8	Municipalities	National Archives	625 wartime municipalities
9	Organization cards	National Archives	ca 500 army units & ca 300 persons & 642 battles
10	National Biography	Finnish Literature Society	ca 500 biographies of wartime persons
11	Wartime events	War history books	1,000 events
12	Persons	War history books, Wikipedia	2,600 persons
13	Army units	War history books	3,200 army units

Conceptual Framework and Model. Since wars are essentially sequences of events, an obvious framework for representing them is event-based modeling. There are many approaches available for this, such as Event Ontology[4], LODE[5], SEM[6], and CIDOC CRM[7] [4]. CIDOC CRM was selected as a commonly used ISO standard (21127:2014). Another reason for the selection was that this conceptual framework is not limited to modeling events only, but can be used for modeling other WarSampo contents as well, such as war diaries, magazine articles, casualty records, and photos.

The core classes used in our event model is represented in Fig. 1 where namespaces crm, dc, and skos refer to CIDOC CRM, Dublin Core, and SKOS standards, respectively. Events are characterized by actors, places, and times that are represented by corresponding CIDOC CRM classes: Actors (crm:E39_Actor) are either persons (crm:E21_Person) or groups (crm:E74_Group). Persons are characterized by the following event types: birth, death, military rank promotion, and getting a medal of honor. Groups have subclasses of military units that may be involved in events where a unit is formed, the unit is renamed, the unit is joined with other units, and a person is joining the unit. There are currently 327,200 events in WarSampo. For Places, the Hipla.fi ontology of Karelian places and historical maps [11] is used, and for times CIDOC CRM time spans. Metadata about documentary objects, such as war diaries, magazine articles,

---

[4] http://motools.sourceforge.net/event/event.html.
[5] http://linkedevents.org/ontology/.
[6] http://semanticweb.cs.vu.nl/2009/11/sem/.
[7] http://cidoc-crm.org.

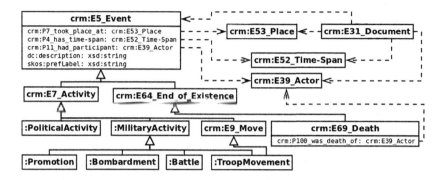

Fig. 1. Core classes of CIDOC CRM used in WarSampo.

casuality records, and photos is represented as instances of crm:E31_Document. For subject matter, the comprehensive Finnish KOKO ontology[8] of over 47,000 keyword concepts is used. Documentation about the data and metadata schemas used are available at the data service homepage[9].

Data Service. WarSampo Data is available as mutually linked open datasets. The data is provided using the "7-star" LD model [10], where the first five stars are equal to the traditional LD 5-star model [6], the 6th star is credited if the data is provided with an explicit schema, and the 7th star if the data has been validated against the schema. WarSampo was given six stars. The idea of the extra stars is to foster reuse of the data. In addition to traditional linked data services, i.e., full dataset download, URI redirection, linked data browsing, and SPARQL querying, the WarSampo Data Service provides the user with a variety of other services for data production, editing, documentation, validation, and visualization available at the hosting Linked Data Finland platform[10] [10]. The service is based on Fuseki[11] with a Varnish Cache[12] front end for serving LOD.

In contrast to the generic LOD Cloud[13], the WarSampo data cloud has a particular application domain in focus. A larger vision behind our work is that by publishing openly shared ontologies and data about WW2 for everybody to use in annotations, future interoperability problems can be prevented before they arise [7].

3   WarSampo Portal

Providing Interlinked Perspectives of War. The WarSampo Portal is not just one application, but a collection of six interlinked applications, and more are

---

[8] https://finto.fi/koko/en/.
[9] http://www.ldf.fi/dataset/warsa/.
[10] See http://www.ldf.fi for more details.
[11] http://jena.apache.org/documentation/serving_data/.
[12] https://www.varnish-cache.org.
[13] http://linkeddata.org.

being designed. The idea is that in order to address different end-user information needs properly, different application perspectives are needed [9,16]. For example, a first user may want to see how the war events evolve in time and geographically, a second one is interested in persons and their stories of the war, and a third one wants to do research on the casualty records of the war. The idea of providing perspectives is different from large monolithic portals like Europeana that may show only one view or search perspective of the data.

An important feature of WarSampo is that the different application perspectives can be supported without modifying the data, which would be costly given the size and complexity of the knowledge graph, but by only modifying the way the data is accessed using SPARQL. In this way new application perspectives to the data can be added more easily and independently without affecting the other perspectives.

WarSampo not only provides multiple perspectives, but also supports their interlinking using a systematic URI referencing policy. While the WarSampo Data Service is able to resolve each WarSampo URI in the traditional LD way, each application perspective is assumed to be able to resolve the URIs of its application domain as domain specific HTML pages for human usage. In a sense, each resource, e.g., a soldier in the "person" perspective, has a kind of homepage, created by the perspective, that can be linked easily to the home pages of the other perspectives, if the URI is known. Each application perspective, and also any application external to WarSampo, is able to use these ready-to-use pages via URLs. For example, an event page describing a battle event, can easily provide more information about the persons involved in the battle or the historical locations where it took place.

Many datasets in Table 1 have their own perspectives, where the user can first search data of interest and then get linked data related to them. The perspectives enrich each other via linked data. The datasets are published in the WarSampo SPARQL endpoint[14] as separate graphs. The URIs of the data resources are minted using the following template: http://ldf.fi/warsa/GRAPH/LOCAL_ID. For example, the URI http://ldf.fi/warsa/events/event_536 identifies the event "Field Marshal Mannerheim inspected the Detachment Sisu consisting of foreign volunteers in Lapua". The WarSampo Data Service documentation page contains further example URIs and SPARQL queries, e.g., one for finding events, photographs, and articles that are situated in the city of Vyborg.

The data service can be used as a basis for Rich Internet Applications (RIA). A demonstration of this is the WarSampo Portal, where *all* functionality is implemented on the client side using JavaScript, only data is fetched from the server side SPARQL endpoints. In below, the six perspectives of the WarSampo portal are presented from the point of view of end-user information needs and technological solutions.

Event-Based Perspective. The WarSampo event-based perspective[15] is aimed towards anyone interested in the course of events of the Winter and Continuation War. The events are visualized using a timeline and a map. Each event has a detailed

---

[14] http://ldf.fi/warsa/sparql.
[15] http://www.sotasampo.fi/events.

description and contextualizing hyperlinks to other perspectives through entities linked to the event.

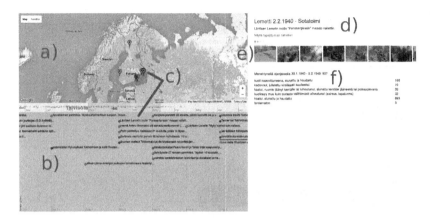

Fig. 2. Event perspective featuring a timeline and map.

Figure 2 illustrates the WarSampo event perspective. Events are displayed on a Google map (a) and on a timeline (b) that shows here events of the Winter War. When the user clicks an event, it is highlighted (c), and the historical place, time span, type, and description for the selected event are displayed (d). Photographs related to the event (e) are also shown. The photographs are linked to events based on location and time. Furthermore, information about casualties during the time span visible on the timeline is shown alongside the event description (f), and the map (a) features a heatmap layer for a visualization of these deaths.

The events can also be found and visualized through other perspectives. For example, in the Army Unit perspective, the events in which a unit participated can be viewed on maps and in time, providing a kind of graphical activity summary of the unit. In the Casualties perspective, military units of the dead soldiers are known, making it possible to sort out and visualize the personal war history of the casualties, e.g., on historical maps that come from a yet another dataset in WarSampo.

The main data sources for events were text books with event lists, including [12,13]. The pages with the lists were scanned, OCR'd, structured as CSV, and transformed into instances of CIDOC CRM event (sub)classes (cf. Fig. 1). In order to keep the visualization comprehensible, the timeline does not show minor events such as troop movements—these are visualized in the unit perspective instead (to be discussed later). The event metadata includes the description, time span, location, and participants of the event, represented using corresponding WarSampo domain ontologies.

The textual event descriptions were annotated using the ARPA automatic annotation service [15]. Automatic linking brings about the issue of name ambiguity.

Military persons mentioned in descriptions mostly have high ranks, which helps identifying them. Approaches to the place name ambiguity problem are discussed later below. Entity recognition for extracting links is still a work in progress, and conditions for it will be tweaked further to achieve a balance between precision, i.e., minimizing the amount of incorrect links, and recall, i.e., extracting as many as links as possible.

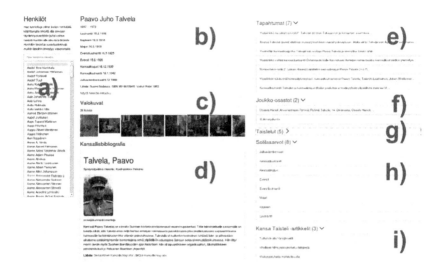

Fig. 3. Person perspective.

Person Perspective. The WarSampo person perspective application[16] is illustrated in Fig. 3. Its typical use case is someone searching for information about a relative who served in the army. On the left, the page has an input field (a) for a search by person's name. The matching names in the triple store are shown in the text field below the input. After making a selection, information about the person is shown at the top of the page (b): name, times and places of birth and death, professions, military ranks and promotions, etc. In the example case, the page shows matching photographs[17] (c), a short biography page from the National Biography[18] (d) and a set of lists linking to related events (e), military units (f), battles (g), military ranks (h), and Kansa Taisteli magazine articles (i) that mention him.

Currently the dataset consists of 96,000 persons. The data has been collected from various sources: lists of generals, lists of commanders in army corps, divisions, and regiments, lists of recipients of honorary medals like the Mannerheim

---

[16] http://sotasampo.fi/persons.
[17] http://sa-kuva.fi/neo?tem=webneoeng.
[18] http://www.ldf.fi/dataset/history.

Cross, casualties database, unit commanders mentioned in Organization Cards, the Finnish National Biography, Wikidata, and Wikipedia. Besides military personnel, an extract of 580 civil persons from the National Biography database and Wikidata was included in WarSampo because of their connections to WarSampo data. This set consists of persons with political or cultural significance during the wartime. The process of producing the data differed a lot depending on the used data source. For example, data lists have been scanned from a variety of documents, OCR'd, converted into CSV, and finally into RDF format. On the other hand, the casualty data of National Archives and the biographies of the National Biography had already been transformed into LOD in our earlier projects.

Some data sources, like the casualties database, provide detailed descriptions of person's life span, places, profession, marital status, etc. In contrast, sources such as the Organization Cards might only mention that, e.g., someone called *Captain Karhunen* has been in command of his unit in a certain battle. Regarding person names, we faced lots of different mentioning practices: a person might be referred to by full name (*Paavo Juho Talvela*), by initials (*P. Talvela*) or by using a combination of rank and family name (*Major General Talvela*, earlier known as *Colonel Talvela*). Recognizing whether such terms refer to the same person or not, often required extra knowledge of the person.

Person instances record only the basic properties, like family name (the only required property), forenames, a description, and provenance data, i.e., a link to the source from which the data was extracted. All other information is modeled as events, such as person's birth, death, promotion, or joining a military unit. Using the event-based approach turned out helpful especially in dealing with changing information. Consider a person's military rank: we may not know it at all, it might be a constant value during the entire wartime, or in the case of a longer military career, the rank is actually defined by a sequence of promotions. In a similar manner a person might be transferred into a different military unit and have a new commanding role in it.

The war diaries[19], data sources[20], and ranks[21] are in separate graphs. The War Diary graph has 13,043 data entries, and there are 10 data sources and 195 entries for ranks. The data includes the full range of ranks used by the Finnish Army added with some ranks used by German and Soviet Armies. Besides the military there are also some civil titles, like the ones used by the women's voluntary association *Lotta Svärd*.

Army Unit Perspective. WarSampo army unit perspective application[22] is illustrated in Fig. 4. A typical use case is someone searching for information about a specific army unit, maybe a unit where an elder relative is known to have served during the Winter War. On the left there is an input field (a) for a search by unit's name.

---

[19] See, e.g., http://digi.narc.fi/digi/hae_ay.ka?sartun=319.SARK.

[20] See, e.g., http://ldf.fi/warsa/actors/source3.

[21] See, e.g., http://ldf.fi/warsa/actors/ranks/Sotamies.

[22] http://sotasampo.fi/units.

The results matching unit labels in the triple store are shown in the text field below the input. The map (b) illustrates the known locations of the unit. The heatmap shows the casualties of the unit and the timeline (c) the events of the unit, e.g., dates of unit foundations, troop movements, and durations of fought battles. On the right there is a list of persons (d) known to have served in that unit. Three lists of related units are shown (e) consisting of (1) larger groups where this unit has been as a member, (2) smaller subunits being parts of this unit, and (3) otherwise related units at the same level in the hierarchy of the Finnish Army. Below this, there are additional information fields for related battles (f) and places (g), and links to entries in War Diaries (h) of the unit. There are also links to Kansa Taisteli magazine articles and photographs if they are related to the unit.

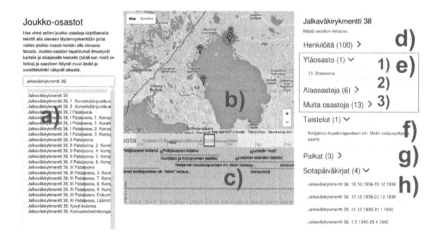

Fig. 4. Army unit perspective.

The data consists of over 3,000 Finnish army units, including Land Forces, Air Forces, Navy and its vessels, Medical Corps, stations of Anti-Aircraft Warfare and Skywatch, Finnish White Guard, and Swedish Volunteer Corps. The main sources of information have been the War Diaries and Organization Cards. The War Diaries provided an excellent starting point with about 3,000 unit labels. Currently only a part of Organization Cards are in the database, including the most important Divisions and Regiments of Infantry—during WW2 most soldiers served in Artillery and Infantry of the Land Forces, which formed the backbone of the Finnish Army.

The data in the Military Unit Ontology has been gathered simultaneously with person data. The event-based data model of a military unit is analogous to the model of a person. Also the problems regarding named entity recognition are similar in many ways. In the data sources, there are several ways of referring to a unit: by full name, e.g., *Jalkaväkirykmentti 11 (11th Infantry Regiment)*, by an abbreviation. e.g., *JR 11*, or in some cases by a nickname, e.g., *Ässärykmentti (Ace Regiment)*.

In addition, during the Winter War many units were renamed in order to confuse the enemy.

Historical Places Perspective. Most datasets used in WarSampo contain references to historical places (crm:E53 Place). If coordinates are available, places can be visualized on maps, providing a yet another perspective[23] to find and view WarSampo contents. Historical places are also essential for interlinking the datasets. For these purposes, a wartime place ontology containing place names with different levels of granularity and types (e.g., counties, municipalities, villages, bodies of water) was created as a pilot implementation of the "Finnish Ontology Service of Historical Places and Maps" [11]. After the creation of the place ontology, the other WarSampo datasets were programmatically linked to its place instances. This made it possible to build a perspective for viewing WarSampo contents on both modern and historical maps.

Figure 5 depicts the main functions of the historical places Perspective. For serendipitous browsing, all places that possess links to other WarSampo datasets can be visualized as markers or polygons on the Google map by pushing the button (a). This gives an overview of all places related to the war. In case the user is searching for a particular place, a tab for federated text search with autocompletion (b) is also provided. The search results are listed below the search field and are dynamically visualized on the map. The user can select a place by clicking on a search result row, or on a marker on the map. In the figure, the user has selected a village with the Finnish place name "Vääräkoski" that is then shown on the map with an infobox (f). By clicking the buttons (g) on the box the user can view and explore the linked events and photographs related to Vääräkoski.

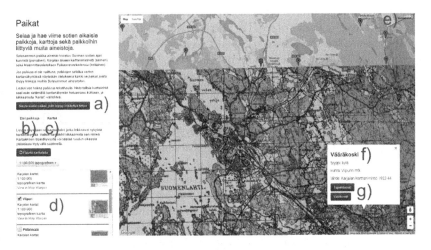

Fig. 5. Historical places perspective.

In addition to the search tab described above, there is also a historical maps tab (c) on the perspective. It provides the user with a list of selectable historical maps that intersect the current Google map view. In the figure, a historical map sheet covering the city of Viipuri and its neighborhoods (d) is selected. The opacity of the historical map sheets can be adjusted with the slider (e), which allows the user to investigate both historical and modern maps at the same time, providing new insight into place names. In this case, she realizes that the place she has selected, the village "Vääräkoski" (f), can be found only from the historical map of Viipuri— obviously the village does not exist anymore.

The historical place ontology was created using four data sources: (1) a map application the National Archives of Finland (612 wartime municipalities), (2) Finnish Spatio-Temporal Ontology (polygon boundaries of the municipalities)[24], (3) a dataset of geocoded Karelian map names (35,000 map names with coordinates and place types), and (4) the current Finnish Geographic Names Registry (800,000 places). The places were modeled with a simple schema used in [11], which contains properties for the place name, coordinates, polygon, place type, and part-of relationship of the place.

The big challenge when working with place names is that place names are highly ambiguous (polysemy). There can be dozens or even hundreds of places around Finland with the same name, which presents problems for automatic annotation of description texts. Utilizing place type information is one partial solution to this problem. When linking place name mentions to the WarSampo place ontology the following order of priority was used: (1) municipality (2) town (3) village (4) body of water. House names were most ambiguous, and they were not used in automatic linking.

Another major difficulty we encountered was that different geographic data sources, such as maps used as the basis for geocoding, are overlapping, producing multiple instances of same places. A partial solution to this issue was to remove duplicate place names in advance, when two places shared a name, were close to each other, and had the same place type. However, in practice there still remained cases where it is not possible to disambiguate multiple place names without manual work.

Casualties Perspective. The casualties perspective[25] is based on the National Archives' dataset of all known Finnish casualties of WW2. The dataset consists of some 95,000 war casualty records from 1939 to 1945. The data has been originally in a relational database, which was then converted into RDF and enriched by linking it to other datasets of WarSampo. In particular, each casualty record is linked to military ranks, units, persons, and wartime municipalities. In addition, there are links to resources within the dataset, such as instances of graveyards around Finland where the deceased are buried. The casualty dataset graph consists of almost 2.5 million triples. As the dataset is large, with links to various kinds of information about each casualty, it is not straightforward to present it in an online service for users to search and browse.

---

[24] http://seco.cs.aalto.fi/ontologies/sapo/.
[25] http://www.sotasampo.fi/casualties.

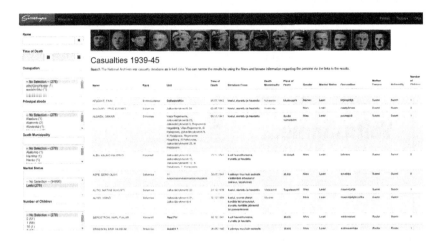

Fig. 6. Casualties perspective with one selected facet.

The casualties perspective, shown in Fig. 6, is a table-like view of the data records that can be filtered using faceted semantic search. Facets associated with the casualties are presented on the left of the interface as hierarchical facets with string search support. The number of hits on each facet category is calculated dynamically and shown to the user, so that selections leading to empty result set can be avoided. In addition, there is a special text search facet for finding persons directly by name, and a date range selector to filter the results by date of death.

In the figure, five facets are open and the other facets are not visible as they don't fit into the browser screen. The user has selected on the marital status facet the category "widow", focusing the search down to 278 killed widows of war that are presented in the table with links to further information.

Faceted search can not only be used for searching but also as a flexible tool for researching the underlying data [18]. In Fig. 6, the hit counts immediately show distributions of the killed widows along the facet categories. For example, the facet "Number of children" shows that one of the deceased had 10 children and most often (in 88 cases) widows had one child. If we next select category "one child" on its facet, we can see that two of the deceased are women and 86 are men in the gender facet.

Our faceted search engine is based purely on SPARQL queries and client side data processing in JavaScript. The system works well even with the large datasets of WarSampo, as pagination is used to limit the amount of results that are queried and displayed to the user.

The casualty records were modeled using the class crm:E31_Document with a distinct property for each facet. The property values are annotation resources selected from the corresponding ontologies, such as places. Record instances refer also to events, e.g., the death events of persons.

Fig. 7. The Contextual Reader interface targeting the Kansa Taisteli magazine articles.

Magazine Article Perspective. This application[26] is for searching and browsing textual articles relating to WW2. Here, the content are the 3,357 Kansa Taisteli magazine articles published by Sotamuisto in 1957–1986, containing mostly memoirs of soldiers related to WW2. The purpose of the perspective is two-fold: (1) to help a user find Kansa Taisteli articles of interest using faceted semantic search and, (2) to provide context to the found articles by extracting links to related WarSampo data from the texts.

The start page of the magazine article perspective is a faceted search browser similar to the one in the casualties perspective (cf. Fig. 6). Here, the facets allow the user to find articles by filtering them based on author, issue, year, related place, army unit, or keyword. Some of the underlying properties, such as the year and issue number of the magazines, are hierarchical and represented using SKOS. The hierarchy is visualized in the appropriate facet, and can be used for query expansion: by selecting an upper category in the facet hierarchy one can perform a search using all subcategories.

After the user has found an article of interest, she can click on it, and the digitized article appears on the screen in the CORE Contextual Reader interface [17]. Depicted in Fig. 7, CORE is able to automatically and in real time annotate PDF and HTML documents with recognized keywords and named entities, such as army units, places, and person names. These are then encircled with colored boxes indicating the linked data source. By hovering the mouse over a box, linked data from the data source is shown to the user, providing contextual information for an enhanced reading experience. In Fig. 7 the user is hovering on the identified place *Ristisalmi*,

---

[26] http://www.sotasampo.fi/articles.

which is then shown on a map for contextualization. If further contextual information is desired, the user can click on an entity to open the WarSampo page for that entity on a pane to the right of the reader interface. In Fig. 7, for example, detailed data are shown about *Raymond August Ericsson*, one of the battalion commanders discussed in the article.

The Kansa Taisteli magazine articles used in the interface have been manually scanned into PDF format by a member of the Association for Military History in Finland, Timo Hakala, and made available on the association's web site[27] in collaboration with the current copyright holder, Bonnier Publications. Our search application additionally makes use of a separate CSV file containing metadata for the 3,357 articles, also manually crafted by Timo Hakala.

After transforming the metadata into instances of documents (crm: E31_Document) and linking it with the WarSampo domain ontologies, the article dataset was further enriched with subject matter keywords by using the ARPA automatic text annotation service in the same way as with the other datasets. The extracted keywords were resources indicating military units, military persons, and places mentioned in the article text. These resources are used as the basis for the keyword facet in searching. The enriched metadata of the articles contains approximately 44,000 triples in total. The metadata is based on Dublin Core, where in addition to some standard properties like *dc:title*, there are object properties corresponding to each search facet, which facilitate the search.

A challenge faced during the linking and annotating of the Kansa Taisteli articles was the quality of the data. For example, because the magazines were manually scanned in a laborious process, full-page advertisements were sometimes not included. However, when locating the articles inside the PDFs based on the metadata, this threw off the reader sometimes even by multiple pages. A more serious concern was errors of the OCR process that caused challenges for the automatic annotation process. For example, unit names as abbreviations are inflected in Finnish by appending a : and the inflection ending. However, in OCR, character : was often read as *i* or *z*. Luckily, being a specialized domain with rigid conventions for writing, e.g., units and ranks, most of these errors could be corrected using a host of 135 regular expression rules.

This still left the problem of semantic disambiguation; in this case this concerned named entity recognition of persons, places, and military units. Formal evaluation on the automatic annotation process has not been made, but based on an informal evaluation, the final outcome is useful for its purpose even if the annotations are incomplete and some errors remain.

4   Related Work, Discussion, and Future Work

There are several projects publishing linked data about the World War I on the web, such as Europeana Collections 1914–1918[28], 1914–1918 Online[29], WW1

---

[27] http://kansataisteli.sshs.fi.

[28] http://www.europeana-collections-1914-1918.eu.

[29] http://www.1914-1918-online.net.

Discovery[30], Out of the Trenches[31], CENDARI[32], Muninn[33], and WW1LOD [14]. There are few works that use the Linked Data approach to WW2, such as [1,2] and Open Memory Project[34] on holocaust victims.

Our results suggest that large heterogeneous datasets of war history can be interlinked with each other through events in ways that provide insightful multiple perspectives for the historians and laymen to the data. Given the wide, deep, and sentimental interest in war history among the public and researchers, we envision that war history will become an important domain for Linked Data applications.

We have also learned that even in the rural northern parts of Europe, massive amounts of WW2 data can be found and opened for public use. We have initially dealt with less than 100,000 people involved in war events. However, there is also data available about hundreds of thousands of soldiers who survived the war only in Finland. Managing the data, and providing it for different user groups, suggests serious challenges when dealing with, e.g., the war events in the central parts of Europe, where the amount of data is orders of magnitude larger than in Finland, multilingual, and distributed in different countries. For example, solving entity resolution problems regarding historical place names and person names can be difficult. However, it seems that Linked Data is a promising way to tackle these challenges.

Future work on WarSampo includes, e.g., end user evaluations, where the portal is compared with existing legacy database services in searching for WW2 materials, and where the usability of the portal is tested in its use cases. We also plan to continue our work on automatic annotation of texts.

Acknowledgements. Jérémie Dutruit created the first RDF version of the casualties data, Jyrki Tiittanen geocoded the Karelian places dataset, Hanna Hyvönen rectified the historical maps on modern ones, Timo Hakala provided the Kansa Taisteli CSV metadata, and Kasper Apajalahti transformed it into RDF. Our work is funded by the Ministry of Education and Culture and Finnish Cultural Foundation. Wikidata Finland project financed rectifying of the historical maps.

References

1. de Boer, V., van Doornik, J., Buitinck, L., Marx, M., Veken, T.: Linking the kingdom: enriched access to a historiographical text. In: Proceedings of the 7th International Conference on Knowledge Capture (KCAP 2013), pp. 17–24. ACM, June 2013
2. Collins, T., Mulholland, P., Zdrahal, Z.: Semantic browsing of digital collections. In: Gil, Y., Motta, E., Benjamins, V.R., Musen, M.A. (eds.) ISWC 2005. LNCS, vol. 3729, pp. 127–141. Springer, Heidelberg (2005)

---

[30] http://ww1.discovery.ac.uk.

[31] http://www.canadiana.ca/en/pcdhn-lod/.

[32] http://www.cendari.eu/research/first-world-war-studies/.

[33] http://blog.muninn-project.org.

[34] http://www.bygle.net/wp-content/uploads/2015/04/Open-Memory-Project_3-1.pdf.

3. Crymble, A., Gibbs, F., Hegel, A., McDaniel, C., Milligan, I., Posner, M., Turkel, W.J. (eds.): The Programming Historian, 2nd edn. (2015). http://programminghistorian. org/
4. Doerr, M.: The CIDOC CRM - an ontological approach to semantic interoperability of metadata. AI Mag. **24**(3), 75–92 (2003)
5. Graham, S., Milligan, I., Weingart, S.: Exploring Big Historical Data: The Historian's Macroscope. Imperial College Press, London (2015)
6. Heath, T., Bizer, C.: Linked Data: Evolving the Web into a Global Data Space. Synthesis Lectures on the Semantic Web: Theory and Technology, 1st edn. Morgan & Claypool, Palo Alto (2011). http://linkeddatabook.com/editions/1.0/
7. Hyvönen, E.: Preventing interoperability problems instead of solving them. Semantic. Web J. **1**(1–2), 33–37 (2010)
8. Hyvönen, E.: Publishing and Using Cultural Heritage Linked Data on the Semantic Web. Synthesis Lectures on the Semantic Web: Theory and Technology. Morgan & Claypool, Palo Alto (2012)
9. Hyvönen, E., Lindquist, T., Törnroos, J., Mäkelä, E.: History on the semantic web as linked data - an event gazetteer and timeline for World War I. In: Proceedings of CIDOC 2012 - Enriching Cultural Heritage, CIDOC, June 2012
10. Hyvönen, E., Tuominen, J., Alonen, M., Mäkelä, E.: Linked data finland: a 7-star model and platform for publishing and re-using linked datasets. In: Presutti, V., Blomqvist, E., Troncy, R., Sack, H., Papadakis, I., Tordai, A. (eds.) ESWC Satellite Events 2014. LNCS, vol. 8798, pp. 226–230. Springer, Heidelberg (2014)
11. Hyvönen, E., Tuominen, J., Ikkala, E., Mäkelä, E.: Ontology services based on crowdsourcing: case national gazetteer of historical places. In: Proceedings of 14th International Semantic Web Conference (ISWC 2015), Posters and Demonstrations Track. CEUR Workshop Proceedings, vol. 1486, October 2015
12. Leskinen, J., Juutilainen, A. (eds.): Jatkosodan Pikkujättiläinen. WSOY, Finland (2005)
13. Leskinen, J., Juutilainen, A. (eds.): Talvisodan pikkujättiläinen, 4th edn. WSOY, Finland (2006)
14. Mäkelä, E., Törnroos, J., Lindquist, T., Hyvönen, E.: World War 1 as Linked Open Data (2015), submitted for review. http://seco.cs.aalto.fi/publications/
15. Mäkelä, E.: Combining a REST lexical analysis web service with SPARQL for mashup semantic annotation from text. In: Presutti, V., Blomqvist, E., Troncy, R., Sack, H., Papadakis, I., Tordai, A. (eds.) ESWC Satellite Events 2014. LNCS, vol. 8798, pp. 424–428. Springer, Heidelberg (2014)
16. Mäkelä, E., Hyvönen, E., Ruotsalo, T.: How to deal with massively heterogeneous cultural heritage data - lessons learned in CultureSampo. Semantic Web - Interoperability, Usability, Applicability **3**(1), 85–109 (2012)
17. Mäkelä, E., Lindquist, T., Hyvönen, E.: CORE - a contextual reader based on linked data. In: Proceedings of Digital Humanities 2016, long papers, July 2016
18. Tunkelang, D.: Faceted Search. Retrieval, and Services, Morgan & Claypool, Palo Alto, CA, USA, Synthesis Lectures on Information Concepts (2009)

Predicting Drug-Drug Interactions Through Large-Scale Similarity-Based Link Prediction

Achille Fokoue[(✉)], Mohammad Sadoghi, Oktie Hassanzadeh, and Ping Zhang

IBM T.J. Watson Research Center, Yorktown Heights, USA
achille@us.ibm.com

Abstract. Drug-Drug Interactions (DDIs) are a major cause of preventable adverse drug reactions (ADRs), causing a significant burden on the patients' health and the healthcare system. It is widely known that clinical studies cannot sufficiently and accurately identify DDIs for new drugs before they are made available on the market. In addition, existing public and proprietary sources of DDI information are known to be incomplete and/or inaccurate and so not reliable. As a result, there is an emerging body of research on in-silico prediction of drug-drug interactions. We present Tiresias, a framework that takes in various sources of drug-related data and knowledge as inputs, and provides DDI predictions as outputs. The process starts with semantic integration of the input data that results in a knowledge graph describing drug attributes and relationships with various related entities such as enzymes, chemical structures, and pathways. The knowledge graph is then used to compute several similarity measures between all the drugs in a scalable and distributed framework. The resulting similarity metrics are used to build features for a large-scale logistic regression model to predict potential DDIs. We highlight the novelty of our proposed approach and perform thorough evaluation of the quality of the predictions. The results show the effectiveness of Tiresias in both predicting new interactions among existing drugs and among newly developed and existing drugs.

1 Introduction

Adverse drug reactions (ADRs) are the 4th leading cause of deaths in United States surpassing complex diseases such as diabetes, pneumonia, and AIDS [8]. ADR risk increases significantly when taking multiple drugs simultaneously, which is often common in the elderly population and for managing chronic diseases. In fact, 3 to 5 % of all in-hospital medication errors are due to "preventable" drug-drug interactions (DDIs) [8]. Unfortunately, most ADRs are not revealed in clinical trials with relatively small sizes (at most tens of thousands of participants) due to the rare toxicity of some drugs and the large number of drug combinations that would need to be tested to detect potential DDIs. As a result, the only practical way to explore the large space of drug combinations in search of interacting drugs is through in-silico DDI predictions.

Recently, there has been a growing interest in computationally predicating potential DDIs [11,14,17–21]. Similar to content-based recommender systems,

© Springer International Publishing Switzerland 2016
H. Sack et al. (Eds.): ESWC 2016, LNCS 9678, pp. 774–789, 2016.
DOI: 10.1007/978-3-319-34129-3_47

the core idea of the predominant similarity-based approach [11,17–21] is to predict the existence of an interaction between a candidate pair of drugs by comparing it against known interacting pairs of drugs. Finding known interacting drugs that are very similar to a candidate pair provides supporting evidence in favor of the existence of a DDI between the two candidate drugs.

In this paper, we introduce Tiresias, a framework that takes in various sources of drug-related data and knowledge as input, and provides as output DDI predictions. In Tiresias, we extend the basic similarity-based DDI prediction framework while addressing the following four significant challenges and shortcomings that are mostly overlooked by prior work:

1. Important Use Case of Newly Developed Drugs: Prior work either (1) are fundamentally unable to make predictions for newly developed drugs (i.e., drugs for which no or very limited information about interacting drugs is available) [18] or (2) could conceptually predict drugs interacting with a new drug, but have not been tested for this scenario [11,17]. Similarity-based approaches (e.g. [11,17]) can clearly be applied to drugs without any known interacting drugs. However, in commonly carried 10-fold cross validation evaluation, prior work using similarity-based approaches have hidden drug-drug interaction associations and not drugs. Thus, the large majority of drugs used at testing are also known during the training phase, which is an inappropriate evaluation strategy to simulate the introduction of a newly developed drug. In our experimental evaluation, we show that the prediction quality of the basic similarity-based approaches drops noticeably when instead of hiding drug-drug associations, we hide drugs. We also show that techniques developed in Tiresias significantly improve the prediction quality for new drugs not seen at training.

2. Skewed Distribution of Interacting Drug Pairs: Contrary to most prior work [11,17,21], we do not assume *a priori* a balanced distribution of interacting drug pairs at training or at testing. There is no reason to believe that the prevalence of pairs of interacting drugs in the set of all the drug pairs is close to 50 % (often assumed in past studies). In fact, in Sect. 6, we present a methodology to estimate a lower and upper bound on the true prevalence of interacting drug pairs in the set of all drug pairs. We show that the true prevalence of DDIs is between 10 % and 30 %.

3. Appropriate Evaluation Metrics and Methodology for Skewed Distribution: Existing work [11,17,21] use mainly the area under the R.O.C curves (AUROC) as the evaluation metric to assess the quality of predictions and often justify their decision to rely on a balanced testing dataset because of the valid observation that AUROC is not too sensitive to the ratio of positive to negative examples. However, as shown in [7] and reinforced in our experimental evaluation section, AUROC is not appropriate for skewed distribution. Metrics designed specifically for skewed distribution such as precision & recall, F-score, or area under Precision-Recall curve (AUPR) should be used instead. Unfortunately, when prior work use these metrics, they do so on a balanced testing data set, which results in artificially high values (e.g., for a trivial classifier that report all pairs of drugs as interacting, recall is 1, precision 0.5 and f-score 0.67).

As shown in our evaluation, on unbalanced testing dataset (with prevalence of drug-drug interacting ranging from 10 % to 30 %), the basic similarity-based prediction produces excellent AUROC values, but mediocre F-score or AUPR.

4. Variety of Data Sources Considered, and Incompleteness of Similarity Measures: Existing techniques have relied on a limited number of data sources (primarily DrugBank) for creating drug similarity measures or drug features. However, in this paper, we exploit information originating from multiple linked data sources (e.g., DrugBank, UMLS, the Comparative Toxicogenomics Database (CTD), Uniprot) to create various drug similarity measures. This poses unique data integration challenges. In particular, since various data sources provide only partial information about a subset of drugs of interest, the resulting drug similarity measures exhibit varying levels of incompleteness. This incompleteness of similarity measures, which has been for the most part overlooked by prior work, is already an issue even when a single data source such as DrugBank is used because not all the attributes needed by a given similarity measure are available for all drugs. Without any additional machine learning features, the learning algorithm cannot distinguish between a low similarity value between two drugs due to incomplete data about at least one of the drugs or real dissimilarity between them. To address this important shortcoming, which affects prediction quality as measured by F-score and AUPR, we introduce a new class of features, called calibration features that captures the relative completeness of the drug-drug similarity measures.

In summary, in this paper, we make the four key contributions. First, we introduce a first of kind semantic integration of a comprehensive set of structured and unstructured data sources including, e.g., DrugBank, UMLS, and CTD (cf. Sect. 3) to construct a knowledge graph. Second, we develop new drug-drug similarity measures based on various properties of drugs including metabolic and signaling pathways, drug mechanism of action and physiological effects (cf. Sect. 5). Third, we build a large-scale and distributed linear regression learning model (in Apache Spark) to predict the existence of DDIs while efficiently coping with skewed distribution of DDIs and data incompleteness through a combination of case control sampling for rare events (cf. Sect. 6) and a new class of calibration features (cf. Sect. 5). Finally, we conduct extensive evaluations with real data to achieve DDI prediction with an average F-Score of 0.74 (vs. 0.65 for the baseline) and area under PR curve of 0.82 (vs. 0.78 for the baseline) using standard 10-fold cross validation for the newly developed drugs scenario (for the existing drug scenario: F-Score of 0.85 vs 0.75 and AUPR of 0.92 vs 0.87). Additionally, we introduce a novel retrospective analysis to demonstrate the effectiveness of our approach to predict correct, but yet unknown DDIs. Up to 68 % of all DDIs found after 2011 were correctly predicted using only DDIs known in 2011 as positive examples in training (cf. Sect. 7).

2   Background: Similarity-Based DDI Predictions

Similar to content-based recommender systems, the core idea of similarity-based approaches [11,17,21] is to predict the existence of an interaction between a candidate pair of drugs by comparing it against known interacting pairs of drugs. These approaches first define a variety of drug similarity measures to compare drugs. A drug similarity measure sim is a function that takes as input two drugs and returns a real number between 0 (no similarity between the two drugs) and 1 (perfect match between the two drugs) indicating the similarity between the two drugs. SIM denotes the set of all drug similarity measures. Entities of interest for drug-drug interaction prediction are not single drugs, but rather pair of drugs. Thus, drug similarity measures in SIM need to be extended to produce drug-drug similarity measures that compare two pairs of drugs (e.g., a pair of candidate drugs against an already known interacting pair of drugs). Given two drug similarity measures sim_1 and sim_2 in SIM, we can define a new drug-drug similarity measure, denoted $sim_1 \otimes sim_2$, that takes as input a two pairs of drugs (a_1, a_2) and (b_1, b_2) and returns the similarity between the two pairs of drugs computed as follows:

$$sim_1 \otimes sim_2((a_1, a_2), (b_1, b_2)) = avg(sim_1(a_1, b_1), sim_2(a_2, b_2))$$

where avg is an average or mean function such as the geometric mean or the harmonic mean. In other words, the first drug similarity measure (sim_1) is used to compare the first element of each pair and the second drug similarity measure (sim_2) is used to compare the second element of each pair. Finally, the results of the two comparisons are combined using, for example, harmonic or geometric mean. The set of all drug-drug similarity measures thus defined by combining drug similarity measures in SIM is denoted $SIM^2 = \{sim_1 \otimes sim_2 | sim_1 \in SIM \wedge sim_2 \in SIM\}$.

Given a set $KDDI$ of known drug-drug interactions, a drug-drug similarity measure $sim_1 \otimes sim_2 \in SIM^2$, and a candidate drug pair (d_1, d_2), the prediction based solely on $sim_1 \otimes sim_2$ that d_1 and d_2 interacts, denoted $predict[sim_1 \otimes sim_2, KDDI](d_1, d_2)$, is computed as the arithmetic mean of the similarity values between (d_1, d_2) and the top-k most similar known interacting drug pairs to (d_1, d_2): $amean(top_k\{sim_1 \otimes sim_2((d_1, d_2), (x, y)) | (x, y) \in KDDI - \{(d_1, d_2)\}\})$ where $amean$ is the arithmetic mean, and, in most cases, k is equal to 1. The power of similarity-based approaches stems from not relying on a single similarity based prediction, but from combining all the individual independent predictions $predict[sim_1 \otimes sim_2, KDDI]$ for all $sim_1 \otimes sim_2 \in KDDI$ into a single score that indicates the level of confidence in the existence of a drug-drug interaction. This combination is typically done through machine learning (e.g., logistic regression): the training is performed using $KDDI$ as the ground truth and, given a drug pair (d_1, d_2), its feature vector consists of $predict[sim_1 \otimes sim_2, KDDI](d_1, d_2)$ for all $sim_1 \otimes sim_2 \in KDDI$.

3   Addressing Data Integration Challenges

One of the salient feature of our Tiresias framework is to leverage many available sources on the Web. More importantly, there is a crucial need to connect these disparate sources in order to create a knowledge graph that is continuously being enriched as ingesting more sources. Notably the life science community has already recognized the importance of the data integration and taken the first step to employ the Linked Open Data methodology for connecting identical entities across different sources. However, most of the existing linkages in the scientific domain are often done statically, which results in many outdated or even non-existent links overtime. Therefore, even when the data is presumably linked, we are forced to verify these links. Furthermore, there are number of fundamental challenges that must be addressed to construct a unified view of the data with rich interconnectedness and semantics — a knowledge graph. For example, we employ entity resolution methodology either through syntactical disambiguation (e.g., cosine similarity, edit distance, or language model techniques [4]) or through semantic analysis by examining the conceptual property of entities [2]. These techniques are not only essential to identify similar entities but also instrumental in designing and capturing similarities among entities in order to engineer features necessary to enable DDIs prediction.

We first begin forming our knowledge graph by ingesting data from variety of sources (including XML, relational, and CSV formats) from the Web. As partially shown in Fig. 1, our data comes from variety of sources such as *DrugBank* [13] that offers data about known drugs and diseases, *Comparative Toxicogenomics Database* [6] that provides information about gene interaction, *Uniprot* [1] that provides details about the functions and structure of genes, *BioGRID* database that collects genetic and protein interactions [5], *Unified Medical Language System* that one is the largest repository of biomedical vocabularies including *NCBI* taxonomy, *Gene Ontology (GO)*, the *Medical Subject Headings (MeSH)* [2], and the *National Drug File - Reference Terminology*

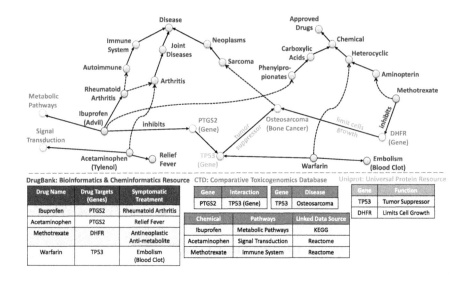

Fig. 1. Semantic curation and linkage of data from variety of sources on the Web.

(NDF-RT) that classifies drug with a multi-category reference models such as cellular or molecular interactions and therapeutic categories [3].

As part of our knowledge graph curation task, we identify which attributes or columns refer to which real world entities (i.e., data instances). Therefore, our constructed knowledge graph possess a clear notion of what the entities are, and what relations exist for each instance in order to capture the data interconnectedness. These may be relations to other entities, or the relations of the attributes of the entity to data values. As an example, in our ingested and curated data, we have a table for *Drug*, and have the columns *Name, Targets, Symptomatic Treatment*. Our knowledge graph has an identifier for a real world drug *Methotrexate*, and captures its attributes such as *Molecular Structure* or *Mechanism of Actions*, as well as relations to other entities including *Genes* that *Methotrexate* targets (e.g., *DHFR*), and subsequently, *Conditions* that it treats such as *Osteosarcoma (bone cancer)* that are reachable through its target genes, as demonstrated in Fig. 1. Constructing a rich knowledge graph is a necessary step before building our predication model as discussed next.

4 Overview

The overview of our similarity-based DDI prediction approach is illustrated in Fig. 2. It consists of five key phases (the arrows in Fig. 2).

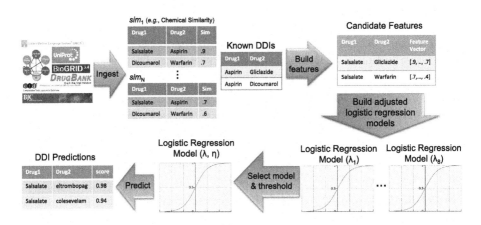

Fig. 2. Overview of similarity-based DDI prediction approach (Color figure online)

Ingestion Phase: In this phase, data originating from multiple sources are ingested and integrated to create various drug similarity measures (represented as blue tables in Fig. 2) and a known DDIs table. Similarity measures are not necessarily complete in the sense that some drug pairs may be missing from the similarity tables displayed in Fig. 2. The known DDIs table, denoted *KDDI*, contains the set of 12,104 drug pairs already known to interact in DrugBank.

In the 10-fold cross validation of our approach, $KDDI$ is randomly split into 3 disjoint subsets: $KDDI_{train}$, $KDDI_{val}$, and $KDDI_{test}$ representing the set of positive examples respectively used in the training, validation and testing (or prediction) phases. Contrary to most prior work, which partition $KDDI$ on the DDI associations instead of on drugs, our partitioning simulates the scenario of the introduction of newly developed drugs for which no interacting drugs are known. In particular, each pair (d_1, d_2) in $KDDI_{test}$ is such that either d_1 or d_2 does not appear in $KDDI_{train}$ or KDD_{val}.

Feature Construction Phase: Given a pair of drugs (d_1, d_2), we construct its machine learning feature vector derived from the drug similarity measures and the set of DDIs known at training. Like previous similarity-based approaches, for a drug candidate pair (d_1, d_2) and a drug-drug similarity measure $sim_1 \otimes sim_2 \in SIM^2$, we create a feature that indicates the similarity value of the known pair of interacting drugs most similar to (d_1, d_2) (see Sect. 5.2). Unlike prior work, we introduce new calibration features to address the issue of the incompleteness of the similarity measures and to provide more information about the distribution of the similarity values between a drug candidate pair and all known interacting drug pairs - not just the maximum value (see Sect. 5.3).

Model Generation Phase: As a result of relying on more data sources, using more similarity measures, and introducing new calibration features, we have significantly more features (1014) than prior work (e.g., [11] uses only 49 features). Thus, there is an increased risk of overfitting that we address by performing L_2-model regularization. Since the optimal regularization parameter is not known a-priori, in the model generation phase, we build 8 different logistic regression models using 8 different regularization values. To address issues related to the skewed distribution of DDIs (for an assumed prevalence DDIs lower than 17 %), we make some adjustments to logistic regression (see Sect. 6).

Model Validation Phase: The goals of this phase are twofold. First, in this phase, we select the best of the eight models (i.e., the best regularization parameter value) built in the model generation phase by choosing the model producing the best F-score on the validation data. Second, we also select the optimal threshold as the threshold at which the best F-score is obtained on the validation data evaluated on the selected model.

Prediction Phase: Let f denote the logistic function selected in the model validation phase and η the confidence threshold selected in the same phase. In the prediction phase, for each candidate drug pair (d_1, d_2), we first get its feature vector v computed in the feature construction phase. $f(v)$ then indicates the probability that the two drugs d_1 and d_2 interact, and the pair (d_1, d_2) is labeled as interacting iff. $f(v) \geq \eta$.

5 Feature Engineering

In this section, we describe the drug similarity measures used to compare drugs and how various machine learning features are generated from them.

5.1 Drug Similarity and Drug-Drug Similarity Measures

Due to space limitation, we describe here only 4 of the 13 similarity measures used to compare two drugs. The other similarity metrics are presented in detail in [9], including physiological effect based similarity, side effect based similarity, two metabolizing enzyme based similarities, three drug target based similarities, chemical structure similarity, MeSH based similarity.

Chemical-Protein Interactome (CPI) Profile Based Similarity: The Chemical-Protein Interactome (CPI) profile of a drug d, denoted $cpi(d)$, is a vector indicating how well its chemical structure docks or binds with about 611 human Protein Data Bank (PDB) structures associated with DDIs [14]. The CPI profile based similarity of two drugs d_1 and d_2 is computed as the cosine similarity between the mean-centered versions of vectors $cpi(d_1)$ and $cpi(d_2)$.

Mechanism of Action Based Similarity: For a drug d, we collect all its mechanisms of action obtained from NDF-RT. To discount popular terms, Inverse Document Frequency (IDF) is used to assign more weight to relatively rare mechanism of actions: $IDF(t, Drugs) = log\frac{|Drugs|+1}{DF(t,Drugs)+1}$ where $Drugs$ is the set of all drugs, t is a mechanism of action, and $DF(t, Drugs)$ is the number of drugs with the mechanism of action t. The IDF-weighted mechanism of action vector of a drug d is a vector $moa(d)$ whose components are mechanisms of action. The value of a component t of $moa(d)$, denoted $moa(d)[t]$, is zero if t is not a known mechanism of action of d; otherwise, it is $IDF(t, Drugs)$. The mechanism of action based similarity measure of two drugs d_1 and d_2 is the cosine similarity of the vectors $moa(d_1)$ and $moa(d_2)$.

Pathways Based Similarity: Information about pathways affected by drugs is obtained from CTD database. The pathways based similarity of two drugs is defined as the cosine similarity between the IDF-weighted pathways vectors of the two drugs, which are computed in a similar way as IDF-weighted mechanism of action vectors.

Anatomical Therapeutic Chemical (ATC) Classification System Based Similarity: ATC [15] is a classification of the active ingredients of drugs according to the organs that they affect as well as their chemical, pharmacological and therapeutic characteristics. The classification consists of multiple trees representing different organs or systems affected by drugs, and different therapeutical and chemical properties of drugs. The ATC codes associated with each drug are obtained from DrugBank. For a given drug, we collect all its ATC code from DrugBank to build a ATC code vector (the most specific ATC codes associated with the drug -i.e., leaves of the classification tree- and also all the ancestor codes are included). The ATC similarity of two drugs is defined as the cosine similarity between the IDF-weighted ATC code vectors of the two drugs, which are computed in a similar way as IDF-weighted mechanism of action vectors.

The set of all drug similarity measures is denoted SIM. As explained in the background Sect. 2, drug similarity measures in SIM need to be extended to produce drug-drug similarity measures that compare two pairs of drugs. SIM^2 denotes the set of all drug-drug similarity measures derived from SIM.

5.2   Top-k Similarity-Based Features

Like previous similarity-based approaches, for a given drug candidate pair (d_1, d_2), a set KDD_{train} of DDIs known at training, and a drug-drug similarity measure $sim_1 \otimes sim_2 \in SIM^2$, we create a new similarity-based feature, denoted $abs_{sim_1 \otimes sim_2}$ and computed as the similarity value between (d_1, d_2) and the most similar known interacting drug pair to (d_1, d_2). In other words,

$$abs_{sim_1 \otimes sim_2}(d_1, d_2) = max(D_{sim_1 \otimes sim_2}(d_1, d_2))$$

where $D_{sim_1 \otimes sim_2}(d_1, d_2)$ is the set of all the similarity values between (d_1, d_2) and all known DDIs:

$$D_{sim_1 \otimes sim_2}(d_1, d_2) = \{sim_1 \otimes sim_2((d_1, d_2), (x, y)) | (x, y) \in KDDI_{train} - \{(d_1, d_2)\}\}$$

5.3   Calibration Features

Calibration of Top-k Similarity-Based Features: For a drug candidate pair (d_1, d_2), a high value of the similarity-based feature $abs_{sim_1 \otimes sim_2}(d_1, d_2)$ is a clear indication of the presence of at least one known interacting drug pair very similar to (d_1, d_2) according to the drug-drug similarity measure $sim_1 \otimes sim_2$. However, this feature value provides to the machine learning algorithm only a limited view of the distribution $D_{sim_1 \otimes sim_2}(d_1, d_2)$ of all the similarity values between (d_1, d_2) and all known DDIs.

For example, with only access to $max(D_{sim_1 \otimes sim_2}(d_1, d_2))$, there is no way to differentiate between a case where that maximum value is a significant outlier (i.e., many standard deviation away from the mean of $D_{sim_1 \otimes sim_2}(d_1, d_2)$) and the case where it is not too far from the mean value of $D_{sim_1 \otimes sim_2}(d_1, d_2)$. Since it would be impractical to have a feature for each data point in D (overfitting and scalability issues), we instead summarize the distribution $D_{sim_1 \otimes sim_2}(d_1, d_2)$ by introducing the following features to capture its mean and standard deviation:

$$avg_{sim_1 \otimes sim_2}(d_1, d_2) = mean(D_{sim_1 \otimes sim_2}(d_1, d_2))$$

$$std_{sim_1 \otimes sim_2}(d_1, d_2) = stdev(D_{sim_1 \otimes sim_2}(d_1, d_2))$$

To calibrate the absolute maximum value computed by $abs_{sim_1 \otimes sim_2}(d_1, d_2)$, we introduce a calibration feature, denoted $rel_{sim_1 \otimes sim_2}$, that corresponds to the z-score of the maximum similarity value of the candidate and a known DDI (i.e., it indicates the number of standard deviations from the mean):

$$rel_{sim_1 \otimes sim_2}(d_1, d_2) = \frac{abs_{sim_1 \otimes sim_2}(d_1, d_2) - avg_{sim_1 \otimes sim_2}(d_1, d_2)}{std_{sim_1 \otimes sim_2}(d_1, d_2)}$$

Finally, for a candidate pair (d_1, d_2), we add a boolean feature, denoted $con_{sim_1 \otimes sim_2}(d_1, d_2)$, that indicates whether the most similar known interacting drug pair contains d_1 or d_2.

Calibration of Drug-Drug Similarity Measures: Features described so far capture similarity values between a drug candidate pair and known DDIs. As such, a high feature value for a given candidate pair (d_1, d_2) does not necessarily indicate that the two drugs are likely to interact. For example, it could be the case that, for a given drug-drug similarity measure, (d_1, d_2) is actually very similar to most drug pairs (whether or not they are known to interact). Likewise, a low feature value does not necessarily indicate a reduced likelihood of drug-drug interaction if (d_1, d_2) has a very low similarity value with respect to most drug pairs (whether or not they are known to interact). In particular, such a low overall similarity between (d_1, d_2) and most drug pairs is often due to the incompleteness of the similarity measures considered. For a drug-drug similarity measure $sim_1 \otimes sim_2 \in SIM^2$ and a candidate pair (d_1, d_2), we introduce a new calibration feature, denoted $base_{sim_1 \otimes sim_2}$, to serve as a baseline measurement of the average similarity measure between the candidate pair (d_1, d_2) and any other pair of drugs (whether or not it is known to interact). The exact expression of $base_{sim_1 \otimes sim_2}(d_1, d_2)$ is as follows:

$$\frac{\displaystyle\sum_{(x,y)\neq(d_1,d_2)\wedge x\neq y} sim_1 \otimes sim_2((d_1, d_2), (x, y))}{|Drugs|(|Drugs| - 1)/2 - 1}$$

The evaluation of this expression is quadratic in the number of drugs $|Drugs|$, which results in a significant runtime performance degradation without any noticeable gain in the quality of the predictions as compared to the following approximation of $base_{sim_1 \otimes sim_2}$ (with a linear time complexity):

$$base_{sim_1 \otimes sim_2}(d_1, d_2) \approx hm(\frac{\displaystyle\sum_{x\neq d_1} sim_1(d_1, x)}{|Drugs| - 1}, \frac{\displaystyle\sum_{y\neq d_2} sim_2(d_2, y)}{|Drugs| - 1})$$

where hm denotes the harmonic mean. In other words, $base_{sim_1 \otimes sim_2}(d_1, d_2)$ is approximated as the harmonic mean of (1) the arithmetic mean of the similarity between d_1 and all other drugs computed using sim_1, and (2) the arithmetic mean of the similarity between d_2 and all other drugs computed using sim_2.

6 Dealing with Unbalanced Data

In evaluating any machine learning system, the testing data should ideally be representative of the real data. In particular, the fraction of positive examples in the testing data should be as close as possible to the prevalence or fraction of DDIs in the set of all pairs of drugs. Although the ratio of positive to negative examples in the testing has limited impact on the area under the ROC curves, as shown in the experimental evaluation, it has significant impact on other key quality metrics more appropriate for skewed distributions (e.g., F-score and area under precision-recall curves). Unfortunately, the exact prevalence of DDIs in the set of all drugs pairs is unknown. Here, we first provide upper and lower bounds on the true prevalence of DDIs in the set of all drug pairs. Then, we discuss logistic regression adjustments to deal with the skewed distribution of DDIs.

Upper Bound: FDA Adverse Event Reporting System (FAERS) is a database that contains information on adverse events submitted to FDA. It is designed to support FDA's post-marketing safety surveillance program for drugs. Mined from FAERS, TWOSIDES [16] is a dataset containing only side effects caused by the combination of drugs rather than by any single drug. Used as the set of known DDIs, TWOSIDES [16] contains many false positives as some DDIs are observed from FAERS, but without rigorous clinical validation. Thus, we use TWOSIDES to estimate the upper bound of the DDI prevalence. There are 645 drugs and 63,473 distinct pairwise DDIs in the dataset. Thus, the upper bound of the DDI prevalence is about 30 %.

Lower Bound: We use a DDI data set from Gottlieb et al. [11] to estimate the lower bound of the DDI prevalence. The data set is extracted from DrugBank [13] and the http://drugs.com website (excluding DDIs tagged as minor). DDIs from this data set are extracted from drug's package inserts (accurate but far from complete). Thus, there are some false negatives in such a data set. There are 1,227 drugs and 74,104 distinct pairwise DDIs in the dataset. Thus the lower bound of the DDI prevalence is about 10 %.

Modified Logistic Regression to Handle Unbalanced Data: For a given assumed low prevalence of DDIs τ_a, it is often advantageous to train our logistic regression classifier on a training set with a higher fraction τ_t of positive examples and to later adjust the model parameters accordingly. The main motivation for this *case-control sampling* approach for rare events [12] is to improve runtime performance of the model building phase since, for the same number of positive examples, the higher fraction τ_t of positive examples yields a smaller total number of examples at training. Furthermore, for an assumed prevalence $\tau_a \leq 0.17$, the quality of the predictions is only marginally affected by the use of a training set with a ratio of one positive example to 5 negative examples (i.e., $\tau_t \sim 0.17$)

A logistic regression model with parameters $\beta_0, \beta_1, \ldots, \beta_n$ trained on a training sample with prevalence of positive examples of τ_t instead of τ_a is then converted into the final model with parameters $\hat{\beta}_0, \hat{\beta}_1, \ldots, \hat{\beta}_n$ by correcting the intercept $\hat{\beta}_0$ as indicated in [12]:

$$\hat{\beta}_0 = \beta_0 + log\frac{\tau_a}{1-\tau_a} - log\frac{\tau_t}{1-\tau_t}$$

The other parameters are unchanged: $\hat{\beta}_i = \beta_i$ for $i \geq 1$.

We have tried more advanced adjustments for rare events discussed in [12] (e.g., weighted logistic regression and ReLogit), but the overall improvement of the quality of our predictions was only marginal.

7   Evaluation

To assess the quality of our predictions, we perform two types of experiments. First, a 10-fold cross validation is performed to assess F-Score and area under Precision-Recall (AUPR) curve. Second, a retrospective analysis shows the ability of our Tiresias framework to discover valid, but yet unknown DDIs.

7.1   10-Fold Cross Validation Evaluation

Data Partitioning: In the 10-fold cross validation of our approach, to simulate the introduction of a newly developed drug for which no interacting drugs are known, 10 % of the drugs appearing as the first element of a pair of drugs in the set $KDDI$ of all known drug pairs are hidden, rather than hiding 10 % of the drug-drug relations as done in [11,17,18]. Since the DDI relation is symmetric, we consider, without loss of generality, only drug candidate pairs (d_1, d_2) where the canonical name of d_1 is less than or equal to the canonical name of d_2 according to the lexicographic order (i.e., $d_1 \leq d_2$). $KDDI$ is randomly split into 3 disjoint subsets: $KDDI_{train}$, $KDDI_{val}$, and $KDDI_{test}$ representing the set of positive examples respectively used in the training, validation and testing (or prediction) phases, and containing respectively about 8/10th, 1/10th and 1/10th of $KDDI$ pairs. In particular, each pair (d_1, d_2) in $KDDI_{test}$ is such that either d_1 or d_2 does not appear in $KDDI_{train}$ or KDD_{val} (more on this partitioning in Sect. 7.1 of [9]). The training data set consists of (1) known interacting drugs in $KDDI_{train}$ as positive examples, and (2) randomly generated pairs of drugs (d_1, d_2) not already known to interact (i.e., not in $KDDI$) such that the drugs d_1 and d_2 appear in $KDDI_{train}$ (as negative examples). The validation data set consists of (1) the known interacting drug pairs in $KDDI_{val}$ as positive examples, and (2) negative examples that are randomly generated pairs of drugs (d_1, d_2) not already known to interact (i.e., not in $KDDI$) such that d_1 is the first drug in at least one pair in $KDDI_{val}$ (i.e., a drug only seen at validation but not at training) and d_2 appears (as first or second element) in at least on pair in $KDDI_{train}$ (i.e., d_2 is known at training). The testing data set consists of (1) the known interacting drug pairs in $KDDI_{test}$ as positive examples, and (2) negative examples that are randomly generated pairs of drugs (d_1, d_2) not already known to interact (i.e., not in $KDDI$) such that d_1 is the first drug in at least one pair in $KDDI_{test}$ (i.e., a drug only seen at testing but not at training or validation) and d_2 appears (as first or second element) in at least on pair in $KDDI_{train} \cup KDDI_{val}$ (i.e., d_2 is known at training or at validation).

Results: Contrary to prior work, in our evaluation, the ratio of positive examples to randomly generated negative examples is not 1 to 1. Instead, the assumed prevalence of DDIs at training and validation is the same and is in the set {10 %, 20 %, 30 %, 50 %}. For a given DDI prevalence at training and validation, we evaluate the quality of our predictions on testing data sets with varying prevalence of DDIs (ranging from 10 % to 30 %). 50 % DDI prevalence at training and validation is used here to assess the quality of prior work (which rely on a balanced distribution of positive and negative examples at training) when the testing data is unbalanced. For a given assumed DDI prevalence at training/validation and a DDI prevalence at testing, to get robust results and show the effectiveness of our calibration-based features, we perform not one, but five 10-fold cross validations with all the features described in Sect. 5 (solid lines in Figs. 3 and 4) and five 10-fold cross validations without calibration features (dotted lines in Figs. 3 and 4). Results reported on Figs. 3 and 4 represent average over the five 10-fold cross validations.

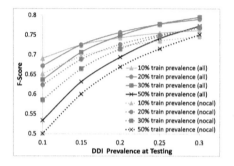
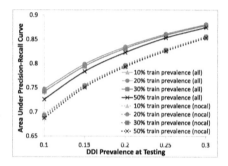

Fig. 3. F-Score for new drugs   **Fig. 4.** AUPR for new drugs

The key results from our evaluation are as follows:

- Regardless of the DDI prevalence used at training and validation (provided that it is between 10 % to 30 % -i.e., the lower and upper bound of the true prevalence of DDIs), our approach using calibration features (solid lines in Figs. 3 and 4) and unbalanced training/validation data (non-black lines) significantly outperforms the baseline representing prior similarity-based approaches (e.g., [11]) that rely on balanced training data without calibration features (the dotted black line with crosses as markers). For an assumed DDI prevalence at training ranging from 10 % to 30 %, the average F-score (resp. AUPR) over testing data with prevalence between from 10 % to 30 % varies from 0.73 to 0.74 (resp. 0.821 to 0.825) when all features are used. However, when the training is done on balanced data without calibration features, the average F-score (resp. AUPR) over testing data with prevalence between from 10 % to 30 % is 0.65 (resp. 0.78)[1]. The difference with the baseline is higher the skewer the testing data distribution is.
- For a fixed DDI prevalence at training/validation, using calibration features is always better in terms of F-Score and AUPR (solid vs. dotted lines in Figs. 3 and 4)

---

[1] Precision (resp. recall) varies from 0.84 to 0.70 (resp. 0.66 to 0.78) with calibration. Precision (resp. recall) is at 0.54 (resp. 0.84) on balanced training w/o calibration.

- As pointed out in prior work, the area under ROC curves (AUROC) is not affected by the prevalence of DDI at training/validation or testing. It remains constant at about 0.92 with calibration features and 0.90 without them.
- Finally, no similarity metric by itself has good predictive power (ATC similarity is the best with 0.58 F-Score and 0.56 AUPR), and removing any given similarity metric has limited impact on the quality of the predictions (the highest decrease was by 1 % in F-Score & AUPR w/o ATC similarity).

Note that we also perform 10-fold cross validation evaluations hiding drug-drug associations instead of drugs. Results presented in [9] show that, even when predictions are made only on drugs with some known interacting drugs, the combination of unbalanced training/validation data and calibration features remains superior to the baseline (F-Score of 0.85 vs 0.75 and AUPR of 0.92 vs 0.87).

7.2   Retrospective Analysis

We perform a retrospective evaluation using as the set of known DDIs ($KDDI$) only pairs of interacting drugs present in an earlier version of DrugBank (January 2011). Figure 5 shows the fraction of the total of 713 DDIs added to Drug-Bank between January 2011 and December 2014 that our approach can discover based only on DDIs known in January 2011 for different DDI prevalence at training/validation. Figure 5 shows that we can correctly predict up to 68 % of the DDI discovered after January 2011, which demonstrates the ability of Tiresias to discover valid, but yet unknown DDIs.

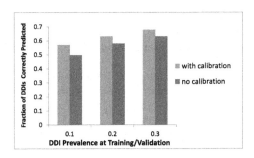

Fig. 5. Retrospective evaluation

8   Conclusion

In this paper, we presented Tiresias, a computational framework that predicts DDIs through large-scale similarity-based link prediction. Experimental results clearly show the effectiveness of Tiresias in both predicting new interactions among existing drugs and among newly developed and existing drugs. The predictions provided by Tiresias will help clinicians to avoid hazardous DDIs in their prescriptions and will aid pharmaceutical companies to design large-scale clinical trial by assessing potentially hazardous drug combinations. We have designed a Web interface and a set of APIs to assist with such use cases [10]. We are currently extending Tiresias to perform link prediction among other entity types in our knowledge graph, turning it into a generic large-scale link prediction system.

References

1. Apweiler, R., Bairoch, A., Wu, C.H., Barker, W.C., Boeckmann, B., Ferro, S., Gasteiger, E., Huang, H., Lopez, R., Magrane, M., et al.: Uniprot: the universal protein knowledgebase. Nucleic Acids Res. **32**(Suppl. 1), D115–D119 (2004)
2. Bodenreider, O.: The unified medical language system (UMLS): integrating biomedical terminology. Nucleic Acids Res. **32**(Suppl. 1), D267–D270 (2004)
3. Brown, S.H., Elkin, P.L., Rosenbloom, S., Husser, C., Bauer, B., Lincoln, M., Carter, J., Erlbaum, M., Tuttle, M.: VA national drug file reference terminology: a cross-institutional content coverage study. Medinfo **11**(Pt. 1), 477–481 (2004)
4. Chandel, A., Hassanzadeh, O., Koudas, N., Sadoghi, M., Srivastava, D.: Benchmarking declarative approximate selection predicates. In: ACM SIGMOD International Conference on Management of Data, SIGMOD 2007, pp. 353–364 (2007)
5. Chatr-aryamontri, A., Breitkreutz, B.J., Oughtred, R., Boucher, L., Heinicke, S., Chen, D., Stark, C., Breitkreutz, A., Kolas, N., O'Donnell, L., et al.: The BioGRID interaction database: 2015 update. Nucleic Acids Res. **43**, D470–D478 (2014). doi:10.1093/nar/gku1204
6. Davis, A.P., Murphy, C.G., Saraceni-Richards, C.A., Rosenstein, M.C., Wiegers, T.C., Mattingly, C.J.: Comparative toxicogenomics database: a knowledgebase and discovery tool for chemical-gene-disease networks. Nucleic Acids Res. **37**(Suppl. 1), D786–D792 (2009)
7. Davis, J., Goadrich, M.: The relationship between precision-recall and roc curves. In: Proceedings of the 23rd International Conference on Machine Learning, pp. 233–240. ACM (2006)
8. Flockhart, D.A., Honig, P., Yasuda, S.U., Rosebraugh, C.: Preventable adverse drug reactions: A focus on drug interactions. Centers for Education & Research on Therapeutics
9. Fokoue, A., Sadoghi, M., Hassanzadeh, O., Zhang, P.: Predicting drug-drug interactions through large-scale similarity-based link prediction. http://researcher.watson.ibm.com/researcher/files/us-achille/adrTechreport.pdf
10. Fokoue, A., Hassanzadeh, O., Sadoghi, M., Zhang, P.: Predicting drug-drug interactions through similarity-based link prediction over web data. In: Proceedings of the 25th International Conference on World Wide Web, WWW 2016. ACM (2016)
11. Gottlieb, A., Stein, G.Y., Oron, Y., Ruppin, E., Sharan, R.: Indi: a computational framework for inferring drug interactions and their associated recommendations. Mol. Syst. Biol. **8**(1), 592 (2012)
12. King, G., Zeng, L.: Logistic regression in rare events data. Polit. Anal. **9**(2), 137–163 (2001)
13. Knox, C., Law, V., Jewison, T., Liu, P., Ly, S., Frolkis, A., Pon, A., Banco, K., Mak, C., Neveu, V., et al.: DrugBank 3.0: a comprehensive resource for 'comics' research on drugs. Nucleic Acids Res. **39**(Suppl. 1), D1035–D1041 (2011)

14. Luo, H., Zhang, P., Huang, H., Huang, J., Kao, E., Shi, L., He, L., Yang, L.: Ddi-cpi, a server that predicts drug-drug interactions through implementing the chemical-protein interactome. Nucleic Acids Res. **42**, W46–W52 (2014). doi:10. 1093/nar/gku433

15. Skrbo, A., Begović, B., Skrbo, S.: Classification of drugs using the atc system (anatomic, therapeutic, chemical classification) and the latest changes. Medicinski arhiv **58**(1 Suppl. 2), 138–141 (2003)

16. Tatonetti, N.P., Patrick, P.Y., Daneshjou, R., Altman, R.B.: Data-driven prediction of drug effects and interactions. Sci. Transl. Med. **4**(125), 125ra31 (2012)

17. Vilar, S., Uriarte, E., Santana, L., Lorberbaum, T., Hripcsak, G., Friedman, C., Tatonetti, N.P.: Similarity-based modeling in large-scale prediction of drug-drug interactions. Nat. Protoc. **9**(9), 2147–2163 (2014)

18. Vilar, S., Uriarte, E., Santana, L., Tatonetti, N.P., Friedman, C.: Detection of drug-drug interactions by modeling interaction profile fingerprints. PLoS ONE **8**(3), e58321 (2013)

19. Zhang, P., Agarwal, P., Obradovic, Z.: Computational drug repositioning by ranking and integrating multiple data sources. In: Blockeel, H., Kersting, K., Nijssen, S., Železný, F. (eds.) ECML PKDD 2013, Part III. LNCS, vol. 8190, pp. 579–594. Springer, Heidelberg (2013)

20. Zhang, P., Wang, F., Hu, J., Sorrentino, R.: Towards personalized medicine: leveraging patient similarity and drug similarity analytics. AMIA Summits Transl. Sci. Proc. **2014**, 132 (2014)

21. Zhang, P., Wang, F., Hu, J., Sorrentino, R.: Label propagation prediction of drug-drug interactions based on clinical side effects. Scientific reports 5 (2015)

PhD Symposium

Semantics Driven Human-Machine Computation Framework for Linked Islamic Knowledge Engineering

Amna Basharat[(✉)]

Department of Computer Science, University of Georgia, Athens, GA 30602, USA
amnabash@uga.edu

Abstract. Formalized knowledge engineering activities including semantic annotation and linked data management tasks in specialized domains suffer from considerable knowledge acquisition bottleneck - owing to the lack of availability of experts and in-efficacy of computational approaches. Human Computation & Crowdsourcing (HC&C) methods successfully advocate leveraging the human processing power to solve problems that are still difficult to be solved computationally. Contextualized to the domain of Islamic Knowledge, my research investigates the synergistic interplay of these HC&C methods and the semantic web and will seek to devise a semantics driven human-machine computation framework for knowledge engineering in specialized and knowledge intensive domains. The overall objective is to augment the process of automated knowledge extraction and text mining methods using a hybrid approach for combining collective intelligence of the crowds with that of experts to facilitate activities in formalized knowledge engineering - thus overcoming the so-called knowledge acquisition bottleneck.

Keywords: Human computation · Semantic web · Task profiles · Crowdsourcing · Islamic knowledge · Quran · Hadith

1 Introduction

Challenges associated with large-scale adoption of semantic web technologies continue to confront the researchers in the field. Researchers have recognized the need for human intelligence in the process of semantic content creation and analytics, which forms the backbone of any semantic application [1,2]. Realizing the potential that human computation, collective intelligence and the fields of the like such as crowdsourcing and social computation have offered, semantic web researchers have effectively taken up the synergy to solve the bottlenecks of human experts and the needed human contribution in the semantic web development processes. This paper presents a novel contribution towards this intersection of the semantic web and human computation paradigm.

© Springer International Publishing Switzerland 2016
H. Sack et al. (Eds.): ESWC 2016, LNCS 9678, pp. 793–802, 2016.
DOI: 10.1007/978-3-319-34129-3_48

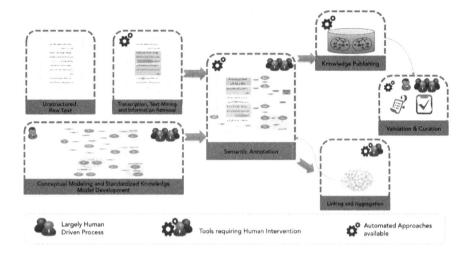

Fig. 1. Challenges in knowledge engineering processes

1.1   Background and Motivation

One of the major challenges hindering successful application of ontology-based approaches to data organization and integration in specialized domains is the so-called *'knowledge acquisition bottleneck'* [3]- that is, the large amount of time and money needed to develop and maintain the formal ontologies. This also includes ontology population and semantic annotation using well established vocabularies. My research primarily is motivated to overcome the inherent knowledge acquisition bottleneck in creating semantic content in semantic applications. We have established how this is particularly true for knowledge intensive domains such as the domain of Islamic Knowledge, which has failed to cache upon the promised potential of the semantic web and the linked data technology; standardized web-scale integration of the available knowledge resources is currently not facilitated at a large scale [4]. To date, only one dataset on the Linked Open Data (LOD) cloud in the domain exists [5].

To understand the knowledge acquisition bottleneck encountered in this domain (and others), consider the knowledge engineering processes illustrated in Fig. 1. Existing methods towards semantic annotation and linked knowledge generation are either (a) computationally driven, employing on text mining and information extraction methods or, (b) expert driven (such as conceptual modelling, annotation and validation). While the computational methods may assist in large-scale knowledge acquisition, however, the lack of formalized and agreed upon knowledge models and sensitivity of the knowledge at hand- primarily obtained from unstructured and multilingual data- makes the knowledge engineering process far from trivial. Islamic domain suffers a great deal from a lack of suitable training data and gold standards. This only adds to the challenge of ensuring the reliability and scalability of these methods. Expert driven methods involve subject specialists however, these are often not scalable (time or cost). It is no wonder that the efforts towards the vision of standardization and formalization of Islamic Knowledge as proposed by [6] have remained futile.

1.2 Research Context and Problem Statement

To address these and similar challenges, researchers have recognized that the realization of the semantic and linked data technologies will require not only computation but also significant human contribution [1,2]. Humans are simply considered indispensable [7] for the semantic web to realize its full potential. Emerging research is advocating the use of Human Computation & Crowdsourcing (HC&C) methods to leverage human processing power to harness the collective intelligence and the wisdom of the crowds by engaging large number of online contributors to accomplish tasks that cannot yet be automated. There has been growing interest to use crowdsourcing methods to support the semantic web and linked open data research by providing means to efficiently create research relevant data and potentially solve the bottleneck of knowledge experts and annotators needed for the large-scale deployment of semantic web and linked data technologies. Recent framework called CrowdTruth proposed by [8] is a step forward that recognizes the challenges in gathering distributed data in crowdsourcing.

Within the realm of (semantic web based) knowledge engineering tasks, several levels of complexity may be encountered. Some tasks are simple, while others are more knowledge intensive. While some may reasonably be amenable to computational approaches, others need domain specific expert annotations and judgements. Therefore, not all tasks are fit for general purpose crowdsourcing. This is specifically true for the domain of islamic knowledge, owing to the specialized nature of the learning needs presented by diverse users, and heterogenous, multilingual knowledge resources. Therefore, I emphasize that specialized domains such as the one that forms the basis of my research, i.e. the Islamic knowledge domain, needs more than just faceless crowdsourcing. Emerging research paradigm recognizes this in the form of nichesourcing [9,10] or Expert-Sourcing [11], as a natural step in the evolution of the crowdsourcing to address the need of solving complex and knowledge intensive tasks, and as means to improve the quality of crowdsourced input.

Based on these ideas, I propose the design and development of a hybrid workflow framework that combines human-machine computation in a manner that allows the ability to compose tasks to a varying degree of granularity and delegating them to generic and specialized crowds depending on the needs of the task fulfillment requirements. In order to achieve this, I propose the utilization of semantics based representations of tasks, workflows and worker profiles.

2 State of the Art

The potential of HC&C has been leveraged by semantic web researches such as Noy et al. [12], Sarsua et al. [13] and others in attempting to solve the bottlenecks of human experts and the needed human contribution in the semantic web development (ontology engineering) processes. Some early efforts that led to the evolution of this approach include Ontogame [14] and inPho [15]. Two major genres of research may be seen emerging in the last few years, in an attempt to bring human computation methods to the semantic web: (1) Mechanized Labour and (2) Games with a Purpose for the Semantic Web. Several recent research prototypes have attempted to use micro-task crowdsourcing for solving semantic web tasks e.g. ontology engineering [12,13] and linked data management [16]. Recent work by Hanika, Wohlgenannt and collegues [17,18] have attempted to provide tool support for integrating crowdsourcing into ontology engineering processes by providing a plugin for the popular ontology development tool Protégé. Other approaches such as [19] and [20] adopted the Lui von Ahn's "games with a purpose" [21] paradigm for creating the next generation of the semantic web. The idea is to tap on the wisdom of the crowds by providing motivation in terms of fun and intellectual challenge.

The evidence of semantic web techniques applied to improve the state of human computation systems is also emerging. Sabou et al. [22] propose the notion of hybrid genre workflows to overcome the limitations of traditional workflows in the crowdsourcing settings. Research also suggests the use of ontologies to improve the human-computation process [23], which provides the motivation for proposing a semantics driven model of human computation for this research.

3   Research Proposition

In this research, I seek to develop a semantics driven, generic and reusable human computation based framework for semantic annotation, knowledge acquisition and generation of Linked Islamic knowledge. The framework utilizes HC&C paradigm incorporated into hybrid knowledge engineering (ontology development and linked data management) workflows to produce semantics based multi-lingual, linked knowledge resources. A hybrid model of human computation, that leverages both generic workers and experts, is utilized to not only validate and verify the findings obtained through computational approaches such as text mining techniques, but also to perform higher level conceptual problem solving, integrative analysis and inter-linking of knowledge sources at hand. The results of the crowd contributions are used to create formalized, shared knowledge models and benchmarks for distributed knowledge sources and to meaningfully link them in primarily the Islamic Knowledge domain.

The proposed framework is validated through contextualized application to the domain of Islamic and Religious texts, where the overall vision is to provide means to enable efficient and reliable knowledge discovery for religious knowledge seekers for a more meaningful knowledge seeking and learning experience. Some of the key research questions are addressed as part of the process are as follows:

RQ-1: What is the amenability of crowdsourcing ontology engineering tasks in knowledge intensive domains? Can the tasks of thematic disambiguation, semantic annotation and thematic interlinking be reliably crowdsourced in the specialized and knowledge intensive domains such as the Islamic knowledge?

RQ-2: Is there significant performance gap between experts and crowd workers when performing such tasks?

RQ-3: Can the contributions from crowd workers and experts be reliably and efficiently combined for the purpose of knowledge engineering, using semantics driven, hybrid and iterative workflows? What methodological considerations would enable such an effective synergy?

4   Research Methodology and Approach

In order to realize the research vision, and to answer the research questions, I undertake the research methodology, a high level view of which is presented in Fig. 2.

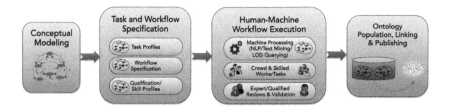

Fig. 2. High level constituent stages of framework design

Conceptual Modeling: The initial step in our approach consists of defining a conceptual model of the domain i.e. an ontology schema. For the initial development, the scope is limited to ontology population tasks. The entities and relations in this conceptual model will then become the candidates for annotation with task profiles, which will define how the instances of the entities and relations will be populated through an iterative workflow of human and machine tasks.

Task and Workflow Specification: As part of this research, I introduce the notion of *Semantics Driven Task Management*, enabled by semantically represented *task profiles*. In this stage, I map the semantic annotation tasks onto associated *task profiles*, which define the data sources, the nature of annotation and disambiguation required, the output mappings and other essential task design parameters. This facilitates modeling tasks to a varying degree of granularity. The provision for customizable and generic task templates is made to encourage reusable tasks templates. The *task profiles* primarily pertain to various common knowledge engineering tasks such as creation of concepts, relations, hierarchies, entity links, annotation and curation to name a few. A key consideration for effective task management is the representation of simple vs. knowledge intensive tasks.

Another key aspect of the framework is enabling *Semantics Driven Workflows*. This is achieved using *workflow profiles*, which contain descriptions of

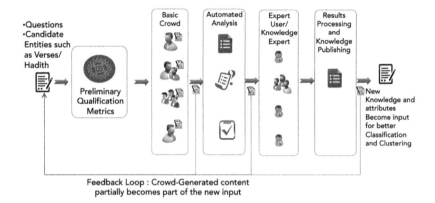

Fig. 3. Illustration of hybrid, iterative workflows

how the tasks that are subject to human vs. machine computation are composed together. The proposed framework utilizes semantics based representations for *workflow specification and management*. This is a unique and novel feature of the framework. Semantic representation of workflows provides the means to dynamically route and delegate suitable tasks to the suitable set of workers based on the skills and knowledge requirements. A workflow may be a composition of machine-task, crowd-task, niche(expert)-task or more depending upon the nature and the complexity of the task subject to the case study. The *workflow profile* captures the composition of tasks, their requirements, processing outcomes and any dependencies amongst other parameters. In relation to the conceptual model, a *workflow profile* for each task will specify the steps required to be performed in order for the entities and relations in a particular triple to be populated. A machine based task in the workflow may include an automated Information Retrieval (IR) task, or a task based on Natural Language Processing (NLP), such as extracting morphological variations of a given word (in Arabic). Text mining techniques are also employed to find portions of semantically relevant texts. Existing knowledge is also retrieved from available LOD sources. For the human driven tasks, a hybrid approach is utilized.

Human-Machine Workflow Execution: Based on the task and workflow specification designed in the previous stage, the system is implemented for executing these workflows. The proposed framework enables *Skill Driven Crowd Management* by utilizing the idea of niche or expert sourcing, managing task allocation and response collection to and from generic and specialized crowds. In the case of Islamic knowledge modeling, niches or experts would essentially be the teachers, scholars/subject-specialists in the domain. As per this idea, distinction is made between the contributors who can only perform simple, atomic tasks, vs. those to whom the knowledge intensive tasks are best delegated to. In addition, an important contribution is the notion of *Iterative and Dynamic*

Workflows, which ensures designing tasks and dynamic workflows, that execute in an iterative manner, suited to the knowledge and the skills of the crowd. Figure 3 shows an illustration of such a workflow, whereby preliminary contributions for a task are gathered from a generic crowd, and based on some analysis and findings, a followup task requires experts or super users to validate or advance upon the contributions obtained to deliver concrete results. Initial efforts aim to bridge and combine the contributions from platforms such as Amazon Mechanical Turk[1] (AMT) and the custom web framework, however, as the framework matures, I consider completely relying on my own framework.

Ontology Population, Linking and Publishing: Once the results are aggregated and validated, the last stage of the framework is designed to automatically populate the ontology under consideration, link with available linked data sources and publish the knowledge base.

5   Ongoing Work and Preliminary Results

My initial work focused on developing a reusable and generic crowdsourcing workflow using the AMT. I have developed and tested this in the context of linked data management tasks, and it has been recently published in our research entitled CrowdLink [24]. Although, this work was limited to linked data management tasks, however based on the experience and findings, I envision that a more mature human-machine computation framework, with hybrid properties, to solve the inherent challenges of knowledge engineering in specialized and knowledge intensive domains may be achieved.

My dwellings into the domain of Islamic knowledge has shown great potential for further undertaking this domain. I have conceptualized a macro-knowledge structure for Islamic knowledge and define macro and micro level links within the Islamic knowledge [4]. I have developed two key case studies to validate the different aspects of the framework. The central theme of the case studies focus on the two primary sources of Islamic knowledge namely: *The Qur'an* and the *Hadith*. The key aspects and the work status of each is summarized:

Case Study A - Thematic Disambiguation, Annotation and Linking of the Quranic Verses: The main focus of this case study includes four tasks: (1) Thematic Disambiguation, (2) Semantic Annotation, (3) Thematic Classification and (4) Semantic Relation Identification for the verses of the Qur'an.

For illustration consider the *Thematic Disambiguation* task. The Arabic is a rich morphological language and a phrase often needs to be disambiguated for its meaning. A typical workflow involves the extraction of the candidate verses that may or may not menifest a particular theme, based on some NLP or information extraction technique. These candidate verses then become input to a micro-task on AMT. I have also implemented a similarity computation framework for the Qur'anic verses [25], for retrieving highly similar and relevant verses. The crowd

---

[1] www.mturk.com.

disambiguates the theme's occurrence in the verse. The responses from the crowd are aggregated and analyzed.

To capture some of this knowledge and links, I have conducted some initial experimentation using the AMT. The results from the AMT have been promising for the simpler tasks, even though the tasks required the workers to have knowledge of the arabic language. However, some tasks such as identifying internal relationships within the verses of the Quran, which require deeper knowledge are not as straight forward.

My ongoing work is focusing on developing decision metrics and getting expert reviews and validations for a subset of the tasks which have been crowdsourced. For this purpose, I have developed a prototype web application, which takes as input the aggregated results of the tasks from the AMT and recommends them to suitable experts for review and validation.

Case Study B - Thematic and Inter-Contextual Linking of the Quran and Hadith: The foremost contribution of this case study is to use the learnersourcing methodology, a specialize form of crowdsourcing, using the custom developed web application to gather annotations, classifications and relationships not only for the verses of the Qur'an but most importantly, the relationships between the Qur'anic verses and the Hadith. This is by far the most knowledge intensive and complex task, since the process may not be automated. The conceptual modeling and the task designs for this case study has been completed. This case study combines all the various aspects of the framework that have been outlined and is currently in progress.

6   Evalution Plan

I evaluate the methodology for my research by contextualizing the domain of application to Islamic knowledge, given its sensitive and knowledge intensive nature. I define task execution workflows pertaining to some key knowledge engineering tasks in the domain primarily focusing on thematic disambiguation and annotation. The purpose of these experiments is two fold. Firstly, to evaluate crowdsourced thematic annotation data in comparison with ground truth obtained from experts. Secondly, to establish the amenability of crowdsourcing for obtaining ground truths in contexts where no prior baseline annotations exist.

In addition, I also plan to design, experiment and analyze other tasks such as semantic relation identification for identifying internal relationships between knowledge units, whereby, a considerable degree of knowledge expertise is required. The aim is to establish the right balance between crowd and expert contribution in achieving reasonably confidence in the acquisition of knowledge.

I will also attempt to carry out an analysis of the extent to which the semantic profiling of tasks and crowds improves the task annotation quality. I expect that some reasonable metrics will evolve in the process. I also aim to investigate to what extent our methodology will be generalizable to other tasks and domains. The results of this analysis will help establish the scalability of my approach.

7   Conclusions

In this paper, I propose a semantics driven framework for combining human and machine computation for the purpose of knowledge engineering. The framework utilizes semantics based tasks, workflow and worker profiles and allows for design and execution of iterative and dynamic workflows based on these profiles. The contextualization that my research aims to tackle, of engineering formalized knowledge models for the purpose of an enhanced knowledge seeking experience in the Islamic knowledge domain, is a high-impact problem, however a non-trivial one. The knowledge at hand is sensitive, and ensuring the credibility and authenticity of knowledge sources is challenging. I therefore believe that a hybrid approach, whereby contributions from crowds and experts, based on skills and knowledge background, combined with automated approaches, will considerably improve the efficiency and reliability of semantic annotation tasks in specialized domains. My initial experiments show favorable results for thematic disambiguation and annotation tasks. For more knowledge intensive tasks, work is underway for obtaining expert contributions and inputs based on aggregated results from crowd inputs. I expect to perform well defined evaluation of our decision metrics, for analyzing and aggregating these annotations. I also envision making the framework generalizable for other knowledge intensive domains.

Acknowledgments. I would like to acknowledge the contributions from my supervisors Dr. Khaled Rasheed and Dr. I. Budak Arpinar for their support and advice.

References

1. Simperl, E., Acosta, M., Flöck, F.: Knowledge engineering via human computation. In: Handbook of Human Computation, pp. 131–151. Springer, New York (2013)
2. Siorpaes, K., Simperl, E.: Human intelligence in the process of semantic content creation. World Wide Web **13**(1–2), 33–59 (2010)
3. Sabou, M., Scharl, A., Michael, F.: Crowdsourced knowledge acquisition: Towards hybrid-genre workflows. Int. J. Semant. Web Inf. Syst. **9**(3), 14–41 (2013)
4. Basharat, A., Rasheed, K., Arpinar, I.B.: Towards linked open islamic knowledge using human computation and crowdsourcing. In: Proceedings of the International Conference on Islamic Applications in Computer Science And Technology (2015)
5. Sherif, M.A., Ngomo, A.C.N.: Semantic Quran - a multilingual resource for natural-language processing. Semant. Web **6**(4), 339–345 (2015)
6. Atwell, E., Brierley, C., Dukes, K., Sawalha, M., Sharaf, A.B.: An artificial intelligence approach to arabic and islamic content on the internet. In: Proceedings of NITS 3rd National Information Technology Symposium (2011)
7. DiFranzo, D., Hendler, J.: The semantic web and the next generation of human computation. In: Handbook of Human Computation, pp. 523–530. Springer, New York (2013)
8. Inel, O., et al.: CrowdTruth: Machine-human computation framework for harnessing disagreement in gathering annotated data. In: Mika, P., et al. (eds.) ISWC 2014, Part II. LNCS, vol. 8797, pp. 486–504. Springer, Heidelberg (2014)

9. de Boer, V., Hildebrand, M., Aroyo, L., De Leenheer, P., Dijkshoorn, C., Tesfa, B., Schreiber, G.: Nichesourcing: Harnessing the power of crowds of experts. In: ten Teije, A., et al. (eds.) EKAW 2012. LNCS, vol. 7603, pp. 16–20. Springer, Heidelberg (2012)

10. Oosterman, J., Bozzon, A., Houben, G.J., et al.: Crowd vs. experts: nichesourcing for knowledge intensive tasks in cultural heritage. In: International WWW Conferences Steering Committee, pp. 567–568 (2014)

11. Retelny, D., Robaszkiewicz, S., To, A., Lasecki, W.S., Patel, J., Rahmati, N., Doshi, T., Valentine, M., Bernstein, M.S.: Expert crowdsourcing with flash teams. In: Proceedings of the 27th Annual ACM Symposium on User Interface Software and Technology, pp. 75–85. ACM (2014)

12. Noy, N.F., Mortensen, J., Musen, M.A., Alexander, P.R.: Mechanical turk as an ontology engineer? Using microtasks as a component of an ontology-engineering workflow. In: Proceedings of the 5th Annual ACM Web Science Conference, WebSci 2013, pp. 262–271. ACM, New York (2013)

13. Sarasua, C., Simperl, E., Noy, N.F.: CROWDMAP: Crowdsourcing ontology alignment with microtasks. In: Cudré-Mauroux, P., et al. (eds.) ISWC 2012, Part I. LNCS, vol. 7649, pp. 525–541. Springer, Heidelberg (2012)

14. Siorpaes, K., Hepp, M.: OntoGame: Weaving the Semantic Web by Online Games. In: Bechhofer, S., Hauswirth, M., Hoffmann, J., Koubarakis, M. (eds.) ESWC 2008. LNCS, vol. 5021, pp. 751–766. Springer, Heidelberg (2008)

15. Niepert, M., Buckner, C., Allen, C.: Working the crowd: Design principles and early lessons from the social-semantic web. In: Proceedings of the Workshop on Web 3.0: Merging Semantic Web and Social Web at ACM Hypertext (2009)

16. Simperl, E., Acosta, M., Norton, B.: A semantically enabled architecture for crowdsourced linked data management. In: CrowdSearch, pp. 9–14. Citeseer (2012)

17. Hanika, F., Wohlgenannt, G., Sabou, M.: The uComp Protégé plugin: crowdsourcing enabled ontology engineering. In: Janowicz, K., Schlobach, S., Lambrix, P., Hyvönen, E. (eds.) EKAW 2014. LNCS, vol. 8876, pp. 181–196. Springer, Heidelberg (2014)

18. Adams, B., Wohlgenannt, G., Sabou, M., Hanika, F., et al.: Crowd-based ontology engineering with the ucomp protégé plugin. Semantic Web (Preprint), pp. 1–20

19. Siorpaes, K., Hepp, M.: Games with a purpose for the semantic web. IEEE Intell. Syst. 23(3), 50–60 (2008)

20. Simko, J., Bielikov, M.: Semantic acquisition games, pp. 35–50 (2014)

21. Von Ahn, L., Dabbish, L.: Designing games with a purpose. Commun. ACM 51(8), 58–67 (2008)

22. Sabou, M., Scharl, A., Fols, M.: Crowdsourced knowledge acquisition: Towards hybrid-genre workflows. Int. J. Semant. Web Inf. Syst. 9(3), 14–41 (2013)

23. Luz, N., Silva, N., Novais, P.: Generating human-computer micro-task workflows from domain ontologies. In: Kurosu, M. (ed.) HCI 2014, Part I. LNCS, vol. 8510, pp. 98–109. Springer, Heidelberg (2014)

24. Basharat, A., Arpinar, I.B., Dastgheib, S., Kursuncu, U., Kochut, K., Dogdu, E.: Semantically enriched task and workflow automation in crowdsourcing for linked data management. Int. J. Semant. Comput. 8(04), 415–439 (2014)

25. Basharat, A., Yasdansepas, D., Rasheed, K.: Comparative study of verse similarity for multi-lingual representations of the qur'an. In: Proceedings on the International Conference on Artificial Intelligence (ICAI), pp. 336–343 (2015)

Towards Scalable Federated Context-Aware Stream Reasoning

Alexander Dejonghe[⊠]

Department of Information Technology (INTEC), Ghent University - iMinds,
Gaston Crommenlaan 8 Bus 201, 9050 Ghent, Belgium
Alexander.Dejonghe@intec.ugent.be

Abstract. With the rising interest in internet connected devices and sensor networks, better known as the Internet of Things, data streams are becoming ubiquitous. Integration and processing of these data streams is challenging. Semantic Web technologies are able to deal with the variety of data but are not able to deal with the velocity of the data. An emerging research domain, called stream reasoning, tries to bridge the gap between traditional stream processing and semantic reasoning. Research in the past years has resulted in several prototyped RDF Stream Processors, each of them with its own features and application domain. They all cover querying over RDF streams but lack support for complex reasoning. This paper presents how adaptive stream processing and context-awareness can be used to enhance semantic reasoning over streaming data. The result is a federated context-aware architecture that allows to leverage reasoning capabilities on data streams produced by distributed sensor devices. The proposed solution is stated by use cases in pervasive health care and smart cities.

Keywords: Semantic stream processing · Stream reasoning · Context awareness · Internet of Things (IoT)

1 Introduction

In recent years we saw an increasing interest in internet connected devices and sensors, also called the Internet of Things (IoT) [18]. This led to a plethora of new applications in different domains like smart cities, traffic monitoring, pervasive healthcare, smart energy grids, smart buildings and environmental sensing. The data generated by these IoT devices is a heterogeneous, voluminous, and a possibly noisy or incomplete set of time-varying data elements called data streams. The combination of these characteristics makes it a challenging task to integrate, interpret and process data streams on the fly. Moreover, to create added value out of these data streams, the data must be combined with domain knowledge.

For example, in smart nursing homes and hospital, rooms are equipped with Wireless Sensor Networks (WSNs) monitoring the environment, and patients' health parameters are monitored by Body Area Networks (BANs). Lots of data

© Springer International Publishing Switzerland 2016
H. Sack et al. (Eds.): ESWC 2016, LNCS 9678, pp. 803–812, 2016.
DOI: 10.1007/978-3-319-34129-3_49

events are generated by these sensor networks. To take advantage of the collected streaming data, integration with domain knowledge containing diseases, symptoms and the patient's Electronic Health Record (EHR) is important. Reasoning on this information can infer new knowledge of the patients' current condition. Well integrated and aggregated information helps to monitor the patients more closely and allows the nursing resources to operate more efficiently.

Semantic Web technologies like the data model RDF[1], the ontology language OWL[2], and the RDF query language SPARQL[3] allow to represent, integrate, query and reason on heterogeneous data. However, these technologies were developed for static or slow changing data sources. On the other end of the spectrum Data Stream Management Systems (DSMS) and Complex Event Processing (CEP) systems allow to query homogeneous streaming data structured according to a fixed data model. They are not able to deal with heterogeneous data sources and lack support for the integration of domain knowledge. To bridge this gap, stream reasoning has emerged as a challenging research area that focuses on the adoption of Semantic Web technologies for streaming data. Della Valle et al. [7] describes stream reasoning as a high-impact research area with a multidisciplinary approach that can provide the abstractions, foundations, methods, and tools required to integrate data streams, the Semantic Web, and reasoning systems.

As a result of stream reasoning research conducted in the past years, different prototypes of RDF Stream Processing (RSP) engines have been presented. These RSP engines can filter and query RDF data streams, but are not able to deal with complex reasoning tasks. Reasoning over large complex ontologies is computationally intensive and slow compared to the velocity of the data streams. Traditional reasoners, like FaCT++ [19], HermiT [15] and Pellet [16], that have the capabilities of performing such complex OWL 2 DL reasoning tasks are designed to process static or slow evolving data, and are not able to manage frequently changing data streams.

Despite all initiatives taken, stream reasoning is not yet mature and there is a need for algorithms, protocols and approaches that support a scalable, efficient and complex stream reasoning [11].

The aim of this research is to present a federated context-aware system that integrates adaptive stream processing and complex reasoning. The federated system should exploit context-awareness to bring together low-level stream reasoning and high-level complex reasoning. The low-level stream reasoning should be performed close to the data stream sources while complex reasoning should be performed deeper in the network.

This approach might especially be useful in large scale sensor applications where an effective processing of the event streams is of high importance. By using context information about the environment, irrelevant sensor data can

---

[1] http://www.w3.org/TR/rdf-syntax-grammar/.

[2] http://www.w3.org/TR/owl-overview/.

[3] http://www.w3.org/TR/sparql11-overview/.

be filtered and only applicable domain knowledge can be take into account for reasoning.

2 State of the Art

Several RSP solutions such as C-SPARQL [4], CQELS [10], EP-SPARQL [2] and SPARQLStream [6], which all mainly focus on stream processing, have been developed in the past years. They extend SPARQL by using proven techniques from DSMSs and CEP systems, namely sliding windows and continuous queries [3]. A continuous query is registered once and produces results continuously over time as the streaming data in the considered window changes.

The prototyped RSP engines enable processing of a continuous flow of data and can provide real-time answers to registered queries. However, each of them has different semantics and targets different scenarios. Steps towards a unifying semantics query model are being taken by the W3C RDF Stream Processing Community Group[4].

Other solutions like Sparkwave [9] and INSTANS [14] make use of extensions of the RETE algorithm [8] for pattern matching. With this approach queries are translated into a RETE network through which the data flows. The resulting network consists of a set of nodes which can memorize partial pattern matches in the streaming data.

All RSP engines, with the exception of INSTANS, support integration of domain knowledge in the querying process, but reasoning capabilities are limited (Table 1). None of the proposed systems is able to perform complex OWL 2 DL reasoning on streaming data. C-SPARQL is the only engine supporting full RDFS reasoning using the Jena Rule Engine. EP-SPARQL, that is build around the Prolog engine ETALIS, also supports RDFS reasoning but the domain ontologies have to be converted into Prolog rules and facts in advance. Sparkwave supports an RDFS subset but as for EP-SPARQL preprocessing is necessary. Both, the domain knowledge and the query conditions have to be compiled into nodes of RETE network in advance. SPARQLStream, CQELS and INSTANS do not provide reasoning features.

Incremental reasoning helps reasoners to handle streaming data by incrementally maintaining the materialization of the knowledge base. By only considering the data that is subject to change, incremental reasoning tries to avoid re-materializing the complete knowledge base [5,13]. However, also this approach is subject to limitation and assumptions.

Despite all effort, reasoning capabilities remain limited due to the gap between the changing frequency of streaming data and the computing time demanded by complex reasoning algorithms. Cascading reasoning is a concept presented by Stuckenschmidt et al. [17] to deal with this problem. The aim is to construct a processing hierarchy by exploiting the trade-off between the complexity of the reasoning method and the frequency of the data stream the reasoner

---

[4] http://www.w3.org/community/rsp/.

Table 1. Reasoning support in state of the art RDF Stream Processing solutions

	Background knowledge	Reasoning capabilities
C-SPARQL	Y	RDFS
SPARQLStream	Y	N
EP-SPARQL	Y	RDFS (in Prolog)
CQELS	Y	N
Sparkwave	Y	RDFS subset
INSTANS	N	N

is able to handle. At the lower levels of the hierarchy, with high frequency data streams, we focus on filtering to reduce the change frequency. The higher in the hierarchy the more complex reasoning can be applied. This approach helps to avoid feeding high frequency data directly to complex reasoners.

StreamRule [12] is a 2-tier approach combining stream processing with rule-based non-monotonic incremental Answer Set Programming (ASP) to enable the ability of reasoning over data streams. The novelty of this approach is that the size of the input stream towards the reasoner is reduced by the stream processor as the reasoning task becomes more computationally intensive. To allow this a feedback loop from the reasoner towards the stream processing is needed.

3   Problem Statement and Contributions

The available RSP engines aim filtering and querying on streaming data but lack support for complex OWL 2 DL reasoning. To perform such complex reasoning we need traditional DL reasoners which are computationally intensive and not capable of handling streaming data. From the analysis of the state of the art, we identified two problems we would like to address with this research:

P1: When integrating an RSP engine together with an OWL reasoning engine using a pipeline architecture, as worked out by Mileo et al. [12] for ASP reasoning, effective stream processing is necessary. Because of the limitations of the reasoners towards streaming data, it is important to neglect irrelevant data streams. Moreover, it is important to choose appropriate window parameters. Today, window parameters and query conditions are defined in advance. There is a lack of adaptive stream reasoning taking into account the changes in stream characteristics and domain knowledge at runtime.

P2: IoT and sensor networks will only bring more streaming data flooding our networks. One of the solutions to deal with this is to (pre-)process data close to its source [11]. This approach fits with IoT architectures proposing edge computing as an intermediate layer between data acquisition and the cloud based processing layer [1]. This intermediate layer allows filtering and aggregation of data, resulting in reduced network congestion, less latency and improved scalability. We believe that this edge computing approach, is in line with the idea of cascading reasoners to reduce the change frequency of the data [17], and can be useful to increase reasoning capabilities on streaming data.

Out of the discussed problems we formulate the following two hypotheses:

H1: When integrating an RSP engine and an OWL reasoner in a pipeline architecture, a context-dependent controller using stream characteristics and domain knowledge can adapt at runtime, the window parameters and filtering conditions of the RSP engine, in such a way it increases the throughput of the pipeline.

H2: A federated reasoning platform, making use of context-awareness, to combine low-level reasoners positioned close the data stream generator, and high-level reasoners positioned deeper in the network, will leverage the reasoning capabilities over streaming data. By considering the context in which the data streams are generated the applicable domain knowledge can be narrowed, streaming frequency can be reduced and reasoning capabilities can be leveraged.

The following research questions are targeted to prove these hypotheses:

Q1: *How can adaptive stream reasoning be implemented?*
The aim is to investigate how the actual stream characteristics and domain knowledge can be used to adapt window parameters and query conditions at runtime.
   Q1.1: *Do there exists relationships between stream characteristics, window parameters and throughput which can be used to adapt the window parameters at runtime?*
   Q1.2: *Can information incorporated in the domain knowledge be used to adapt the window parameters at runtime?*
   Q1.3: *Can information incorporated in the domain knowledge be used to change query conditions at runtime?*

Q2: *How can context-awareness be exploited to manage domain knowledge among distributed reasoning systems?*
Low-level reasoning should be performed close to the network edges and the data stream generators. To reduce the size and complexity of the knowledge bases among the different systems context information should be used.
   Q2.1: *How can context-awareness be used to distribute domain knowledge among different reasoning systems?*
   Q2.2: *How to manage the distributed domain knowledge when the context changes?*
   Q2.3: *How to deal with changes in the domain knowledge that is shared among different reasoning systems?*

Q3: *How to build a federated context-aware reasoning system in which distributed adaptive stream reasoning is combined with a global reasoning system using the results of Q1 and Q2?*
Adaptive RSP engines (Q1) should be placed close to the data stream generators and being supervised by a more complex reasoning system for supporting global decision making (Q2).

4   Research Methodology and Approach

The different steps and approaches we will use to answer the research questions are discussed in the following section making use of Fig. 1.

We start our research by creating an integrated reasoning environment combining both an RSP engine and a OWL DL reasoner (Fig. 1 Q1). The output of the RSP engine will be forwarded to and processed by an OWL DL reasoner. The resulted framework allows us to perform experiments with different RSP engines and OWL DL reasoners to process sensor data streams. The tests will focus on throughput and latency and keep into account different semantics of current RSP engines. The outcome of the conducted experiments will lead to the selection of an RSP engine and OWL reasoner pair which will be used in later research.

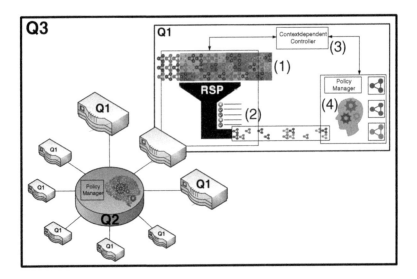

Fig. 1. Research approach

In a second phase we look how adaptive windows (Fig. 1 Q1(1)) and adaptable query conditions (Fig. 1 Q1(2)) can be integrated in the RSP engine. We investigate how these features can be used to increase overall querying performance. More specific, we want to know the impact of different window parameters on throughput, latency, network load and resource usage.

Using the outcome of the previous phase, we introduce a context-dependent controller (Fig. 1 Q1(3)) that controls the RSP engine based on available domain knowledge. We investigate how context information included in the knowledge base, can be used by the context-dependent controller to determine windows parameters and query conditions used by the RSP engine. Results of the previous phase, about the impact of window parameters, should be taken into account by the controller when determining window parameters.

Next, we focus on the OWL DL reasoning (Fig. 1 Q1(4)). We investigate how modularization of the ontology in combination with context-awareness can be used to improve reasoning performance. For making this possible, we think about the introduction of semantic policies and rules which allow to neglect or deactivate certain modules based on context information available in the knowledge base.

In the next phase, we focus on the distribution of the domain knowledge among reasoners positioned in different locations (Fig. 1 Q2). A centralized node will be the only reasoning system dealing with the complete knowledge base. It will manage and distribute knowledge over the edge reasoners depending on their context. To apply context-awareness we have to identify which context information is available, how we can access it, and how we will manage it. Once the distributed system is running, we have to deal with synchronization of domain knowledge and changes in the context of the different systems.

The latter part of the research consist of putting together the solutions found in Q1 and Q2, in such a way we get a federated context-aware stream reasoning platform (Fig. 1 Q3). Success of this step is highly dependent of previous results.

5 Initial Investigation

As initial step a literature study on the state of the art of stream reasoning and the RSP engines has been conducted. The most mature technologies where presented in Sect. 2. At the moment we are working on a test environment for RSP engines.

To perform credible tests with the RSP and reasoning engines, it is important to do some tests in a real-life context, and not only in a simulated environment. For this we can make use of the iMinds iLab.t[5] testing facilities. These include: the Virtual Wall, the w-ilab.t testbed and the iMinds Homelab. The Virtual Wall, a fully configurable set of servers, can be used to test distributed reasoning in the future. The w-ilab.t testbed consisting of 200 sensor nodes can be used to generate and test stream reasoning and RSP engines.

6 Evaluation Plan

The researched algorithms will be evaluated using two cases. The first one consists of a pervasive healthcare use case based on a continuous care OWL DL ontology. It will be used to evaluate the performance of the solutions presented for the different research questions. On the one hand, we are interested in throughput and latency to verify processing speed of the solutions. We expect the reasoning component on the gateway (Fig. 1 Q1(1)) to be the bottleneck but we strive to prevent queuing of events. On the other hand, we want to see the impact of the choices we made on network traffic. Using local reasoning at the gateways should result in a decrease of the number of events arriving at the central reasoner component.

---

[5] http://ilabt.iminds.be/testbeds/.

Pervasive Healthcare. The proposed solution is explained using Fig. 2. The environment is a nursing home equipped with sensors monitoring both, the patients' room (yellow box) and the patients' health parameters (green box). The sensor data of a single room are captured by a smart gateway, presented as Q1 in Fig. 1. On their turn all these gateways are connected to a central reasoning system, presented as Q2 in Fig. 1. The continuous care ontology used in this use case models among others sensors, rooms, patients, medical diagnoses, nurses and nursing activities.

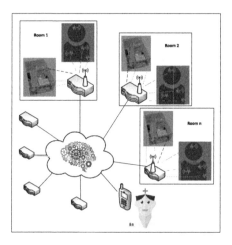

Fig. 2. Patient monitoring (Color figure online)

Suppose a patient, suffering from a concussion, stays in room 2 (Fig. 2). According to the domain ontology such a patient is sensitive to light and noise, hence has to recover in a dark and quiet environment. The patient and its room are monitored with different sensors including light, sound, temperature, blood pressure, heart rate and body temperature sensors.

Sensor values obtained on the patient and in the patient's room are collected by the smart room gateway. The gateway also receives the patients' EHR. Based on the medical diagnose in the EHR, the sense rate of the sensors, which is part of the domain knowledge, and the actual stream characteristics, the data streams can be filtered in an intelligent way. Light and sensor values, which are responsive values, should be monitored more closely than room temperature, room temperature, blood pressure and others which are of less significance.

When increasing light or sound values are registered, the gateway reasoner detects the possible unpleasant situation for the patient. A warning event will be sent to the central nursing reasoning system. Based on the current locations, competences and activities of the nurses the system decides who can visit the patient. A nursing task is assigned and sent to the mobile device of the nurse.

A second use case, about smart cities, will be used in a later part of the research to show applicability in an IoT domain different from healthcare.

Smart Cities. This use case is about parking and traffic information in smart cities. Consider a city where roads are equipped with traffic sensors and parking spot are monitored on their availability. In stead of collecting all sensor data in one single place, smart gateways capturing the data are distributed over the city. Each gateway monitors a district, a couple of streets or parking site.

Based on their location, gateways are informed about events taking place in their environment by a central reasoning system. For example public markets, manifestations or road works will lead to unaccessible roads and parking spots. Using this knowledge, smart gateways can filter out irrelevant sensor data about unaccessible places that would lead to false positives. As a result, more accurate information about parking spots can be forwarded to the central system.

In another situation, a sudden increase or decrease of the traffic intensity in a certain area, without any apparent reason, can be an indication of an unexpected incident. In this case a warning event should be sent to the central system.

In the central reasoning system, information of the different regions will be merged and used for traffic routing towards parking spots or for the creation of heat maps with information about parking spots, traffic and incidents.

7   Conclusions

This paper presented our vision on how adaptive stream processing and context-awareness can help to deal with challenges in stream reasoning. The aim is to exploit context-awareness in an attempt to bridge the gap between stream processing and complex reasoning. We described our approach and discussed it using use cases in pervasive healthcare en smart cities. With this approach we intend to contribute to future federated stream reasoning solutions.

Acknowledgments. This research was partly funded by the strategic research project DiSSeCt funded by the IWT and the CAPRADS Project co-funded by the IWT, iMinds, Luciad, Televic and JForce.

References

1. Al-Fuqaha, A., Guizani, M., Mohammadi, M., Aledhari, M., Ayyash, M.: Internet of things: a survey on enabling technologies, protocols and applications. IEEE Commun. Surv. Tutorials **99**, 1 (2015)
2. Anicic, D., Fodor, P., Rudolph, S., Stojanovic, N.: EP-SPARQL: a unified language for event processing and stream reasoning. In: Proceedings of the 20th International Conference on World Wide Web, pp. 635–644 (2011)
3. Arasu, A., Babu, S., Widom, J.: The CQL continuous query language: semantic foundations and query execution. VLDB J. **15**(2), 121–142 (2006)
4. Barbieri, D.F.: C-SPARQL: SPARQL for continuous querying. In: Proceedings of the 18th International Conference on WWW. vol. 427, pp. 1061–1062 (2009)
5. Barbieri, D.F., Braga, D., Ceri, S., Della Valle, E., Grossniklaus, M.: Incremental reasoning on streams and rich background knowledge. In: Aroyo, L., Antoniou, G., Hyvönen, E., ten Teije, A., Stuckenschmidt, H., Cabral, L., Tudorache, T. (eds.) ESWC 2010, Part I. LNCS, vol. 6088, pp. 1–15. Springer, Heidelberg (2010)

6. Calbimonte, J.-P., Corcho, O., Gray, A.J.G.: Enabling ontology-based access to streaming data sources. In: Patel-Schneider, P.F., Pan, Y., Hitzler, P., Mika, P., Zhang, L., Pan, J.Z., Horrocks, I., Glimm, B. (eds.) ISWC 2010, Part I. LNCS, vol. 6496, pp. 96–111. Springer, Heidelberg (2010)

7. Valle, E.D., Ceri, S., Harmelen, F.V., Fensel, D.: It's a streaming world! Reasoning upon rapidly changing information. IEEE Intell. Syst. **6**, 83–89 (2009)

8. Forgy, C.L.: Rete: a fast algorithm for the many pattern/Many object pattern match problem. Artif. Intell. **19**(1), 17–37 (1982)

9. Komazec, S., Cerri, D., Fensel, D.: Sparkwave: continuous schema-enhanced pattern matching over RDF data streams. In: Proceedings of the 6th ACM International Conference on Distributed Event-Based Systems, pp. 58–68 (2012)

10. Le-Phuoc, D., Dao-Tran, M., Xavier Parreira, J., Hauswirth, M.: A native and adaptive approach for unified processing of linked streams and linked data. In: Aroyo, L., Welty, C., Alani, H., Taylor, J., Bernstein, A., Kagal, L., Noy, N., Blomqvist, E. (eds.) ISWC 2011, Part I. LNCS, vol. 7031, pp. 370–388. Springer, Heidelberg (2011)

11. Margara, A., Urbani, J., van Harmelen, F., Bal, H.: Streaming the web: reasoning over dynamic data. Web Seman. Sci. Serv. Agents World Wide Web **25**, 24–44 (2014)

12. Mileo, A., Abdelrahman, A., Policarpio, S., Hauswirth, M.: StreamRule: a non-monotonic stream reasoning system for the semantic web. In: Faber, W., Lembo, D. (eds.) RR 2013. LNCS, vol. 7994, pp. 247–252. Springer, Heidelberg (2013)

13. Ren, Y., Pan, J., Zhao, Y.: Towards scalable reasoning on ontology streams via syntactic approximation. In: Proceedings of IWOD (2010)

14. Rinne, M., Nuutila, E., Törmä, S.: INSTANS: High-performance event processing with standard RDF and SPARQL. In: 11th International Semantic Web Conference ISWC 2012, vol. 914, pp. 101–104 (2012)

15. Shearer, R., Motik, B., Horrocks, I.: HermiT: A Highly-efficient OWL reasoner. In: OWLED, vol. 432 (2008)

16. Sirin, E., Parsia, B., Grau, B.C., Kalyanpur, A., Katz, Y.: Pellet: a practical OWL-DL reasoner. Web Seman. Sci. Serv. Agents World Wide Web **5**, 51–53 (2007)

17. Stuckenschmidt, H., Ceri, S., Della Valle, E., van Harmelen, F.: Towards expressive stream reasoning. In: Proceedings of the Dagstuhl Seminar on Semantic Aspects of Sensor Networks, pp. 1–14 (2010)

18. Sundmaeker, H., Guillemin, P., Friess, P.: Vision and Challenges for Realising the Internet of Things. Publications Office of the European Union, Luxembourg (2010)

19. Tsarkov, D., Horrocks, I.: FaCT++ description logic reasoner: system description. In: Proceedings of the Third International Joint Conference pp. 292–297 (2006)

Machine-Crowd Annotation Workflow for Event Understanding Across Collections and Domains

Oana Inel[1,2]([⊠])

[1] Vrije Universiteit Amsterdam, Amsterdam, The Netherlands
oana.inel@vu.nl
[2] IBM Center for Advanced Studies Benelux, Amsterdam, The Netherlands
oana.inel@nl.ibm.com

Abstract. People need context to process the massive information online. Context is often expressed by a specific event taking place. The multitude of data streams used to mention events provide an inconceivable amount of information redundancy and perspectives. This poses challenges to both humans, *i.e.*, to reduce the information overload and consume the meaningful information and machines, *i.e.*, to generate a concise overview of the events. For machines to generate such overviews, they need to be taught to understand events. The goal of this research project is to investigate whether combining machines output with crowd perspectives boosts the event understanding of state-of-the-art natural language processing tools and improve their event detection. To answer this question, we propose an end-to-end research methodology for: machine processing, defining experimental data and setup, gathering event semantics and results evaluation. We present preliminary results that indicate crowdsourcing as a reliable approach for *(1)* linking events and their related entities in cultural heritage collections and *(2)* identifying salient event features (*i.e.*, relevant mentions and sentiments) for online data. We provide an evaluation plan for the overall research methodology of crowdsourcing event semantics across modalities and domains.

Keywords: Crowdsourcing · Event extraction · Machine-human computation · Information extraction · Event semantics annotation

1   Introduction/Motivation

With the progress on the Web, significant amounts of information are made available online. The information ranges from different data types such as tweets, news, cultural heritage and news archives and across various distribution channels such as traditional or social media. This poses a lot of challenges for search engines and information retrieval systems as they need *(1)* to extract meaningful information from any modality (*i.e.*, text, image, video) and *(2)* to synthesize streams from various channels in order to provide succinct pieces of information

© Springer International Publishing Switzerland 2016
H. Sack et al. (Eds.): ESWC 2016, LNCS 9678, pp. 813–823, 2016.
DOI: 10.1007/978-3-319-34129-3_50

that answer the end user needs. Thus, there is a challenge to interpret the information gain of each data stream, identify the meaningful pieces of information and generate a concise and complete summary of all the information requested.

Events are by definition complex entities, essential for querying, perceiving and consuming the meaning of the information we are surrounded by. We need to understand what an event is, how to describe an event and to what extent an event is useful for searching on a given topic. Usually, events create context by introducing related entities such as participants involved, locations where the event takes place or the time period when the event takes place. For everyday events, the event space is represented in the different data streams and channels. Hence, besides relevance, we need to extend the event understanding with salience, novelty features, thus, minimized redundancy, multitude of perspectives and subjective semantics such as sentiments and sentiment intensities.

The natural language processing (NLP) community recognizes the importance of events [1,2]. While the accuracy of the NLP tools for extracting named entities (NE) is continuously improving, their performance in detecting events is still poor. The reasons are three-fold: *(1)* events are vague, *(2)* events carry multiple perspectives and *(3)* events have different granularity. The mainstream procedure for event annotation is by means of experts. However, even experts disagree a lot. To overcome this, people create strict annotation guidelines which instead, make the task rigid and hardly adaptable to other domains. This over-generalization does not deal with the intrinsic ambiguity, the multitude of interpretations and perspectives of the language. Thus, many NLP tools suffer from lack of training and evaluation data [3], as well as understanding the ambiguity. Setting up annotation tasks is also time and cost consuming due to both the length of the process and the costs associated with the experts. The constant lack of training data is also a downfall for increasing the performance of tools to automatically assess event novelty [4] and event clustering (*e.g.*, Google News[1]).

Crowdsourcing has emerged as a reliable, time and cost efficient approach for gathering semantic annotations. Typical solutions for assessing the quality of crowdsourced data are based on the hypothesis [5] that there is only one right answer. However, this contradicts with the three angles of events, *i.e.*, vagueness, multiple perspectives and granularities and with the natural language ambiguity. Recent work [6,7] has shown that disagreement between workers is a signal for identifying low quality workers and provides better understanding of the data ambiguity. A major crowdsourcing bottleneck is that most practices are not systematic and sustainable, while state-of-the-art methods are only developed for a specific domain or input. Crowdsourcing became an efficient way of gathering ground truth data for active learning systems [8] as well. However, this did not change the assumption that there is only one correct answer. Thus, the variety of perspectives, interpretations and language ambiguity are still not considered.

The current research defines events as *something that happened, is happening or will happen*, thus, using minimum restrictions. The primary purpose and focus is to gain event understanding by exploring event streams with regard to

---

[1] https://news.google.com.

(1) surface form, *i.c.*, event granularity; *(2)* space, *i.e.*, actors, location, time period; *(3)* relevance, *i.e.*, the most representative entities or phrases; *(4)* subjective perspectives, *i.e.*, the sentiment an event or entity triggers; *(5)* novelty and salience, *i.e.*, new or notable event features. Our experimental workflow builds on *(1)* machine-optimized stages where the data is pre processed and pre-annotated with semantics and *(2)* crowd-driven gathering of event semantics ground truth. The novelty of the research comes from the event-centric approach of generating a ground truth of events, *i.e.* dealing with various concepts around events.

Our aim is to investigate how events are perceived and represented across data modalities (*e.g.*, text, image, video), sources (*e.g.*, news articles, tweets, video broadcasts) and languages. We want to analyze to what extent the crowd can help the NLP tools to understand events and improve their event detection. In summary, our aim is to improve event understanding and translate the data that we gather through experiments into a ground truth that can be used for training machines. Further, we aim to integrate machines and humans in a systematic way, *i.e.*, with focus on experimental methodologies and replicability and a sustainable way, *i.e.*, with focus on reusability of data, code and results.

2   State of the Art

In the literature, the process of detecting and extracting events presents high interest. The research focussed on this topic covers multiple perspectives, among others: event detection, related entities extraction, sentiments and novelty. Many NLP tools deal with extracting NE [1], but only a few with event detection such as, *e.g.*, OpenCalais[2], FRED[3]. One drawback of the supervised machine learning tools is the need of manually annotated data. For each new domain or data type new annotation guidelines need to be created. Preliminary research has been done on extracting earthquake-related events from tweets using distant supervision [9]. Their results show a performance of 88 % compared to a system that was trained on manually annotated microblogs. Using the sentence dependency tree provides F1-score of 53 % for event recognition on biomedical data [2].

For extracting event-related concepts such as people, locations, NER tasks are envisioned. In [10], the authors extract such concepts from video synopsis. State-of-the-art NE extractors [1] are developed using different algorithms and training data, making each targeted for specific NE recognition and classification tasks or more reliable on particular data [11]. [12] shows that evaluating agreement among extractors is effective: entities missed by one extractor can be found by others. Annotation of texts with heterogeneous topics and formats benefits from integrating extractors [12,13]. Nevertheless, semantic annotation of texts with heterogeneous topics, like news articles or TV-news bulletins is challenging, due to difficulties in training a single extractor to perform well across domains.

Agreement among NER tools is well captured by majority vote systems [14]. However, this could cut off relevant information such as, information supported

---

[2] http://viewer.opencalais.com/.
[3] http://wit.istc.cnr.it/stlab-tools/fred.

by only one extractor and cases with more than one solution. The evaluation of different extractor results will show disagreement, thus, aggregation of different NER tool results does not always solve the problem. Capturing events and other keywords from videos is even more challenging. A mainstream approach to tackle this problem is through automatic enrichment of the metadata [15]. This can be done either through machine processing of textual documents related to the videos [10,16], or through crowdsourcing descriptive tags for the media itself [17] and the media description [18]. Each of those approaches achieves reasonable results, however, each of them processes the videos from only one perspective.

Extensive research is also performed on event summarization, novelty and sentiment analysis. In [19], the authors perform single document summarization for creating news highlights by combining news articles with microblogs. However, this method has a very restrictive set of tweets that are considered relevant, *i.e.*, only the tweets that are linked to the article. An extensive literature study of automated novelty detection in texts is presented in [20]. Crowdsourcing proved to be a useful method for gathering ground truth on temporal events ordering in [21]. Another dimension of events is given by the sentiment [22,23]. Crowdsourcing has been also used on annotating NE in tweets [24], but its value on event annotation has not been thoroughly tackled until now.

We perform the crowdsourcing experiments in the context of the CrowdTruth [25] approach and methodology [6,7]. Using this approach the crowd annotations are stored in a vectorized fashion on which we apply cosine measures to identify low quality workers, unclear and ambiguous input units and annotation labels.

3   Problem Statement and Contributions

Current research builds upon the limitations of existing approaches on gathering and detecting event semantics presented in Sect. 2. Furthermore, we define the main research question: *can event detection tools benefit and gain event understanding by employing hybrid machine-optimized crowd-driven event semantics?*. The research novelty is two-fold: *(1)* the approach of gathering a ground truth by studying how events are represented in different modalities; *(2)* presenting the results in a machine readable form for improving NLP tools performance. To answer the main research question, we focus on the following sub-questions:

1. *How can we take advantage of existing natural language processing tools in order to ease and optimize the process of understanding events?*
   Many NLP tools deal with extracting NE that usually define the event space, but, they all have different precision. We aim to identify key features of the data for which different tools perform better than the rest and thus, take an informed decision of which tool should be used for the case at hand.
2. *How can we employ an optimized, replicable across data types, hybrid machine-crowd workflow for event understanding?*
   We focus on developing a workflow that combines machine processing and crowd perspectives for understanding event-related features across data types.

3. *How can we provide reliable crowdsourced training data to automated tools?*
   The multitude of crowdsourcing experiments that we need to perform to understand events requires us to validate the existing CrowdTruth metrics. Moreover, we need to adapt or define new disagreement metrics in the context of the CrowdTruth methodology to provide reliable crowdsourced data.
4. *How can we improve existing event and event-related feature annotation tools with machine-optimized and crowd-generated data?*
   Throughout the entire research process we focus on event exploration and understanding. As a final goal, we aim to grasp the expertise and semantics gained and ingest them as training data in specialized, existing NLP tools.

4 Research Methodology and Approach

We propose to answer the research questions defined in Sect. 3 and tackle some of the limitations presented in Sect. 2 through hybrid machine-crowd workflows. Our research methodology is three-fold, as shown in Fig. 1: *(i) data enrichment* layer where the data is enriched by machines and humans throughout a series of tasks; *(ii) analytics* layer that evaluates the results of the data enrichment layer and generates feedback; *(iii) feedback* layer that creates a continuous feedback loop throughout the methodology.

4.1 Data Enrichment Layer

The first step of our methodology is to apply relevant *machine processing and annotation* for the data input type and language at hand, while determining the *suitable input data*. In general, events are characterized by many features and thus, their detection and interpretation need multiple iterations and various annotation goals. Thus, the next step is *defining the annotation task*. At this stage we decide over the features that we want to annotate, implement the *annotation template*, choose the *crowdsourcing platform* and setup the *crowd-specific settings*. For every crowdsourcing experiment we have a *pilot run* that helps us identify the proper setup of the aforementioned methodology steps.

A couple of aspects make our methodology an efficient approach for gathering event understanding. On the one hand, our crowdsourcing tasks are replicable across data types (*e.g.*, running sentiment analysis and novelty ranking for news and tweets). This reduces the time spent to setup a proper crowdsourcing experiment for a new data type and generalizes the approach. One the other hand, we perform a gradual workflow of crowdsourcing experiments. The output of one crowdsourcing task becomes the input for a following task (*i.e.* data units that are not relevant for our event understanding goal are immediately discarded).

4.2 Analytics Layer

Human annotation is a process of semantic interpretation, often described by the triangle of reference [6]. Thus, disagreement signals low quality that can be measured at any corner of the triangle: input, workers or annotation labels. In the **Analytics** layer we deal with the assessment of the data produced in the **Data Enrichment** layer. In order to evaluate the results of the crowd we first use the *vector representation* to measure the quality of the annotations in the three corners of the triangle. Thus, each worker annotation is stored in a vectorized data structure which eases the overall evaluation of the results. Second, we apply CrowdTruth metrics that provide useful insights for: *(1)* spam or low-quality *crowd workers*; *(2)* ambiguous or unclear *input units*; *(3)* unclear *annotation labels*. We iterate this process based on the input provided by the **Feedback Layer**, until we have a set of clear, reliable and correct results.

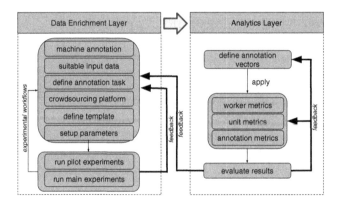

Fig. 1. Research methodology

4.3   Feedback Layer

The final part of our research methodology integrates various feedback loops that aim to improve the overall system. The feedback loops take advantage of the analysis generated in the **Analytics** layer and applies the necessary changes. Based on the current set of results, decisions are evaluated as follows. We assess the suitability of the data in the **Data Enrichment** layer and apply additional pre-processing or filtering steps. Similarly, based on the analysis of the *pilot experiments*, *i.e.*, the overall task clarity and crowd workers performance, we re-design the annotation task and template and tune the parameters. We analyze *(1)* the outcome of the *disagreement metrics* and improve the spam filtering (*i.e,* tune the worker metrics or perform additional filtering) and *(2)* the evaluation of the crowdsourcing outcomes (*i.e,* tune the worker and unit vector space).

5   Preliminary Results

In the initial stage of this PhD project we performed several experiments, across data streams and languages. We report on *(1)* event extraction from Dutch video synopsis and *(2)* event space and sentiment analysis on English news articles and tweets. The crowdsourcing experiments were run on the CrowdFlower[4] platform and were analyzed using the CrowdTruth methodology and metrics [7,26].

The results of the first experiment are integrated in the DIVE[5] demonstrator [27], a linked-data digital cultural heritage collection browser for collections interpretation and navigation by means of events and related entities. The experiments were performed on a dataset of 300 video synopsis of news broadcasts from The Netherlands Institute for Sound and Vision. We use various NER tools to extract NE from video synopses, but their accuracy vary significantly. Empirical analysis of the results indicated that combining the output of several NER tools provides better results than using a single extractor (*e.g.*, a richer set of entities, types, surface forms). However, a thorough evaluation is necessary to answer the first sub-question. Section 6 provides a plan for this. Next, we designed crowdsourcing experiments for tasks were machines under-perform: *(i)* extracting events, *(ii)* linking events to their participating entities.

For the second set of experiments we identify relevant news snippets and tweets for the event of whaling. We performed crowdsourcing experiments to identify relevant texts, word phrases and sentiments that are expressed in news and tweets. The results of these experiments[6] present preliminary insights for identifying salient features across data sources. Further experiments will be performed for novelty assessment and determining the event features saturation, *i.e.*, compiling, over time, a complete set of side-events, participating actors, locations, time frames and subjective perspectives such as sentiment changes.

Overall, the experiments performed gave us an important head start for reasoning on the third research question. We gain understanding of various crowdsourcing tasks and we are able to analyze the crowd workers behavior in order to assess their work. We applied existing CrowdTruth disagreement metrics and defined new metrics and filters that identify with high precision, recall and accuracy the spam and low-quality workers. The metrics proved to support us in providing reliable crowdsourced event semantics annotations. However, we only performed manual evaluation of the data, which is not scalable over time. We plan to extend this evaluation with automated methods and ground truth data.

6   Evaluation Plan

Our modular and multi-layer research methodology with combined machine and human generated data stages, needs various types of evaluation. To answer the first sub-question from Sect. 3 we need to perform extensive automated annotations by means of NLP tools. Benchmarking such tools helps us to gain enough empirical evidence for choosing the most accurate tool based on the input (*e.g.* text dimension, language, domain) and the task (*e.g.* identification of locations,

---

[4] http://www.crowdflower.com/.
[5] Available at http://diveplus.beeldengeluid.nl/.
[6] http://data.crowdtruth.org/salience-news-tweets/.

actors, times) at hand. The "Open Knowledge Extraction" challenge at ESWC[7] and frameworks such as GERBIL [28] are good systems to validate our approach.

In the context of sub-question 3, we will perform various crowdsourcing tasks (*e.g.* event detection, sentiment analysis, novelty ranking). For each task we need to validate the correctness of the crowdsourced data. Simply applying the CrowdTruth disagreement metrics may not always correctly identifying the low-quality workers. The main component of the crowdsourcing task is the *task template*. Depending on its fields (*e.g.* multiple or single choice, free input text), low-quality workers can exhibit particular behaviors, such as, always choosing the same answer, taking the shortest path to solve the task. In order to guarantee reliable data, additional behavioral and disagreement metrics are applied. Further, curated annotations can be compared with existing ground truth datasets.

To evaluate our hybrid machine-crowd workflow we start by instantiating it with a specific input type and use case. This gives us insights to answer the inter-related sub-questions 1 and 3 and finally sub-question 2. Next, we identify similarities between input types and tasks, as well as their independent features, which helps us investigating to what degree our methodology is data agnostic. The semantic annotations generated throughout the various annotations tasks will be used to answer sub-question 4 and ultimately our main research question. We investigate here whether the event-specific generated data (*i.e.*, events, concepts, sentiments, salient features) can improve understanding of events and whether this ground truth can be used in improving machines performance.

Overall, the task of understanding events is tackled and investigated by many venues. TREC[8] is focusing on the Temporal Summarization Track, a track aiming to develop a system that is able to provide concise and non-redundant information with regard to a given event. The structure of this challenge is very close to our novelty approach: identifying relevant and novel texts that describe an event. In the past years, main conferences focused on sentiment analysis as well. For example, our crowd-annotated corpus could be evaluated in semantic web challenges, such as, "Concept-Level Sentiment Analysis" during ESWC.

7   Conclusions

The primary focus of this research proposal is to gain event understanding through employing automated tools and collecting diverse crowd semantic interpretations on different data modalities, sources and event-related tasks. Our main hypothesis is that existing tools for detecting events and event-related features can acquire event semantics and understanding by employing such additional training data or by validating their current results with crowd-empowered semantics. In order to answer our main research question, we defined a set of 4 related research questions to investigate: *(1)* how can we harness the semantics of existing tools; *(2)* how can we create data agnostic annotation workflows;

---

[7] http://eswc-conferences.org.
[8] http://trec.nist.gov.

(3) how can we guarantee reliable crowdsourced semantics; and *(4)* how can we improve existing tools with our machine-optimized and crowd-generated data.

The research methodology consists of a continuous experimental loop of interconnected components for providing the event space, context, semantics and perspectives. The methodology allows for: *(1)* running combined crowd and machine workflows across different data types; *(2)* analysis of the crowdsourced data; *(3)* extensive feedback layer for improving existing or future results. We presented preliminary results of in-use crowd-generated linked data and an example of an optimized and generic annotation workflow for identifying salient event features. Our evaluation plan is two-fold: *(1)* validating our methodology by testing it with new data types and *(2)* assessing our methodology by ingesting the semantics collected for training and testing existing tools.

Acknowledgements. We thank Lora Aroyo for helping in this research proposal, Robert-Jan Sips, Victor de Boer, Tommaso Caselli for assistance in performing the experiments.

References

1. Gangemi, A.: A comparison of knowledge extraction tools for the semantic web. In: Cimiano, P., Corcho, O., Presutti, V., Hollink, L., Rudolph, S. (eds.) ESWC 2013. LNCS, vol. 7882, pp. 351–366. Springer, Heidelberg (2013)
2. McClosky, D., Surdeanu, M., Manning, C.D.: Event extraction as dependency parsing. In: Proceedings of the 49th Annual Meeting of the Association for Computational Linguistics: Human Language Technologies, vol. 1, pp. 1626–1635 (2011)
3. Kim, S.M., Hovy, E.: Automatic detection of opinion bearing words and sentences. In: Companion Volume to the Proceedings of the International Joint Conference on Natural Language Processing (IJCNLP), pp. 61–66 (2005)
4. Soboroff, I., Harman, D.: Novelty detection: the TREC experience. In: Proceedings of the Conference on Human Language Technology and Empirical Methods in Natural Language Processing, pp. 105–112. ACL (2005)
5. Nowak, S., Rüger, S.: How reliable are annotations via crowdsourcing: a study about inter-annotator agreement for multi-label image annotation. In: Proceedings of the International Conference on Multimedia IR, pp. 557–566. ACM (2010)
6. Aroyo, L., Welty, C.: Truth is a lie: CrowdTruth and the seven myths of human annotation. AI Mag. **36**(1), 15–24 (2015)
7. Aroyo, L., Welty, C.: The three sides of CrowdTruth. J. Hum. Comput. **1**, 31–34 (2014)
8. Yan, Y., Fung, G.M., Rosales, R., Dy, J.G.: Active learning from crowds. In: Proceedings of the 28th International Conference on Machine Learning (ICML 2011), pp. 1161–1168 (2011)
9. Intxaurrondo, A., Agirre, E., de Lacalle, O.L., Surdeanu, M.: Diamonds in the rough: event extraction from imperfect microblog data. In: Proceedings of the Conference of the North American Chapter of the Association for Computational Linguistics - Human Language Technologies (NAACL HLT) (2015)
10. Li, Y., Rizzo, G., Redondo García, J.L., Troncy, R., Wald, M., Wills, G.: Enriching media fragments with named entities for video classification. In: Proceedings of the 22nd International Conference on World Wide Web Companion, pp. 469–476 (2013)

11. Rizzo, G., van Erp, M., Troncy, R.: Benchmarking the extraction and disambiguation of named entities on the semantic web. In: Proceedings of the 9th International Conference on Language Resources and Evaluation, pp. 4593–4600 (2014)

12. Chen, L., Ortona, S., Orsi, G., Benedikt, M.: Aggregating semantic annotators. Proc. VLDB Endowment **6**(13), 1486–1497 (2013)

13. Hellmann, S., Lehmann, J., Auer, S., Brümmer, M.: Integrating NLP using linked data. In: Alani, H., Kagal, L., Fokoue, A., Groth, P., Biemann, C., Parreira, J.X., Aroyo, L., Noy, N., Welty, C., Janowicz, K. (eds.) ISWC 2013, Part II. LNCS, vol. 8219, pp. 98–113. Springer, Heidelberg (2013)

14. Kozareva, Z., Ferrández, Ó., Montoyo, A., Muñoz, R., Suárez, A., Gómez, J.: Combining data-driven systems for improving named entity recognition. Data Knowl. Eng. **61**(3), 449–466 (2007)

15. Schreiber, G., Amin, A., Aroyo, L., van Assem, M., de Boer, V., Hardman, L., Hildebrand, M., Omelayenko, B., et al.: Semantic annotation and search of cultural-heritage collections: the MultimediaN E-Culture demonstrator. Web Seman. Sci. Serv. Agents WWW **6**(4), 243–249 (2008)

16. Oomen, J., Belice Baltussen, L., Limonard, S., van Ees, A., Brinkerink, M., Aroyo, L., Vervaart, J., Asaf, K., Gligorov, R.: Emerging practices in the cultural heritage domain-social tagging of audiovisual heritage. In: Proceedings of the WebSci 2010: Extending the Frontiers of Society On-Line (2010)

17. Oosterman, J., Nottamkandath, A., Dijkshoorn, C., Bozzon, A., Houben, G.J., Aroyo, L.: Crowdsourcing knowledge-intensive tasks in cultural heritage. In: Proceedings of the 2014 ACM Conference on Web Science, pp. 267–268. ACM (2014)

18. Maccatrozzo, V., Aroyo, L., Van Hage, W.R., et al.: Crowdsourced evaluation of semantic patterns for recommendation. In: UMAP Workshops (2013)

19. Wei, Z., Gao, W.: Utilizing microblogs for automatic news highlights extraction. In: COLING (2014)

20. Verheij, A., Kleijn, A., Frasincar, F., Hogenboom, F.: A comparison study for novelty control mechanisms applied to web news stories. In: 2012 IEEE/WIC/ACM International Conferences on Web Intelligence and Intelligent Agent Technology (WI-IAT), vol. 1, pp. 431–436. IEEE (2012)

21. Snow, R., O'Connor, B., Jurafsky, D., Ng, A.Y.: Cheap and fast–but is it good?: evaluating non-expert annotations for natural language tasks. In: Proceedings of the Conference on Empirical Methods in NLP, pp. 254–263 (2008)

22. Rao, Y., Lei, J., Wenyin, L., Li, Q., Chen, M.: Building emotional dictionary for sentiment analysis of online news. World Wide Web **17**(4), 723–742 (2014)

23. Balahur, A., Steinberger, R., Kabadjov, M., Zavarella, V., Van Der Goot, E., Halkia, M., Pouliquen, B., Belyaeva, J.: Sentiment analysis in the news. In: Proceedings of the 7th International Conference on Language Resources and Evaluation, pp. 2216–2220 (2010)

24. Finin, T., Murnane, W., Karandikar, A., Keller, N., Martineau, J., Dredze, M.: Annotating named entities in twitter data with crowdsourcing. In: Proceedings of the NAACL HLT 2010 Workshop on Creating Speech and Language Data with Amazon's Mechanical Turk, pp. 80–88. ACL (2010)

25. Inel, O., Khamkham, K., Cristea, T., Dumitrache, A., Rutjes, A., van der Ploeg, J., Romaszko, L., Aroyo, L., Sips, R.-J.: CrowdTruth: machine-human computation framework for harnessing disagreement in gathering annotated data. In: Mika, P., et al. (eds.) ISWC 2014, Part II. LNCS, vol. 8797, pp. 486–504. Springer, Heidelberg (2014)

26. Soberón, G., Aroyo, L., Welty, C., Inel, O., Lin, H., Overmeen, M.: Measuring crowd truth: disagreement metrics combined with worker behavior filters. In: Proceedings of CrowdSem 2013 Workshop, ISWC (2013)
27. de Boer, V., Oomen, J., Inel, O., Aroyo, L., van Staveren, E., Helmich, W., de Beurs, D.: Dive into the event-based browsing of linked historical media. Web Semant. Sci. Serv. Agents WWW 35(3), 152–158 (2015)
28. Usbeck, R., Röder, M., Ngonga Ngomo, A.C., Baron, C., Both, A., Brümmer, M., Ceccarelli, D., Cornolti, M., Cherix, D., Eickmann, B., et al.: Gerbil: general entity annotator benchmarking framework. In: Proceedings of the 24th International Conference on World Wide Web, pp. 1133–1143 (2015)

Distributed Context-Aware Applications by Means of Web of Things and Semantic Web Technologies

Nicole Merkle[✉]

FZI Forschungszentrum Informatik am KIT, Information Process Engineering,
Haid-und-Neu-Str. 10-14, 76131 Karlsruhe, Germany
merkle@fzi.de

Abstract. Ambient Assisted Living aims for providing context-aware and adaptive applications to assist elderly and impaired people in their everyday living environment. This requires the recognition of user intentions and activities by means of multi-modal and heterogeneous sensing devices. An unresolved problem is the lack of interoperability and extendibility of the setting. Moreover, to achieve adaptivity, a context-aware environment requires to consider user impairments as well as capabilities and to monitor non-stop user activities. This complicates an on the fly integration of new sensing devices and applications. Furthermore, a flexible and expressive domain model for describing and processing user profiles, intentions and activities, is required. Our approach to overcome these integration and modeling problems, is to use the Web of Things and Semantic Web technologies. Another unresolved problem concerns the security of the collected sensitive data. To avoid the manipulation of applications by an unauthorized access, we introduce ontology based security policies for context-aware applications, considering their managed context data.

1 Introduction/Motivation

Context-aware platforms and assisting applications -summarized under the term Ambient Intelligence (AmI)- have mainly the goal to recognise the user intention, to achieve adaption. The objective is that users are not required to learn how to use the assisting system, but the assisting system does have to learn and to adapt itself to the user and his/her needs [2]. The environmental data have to be sensed and to be mapped into a machine interpretable domain model, enriched usually with semantics so that applications can process these data. In AmI use cases this is a common applied approach to allow context-awareness and adaption. A subset of AmI is Ambient Assisted Living (AAL). The target group of AAL are elderly and impaired people who have problems to overcome the complexity of daily flows. For this reason, a context-aware application must be able to **recognise the user activity** and to **infer** by means of the activity and context recognition the **user intention** to provide an adequate assistance by means of actions. The user intention specifies the current need and objective

© Springer International Publishing Switzerland 2016
H. Sack et al. (Eds.): ESWC 2016, LNCS 9678, pp. 824–833, 2016.
DOI: 10.1007/978-3-319-34129-3_51

of the user in an appropriate context while actions enable changes in the environment. The challenge here is to provide a domain model of user intentions and activities and to enable, a reasoner by this model to deduce the matching actions. Currently there seems to be no approach which is addressing adaption by means of user intention and activity recognition. Moreover, a context-aware environment consists of multi-modal and heterogeneous devices for sensing user-specific and environmental data. The concerned devices can be separated into wearable and stationary devices. A wearable device is worn in common by the user to sense for instance his/her vital parameters and to monitor his/her health state while stationary devices are sensing state changes in the environment. For this reason, an additional requirement of a context-aware environment is to provide a standardised description of these devices to enable their **integration**. But current approaches are lacking of mechanisms for a simplified integration and interoperability. Our approach to use Web of Things (WoT) and Semantic Web allows to integrate devices on the fly. However not just devices have to be integrated on the fly, but also context-aware applications.

In this work, we consider context-aware applications as a composition of services which execute appropriate tasks to assist according to the user characteristics, activities and intentions. It is self-explanatory that applications impact the user context by **actions** which they initiate to accomplish state changes in the environment. For this reason, we need an approach to express their functionality by means of action descriptions and to relate them with the device functionality as well as with the user intention. From the perspective of a context-aware application, a user intention has always to lead to at least one action which is satisfying the user intention. In our previous work [5], we present a service description which is linking user intentions to matching actions. The advantage is that the knowledge and functionality of context-aware applications can be increased by new incorporated service descriptions, serving for appropriate use cases.

We also have to consider that context-aware applications collect the entire time data about the user and his/her context. These data are sensitive data as they allow derivations about the user and his/her current activity and situation or even about his/her health state. This implies that we need a mechanism to protect these data from unauthorized access and manipulation. Current attempts to provide ontology-based security policies do not satisfy the requirements of an AmI environment. The overall objective of this work is to overcome the mentioned problems and challenges and to enhance state of the art approaches.

Section 2 gives an overview of state of the art techniques to distinguish this work from current approaches. Section 3 deals with the general problems of context-aware environments and introduces our contribution for developing distributed context-aware applications by means of WoT and Semantic Web technologies. Section 4 presents the research methodology and the technical approach of this proposal. Section 5 gives a brief overview of the current work state. Section 6 introduces the evaluation plan of our work and Sect. 7 summarizes the discussed topics and gives an outlook to further open issues and research questions.

2   State of the Art

There are a lot of efforts in developing context-aware AAL environments. For instance, Hristova et al. introduce in [4] an Ambient Home care system which implements context-aware services with reasoning functionalities. The framework uses no ontology for modelling the AAL environment and its participating instances (users, devices, services, etc.). The problems of this approach are that every application developer does have to learn different APIs to implement their application logic. It would be more comfortable if every developer just needs to follow one standarised API, which is extendible in a semantic way. The presented use cases of [4] do not evaluate user capabilities and impairments. The reasoning is done by a rule-based engine, modelling the context with key-value pairs. The engine in [4] is very restrictive because the context of a user is more complex than simple rules can cover by key-value pairs. Hristova et al. also state that privacy and security are not supported by their framework. We want to address this issue with our approach to model ontology-based security policies.

Bacciu et al. discuss in their work [1] an AAL platform for prohibiting sedentariness and unhealthy dietary habits by means of activity recognition. The sensed data is aggregated to allow the platform to recognise social and physical activities [1]. However the mentioned platform does not consider user intentions and profiles. An adaption mechanism is not given. In contrast to our approach, the given activity recommendations are restrictive because they are based on general expert assumptions and do not consider the individuality of the user. The DemaWare platform in [8] aims for supporting people with dementia, i.e. the platform and its ontology is addressing a very specific domain of AAL. According to [8] the platform is restricted in terms of the used hardware and includes a weak provision of context information. The used ontology is just matching to people with Dementia. So the field of application is very specific and restricted while our approach aims at being applicable in the entire field of AAL.

Regarding security policies, the work in [7] presents an approach to compose privacy policies, based on Semantic Web technologies. The engine in [7] generates composition rules and deduces implicit terms by means of the data usage context. However the mentioned approach provides no rules for other contexts as for the presented one. Moreover their approach provides no storage location regarding the appropriate data [7]. Considering our AAL use case, it is necessary to enable storage location information, because every context-aware application is distributed due to security aspects and provides and manages its context data itself. If an application wants to access the data of a device or another application, the access must be regulated. The approach in [7] provides no user feedback if the policy composition fails. As a consequence to the mentioned problem, our approach aims at giving the user an appropriate feedback, if the security policies are not obeyed.

3   Problem Statement and Contributions

There are criteria and aspects for implementing a context-aware infrastructure in the AAL domain. **Security** is such an aspect. If the platform is attacked, the entire system and the privacy of the user is concerned and in danger. An intended manipulation of these sensitive data can have serious matters for the user. An offender could infer by means of the gathered data the activities and habits of the user and exploit this knowledge to mean ill. The contribution of this proposal is to devise semantical security policies, for the handling of collected and distributed context data. Another aspect is that an assistive environment consists of different sensing devices and user interfaces which can be used through different communication channels like voice, gaze and haptic interaction. The usage of these devices and user interfaces can require different abilities which the user might not have because of his/her impairments. Therefore the **recognition of and adaption to user intentions and activities** requires to happen in a natural and intuitive way by means of sensed activities and user profiles. Considering these aspects and challenges as well as our work in [5], this work answers the following research questions:

1. Do ontology-based security policies for context-aware applications in the AAL domain prevent misuse and manipulation of sensitive context data?
2. Do lightweight ontology languages such as RDF(S) provide the necessary expressivity for representing devices and their functionalities in AAL environments? See also [5]
3. How can user intentions, user activities as well as environmental aspects such as device functionalities be semantically linked together using lightweight ontological semantics to improve the integration of devices and the user intention and activity recognition? See also [5]

With respect to these research questions, we present a platform that facilitates the integration and interoperability of heterogeneous devices by means of the WoT and Semantic Web technologies. Furthermore, we devise a context-aware application called `Sherlock` that recognises the user intention and activity by means of the mentioned `Service Description`. Moreover, an objective is to enable the `Sherlock` application to implement ontology based security policies.

4   Research Methodology and Approach

The methodology of this work comprises to define the target group and to understand by means of workshops and interviews with the user group and domain experts, the lifecycle of AAL environments and use cases. Moreover, we analyse *Ethical Legal Social Implications* (ELSI) to infer general user requirements of assistive systems. Considering ELSI and state of the art approaches, we determine which aspects an assisting system needs to satisfy and which of these aspects are not yet achieved by current approaches. In a next step, we create a domain model by means of a light-weight ontology. For this reason, we introduce -according to the WoT concept- the term *Things of Interest*. *Things of Interest* are physical or abstract objects of the intended use cases. The semantical description of these *Things* is the basis for modelling an AAL domain ontology which is considering the impairments and capabilities of the user as well as the functionality of applications and devices. For a future evaluation of our approach, we implement a WoT server and create by means of the proposed domain model a machine-processable AAL environment. We cite the Light Switching use case as an example use case. The implementation of this use case requires a Light Service description, which is used by the Sherlock application to deduce, if the user does have the intention to turn a lamp on or off. The objective is to enhance the functionality of Sherlock by rule description languages (such as SWRL[1] or SPIN[2]). For this reason, we need to consider and analyse existing rule description languages, regarding their expressivity. There might be sometimes fuzzy situations, i.e. rule-based languages are not suitable in this cases. So we have to investigate in fuzzy logics and in pattern recognition methodologies. But before we can apply these rules, we require to model sequences of user activities. Therefore, we have to consider different models such as *Donald Norman's Seven Stage Model*, which are describing established user activities.

The planned security policies also require to follow rules. These rules enable the Sherlock application to recognise security violations.

We devised a first version of a semantical domain model, describing *Things of Interest*. For this reason, we considered different existing ontologies such as the SSN[3] and DUL[4] ontology. But we noticed that these ontologies do not cover all aspects which are necessary for the considered AAL application domain.

Figure 1 depicts our approach to model a Lamp device and the linking of its inherent functionalities and location. The location of Lamp1 references further relevant devices (AmbientLightSensor1), which are sensing relevant context data for the participating applications. The composition of the sensed observations allows the derivation of the user context.

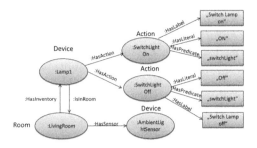

Fig. 1. A snippet of the device ontology as graph representation [5]

For the sake of simplicity, it is necessary to provide a tool for generating ontology instances which can be accessed immediately during runtime. The approach in this proposal is to use Semantic MediaWiki (SMW) because it provides forms and templates for creating semantically annotated data. Context-aware

---

[1] Semantic Web Rule Language.
[2] SPARQL Inferencing Notation.
[3] Semantic Sensor Network.
[4] Descriptive Ontology for Linguistic and Cognitive Engineering.

applications get the possibility to request and subscribe for context data depending on their need and interest without knowing the (details of the) different context resources. The knowledge of the domain is distributed by different context-aware applications and devices, managing their own context and information. To accomplish this, every service requires to provide at least one `Service Description`, which the applications can use for reasoning. This `Service Description` is based on the WoT recommendation. Every *Thing* whether physical or abstract in the WoT recommendation is described by [6]:

(i) **properties**, describing the `Thing` composition
(ii) **events**, the Thing is interested in or which can be triggered by a Thing
(iii) **actions**, which can be invoked on a Thing, if matching events are occurring

Furthermore, the WoT approach allows the exchange of data by open Web standards. The goal of this approach will be to use the WoT for describing and managing *Things of Interest* to make them *findable, sharable, accessible and composable* [3]. The knowledge of the AAL environment is distributed in this way by a net of smart applications, managing and sharing their acquired context data. Figure 2 depicts the `Light Service` description. The `Light Service` consists of `Capabilities`, the service provides and `Intentions` the application is interested in. The service description expresses by linking `Capabilities` and `Intentions`, the ability of a service to serve for an appropriate task and that this task is depending on the linked user intention. The advantage of the mentioned approach is, that an application can request various service descriptions to expand its knowledge about the domain and context. A user `Intention` is described by `Actions` and `Rules`. Every `Rule` consists of pre-conditions (triples) which are denoting possible events. In Fig. 3, the `SwitchLightOff` intention references the `SwitchLightOff` action and three triples describing the context conditions to recognise the presented `Intention`. Moreover, our approach aims at achieving the security and privacy of this context-aware infrastructure by distributing the acquired data to various context-aware applications which are managing and creating their own context data. Therefore every application requires to follow semantical security policies. For the evaluation of our approaches, we extend the `Sherlock` application with the proposed functionality. Moreover, in future

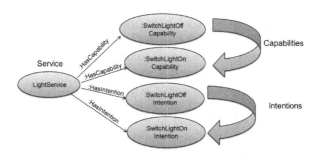

Fig. 2. The `Light Service` description linking user intentions with actions [5]

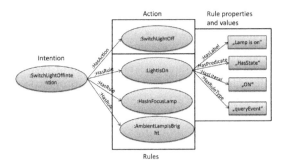

Fig. 3. The `Light Switching` user intention with its rules. See also [5]

steps, we adapt and enhance the work of [7] so that `Sherlock` might follow the devised semantical security policies and warn the user if a security policy was violated by external influences. Furthermore, we want to extend our domain ontology to allow the semantical description and evaluation of `user profiles`. A user profile constitutes of relevant user characteristics, such as user capabilities and disabilities. One possibility is to define and integrate semantical user profiles based on the *WHO International Classification of Diseases Guidelines*[5] to derive compensating actions in appropriate situations. Our objective is to enable context-aware applications to include user profiles in their reasoning process.

5   Prelimenary Results

The target group and involved people were interviewed in two workshops to determine by means of ELSI the technical requirements and to specify useful use cases. Three use cases are defined.

1. a `Light Switching` use case, controlling lamps by the user intention and activity.
2. the recognition of emergency situations (for instance fire in the living environment).
3. the controlling of an autonomous wheelchair by gaze patterns and user context.

A first prototype of the `Sherlock` application and the components in Fig. 4 were already implemented and have to be extended. Furthermore, the `Light Switching` use case was implemented to demonstrate our approach. The `Sherlock` engine is running on a mobile phone and is communicating with the user. In a first step, `Sherlock` requests the `Light Service` description from a `Query Generator` component. This component uses the vocabulary of our introduced light-weight ontology and generates from the application requests SPARQL queries. The `Triple Store` imports the RDF instances of the SMW. The `Configurator` transforms created device descriptions from the SMW into

---

[5] For more details see here: http://www.who.int/classifications/icd/en/.

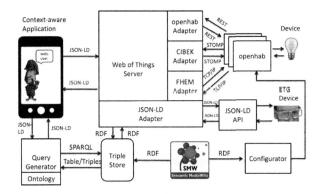

Fig. 4. The proposed architecture of our approach. See also [5]

appropriate IoT system configurations to introduce the devices. After `Sherlock` has received its service description, it registers at the WoT Server for appropriate sensor events. An Eye Tracking Glass (ETG) is measuring the gazes of the user who is sitting in an autonomous wheelchair. If the user focusses for instance a Lamp, this sensor event is sent to the WoT. The WoT uses different `Adapters` to transform events into a standard structure, which the `Sherlock` engine can understand. `Sherlock` requests after the sensor event from the WoT server further context information according to its `Light Service` description and the contained intention rules. By means of this context information, Sherlock derives the current intention of the user and asks the user, if the inference was right. The user confirms Sherlocks conclusion and Sherlock triggers the appropriate action, which is forwarded to the WoT server for switching the appropriate devices, i.e. `Lamp1`.

A first version of the presented domain ontology with the service description was defined and inserted in SMW. A general message structure was already defined for exchanging messages between different components to abstract the different resources and to achieve a common understanding. The message contains all relevant information of the underlying ontology.

6   Evaluation Plan

The evaluation will be conducted as part of the AICASys[6] project. We plan to make lab and field trials with people of the target group in real living environments. An objective is to evaluate the usability, feasibility and reliability of the context-aware `Sherlock` application by measuring and statistically evaluating the success factor of the user intention and context recognition. We want to show in our evaluation that (a) the semantically specified security policies prevent the misuse of sensitive data, (b) the user intention detection is improved (at least

---

[6] A national funded project for supporting impaired people in their living environment by means of an autonomuous wheelchair, an ETG and a context-aware environment.

80 % reliability) by means of introducing user profiles and (c) the integration of various devices and applications is simplified for not technically minded persons compared to other related work approaches. It is planned to compare the improvements or degradations with other similar approaches to show that the presented approach was successful or not.

7   Conclusion

We introduced service descriptions which are providing the knowledge of our context-aware applications. Context-aware applications need these service descriptions to link user intentions to appropriate actions. We discussed that the detection of user intentions and activities needs also to consider user characteristics by means of user profiles. In our approach context-aware applications are distributed by running independent on different devices and managing their own context data. The exchanged context and user data is sensitive and requires to be protected from unauthorized access. For this reason, we proposed to integrate security policies in our ontology, so that every application needs to follow these policies. The utilisation of the WoT concept offers the solution for overcoming integration problems of heterogeneous data sources. Moreover, the Semantic Web opens a way to map the real world into a machine understandable format so that reasoning techniques can be applied to infer the context and intention of an appropriate user. For demonstration purposes we introduced our context-aware and extendible `Sherlock` application. However, this proposal does not address to solve problems concerning the realtime reaction of applications to different events. Furthermore, it is was not discussed how to enhance the quality of context data. In an AAL environment it is necessary to evaluate and assure the trustworthiness of data sources to warrant a reliable context-awareness. In future work we have to consider what happens if fuzzy and uncertain situations are detected. A conceivable approach is to apply machine learning and pattern recognition to combine it with rule-based approaches. One enhancement can be the implementation of a mood classifier. This would allow the Sherlock engine to learn from context data to motivate the user to do some activities. But the mentioned challenges remain for future research work as part of this Ph.D.

Acknowledgement. I want to thank Prof. Dr. Rudi Studer, Prof. Dr. Sören Auer, Dr. Stefan Zander, Ignacio Traverso Ribon and Prof. Dr. Maria-Esther Vidal for their support in accomplishing this proposal. Furthermore, this work is supported by the German Federal Ministry of Education and Research (BMBF) under the AICASys project.

References

1. Bacciu, D., et al.: Smart environments and context-awareness for lifestyle management in a healthy active ageing framework. In: Pereira, F., Machado, P., Costa, E., Cardoso, A. (eds.) EPIA 2015. LNCS, vol. 9273, pp. 54–66. Springer, Heidelberg (2015)

2. Grguric, A.: ICT towards elderly independent living, Croatia (2012)
3. Guinard, D.: A Web of Things Application Architecture - Integrating the Real World into the Web, Switzerland (2011)
4. Hristova, A., et al.: Context-aware services for ambient assisted living: a case-study, Spain (2008)
5. Merkle, N., Zander, S., (n,d); Representing and Reasoning over User Intentions and Actions in Adaptive Ambient Assisted Living, Germany, (i), 1124
6. Raggett, D.: The Web of Things: Challenges and Opportunities, May 2015
7. Soto-Mendoza, V., et al.: Policies composition based on data usage context. In: COLD (2015)
8. Stavropoulos, T.G., et al.: The DemaWare Service-Oriented Platform for AAL of Patients with Dementia, Czech Republic (2014)

On Learnability of Constraints from RDF Data

Emir Muñoz[1,2(✉)]

[1] Fujitsu Ireland Limited, Dublin, Ireland
[2] Insight Centre for Data Analytics, National University of Ireland,
Galway, Ireland
emir.munoz@insight-centre.org

Abstract. RDF is structured, dynamic, and schemaless data, which
enables a big deal of flexibility for Linked Data to be available in an
open environment such as the Web. However, for RDF data, flexibility
turns out to be the source of many data quality and knowledge representation issues. Tasks such as assessing data quality in RDF require
a different set of techniques and tools compared to other data models.
Furthermore, since the use of existing schema, ontology and constraint
languages is not mandatory, there is always room for misunderstanding the structure of the data. Neglecting this problem can represent a
threat to the widespread use and adoption of RDF and Linked Data.
Users should be able to *learn* the characteristics of RDF data in order to
determine its fitness for a given use case, for example. For that purpose,
in this doctoral research, we propose the use of constraints to inform
users about characteristics that RDF data naturally exhibits, in cases
where ontologies (or any other form of explicitly given constraints or
schemata) are not present or not expressive enough. We aim to address
the problems of defining and discovering classes of constraints to help
users in data analysis and assessment of RDF and Linked Data quality.

Keywords: RDF constraints · Linked data mining · Data quality · Data
semantics

1 Introduction

Background. The flexibility of RDF comes from the Web Ontology Language (OWL): Open World Assumption (OWA), missing information treated
as unknown; and Unique Name Assumption (UNA), individuals may have more
than one name. These characteristics difficult data validation [28] and data quality assessment. The notion of Data Quality (DQ) is related to individual use
cases and cannot be assessed independently from the user. Yet there is a lack of
methodologies tailored to Linked Data (LD) that consider users' requirements
and users without previous experience on the data. In this doctoral research,
we explore the use of constraints as a tool for users to identify the modeling
behind data represented using RDF. Constraints are limitations incorporated
on the data that are supposed to be satisfied all the time by instances of the

H. Sack et al. (Eds.): ESWC 2016, LNCS 9678, pp. 834–844, 2016.
DOI: 10.1007/978-3-319-34129-3_52

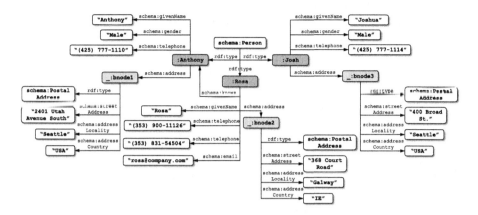

Fig. 1. RDF data describing information about people.

database [1]. They are useful for users to understand data as they represent characteristics that data naturally exhibits [15]. However, a deeper study of constraints and the benefits that they bring to the RDF model, especially for use cases like quality assessment, has not yet been made.

Motivation. As of April 2014, an amount of 1014 Linked Datasets were registered in the Linked Open Data (LOD) cloud [19], containing billions of RDF statements. Due to the heterogeneity of the schema(s) and modeling used in each one of the sources, the task of determining whether a dataset is relevant for a given use case becomes a daunting and non-trivial problem. A way to understand data is by understanding the model behind that all facts follow. However, under the OWA multiple possible models can be satisfied by the facts. Considering the data as complete can allow us to adopt a Closed World Assumption (OWA) with UNA (sacrificing flexibility), which is equivalent to a relational database [16,28], where a single model contains all and only the facts assessed. Although advances in the Semantic Web and Linked Data have been made in standards to communicate the semantics behind data (*e.g.*, ontologies), they most of the time fall short in providing expressive schemas. Therefore, there is a clear need for methods and tools to help users to comprehend the structure of RDF data in the presence of inconsistencies and poor or absent schema.

The problem of lacking structure of RDF data can be addressed by extracting constraints. In the RDF model, they can help to cope with this lack of schema, representing some missing meta-data, such as identifiers, and cardinalities. Though we sacrifice some of the greater flexibility and interoperability.

Example 1. *In the following we consider the RDF data in Fig. 1, which contains information about people. Each entity is defined as a member of the class* schema:Person, *thus they can have all properties of this class. We can describe some characteristics of the data (in Fig. 1) by means of constraints:*

C1. Every person contains exactly one value for the schema:`givenName` *and* schema:`address` *properties.*

C2. The combined properties schema:`givenName` *and* schema:`address` *uniquely identify each person in the data, every person has a different pair of values.*

C3. Each person is connected to at least one value for the schema:`telephone` *property, and at most two values.*

C4. The value of the property schema:`knows` *in the person* :`Rosa` *makes direct reference to the person* :`Anthony`.

C5. The property schema:`email` *is uncommon in the data and, in this dataset, has a probability of $\frac{1}{3}$ to appear once.*

C6. All values of the property schema:`telephone` *follow the same '(NUMBER) NUMBER-NUMBER' syntactic pattern.*

C7. Entities with a schema:`givenName` *and* schema:`address` *must be instances of the class* schema:`Person`.

C8. Values for property schema:`gender` *are expected to be of type text (string).*

By looking at the constraints in the example, we can see that they partially uncover the structure and schema of the data. Formally, constraints *C2* and *C4* are known as *integrity constraints* (used to ensure accuracy and consistency of data), whilst constraints *C1* and *C3* are known as *cardinality constraints* (used to specify a minimum and maximum bound for relationships with properties that an entity can have [15, 29]). *C5* is special in the sense it expresses the marginal probability by which the constraint holds in the dataset [4, 6]. *C6* is known as *syntactic pattern constraint* [17], and restrict the syntax that the values of a property can have. Constraints *C7* and *C8* are known as *domain and range constraints*, respectively; they restrict the types that entities (in the domain) and values (in the range) of relations for a given property can have, respectively.

In databases, applications of constraints include: *data cleaning*, constraints verify that the data conforms to a basic level of data consistency and correctness, preventing the introduction of dirty data; *integration, modeling, retrieval*, among others [1]. Although other classes of constraints can be defined for RDF, in this work, we will limit ourselves to mainly study the ones described above because of: (a) their potential for DQ analysis; and (b) the lack of existing research on learning such constraints from the data. Most of the actions involved in LD quality assessment, in order to be performed, require users to have knowledge on the underlying structure of the data. Here, we aim to provide such a notion to the users in the absence of explicit programming, by means of identifying constraints in the data. In this thesis work, we refer to this problem as *"learnability of constraints"*. During this research our goal is three-fold: (1) increase the expressiveness of RDF constraints to enrich the user's understanding of the RDF data modeling; (2) present algorithms to automatically extract such constraints; and (3) study their effects during the assessment of quality in RDF data.

2   State of the Art

The concept of constraints has been present for long time in databases [1], and only recently has been introduced in RDF [14]. Lausen et al. [14] presents an approach for translating constraints while converting a relational database to RDF without losing semantic information. The authors extended the RDF vocabulary in order to encode integrity constraints such as keys and foreign keys. Theoretical aspects of integrity constraints in RDF and RDFS were studied in [2,7], introducing a mechanism to express functional constraints in terms of equality. While constraints like keys are indispensable for data consistency in relational databases, in the RDF model they are still not considered first-class citizens.

OWL 2[1] is the most straightforward way to represent some of the constraint here introduced. Key constraints can be defined using the axiom `owl:hasKey` based upon object and data properties. Cardinality constraints can also be expressed in OWL 2 by means of three expressions: `owl:minCardinality`, `owl:maxCardinality`, and `owl:exactCardinality`. However, the semantics of OWL 2 considers OWA and does not make the UNA, making it hard to evaluate consistency when same or different individual relations are not explicitly stated (very common in lightweight ontologies). Tao *et al.* [28] also shows that the expressiveness allowed by OWL 2 is limited to express Integrity Constraints.

Due to the huge effort involved in generating full ontologies, several works aim to (semi)automatically learn or enrich ontologies from text or instance data. Property domain/range constraints have been studied in [30], and used to acquire class disjointness axioms to detect inconsistencies in DBpedia. Völker et al. [31] uses two approaches for the same problem: extensional (relying solely on instances) and intensional (based on logical and lexical description of classes). Key constraints have been explored by [23,27] for link discovery and data integration purposes. And our previous work [17] introduces the concept of syntactic constraints in RDF. (For a recent survey on methods in this line see [18].)

Languages for expressing constraints in RDF data intended to validate RDF documents, and communicate expected patterns exists. Among the most populars is Shape Expressions (ShEx)[2], which is more focused on type inference than on verification, unlike the RDF Data Descriptions (RDD) [20]. RDD uses a compact special-purpose syntax which is independent of a specific inference machinery. Shapes Constraint Language (SHACL)[3] is the recent output of the W3C RDF Data Shapes Working Group, which provides a high-level vocabulary to identify predicates and their associated cardinalities, datatypes and other constraints (by using SPARQL). SPARQL Infering Notation (SPIN)[4] is a low-level language that allows users to use SPARQL to specify rules and logical constraints. All of them, namely, ShEx, RDD, SHACL, and SPIN, are aimed to validate RDF data, and communicate data semantics among users. They do cover constraints such as keys and cardinality; however, their expressivity is limited compared to our proposal and they do not consider a probabilistic notion.

Multiple applications of constraints can be envisioned in RDF based on their success in the relational model, including: data cleaning, integration, modeling,

---

[1] https://www.w3.org/TR/owl2-syntax/.

[2] https://www.w3.org/2013/ShEx/Primer.

[3] https://www.w3.org/TR/shacl/.

[4] http://spinrdf.org/.

processing, and retrieval [21]. In terms of quality assurance, keys were recently used in [24] to determine discriminability of resources, *i.e.*, determine if datasets contain indistinguishable resources *w.r.t.* a given set of properties. Likewise, we plan the application of constraints as a tool that serves users to understand the structure of the data and help them to assess RDF data quality.

The discovery task of constraints from RDF data is a new topic and has been focused solely on keys and under a particular and limited view of RDF, which is known as *Concise Bounded Description* (CBD)[5]. CBD limits the (over)use of blank nodes (*e.g.*, the ones generated in the `schema:address` property shown in Example 1), the path length, total number of statements, and reifications. It also adds an unexpected complexity to the discovery problem.

3   Problem Statement and Contributions

The benefits that the definition of expressive classes of constraints for RDF data brings to the table are manifold, *e.g.*, data cleaning, integration, modeling, processing, and retrieval, akin to constraints in relational databases. However, they have not been studied in depth by the Linked Data community. Although various of the existing technologies can be used to define new classes of constraints, there is no practical or theoretical framework that connects them yet. In this Ph.D. research, we study solutions to implement constraints in RDF using current theoretical foundations. We also aim to show how users can benefit from the presence of constraints while assessing quality of RDF datasets.

Unlike previous works on key constraints and linked data quality, we acknowledge the volume, variety, veracity, and velocity of data. Therefore, we rely on the following assumptions: (i) we do not assume the existence of a full/complete/rich schema or ontology; (ii) RDF data is not always formatted in a specific way (*e.g.*, CBD), this is more unlikely if we consider the Web of Data; (iii) RDF data contain plenty of RDF Blank Nodes [12,13], which radically change the current view of constraints solely based on *sets of properties* to a view of *sets of property paths*; and (iv) we assume that RDF data are error prone, and incomplete, which requires that these techniques work under uncertainty. The problem and assumptions stated above lead to the following research questions:

RQ1. *Can we define more expressive and novel constraints for RDF data?*
We observed that RDF key constraints are defined as a set of single properties and under the assumption of CBD [25]. We thus see three drawbacks of this approach: (a) CBD is application dependent and complex to compute, and (b) it does not consider complex values (*e.g.*, `schema:address` in Example 1), (c) it does not take into account that RDF data on the Web of Data contains plenty of blank nodes [12,13]. For example, considering Example 1 and the state-of-the-art, we have that constraint *C5* cannot be expressed with current approaches; and, constraint *C2* cannot be expressed under the CBD assumption, and under the assumption

---

[5] https://www.w3.org/Submission/CBD/.

that two blank nodes are always considered to be equal. Nevertheless, a relaxed version of *C2* including only the schema:givenName property can be expressed with current approaches. Here, we aim to increase the coverage and expressivity of RDF constraints with the so-called SPARQL Property Paths [22] and a value similarity definition that can help to cope with complex values.

RQ2. *Can constraints be automatically extracted under a non-CBD assumption?* So far the task of defining constraints for RDF data has been relegated to users mainly with the help of high/low-level languages. ROCKER [23] was the first machine-learning-based approach for key discovery. We expect to determine whether the methods behind ROCKER can be extended to identify more expressive keys and our new classes of constraints. Regardless of the solution, we aim to account the existence of blank nodes, and to consider scalability as a key feature of our framework.

RQ3. *What is the impact of constraints in the assessment of RDF data quality?* Constraints in RDF have been used for data validation mainly. Specific applications of RDF keys have been investigated in data linkage by [5,23, 27]. But the semantics of constraints can have several implications in data quality assessment that have not been investigated yet. In [24] the authors present an approach to discover redundant entities in RDF data using a key discovery algorithm. However, we want to investigate further how to exploit the rich semantics of constraints in RDF, especially considering different DQ scenarios and dimensions.

4   Research Methodology and Approach

The expressivity of current RDF constraints definitions is limited, and do not deal with complex data values. Currently, the definition of constraints is based only on sets of properties and does not consider the graph structure of RDF data. This is partially due to certain intractabilities when working with complex data values. For instance, to determine whether two blank nodes are equal (*e.g.*, _:bnode1 and _:bnode3 in Example 1), it is needed to determine if there exists an isomorphism between the RDF graphs — which results to be an NP (GI-complete) problem [12]. In order to cope with these limitations, our proposal unfolds in the following subsections.

4.1   Definition of Constraints for RDF

The semantics of constraints in Example 1 is similar across different vocabularies, and similar to their pairs in relational databases. However, current vocabularies/ontologies do not consider complex data values or blank nodes in the constraints definition. In order to cope with this lack of expressivity, our approach includes the definition of constraints using existing standards like SPARQL property paths [22]. In turn, an RDF key could be defined as a set of property paths, instead of single properties. For example, we could express that the path `schema:address/schema:streetAddress` is used as a key. This will also require us to deal with blank nodes and their equivalence. For this we plan to use the skolemization algorithm proposed by Hogan [12].

RDF data may also contain uncertain data, which usually lead to inconsistencies in later processing. In these cases, a strict consideration of constraints would lead to data loss. Allowing some exceptions can prevent applications from losing data [11]. For that we believe that it is missing a definition of RDF constraints with an associated probability of occurrence. Constraint *C5* in Example 1 is an example of this case.

4.2   Discovery of Constraints

Together with the definition of more expressive classes of constraints, we will propose algorithms to discover them from RDF data. Existing approaches only discover key constraints from RDF data in a CBD form, and domain/range constraints. Here, we can follow two non-exclusive possible paths: (i) determine whether existing approaches, such as ROCKER or SAKey, can be extended to identify more expressive keys, and if they are suitable for other classes of constraints (*e.g.*, cardinality, probabilistic, syntactic); (ii) analyse the translation of approaches used to extract constraints in other data models, for example, the XML model [3,10]. (XML is meant for data serialization as a tree where order is important, RDF is a knowledge graph and order is not important.)

4.3   Constraints and Data Quality

We argue that constraints will allow users to efficiently understand the structure and nature of the data. This empowers users to perform different analytic tasks, being DQ one of them. Constraints could be related with several of the DQ dimensions defined in [26,32], such as, relevancy (data helps to know what you want), completeness (data do not leave any open questions), amount of data, interpretability, concise representation, consistent representation, to name just a few. These dimensions are categorized as contextual or representational DQ characteristics of high-quality data [26]. Constraints as the ones listed in Example 1 could be defined and extracted from RDF data to help users in the identification and analysis of those dimensions. A more practical study on how users can benefit from these constraints should be done to derive relations between constraints and DQ dimensions.

5   Preliminary Results

Our first step into the path of learning the structure of RDF data was at property values level. In [17] we provide an unsupervised approach to extract syntactic patterns from property values used in RDF data. These patterns attempt to address the lack of `rdfs:range` definitions. A set of patterns extracted for a given property enables: (i) human-understanding of syntactic patterns that each property follows, (ii) a structural description of properties, (iii) the detection of data inconsistencies, and (iv) the validation and suggestion of new values for a property. In the learning process, first, a lexical analysis is applied over values of properties to generate a sequence of tokens. From these tokens, lexico-syntactic rules are extracted to generate content patterns. For instance, a pattern for the values of property `schema:telephone` in Example 1 could be as '`(NUMBER) NUMBER-NUMBER`'. We experimented using DBpedia v3.9 to extract content patterns for all properties in the ontology. The extraction generated a database with *ca.* 500,000 content patterns.

The extraction of content patterns could be used to enrich the definition of `rdfs:range` properties. Interestingly, a content pattern goes beyond the current definition of a RANGETYPECONSTRAINT in RDD, which is limited to indicate that a property points either a URI, BlankNode, Resource, or a Literal.

6   Evaluation Plan

In order to evaluate the outputs of this Ph.D. research, we plan the following segmented evaluation, for the three main parts of this work.

6.1   Definition of Constraints for RDF

A definition for the types of constraints we propose here can be evaluated in terms of the expressivity and applicability. The expressivity of our definition of RDF keys can be compared against OWL 2, ShEx, SHACL, RDD, and [23]. While for the rest of our definitions, to the best of our knowledge, there is no work to compare with. Conversely, we plan to use related works performed in XML [8,9] and relational databases [6] to define semantically similar constraints and applications.

6.2   Discovery of Constraints

The quality of the discovery part can be evaluated against evaluation datasets, measuring the number of retrieved constraints, the time and memory complexity, precision, recall, correctness, and completeness. In the state-of-the-art, we could find only the datasets used in [23], which can be used to partially evaluate our extracted RDF keys. Since there is no other work that considers SPARQL property paths nor probabilistic constraints, there is a lack of manually-annotated gold standards that we can use. We will generate new evaluation datasets to show the effectiveness of our approach, considering different data sources such as Web Data Commons[6]. In order to measure scalability, these datasets must be of different sizes (from thousands up to millions).

6.3   Constraints and Data Quality

The evaluation of this interaction between constraints and data quality will be measured in two ways. First, we can express our constraints in ShEx or RDD, and apply existing tools to validate a dataset against a set of constraints. Second, we plan to perform a user study to determine how useful are the extracted constraints for several scenarios and tasks of quality assessment.

---

[6] http://webdatacommons.org/.

7 Conclusions and Future Work

From the beginning, the definition of constraints in RDF has been limited by their mapping from relational databases. Current constraints are based on single property names to, for example, uniquely identify entities in a dataset. Thereby, current approaches do not consider the existence of complex property values. Therefore there is a lack of expressiveness and applications for RDF constraints.

In this paper, we present a doctoral research proposal that identifies research questions in the area of constraints for RDF and data quality. This research proposes the definition of more expressive classes of RDF constraints, and the development of methodologies where constraints can help users to understand the structure behind RDF data. We believe this kind of tools will help users in the assessment of quality. Also it will unlock further applications in data cleaning, integration, modeling, processing, and retrieval, akin to constraints in relational databases.

Acknowledgments. This thesis is supervised by Dr. Matthias Nickles. The author would like to thank Prof. Dr. Heiko Paulheim for his valuable comments and suggestions. The work presented in this paper has been supported by TOMOE project funded by Fujitsu Laboratories Limited and Insight Centre for Data Analytics at NUI Galway.

References

1. Abiteboul, S., Hull, R., Vianu, V.: Foundations of Databases: The Logical Level, 1st edn. Addison-Wesley, Boston (1995)
2. Akhtar, W., Cortés-Calabuig, A., Paredaens, J.: Constraints in RDF. In: 4th International Workshops on Semantics in Data and Knowledge Bases, SDKB, pp. 23–39 (2010)
3. Arenas, M., Daenen, J., Neven, F., Ugarte, M., den Bussche, J.V., Vansummeren, S.: Discovering XSD keys from XML data. ACM Trans. Database Syst. **39**(4), 28:1–28:49 (2014)
4. Atencia, M., et al.: Defining key semantics for the RDF datasets: experiments and evaluations. In: Hernandez, N., Jäschke, R., Croitoru, M. (eds.) ICCS 2014. LNCS, vol. 8577, pp. 65–78. Springer, Heidelberg (2014)
5. Atencia, M., David, J., Scharffe, F.: Keys and pseudo-keys detection for web datasets cleansing and interlinking. In: ten Teije, A., Völker, J., Handschuh, S., Stuckenschmidt, H., d'Acquin, M., Nikolov, A., Aussenac-Gilles, N., Hernandez, N. (eds.) EKAW 2012. LNCS, vol. 7603, pp. 144–153. Springer, Heidelberg (2012)
6. Brown, P., Link, S.: Probabilistic keys for data quality management. In: Zdravkovic, J., Kirikova, M., Johannesson, P. (eds.) CAiSE 2015. LNCS, vol. 9097, pp. 118–132. Springer, Heidelberg (2015)
7. Cortés-Calabuig, A., Paredaens, J.: Semantics of constraints in RDFS. In: Proceedings of the 6th Alberto Mendelzon International Workshop on Foundations of Data Management, pp. 75–90 (2012)

8. Ferrarotti, F., Hartmann, S., Link, S., Marin, M., Muñoz, E.: The finite implication problem for expressive XML keys: foundations, applications, and performance evaluation. In: Hameurlain, A., Küng, J., Wagner, R., Liddle, S.W., Schewe, K.-D., Zhou, X. (eds.) TLDKS X. LNCS, vol. 8220, pp. 60–94. Springer, Heidelberg (2013)
9. Ferrarotti, F., Hartmann, S., Link, S., Marin, M., Muñoz, E.: Soft cardinality constraints on XML data. In: Lin, X., Manolopoulos, Y., Srivastava, D., Huang, G. (eds.) WISE 2013, Part I. LNCS, vol. 8180, pp. 382–395. Springer, Heidelberg (2013)
10. Grahne, G., Zhu, J.: Discovering approximate keys in XML data. In: Proceedings of the 2002 ACM CIKM, pp. 453–460 (2002)
11. Hartmann, S.: Soft constraints and heuristic constraint correction in entity-relationship modelling. In: Bertossi, L., Katona, G.O.H., Schewe, K.-D., Thalheim, B. (eds.) Semantics in Databases 2001. LNCS, vol. 2582, pp. 82–99. Springer, Heidelberg (2003)
12. Hogan, A.: Skolemising blank nodes while preserving isomorphism. In: Proceedings of the 24th WWW, pp. 430–440 (2015)
13. Hogan, A., Arenas, M., Mallea, A., Polleres, A.: Everything you always wanted to know about blank nodes. Web Semant. Sci. Serv. Agents World Wide Web 27–28, 42–69 (2014). Semantic Web Challenge 2013
14. Lausen, G., Meier, M., Schmidt, M.: SPARQLing constraints for RDF. In: Proceeding of the 11th EDBT, pp. 499–509 (2008)
15. Liddle, S.W., Embley, D.W., Woodfield, S.N.: Cardinality constraints in semantic data models. Data Knowl. Eng. 11(3), 235–270 (1993)
16. Motik, B., Horrocks, I., Sattler, U.: Bridging the gap between OWL and relational databases. Web Semant. Sci. Serv. Agents World Wide Web 7(2), 74–89 (2009)
17. Muñoz, E.: Learning content patterns from linked data. In: Proceedings of the Linked Data for Information Extraction (LD4IE) Workshop, ISWC, CEUR Workshop Proceedings, vol. 1267, pp. 21–32. CEUR-WS.org (2014)
18. Paulheim, H.: Knowledge graph refinement: a survey of approaches and evaluation methods. Semant. Web - Interoperability Usability Appl. IOS Press J. (2016, to appear). http://www.semantic-web-journal.net/
19. Schmachtenberg, M., Bizer, C., Paulheim, H.: Adoption of the linked data best practices in different topical domains. In: Mika, P., et al. (eds.) ISWC 2014, Part I. LNCS, vol. 8796, pp. 245–260. Springer, Heidelberg (2014)
20. Schmidt, M., Lausen, G.: Pleasantly consuming linked data with RDF data descriptions. In: Proceedings of the 4th COLD Workshop (2013)
21. Schmidt, M., Meier, M., Lausen, G.: Foundations of SPARQL query optimization. In: Proceedings of the 13th ICDT, pp. 4–33. ACM (2010)
22. Seaborne, A.: SPARQL 1.1 Property Paths (2010). http://www.w3.org/TR/sparql11-property-paths/. Accessed Nov 2015
23. Soru, T., Marx, E., Ngomo, A.N.: ROCKER: a refinement operator for key discovery. In: Proceedings of the 24th WWW, pp. 1025–1033 (2015)
24. Soru, T., Marx, E., Ngonga Ngomo, A.-C.: Enhancing dataset quality using keys. In: Proceedings of the 14th ISWC, Posters & Demonstrations Track (2015)
25. Stickler, P.: CBD - Concise Bounded Description (2005). http://www.w3.org/Submission/CBD/. Accessed Oct 2015
26. Strong, D.M., Lee, Y.W., Wang, R.Y.: Data quality in context. Commun. ACM 40(5), 103–110 (1997)
27. Symeonidou, D., Armant, V., Pernelle, N., Saïs, F.: SAKey: scalable almost key discovery in RDF data. In: Mika, P., et al. (eds.) ISWC 2014, Part I. LNCS, vol. 8796, pp. 33–49. Springer, Heidelberg (2014)

28. Tao, J., Sirin, E., Bao, J., McGuinness, D.L.: Extending OWL with integrity constraints. In: Description Logics, CEUR Workshop Proceedings, vol. 573. CEUR-WS.org (2010)
29. Thalheim, B.: Fundamentals of cardinality constraints. In: Pernul, G., Tjoa, A.M. (eds.) ER 1992. LNCS, vol. 645, pp. 7–23. Springer, Heidelberg (1992)
30. Töpper, G., Knuth, M., Sack, H.: DBpedia ontology enrichment for inconsistency detection. In: Proceedings of the 8th International Conference on Semantic Systems, I-SEMANTICS 2012, pp. 33–40. ACM, New York (2012)
31. Völker, J., Fleischhacker, D., Stuckenschmidt, H.: Automatic acquisition of class disjointness. Web Semant. Sci. Serv. Agents World Wide Web 35(Part 2), 124–139 (2015). Machine Learning and Data Mining for the Semantic Web (MLDMSW)
32. Wang, R.Y., Strong, D.M.: Beyond accuracy: what data quality means to data consumers. J. Manage. Inf. Syst. 12(4), 5–33 (1996)

A Knowledge-Based Framework for Events Representation and Reuse from Historical Archives

Marco Rovera[✉]

Department of Computer Science, Università di Torino, Turin, Italy
rovera@di.unito.it

Abstract. Thanks to the digitization techniques, historical archives become source of a considerable amount of biographical, factual and geographical data, that need to be structured in order to be usable in higher-level applications. In this paper we present the project of an ontology-based framework aimed at formally representing and extracting historical events from archives; this process should serve to the purpose of (semi)-automatically building narratives that allow users to explore the archives themselves. This proposal refers to a Ph.D. research at early stage and is part of a wider project, Harlock'900, established between the Computer Science department of the University of Turin and the Istituto Gramsci, a cultural foundation promoting research in contemporary history.

Keywords: Historical events · Ontology modeling · Events extraction

1 Introduction and Motivation

The progressive digitization process of historical archives makes documentary and archival contents spread beyond their traditional boundaries – the archives themselves – and opens new possibilities of employment for a significant amount of biographical, factual and geographical data available in the archival resources. On the other hand, the Semantic Web and Linked Open Data paradigm provides a stable technological layer for representing such information and for making it reusable in a wide range of applications. In such a context, how to extract and represent this knowledge, in order to make it accessible and usable for the end user, becomes an issue.

In this paper we propose to use a knowledge-based approach for extracting and representing historical events gathered from archives and to make them available for higher-level applications in the form of narrative chains. The resulting framework should provide a reusable ICT solution for the historical domain.

2 State of the Art

This project is situated at the intersection of different research areas; in this section we give a concise account of the most relevant works in such domains, highlighting interesting results that have been achieved, according to our purpose and scope.

© Springer International Publishing Switzerland 2016
H. Sack et al. (Eds.): ESWC 2016, LNCS 9678, pp. 845–852, 2016.
DOI: 10.1007/978-3-319-34129-3_53

Historical events and cultural heritage. In general, a comprehensive survey on the adoption of Semantic Web technologies in the historical research can be found in [20]. Our project shares some of the goals and, partially, the application domain of the Agora project, developed at the VU Amsterdam, aimed at providing a context for cultural heritage collections (in particular, museum objects), connecting them to relevant historical events [26]. The Agora project, in turn, was tightly related to the Semantics of History project, aimed at the extraction of historical events from textual resources. Further results of the two projects, particularly in providing semantically-based access to museum resources, can be found in [7]. An example of semantic, event-based access to cultural resources is depicted in [8] while in [4] events are first class concept in an ontology-based application that employs narratives for museums and exhibitions.

Event ontologies and ontology modeling. For a synthetic introduction to the concept of "event" in the philosophical thought, the "Events" item in the Stanford Encyclopedia of Philosophy can be taken as a reference [2]. As a first step of this Ph.D. project, an analysis of the existing ontologies has been carried out (and it is still ongoing) to assess how each model covers the concept of "event"; among the existing ontologies and models, the Simple Event Model (SEM) [27], the Linked Open Description of Events (LODE) [24], the Event Ontology [10] and the F Event Model [22] are explicitly designed for representing events, though they show different purposes and levels of expressiveness and reusability. SEM is a domain-independent model, aimed at representing events on the web. It explicitly models the concepts of *event, actor, time, space, role* and *authority* and it builds on a loose (inclusive) definition of "event". LODE is a lightweight and property-based ontology and it is the result of an alignment between different ontologies, namely ABC, CIDOC-CRM, Event Ontology and DOLCE + DnS Ultralite (DUL). LODE aims at describing the aspects of an event that result from the answer to the four w-questions: what is happened, where, when and who is involved. The Event Ontology (EO) is a minimal event model, it represents the basic features of an event and, though it can be considered domain-independent, it was designed having in mind the description of music performances (like concerts). The F-Event Model is an upper ontology extending DOLCE + DnS Ultralite (DUL); by means of a rich conceptualization and a quite complex formalization, it allows to describe events in time and space, participation, structural relations between events (like mereology, causality, correlation), documentary support about an event and interpretation.

Other ontologies, like CIDOC-CRM [19] and the Europeana Data Model [9], provide a modeling of the concept of event within the wider framework represented by Cultural Heritage-related resources. An interesting classification of events (at upper level) can be found in [18], where the authors present a formal characterization of events, based on order-sorted logic. Among the foundational ontologies, DOLCE [1] implements in its backbone taxonomy one of the possible classifications of events discussed in [2], distinguishing between *activities, accomplishments, achievements* and *states*. The analysis also suggested that there are ontological problems yet not tackled in the representation of historical events [See Sect. 3].

Semantics for diachronical geography. As illustrated in Sect. 3, one of the goals of this Ph.D. proposal is to support a diachronical perspective on the geographical knowledge involved in the representation of historical events and concepts; in this field, valuable experience can be provided by projects like Pleiades [21], a gazetteer of places of ancient times, used for educational purposes, and GeoLat [3], a system allowing users to explore the geography of ancient times in Latin literature texts.

Event extraction. The task of extracting events from textual corpora has been performed according to different approaches, that can be roughly classified in data-driven or knowledge-driven [16]. While the former employ quantitative, statistical techniques and require to be trained on large textual corpora, the latter rely on domain knowledge and perform text mining by means of rules and patterns. A set of publicly available Information Extraction tools (some of which including event extraction), combining NLP and Semantic Web technologies, has been reviewed in [11]. However, as far as historical events are concerned, few examples are available (see [6, 23]), but there seems to be no benchmark to refer to. An interesting result has been achieved in [15], though in that case the input was represented by Wikipedia pages (i.e., a single type of source, providing semi-structured content) rather than by heterogeneous plain texts.

3   Problem Statement and Contributions

The Ph.D. proposal partially originates within Harlock'900, a three-year project established in December 2015 and involving the Department of Computer Science of the University of Turin and the Istituto Piemontese Antonio Gramsci, a non-profit institute promoting research on contemporary history topics. An overview of the approach that will be adopted in Harlock'900 can be found in [13]. The project's main goal is that of employing a set of resources from the Istituto Gramsci's archive and enriching the existing metadata with information describing the content of the resource itself; such information will be expressed using formal representations based on ontologies. The goal of the enrichment process is that of allowing the reuse and exploration of information for application (e.g. educational, touristic) purposes. The enriched metadata represent a semantic layer that will ensure a content-based access to archival resources.

The semantic layer should take into account the following aspects:

- supporting a diachronical perspective on geography, connecting the representation of a place in history with the actualized representation of the same place;
- connecting the representation of factual/historical entities at different granularity levels, allowing, for example, both a biographical and a general history perspective to be mutually connected and aligned;
- supporting the use – at application level – of narrative structures by giving a formal representation of events, their factual components and their complex relations.

This Ph.D. research will become part of such a wider project, taking on specific responsibility for the following **research questions**:

- what are the full requirements of the semantic layer (from an ontological point of view), accordingly to the previously stated aspects?
- Are the currently available event models adequate to fulfill the mentioned aspects?

- If not, what extensions to existing ontologies or integrations between different ontology modules are required for the purpose?
- Which tools can be used to automatically or semi-automatically perform event mining from text and annotation based on the model?
- What contribution can be given by Linked Open Data (LOD) (as input) in such a process?
- What contribution can the resulting framework give to the LOD (as output)?

The **research problem** at the basis of this proposal can be further decomposed into three main sub-problems:

- from the analysis performed so far it seems that the current available event models are not able to accurately account for all the conceptual requirements in the historical domain: for example, the reviewed models do not seem to provide the needed means to accurately represent collective entities (e.g. a political party); secondly, it seems difficult, using such models, to give a diachronical representation of places, asserting for example that a certain place, named in a certain way in 1920 and having a certain role at that time, was named differently in 2005 and had a different purpose (a military base became a school, a private house became an hotel, and so on). Third, in the event extraction process it could be beneficial to have an articulated taxonomy of historical events available, to classify the extracted events. In wider terms, it seems that, when individually taken, none of the available event models is able to cover all the conceptual aspects involved in a rich representation of (historical) events.
- an event-based framework supporting narrative is needed in order to put the domain information enclosed in historical archives at the service of different sorts of applications; this problem could impose backward requirements, especially on the event model;
- despite remarkable improvements in Natural Language Processing (NLP) and Semantic Web methodologies during the last decade, and even if some state-of-the-art tools are available, there still is no standard for event extraction from text; this project does not have as main goal that of improving NLP methodologies themselves; nevertheless, the availability of a domain event model, designed for the purpose of representing historical events, could allow us to treat our archival resources as a valuable test bed for such technologies.

The Ph.D. proposal aims at providing an event-based framework able to make contents provided by historical archives reusable for original applications. The underlying **research hypothesis** is that none of the available event models, individually taken as they are, provides a covering of all the aspects that characterize the concept of (historical) "event", while their integration (and their enrichment, where necessary) would allow us to succeed both in extracting and in representing the semantics of events in historical archives in a form that is suitable for their exploitation in novel contexts (such as tourism or education applications) provided by the Harlock'900 project.

4   Research Methodology and Approach

In order to achieve our research goals, we will adopt an iterative methodology, involving the following steps: (*a*) analysis of the existing event ontologies and, simultaneously, (*b*) requirements and data analysis (based on an appropriate selection of texts from the historical archives); (*c*) design hypothesis of the model and (*d*) resources annotation based on the model. "Outside" the iterative cycle, once the model will have become stable enough, the connection with the LOD will be established by contributing with the extracted data and, if necessary, by providing mappings of our model.

It is relevant to point out that, while the Harlock'900 project aims at covering different resource types, the Ph.D. proposal will focus only on textual digitized resources (hence images, posters, handwritten papers and similar are excluded from its scope). In particular, given the traditional distinction between primary and secondary historical sources (as recalled in [20]), the considered input data at application level will consist of primary ("narrative") sources like biographies, newspaper articles, chronicles. In the analysis stage, also secondary resources (e.g., history books) will be employed, as well as digital historical datasets (see for instance the one cited in [15] and available as API at http://www.vizgr.org/historical-events/) for hands-on testing of the existing event models and for creating a corpus of semantic representations of historical events to be used by the system. Since the Harlock'900 project aims at making accessible archival resources, which enclose much information about historical events (that are, in that form, often barely accessible), primary narrative sources are more interesting for the purpose of this work.

The described methodology will be instantiated as follows:

- requirements: a textual corpus from the archive will be selected and manually analyzed, extracting the ontological design requirements; simultaneously, the available event models are analyzed in order to identify the requirements fulfilled by each model, design patterns, lacks in coverage and differences between the models;
- hypothesis on the model: based on the previous step, an ontological analysis of the event domain will be carried out, in order to define the intended model; in doing so, particular attention will be paid to the reuse of existing (parts of) ontologies, integrating them as needed; in fact, the analysis will not necessarily produce a new ontological model (i.e., it is not aimed at creating a new ontology from scratch), but it should identify the classes and relations needed for the purpose of the project and the existing models that provide them. Only if needed, an extension in terms of classes or properties will be realized, but the current work is mainly conceived as an integration between different models. This approach will also guarantee a high degree of interoperability with existing ontologies. Furthermore, in the design process, state-of-the-art design methodologies will be employed like OntoClean [14];
- event extraction and annotation: based on the designed model, historical events must be extracted from textual resources and annotated; a set of tools will be implemented/employed that make use of extraction techniques [11]; a second means for enriching resources (that will be taken into account in Harlock'900 but not in this Ph.D. project)

is that of allowing "trusted users" – i.e. domain professionals – to contribute by editing metadata in order to participate to their enrichment.

5 Current Work

Since the presented project refers to a Ph.D. at early stage (beginning: 1 October 2015), no significant result has been achieved so far. Current work is focused on the analysis of the literature and of the existing event ontologies, in order to identify requirements and design choices. Also, the ontology design and annotation experience gained in previous projects, like the one documented in [12], will represent a valuable starting point. The next steps in the analysis will be (a) a hands-on test of the models aimed at understanding their strength and weaknesses in the annotation of historical texts, and (b) an investigation of the LOD cloud to verify which models are employed to describe events, how many and what kind of events are represented.

6 Evaluation Plan

Within the described proposal, three main phases will be subject to evaluation. The *semantic event model* will be evaluated during each iteration cycle, identifying a suitable corpus of texts in order to asses the coverage and expressiveness of the designed model. The *extraction and annotation tool(s)* will be evaluated using traditional statistical performance measures (precision, recall, accuracy, for instance). Lastly, once the whole *framework* is available and operative, a user study will be set up to test its main functionalities with end users.

7 Conclusions

In this paper I outlined my Ph.D. proposal, aimed at extracting semantic knowledge from historical archives and making it available for application purposes. To this end, I stated the main research problem I intend to tackle, together with the approach and research methodology to be followed. Also, the steps to be evaluated and the evaluation approach have been sketched. Due to the early stage of this Ph.D., it has not been possible to provide achieved results yet.

References

1. Borgo, S., Masolo, C.: Foundational choices in DOLCE. Handbook on Ontologies. International Handbooks on Information Systems, pp. 361–381. Springer, Heidelberg (2009)
2. Casati, R., Varzi, A.: Events. Stanford Encyclopedia of Philosophy (2002). http://plato.stanford.edu/entries/events (Substantive revision 2014). Accessed 10 Dec 2015
3. Ciotti, F., Lana, M., Tomasi, F.: TEI, ontologies, linked open data: geolat and beyond. J. Text Encoding Initiative (8) (2014)

4. Collins, T., Mulholland, P., Wolff, A.: Web supported employment: using object and event descriptions to facilitate storytelling online and in galleries. In: Proceedings of 4th Annual ACM Web Science Conference, pp. 74–77. ACM, June 2012

5. Cybulska A., Vossen P.: Event models for historical perspectives: determining relations between high and low level events in text, based on the classification of time, location and participants. In: LREC (2010)

6. Cybulska, A., Vossen, P.: Historical event extraction from text. In: Proceedings of 5th ACL-HLT Workshop on Language Technology for Cultural Heritage, Social Sciences, and Humanities. Association for Computational Linguistics (2011)

7. de Boer, V., Oomen, J., Inel, O., Aroyo, L., van Staveren, E., Helmich, W., de Beurs, D.: DIVE into the event-based browsing of linked historical media

8. Den Akker, C., van Aroyo, L., Cybulska, A., Van Erp, M., Gorgels, P., Hollink, L., Jager, C., Legene, S., van der Meij, L., Oomen, J., van Ossenbruggen, J., Wielinga, B.: Historical event-based access to museum collections. In: Proceedings of 1st International Workshop on Recognising and Tracking Events on the Web and in Real Life (EVENTS2010), Athens, Greece, May 2010

9. EDM Definition v. 5.2.6. http://pro.europeana.eu/page/edm-documentation

10. Event Ontology (EO). http://motools.sourceforge.net/event/event.html

11. Gangemi, A.: A comparison of knowledge extraction tools for the semantic web. In: Cimiano, P., Corcho, O., Presutti, V., Hollink, L., Rudolph, S. (eds.) ESWC 2013. LNCS, vol. 7882, pp. 351–366. Springer, Heidelberg (2013)

12. Goy, A., Magro, D., Petrone, G., Rovera, M., Segnan, M.: A semantic framework to enrich collaborative tables with domain knowledge. In: Proceedings of the 7th International Joint Conference on Knowledge Discovery, Knowledge Engineering and Knowledge Management, vol. 3, pp. 371–381. KMIS (2015)

13. Goy, A., Magro, D., Rovera, M.: Ontologies and historical archives: a way to tell new stories. Appl. Ontol. (2015, in press)

14. Guarino, N., Welty, C.A.: An overview of OntoClean. Handbook on ontologies. International Handbooks on Information Systems, pp. 201–220. Springer, Heidelberg (2009)

15. Hienert, D., Wegener, D., Paulheim, H.: Automatic classification and relationship extraction for multi-lingual and multi-granular events from Wikipedia. In: Detection, Representation, and Exploitation of Events in the Semantic Web (DeRiVE 2012), vol. 902, pp. 1–10 (2012)

16. Hogenboom, F., Frasincar, F., Kaymak, U., De Jong, F.: An overview of event extraction from text. In: Workshop on Detection, Representation, and Exploitation of Events in the Semantic Web (DeRiVE 2011) at 10th International Semantic Web Conference (ISWC 2011), vol. 779, pp. 48–57, October 2011

17. Hyvönen, E.: Semantic portals for cultural heritage. Handbook on Ontologies. International Handbooks on Information Systems, pp. 757–778. Springer, Heidelberg (2009)

18. Kaneiwa, K., Iwazume, M., Fukuda, K.: An upper ontology for event classifications and relations. In: Orgun, M.A., Thornton, J. (eds.) AI 2007. LNCS (LNAI), vol. 4830, pp. 394–403. Springer, Heidelberg (2007)

19. Le Boeuf, P., Doerr, M., Ore, C.E., Stead, S. (eds.): Definition of the CIDOC Conceptual Reference Model (Version 6.1). ICOM/CIDOC CRM Special Interest Group (2015)

20. Meroño-Peñuela, A., Ashkpour, A., Erp, M., Mandemakers, K., Breure, L.: Semantic technologies for historical research: a survey. Semant. Web J. **6**(6), 539–564 (2015)

21. PLEIADES project. http://pleiades.stoa.org/. Accessed 10 Dec 2015

22. Scherp, A., Franz, T., Saathoff, C., Staab, S.: F–a model of events based on the foundational ontology DOLCE + DnS ultralight. In: Proceedings of 5th International Conference on Knowledge Capture (K-CAP 2009). pp. 137–144. ACM, New York, NY, USA (2009)

23. Segers, R., Van Erp, M., Van Der Meij, L., Aroyo, L., van Ossenbruggen, J., Schreiber, G., Wielinga, B., Oomen, J., Jacobs, G.: Hacking history via event extraction. In: Proceedings of 6th International Conference on Knowledge Capture, pp. 161–162. ACM (2011)
24. Shaw, R., Troncy, R., Hardman, L.: LODE: linking open descriptions of events. In: Gómez-Pérez, A., Yu, Y., Ding, Y. (eds.) ASWC 2009. LNCS, vol. 5926, pp. 153–167. Springer, Heidelberg (2009)
25. Van Den Akker, C., Aroyo, L., Cybulska, A., Van Erp, M., Gorgels, P., Hollink, L., Wielinga, B.: Historical event-based access to museum collections. In: Proceedings of 1st International Workshop on Recognising and Tracking Events on the Web and in Real Life (EVENTS 2010), Athens, Greece (2010)
26. Van Den Akker, C., Legêne, S., Van Erp, M., Aroyo, L., Segers, R., van Der Meij, L., Van Ossenbruggen, J., Schreiber, G., Wielinga, B., Oomen, J., Jacobs, G.: Digital hermeneutics: agora and the online understanding of cultural heritage. In: Proceedings of 3rd International Web Science Conference, p. 10. ACM, June 2011
27. Van Hage, W.R., Malaisé, V., Segers, R., Hollink, L., Schreiber, G.: Design and use of the simple event model (SEM). Web Semant.: Sci. Serv. Agents World Wide Web **9**(2), 128–136 (2011)

Unsupervised Conceptualization and Semantic Text Indexing for Information Extraction

Eugen Ruppert[✉]

FG Language Technology, Technische Universität Darmstadt,
Darmstadt, Germany
ruppert@lt.informatik.tu-darmstadt.de

Abstract. The goal of my thesis is the extension of the Distributional Hypothesis [13] from the word to the concept level. This will be achieved by creating data-driven methods to create and apply conceptualizations, taxonomic semantic models that are grounded in the input corpus. Such conceptualizations can be used to disambiguate all words in the corpus, so that we can extract richer relations and create a dense graph of semantic relations between concepts. These relations will reduce sparsity issues, a common problem for contextualization techniques. By extending our conceptualization with named entities and multi-word entities (MWE), we can create a Linked Open Data knowledge base that is linked to existing knowledge bases like Freebase.

1 Motivation

The current NLP research is moving from the linear word/sentence/discourse representation towards semantic representations, where entities and their relations are made explicit. Even though NLP components are still being improved by emerging techniques like deep learning, the quality of existing components is sufficient to work on the semantic level – one level of abstraction up from surface text. We want to semantify text by assigning word sense IDs to the content words in the document. Working on the semantic level does not only provide us with entities like nouns, but also with their relations between each other. After semantification, we can use this representation to grasp the meaning of the document, e.g. what are the subjects in the document and to which class do they belong.

Semantic Web (SW) applications use entities, e.g. to disambiguate which *Turkey* the text refers to: the country or the animal[1]. However, knowledge bases like Freebase [6] or DBpedia [8] only relate concepts, not necessarily disambiguating senses. While linking words to such knowledge bases is useful in their current state, we propose an all-word conceptualization where all content words are identified by their senses.

---

[1] Throughout this proposal, we are using *italics* for text examples, underscores for hypernyms and `monospace text` for technical details, e.g. explicit context features.

© Springer International Publishing Switzerland 2016
H. Sack et al. (Eds.): ESWC 2016, LNCS 9678, pp. 853–862, 2016.
DOI: 10.1007/978-3-319-34129-3_54

With our symbolic conceptualizations we are able to identify concepts in a text (contextualization). The contextualization of a document text will improve the performance of entity linking for the 'classic Semantic Web' because we assume that concepts/entities also follow the Distributional Hypothesis – concepts co-occur with similar concepts in a document. Identification of all content words in a document will enable semantic applications like semantic indexing.

2   State of the Art

2.1   Conceptualizations

The creation of conceptualizations is related to Ontology Learning [2], which uses supervised and unsupervised methods. Supervised approaches use Word-Net [11] (e.g. [32]), Wikipedia, DBpedia and Freebase for ontology induction or entity extraction [8,21,35]. DBpedia Spotlight [8] relies on explicit links between entities in Wikipedia. Entity extraction is performed by a pre-trained prefix tree, and the disambiguation is done with a language model (LM). Sense embeddings can also be trained for conceptualizations [15]. The proposed approach is related to [7] but does not use knowledge bases.

Unsupervised methods often rely on relation extraction (cf. Sect. 2.3). OntoUSP [27] extracts a probabilistic ontology for the medical domain using dependency path features. OntoGain [9] relies on multi-word expressions (MWEs) to construct a taxonomic graph using hierarchical clustering. This graph is expanded with non-taxonomic relations. The hybrid approach of [35], which accesses search engines, can also identify novel MWEs, which signify entities. Local taxonomic relations can be extracted from text by identifying certain syntactic patterns in a text [14,16]. There are also approaches that directly work on the SW graph and utilize distributional methods to establish concept similarities, e.g. [24].

2.2   Contextualization

The two most recent contextualization shared tasks are the Word Sense Disambiguation (WSD) tasks of SemEval 2010 [20] and SemEval 2013 [23]. The participating systems often use knowledge bases like YAGO [19], WordNet or other ontologies to assign sense identifiers to target words (usually nouns) in a sentence. Knowledge-free systems employ co-occurrence and distributional similarities together with language models.

2.3   Relation Extraction

TextRunner [36] extracts explicit relationship tuples (R, T_1, T_2) from POS-tagged text. These relations can be seen as 'facts' and aid question answering. GraBTax [34] can build taxonomies by utilizing co-occurrence and lexical similarity of n-gram topics from document titles.

OntoUSP [27] focuses on identifying specific relations in the medical domain. It can identify nominal MWEs and group different spellings into clusters. Also, it performs hierarchical clustering on verbs. [31] present a relation learning approach. By extracting known facts, they create triggers to extract similar relations of different entities. [28] uses distributional statistics to extract relations between nouns. This produces highly precise relations, but the recall is quite low. We believe that we can alleviate this problem by extracting relations not on the word level (leaf in the taxonomy graph) but between taxonomic concepts (nodes).

2.4 Linked Open Data

There are many useful data collections available on the Web. Freebase [6] or DBpedia [17] offer a large number of relations, usually as RDF triples, that can be queried using APIs. On top of that, there are applications like DBPedia Spotlight [8] or Babelfy [22] that can annotate texts with e.g. DBPedia entities, which in turn can link to Wikipedia.

We plan to extract and display information similar to the Weltmodell [1] or ConceptNet [33]. By utilizing disambiguated concepts, we believe that we can extract more (higher recall on concept level vs. word level) and more precise relations (handling polysemy).

3   Problem Statement and Contributions

The most trivial sentences and phrases can be difficult to understand and process for computers. Supervised ontologies and dictionaries help to a large extent, however, often they do not fit the textual domain to which they are applied. Search engines can be viewed as semantic applications, as they are able to identify word senses by using the provided keywords, e.g. *throw a ball* vs. *attend a ball*. However, it is not shown how many senses of the provided query terms the search engine knows about. We believe that this is important information and making this information available would help many users. Semantic resources like Word-Net are often too fine-grained, which reduces usability in semantic applications[2].

To alleviate such problems, we plan to investigate the following research questions:

- How can we construct a semantic model that improves WSD performance? In this task we are going to use Distributional Semantics (DS) methods [13] to create conceptualizations from an input corpus. Afterwards, we are going to structure the concepts in a global taxonomy graph by utilizing the hypernymy structure.
- Which unsupervised, knowledge-free methods can we use to obtain state-of-the-art WSD? This task involves identifying concepts in their context to obtain an explicit semantic representation of the corpus. To the best of our knowledge, there have been no attempts to create a fully disambiguated corpus.

---

[2] WordNet identifies 12 senses for *ball*, some are highly domain-specific, like sense 12, "a pitch that is not in the strike zone." Humans intuitively identify fewer senses.

– How can we identify significant relations between concepts to enrich our conceptualization? Using the contextualized corpus, we can extract relations on the concept and hypernym level, allowing us to extract more relations. We need to make sure that we only add significant relations to our concept graph.

Furthermore, we plan to create demonstrator applications based on the conceptualizations to exemplify the semantic annotation capability. We are also going to release the created applications as free, open-source applications with a focus on usability.

4   Research Methodology and Approach

4.1   Conceptualizations

To be able to annotate and semantify text, we need a knowledge model. Therefore, we are going to use the JoBimText framework [5] to create symbolic conceptualizations. We believe that having an explicit symbolic representation is an advantage to vector-based models like deep learning because of direct interpretability. We are going to create JoBimText models [30] and extend those to interconnected graphs, where we introduce new semantic relations between the nodes. A JoBimText model consists of a Distributional Thesaurus (DT) with sense-disambiguated entries. We induce word senses using Chinese Whispers [3], a knowledge-free graph clustering algorithm. The senses are labeled with the most frequent hypernym terms that were obtained using lexico-syntactic patterns [14,16], producing local taxonomies. In addition, the models contain significant context features for each word and a DT of such context features. The combination of entities like named entities and multi-word expressions [29] with common words will create an all-word knowledge base. Since it is fully based on the input corpus, there is no need for domain adaptation.

By utilizing the hypernymy structure, we can aggregate context features on concept levels. E.g. *jaguars*, *tigers* and *wolves* are all <u>animals</u>, but in our corpus, we only find sentences where *tigers* and *wolves* **hunt**. From this information, we can infer that *jaguars* probably can **hunt** as well, thus projecting contextual information through aggregation into the <u>animal</u> concept.

We apply contextualization to obtain a sense-disambiguated corpus. Then we compute the similarity graph once again, this time using word sense IDs instead of words. This should result in a DT, where each entry is fully disambiguated, allowing us to create a more detailed and more precise model (more entries per word sense, disambiguated context features).

4.2   Contextualization

With our sense-disambiguated semantic models, we can perform semantic text annotation. We put a strong focus on the contextualization technique, since it is going to connect the conceptualized knowledge to text. Using word sense disambiguation on the input text, we are going to infer the senses of the words.

By using similar context features from the model, we are able to identify the word sense, even if the term–context feature combination has never been observed in the corpus. Preliminary experiments have shown that utilizing similar features improves recall, with a slight decline in precision. To further increase recall, we will use co-occurrence features and a language model.

Once we have established a conceptualization with relations between concepts and aggregated context features per concept, we can even infer the concept for yet unseen words – a zero-shot contextualization, e.g. by matching the context features of concepts to the unseen word, we can assume that X in the phrase X *hunts its prey* is an <u>animal</u>.

4.3 Relation Extraction

The conceptualization yields a taxonomy graph with sense-labeled leaf nodes. We want to extend such a 'taxonomic skeleton' into a dense graph with many types of relations. Our semantic model already contains a large number of facts, like *jaguar* is-an <u>animal</u> or *cars* can be *driven*. While such facts seem trivial, in our model there are many facts that are not considered common knowledge, e.g. *impala* is-a <u>car</u> (indicating a Chevrolet Impala). We employ Open Information Extraction (OIE) techniques [26] to extract additional semantic relations.

Using the disambiguated corpus, we propagate the dependency relations in the hypernym graph to extend our conceptualization. If we take the input phrase *jaguar kills deer*, we can extract a multitude of facts:

basic *jaguar kills deer*
agent expansion *tiger kills deer*
object expansion *jaguar kills prey*
verb expansion *jaguar wounds deer*
hypernymies <u>animal</u> kills <u>animal</u>
combinations <u>animal</u> *kills deer, jaguar wounds* <u>animal</u>, etc.

Especially to identify relations between named entities [12], we need a larger, richer set of relations. Therefore, we are going to use supervised resources like WordNet to extract examples of a relation tuple (R, T_1, T_2) and – based on this input relation – find patterns that can be used to extract such relations. This bootstrapping method is similar to [31].

4.4 Linked Open Data

To make our approach usable to other researchers, as well as to incorporate our conceptualizations into the SW, we are going to publish the models as Linked Open Data. We are going build a semantic network similar to the hyper graph that is available in JoBimViz [30] and extend it with the concept taxonomies and contextualization techniques. This will allow to browse our conceptualizations with unique identifiers.

We are going to offer our contextualization technique through an open API. The annotated text would consist of a sequence of sense IDs, e.g. *jaguar#NN_2 hunt#VB_1 deer#NN_1* for the source sentence *jaguar hunts deer*. To bridge the gap between our inferred knowledge base to existing knowledge bases, we can create an alignment using Lesk [18]. This will increase the usability, since users can obtain identifiers of established SW resources and use this information to semantify their texts.

5   Preliminary Results

5.1   Conceptualization

Extending [5], we are now able to create semantic models that contain word senses and (unstructured) hypernyms. We call these models JoBimText models [30] and already use them for contextualization. The next step is to create a concept taxonomy with aggregated context features.

5.2   Contextualization

We have implemented a contextualization system that we are now extending with new features for a publication in the near future. Currently, it performs sense annotation based on a context feature extractor, e.g. trigram or dependency features. Using large language model with and word co-occurrences, we achieve a performance comparable to the systems in SemEval 2013, task 13 [23].

5.3   Relation Extraction

This task has not yet started, because it relies on a contextualized corpus.

5.4   Web Demonstrators/LOD

Our JoBimViz[3] web application is used to exemplify our semantic models [30]. It already features word identifiers, consisting of the model and a word representation (e.g. `lemma#POS`). It can be browsed as a semantic network, by following the links, similarly to LOD repositories. Furthermore, it features a transparent Java API for machine access. As a demonstrator for contextualized corpora, we have created a semantic search demo based on Apache Solr and PHP. It incorporates keyword search as well as search for concepts and displays possible MWE expansions.

6   Evaluation Plan

6.1   Conceptualizations

Using a path based measure [25], we can assess the structural similarity of our conceptualizations with WordNet. The sense clustering method is flexible and

---

[3] http://maggie.lt.informatik.tu-darmstadt.de/jobimviz/.

allows for different granularities. We want to identify the best granularity settings extrinsically, by evaluating the performance of several clusterings (with different granularities) using the contextualization technique. We use the Turk Bootstrap Word Sense Inventory (TWSI) 2.0 dataset [4]. It contains sense-annotated sentences from Wikipedia and a crowdsourced sense inventory with substitutions for about 1,000 nouns.

To verify our intuition that a model computed on a domain-specific corpus outperforms general or foreign-domain models, we plan to compute several models and cross-evaluate them.

6.2 Contextualization

To evaluate the performance of the contextualization system, we are going to use the TWSI dataset [4] here as well. It contains contextualized substitutions for about 150,000 sentences, a larger collection than used for SemEval WSD tasks. The TWSI dataset is mostly used for parameter tuning and determining the best feature configuration. Once the best feature set is established, we are going to evaluate our contextualization on the SemEval 2010 [20] and SemEval 2013 [23] datasets. This allows us to compare our unsupervised contextualization technique to state-of-the-art techniques, and possibly to participate in a future WSD challenge.

To evaluate the zero-shot contextualization, we can remove sentences with certain (even polysemous) target terms from the input corpus and create the conceptualization. Then we can input the sentences with the "unknown" words and evaluate the concept identification. To demonstrate improvements of the complex structured semantic model, we compare it with a simple distributional model.

6.3 Relation Extraction

The evaluation of relation extraction is challenging. To evaluate our approach, we are going to apply the relation extraction on a slightly different task. We are going to extract named entities like politicians (news data) and use our conceptualization to identify the events and relations in which they are involved. Most other open information systems rely on manually created datasets [10] to evaluate their systems.

7 Conclusion

In this proposal we have presented a framework for unsupervised conceptualizations based on unstructured text collections. Its advantage is that the resulting models are tied to the input text, thus allowing for applications without domain adaptation. The generated models are data-driven and can therefore be created for every domain where large amounts of texts are available. Using a contextualization technique, the framework creates a fully semantified sentence and document representation. This representation is tied to Linked Open Data resources.

Acknowledgments. This work has been supported by the German Federal Ministry of Education and Research (BMBF) within the context of the Software Campus project LiCoRes under grant No. 01IS12054. The author would like to thank his mentor Simone Paolo Ponzetto, his advisers Chris Biemann and Martin Riedl, and the reviewers for their valuable feedback.

References

1. Akbik, A., Michael, T.: The weltmodell: a data-driven commonsense knowledge base. In: Proceedings of LREC 2014, Reykjavik, Iceland, pp. 3272–3276 (2014)
2. Biemann, C.: Ontology learning from text: a survey of methods. LDV forum **20**(2), 75–93 (2005)
3. Biemann, C.: Chinese whispers - an efficient graph clustering algorithm and its application to natural language processing problems. In: Proceedings of TextGraphs-1, New York City, NY, USA, pp. 73–80 (2006)
4. Biemann, C.: Turk bootstrap word sense inventory 2.0: a large-scale resource for lexical substitution. In: Proceedings of LREC 2012, Istanbul, Turkey, pp. 4038–4042 (2012)
5. Biemann, C., Riedl, M.: Text: now in 2D! a framework for lexical expansion with contextual similarity. J. Lang. Model. **1**(1), 55–95 (2013)
6. Bollacker, K., Evans, C., Paritosh, P., Sturge, T., Taylor, J.: Freebase: a collaboratively created graph database for structuring human knowledge. In: Proceedings of ACM SIGMOD 2008, Vancouver, Canada, pp. 1247–1250 (2008)
7. Chen, X., Liu, Z., Sun, M.: A unified model for word sense representation and disambiguation. In: Proceedings of EMNLP 2014, Doha, Qatar, pp. 1025–1035 (2014)
8. Daiber, J., Jakob, M., Hokamp, C., Mendes, P.N.: Improving efficiency and accuracy in multilingual entity extraction. In: Proceedings of I-SEMANTICS 2013, Graz, Austria, pp. 121–124. ACM (2013)
9. Drymonas, E., Zervanou, K., Petrakis, E.G.M.: Unsupervised ontology acquisition from plain texts: the *OntoGain* system. In: Hopfe, C.J., Rezgui, Y., Métais, E., Preece, A., Li, H. (eds.) NLDB 2010. LNCS, vol. 6177, pp. 277–287. Springer, Heidelberg (2010)
10. Etzioni, O., Fader, A., Christensen, J., Soderland, S., Mausam, M.: Open information extraction: the second generation. In: IJCAI, Barcelona, Spain, vol. 11, pp. 3–10 (2011)
11. Fellbaum, C.: Wordnet. An Electronic Lexical Database. MIT Press, Cambridge (1998)
12. Feuerbach, T., Riedl, M., Biemann, C.: Distributional semantics for resolving bridging mentions. In: Proceedings of RANLP 2015, Hissar, Bulgaria, pp. 192–199 (2015)
13. Harris, Z.S.: Methods in Structural Linguistics. University of Chicago Press, Chicago (1951)
14. Hearst, M.A.: Automatic acquisition of hyponyms from large text corpora. In: Proceedings of COLING-1992, Nantes, France, pp. 539–545 (1992)
15. Iacobacci, I., Pilehvar, M.T., Navigli, R.: Sensembed: learning sense embeddings for word and relational similarity. In: Proceedings of ACL 2015, Beijing, China, pp. 95–105 (2015)
16. Klaussner, C., Zhekova, D.: Lexico-syntactic patterns for automatic ontology building. SRW at RANLP **2011**, 109–114 (2011)

17. Lehmann, J., Isele, R., Jakob, M., Jentzsch, A., Kontokostas, D., Mendes, P.N., Hellmann, S., Morsey, M., van Kleef, P., Auer, S., Bizer, C.: DBpedia - a large-scale, multilingual knowledge base extracted from wikipedia. Semant. Web J. **6**(2), 167–195 (2015)

18. Lesk, M.: Automatic sense disambiguation using machine readable dictionaries: how to tell a pine cone from an ice cream cone. In: Proceedings of SIGDOC 1986, pp. 24–26. ACM, Toronto, Ontario, Canada (1986)

19. Mahdisoltani, F., Biega, J., Suchanek, F.: YAGO3: a knowledge base from multi-lingual wikipedias. In: Proceedings of CIDR 2015, Asilomar, CA, USA (2015)

20. Manandhar, S., Klapaftis, I.P., Dligach, D., Pradhan, S.S.: SemEval-2010 task 14: word sense induction & disambiguation. In: Proceedings of SemEval-2010, Uppsala, Sweden, pp. 63–68 (2010)

21. Medelyan, O., Manion, S., Broekstra, J., Divoli, A., Huang, A.-L., Witten, I.H.: Constructing a focused taxonomy from a document collection. In: Cimiano, P., Corcho, O., Presutti, V., Hollink, L., Rudolph, S. (eds.) ESWC 2013. LNCS, vol. 7882, pp. 367–381. Springer, Heidelberg (2013)

22. Moro, A., Raganato, A., Navigli, R.: Entity linking meets word sense disambigua-tion: a unified approach. TACL **2**, 231–244 (2014)

23. Navigli, R., Vannella, D.: SemEval-2013 task 11: word sense induction and disam-biguation within an end-user application. In: Proceedings of *SEM 2013, Atlanta, GA, USA, vol. 2, pp. 193–201 (2013)

24. Nováček, V., Handschuh, S., Decker, S.: Getting the meaning right: a complemen-tary distributional layer for the web semantics. In: Aroyo, L., Welty, C., Alani, H., Taylor, J., Bernstein, A., Kagal, L., Noy, N., Blomqvist, E. (eds.) ISWC 2011, Part I. LNCS, vol. 7031, pp. 504–519. Springer, Heidelberg (2011)

25. Pedersen, T., Patwardhan, S., Michelizzi, J.: Wordnet::similarity: measuring the relatedness of concepts. In: Demonstration Papers at HLT-NAACL 2004, Boston, MA, USA, pp. 38–41 (2004)

26. Piskorski, J., Yangarber, R.: Information extraction: past, present and future. In: Poibeau, T., Saggion, H., Piskorski, J., Yangarber, R. (eds.) Multi-source, Mul-tilingual Information Extraction and Summarization. Theory and Applications of Natural Language Processing, pp. 23–49. Springer, Heidelberg (2013)

27. Poon, H., Domingos, P.: Unsupervised ontology induction from text. In: Proceed-ings of ACL 2010, Uppsala, Sweden, pp. 296–305 (2010)

28. Remus, S.: Unsupervised relation extraction of in-domain data from focused crawls. In: SRW at EACL 2014, Gothenburg, Sweden, pp. 11–20 (2014)

29. Riedl, M., Biemann, C.: A single word is not enough: ranking multiword expressions using distributional semantics. In: Proceedings of EMNLP 2015, Lisboa, Portugal, pp. 2430–4440 (2015)

30. Ruppert, E., Kaufmann, M., Riedl, M., Biemann, C.: JoBimViz: a web-based visu-alization for graph-based distributional semantic models. In: System Demonstra-tions at ACL 2015, Beijing, China, pp. 103–108 (2015)

31. Shinyama, Y., Sekine, S.: Preemptive information extraction using unrestricted relation discovery. In: Proceedings of HLT-NAACL 2006, New York, NY, USA, pp. 304–311 (2006)

32. Snow, R., Jurafsky, D., Ng, A.Y.: Semantic taxonomy induction from heterogenous evidence. In: Proceedings of COLING/ACL 2006, Sydney, Australia, pp. 801–808 (2006)

33. Speer, R., Havasi, C.: Conceptnet 5: a large semantic network for relational knowl-
    edge. In: Gurevych, I., Kim, J. (eds.) The Peoples Web Meets NLP. Theory and
    Applications of Natural Language Processing, pp. 161–176. Springer, Heidelberg
    (2013)
34. Treeratpituk, P., Khabsa, M., Giles, C.L.: Graph-based approach to automatic
    taxonomy generation (grabtax). CoRR abs/1307.1718 (2013)
35. Wong, W., Liu, W., Bennamoun, M.: Acquiring semantic relations using the web
    for constructing lightweight ontologies. In: Theeramunkong, T., Kijsirikul, B.,
    Cercone, N., Ho, T.-B. (eds.) PAKDD 2009. LNCS, vol. 5476, pp. 266–277.
    Springer, Heidelberg (2009)
36. Yates, A., Cafarella, M., Banko, M., Etzioni, O., Broadhead, M., Soderland, S.:
    Textrunner: open information extraction on the web. In: System Demonstrations
    at NAACL 2007, Rochester, NY, USA, pp. 25–26 (2007)

Continuously Self-Updating Query Results over Dynamic Heterogeneous Linked Data

Ruben Taelman[(⊠)]

Data Science Lab (Ghent University - iMinds),
Sint-Pietersnieuwstraat 41, 9000 Ghent, Belgium
`ruben.taelman@ugent.be`

Abstract. Our society is evolving towards massive data consumption from heterogeneous sources, which includes rapidly changing data like public transit delay information. Many applications that depend on dynamic data consumption require highly available server interfaces. Existing interfaces involve substantial costs to publish rapidly changing data with high availability, and are therefore only possible for organisations that can afford such an expensive infrastructure. In my doctoral research, I investigate how to publish and consume real-time and historical Linked Data on a large scale. To reduce server-side costs for making dynamic data publication affordable, I will examine different possibilities to divide query evaluation between servers and clients. This paper discusses the methods I aim to follow together with preliminary results and the steps required to use this solution. An initial prototype achieves significantly lower server processing cost per query, while maintaining reasonable query execution times and client costs. Given these promising results, I feel confident this research direction is a viable solution for offering low-cost dynamic Linked Data interfaces as opposed to the existing high-cost solutions.

Keywords: Linked Data · Triple Pattern Fragments · SPARQL · Continuous querying · Real-time querying

1 Introduction

The Web is an important driver of the increase in data. This data is partially made up of *dynamic* data, which does not remain the same over time, like for example the delay of a certain train or the currently playing track on a radio-station. Dynamic data is mostly published as *data streams* [3], which tend to be offered in a push-based manner. This requires data providers to have a persistent connection with all clients who consume these streams. On top of that, queries over real-time data are expected to be *continuous*, because the data are now continuously updating streams instead of just finite stored datasets. At the same time, this dynamic data also leads to the generation of *historical* data, which may be useful for data analysis.

R. Taelman—Supervised by Ruben Verborgh and Erik Mannens.

H. Sack et al. (Eds.): ESWC 2016, LNCS 9678, pp. 863–872, 2016.
DOI: 10.1007/978-3-319-34129-3_55

In this work, I investigate how to publish and consume non-high frequency real-time and historical Linked Data. This real-time data for example includes sensor results which update at a frequency in the order of seconds, use cases that require updates in the order of milliseconds are excluded. The focus lies at low-cost publication, so that large scale consumption of this data becomes possible without endpoint availability issues.

In the next section, the existing work in the area will be discussed. Section 3 will explain the problem I am trying to solve, after which Sect. 4 will briefly explain the methodology for solving this problem. Section 5 will discuss the evaluation of this solution after which Sect. 6 will present some preliminary results. Finally, in Sect. 7 I will explain the desired impact of this research.

2   State of the Art

Current solutions for querying and publishing dynamic data is divided in the two generally disjunct domains of *stream reasoning* and *versioning*, which will be explained hereafter. After that, a low cost server interface for static data will be explained.

Stream reasoning is defined as "the logical reasoning in real time on gigantic and inevitably noisy data streams in order to support the decision process of extremely large numbers of concurrent users" [4]. This area of research integrates data streams with traditional RDF reasoners. Existing SPARQL extensions for stream processing solutions like C-SPARQL [5] and CQELS [10] are based on *query registration* [4,7], which allows clients to register their query at a streaming-enabled SPARQL endpoint that will continuously evaluate this query. These data streams consist of triples that are *annotated* with a timestamp, which indicates the moment on which the triple is valid. These querying techniques can for example be used to query semantic sensor data [13]. C-SPARQL is a first approach to querying over both static and dynamic data. This solution requires the client to register a query in an extended SPARQL syntax which allows the use of *windows* over dynamic data. The execution of queries is based on the combination of a traditional SPARQL engine with a *Data Stream Management System* (DSMS) [2]. The internal model of C-SPARQL creates queries that distribute work between the DSMS and the SPARQL engine to respectively process the dynamic and static data. CQELS is a "white box" approach, as opposed to the "black box" approaches like C-SPARQL. This means that CQELS natively implements all query operators, as opposed to C-SPARQL that has to transform the query to another language for delegation to its subsystems. This native implementation removes the overhead that black box approaches like C-SPARQL have. The syntax is very similar as to that of C-SPARQL, also supporting query registration and time windows. According to previous research [10], this approach performs much better than C-SPARQL for large datasets, for simple queries and small datasets the opposite is true.

Offering historical data can be achieved by versioning entire datasets [15] using the Memento protocol [14] which extends HTTP with content negotiation in the datetime dimension. Memento adds a new link to resources in the

HTTP header, named the *TimeGate*, which acts as the datetime dimension for a resource. It provides a list of timely versions of the resource which can be requested. Using Memento's datetime content negotiation and TimeGates, it is possible to do *Time Travel* over the web and browse pages at a specific point in time. R&WBase [17] is a triple-store versioning approach based on delta storage combined with traditional snapshots. It offers a method for querying these versioned datasets using SPARQL. The dataset can be retrieved as a virtual graph for each delta revision, thus providing Memento-like time travel without an explicity time indication. TailR [11] provides a platform through which datasets can be versioned based on a combination of snapshot and delta storage and offered using the Memento protocol. It allows queries to retrieve the dataset version at a given time and the times at which a dataset has changed.

Triple Pattern Fragments (TPFS) [18] is a Linked Data publication interface which aims to solve the issue of low availability and performance of existing SPARQL endpoints for static querying. It does this by moving part of the query processing to the client, which reduces the server load at the cost of increased data transfer and potentially increased query evaluation time. The endpoints are limited to an interface with which only separate triple patterns can be queried instead of full SPARQL queries. The client is then responsible for carrying out the remaining work.

3   Problem Statement

Traditional public static SPARQL query endpoints have a major availability issue. Experiments have shown that more than half of them only reach an availability of less than 95 % [6]. The unrestricted complexity of SPARQL queries [12] combined with the public character of SPARQL endpoints requires an enormous server cost, which can lead to a low server availability. Dynamic SPARQL streaming solutions like C-SPARQL and CQELS offer combined access to dynamic data streams and static background data through continuously executing queries. Because of this continuous querying, the cost of these servers can become *even bigger* than with static querying for similarly sized datasets.

The definition of stream reasoning [4] states that it requires reasoning on data streams for "an extremely large number of concurrent users". If we can not even reach a large number of concurrent static SPARQL queries against endpoints without overloading them, how can we expect to do this for dynamic SPARQL queries? Because evaluating these queries put an even greater load on the server if we assume that the continuous execution of a query requires more processing than the equivalent single execution of that query.

The main research question of our work is:

Question 1: How can we combine the low cost publication of non-high frequency real-time and historical data, such that it can efficiently be queried together with static data?

To answer this question, we also need to find an answer to the following questions:

Question 2: How can we efficiently store non-high frequency real-time and historical data and allow efficient transfer to clients?

Question 3: What kind of server interface do we need to enable client-side query evaluation over both static and dynamic data?

These research questions have lead to the following hypotheses:

Hypothesis 1: Our storage solution can store new data in linear time with respect to the amount of new data.

Hypothesis 2: Our storage solution can retrieve data by time or triple values in linear time with respect to the amount of retrieved data.

Hypothesis 3: The server cost for our solution is lower than the alternatives.

Hypothesis 4: Data transfer is the main factor influencing query execution time in relation to other factors like client processing and server processing.

4   Research Approach

As discussed in Sect. 2, TPF is a Linked Data publication interface which aims to solve the high server cost of static Linked Data querying. This is done by partially evaluating queries client-side, which requires the client to break down queries into more elementary queries which can be solved by the limited and low cost TPF server interface. These elementary query results are then locally combined by the client to produce results for the original query.

We will extend this approach to *continuously updating* querying over *dynamic* data.

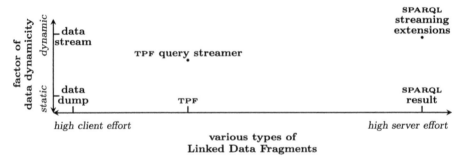

Fig. 1. LDF axis showing the server effort needed to publish different types of interface together with a vertical axis showing the factor of data dynamicity an interface exposes.

Figure 1 shows this shift to more static data in relation to the *Linked Data Fragments (*ldf*)* [19] axis. LDF is a conceptual framework to compare Linked Data publication interface in which TPF can be seen as a trade-off between high server and client effort for data retrieval. SPARQL streaming solutions like C-SPARQL and CQELS can handle high frequency data and they require a high server effort because they are at least as expressive as regular SPARQL. Data streams on the other hand expose high frequency data as well, but here it is the client that has to do most of the work when selecting data from those streams. Our TPF query streaming extension focuses on non-high frequency data and aims to lower the server effort for more efficient scaling to large numbers of concurrent query executions.

We can split up this research in three parts, which are shown in Fig. 2. First, the server needs to be able to efficiently store dynamic data and publicly offer it. Second, this data must be transferable to the client. Third, a query engine at the client must be able to evaluate queries using this data and keep its answers up to date.

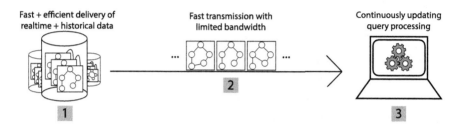

Fig. 2. A client must be able to evaluate queries by retrieving data from multiple heterogeneous datasources.

The storage of historical and real-time data requires a delicate balance between storage size and lookup speed. I will develop a method for this storage and lookup with a focus on the efficient retrieval and storage of versions, the dynamic properties of temporal data and the scalability for historical data. This storage method can be based on the differential storage concept TailR uses, combined with HDT [8] compression for snapshots. The interface through which data will be retrieved could benefit a variant of Memento's *timegate* index to allow users to evaluate historical queries.

For enabling the client to evaluate queries, the client needs to be able to access data from one or more data providers. I will develop a mechanism that enables efficient transmission of temporal data between server and client. By exploiting the similarities between and within temporal versions, I will limit the required bandwidth as much as possible.

To reduce the server cost, the client needs to help evaluating the query. Because of this, we assume that our solution will have a higher client processing cost than streaming-based SPARQL approaches for equivalent queries. For this last goal, I will develop a client-side query engine that is able to do federated querying of temporal data combined with static data against heterogeneous datasources. The engine must keep the real-time results of the query up to date. We can distinguish three requirements for this solution:

- Allowing queries to be declared using a variant of the SPARQL language so that it becomes possible for clients to declare queries over dynamic data. This language should support the RDF stream query semantics that are being discussed within the RSP Community Group[1]. We could either use the C-SPARQL or CQELS query language, or make a variant if required.

- Building a client-side query engine to do continuous query evaluation, which means that the query results are updated when data changes occur.
- Providing a format for the delivery of continuously updating results for registered queries that will allow applications to handle this data in a dynamic context.

5   Evaluation Plan

I will evaluate each of the three major elements of this research independently: the storage solution for dynamic data at the server; the retrieval of this data and its transmission; and the query evaluation by the client.

5.1   Storage

The evaluation of our storage solution can be done with the help of two experiments.

First, I will execute a large number of insertions of dynamic data against a server. I will measure its CPU usage and determine if it is still able to achieve a decent quality of service for data retrieval. I will also measure the increase in data storage. By analyzing the variance of the CPU usage with different insertion patterns we should be able to accept or reject Hypothesis 1, which states that data can be added in linear time.

The second experiment will consist of the evaluation of data retrieval. This experiment will consist of a large number of lookups against a server by both triple contents and time instants. Doing a variance analysis on the lookup times over these different lookup types will help us to determine the validity of Hypothesis 2, which states that data can be retrieved in linear time.

These two experiments can be combined to see if one or the other demands too much of the server's processing power.

5.2   Retrieval and Transmission

To determine the retrieval cost of data from a server and its transmission, we need to measure the effects of sending a large amount of lookup requests. One of the experiments I performed on the solution that was built during my master's thesis [16] was made up of one server and ten physical clients. Each of these clients could execute from one to ten concurrent unique queries. This results in a series of 10 to 200 concurrent query executions. This setup was used to test the client and server performance of my implementation compared to C-SPARQL and CQELS.

Even though this experiment produced some interesting results, as will be explained in the next section, 200 concurrent clients are not very representative

---

[1] https://www.w3.org/community/rsp/.

for large scale querying on the public Web. But it can already be used to partially answer Hypothesis 3 that states that our solution has a lower server cost than the alternatives. To extend this, I will develop a mechanism to simulate thousands of simultaneous requests to a server that offers dynamic data. The main bottleneck in the current experiment setup are the query engines on each client. If we were to detach the query engines from the experiment, we could send much more requests to the server and this would result in more representative results. This could be done by first collecting a representative set of HTTP requests that these query engines send to the server. This set of requests should be based on real non-high frequency use cases where it makes sense to have a large number of concurrent query evaluations. These requests can be inspired by existing RSP benchmarks like SRBench [20] and CityBench [1]. Once this collection has been built, the client-CPU intensive task is over, and we can use this pool of requests to quickly simulate HTTP requests to our server. By doing a variance analysis of the server CPU usage for my solution compared to the alternatives, we will be able to determine the truth of Hypothesis 3.

5.3 Query Evaluation

The evaluation of the client side query engine can be done like the experiment of my master's thesis, as explained in the previous section. In this case, the results would be representative since the query engine is expected to be the most resource intensive element in this solution. The CityBench [1] RSP benchmark could for example be used to do measurements based on datasets from various city sensors. By doing a variance analysis of the different client's CPU usage for my solution compared to the alternatives, we will be able to determine how much higher our client CPU cost is than the alternatives. The alternatives in this case include server-side RSP engines like C-SPARQL and CQELS, but also fully client-side stream processing solutions using stream publication techniques like Ztreamy [9]. This way, we test compare our solution with both sides of the LDF axis, on the one hand we have the cases where the server does all of the work while evaluating queries, while on the other hand we have cases where the client does all of the work. For Hypothesis 4, which assumes that data transfer is the main factor for query execution time, we will do a correlation test of bandwidth usage and the corresponding query's execution times.

6 Preliminary Results

During my master's thesis, I did some preliminary research on the topic of continuous querying over non-high frequency real-time data. My solution consisted of *annotating* triples with *time* to give them a timely context, which allowed this dynamic data to be stored on a regular *static* TPF server. An extra layer on top of the TPF client was able to interpret these time-annotated triples as dynamic versions of certain facts. This extra software layer could then derive the exact moment at which the query should be *re-evaluated* to keep its results up to date.

The main experiment that was performed in my master's thesis resulted in the output from Fig. 3. We can see that our approach significantly reduced the server load when compared to C-SPARQL and CQELS, as was the main the goal. The client now pays for the largest part of the query executions, which is caused by the use of TPF. The client CPU usage for our implementation spikes at the time of query initialization because of the rewriting phase, but after that it drops to around 5 %.

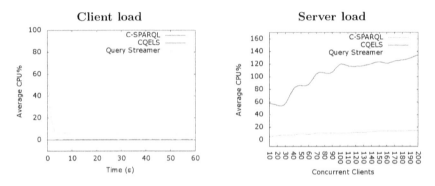

Fig. 3. The client and server CPU usages for one query stream for C-SPARQL, CQELS and our preliminary solution. Our solution has a very low server cost and a higher average client CPU usage when compared to the alternatives.

7   Conclusions

Once we can publish both non-high frequency real-time and historical data at a low server cost, we can finally allow many simultaneous clients to query this data while keeping their results up to date, so this dynamic data can be used in our applications with the same ease as we already do today with static data.

The Semantic Sensor Web already promotes the integration of sensors in the Semantic Web. My solution would make medium to low frequency sensor data queryable on a web-scale, instead of just for a few machines in a private environment for keeping the server cost maintainable.

Current Big Data analysis techniques are able to process data streams, but combining them with other data by discovering semantic relations still remains difficult. The solution presented in this work could make these Big Data analyses possible using Semantic Web techniques. This would make it possible to perform these analyses in a federated manner over heterogeneous sources, since a strength of Semantic Web technologies is the ability to integrate data from the whole web. These analyses could be executed by not only one entity, but all clients with access to the data, while still putting a reasonable load on the server.

References

1. Ali, M.I., Gao, F., Mileo, A.: CityBench: a configurable benchmark to evaluate RSP engines using smart city datasets. In: Arenas, M., et al. (eds.) ISWC 2015. LNCS, vol. 9367, pp. 374–389. Springer, Heidelberg (2015). doi:10.1007/978-3-319-25010-6_25

2. Arasu, A., Babcock, B., Babu, S., Cieslewicz, J., Datar, M., Ito, K., Motwani, R., Srivastava, U., Widom, J.: STREAM: the Stanford data stream management system. Book chapter (2004). http://ilpubs.stanford.edu:8090/641/1/2004-20.pdf

3. Babu, S., Widom, J.: Continuous queries over data streams. ACM Sigmod Rec. **30**(3), 109–120 (2001). http://dl.acm.org/citation.cfm?id=603884

4. Barbieri, D., Braga, D., Ceri, S., Della Valle, E., Grossniklaus, M.: Stream reasoning: where we got so far. In: Proceedings of the NeFoRS2010 Workshop, Co-located with ESWC 2010 (2010). http://wasp.cs.vu.nl/larkc/nefors10/paper/nefors10_paper_0.pdf

5. Barbieri, D.F., Braga, D., Ceri, S., Valle, E.D., Grossniklaus, M.: Querying RDF streams with C-SPARQL. SIGMOD Rec. **39**(1), 20–26 (2010)

6. Buil-Aranda, C., Hogan, A., Umbrich, J., Vandenbussche, P.-Y.: SPARQL web-querying infrastructure: ready for action? In: Alani, H., et al. (eds.) ISWC 2013, Part II. LNCS, vol. 8219, pp. 277–293. Springer, Heidelberg (2013)

7. Della Valle, E., Ceri, S., van Harmelen, F., Fensel, D.: It's a streaming world! Reasoning upon rapidly changing information. IEEE Intell. Syst. **24**(6), 83–89 (2009). http://www.few.vu.nl/ frankh/postscript/IEEE-IS09.pdf

8. Fernández, J.D., Martínez-Prieto, M.A., Gutiérrez, C., Polleres, A., Arias, M.: Binary RDF representation for publication and exchange (HDT). Web Semant. Sci. Serv. Agents World Wide Web **19**, 22–41 (2013). http://www.sciencedirect.com/science/article/pii/S1570826813000036

9. Fisteus, J.A., Garcia, N.F., Fernandez, L.S., Fuentes-Lorenzo, D.: Ztreamy: a middleware for publishing semantic streams on the web. Web Semant. Sci. Serv. Agents World Wide Web **25**, 16–23 (2014)

10. Le-Phuoc, D., Dao-Tran, M., Xavier Parreira, J., Hauswirth, M.: A native and adaptive approach for unified processing of linked streams and linked data. In: Aroyo, L., et al. (eds.) ISWC 2011, Part I. LNCS, vol. 7031, pp. 370–388. Springer, Heidelberg (2011)

11. Meinhardt, P., Knuth, M., Sack, H.: TailR: a platform for preserving history on the web of data. In: Proceedings of the 11th International Conference on Semantic Systems, pp. 57–64. ACM (2015). http://dl.acm.org/citation.cfm?id=2814875

12. Pérez, J., Arenas, M., Gutierrez, C.: Semantics and complexity of SPARQL. In: Cruz, I., Decker, S., Allemang, D., Preist, C., Schwabe, D., Mika, P., Uschold, M., Aroyo, L.M. (eds.) ISWC 2006. LNCS, vol. 4273, pp. 30–43. Springer, Heidelberg (2006)

13. Sheth, A., Henson, C., Sahoo, S.: Semantic sensor web. IEEE Internet Comput. **12**(4), 78–83 (2008). http://corescholar.libraries.wright.edu/cgi/viewcontent.cgi?article=2125&context=knoesis

14. de Sompel, H.V., Nelson, M.L., Sanderson, R., Balakireva, L., Ainsworth, S., Shankar, H.: Memento: time travel for the web. CoRR abs/0911.1112 (2009). http://arxiv.org/abs/0911.1112

15. de Sompel, H.V., Sanderson, R., Nelson, M.L., Balakireva, L., Shankar, H., Ainsworth, S.: An HTTP-based versioning mechanism for linked data. CoRR abs/1003.3661 (2010). http://arxiv.org/abs/1003.3661

16. Taelman, R.: Continuously updating queries over real-time linked data. Master's thesis, Ghent University, Belgium (2015). http://rubensworks.net/raw/publications/2015/continuously_updating_queries_over_real-time_linked_data.pdf
17. Vander Sande, M., Colpaert, P., Verborgh, R., Coppens, S., Mannens, E., Van de Walle, R.: R&Wbase: git for triples. In: LDOW (2013). http://events.linkeddata.org/ldow2013/papers/ldow2013-paper-01.pdf
18. Verborgh, R., Hartig, O., De Meester, B., Haesendonck, G., De Vocht, L., Vander Sande, M., Cyganiak, R., Colpaert, P., Mannens, E., Van de Walle, R.: Querying datasets on the Web with high availability. In: Proceedings of the 13th International Semantic Web Conference (2014). http://linkeddatafragments.org/publications/iswc2014.pdf
19. Verborgh, R., Vander Sande, M., Colpaert, P., Coppens, S., Mannens, E., Van de Walle, R.: Web-scale querying through Linked Data Fragments. In: Proceedings of the 7th Workshop on Linked Data on the Web (2014). http://events.linkeddata.org/ldow2014/papers/ldow2014_paper_04.pdf
20. Zhang, Y., Duc, P.M., Corcho, O., Calbimonte, J.-P.: SRBench: a streaming RDF/SPARQL benchmark. In: Heflin, J., et al. (eds.) ISWC 2012, Part I. LNCS, vol. 7649, pp. 641–657. Springer, Heidelberg (2012)

Exploiting Disagreement Through Open-Ended Tasks for Capturing Interpretation Spaces

Benjamin Timmermans$^{(\boxtimes)}$

VU University, Amsterdam, The Netherlands
b.timmermans@vu.nl

Abstract. An important aspect of the semantic web is that systems have an understanding of the content and context of text, images, sounds and videos. Although research in these fields has progressed over the last years, there is still a semantic gap between data available of multimedia and metadata annotated by humans describing the content. This research investigates how the complete interpretation space of humans about the content and context of this data can be captured. The methodology consists of using open-ended crowdsourcing tasks that optimize the capturing of multiple interpretations combined with disagreement based metrics for evaluation of the results. These descriptions can be used meaningfully to improve information retrieval and recommendation of multimedia, to train and evaluate machine learning components and the training and assessment of experts.

Keywords: Semantic interpretation · Multimedia · Crowdsourcing · Disagreement

1 Introduction

The semantic web has contributed to improving the usability of websites through semantic metadata, but this has not proven effective for media such as images, sounds and videos. The problem is that although every day more and more media content is shared online, the amount of metadata available is still limited [1]. Often some metadata is made available by the author, there is still a semantic gap between the available metadata and how it is perceived by humans. As a result, the metadata does not give a good representation of the actual content and context. The underlying systems need to have an understanding of this context, the human perspectives and opinions in order to provide meaningful search and discovery of information.

Recent work on video search engines has shown the existence of a semantic gap in video content [2]. In order to bridge this gap, it is essential for these systems to gain an understanding of the actual content. Search in social platforms like YouTube is limited by the annotations made by the uploader of a video. This means that the content you are looking for can be there, but that there is no representation of it in the metadata. The current improvements in information extraction methods will contribute to solving this problem, but the problem

© Springer International Publishing Switzerland 2016
H. Sack et al. (Eds.): ESWC 2016, LNCS 9678, pp. 873–882, 2016.
DOI: 10.1007/978-3-319-34129-3_56

remains that sounds, images and videos are prone to have multiple interpretations because they can be represented in multiple ways [3]. This means more interpretations of what can be heard are needed, in order to form a full spectrum of meaningful representations. In addition, gold standards for sounds are limited in size, and are often homogeneous because they only contain single interpretations [4]. Rich information can be obtained by aggregating multiple of these interpretations, which can be seen as collective intelligence [5].

In crowdsourcing, annotations are obtained from a large crowd of people using small microtasks. Often these tasks consist of the crowd workers selecting the right answer to a question. Capturing the complete answer space of possible interpretations has proven to be difficult with annotation tasks. Guidelines in tasks often focus on only one semantic interpretation in order to increase the inter-annotator agreement [6]. Also, no common reference system for defining the answers may be available, because for many tasks the answers simply cannot be predefined [7]. It can also be that there simply is no standardization in the categorization available [4]. Experts can help to define the reference system, but their views are biased and often it is difficult to define who the "experts" are in tasks such as general question-answering, interpretations of multimedia or sentiment analysis of texts. It may even be possible that the answer space is infinite, because the boundaries of the space are unknown or cannot be known.

The CrowdTruth[1] initiative [8–11] has been investigating how annotations can be collected on a large scale using crowdsourcing based on disagreement, rather than artificially forcing agreement through the inter-annotator agreement. The results have shown that crowdsourcing is a quick and cheaper way to solve the problem of scale, lack of experts and lack of interpretations [12]. Gathering many annotations from a lay crowd rather than few annotations from experts gives a closer representation of reality. Crowdsourcing has proven to be a successful tool for bridging the semantic gap [13]. Yet the question remains how crowdsourcing tasks can be designed if the interpretation space of possible answers is unknown, such that interpretations are not limited by the design of a task. Such open-ended tasks in which the user is less restricted in the answer that can be given can help gather annotations that better represent what is heard in a sound or what can be seen in a video. For instance for improving search or recommendation through content-based indexing.

In Sect. 2 the state of the art is described. Next, the problem statement and contributions of this research are formulated. In Sect. 4, the approach of this research is presented, followed by the preliminary results in Sect. 5. Last follow the evaluation plan for the approach and the conclusions of this paper.

2   State of the Art

There are many questions to which there is no single answer, or the answer is unknown. In such case an open ended task can be used to allow a multitude of possible answers to be annotated, without there being one correct answer. This

---

[1] http://crowdtruth.org.

does not restrict an annotator in giving a predefined answer, but has shown to give the worker freedom to answer according to their own interpretation [14]. This makes these tasks efficient for gathering a wide range of human interpretations, and that the resulting ground truth has a low influence of the reference system or answer options.

The problem of interpretation and answer spaces of unknown or large size is addressed in a study on object localization in images [15]. It is claimed that probabilistic inference on the ability to solve a task is necessary, because there is a significant difference in the ability of a person to perform a cognitive task. Related, another study found that if the data spans multiple domains, it is more likely that the answer space changes across different tasks [16]. If there is agreement between different workers in an open-ended task, it can be seen as a stronger signal because the answer space is much larger [17]. The chance of workers selecting the same answer in a small answer space is much higher, and having no overlap increases the difficulty of evaluating the answers.

Designing a clear and more detailed crowdsourcing task is needed in order to avoid misunderstanding and result in more substantial answers [18]. For open-ended tasks this is more important, because closed tasks have better defined rules and guidelines than open ended tasks. As a result, the interpretation of every aspect of an open-ended task plays a more important role. This feature of open-ended tasks is often exploited for teaching purposes, because the task can be approached in more ways where the difficulty plays less of a role. This has been found to result in a more diverse and larger group of answers that can be given than with closed questions, while providing valuable feedback to the teacher about the level of understanding of the students [19]. In a study on describing images through crowdsourcing [20] it was found that for efficient task designs, the resulting annotations should be as simple as possible while being as meaningful as possible. The simple annotations can also improve the detection of low quality annotations, because they have a more unified structure. This combination of an open-ended design with clear guidelines and simple meaningful annotations can be used to gather the interpretations.

The multitude of interpretations resulting from open-ended crowdsourcing tasks has also shown to be more difficult to evaluate [21], because these tasks can be highly subjective. Furthermore, if multiple answers can be correct majority voting as quality measure is not appropriate. Instead a peer-review approach should be used where workers verify each others work [22]. In [23] an open-ended crowdsourcing task was used to describe in words the differences between two images. The resulting lexicon was found to be comparable in quality to one created by experts using the different interpretations. For spam detection, open-ended questions have also been found to have the advantage that the response time is an indicator for lies and deception [24]. A study has been done on the use of open tasks [7] using free-response formulation with infinite outcome spaces. However, it was shown that although there are many crowdsourcing tasks that can match the quality of expert annotations, tasks with an infinite interpretation space where the boundaries of possible answers were unknown are not yet used in practice. Although these results have been found to be more difficult to evaluate, the design can result in high quality annotations.

There has been an increasing interest in research on the nature of sounds, and how people perceive sounds. The content that is perceived in sounds can be described in three categories: the source of the sound, attributes of the sound and the environment of the sound [25]. The perception differs from visual interpretation, because the awareness of sound sources are not as direct as when the source can be visually identified. Also, visible entities such as images differ from sound because they fully exist over time, while a sound always begins and ends like an event does. During this, its audible features change over time and do not have to fully exist at any given time [26]. This makes the sounds prone to have multiple interpretations, which makes the inter-annotator agreement a difficult metric to for instance measure the emotion in a sound [3]. Although our recent work on CrowdTruth has shown that annotator disagreement should be used as a signal, it is still considered a consensus problem [27,28]. This is specifically present in the ambiguity of sounds, which is why it is used in this study.

In music information retrieval, the features that influence how humans perceive music can be categorized into four factors [29]: (1) The content of what is heard in the music, discriminated using low to high level features like harmony or rhythm. (2) The context of the music such as lyrics, video clips, artist info and semantic labels. (3) The user listening to music, such as demographics and experience. (4) The context the user is in, such as mood and temporal context. The need for common representations in music using linked open data is discussed in [30], where they developed a semantic audio analysis interface for extracting content features of music. Another automatic tag classification of music was found to be improved in [31] by using the Music Information Retrieval Evaluation eXchange[2] datasets. An example of these datasets is majorminer.org, which is created through its own music labeling game. The goal of the game is to label music with words that people agree on. Several games with a purpose exist for annotating music, such as Listen Game and TagATune. The latter resulted in the Magnatagatune dataset, which according to [31] contains 21642 songs annotated with 188 unique tags. These findings from music information retrieval can be applied in this study for sound and video annotations.

3   Problem Statement and Contributions

The identified problem is that multimedia annotations are sparse, homogeneous, do not represent everything that can be heard or seen. Furthermore, crowdsourcing tasks are designed to stimulate agreement, while an open-ended approach is necessary to capture the full spectrum of subjectivity in human interpretations. Here it does not matter whether these are subjective or objective annotations, as they are both relevant interpretations. Based on this, the main research questions is: *Are open-ended crowdsourcing tasks a feasible method for capturing the*

---

[2] http://www.music-ir.org/mirex/.

interpretation space of multimedia? In order to answer this, we need to investigate how to evaluate the quality of the results, how to efficiently design such tasks and how this can solve the growing need for improved classification and semantic understanding of the content of multimedia.

1. The first step is to assess the quality of annotations captured using open-ended tasks. This is done under the hypothesis that open-ended tasks contribute to a larger interpretation space than closed tasks. The question to be answered is: *How can the quality of results of open-ended tasks be measured?* In order to answer this, the following sub questions are asked:
   - How can crowdsourcing quality measures be improved for open-ended crowdsourcing tasks which have no clear answer? The quality of open-ended tasks has shown to be more difficult to measure because these tasks can be highly subjective [21].
   - How can the confidence that people performing crowdsourcing tasks have in their answers be measured? With a multitude of answers that people provide, it is expected that they are more confident in some answers than others.

2. The second step is to assess the design of open-ended tasks for gathering large interpretation spaces. This is done under the hypothesis that design features such as pre-selected answers or visual clues have a positive effect on the efficiency of open-ended crowdsourcing tasks while maintaining the openness of the task. The research question asked is: *Can open-ended crowdsourcing tasks efficiently generate reliable ground-truth data?* This is investigated using the following sub questions:
   - How can constraints be used in open-ended tasks to improve the detection of low quality results? By constraining the tasks, the results may be able to be validated better.
   - How can existing automated feature extraction methods be used to optimize the use of the crowdsourcing tasks?
   - How can the threshold be measured for gaining a clear distribution of answers? If the interpretation space is unknown, a measure has to indicate when there is a clear distribution.

3. The third step is to assess the usability of the captured interpretation spaces. The hypothesis is that a larger interpretation space generated through open-ended tasks results in improved descriptions of the content, which lead to better search and discovery of the multimedia. The research question is: *How can a ground truth with a large interpretation space of what can be heard or seen in multimedia improve their search and discovery?*
   - How can the annotated human interpretation space of multimedia be combined with the context and content factors of both the entity and user which influence the human perception?
   - Can the novel ground truth data improve the indexing of multimedia on existing platforms? By analyzing the ground truth improvements, we can measure if and how the quality of search results can be improved.

The main contribution of this paper is to investigate how open-ended crowd-sourcing tasks can be used for gathering training data on things for which the interpretation space is unknown. Because it is unclear what the classifications are, designing crowdsourcing tasks becomes more difficult. The results should lead to more efficient crowdsourcing tasks resulting in higher quality representations of what can be heard in a sound or seen in a video. This will lead towards systems that have a more human friendly interaction through improved search and discovery of multi-representational entities.

4   Research Methodology and Approach

First, the quality of annotations captured through open-ended tasks has to be evaluated. In order to test this experiments will be performed with different open and closed approaches that represent different gradations of constraints. The least constrained design is free-text input The results will be evaluated using the CrowdTruth disagreement-based metrics [10] by transforming the annotated answers into a vector space representation. This allows for the evaluation of the quality of the annotations, the task, and the people performing annotations [32]. The metrics will have to be improved for these open-ended tasks in order to support interpretation spaces for which the size is unknown. As a result the vector space representation may be infinite, but an estimation of its size may be possible to deal with the unknown dimensions. The confidence people have in their answers can be tested through several approaches such as asking people to order their answers by level of confidence or using peer-reviews as done in [19]. This can then be compared to measures such as time spent on the task, the amount of times listened to a sound or video and the order the annotations were made in.

The second step of the approach is to find the optimal design of open-ended tasks for gathering large interpretation spaces. The task should have a clear design to avoid misunderstandings [18], but several features such as free text input, auto-completion or gamification can be used to reach the most efficient task in terms of total cost and time spent to complete it. Another method is to present an answer choice of pre-defined probable answers, but allow users to add more options. This can be extended by showing the options other users have added. By adding constraints to the open-ended tasks, the detection of low quality workers can be improved. This can again be done by testing and comparing features that are normally present in closed tasks. For instance a two-step task like presented in [6], where the user is allowed to provide self-contradicting answers. This has proven a useful measure, but requires at least two steps.

Another method of optimizing the tasks is by using automated feature extraction. Several studies have shown that high level descriptions such as abstract can be extracted [30,31]. By giving predefined descriptions through distant supervision, the annotator is forced to focus on making low-level annotations. Experiments have to show whether this approach works and does not bias the users into providing only high level descriptions that are similar to the predefined descriptions. Finally, an experiment has to be performed to compare measurements that indicate when there is a clear distribution of answers. In other words: how do you know when you have enough annotations?

For instance, short clear sounds may need less descriptions than long ambiguous sounds. This means that instead of requiring a fixed number of annotators, it can be dynamic depending on the complexity or ambiguity of the thing to annotate. This can not only help save cost by reducing the amount of needed annotations, but also increases the captured annotations for the most ambiguous examples. Content descriptive features such as length, pitch and melody can be extracted can be tested for determining the optimal threshold.

The third step in the approach is to test the usefulness of the captured interpretations. First, the annotations of sounds and videos are placed within the context and content the users and items. For instance, some annotations may describe sound content while other describe the context, which can be categorized following the research described in [29]. Combining the different factors with the context and content of the user is expected to improve the performance of search results. These are to be further investigated by testing whether the indexing of sounds with the ground truth built through crowdsourcing in this research improves the results. This requires an information retrieval evaluation that can be performed using measures such as precision and recall and by assessing users.

5   Preliminary or Intermediate Results

In [6] we describe our initial work on crowdsourcing semantic interpretations for open-domain questions answering. The goal of this study was to gather training data on open-domain questions for IBM Watson, as part of the Crowd-Watson collaboration between IBM Research and the VU University Amsterdam. The problem with open-domain questions is that it is difficult to define who experts are, and both the question and the answer can be highly ambiguous. For Watson to better understand why a text passage justifies the answer to a question, we used multiple crowdsourcing tasks to map question and answer pairs and disambiguate terms. The results showed that CrowdTruth is an efficient approach for gathering ground truth data on open-domain questions that can have multiple interpretations and answers. By designing the crowdsourcing tasks with limited constraints, we found that self-contradicting answers were an effective measure for identifying low quality annotations.

In [14] we describe and publish the VU sound corpus[3], which is a continuation on the work of [33]. The gathered corpus consists of fine-grained annotations of 2000 short sound effects in the Freesound database[4]. The annotations were obtained through a simple free-text input form, where per sound 10 crowd workers were asked to describe with keywords what they heard in a given sound. Due to the open-ended design of the task there was high disagreement between

---

[3] http://dx.doi.org/10.5281/zenodo.35508.
[4] http://freesound.org.

the annotators, which increased the difficulty of detecting low quality results. The perspectives of crowd annotators proved to be important, because there was a large distinction in descriptions made by the author of a sound and the crowd. These descriptions are essential for having effective search and discovery of sounds.

6   Evaluation Plan

The approach presented in the previous chapter is assessed through the three hypotheses and associated research questions. The first evaluation is the quality of the gathered crowdsourcing annotations. The use of open-ended tasks should result in a larger interpretation space than if closed tasks were used. The second evaluation is of the design of the open-ended tasks. Alternating between different features should conclude which features result in a higher quality of results. Furthermore, the quality of the task can be evaluated through the cost and time of the crowdsourcing task to complete. The third evaluation is of the performance of the captured interpretations, which can be measured through precision and recall and by assessing users. This will show the effectiveness of the larger interpretation space of sounds.

7   Conclusions

Information systems should always have a semantic understanding of their content. This is exemplified in this study through sounds and videos, because their annotations are sparse and homogeneous. Furthermore, there is a semantic gap between the available descriptions and what can be heard or seen in those sounds and videos. The existing approaches can be improved because they stimulate agreement between annotators or do not deal with the fact that the interpretation space is unknown.

This study aims to investigate how open-ended crowdsourcing tasks and disagreement-based metrics can be used to capture the complete human interpretation space of multimedia. The approach is to first investigate how the quality of results for open-ended crowdsourcing tasks can be measured. Next, the design of these open-ended tasks is assessed for gathering large interpretation spaces, and their usability for improving the search and discovery of multimedia. The preliminary experiments have shown that the approach cam be feasible, but evaluation through applying the ground truth is necessary.

Acknowledgements. I would like to thank Dr. Lora Aroyo for her supervision, Dr. Matteo Palmonari for his guidance, Emiel van Miltenburg for our collaborative work and Robert-Jan Sips for his support.

References

1. Nixon, L., Troncy, R.: Survey of semantic media annotation tools for the web: towards new media applications with linked media. In: Presutti, V., Blomqvist, E., Troncy, R., Sack, H., Papadakis, I., Tordai, A. (eds.) ESWC Satellite Events 2014. LNCS, vol. 8798, pp. 100–114. Springer, Heidelberg (2014)
2. Jiang, L.: Web scale multimedia search for internet video content. In: Proceedings of the Ninth ACM International Conference on Web Search and Data Mining, WSDM 2016, p. 701. ACM, New York (2016)
3. Aljanaki, A., Wiering, F., Veltkamp, R.C.: Emotion based segmentation of musical audio. In: Proceedings of the 15th Conference of the International Society for Music Information Retrieval (ISMIR 2014) (2015)
4. Campos, G., Quintas, J.: On the validation of computerised lung auscultation. In: Proceedings of the International Conference on Health Informatics (BIOSTEC 2015), pp. 654–658 (2015)
5. Singh, P., Lasecki, W.S., Barelli, P., Bigham, J.P.: Hivemind: A framework for optimizing open-ended responses from the crowd. Technical report, URCS Technical Report (2012)
6. Timmermans, B., Aroyo, L., Welty, C.: Crowdsourcing ground truth for question answering using crowdtruth. In: WebSci (2015)
7. Lin, C.H., Mausam, M., Weld, D.S.: Crowdsourcing control: moving beyond multiple choice. In: Workshops at the Twenty-Sixth AAAI Conference on Artificial Intelligence (2012)
8. Inel, O., et al.: CrowdTruth: machine-human computation framework for harnessing disagreement in gathering annotated data. In: Mika, P., et al. (eds.) ISWC 2014, Part II. LNCS, vol. 8797, pp. 486–504. Springer, Heidelberg (2014)
9. Aroyo, L., Welty, C.: Measuring crowd truth for medical relation extraction. In: AAAI 2013 Fall Symposium on Semantics for Big Data (2013)
10. Soberón, G., Aroyo, L., Welty, C., Inel, O., Lin, H., Overmeen, M.: Measuring crowdtruth: disagreement metrics combined with worker behavior filters. In: Proceedings of 1st International Workshop on Crowdsourcing the Semantic Web (CrowdSem), ISWC, pp. 45–58 (2013)
11. Inel, O., Aroyo, L., Welty, C., Sips, R.-J.: Domain-independent quality measures for crowd truth disagreement. J. Detect. Representation Exploit. Events Semant. Web, 2–13 (2013)
12. Aroyo, L., Welty, C.: Truth is a lie: 7 myths about human annotation. AI Mag. **36**(1), 15–24 (2015)
13. Macanas, J., Ouyang, L., Bruening, M.L., Muñoz, M., Remigy, J.C., Lahitte, J.F.: Development of polymeric hollow fiber membranes containing catalytic metal nanoparticles. Catal. Today **156**(3), 181–186 (2010). doi:10.1016/j.cattod.2010.02.036
14. van Miltenburg, E., Timmermans, B., Aroyo, L.: The VU sound corpus: adding more fine-grained annotations to the freesound database. In: LREC 2016 (2016)
15. Salek, M., Bachrach, Y., Key, P.: Hotspottinga probabilistic graphical model for image object localization through crowdsourcing. In: Twenty-Seventh AAAI Conference on Artificial Intelligence (2013)
16. Kurve, A., Miller, D.J., Kesidis, G.: Multicategory crowdsourcing accounting for variable task difficulty, worker skill, and worker intention. IEEE Trans. Knowl. Data Eng. **27**(3), 794–809 (2015)

17. Lasecki, W.S., Homan, C., Bigham, J.P.: Architecting real-time crowd-powered systems. Human Comput. **1**(1), 69 (2014)
18. Liu, D., Bias, R.G., Lease, M., Kuipers, R.: Crowdsourcing for usability testing. Proc. Am. Soc. Inf. Sci. Technol. **49**(1), 1–10 (2012)
19. Sullivan, P., Clarke, D., Clarke, B.: Using content-specific open-ended tasks. In: Sullivan, P., Clarke, D., Clarke, B. (eds.) Teaching with Tasks for Effective Mathematics Learning, vol. 104, pp. 57–70. Springer, New York (2013)
20. Ooi, W.T., Marques, O., Charvillat, V., Carlier, A.: Pushing the envelope: solving hard multimedia problems with crowdsourcing. MMTC e-letter **8**(1), 37–40 (2013)
21. Deng, J., Krause, J., Fei-Fei, L.: Fine-grained crowdsourcing for fine-grained recognition. In: Proceedings of the IEEE Computer Society Conference on Computer Vision and Pattern Recognition, pp. 580–587 (2013)
22. Schulze, T., Nordheimer, D., Schader, M.: Worker perception of quality assurance mechanisms in crowdsourcing and human computation markets. In: Proceedings of 19th Americas Conference on Information Systems, AMCIS 2013, pp. 1–11 (2013)
23. Maji, S.: Discovering a lexicon of parts and attributes. In: Fusiello, A., Murino, V., Cucchiara, R. (eds.) ECCV 2012 Ws/Demos, Part III. LNCS, vol. 7585, pp. 21–30. Springer, Heidelberg (2012)
24. Walczyk, J.J., Roper, K.S., Seemann, E., Humphrey, A.M.: Cognitive mechanisms underlying lying to questions: response time as a cue to deception. Appl. Cogn. Psychol. **17**(7), 755–774 (2003)
25. Nudds, M., O'Callaghan, C.: Sounds and Perception: New Philosophical Essays. Oxford University Press, Oxford (2009)
26. O'Callaghan, C.: Objects for multisensory perception. Philos. Stud. **173**(5), 1269–1289 (2016). doi:10.1007/s11098-015-0545-7
27. Ekeroma, A., Kenealy, T., Shulruf, B., Hill, A.: Educational and wider interventions that increase research activity and capacity of clinicians in low to middle income countries: a systematic review and narrative synthesis. IBM J. Res. Dev. **3**, 120 (2015)
28. Boland, M.R., Miotto, R., Gao, J., Weng, C.: Feasibility of feature-based indexing, clustering, and search of clinical trials. Methods Inform. Med. **52**(5), 382–394 (2013). doi:10.3414/ME12-01-0092
29. Schedl, M., Widmer, G., Knees, P., Pohle, T.: A music information system automatically generated via web content mining techniques. Inform. Process. Manage. **47**(3), 426–439 (2011). dx.doi.org/10.1016/j.ipm.2010.09.002
30. Allik, A., Fazekas, G., Dixon, S., Sandler, M.: Facilitating music information research with shared open vocabularies. In: Cimiano, P., Fernández, M., Lopez, V., Schlobach, S., Völker, J. (eds.) ESWC 2013. LNCS, vol. 7955, pp. 178–183. Springer, Heidelberg (2013)
31. Seyerlehner, Klaus, Schedl, Markus, Sonnleitner, Reinhard, Hauger, David, Ionescu, Bogdan: From Improved Auto-Taggers to Improved Music Similarity Measures. In: Nürnberger, Andreas, Stober, Sebastian, Larsen, Birger, Detyniecki, Marcin (eds.) AMR 2012. LNCS, vol. 8382, pp. 193–202. Springer, Heidelberg (2014). doi:10.1007/978-3-319-12093-5_11
32. Aroyo, L., Welty, C.: The three sides of CrowdTruth. J. Human Comput. **1**, 31–34 (2014)
33. Lopopolo, A., van Miltenburg, E.: Sound-based distributional models. In: IWCS 2015, p. 70 (2015)

A Semantic Approach for Process Annotation and Similarity Analysis

Tobias Weller[✉]

Institute AIFB, Englerstr. 11, 76128 Karlsruhe, Germany
tobias.weller@kit.edu
http://www.aifb.kit.edu

Abstract. Research in the area of process modeling and analysis has a long-established tradition. There are quite few formalism for capturing processes, which are also accompanied by a number of optimization approaches. We introduce a novel approach, which employs semantics, for process annotation and analysis. In particular, we distinguish between target processes and current processes. Target process models describe how a process should ideally run and define a framework for current processes, which in contrast, capture how processes actually run in real-life use cases. In some cases, current processes do not match the target process models and can even overhaul them. Therefore, one is interested in the similarity between the defined target process model and current processes. The comparisons can consider different characteristics of processes such as service quality measures and dimensions. Current solutions perform process mining methods to discover hidden structures or try to infer knowledge about processes by using specific ontologies. To this end, we propose a novel method to capture and formalize processes, employing semantics and devising strategies and similarity measures that exploit the semantic representation to calculate similarities between target and current processes. As part of the similarity analysis, we consider different service qualities and dimensions in order to determine how they influence the target process models.

Keywords: Process annotation · Similarity analysis · Process analysis · Quality of Service · Semantic process modelling

1 Introduction

Process modeling and analysis has multiple application domains. I.e. clinical pathways are an evidence-based response to specific problems and care needs in clinics. They support physicians by providing recommendations on the sequence and timing of actions necessary to achieve an efficient treatment of patients [1,2]. Each clinic has their own pathways based on their individual evidence and experience. Therefore, there are multiple pathways that target different problems and care needs [3–5].

© Springer International Publishing Switzerland 2016
H. Sack et al. (Eds.): ESWC 2016, LNCS 9678, pp. 883–893, 2016.
DOI: 10.1007/978-3-319-34129-3_57

However, physicians are not strictly restricted to the published pathways. Therefore, the process, defined in the pathway (target process model), can differ from the actually performed workflow (current process). As a result, there might be discrepancies between the published clinical pathways and the actually performed workflow, which is based on the decisions of the physician on how to treat the patient.

This situation is aggravated by the fact that there is a lot of data, generated and used during the treatment of patients, that needs to be managed and interpreted. In order to ease this task, the information can be captured semantically and used for comparisons. For this purpose, there are already many ontologies in the medical domain, which can be used to structure the semantic information – the Disease Ontology[1] that provides descriptions and related medical terms about human diseases and the Foundational Model of Anatomy (FMA)[2], which describes classes, structures and relationships of all parts of the human body. In addition, processes can be compared based on different service qualities and dimensions, such as complexity, runtime, outcome or costs.

The problem of having current processes that diverge from the defined target process models does not occur only in the medical domain. The same difficulties arise also in enterprises and in the domain of Internet of Things (IoT) applications, in which the actual communication flow between devices can diverge from a defined target process model. This is precisely the topics that we want to explore. One aspect is that it is debatable whether the current process performs better, in terms of certain service qualities and dimensions, than the defined target process model. Knowing the deviation and different outcome of the service qualities and dimensions could lead to the incentive of adapting the target process model.

Given a set of current processes and its target process model, we are interested in calculating the similarity between them, in order to be able to quantify the variety and see how different processes behave in terms of different service quality aspects and dimensions. The revelation of the effect of service qualities and dimensions can be used for different aspects. One aspect is to provide a confidence interval for a service quality variable. Another aspect could be the hint of adapting the target process model, if the current process instances diverge too much from the target process model. The adaption of the target process model can than be performed in respect to certain service qualities and dimensions.

2   State of the Art

An important aspect, in order to have a common point of view on processes, is to define the term *process*. We use the process definition from ISO 9000:2015 [6], which is given in the following.

---

[1] http://disease-ontology.org.
[2] http://sig.biostr.washington.edu/projects/fm/index.html.

Definition 1. *ISO 9000:2015 Process: Set of interrelated or interacting activities that use inputs to deliver an intended result.*

Note 1 to entry: Whether the "intended result" of a process is called output, product or service depends on the context of the reference.

Note 2 to entry: Inputs to a process are generally the outputs of other processes and outputs of a process are generally the inputs to other processes.

This definition is specifically related to quality management systems, but we aim to use it in a broader way. We do not focus on quality management systems in particular but rather on processes in general.

Process semantic annotations: There are already widely used ontologies, such as Dublin Core Schema[3] that provide a set of metadata that can be used to annotate resources. We can use these ontologies to annotate process elements like e.g. tasks and gateways of the target process model. The advantage of such schemata is that they can be integrated easily in order to annotate resources and provide interoperability with further datasets.

Semantic process-based formalizations and conformance checking: Business Process Abstract Language (BPAL) provides a formal semantic to process modeling languages [7,8] and allows enriching it with semantic annotations. The formal definition allows a verification of the used properties and the ontology-based annotations. Thus, this approach can be used to semi-automatically map current process instances to its target process model and verify the mapping according to a correct semantic annotation [9–12]. There are also some ontology-based annotations for process models available that can be reused [13,14]. In addition to semantic annotations, there are also ontologies available to describe the components of a process and the relationships between them such as SUPER [15,16] and the Process Specification Language[4], which has been approved as an international standard [17].

Service qualities and dimensions: Existing approaches describe how service qualities and dimensions can be captured [18]. Thereby, frameworks like SERVQUAL can be used to measure the quality of processes [19]. Service qualities from e-services [20] or other process performance indicators [21] can also be used as metrics to measure the performance of processes.

Process matching: There are a number of different process similarity measurements for comparing processes. Some uses node similarity, structural similarity, behavioral similarity and language based matching [22,23]. However, most of them focus on business processes [24,25] and do not distinguish between target and current process models. The similarity of processes is, among others, used to cluster processes [26].

Adaption of target process models: Approaches like e.g. Process Mining try to reveal hidden structures and create a target process by using i.e. log files

---

[3] http://dublincore.org.
[4] http://www.mel.nist.gov/psl/.

or other data produced by process instances [27, 28]. These approaches reveal hidden structures but not the influence of processes on different service qualities and dimensions. However, process mining techniques can be used to discover new insights based on a created reference model from the current process data [29].

3   Problem Statement and Contributions

We focus on performing similarity analysis between target and current processes by exploiting the semantics of processes. The semantics that we use to compare process models consist, among others, of the semantic annotations (like labels and descriptions) that we add to the process models, a domain hierarchy of the process elements, the user roles (for example, only specific users are allowed to perform a task or a decision) and rules that define the workflow of processes. Based on the presented motivation and the current state-of-the-art, we formulate the following research question and its subquestions:

How do we benefit from the combination of process models with semantics in order to improve processes by performing similarity analysis?

RQ1 How can we formally specify process data with semantics?
RQ2 Which service qualities and dimensions can we use to compare processes?
RQ3 Which methods can we use to perform similarity analysis of target processes and current process data?

During the PhD we will develop an approach to annotate process data with semantic information and perform similarity analysis of target process models and current processes. This approach will be modeled in a common way, so it is generally applicable. In the following, we discuss the subquestions in more detail.

(RQ1) How can we formally specify process data with semantics?
There are already established formal representations for modeling languages e.g. for BPMN 2.0, the standard language BPMN 2.0 XML published by OMG[5] or the Petri Net Markup Language [30] for representing petri nets. However, while the execution semantics of processes is partly covered, there is a lack of semantics for the inputs/outputs used in the processes, annotations and the terminology of process elements. Therefore, we will show how to combine formally specified process models with semantics that can be queried and processed. The enriched current process instances can be used for comparisons and analysis.

(RQ2) Which service qualities and dimensions can we use to compare processes?
Processes can be compared based on different service qualities and dimensions such as runtime, outcome, costs or reliability. Capturing these service qualities and dimensions is a first step towards being able to compare the defined target

---

[5] http://www.omg.org/spec/BPMN/2.0/.

process model and the current processes. We will analyze existing frameworks (see Sect. 2) according to their extent and their usability in different domains and maybe extend them. As possible output, we may propose a new framework.

(RQ3) Which methods can we use to perform similarity analysis of target processes and current process data?
We will show which methods can be used to compare a target process model with a set of current processes. During the use of different similarity methods, we will exploit semantics such as the hierarchical arrangements of process elements, as well as domain semantics, and user roles, linked to tasks, and rules, which influence the process flow. Figure 1 shows the comparisons of a target process model with current process instances.

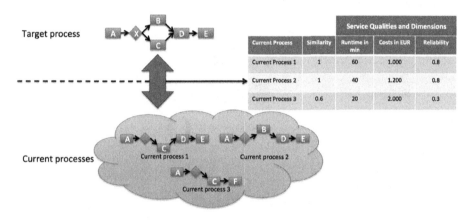

Fig. 1. Determining the similarity between target process model and current process instances

The research questions aim to result in multiple contributions. The first contribution is the introduction of an approach that integrates processes with semantic information that can be queried and processed. We would like to integrate as much semantic information as possible to allow, in a later step, enhanced similarity analysis that considers all these aspects. Another contribution is a set of service qualities and dimensions that can be used to compare processes. We will show different metrics and how they can be used in multiple domains. The last contribution is the similarity analysis between target process models and current processes. Thereby, we will use methods that exploit the semantics, captured in the previous step, such as the hierarchy of activities and process flows, to quantify the similarity.

4 Research Methodology and Approach

The structure of the research methodology and approach is directly derived based on the research questions (Sect. 3). Research methodologies can be classified as quantitative, qualitative and mixed research methodologies. Quantitative research methods collect numerical data and use it to analyze and explain a circumstance [31]. We will apply quantitative methodologies to plan and approach the research problems. In particular we will collect a sample of process instances, formalize the data, map them to the target process model, calculate the similarity between them and reveal the effect of current processes to service qualities and dimensions. We will investigate how semantics affects the analysis and comparisons of processes, and test different methods to compare processes.

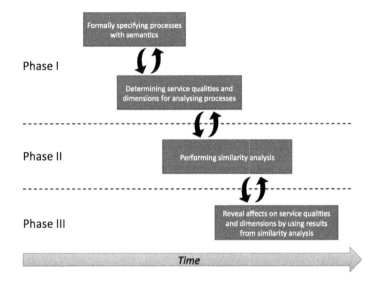

Fig. 2. Research approach – divided into three phases. Each tackles another aspect and influences tasks in other phases.

Figure 2 shows the planned thesis approach, divided into three main phases. Each phase tackles a specific part of the thesis, consists of performed activities and influences or is influenced by other activities. In the following, we will explain each phase in more detail.

Phase I: We assume that semantics provide a huge potential for similarity analysis and in revealing the effects on service qualities and process performance indicators. In order to exploit semantics in processes, we first have to annotate current process instances by mapping them to the semantically enriched target model.

The definition of dimensions partially overlaps with the formal specification of processes with semantics. Both activities are performed in phase I.

Phase II: This phase focuses on performing similarity analysis of target process models and current processes. We will use different similarity methods e.g. node similarity, structural similarity and behavioral similarity. Among others, we will also use methods that do not exploit semantics and compare them to methods that exploit semantics in order to show the advantages of having semantic annotations. We will also consider combining different methods for similarity analysis, resulting in a hybrid approach.

Phase III: The last phase uses the similarity analysis as an input in order to reveal the effect on service qualities and dimensions. We will evaluate whether current processes have an influence on the service qualities and dimensions. In addition, during this phase, we can also discover new insights that motivate to capture additional service qualities and dimensions. Therefore, this activity influences in turn phase II.

We will show that the methods are not constrained to a single domain by applying them to different domains (Sect. 6).

5   Preliminary Results

Currently, we are facing the first phase (see Sect. 4), which is about formally specifying process models with semantics. To this end, we analyzed different tools that allow to model processes. However, existing tools do not allow to enrich process data with semantics. In addition, we aim to follow the Linked Data Principles[6] for publishing data.

In order to combine processes with semantic information, we created a tool that captures processes and allow users to enrich them with semantic information. We used bpmn-io as web modeler and extended it with further functionalities. bpmn-io[7] is a JavaScript renderer that allows modeling and checking the syntax of BPMN processes. We embedded our developed tool into a Semantic MediaWiki[8]. Thus, Semantic MediaWiki, in combination with our developed tool, serves as platform to capture, annotate, query and process the information in a structured way and publishing it as Linked Data.

With this tool, we can integrate processes, stored in the standard format BPMN 2.0 XML[9] into Semantic MediaWiki and enrich them with semantics. The integrated and semantically enriched processes can in turn be exported into BPMN 2.0 XML format, allowing for exchange and reuse of the modeled processes.

As the next step, we will determine service quality measures and dimensions for comparing processes but also measuring the efficiency of a target process such as runtime, outcome or costs and study approaches to map current process instances to the target process model. In addition, we will consider different similarity methods to quantify the similarity between target process and current processes and necessary information that will improve the calculation of similarity. These considerations will influence the enrichment of semantic information, since we have to capture it during the annotation of the processes.

---

[6] http://www.w3.org/DesignIssues/LinkedData.html.
[7] https://github.com/bpmn-io/bpmn-js.
[8] https://semantic-mediawiki.org.
[9] http://www.omg.org/spec/BPMN/2.0/.

6   Evaluation Plan

For validating our solution, we will implement the designed approach and methods in different use-case scenarios. This ensures on the one hand that our approach and methods abstract from the used domain and on the other hand to capture independent results that can be evaluated.

We plan to use the following two domains to evaluate our approach:

(1) Medical Domain: Current processes in clinics differ from target process models. This is caused by latest insights and developments in the medical domain and the slow adoption of clinical pathways. In addition, there are many ontologies i.e. Foundational Model of Anatomy ontology (FMA)[10] or Gene Ontology[11] that can be used to structure processes with semantic information. Therefore, we will use our approach to calculate the similarity between target and current processes and show the influences of processes on different service qualities and dimensions.

(2) Internet of Things: Another field of application is the domain Internet of Things. In this domain, the communication and data flow between devices is not strictly given. Hence, there are more ad-hoc processes, which makes it hard to get an overview of the processes in general. Although this domain is rather new, there are already some ontologies available [32,33]. We will use data from devices (i.e. communication data and process data) and annotate the tasks with semantic information. This allows for enhanced analysis of communication workflows, and allows us to see the deviation of current processes from target process models.

For evaluating the first research question, we will validate the formalized process data, enriched with semantics, by comparing the usability of the provided methods with different approaches and the expressiveness of the formally specified processes. The formalization of data should not be focused on a specific scenario or domain, which is shown by applying our approach and methods in multiple scenarios and domains. Mapping the current process instances to its target process model is validated by comparing it to different methods.

To evaluate the second and third research question, we will start with comparing very simple target and current process models and gradually extend the process with further details and expressiveness. Hence, we will start performing similarity analysis and revealing the effect on different service qualities and dimensions in each applied domain with a sequential process and then successively extend the expressiveness of the process and the used service qualities and dimensions.

7   Conclusions

We aim to develop an approach to annotate and perform similarity analysis between target and current processes.

---

[10] http://sig.biostr.washington.edu/projects/fm/.
[11] http://geneontology.org.

We will consider the similarity in relation to service qualities and dimensions in order to (1) provide confidence intervals for service qualities and dimensions so one can estimate which values will be assigned by a process variable of the target process model (2) reveal weak spots, which has influence on different service qualities and dimensions and (3) motivate to adapt the target process model if its current process instances diverge too much from it.

In addition, the knowledge from this approach can also be used to support people with process optimization and improvements of the target process models.

References

1. Panella, M., Marchisio, S., Di Stanislao, F.: Reducing clinical variations with clinical pathways: do pathways work? Int. J. Qual. Health Care **15**(6), 509–521 (2003)
2. Kinsman, L., Rotter, T., James, E., Snow, P., Willis, J.: What is a clinical pathway? Development of a definition to inform the debate. BMC Medicine **8**, 31 (2010)
3. Zand, E.K.: Integrated care pathways: eleven international trends. Int. J. Care Coord. **6**(3), 101–107 (2002)
4. Vanhaecht, K., Bollmann, M., Bower, K., Gallagher, C., Gardini, A., Guezo, J., Jansen, U., Massoud, R., Moody, K., Sermeus, W., Zelm, R., Whittle, C., Yazbeck, A.M., Zander, K., Panella, M.: Prevalence and use of clinical pathways in 23 countries - an international survey by the European pathway association. Int. J. Care Coord. **10**(1), 28–34 (2006)
5. Hindle, D., Yazbeck, A.: Clinical pathways in 17 European union countries: a purposive survey. Aust. Health Rev. **29**(1), 94–104 (2005)
6. European Committee for Standardization, Quality management systems - Fundamentals and vocabulary (ISO 9000:2015), September 2015
7. de Nicola, A., Lezoche, M., Missikoff, M.: An ontological approach to business process modeling. In: 3rd Indian International Conference on Artificial Intelligence 2007, pp. 1794–1813, December 2007
8. Smith, F., De Sanctis, D., Proietti, M.: A platform for managing business process knowledge bases via logic programming. In: CILC 2013, pp. 247–251 (2013)
9. van der Aalst, W.M.P.: Process mining in the large: a tutorial. In: Zimányi, E. (ed.) eBISS 2013. LNBIP, vol. 172, pp. 33–76. Springer, Heidelberg (2014)
10. Di Francescomarino, C., Ghidini, C., Rospocher, M., Serafini, L., Tonella, P.: Reasoning on semantically annotated processes. In: ICSOC 2008, pp. 132–146 (2008)
11. Rozinat, A., van der Aalst, W.M.P.: Conformance checking of processes based on monitoring real behavior. Inf. Syst. **33**(1), 64–95 (2008)
12. van der Aalst, W.M.P., Adriansyah, A., van Dongen, B.F.: Replaying history on process models for conformance checking and performance analysis. Wiley Interdisc. Rev. Data Min. Knowl. Disc. **2**(2), 182–192 (2012)
13. Lin, Y., Ding, H.: Ontology-based semantic annotation for semantic interoperability of process models. In: Computational Intelligence for Modelling, Control and Automation, 2005 and International Conference on Intelligent Agents, Web Technologies and Internet Commerce, vol. 1, pp. 162–167, November 2005. doi:10.1109/CIMCA.2005.1631259
14. Lin, Y., Strasunskas, D.: Ontology-based semantic annotation of process templates for reuse. In: 10th International Workshop on Exploring Modeling Methods in System Analysis and Design (EMMSAD 2005), Porto, Portugal (2005)

15. Dimitrov, M., Simov, A., Stein, S., Konstantinov, M.: A BPMO based semantic business process modelling environment, semantic business process and product lifecycle management. In: Proceedings of the Workshop SBPM 2007, Innsbruck, April 2007
16. Hepp, M., Roman, D.: An ontology framework for semantic business process management. In: Wirtschatsinformatik Proceedings (2007)
17. European Committee for Standardization, Industrial automation systems and integration – Process specification language (ISO 18629-1:2004), November 2004
18. Bauer, H.H., Falk, T., Hammerschmidt, M.: eTransQual: a transaction process-based approach for capturing service quality in online shopping. J. Bus. Res. **59**(7), 866–875 (2006). ISSN: 0148-2963
19. Gawyar, E.T.H., Ehsani, M., Kozehchian, H.: Measuring service quality of state-clubs in Lorestan Province using SERVQUAL model. Int. J. Sport Stud. **4**(2), 233–237 (2014). ISSN: 2251-7502
20. Collier, J.E., Bienenstock, C.C.: A conceptual framework for measuring e-service quality. In: Proceedings of the Academy of Marketing Science (AMS) Annual Conference, pp. 158–162 (2003). ISSN: 2363-6165
21. del-Río-Ortega, A., Resinas, M., Ruiz-Cortés, A.: Defining process performance indicators: an ontological approach. In: Meersman, R., Dillon, T.S., Herrero, P. (eds.) OTM 2010. LNCS, vol. 6426, pp. 555–572. Springer, Heidelberg (2010). ISSN: 0302-9743
22. Weidlich, M., Sheetrit, E., Branco, M.C., Gal, A.: Matching business process models using positional passage-based language models. In: Ng, W., Storey, V.C., Trujillo, J.C. (eds.) ER 2013. LNCS, vol. 8217, pp. 130–137. Springer, Heidelberg (2013)
23. Dijkman, R., Dumas, M., García-Bañuelos, L.: Graph matching algorithms for business process model similarity search. In: Dayal, U., Eder, J., Koehler, J., Reijers, H.A. (eds.) BPM 2009. LNCS, vol. 5701, pp. 48–63. Springer, Heidelberg (2009)
24. Dijkman, R., Dumas, M., van Dongen, B., Käärik, R., Mendling, J.: Similarity of business process models: metrics and evaluation. Inf. Syst. **36**(2), 498–516 (2011). ISSN: 0306-4379
25. Zhang, Y., Liu, J., Wang, L.: Product manufacturing process similarity measure based on attributed graph matching. In: 3rd International Conference on Mechatronics, Robotics and Automation (ICMRA 2015), June 2015
26. Jung, J., Bae, J.: Workflow clustering method based on process similarity. In: Gavrilova, M.L., Gervasi, O., Kumar, V., Tan, C.J.K., Taniar, D., Laganá, A., Mun, Y., Choo, H. (eds.) ICCSA 2006. LNCS, vol. 3981, pp. 379–389. Springer, Heidelberg (2006)
27. van der Aalst, W.M.P., Reijers, H.A., Weijters, A.J.M.M., van Dongen, B.F., Alves de Medeiros, A.K., Song, M., Verbeek, H.M.W.: Business process mining: an industrial application. Inf. Syst. **32**(5), 713–732 (2007)
28. van Dongen, B.F., de Medeiros, A.K.A., Verbeek, H.M.W., Weijters, A.J.M.M., van der Aalst, W.M.P.: The ProM framework: a new era in process mining tool support. In: Ciardo, G., Darondeau, P. (eds.) ICATPN 2005. LNCS, vol. 3536, pp. 444–454. Springer, Heidelberg (2005)
29. Maggi, F.M., Mooij, A.J., van der Aalst, W.M.P.: Analyzing vessel behavior using process mining. In: van de Laar, P., Tretmans, J., Borth, M. (eds.) Situation Awareness with Systems of Systems, pp. 133–148. Springer, New York (2013)

30. Billington, J., Christensen, S., van Hee, K.M., Kindler, E., Kummer, O., Petrucci, L., Post, R., Stehno, C., Weber, M.: The Petri Net markup language: concepts, technology, and tools. In: van der Aalst, W.M.P., Best, E. (eds.) ICATPN 2003. LNCS, vol. 2679, pp. 483–505. Springer, Heidelberg (2003)

31. Muijs, D.: Doing Quantitative Research in Education with SPSS, 2nd edn. SAGE Publications, London (2010)

32. Hachem, S., Teixeira, T., Issarny, V.: Ontologies for the Internet of Things. In: ACM/IFIP/USENIX 12th International Middleware Conference, Lisbon, Portugal, December 2011. Springer (2011)

33. Wang, W., De, S., Toenjes, R., Reetz, E., Moessner, K.: A comprehensive ontology for knowledge representation in the Internet of Things. In: 2012 IEEE 11th International Conference on Trust, Security, Privacy in Computing, Communications (TrustCom), pp. 1793–1798 (2012). doi:10.1109/TrustCom.20

Author Index

Printed in the United States
By Bookmasters